D1189485

Partial Stability and Control

V. I. Vorotnikov

1998
Birkhäuser
Boston • Basel • Berlin

V. I. Vorotnikov
Ural State Technical University
Nizhny Tagil Campus
59 Krasnogvardeiskaya ul.
Nizhny Tagil 622031
Russia

Library of Congress Cataloging-in-Publication Data

Vorotnikov, V. I. (Vladimir Il ' ich)
Partial stability and control / V.I. Vorotnikov
p. cm.
Includes bibliographical references.
ISBN 0-8176-3917-9 (hc : alk. paper). -- ISBN 3-7643-3917-9 (alk. paper)
1. Automatic control. 2. Stability. 3. Control theory.
I. Title
TJ213.V63 1998
629.8--dc21 98-181
CIP

Printed on acid-free paper

Birkhäuser 

ISBN 0-8176-3917-9
ISBN 3-7643-3917-9
Camera-ready text provided by the author in TEX.
Printed and bound by Hamilton Printing Company, Rensselear, NY.
Printed in the U.S.A.

9 8 7 6 5 4 3 2 1

Preface

Unlike the conventional research for the general theory of stability, this monograph deals with problems on stability and stabilization of dynamic systems with respect *not to all* but just to a *given part* of the variables characterizing these systems. Such problems are often referred to as the problems of *partial stability* (*stabilization*). They naturally arise in applications either from the requirement of proper performance of a system or in assessing system capability. In addition, a lot of actual (or desired) phenomena can be formulated in terms of these problems and be analyzed with these problems taken as the basis. The following multiaspect phenomena and problems can be indicated:

- "Lotka–Volterra ecological principle of extinction;"
- focusing and acceleration of particles in electromagnetic fields;
- "drift" of the gyroscope axis;
- stabilization of a spacecraft by specially arranged relative motion of rotors connected to it.

Also very effective is the approach to the problem of stability (stabilization) *with respect to all* the variables based on *preliminary* analysis of partial stability (stabilization).

A.M. Lyapunov, the founder of the modern theory of stability, was the first to formulate the problem of partial stability. Later, works by V.V. Rumyantsev drew the attention of many mathematicians and mechanicians around the world to this problem, which resulted in its being intensively worked out. The method of Lyapunov functions became the key investigative method which turned out to be very effective in analyzing both theoretic and applied problems.

The method of Lyapunov functions allows one to derive strict and easily interpretable conditions for partial stability and stabilizability in various practically important problems. However, the tasks accompanying constructive construction of Lyapunov functions are as yet imperfectly studied. This situation has awakened great interest both in the further elaboration of the method as regards weakening the requirements imposed on Lyapunov functions and in the development of constructive ways of their construction. It is also reasonable to extend other approaches to solving problems on partial stability and stabilization.

In this connection the monograph suggests a new method for solving the indicated problems. It is based on transformations of original systems of *ordinary* differential equations (in the case of controlled systems, combined with a special choice of a feedback). Such transformations make it possible to construct some auxiliary system that is more convenient for investigation. Then, the analysis of the constructed system for stability (stabilizability) with respect to all the variables allows us to judge whether the original system is stable (stabilizable) with resect to a given part of the variables. As a result, the two kinds of stability (with respect to all and to part of the variables) appear not only closely interrelated but also supplement each other in the course of investigation.

Though not a key to the solution of all the problems, this method makes it possible in many cases to obtain constructively verifiable conditions for partial stability and rules for design of stabilizing controls. Another advantage of this method is that it strengthens a number of familiar results and solves certain new problems in the sphere concerned. In addition, in the context of this approach new results can be obtained through the development of the method of Lyapunov functions itself.

The problems of *control with respect to part of the variables*, including *game-theoretic* control problems with respect to part of the variables in the presence of uncontrollable interference are to the same (and even greater) extent natural for the theory and applications. Scientifically and methodically, it is not devoid of interest to treat these problems together with the problems of partial stability and stabilization. The proposed approach of investigation is general enough to attain the overall objective: it can serve as a basis for systematic study (from unified grounds) of both the problems of partial stability and stabilization and the problems of control with respect to part of the variables.

This book is intended to achieve the objective outlined. It comprises the introduction that analyzes the state of the art in the theory of stability and control considered and seven chapters. The references include 398 entries.

The contents of the book can be divided into *three* parts.

The *first part* (Chapters 1–4, excluding Subsection 1.6, 4.5, and 4.6) deals with the development of this method and its applications to theoretic and applied problems of partial stability and stabilization.

This method made it possible to

- obtain the necessary and sufficient conditions for partial stability and stabilizability for linear systems with constant as well as sufficiently smooth periodic coefficients;
- expand the class of nonlinear systems for which the problem of partial stability is solved through the analysis for stability of a linear (specially constructed) system;
- introduce a new class of Lyapunov functions that extend the possibility of their usage in nonlinear problems of partial stability;
- reduce the solution of nonlinear problems of partial stabilization and, based on them, the nonlinear problems of stabilization with respect to all the variables to the solution of the simplest problems of stabilization for linear (specially constructed) systems.

The efficiency of the method is illustrated by solving a number of problems of controlling the angular motion of an asymmetric solid in various formulations as well as problems of stabilizing an artificial satellite in circular and geostationary orbits.

The *second part* deals with linear (Section 1.6) and nonlinear (Sections 4.5 and 4.6) problems of control with respect to part of the variables as well as with nonlinear *game-theoretic* problems of control with respect to part of the variables in the presence of *uncontrollable interference* (Chapter 5). Allowance is made for given constraints on controls. The solution of these problems is also reduced to the corresponding linear control problems (under the action of interference—to game-theoretic problems). In addition, using a preliminary solution of the problems of control with respect to part of the variables, we solve the problems of control (the game-theoretic ones included) with respect to *all* the variables.

Great attention is paid to solving nonlinear game-theoretic problems of reorientation of an asymmetric solid in the presence of uncontrollable interference posed in a variety of ways. Constructive solutions of these problems are, evidently, obtained for the first time in scientific literature. Results of computer simulation are cited that confirm the effectiveness of the proposed approach.

The *third part* of the book (Chapters 6 and 7) extends the devised method to the problems of partial stability and stabilization embracing more general classes of systems of differential equations:

- *functional-differential* equations (Chapter 6);
- *stochastic* equations (Chapter 7).

In addition, the method of Lyapunov functions and functionals is developed as applied to these problems.

Everywhere when possible we emphasize the distinguishing features and potentialities of both the problems of partial stability and stabilization and problems of control with respect to part of the variables. The theoretic line is accompanied with numerous illustrative examples whose objective is not only to work through the technique of the proposed approach in various situations a researcher encounters but also to demonstrate the peculiarities and potentialities mentioned.

The works appearing in the references can be divided into three groups connected with the problems of:

- partial stability and stabilization as well as control with respect to part of the variables;
- control of the angular motion of a solid;
- control in the presence of uncontrollable interference.

We also list some fundamental monographs concerning the general theories of stability, control, and differential games results of which are used in the book.

We should note that we managed to reflect adequately only those works which are concerned with the problem of partial stability and stabilization. As to other lines of research, mainly those works were cited which have much in common with the approach proposed in the book. For this reason a lot of outstanding publications and monographs also dealing with the issues discussed in the book were left unmentioned. So, I express my sincere regrets to the authors of the works which were not cited.

In the last forty years the problems of partial stability and stabilization were quite intensively developed in many scientific centers. We can expect that these problems as well as the problems of control with respect to part of the variables will remain in the focus of research and will gain in popularity. Naturally, it is necessary to formulate a kind of "*ideology*" for such problems.

In an effort to conceive such an ideology, I held mainly to my own vision of the subject which was molded in the scientific school of Prof. V.V. Rumyantsev. Being the first, this attempt cannot be devoid of drawbacks. Any remarks and suggestions regarding the essence of the issues considered in the book will be appreciated.

The monograph is based on my own results published in the last two decades. It is intended for a rather wide scientific and engineering community engaged in research involving differential equations, general theories of stability, control, and differential games, as well as for the students and postgraduates specializing in these areas.

The concept of the book has grown from conversations and discussions with Prof. V.V. Rumyantsev, to whom I am very much obliged for his valuable advice, support, and understanding. The work on material of the book concerning control problems was encouraged to a great extent by Prof. F.L. Chernous'ko, to whom I am thankful for his worthwhile advice and help. It is hard to overestimate the contribution made by Prof. L.D. Akulenko and Prof. V.P. Prokop'ev, who drew my attention to the problem of partial stability. Certain issues touched upon in the book were fruitfully discussed with Prof. V.V. Beletskii, Prof. V.V. Kozlov, Prof. A.A. Krasovskii, Prof. V.I. Maksimov, Prof. C. Risito, Prof. Yu.S. Osipov, Prof. V.I. Zubov, Dr. I.M. Anan'evskii, Dr. V.A. Feofanova, Dr. I.N. Isayev, Dr. S.Ya. Stepanov, and with the late Dr. A.S. Oziraner. Mr. S.V. Abramov has greatly assisted me with computer simulation. I express my gratitude to all these people.

The book incorporates certain results obtained with the financial support of the Russian Foundation for Fundamental Research, International Scientific Foundation, Russian Government, State Committee of the Russian Federation on Higher Education, and CNR (Italy). I am very appreciative to these funds for their aid.

This project has benefitted from the comments of Prof. C. Corduneaunu and has been promoted with the financial support of

- Birkhauser Boston;
- International Academy and Russian Academy of Engineering.

The author is deeply grateful to Prof. C. Corduneanu, these organizations, and Prof. E.G. Zudov, the head of the Department of Economics, Law, and Management of the Ural branch of Russian Academy of Engineering for the interest in his research and for the support rendered.

The author thanks Mrs. Svetlana V. Rizzo, Acquiring Editor, for her helpful efforts at the stage of launching the project and preparation of the book. Mr. Igor E. Merinov was engaged in translation and preparation of the camera-ready copy. He also made helpful comments and performed some calculations. Finally, the author sincerely thanks the staff of Birkhauser Boston for effective collaboration and professionalism.

Ural State Technical
University, Russia
February 1997

V.I. Vorotnikov

Contents

Introduction

Stability, Stabilization, and Control with Respect to Part of the Variables: Objectives and Methods

The introduction discusses approaches to research within the scope of problems of stability and stabilization of motion with respect to part of the variables (part of the coordinates of the phase vector of a dynamic system). The basic methods are analyzed, and the distinguishing features and possibilities of such problems are given much attention. A detailed analysis of the general situations and concrete problems that led to research in this area is presented. Also considered is the problem of control with respect to part of the variables in a finite time interval.

0.1 Preliminary Remarks

The foundations of the theory of stability of motion were laid down by the great Russian mathematician and mechanician Lyapunov [1892], [1893]. By now, having been developed in the works of many scientists, this theory is generally accepted throughout the world and is widely used in numerous fields of science and engineering. As treated by Lyapunov, the problem of stability of a solution (unperturbed motion) of an original system of ordinary differential equations is reduced to the problem of stability of the *zero equilibrium position* for another system—the *system of perturbed motion*. This reduction makes it possible, in the theory of Lyapunov stability, to consider a very general uniform problem of stability of the zero equilibrium position under most general assumptions concerning the right-hand side of the system under investigation, even though such a system is constructed anew for each concrete unperturbed motion.

The stability of the equilibrium position $x_1 = \cdots = x_n = 0$ of the perturbed-motion system is considered *with respect to all the variables* $x_1, \ldots, x_n$ determining the state of the system: for any $t \geq t_0$ the values of all the variables $x_i(t)$ $(i = 1, \ldots, n)$ must remain small, provided the initial values $x_i(t_0)$ of the very variables are chosen sufficiently small. The results of Lyapunov have been sufficiently completely elucidated in the literature, and there is no need to dwell on them in further detail here.

However, as was noted by Lyapunov [1893], a *more general* problem can be considered: the problem of stability of the same motion, *not with respect to all* of the quantities determining it, but only *with respect to some* of them. As applied to a perturbed-motion system, this would mean stability of the equilibrium position $x_1 = \cdots = x_n = 0$ relative only to some of the variables $x_1, \ldots, x_n$, rather than to all of them, for example, with respect to variables $x_1, \ldots, x_m$ $(m < n)$. In this case for all $t \geq t_0$ *only the values of variables* $x_i(t)$ $(i = 1, \ldots, m)$, and not the values of *all of the variables* $x_i(t)$ $(i = 1, \ldots, n)$, must remain small, provided the initial values $x_i(t_0)$ $(i = 1, \ldots, n)$ are chosen sufficiently small. However, Lyapunov himself did not investigate this problem.

Malkin [1938] later stated (without proof) certain conditions permitting Lyapunov's theorems to be extended to the case of stability with respect to part of the variables.

The fundamental results for the partial stability problem for *finite-dimensional* systems of ordinary differential equations with continuous right-hand sides were obtained by Rumyantsev [1957a]:

(1) the formulation of the problem of stability with respect to part of the variables was made more specific;

(2) the notion (crucial for this problem) of a function $V = V(t, x_1, \ldots, x_n)$ was introduced that is sign-definite relative to part of the variables, namely, with respect to variables $x_1, \ldots, x_m$ $(m < n)$;

(3) theorems on stability and asymptotic stability with respect to part of the variables were proved that generalize the classical theorems of Lyapunov;

(4) examples were cited that apply the results obtained to problems of mechanics.

The problem of stability of motion with respect to part of the variables, which is also frequently called the *partial stability* problem (or, briefly, the *PSt-problem*), was later intensively studied.

Peiffer and Rouche [1969] noted that the PSt-problem in the Lyapunov–Rumyantsev formulation generally *is not reduced* to problems of some kind of *stability of sets* (*trajectories*), which were thoroughly studied beginning with publications by Barbashin [1951] and Zubov [1957].* This result necessitates separate development of appropriate methods for investigating PSt-problems, in which interest is "heating up" by the needs of applications in very diverse fields. Naturally, after appropriate modification, such methods include previously known methods from Lyapunov's theory of stability. However, this path

*From here on, the references are, as a rule, arranged *chronologically*.

is fraught with difficulties of a general nature which can be overcome only by employing certain new ideas that appeared in a number of publications.

Also typical of many practically important situations is the PSt-problem *with respect to that part of the variables* of the original system (or the corresponding perturbed-motion system) *for which this system has an equilibrium position.* In such a case, for all $t \geq t_0$, the values of the specified variables must remain small, provided that their initial values are chosen sufficiently small. The indicated PSt-problem with rather general assumptions was analyzed by Zubov [1980], [1982], Khapaev [1986], and Rumyantsev and Oziraner [1987]. *"Partial" equilibrium positions* of this kind (also termed a *"balanced motion"*) are, as it were, *"indifferent"* to the remaining variables and are (assuming uniqueness of the solutions of the original system) *invariant sets* of such systems. As a result, it is actually the problem of stability of sets that is analyzed in this case. In a more general formulation of the problem, we obtain the PSt-problem *with respect to part of that (major) part of the variables relative to which the original system has an equilibrium position*—a PSt-problem of the Lyapunov–Rumyantsev type.

This type of PSt-problem includes the PSt-problem of *stationary motions of mechanical systems with cyclic coordinates* (Rumyantsev [1967], Rumyantsev and Oziraner [1987]).

All of the above noted general formulations of the PSt-problem appear so frequently in applications that they can be considered *typical.*

The formulation of the PSt-problem was mathematically meaningful and adequate for needs of applications, and further progress was made in the study of this problem. New general concepts of stability arose centered around the problem, among which are

(1) *stability with respect to two measures*, the property in which the closeness of the unperturbed and perturbed motions is estimated in terms of different measures at the initial time and at subsequent times (Movchan [1960]);
(2) *polystability* (*polystability with respect to part of the variables*), the property in which different groups of the variables (different groups of a given part of the variables) possess different types of stability (Aminov and Sirazetdinov [1987]);
(3) *stability with respect to a specified number of the variables*, with their composition not fixed (Vorotnikov [1986]).

In addition, the PSt-problem was extended to include the case of *controlled* systems (Rumyantsev [1970], [1971b]). This made it possible to pose the *problem of stabilization of motion with respect to part of the variables*, which is also referred to as the problem of *partial stabilization* (or, briefly, the *PSb-problem*).

The literature on the subject includes a number of surveys (Oziraner and Rumyantsev [1972], Rumyantsev [1972b], [1987], Hatvani [1979a], [1991], Muller [1982], Vorotnikov [1993]) and monographs (Rumyantsev and Oziraner [1987], Vorotnikov [1991a]) that provide analyses of numerous publications. In the present section, these publications, as well as others that are not mentioned as sources (some of the more recent publications included) are not only systematized in terms of research approaches, but are also discussed from the viewpoint of interpreting the scientific potential they have accumulated. In so doing, it is useful, from the standpoint of applications, to distinguish the part of the published research that "works," i.e., the part that has already found practical application. Also important is to alert one to the *pitfalls* along the path to practical application of other possibilities of PSt-problems that seem inviting at first sight.

The subsequent analysis is conceived as an attempt to meet the goal set. The following issues are addressed:

(1) general situations and specific problems leading to the investigation of PSt- and PSb-problems;
(2) formulation of PSt- and PSb-problems, stages, and approaches to their investigation;
(3) application of Lyapunov functions and vector functions method;
(4) the PSt-problem for linear systems, the PSt-problem in the linear approximation, and the PSt-problem in critical cases;
(5) distinguishing features and possibilities of the PSt-problem;
(6) problems of motion control with respect to part of the variables.

Of course, for reasons of space the analysis of the issues listed cannot be complete. Besides, the subsequent discussion itself cannot be devoid of faults common for digesting each stage of investigating any problem. However, the author hopes that the analysis of the issues involved will provide an insight into the subject of PSt-problems for a rather wide readership.

0.2 General Situations and Specific Problems Leading to Investigation of Problems of Stability and Stabilization with Respect to Part of the Variables

0.2.1 General Classification

A number of general situations and specific problems are presented in publications by Rumyantsev [1957a], [1972b], [1987], Ehricke [1962], Halanay [1963], Corduneanu [1964], Demin [1968], Furasov [1977], Rouche et al. [1977], Zubov [1980], Muller [1982], Phillis [1984], Rumyantsev and Oziraner [1987], Aminov and Sirazetdinov [1987], and Vorotnikov [1991a], [1993].

Summing up the approaches to research within the scope of the PSt-problem outlined and weighing the prospective ones, we shall attempt to distinguish the basic motives for investigating the problem. We consider it reasonable that the following general problems should be primarily numbered among them.

(1) Investigation of the stability of systems with so-called "*superfluous*" variables. This problem is closely related to the practically always inevitable

problem of finding "essential" variables for a dynamic system under investigation.

(2) Investigation of those systems whose PSt-property is *sufficient* for them to operate adequately.

(3) Analysis of situations in which the PSt-property is either *natural*, *desirable*, or even *necessary*.

(4) Investigation of the PSt-problem when a system is *inherently unstable* with respect to *all* the variables. *Assessment of capabilities* of a system being designed in "*non-nominal*" situations is often meant in this case.

(5) Solution of the PSt-problem when investigation of stability with respect to *all* the variables *presents difficulties* or when it is necessary to estimate (with respect to certain variables) the transient of a system stable with respect to all the variables.

(6) Solution of the PSt-problem as an *auxiliary* one. Here, stability with respect to certain variables often implies stability with respect to the remaining variables. In other cases, stability with respect to all the variables may turn out to be conveniently proved through successive investigation (possibly by different methods) of several PSt-problems.

Similar reasons also cause the investigation of the PSb-problem.

On these grounds a number of applied investigations of PSt- and PSb-problems for specific classes of systems were carried out which include, for instance:

(1) *mechanical holonomic and nonholonomic systems* (Rumyantsev [1957a], [1972a], [1972b], [1987], Risito [1970], Magnus [1971], [1980a], [1980b], Lilov [1972], Rumyantsev and Karapetyan [1976], Andreev [1979], [1987], [1991], [1996], Zubov [1980], [1982], Hagedorn [1982], Karapetyan [1981], Terjeki and Hatvani [1981], [1982], [1985], Muller[1982], Karapetyan and Rumyantsev [1983], Hatvani [1983a], [1984], [1985a], [1985b], [1991], Hatvani and Terjeki [1985], Rumyantsev and Oziraner [1987], Smirnov and Ermolina [1988], Kozlov [1991a], Vorotnikov [1991a], Zhu Hai-ping and Mei Fang-xiang [1995]);

(2) *complex mechanical systems*, including models of spacecraft, artificial satellites, and orbital complexes with allowance for elastic elements and tanks filled with liquid (Anchev and Rumyantsev [1979], Rubanovskii and Rumyantsev [1979], Rubanovskii [1982a], [1982b], Nabiullin [1990], Kholostova [1992], and Anapolskii and Chaikin [1993]);

(3) systems describing the dynamics of a *controlled solid*, i.e., *spacecraft* (Rumyantsev [1967], Zubov [1975], [1980], [1982], Krementulo [1977], Vorotnikov [1985], [1986a]–[1986c], [1988b], [1990], [1991a], [1991b], [1997b], [1997d], Rumyantsev and Oziraner [1987]), including systems describing the dynamics of a solid under the action of *uncontrollable interference* (Vorotnikov [1994a], [1994c], [1995c], [1997c], [1997e]);

(4) systems that describe *focusing and acceleration of particles in electromagnetic fields* and the *dynamics of particle collisions* (Zubov [1980], [1982]);

(5) *Lotka–Volterra ecological systems* (Rouche et al. [1977]);

(6) certain classes of *adaptive self-tuning systems* and *parametric identification systems* (Furasov [1977], Petrov et al. [1980], Kolmanovskii and Nosov [1981]).

We should also note cases in which the PSt-problem served as a basis for investigating problems of *celestial mechanics* (Demin [1964], [1968], Khapaev [1986]), stability of a *regular vortex n-gon* (Kurakin [1994]), convergence of certain *optimization algorithms* (Nosov and Furasov [1979]), and *approximate synthesis of optimal motion control with respect to part of the variables* (Akulenko [1980], [1987], [1994a]).

We shall consider these problems in more detail, focusing attention on research that has already found definite practical application. We note that in a number of cases, the examples given below highlight various aspects of the indicated motives for research of PSt- and PSb-problems; such examples, once discussed, may appear elsewhere as appropriate.

0.2.2 Analysis of PSt- and PSb-Problems for Systems with "Superfluous" Variables

When constructing mathematical models of real phenomena or when forming systems of differential equations of perturbed motion in *Lyapunov's problem of stability with respect to certain given functions of state*, one often uses the so-called "superfluous" variables in addition to the "basic" variables. As a rule, the behavior of the "superfluous" variables is of no interest to the researcher. In such cases it is *justifiable* and even *reasonable* to pose a PSt-problem *relative to the "basic" phase variables*, although in such cases we should bear in mind essential limitations associated with "noncoarseness" of such a problem (this is considered in more detail in Section 0.6).

One also encounters a similar situation when compiling systems of differential equations of perturbed motion for *controlled systems.*

A material example of this situation is provided by the problem of analyzing *stability "in output"* in control systems, which is a fundamental problem in modern applied automatic control theory. In solving these problems, it may be advantageous to employ methods for analyzing PSt-problem (appropriately specifying the notion of stability), which find application here to the extent they have been developed (see, for example, Furasov [1982]).

At the same time, given functions of state (given functions of the phase coordinates of a system) or "output coordinates" of a system *may not be independent* functions in the corresponding domain of definition and cannot always be treated as independent variables (Kozlov [1991b]). In this regard, the problem of stability with respect to given functions of state is more general than the PSt-problem. However, even in the case indicated, the scientific potential of the PSt-approach may be useful for dealing with some aspects of the problem indicated.

0.2.3 Cases in Which Analysis of PSt- and PSb-Problems is Sufficient

A mechanical system consisting of a *solid (spacecraft) with cavities (tanks) partially or completely filled with liquid* has an infinite number of degrees of freedom. Its motion is described by a system of simultaneous ordinary and partial differential equations. The problem of the stability of motion of such a system presents considerable difficulty. However, one can avoid these difficulties to a certain extent by stating the problem somewhat differently. Although, of course, the motion of the body depends on the motion of the liquid and vice versa, in applied problems we are mostly interested in the stability of a solid. The problem of the stability of the liquid's motion is of interest only to the extent that it influences the stability of the solid. As a result, Rumyantsev posed and solved the following PSt-problem for a number of principal cases (see Moiseev and Rumyantsev [1965]): analyze the stability of the mechanical system under consideration with respect to the variables determining the motion of the solid and with respect to certain quantities that *integrally* characterize the liquid's motion. Here it is important that the stability of the system with respect to the indicated (*finite*) number of variables ensures the stability of the solid's motion.

The PSt-problem of the stability of motion of a *solid with elastic components (elastic components and liquid-filled cavities)* is treated analogously, and so is the problem of a *solid with an elastic hull containing liquid.* The reader can get familiar with this research in the surveys by Rubanovskii and Rumyantsev [1979], Rubanovskii [1982a] and the monograph by Nabiullin [1990].

Also frequently considered in spacecraft dynamics are the models of a solid with *viscoelastic* elements for which problems of investigating stationary motions and their stability are relevant. As compared to the conditions of the stability of a solid, the domain of stability of the model indicated shrinks to transfer to the domain of *asymptotic* stability with respect to part of the variables of the corresponding system (Kholostova [1992]). We note that the emergence of asymptotic stability with respect to part of the variables is also characteristic of solids with cavities filled with highly viscous liquid (Rumyantsev [1967]).*

Developing the Lyapunov–Rumyantsev PSt-problem (Rumyantsev [1957a]) and the researches by Zubov [1957] on stability of sets, Movchan [1960] suggested the formulation of the problem of *stability with respect to two measures* for *distributed systems* (in particular, for *elastic systems*). This problem was elaborated in the work of many scientists and permitted solving a number of complicated practical problems. Further wide usage and development of the approaches of Rumyantsev, Zubov, and Movchan can be expected in numerous problems involving analysis and design of various complex systems, which are finding ever widening application.

*Asymptotic stability with respect to part of the variables in the absence of active external dissipative forces is also an essential feature of the problem on stability of permanent rotations of a solid in an absolutely rough plane (Karapetyan [1981], Markeev [1992]).

For many important classes of mechanical systems with a *finite* number of degrees of freedom, it is also sufficient to ensure stability or asymptotic stability with respect to only part of the variables.

For example, this is the case with *stabilizing a solid (artificial satellite) in a circular orbit.* Together with the so-called "restricted formulation" of the problem in which only perturbations of the satellite itself in the orbit are taken into account (orbital perturbations excluded), we consider a more general formulation which allows for orbital perturbations.

In this case, for the aggregate system of differential equations of perturbed motion of a satellite, we first solve the problem of stabilization with respect to the variables that determine its position in an orbital coordinate system. This is done by considering a "*truncated*" controlled system (which is obtained from the original aggregate system by setting the variables that are not being controlled in this stage of the solution equal to zero). Then we use the Lyapunov–Malkin theorem to prove that the stabilization carried out actually ensures not only asymptotic stability with respect to the specified part of the variables, but also (nonasymptotic) stability with respect to all the variables of the unperturbed motion under study for the aggregate system. The stabilization of this type guarantees a satellite's proper functioning in orbit.

0.2.4 Cases in Which Analysis of PSt- and PSb-Problems is Natural, Desirable, or Even Necessary

One of these situations has drawn the attention of Ehricke [1962], a well-known spaceflight specialist. The point is that to decrease the structural load and increase the mass ratio, it is *desirable* (*beneficial*) that there be a weak instability in some planes on certain parts of spacecraft trajectories. In other words, it is desirable for the spacecraft to be stable with respect to part of the variables.

An example of another applied problem in which analysis within the scope of PSt-problem is by all means *necessary* is a *simulation study of focusing and acceleration of particles in electromagnetic fields.* An approach to solving this problem was proposed by Zubov [1980], [1982]. Here the necessity of the PSt-property (being made appropriately precise) is *dictated by the very nature of the process* of focusing and accelerating particles, since, for particles located in a manifold in the space of coordinates $Ox_1x_2x_3$ and velocities at the initial time which remain in this manifold in the sequel, we require, over the course of time (with focusing along the Ox_1-axis, for example), the relationships $x_1 \to \infty$, $x_2 \to 0$, $x_3 \to 0$. In the case indicated we have the PSt-problem of *conditional*[*] stability with respect to x_2, x_3 for the balanced motion $x_2 = x_3 = 0$ of the system modeling the phenomenon in question.

[*]The situation of *conditional* stability with respect to part of the variables also arises in the study of *inertial navigation systems* (see Sinitsyn, V.A. [1991] On Stability of Solution in Inertial Navigation Problem, in *Certain Problems on Dynamics of Mechanical Systems*, 46–50, Moscow: Izd. MAI (in Russian)).

Some research is devoted to the study of the joint motion of two mechanical systems, one of which is a *carrying* (*controlled*) system and the other is a system *being carried* (*controlling* system). The technique is given for determining laws of variation for variables of the controlling system which ensure stable motion of the carrying system in a specified mode. Here we have partial stability of the entire system with respect to the variables characterizing the motion of the "carrier." At the same time, stability of the entire system with respect to all the phase variables is not only undesirable, but also inadvisable.

Thus, for example, the stabilization of an equilibrium position of a solid can be ensured by flywheels attached to it which are controlled by special engines. The flywheels "take up" the perturbations of the angular momentum of the system and, when a stabilization session is completed, keep on rotating by inertia; as a result, stabilization of the initial state of the mechanical system under study (gyrostat) is carried out only with respect to part of the variables, those determining the gyrostat's main body.

The problem of stabilizing an equilibrium position of a solid by a gimballed gyroscope rests on the same idea. However, as distinct from stabilization with flywheels, in this case one must not only ensure asymptotic stability of an equilibrium position of the main body of the system under consideration, but also guarantee (nonasymptotic) stability of the axis of the gyroscope relative to the body. (For more details, see Section 4.3.)

The PSt-problem of asymptotic stability can also be used in designing *dynamic dampers* whose application sphere is rather wide. In this case the damper's motion results in damping oscillations of the main body (or the components required) of the construction in question; in other words, the "entire" system must admit an asymptotically stable "partial" (with respect to the variables determining the dynamics of the main body) equilibrium position.

However, it would be erroneous to assume that only mechanics and engineering problems stimulate the progress of the PSt-problem. Thus, for example, the monograph by Rouche et al. [1977] analyzes a number of problems that appear in *biocenose dynamics* in the framework of PSt-problem. Here the statement of the PSt-problem is *natural* and reflects the very essence of a series of processes occurring in such systems. In particular, with the use of theorems on the method of Lyapunov functions modified for the PSt-problem, certain phenomena appearing in the context of the "*Lotka–Volterra ecological principle of extinction*" are analyzed.

It is clear from the examples given in this subsection that the essence of the PSt-notion is that it provides an appropriate mathematical formulation of a "*measure of robustness*" of a process under study (or an equilibrium position) and also the possibility of giving a rigorous mathematical formulation of some phenomena that appear in dynamic systems. In this regard, the sense-aspects of the PSt-problem, compared with the problem of stability with respect to all the variables, are *more diversified* and, in general, the PSt-problem can be interpreted as a problem of analyzing systems with phase variables having *various* behavior.

0.2.5 PSt-Problem and Assessment of System Capabilities (Analysis of Systems that are Inherently Lyapunov Unstable)

The possibility of "partial" equilibrium positions that are "indifferent" to the remaining variables is characteristic of various mechanical systems with a finite number of degrees of freedom. It is sufficient to point to holonomic systems with cyclic and quasi-cyclic coordinates and, more specifically, to the situations that occur in studies of the motion of the *gimballed gyroscope* (Magnus [1955], [1971], Khapaev [1986], Andreev [1991]). In such cases, it is reasonable to investigate the corresponding PSt-problems *to determine the possible operating modes of a system.*

The formulation of the issue presented is closely related to the practically frequent need to *assess the capabilities* of a system in emergency ("nonnominal") situations. One of these situations is associated with dampening of the angular rotation of a solid—spacecraft (Vorotnikov [1986a], [1991a]) and is considered in Section 2.3.

In some cases the system can be stable with respect to part of the variables, but the stability with respect to the remaining part of the variables is physically impossible. Thus, for example, in the system describing the *dynamics of a microwave oscillator*, a phenomenon of "phase drift" exists, and this system is partially stable because "complete stability" is physically impossible (see Phillis [1984]).

Also frequently encountered in applications is another example of investigating the PSt-problem for a system which is inherently unstable with respect to all the variables. This is a system for *direct automatic control in the Lur'e form* (Lur'e [1951]) which we consider here for the critical case of two zero roots:

$$\begin{gathered}\dot{\boldsymbol{y}} = A\boldsymbol{y} + \boldsymbol{b}f(\sigma), \quad \sigma = \boldsymbol{c}^{\mathsf{T}}\boldsymbol{y} + \beta_1 z_1 + \beta_2 z_2, \\ \dot{z}_1 = \gamma_1 f(\sigma), \quad \dot{z}_2 = \gamma_2 f(\sigma), \\ f(0) = 0, \quad \sigma f(\sigma) > 0 \quad (\sigma \neq 0).\end{gathered} \tag{0.2.1}$$

Here $\boldsymbol{y} \in R^m$, $m \geq 1$, $z_i \in R^1$ $(i = 1,2)$ are the phase variables, A, $\boldsymbol{b}$, $\boldsymbol{c}$ are a constant matrix and vectors of appropriate sizes, γ_i, β_i $(i = 1,2)$ are constants, and $f(\sigma)$ is the nonlinear characteristic of a servomotor.

The inherent absence of *absolute stability* (global asymptotic stability) with respect to all the variables of the equilibrium position $\boldsymbol{y} = \boldsymbol{0}$, $z_i = 0$ $(i = 1,2)$ of system (0.2.1) is connected with the fact that the equation $\sigma = 0$ can hold at points of the form $\boldsymbol{y} = 0$, $z_i = z_i^*$ $(i = 1,2)$, where z_i^* satisfy the condition $\beta_1 z_1^* + \beta_2 z_2^* = 0$; as a result, points of the indicated form are nonzero equilibrium positions of system (0.2.1).

Under these conditions the PSt-problem of absolute stability of system (0.2.1) is considered with respect to the variables that appear in vector $\boldsymbol{y}$ (Vorotnikov [1979a], [1979b], [1991a]). (Regarding this issue, see Section 2.2.)

From the standpoint of applications, an interesting and important class of mechanical systems that admits no asymptotic stability with respect to all the variables is the *class of "asymptotically stopping" systems.* The study of these systems was initiated by the work of Lyapunov [1893], Merkin, and

Matrosov.* Along the motions of these systems, their generalized coordinates (or some of them) approach constants, and the generalized velocities tend to zero. (Asymptotic stability and asymptotic stability with respect to part of the variables are special cases of this phenomenon: the generalized coordinates (or some of them) tend to the values they attain at an equilibrium position.) Terjeki and Hatvani [1981] used a modification of the method of Lyapunov functions to obtain conditions sufficient for the zero solution of a system of ordinary differential equations to possess the PSt-property and for the stable variables to converge to constants as time tends to infinity. These conditions were further developed by Terjeki [1983b].

Consider, for example, the motion of a *mathematical pendulum of constant length* under the action of a damping force $F = \alpha(t)\dot{\varphi}$ that is proportional to the velocity. This motion is described by the equation

$$\ddot{\varphi} + a(t)\dot{\varphi} + b\sin\varphi = 0, \quad a = \alpha/m, \quad b = g/l$$

(m is mass, g is the acceleration of gravity, l is the length of the pendulum).

It is known that in the case $\alpha = \text{const} > 0$, the equilibrium position $\varphi = \dot{\varphi} = 0$ of the pendulum is asymptotically stable with respect to φ, $\dot{\varphi}$. However, in the case of a positive but *rapidly increasing* damping factor, the pendulum *may* "*deviate*" from the equilibrium position. Thus, for example, Hale [1969] showed that the equilibrium position $\varphi = \dot{\varphi} = 0$ of the equation $\ddot{\varphi} + (2 + e^t)\dot{\varphi} + \sin\varphi = 0$ is not asymptotically stable. Ballieu and Peiffer [1978] continued the analysis in this direction and showed that a similar result holds for the case $a(t) = (1+t)^\alpha$, $\alpha > 1$; in this case, along each motion, the function $\varphi(t)$ approaches a finite, possibly nonzero, limit (Terjeki [1983b]), and the velocity tends to zero (Hatvani [1984]).

0.2.6 Analysis of PSt- and PSb-Problems when Investigation of Stability with Respect to All the Variables is Difficult

In certain cases, the design of automatic control systems can, in principle, be based on either a PSt-problem or a problem of stability with respect to all the variables (it is the designer who makes the final decision).

Thus, for example, the goal of the *design of a self-tuning system* (*STS*) *with a reference model* is construction of such a self-tuning circuit which ensures that some performance criterion of the material object is close to that of the reference model. The most simple way in this situation is to require stability of the STS with a reference model relative only to the "*phase mismatch*". This problem is solved (Petrov et al. [1980], Kolmanovskii and Nosov [1981]) as a PSt-problem of the Lyapunov stability and simultaneous asymptotic stability of the zero solution of the corresponding perturbed motion system with respect to one group of variables (characterizing the indicated

---

*Matrosov, V.M. [1959] On Stability of Gyroscopic Systems with Dissipation, *Trudy Kazan. Aviatsion. Inst.*, **45**, 63–76; Matrosov, V.M. [1959] On Stability of Gyroscopic Systems, *Trudy Kazan. Aviatsion. Inst.*, **49**, 3–24; Merkin, D.R. [1974] *Gyroscopic Systems*, 2nd ed., Moscow: Nauka, (1st ed.—1956, both in Russian).

phase mismatch). At the same time, a stronger requirement is that of asymptotic stability not only with respect to phase mismatch, but with respect to "*parametric mismatch*" between the object and the model, which is a more complicated problem of asymptotic stability with respect to all the variables.

Consider, for example, the problem of constructing a self-tuning controller for stabilizing an object with undetermined parameters whose dynamics are described by the linear system of ordinary differential equations

$$\dot{\boldsymbol{y}} = A\boldsymbol{y} + B(\boldsymbol{u} + C\boldsymbol{y}). \tag{0.2.2}$$

Here $\boldsymbol{y}$ is the vector of the phase coordinates of the system relative to which the stabilization problem is solved, $\boldsymbol{u}$ is the vector of controls, A and B are constant matrices determining the known parameters of the object, and C is a matrix of unknown parameters, which, for simplicity, we assume to be constant.

Let $\boldsymbol{u}^* = K\boldsymbol{y}$ (K is a constant matrix) be linear control rules solving the stabilization problem for the system $\dot{\boldsymbol{y}} = A\boldsymbol{y} + B\boldsymbol{u}$. Here, as is known, for any positive-definite matrix Q there is a corresponding unique positive-definite matrix L that satisfies the Lyapunov matrix equation. Using principles of design for self-tuning systems (Petrov et al. [1980]), we define the structure of the self-tuning controller as follows:

$$\boldsymbol{u} = (K + \Delta K)\boldsymbol{y}, \quad \frac{d}{dt}(\Delta K) = \Theta(\boldsymbol{y}). \tag{0.2.3}$$

Thus we obtain *nonlinear* system (0.2.2), (0.2.3). Using a V-function of the form (Petrov et al. [1980])

$$V = \alpha \boldsymbol{y}^{\mathsf{T}} L \boldsymbol{y} + \operatorname{tr}(Z^{\mathsf{T}} Z), \tag{0.2.4}$$

where $Z = C - \Delta K$, $\alpha = \text{const} > 0$, and $\operatorname{tr}(Z^{\mathsf{T}} Z)$ is the trace of the matrix $Z^{\mathsf{T}} Z$, we construct algorithms $\Theta(\boldsymbol{y})$ for tuning coefficients ΔK of the controller that compensates for the effects of unknown parameters on the dynamical properties of closed system (0.2.2), (0.2.3) so that the derivative $\dot{V}$ of function V by virtue of system (0.2.2), (0.2.3) has the form $\dot{V} = -\alpha \boldsymbol{y}^{\mathsf{T}} Q \boldsymbol{y}$.

Analysis (Petrov et al. [1980]) of stability of the zero solution $\boldsymbol{y} = \boldsymbol{0}$, $\Delta K = 0$ of *nonlinear* system (0.2.2), (0.2.3) shows that this solution is not only Lyapunov stable, but globally asymptotically $\boldsymbol{y}$-stable. It is this fact that ensures stabilization of the original system relative to the phase variables $\boldsymbol{y}$ in the class of self-tuning controllers. We note that V-function (0.2.4) thus chosen satisfies the conditions of a Marachkov-type theorem on asymptotic $\boldsymbol{y}$-stability (Peiffer and Rouche [1969]).

Also naturally appearing in the theory of STS is the PSt-problem in the presence of constantly acting perturbations; this occurs in analyzing stability with respect to the phase mismatch of nonstationary objects (in contrast to objects with constant but varying, within definite limits, parameters) (Furasov [1977], Petrov et al. [1980], Kolmanovskii and Nosov [1981]).

Also important from the standpoint of applications are examples of multidimensional dynamic systems, where, under unpredictable operating conditions and with control of all the variables being difficult, we *require* asymptotic stability with respect to *some part* of the variables, while stabilization with respect to their *remaining* part *is to be carried out.* We use this approach to achieve ultimate asymptotic stability with respect to all the variables. The situation indicated appears, for example, in the *dynamics of aircraft* (see Haker [1960], Halanay [1963]). This approach may turn out to be very effective for supporting normal operation of objects and deserves thorough study in the general context of analyzing (synthesizing) multidimensional systems.

When synthesizing trajectories for different kinds of aircraft, one is frequently faced with the *necessity of imposing different requirements on different groups of variables* (the problem of *polystability of motion*).

Thus, for example (Aminov and Sirazetdinov [1987]), in considering *spatial maneuvers of aircraft* with a constant load factor, it is important to ensure asymptotic stability with respect to angles of attack and glide, whereas the angular velocities of pitch, yaw, and roll need only be uniformly stable. Here it is not necessary to have stable behavior relative to the angles of pitch, yaw, and roll.

Often it is possible to use available information on the properties of solutions of an original system (an example of which is knowledge of some of its first integrals) to study a PSt-problem with respect to some part of the variables characterizing the system. Then the information thus obtained can be used in a different way:

(1) as a *condition for one of possible operating modes of the system under study* (this problem must be solved separately by the designer in each particular case);

(2) as *auxiliary information for further analysis* of stability with respect to the remaining variables or for analysis of other properties of the system.

Let us consider both of these aspects of PSt-problems in more detail.

1. As one example of the first of the situations indicated, we note Rumyantsev's research [1956], [1957b] on *stability of permanent rotations of a solid.* Construction of the corresponding Lyapunov function with Chetayev's method of bundle of first integrals was used to demonstrate that, for arbitrary parameters of the body, its permanent rotations in the Lagrange case are stable with respect to the angle of nutation and to the projection of the angular velocity vector onto the axis of dynamic symmetry.

This result can be treated as a condition for an admissible operating mode of the permanent rotation of the solid. At the same time, by restricting the domain of variation of the object parameters, it is possible, with a more complete analysis, to find conditions for stability of permanent rotations with respect to all the variables.

Another related example is Rumyantsev's [1957a] generalization of the classical *Lagrange–Dirichlet theorem* on the stability of an equilibrium position of a conservative holonomic mechanical system. In Lagrangian coordinates, the motion of this system is described by the equations

$$\frac{d}{dt}\left(\frac{\partial T}{\partial \dot{q}_i}\right) - \frac{\partial T}{\partial q_i} = -\frac{\partial P}{\partial q_i} \qquad (i = 1, \ldots, n),$$

where T is the kinetic energy, which, in the case of nonstationary constraints, has the form $T = T_2 + T_1 + T_0$ (T_s is a form of degree s in $\dot{q}_i$), and $P(\boldsymbol{q})$ is the potential energy, $P(\mathbf{0}) = 0$.

Suppose that $\boldsymbol{q} = \dot{\boldsymbol{q}} = \mathbf{0}$ is an equilibrium position of the system and T does not explicitly depend on time. In this case we have the generalized energy integral $H = T = T_2 - T_0 + P = \text{const}$.

If T_2 is positive-definite with respect to part of the generalized velocities $\dot{q}_r$ $(r = 1, \ldots, k < n)$, and $P - T_0$ is positive-definite with respect to part of the generalized coordinates q_s $(s = 1, \ldots, m < n)$, then the equilibrium position $\boldsymbol{q} = \dot{\boldsymbol{q}} = \mathbf{0}$ of the system in question is stable with respect to part of the variables q_s, $\dot{q}_r$ $(s = 1, \ldots, m;\ r = 1, \ldots, k)$.

2. As an example of the second of the situations indicated, consider a mechanical system whose motion is described by a nonlinear finite-dimensional system of an ordinary differential equation (in the vector form)

$$\ddot{\boldsymbol{x}} = \boldsymbol{X}(t, \boldsymbol{x}, \dot{\boldsymbol{x}}). \tag{0.2.5}$$

Suppose that we have somehow established *exponential* asymptotic stability of the unperturbed motion $\boldsymbol{x} = \dot{\boldsymbol{x}} = \mathbf{0}$ of system (0.2.5) with respect to $\dot{\boldsymbol{x}}$ (*with respect to velocities*). Because the asymptotic stability with respect to $\dot{\boldsymbol{x}}$ is exponential, we can automatically use this fact to conclude that the unperturbed motion $\boldsymbol{x} = \dot{\boldsymbol{x}} = \mathbf{0}$ is (*nonasymptotically*) stable *with respect to* $\boldsymbol{x}$ (*with respect to coordinates*). In particular, when the forces acting on the system depend only on velocities (i.e., when $\boldsymbol{X}(t, \boldsymbol{x}, \dot{\boldsymbol{x}}) \equiv \boldsymbol{X}(t, \dot{\boldsymbol{x}})$), analysis of the exponential asymptotic stability of the unperturbed motion of system (0.2.5) *with respect to* $\dot{\boldsymbol{x}}$ reduces to analysis of the exponential asymptotic stability of the zero solution $\boldsymbol{w} = \mathbf{0}$ of the system $\dot{\boldsymbol{w}} = \boldsymbol{X}(t, \boldsymbol{w})$ *with respect to all the variables.*

Let us dwell on some aspects of the auxiliary function of PSt- and PSb-problems, other than those already mentioned.

0.2.7 Auxiliary Function of PSt- and PSb-Problems

This function of the PSt(PSb)-problem is important in studies of both theoretical and applied problems of system analysis and synthesis.

For example, the PSt-problem holds an appreciable position among the contemporary analytic tools of the method of Lyapunov functions. Thus, in the *method of Lyapunov vector functions*, the fact of stability of the zero solution of the comparison system with respect to part of the variables was employed (Bellman [1962], Matrosov [1962b]) to draw a conclusion about Lyapunov stability or stability with respect to part of the variables of the unperturbed motion of the original system.

In the automatic control theory we note *Pyatnitskii's method* [1970] of using an auxiliary PSt-problem for analyzing absolute stability with respect to all the variables of nonlinear nonstationary two-dimensional systems in the Lur'e form. The approach is based on the use of a variational method to construct a particular auxiliary limiting system which is the worst in the sense of being closest to the boundary of the domain of absolute stability. The asymptotic stability with respect to part of the variables of this limiting system implies absolute stability with respect to all the variables of the original system. In the work of Pyatnitskii and Rapoport [1991], this approach was extended to more general cases.

A *stepwise strategy* is often used to evade difficulties that accompany solving numerous control problems. Here PSt- and PSb-problems naturally appear in individual stages.

As an example, consider the problem of *damping of rotations of an asymmetric spacecraft with a single flywheel* having its axis of rotation fixed along the major axis of inertia ("Dual Spinner" of the Intelsat VI type, etc.). The following stepwise technique was suggested for solving this problem (Guelman [1989]):

(1) the (*variable*) controlling moment applied to the flywheel is first constructed to bring the system into one of the stable equilibrium states of the spacecraft;

(2) then, to attain the objective of control, the complete dampening of rotations, it suffices to apply a *constant* (in magnitude) torque.

Here the analysis of the first stage of control is carried out with the Lyapunov functions method by analyzing the corresponding PSb-problem.

In investigating various applied problems, stability (stabilization) with respect to one part of the variables often implies stability (stabilization) with respect to either their major part or even to all the remaining variables. In particular, this idea was employed to devise a constructive scheme (Vorotnikov [1985], [1986a]–[1986c], [1988b], [1990], [1991a], [1991b], [1994a], [1994c], [1995c], [1997b]–[1997e]) for solving numerous problems of controlling the angular motion of a solid (spacecraft, artificial satellite, manipulator), which will be considered later in this book.

0.3 Formulation of the Problems of Stability and Stabilization with Respect to Part of the Variables. Lines and Stages of Their Research

0.3.1 General Classification of PSt-Problems

As was already noted, the following basic types of PSt-problems are presently considered:

(1) the Lyapunov–Rumyantsev PSt-problem;

(2) the problem of stability of "partial" equilibrium positions (with respect to all the variables);

(3) the PSt-problem of stability of "partial" equilibrium positions;

(4) the problem (PSt-problem) of polystability;

(5) the problem of stability with respect to a specified number of the variables.

0.3.2 Class of Systems Dealt with in the Theory of Partial Stability

Let there be given a nonlinear system of ordinary differential *equations of perturbed motion* (Lyapunov [1892])

$$\begin{gathered}\dot{\boldsymbol{x}} = \boldsymbol{X}(t, \boldsymbol{x}), \qquad \boldsymbol{X}(t, \boldsymbol{0}) \equiv \boldsymbol{0}, \\ \boldsymbol{x}^{\mathsf{T}} = (y_1, \dots, y_m, z_1, \dots, z_p) = (\boldsymbol{y}^{\mathsf{T}}, \boldsymbol{z}^{\mathsf{T}}), \\ m > 0, \quad p \geq 0, \quad n = m + p,\end{gathered} \tag{0.3.1}$$

where "T" is the symbol of transposition.*

The variables constituting the phase vector $\boldsymbol{x}$ of system (0.3.1) are divided into two groups:

(1) the variables $y_1, \ldots, y_m$ with respect to which the stability of the *unperturbed motion* $\boldsymbol{x} = \mathbf{0}$ is to be investigated;

(2) the remaining variables $z_1, \ldots, z_p$.

Specifically, this partitioning depends on the nature of the problem under study. As a rule, we assume that the choice of variables $y_1, \ldots, y_m$ *has already been made* by the time that a PSt-problem should be analyzed. This means that the PSt-problem is a problem of stability with respect to a *prescribed* part of the variables ("*with respect to named coordinates*," in the terminology of Zubov [1959]). Variables $z_1, \ldots, z_p$ are correspondingly called the "*uncontrollable*" variables.

The behavior of variables $z_1, \ldots, z_p$ of system (0.3.1) is, in principle, of no interest in the study of a PSt-problem. However, the dynamics of the "basic" variables $y_1, \ldots, y_m$, *to a great extent*, are related to the dynamics of variables $z_1, \ldots, z_p$. As a result, in one way or another the analysis of a PSt-problem will principally require a definite analysis of the behavior of all the variables of system (0.3.1). Of course, the specifics of such a "complete" analysis stem from the desire to study only "partial" properties of the system.

We denote by $\boldsymbol{x}(t) = \boldsymbol{x}(t; t_0, \boldsymbol{x}_0)$ the solution of system (0.3.1) subject to the initial conditions $\boldsymbol{x}_0 = \boldsymbol{x}(t_0; t_0, \boldsymbol{x}_0)$.

In the theory of stability with respect to part of the variables, the following assumptions are usually made (Rumyantsev and Oziraner [1987]):

(*a*) the right-hand sides of system (0.3.1) in the domain

$$t \geq 0, \quad \|\boldsymbol{y}\| \leq H, \quad \|\boldsymbol{z}\| < \infty,$$

$$\|\boldsymbol{y}\| = \Big(\sum_{i=1}^{m} y_i^2\Big)^{1/2}, \quad \|\boldsymbol{z}\| = \Big(\sum_{j=1}^{p} z_j^2\Big)^{1/2}, \quad \|\boldsymbol{x}\| = \big(\|\boldsymbol{y}\|^2 + \|\boldsymbol{z}\|^2\big)^{1/2} \tag{0.3.2}$$

are *continuous* and satisfy *conditions of uniqueness* of solutions (for example, the local *Lipschitz condition*).

(*b*) solutions of system (0.3.1) are *z-continuable* (Corduneanu [1964]), i.e., any solution $\boldsymbol{x}(t)$ is defined for all $t \geq 0$ for which $\|\boldsymbol{y}(t)\| \leq H$. From a practical viewpoint, $\boldsymbol{z}$-continuability of solutions of system (0.3.1) means that the $\boldsymbol{z}$-component of any solution $\boldsymbol{x} = \boldsymbol{x}(t; t_0, \boldsymbol{x}_0)$ of this system that originates from a sufficiently small neighborhood of the unperturbed motion cannot "*become infinite*" in a finite time.

It should be stressed that in the PSt-problem (as differed from the problem of stability with respect to all the variables), $\boldsymbol{z}$-continuability of solutions of system (0.3.1) *is not*, generally speaking, a consequence of $\boldsymbol{y}$-stability of its unperturbed motion $\boldsymbol{x} = \mathbf{0}$. We should bear this circumstance in mind when posing and investigating a PSt-problem. In particular, $\boldsymbol{y}$-stability implies $\boldsymbol{z}$-continuability if a

*From here on, except for Sections 4.2, 4.5, 5.1, and 5.3, we define vectors $\boldsymbol{x}$, $\boldsymbol{y}$, $\boldsymbol{z}$ as *column* vectors of the corresponding variables, which is required for the subsequent representation of systems (0.5.1), (1.1.1), etc. in the vector form.

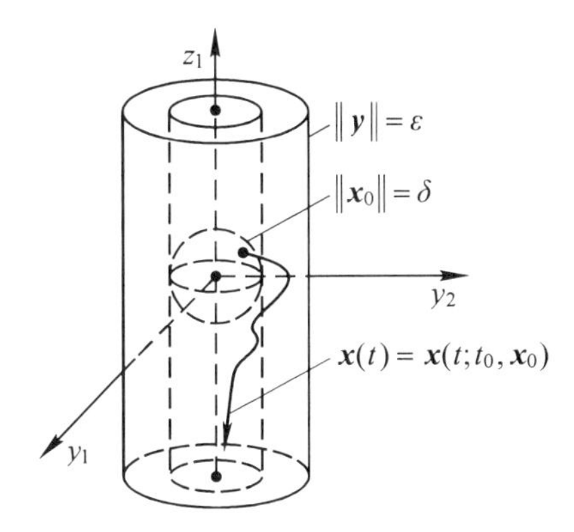

Figure 0.3.1. Geometric interpretation of Definition 0.3.1, where $m = 2$, $p = 1$, $\lim\|\boldsymbol{y}(t; t_0, \boldsymbol{x}_0)\| = 0$, as $t \to \infty$.

PSt-problem is considered in conjunction with the problem on *boundedness* of the corresponding solutions of system (0.3.1) with respect to all the variables. This approach is of importance in terms of "robustness" of the unperturbed motion being studied.

0.3.3 Definitions of Stability with Respect to Part of the Variables

Definition 0.3.1. (Lyapunov [1893], Rumyantsev [1957a]) The unperturbed motion $\boldsymbol{x} = \boldsymbol{0}$ of system (0.3.1) is said to be

(1) *stable with respect to* $y_1, \ldots, y_m$ (or, briefly, $\boldsymbol{y}$-*stable* ($\boldsymbol{y}$-*St*)), if for any numbers $\varepsilon > 0$, $t_0 \geq 0$, there is a number $\delta(\varepsilon, t_0) > 0$ such that from $\|\boldsymbol{x}_0\| < \delta$ it follows that $\|\boldsymbol{y}(t; t_0, \boldsymbol{x}_0)\| < \varepsilon$ for all $t \geq t_0$;

(2) *asymptotically* $\boldsymbol{y}$-*stable* ($\boldsymbol{y}$-*ASt*), if it is $\boldsymbol{y}$-St and, besides, for each $t_0 \geq 0$, there is a number $\Delta(t_0) > 0$ such that each solution $\boldsymbol{x}(t; t_0, \boldsymbol{x}_0)$ with $\|\boldsymbol{x}_0\| < \Delta$ satisfies the condition

$$\lim\|\boldsymbol{y}(t; t_0, \boldsymbol{x}_0)\| = 0, \quad t \to \infty, \tag{0.3.3}$$

domain $\|\boldsymbol{x}_0\| < \Delta$ being contained in the *domain of* $\boldsymbol{y}$-*attraction* of the point $\boldsymbol{x} = \boldsymbol{0}$ for the initial time t_0.

Geometrically, the definitions introduced mean the following. In the case of $\boldsymbol{y}$-St, inside any ε-cylinder $\|\boldsymbol{y}\| < \varepsilon$ is a δ-sphere $\|\boldsymbol{x}_0\| = \delta$ such that any solution $\boldsymbol{x}(t; t_0, \boldsymbol{x}_0)$ of system (0.3.1), once originated from the δ-sphere at $t = t_0$, will remain inside the ε-cylinder for all $t \geq t_0$. In the case of $\boldsymbol{y}$-ASt, the solution $\boldsymbol{x}(t)$ will, in addition, asymptotically approach the ε-cylinder axis (see Figure 0.3.1). (The case when the solution $\boldsymbol{x}(t)$ will remain inside the ε-sphere for all $t \geq t_0$ and, in addition, even asymptotically approach point $\boldsymbol{x} = \boldsymbol{0}$ is not excluded from consideration; in this case the unperturbed motion is Lyapunov stable and Lyapunov asymptotically stable, respectively).

Definition 0.3.1 was subsequently made more specific and generalized in many works. As in the general theory of stability at large, this was motivated both by the needs of applications and by the quest for a comprehensive theory. In particular, the motion $\boldsymbol{x} = \mathbf{0}$ of system (0.3.1) is said to be

(3) *uniformly $\boldsymbol{y}$-stable* ($\boldsymbol{y}$-*USt*) (Halanay [1963], Corduneanu [1964]), if in definition (1) number δ does not depend on t_0;

(4) *$\boldsymbol{y}$-stable on the whole with respect to $\boldsymbol{z}_0$* (Rouche and Peiffer [1967], Rumyantsev [1972]), if in definition (1) the inequalities $\|\boldsymbol{y}_0\| < \delta$, $\|\boldsymbol{z}_0\| < \infty$ take place instead of $\|\boldsymbol{z}_0\| < \delta$, and $\boldsymbol{y}$-stability is uniform if δ does not depend on t_0;

(5) *asymptotically $\boldsymbol{y}$-stable uniformly with respect to $\boldsymbol{x}_0$* ($\boldsymbol{y}$-*UASt*($\boldsymbol{x}_0$)) (Peiffer [1968], Peiffer and Rouche [1969]), if in definition (2) relationship (0.3.3) holds uniformly with respect to $\boldsymbol{x}_0$ from the domain $\|\boldsymbol{x}_0\| < \Delta$;

(6) *uniformly asymptotically $\boldsymbol{y}$-stable* ($\boldsymbol{y}$-*UASt*) (Halanay [1963], Corduneanu [1964]), if in definition (2) number Δ does not depend on t_0 and relationship (0.3.3) holds uniformly with respect to t_0, $\boldsymbol{x}_0$ from the domain $t_0 \geq 0$, $\|\boldsymbol{x}_0\| < \Delta$;

(7) *globally asymptotically $\boldsymbol{y}$-stable* (Peiffer and Rouche [1969]), if it is $\boldsymbol{y}$-St and condition (0.3.3) holds for arbitrary $t_0 \geq 0$, $\boldsymbol{x}_0$ (in this case it is assumed that conditions (a), (b) are valid for system (0.3.1) in the domain $t \geq 0$, $\|\boldsymbol{x}\| < \infty$);

(8) *exponentially asymptotically $\boldsymbol{y}$-stable* ($\boldsymbol{y}$-*EASt*) (Corduneanu [1971]), if there exist constants $\Delta > 0$, $M > 0$, and $\alpha > 0$ such that each solution $\boldsymbol{x}(t; t_0, \boldsymbol{x}_0)$ of system (0.3.1) satisfies the inequality

$$\|\boldsymbol{y}(t; t_0, \boldsymbol{x}_0)\| \leq M\big(\|\boldsymbol{x}_0\|\big)e^{-\alpha(t-t_0)}, \quad t \geq t_0.$$

The literature on the PSt-problem confirms that the definitions listed are universally accepted. At the same time, also engaged are the PSt-notions such as, for example, *global $\boldsymbol{y}$-UASt* (Terjeki and Hatvani [1985]), *$\boldsymbol{y}$-UASt* and *$\boldsymbol{y}$-EASt on the whole with respect to $\boldsymbol{z}_0$* (Rumyantsev and Oziraner [1987]), *$\boldsymbol{y}$-St* and *$\boldsymbol{y}$-USt for large $\boldsymbol{z}_0$* (Vorotnikov [1988a], [1991a]).

In studying the stability with respect to all the variables, the familiar examples by Krasovskii and Vinograd (see Hahn [1967]) resulted in the notion of asymptotic stability being divided into stability and attraction. Indeed, even in the case of autonomous systems, attraction does not, generally, imply stability. Similarly, in the case of a PSt-problem it is reasonable to divide (Peiffer and Rouche [1969]) *$\boldsymbol{y}$-ASt* into *$\boldsymbol{y}$-St* and *$\boldsymbol{y}$-attraction* ($\boldsymbol{y}$-*At*). At the same time, as in the classical case, the following relationships hold (Rumyantsev and Oziraner [1987]):

$$\boldsymbol{y}\text{-UAt}(\boldsymbol{x}_0) \Rightarrow \boldsymbol{y}\text{-St}, \quad \boldsymbol{y}\text{-UASt}(\boldsymbol{x}_0) \Rightarrow \boldsymbol{y}\text{-USt}.$$

However, one *fails to substantiate* the analog of the familiar Massera relationship [1949] $\boldsymbol{y}$-ASt $\Rightarrow$ $\boldsymbol{y}$-UASt for autonomous systems. This leads to the *necessity of separately proving* the property of uniformity of $\boldsymbol{y}$-ASt, desirable from the standpoint of applications, *even in the case of autonomous systems.*

0.3.4 PSt-Problem and Stability of Sets (Trajectories)

The requirement of Lyapunov stability of unperturbed motion with respect to all the variables means, among other things, the *closeness* of the unperturbed and perturbed *trajectories* (which correspond to unperturbed and perturbed motions). The converse is not always true. Indeed, because the velocities of a representative point can differ along each trajectory, the closeness of trajectories may occur in the absence of Lyapunov stability of unperturbed motion. Therefore, along with Lyapunov stability, a weaker form of stability, namely, the *orbital stability* of unperturbed motion is often invoked. More generally, this form of stability is termed the *stability of invariant sets* (Zubov [1957]).

As regards the Lyapunov–Rumyantsev PSt-problem, the "partial stability" of unperturbed motion *does not mean*, generally, the *closeness of trajectories* corresponding to unperturbed and perturbed motions. That is why, as was noted by Peiffer and Rouche [1969], the Lyapunov–Rumyantsev PSt-problem does not reduce to any kind of problems of stability of sets (trajectories). In this connection, the attempt to relate the concept of partial stability to the stability of sets led Habets and Peiffer [1973] to consider the *stability of a set with respect to another set.* This problem was analyzed by Michel [1969] somewhat earlier.

0.3.5 Formulation of the Problem of Stabilization with Respect to Part of the Variables

The *problem of stabilization of motion (processes) with respect to part of the variables* (or, in brief, the *PSb-problem*) is a further extension of the PSt-problem to the class of *controlled systems.*

Let there be given equations of the *perturbed motion of a controlled system* (Krasovskii [1966])

$$\begin{gathered}\dot{\boldsymbol{x}} = \boldsymbol{X}(t, \boldsymbol{x}, \boldsymbol{u}), \qquad \boldsymbol{X}(t, \mathbf{0}, \mathbf{0}) \equiv \mathbf{0}, \\ \boldsymbol{x}^{\mathsf{T}} = (y_1, \ldots, y_m, z_1, \ldots, z_p) = (\boldsymbol{y}^{\mathsf{T}}, \boldsymbol{z}^{\mathsf{T}}), \quad \boldsymbol{u}^{\mathsf{T}} = (u_1, \ldots, u_r), \\ m \geq 1, \quad p \geq 0, \quad n = m + p, \quad r \geq 1,\end{gathered} \tag{0.3.4}$$

with right-hand sides defined and continuous in the domain

$$t \geq 0, \quad \|\boldsymbol{y}\| \leq H, \quad \|\boldsymbol{z}\| < \infty, \quad \|\boldsymbol{u}\| < \infty. \tag{0.3.5}$$

In system (0.3.4) $\boldsymbol{x}$ is the vector of phase variables, and $\boldsymbol{u}$ is the vector of controls, which we shall seek in the form $\boldsymbol{u} = \boldsymbol{u}(t, \boldsymbol{x})$. Vector function $\boldsymbol{u} = \boldsymbol{u}(t, \boldsymbol{x})$ satisfying the condition $\boldsymbol{u}(t, \mathbf{0}) = \mathbf{0}$ is assumed to be defined and continuous in domain (0.3.5). System (0.3.4) satisfies the conditions imposed on system (0.3.1) at $\boldsymbol{u} = \boldsymbol{u}(t, \boldsymbol{x})$. (In this case we further write $\boldsymbol{u} \in U$.)

The performance criterion to be minimized is the integral of the form

$$I = \int_{t_0}^{\infty} \omega(t, \boldsymbol{x}[t], \boldsymbol{u}[t]) dt,$$

where $\omega = \omega(t, \boldsymbol{x}, \boldsymbol{u})$ is a scalar function continuous in domain (0.3.5), $\boldsymbol{x}[t]$ is a solution of system (0.3.4) for $\boldsymbol{u} = \boldsymbol{u}(t, \boldsymbol{x})$, $\boldsymbol{u}[t] = \boldsymbol{u}(t, \boldsymbol{x}[t])$.

Problem 0.3.1. (*on optimal* $\boldsymbol{y}$*-stabilization*) (Rumyantsev [1970], [1971b]). *Find vector function* $\boldsymbol{u} = \boldsymbol{u}^0(t, \boldsymbol{x}) \in U$ *that guarantees the asymptotic* $\boldsymbol{y}$*-stability of the unperturbed motion* $\boldsymbol{x} = \mathbf{0}$ *of system* (0.3.4). *For any other vector function* $\boldsymbol{u} = \boldsymbol{u}^*(t, \boldsymbol{x}) \in U$ *satisfying this condition, the following inequality should hold*:

$$\int_{t_0}^{\infty} \omega(t, \boldsymbol{x}^0[t], \boldsymbol{u}^0[t])dt \leq \int_{t_0}^{\infty} \omega(t, \boldsymbol{x}^*[t], \boldsymbol{u}^*[t])dt,$$

$$t_0 \geq 0, \quad \boldsymbol{x}^0[t_0] = \boldsymbol{x}^*[t_0] = \boldsymbol{x}_0, \quad \|\boldsymbol{x}_0\| < \delta = \text{const} > 0.$$

Number δ *is either preassigned or is taken in accordance with Definition* 0.3.1.

An extension of Problem 0.3.1 is the problem of synthesizing a system whose unperturbed motion is not only asymptotically stable with respect to a given part of the variables, but also (nonasymptotically) Lyapunov stable (Demin and Furasov [1976], Vorotnikov [1990]). This problem can be treated as a version of the *polystabilization* problem.

In selecting definite functionals I to be minimized as performance criteria, one often follows principles stated by Krasovskii [1966]:

(1) the integrand ω of the functional I must ensure an acceptable compromise between sufficiently rapid damping of the transient process (in the PSb-problem, with respect to variables $y_1, \ldots, y_m$) and the resources for generating the controls;

(2) the solution of the problem must be as simple as possible, and so, more convenient from the viewpoint of practical implementation of the control rules obtained.

However, the formalization of these requirements for the majority of nonlinear systems proves possible only in an *iterative mode*, in practice, reduced to the *cut-and-try method* which implies that the minimized functional will somehow be corrected to obtain an acceptable (optimal in some sense) solution.

0.3.6 Stages and Approaches in Analysis of PSt- and PSb-Problems

The research conducted so far can be divided into *three stages.*

The first stage (the end of the 50s to the beginning of the 70s) is associated almost exclusively with development of the method of Lyapunov functions (MLF) and was summarized in the survey by Oziraner and Rumyantsev [1972]. This article provided an important stimulus for further research on PSt- and PSb-problems.

Research by a number of scientists, of which the most important was Rumyantsev [1957a], [1960], [1971a], [1971b], [1972a], [1972b], Zubov [1959],

Halanay [1963], Corduneanu [1964], [1971], Matrosov [1965], Rouche and Peiffer [1967], Peiffer [1968], Peiffer and Rouche [1969], Risito [1970], Fergola and Moauro [1970], Oziraner [1971], [1972a]–[1972c], [1973], and Salvadori [1972], led to modifications of the basic theorems of the MLF as applied to the PSt-problem. The results obtained found important practical application not only for systems with a finite number of degrees of freedom, but also in continuous media (Rumyantsev [1957a], [1960], [1968], [1972a], [1972b], Movchan [1960], Demin [1964], [1968], Moiseev and Rumyantsev [1965]). The *problem of stability with respect to two measures* was formulated for distributed systems (Movchan [1960]). During this period the PSt-problem was also extended to controlled processes (Rumyantsev [1970], [1971b]); here the *problem of stabilization of motion (processes) with respect to part of the variables* (in short, the *PSb-problem*) appeared naturally.

The second stage of research (beginning with the middle of the 70s) is characterized by the range of issues solved within the framework of the PSt- and PSb-problems significantly extended. The following research approaches proved to be fundamental.

(1) *Further development of the MLF in the context of the PSt-problem for systems of ordinary differential equations.* This was necessitated, in particular, by a number of substantial difficulties which arose in the first stage of research when attempts were made to extend the basic theorems of the MLF to the PSt-problem. Among these difficulties, we note two that concern a large group of problems.

(*a*) As noted above, *separate* proof is required for the practically desirable *property of uniformity* of asymptotic stability with respect to part of the variables, even in the *case of autonomous systems.*

(*b*) Forced introduction of the additional requirement of boundedness of the solutions of systems under investigation to extend the widely used scheme for proof of the Barbashin–Krasovskii type theorems to the PSt-problem.

In addition, for the PSt-problem (and to an equal degree for the problem of stability with respect to all the variables), further development of the MLF in the direction of weakening the requirements imposed on Lyapunov functions is, to a considerable extent, dictated by the desire to extend the possibilities for their constructive construction. Thus, the applicability of the MLF to the PSt-problem has been considerably extended along the following lines:

(*a*) by introducing various types of "*limiting*" *systems* (Andreev [1979], [1984], [1987], [1991], Hatvani [1983a], [1983b], [1985a], [1985b], [1991], Bondi et al. [1981], Kosov [1988], Rusinov [1988], Jia-xiang [1990], Mamedova [1995]);

(*b*) by constructing various types of *auxiliary systems* (Zubov [1959], Vorotnikov [1991a]);

(*c*) by refining the notion of a V-function sign-definite with respect to part of the variables (Vorotnikov [1993], [1995a], [1995b]);

(*d*) by using the MLF in conjunction with *asymptotic averaging* (Khapaev [1986]).

The *method of Lyapunov vector functions* provides an effective means of solving the PSt-problem (Bellman [1962], Matrosov [1962b], Lakshmikantham et al. [1991]); this applies especially to research on high-dimensional complex systems.

In publications by Corduneanu [1964], Karimov [1973], Oziraner [1981], Miki et al. [1985], and Ignat'ev [1989a], [1989b] the MLF is used for solving the PSt-problems of *stability in the presence of constantly acting perturbations* (*CAP*) and of *preservation of stability.* We note that one of the peculiarities of the PSt-problem in the presence of CAP is the different nature, as compared to the case of stability with respect to all the variables in the presence of CAP, of the interrelationship with the PSt-problem of *stability in the presence of structural* (*parametric*) *perturbations* (for more detail, see Section 0.6).

The method of Lyapunov functions is also employed in solving the PSt-problem for the following classes of systems:

(*a*) *systems with impulse effect* (Simenov and Bainov [1986]);

(*b*) *semistate systems* of the form $A(t)\dot{\boldsymbol{x}} = \boldsymbol{X}(t, \boldsymbol{x})$, $\operatorname{rank} A(t) < \dim \boldsymbol{x}$ (Bajic [1986]);

(*c*) *systems with multivalued right-hand sides* (Talpalary and Stefancu [1987]).

Besides, there is a number of publications (Miki [1990], Shoichi [1990]) in which the MLF is used to analyze the PSt-problem of *integral stability.*

The Lyapunov–Rumyantsev problem of stability with respect to part of the variables can be interpreted as a *problem of stability with respect to two measures.* One of two measures characterizes the closeness of the unperturbed and perturbed motions at the initial time, and the other characterizes their closeness at subsequent times. This formulation of the problem was developed by Movchan [1960] for the case of stability of *infinite-dimensional processes* described by *distributed systems.* There are versions of the problem of stability with respect to two measures which involve systems of ordinary differential and functional-differential equations (Lakshmikantham et al. [1989]). Details regarding the present state of the theory of stability with respect to two measures and bibliographies on this subject may be found in Sirazetdinov [1987], Lakshmikantham et al. [1989], Matrosov [1989], and Shestakov [1990].

Beginning with the work by Aminov and Sirazetdinov [1987], the MLF is applied to analyzing the *problem of polystability of motion.* Among the authors contributed to the subject is also Martynyuk [1994].

(2) *Development of methods for investigation of the PSt-problem for general classes of linear systems, in the linear approximation, and in critical cases* (Zubov [1959], Matrosov [1965], Corduneanu [1971], Lutsenko and Stadnikova [1973], Oziraner [1973], [1975], [1986], Prokop'ev [1975], Muller [1977], [1984], Vorotnikov and Prokop'ev [1978], Vorotnikov [1979a], [1979b], [1983a], [1988a], [1991a], Krivosheev and Lutsenko [1980], Vu Tuan [1980], Shchennikov [1985], and Lizunova [1991]). Than is the case in investigating stability with respect to all the variables, the much stronger dependence of the PSt-property both on the coefficients of the linear part of the system being studied and on the structure of its nonlinear terms leads to major difficulties in solving the PSt-problem in the cases indicated.

In connection with application of *Lyapunov's first method* to the PSt-problem, we note Hatvani's work [1976], which presents rather general conditions for the absence of asymptotic stability with respect to part of the variables for nonlinear systems, and publications by Kozlov [1991b] and by Vuyichich and Kozlov [1991], which analyze *stability with respect to given state functions.* In general, as Kozlov noted, for further development of the PSt-problem it is reasonable to make broader use of Lyapunov's first method in conjunction with recent advances in the qualitative theory of differential equations.

(3) *Investigation of the PSb-problem.* The fundamental problem is development of constructive techniques for designing stabilizing controllers. Various methods are presented in Rumyantsev [1970], [1971b], [1972a], Demin and Furasov [1976], Furasov [1977], Oziraner [1978], Vorotnikov [1979a], [1982a], [1982b], [1988b], [1990], [1991a], [1991b], [1997b], [1997d], Zubov [1980], [1982], Akulenko [1980], [1987], [1994a], Smirnov et al. [1985], Smirnov and Ermolina [1988], Yurkov [1988], and Pyatnitskii [1993] for both relatively general classes and particular systems.

(4) *Investigation of the PSt-problem based on modifications of Lyapunov's treatment of the notion of stability*, for example:

(*a*) studies on *engineering (practical) stability with respect to part of the variables*, including *stability with respect to part of the variables in a finite time interval* (Zubov [1959], Weiss and Infante [1967], Martynyuk [1972], [1983], Bajic [1986], Khapaev [1986]);

(*b*) studies on the *stiffness of motion* in problems of *approximate integration with respect to part of the variables* of ordinary differential equations (Skimel' [1978], [1992]).

(5) *Investigation of PSt-problem for systems of other than ordinary differential equations*, among which are

(*a*) *functional-differential systems (systems with time delay)* (Halanay [1963], Corduneanu [1975], Yudaev [1975], Vorotnikov [1979a], [1980a], [1980b], [1991a], Terjeki [1983a], Il'yasov [1984], Wei-Xuan [1984], Kalistratova [1986], Savchenko and Ignat'ev [1989]);

(*b*) *stochastic systems* (Yudaev [1977], Sharov [1978], Khapaev [1986], Vorotnikov [1983b], [1991a]);

(*c*) *discrete models* (Silakov and Yudaev [1975], Nosov and Furasov [1979], Furasov [1982], Il'yasov [1984]);

(*d*) *stochastic discrete models* (Phillis [1984]).

(6) *Investigation of the problem of boundedness of solutions with respect to part of the variables*, which is closely related to the PSt-problem (Yoshizawa [1966], Peiffer [1968], Oziraner [1972a], Rumyantsev and Oziraner [1987], Risito [1976], Cantarelli and Risito [1992], Cantarelli [1995]), and investigation of the problem of *boundedness of invariant sets with respect to part of the variables* (Zaitsev [1993]).

The investigations carried out along the lines indicated exhibited a definite difference in the study of the above listed problems with respect to part of and to all the variables. Apparently, the difficulties accompanying the analysis of some versions of the PSt-problem (for example, investigating stability in the linear approximation) cannot be easily avoided. Also revealed were the fundamental peculiarities of the PSt-problem, in particular, its *greater "non-coarseness"* compared with the problem of stability with respect to all the variables. This last circumstance is not surprising, since the PSt-problem is more refined and general.

It has also turned out that the PSt-problem and the problem of stability with respect to all the variables are closely related and complement each other in solving practical problems. The PSt-property is often sufficient for normal operation of a system, and even desirable. As we noted in Section 0.2, there are also other reasons for studying the PSt-problem.

The third stage of research of the PSt- and PSb-problems is being crystallized. Apart from continuing the study of the problems indicated above and making it more thorough, this stage is characterized by the following activities.

(1) An attempt is made to conceptualize the results obtained in the context of the PSt-problem, to get a deeper insight into the PSt-problem, and to elucidate the laws underlying the operation of PSt-systems and the dangers on the way to implementing the attractive results of the PSt-theory. The first steps have been taken in this direction (Vorotnikov [1993], [1995a], [1995b]), which can be viewed as the inception of a kind of "physical" PSt-theory.

(2) Methods developed in the course of investigating PSt-problems are applied to other problems of the system dynamics, for example, to the problems of motion control (with respect to part of or to all the variables) in a *finite* time interval (Vorotnikov [1994a]–[1994c], [1995c], [1997b]–[1997e]). This allows one to extend the potential of the approaches worked out earlier and to understand the role and place of the PSt-problems among the principal problems of the system theory.

0.4 The Method of Lyapunov Functions in Problems of Stability and Stabilization with Respect to Part of the Variables

0.4.1 Basic Notions

It is known that the essence of the MLF (also called Lyapunov's *second* or *direct method*) consists in finding auxiliary V-functions that make it possible to estimate the deviation of the solutions $\boldsymbol{x} = \boldsymbol{x}(t; t_0, \boldsymbol{x}_0)$ of system (0.3.1) from its equilibrium position $\boldsymbol{x} = \mathbf{0}$, V-functions being a generalized measure of the indicate deviation.

To state certain fundamental results, we introduce two classes of auxiliary functions:

(1) functions $V(t, \boldsymbol{x})$ that are single-valued, continuous, satisfy the condition $V(t, \mathbf{0}) = 0$, have continuous partial derivatives with respect to t and $\boldsymbol{x}$

in domain (0.3.2), and have total derivatives with respect to time

$$\dot{V}(t, \boldsymbol{x}) = \frac{\partial V}{\partial t} + \sum_{i=1}^{n} \frac{\partial V}{\partial x_i} X_i(t, \boldsymbol{x})$$

by virtue of system (0.3.1). Here X_i $(i = 1, \ldots, n)$ are the components of vector function $\boldsymbol{X}$, which defines the right-hand side of system (0.3.1).

(2) continuous functions $a(r)$, $b(r)$, and $c(r)$ strictly monotonically increasing for $r \in [0, H]$ such that $a(0) = b(0) = c(0) = 0$.

0.4.2 Notion of a Function Sign-Definite with Respect to Part of the Variables

In the study of the PSt-problems, V-function generally depends on the variables relative to which stability is being analyzed, and also on the remaining variables, and it does not satisfy the conditions of the classic Lyapunov theorem. In this connection, Rumyantsev [1957a] introduced the notion of V-function which is *sign-definite with respect to part of the variables*, and proved the Lyapunov theorem for the case of the PSt-problem.

Definition 0.4.1. (Rumyantsev [1957a]) A function $V = V(t, \boldsymbol{y}, \boldsymbol{z})$ is said to be *$\boldsymbol{y}$-positive-definite*, if there exists continuous function $W(\boldsymbol{y})$, $W(\boldsymbol{0}) = 0$ positive for $||\boldsymbol{y}|| \leq H$ such that the following inequality is valid in domain (0.3.2):

$$V(t, \boldsymbol{y}, \boldsymbol{z}) \geq W(\boldsymbol{y}). \tag{0.4.1}$$

Corduneanu [1964] showed that V-function is $\boldsymbol{y}$-positive-definite if and only if the following inequality is satisfied in domain (0.3.2):

$$V(t, \boldsymbol{y}, \boldsymbol{z}) \geq a(||\boldsymbol{y}||),$$

which is more convenient for proving MLF theorems.

Along with the notion of the $\boldsymbol{y}$-sign-definite Rumyantsev V-function, the notion of the $\boldsymbol{y}$-positive-definite $\boldsymbol{\mu}$-function V can also be helpful in applications.

Definition 0.4.2. (Vorotnikov [1979a]) A function $V = V(t, \boldsymbol{y}, \boldsymbol{z})$ is said to be *$\boldsymbol{y}$-positive-definite $\boldsymbol{\mu}$-function V*, if there can be found a single-valued continuous vector function $\boldsymbol{\mu} = \boldsymbol{\mu}(\boldsymbol{x})$ such that $\boldsymbol{\mu}(\boldsymbol{0}) = \boldsymbol{0}$ and in the domain

$$t \geq 0, \quad ||\boldsymbol{\xi}|| \leq H, \quad ||\boldsymbol{z}|| < \infty \tag{0.4.2}$$

the following inequality is satisfied

$$V(t, \boldsymbol{y}, \boldsymbol{z}) \geq W(\boldsymbol{\xi}), \quad \boldsymbol{\xi} = (\boldsymbol{y}, \boldsymbol{\mu}). \tag{0.4.3}$$

Here W is a single-valued continuous function of variables $\boldsymbol{y}$, $\boldsymbol{\mu}$ such that $W(\boldsymbol{\xi}) > 0$, $W(\boldsymbol{0}) = 0$.

It was shown (Vorotnikov [1993], [1995a], [1995b]) that even in the case $m = p = 1$, one can choose a function $V(y_1, z_1)$, as a $\boldsymbol{\mu}$-function V, that is not sign-definite both with respect to y_1, z_1 (in the sense of Lyapunov [1892]) and with respect to y_1 (in the sense of Definition 0.4.1). This issue is considered in detail in Section 3.1.

Table 0.4.1.

V-function	Lyapunov V-function		Rumyantsev V-function		$\boldsymbol{\mu}$-function V		
	Locally	Globally	Locally	Globally	Locally	Globally	μ_1
$V_1 = y_1^2 + z_1^2$	Yes	Yes	Yes	Yes	Yes	Yes	z_1
$V_2 = \dfrac{y_1^2}{1+y_1^2} + \dfrac{z_1^2}{1+z_1^2}$	Yes	No	Yes	No	Yes	No	z_1
$V_3 = y_1^2 + \dfrac{z_1^2}{1+z_1^2}$	Yes	No	Yes	Yes	Yes	No	z_1
$V_4 = \dfrac{y_1^2 + z_1^2}{1+z_1^4}$	Yes	No	No	No	Yes	No	z_1
$V_5 = y_1^2 + \dfrac{(y_1 z_1)^2}{1+(y_1 z_1)^2}$	No	No	Yes	Yes	Yes	No	$y_1 z_1$
$V_6 = y_1^2 + y_1^2 z_1^2$	No	No	Yes	Yes	Yes	Yes	$y_1 z_1$
$V_7 = \dfrac{y_1^2}{1+y_1^2} + \dfrac{y_1^2 + y_1^2 z_1^2}{1+(y_1^2 z_1^2)^2}$	No	No	Yes	No	Yes	No	$y_1 z_1$
$V_8 = \dfrac{y_1^2 + y_1^2 z_1^2}{1+(y_1^2 z_1^2)^2}$	No	No	Yes	No	Yes	No	$y_1 z_1$

For the case $m = p = 1$, Table 0.4.1 presents eight V-functions which demonstrate possible interrelationships among the notions introduced. Some of these functions are well known in the literature on the theory of stability.

0.4.3 Conditions for Stability with Respect to Part of the Variables

The notion of $\boldsymbol{y}$-sign-definite V-function introduced made it possible to extend the classic Lyapunov theorem [1892] on stability of motion to the case of the PSt-problem.

Theorem 0.4.1. (Rumyantsev [1957a]) *If for system* (0.3.1) *one can find a V-function satisfying conditions*

$$V(t, \boldsymbol{x}) \geq a(\|\boldsymbol{y}\|), \quad \dot{V}(t, \boldsymbol{x}) \leq 0 \tag{0.4.4}$$

in domain (0.3.2), *then the unperturbed motion* $\boldsymbol{x} = \mathbf{0}$ *of system* (0.3.1) *is* $\boldsymbol{y}$*-stable.*

Halanay [1963] and Corduneanu [1964] noted that if, together with conditions of Theorem 0.4.1, the inequality $V \leq b(\|\boldsymbol{x}\|)$ is satisfied, then, as in the classic case, $\boldsymbol{y}$-stability is *uniform*.

The analysis of the proof of Theorem 0.4.1 made it possible to draw a number of conclusions valid exclusively within the context of the PSt-problem. Thus, Rouche and Peiffer [1967], Fergola and Moauro [1970] considered $\boldsymbol{y}$-stability on the whole with respect to $\boldsymbol{z}_0$. This property occurs if, in addition to (0.4.4), $V \leq b(\|\boldsymbol{y}\|)$. Further analysis in this direction (Vorotnikov [1988a], [1991a]) demonstrated that conditions (0.4.4) together with

$$V(t; \mathbf{0}, \boldsymbol{z}) \equiv 0 \quad (V(t; \mathbf{0}, \boldsymbol{z}) \equiv 0, \quad V \leq b(\|\boldsymbol{x}\|))$$

yield $\boldsymbol{y}$-stability (uniform $\boldsymbol{y}$-stability) for large $\boldsymbol{z}_0$.

We note that these conditions are weaker than the condition $V \leq b(\|\boldsymbol{y}\|)$.

Further development of the Rumyantsev theorem [1957a] is associated with making the notion of function sign-definite with respect to part of the variables more specific (Vorotnikov [1995a], [1995b]). (For more detail, see Section 3.1.)

0.4.4 Conditions for Asymptotic y-Stability Using V-Functions with a Sign-Definite Derivative

A number of results in this direction that generalize the classic Lyapunov [1892] and Marachkov [1940] theorems were obtained by Rumyantsev [1957a], [1970], Peiffer and Rouche [1969], and Rumyantsev and Oziraner [1987].

Theorem 0.4.2. (Rumyantsev [1957a], [1970], Rumyantsev and Oziraner [1987]) *If for system* (0.3.1) *one can specify a V-function that in domain* (0.3.2) *satisfies the conditions*

$$a(\|\boldsymbol{y}\|) \leq V(t, \boldsymbol{x}) \leq b(\|\boldsymbol{w}\|), \quad \dot{V}(t, \boldsymbol{x}) \leq -c(\|\boldsymbol{w}\|)$$
$$\|\boldsymbol{w}\| = (x_1^2 + \cdots + x_k^2)^{1/2}, \quad m \leq k \leq n,$$

then the unperturbed motion $\boldsymbol{x} = \mathbf{0}$ of system (0.3.1) *is uniformly asymptotically $\boldsymbol{y}$-stable* (*on the whole with respect to $x_{k+1,0}, \ldots, x_{n,0}$*).

Also appearing in the literature are the following conditions for asymptotic $\boldsymbol{y}$-stability:

$$(1) \qquad a(\|\boldsymbol{y}\|) \leq V(t, \boldsymbol{x}) \leq b(\|\boldsymbol{y}\|), \quad \dot{V} \leq -c(\|\boldsymbol{y}\|),$$
$$(2) \qquad a(\|\boldsymbol{y}\|) \leq V(t, \boldsymbol{x}) \leq b(\|\boldsymbol{x}\|), \quad \dot{V} \leq -c(\|\boldsymbol{x}\|),$$

which present particular cases of the conditions of Theorem 0.4.2. We note that the first group actually establishes the condition for uniform asymptotic stability of the invariant set $\boldsymbol{y} = \mathbf{0}$, such conditions being, under certain additional assumptions as to the right-hand side of system (0.3.1), not only sufficient but also necessary (Oziraner [1971], Rumyantsev and Oziraner [1987]). The second group of conditions will be beneficial in terms of the PSt-problem (i.e., will not reduce to the classic Lyapunov conditions) provided that system (0.3.1) and the corresponding V-function are "essentially" dependent on t (Rumyantsev and Oziraner [1987]).

Conditions $a(\|\boldsymbol{y}\|) \leq V(t, \boldsymbol{x}) \leq b(\|\boldsymbol{x}\|)$, $\dot{V} \leq -c(\|\boldsymbol{y}\|)$ are sometimes wrongly taken for Rumyantsev's conditions for asymptotic $\boldsymbol{y}$-stability. (Evidently, this is due to the fact that some inaccuracy slipped into the formulation of the corresponding theorem in the work by Rumyantsev [1957]; see the refinement in Rumyantsev [1970].) However, in the theory of stability with respect to part of the variables, these conditions (under the general assumptions concerning the right-hand sides of system (0.3.1) made in Subsection 0.3.2) are *not justified.* We point out that if conditions $a(\|\boldsymbol{y}\|) \leq V(t, \boldsymbol{x}) \leq b(\|\boldsymbol{x}\|)$, $\dot{V} \leq 0$ are satisfied, we get uniform $\boldsymbol{y}$-stability and, by virtue of $V \geq a(\|\boldsymbol{y}\|)$, $\dot{V} \leq -c(\|\boldsymbol{y}\|)$, we have weak $\boldsymbol{y}$-attraction, and asymptotic stability is guaranteed by these properties considered in the aggregate when stability with respect to all the variables is analyzed (Rouche et al. [1977]).

Let us consider one of possible modifications of Theorem 0.4.2.

Corollary 0.4.1. (Vorotnikov [1979a]) *If for system* (0.3.1), *together with a V-function, it is possible to specify a vector $\boldsymbol{\mu}(\boldsymbol{x})$-function, $\boldsymbol{\mu}(\boldsymbol{0}) = \boldsymbol{0}$ that in domain* (0.3.2) (*in domain* (0.4.2)) *satisfies the conditions*

$$a(\|\boldsymbol{\eta}\|) \leq V(t, \boldsymbol{x}) \leq b(\|\boldsymbol{\xi}\|), \quad \dot{V}(t, \boldsymbol{x}) \leq -c(\|\boldsymbol{\xi}\|),$$
$$\boldsymbol{\eta} = \boldsymbol{y} \quad (\boldsymbol{\eta} = \boldsymbol{\xi}, \textit{ respectively}), \quad \boldsymbol{\xi} = (\boldsymbol{y}, \boldsymbol{\mu}),$$

then the unperturbed motion $\boldsymbol{x} = \boldsymbol{0}$ of system (0.3.1) *is uniformly asymptotically $\boldsymbol{y}$-stable.*

The conditions of Corollary 0.4.1 are *weaker* than those of Theorem 0.4.2.

Now let us turn to discussing the conditions for asymptotic $\boldsymbol{y}$-stability which develop the conditions of the Marachkov theorem [1940]. To this end we represent system (0.3.1) in the form

$$\dot{\boldsymbol{y}} = \boldsymbol{Y}(t, \boldsymbol{y}, \boldsymbol{z}), \quad \dot{\boldsymbol{z}} = \boldsymbol{Z}(t, \boldsymbol{y}, \boldsymbol{z}).$$

Theorem 0.4.3. (Peiffer and Rouche [1969], Rumyantsev and Oziraner [1987]) *The unperturbed motion $\boldsymbol{x} = \boldsymbol{0}$ of system* (0.3.1) *is asymptotically $\boldsymbol{y}$-stable if in domain* (0.3.2) *the following conditions hold*:

$$V(t, \boldsymbol{x}) \geq a(\|\boldsymbol{y}\|), \quad \dot{V} \leq -c(\|\boldsymbol{y}\|),$$

and, in addition, for any $t_0 \geq 0$, there exists $\delta'(t_0) > 0$ such that, for any point $\boldsymbol{x}_0$ with $\|\boldsymbol{x}_0\| < \delta'$, each component Y_i $(i = 1, \dots, m)$ of vector function $\boldsymbol{Y}$ is upper or lower bounded by a number $N = N(t_0, \boldsymbol{x}_0)$ for all $t \geq t_0$.

As in the classic case, the conditions of Theorem 0.4.3 do not require *decrescency* of either all ($V \leq b(\|\boldsymbol{x}\|)$) or part of the variables ($V \leq b(\|\boldsymbol{y}\|)$) or, in Russian terminology, the variables are not required to admit an *infinitely small upper bound*; Marachkov [1940] was the first to obtain such conditions. Here (the first group of conditions), in contrast to the classic case, boundedness only of vector function $\boldsymbol{Y}$ is required, not of the entire vector function $\boldsymbol{X}^{\mathsf{T}} = (\boldsymbol{Y}^{\mathsf{T}}, \boldsymbol{Z}^{\mathsf{T}})$. (Apparently, the verification of boundedness of $\boldsymbol{Y}$ presents greater difficulty in the case of the PSt-problem). Further development of conditions for asymptotic $\boldsymbol{y}$-stability originating from the Marachkov theorem can be found in the works by Hatvani [1984], Andreev [1991], and Section 3.1 of the present book.

0.4.5 Conditions for Asymptotic y-Stability Using V-Functions with a Sign-Constant Derivative

In most cases, as experience shows, it is possible to construct a V-function with only a *sign-constant* $\dot{V} \leq 0$ (and not a *sign-definite*) derivative. For more detailed consideration of this issue we shall separate cases of *autonomous* and *nonautonomous* systems.

1. The case of autonomous systems. In this class of systems the conditions of asymptotic stability with respect to all the variables for the case of $\dot{V} \leq 0$ are obtained by Barbashin and Krasovskii [1952] under an additional requirement to the set $M = \{\boldsymbol{x} : \dot{V}(\boldsymbol{x}) = 0\}$. The Barbashin–Krasovskii method as applied to the PSt-problem for autonomous systems has been developed in Risito [1970], Rumyantsev [1971a], [1971b], Oziraner [1973], Vorotnikov [1979a], [1997a], Hatvani [1983a], and Andreev [1987], [1991]. (In the works by Andreev [1987], [1991], a more general class of nonautonomous systems is considered. However, the results are of sufficient value for the case of autonomous systems proper.)

Initially, the conditions were found which made it possible to apply the classic scheme of proving the theorems of the Barbashin–Krasovskii type to the PSt-problem. The factor that kept this scheme working was the requirement of *z-boundedness* of solutions of system (0.3.1) originating from a sufficiently small neighborhood of the point $\boldsymbol{x} = \mathbf{0}$. This requirement is basic in Theorems 0.4.4–0.4.6 to be formulated below.

Theorem 0.4.4. (Risito [1970], Rumyantsev [1971a]) *Suppose that for autonomous system* (0.3.1) *one can specify a V-function satisfying the conditions*
(1) $V(\boldsymbol{x}) \geq a(\|\boldsymbol{y}\|)$,
(2) $\dot{V}(\boldsymbol{x}) = 0$ $(\boldsymbol{x} \in M)$, $\dot{V} < 0$ $(\boldsymbol{x} \,\overline{\in}\, M)$, *and*
(3) *M is the set of $\{\boldsymbol{x}\}$ that contains no entire semitrajectories for $t \in [t_0, \infty)$ in the domain $t \geq 0$, $\|\boldsymbol{x}\| \leq H$.*

Then the unperturbed motion $\boldsymbol{x} = \mathbf{0}$ is uniformly asymptotically $\boldsymbol{y}$-stable.

Theorem 0.4.5. (Risito [1970]) *Suppose that for autonomous system* (0.3.1) *one can specify a V-function such that in the domain $t \geq 0$, $\|\boldsymbol{x}\| \leq H$*
(1) *conditions* (1) *and* (2) *of Theorem* 0.4.4 *are satisfied,*
(2) *the set $M_1 = \{\boldsymbol{x} : \boldsymbol{y} = \mathbf{0}\}$ is invariant, and*
(3) *$M \setminus M_1$ contains no entire semitrajectories for $t \in [t_0, \infty)$.*

Then the unperturbed motion $\boldsymbol{x} = \mathbf{0}$ is asymptotically $\boldsymbol{y}$-stable.

Theorem 0.4.6. (Oziraner [1973]) *Suppose that for autonomous system* (0.3.1) *one can specify a V-function such that in the domain $t \geq 0$, $\|\boldsymbol{x}\| \leq H$*
(1) *conditions* (1) *and* (2) *of Theorem* 0.4.4 *are satisfied and*
(2) *the set $M \cup \{\boldsymbol{x} : V(\boldsymbol{x}) > 0\}$ contains no entire semitrajectories for $t \in [t_0, \infty)$.*

Then the unperturbed motion $\boldsymbol{x} = \mathbf{0}$ is uniformly asymptotically $\boldsymbol{y}$-stable.

Further development of the theorems of type 0.4.4–0.4.6 rests on introducing a second (vector) auxiliary function (Vorotnikov [1979a], [1997a]); for more detail, see Chapter 3.

Oziraner [1973] provided an example demonstrating that, without the requirement of z-boundedness of solutions, Theorems 0.4.4–0.4.6 are not generally true. Removing this restriction was the objective of the works by Hatvani [1983a], [1983b] and Andreev [1987], [1991].

Thus, to solve this problem Hatvani [1983a], [1983b] proposed two approaches which respectively imply

(a) additional requirements to the V-function for $\|z\| \to \infty$ and

(b) the possibility of constructing some *"limiting" system*

$$\dot{\boldsymbol{y}} = \boldsymbol{Y}_*(\boldsymbol{y}) \tag{0.4.5}$$

(where $\boldsymbol{Y}_* = \lim \boldsymbol{Y}(\boldsymbol{y}, \boldsymbol{z})$ for $\|\boldsymbol{z}\| \to \infty$ uniformly with respect to $\boldsymbol{y}$, $\|\boldsymbol{y}\| \leq H' < H$) obtained from the first m equations of the original system (0.3.1) as $\|\boldsymbol{z}\| \to \infty$.

If conditions (1) or (2) of Theorem 0.4.6 (without assuming that the solutions are z-bounded) are satisfied, the following alternative arises (Hatvani [1983a]): either $V \to 0$ (and, consequently, the unperturbed motion $\boldsymbol{x} = \mathbf{0}$ is asymptotically $\boldsymbol{y}$-stable) or $\|\boldsymbol{z}\| \to \infty$, $t \to \infty$. Taking this circumstance into account, Hatvani [1983a], [1983b] proved a number of conditions for asymptotic $\boldsymbol{y}$-stability which *do not imply* z-boundedness of solutions. These conditions allow both of the above approaches to achieve the objective to be implemented. We cite some of these conditions.

Theorem 0.4.7. (Hatvani [1983a]) *Suppose that for autonomous system* (0.3.1) *one can specify a V-function such that in domain* (0.3.2)

(1) $V(\boldsymbol{x}) \geq a(\|\boldsymbol{y}\|)$,

(2) *condition* (2) *of Theorem* 0.4.4 *and condition* (2) *of Theorem* 0.4.6 *are satisfied, and*

(3) $V \to 0$ *uniformly with respect to* $\boldsymbol{y}$ ($\|\boldsymbol{y}\| \leq H' < H$) *as* $\dot{V} \to 0$ *and* $\|\boldsymbol{z}\| \to \infty$.

Then the unperturbed motion $\boldsymbol{x} = \mathbf{0}$ *is uniformly asymptotically* $\boldsymbol{y}$*-stable.*

Following Hatvani [1983b], we denote by $V_m^{-1}[c, \infty]_0$ the set of $\{\boldsymbol{y}\}$ for which there exists a sequence $\{(\boldsymbol{y}_i, \boldsymbol{z}_i)\}$ such that $\boldsymbol{y}_i \to \boldsymbol{y}$, $\|\boldsymbol{z}_i\| \to \infty$, $V(\boldsymbol{y}_i, \boldsymbol{z}_i) \to c$, and $\dot{V}(\boldsymbol{y}_i, \boldsymbol{z}_i) \to 0$ as $i \to \infty$.

Theorem 0.4.8. (Hatvani [1983a]) *Suppose that for autonomous system* (0.3.1) *one can specify a V-function such that in domain* (0.3.2)

(1) *conditions* (1) *and* (2) *of Theorem* 0.4.7 *are satisfied and*

(2) *for each* $c > 0$ *the set* $V_m^{-1}[c, \infty]_0$ *contains no entire semitrajectories of the limiting system* (0.4.5) *except for* $\boldsymbol{y} = \mathbf{0}$.

Then the unperturbed motion $\boldsymbol{x} = \mathbf{0}$ *is asymptotically* $\boldsymbol{y}$*-stable.*

Another important advance toward solving the problem of eliminating the requirement of z-boundedness of solutions was made by Andreev [1987], [1991], who put forward the new idea of *z-limiting systems*, which, in contrast to Hatvani's limiting systems, make it possible *to take into account the z-properties* of the solutions of the original system.

The condition for existence of a system that is z-limiting in the sense of Andreev for autonomous system (0.3.1) is *z-precompactness* of its translations:

for any sequence $\eta_n \to \infty$ there must exist a subsequence $\{\eta_{nk}\} \in \{\eta_n\}$ and a vector function $\boldsymbol{X}^*(\boldsymbol{x})$ such that

$$\boldsymbol{X}(\boldsymbol{y}, \boldsymbol{z} + \eta_{nk}) \to \boldsymbol{X}^*(\boldsymbol{x})$$

uniformly on each compactum $\{\|\boldsymbol{y}\| \le H' < H, \|\boldsymbol{z}\| \le Q\}$.

Then the system

$$\dot{\boldsymbol{x}} = \boldsymbol{X}^*(\boldsymbol{x}) \tag{0.4.6}$$

is *$\boldsymbol{z}$-limiting* for (0.3.1).

In turn, (0.3.1) is $\boldsymbol{z}$-precompact if and only if the function $\boldsymbol{X}(\boldsymbol{x})$ is bounded and uniformly continuous on each set $S = \{\|\boldsymbol{y}\| \le H_1 < H, \|\boldsymbol{z}\| < \infty\}$. Similar conditions are required for the existence of *limiting V-functions.*

Theorem 0.4.9. (Andreev [1987]) *Suppose that for $\boldsymbol{z}$-precompact autonomous system* (0.3.1) *in domain* (0.3.2) *one can find a V-function such that*

(1) $V(\boldsymbol{x}) \ge a(\|\boldsymbol{y}\|)$, $\dot{V}(t, \boldsymbol{x}) \le -W(\boldsymbol{x}) \le 0$ *and*

(2) *functions V, W are $\boldsymbol{z}$-precompact and for any limiting set $(\boldsymbol{X}^*, V^*, W^*)$ that is defined for the same sequences $\eta_n \to \infty$ the set $\{V^*(\boldsymbol{x}) = \mathrm{const} > 0\} \cup \{W^*(\boldsymbol{x}) = 0\}$ contains no entire semitrajectories of the limiting system* (0.4.6).

Then the unperturbed motion $\boldsymbol{x} = \boldsymbol{0}$ is uniformly asymptotically $\boldsymbol{y}$-stable.

2. The case of nonautonomous systems. Matrosov [1962a] showed that the Barbashin–Krasovskii theorem cannot be generally applied to *nonautonomous* systems (periodic systems being an exception), which gave rise to two approaches in the theory of stability:

(1) one which rests on the use of *two* or *several* V-functions;

(2) the other is associated with the use of *limiting systems* and *limiting V-functions.*

Within the confines of these approaches analysis of nonautonomous systems no longer uses the condition that the set $M = \{\dot{V}(t, \boldsymbol{x}) = 0\}$ contains no entire semitrajectories for $t \in [t_0, \infty)$, which is caused, generally speaking, by the impossibility of relying on some *invariance* property of limiting sets of solutions in the nonautonomous case.

The difficulty can be overcome (Matrosov [1962a]) by introducing a *second auxiliary W-function* that "does not permit" solutions of a nonautonomous system to remain too long close to M. The concept of *vector function* V, which later led to the development of a general *principle of comparison with vector function* V, still greatly extends our ability to analyze the stability of nonautonomous systems. An overview of works in this area, connected with the PSt-problem, is presented in Subsections 0.4.9 and 0.4.10.

The other analytical approach associated with constructing limiting systems and V-functions considers those general classes of nonautonomous systems for which the limiting sets of solutions possess some *invariance-type* properties (that are characteristic of autonomous systems) with respect to the family of limiting systems. This approach uses methods of *topological dynamics* and, in particular, results of Sell [1967] and Artstein [1977].

Thus, having suggested an approach to constructing *t-limiting systems* $\dot{\boldsymbol{x}} = \boldsymbol{X}^*(t, \boldsymbol{x})$ and *t-limiting V-functions* for the original nonautonomous system (0.3.1) (under certain additional assumptions as to its right-hand sides), Andreev [1979], [1984] obtained a number of theorems on asymptotic (including uniform) $\boldsymbol{y}$-stability based on a single V-function with a sign-constant derivative. These theorems include conditions for asymptotic $\boldsymbol{y}$-stability that were obtained earlier for autonomous systems and are based on the *assumption of* $\boldsymbol{z}$*-boundedness* of the solutions of the original system.

Another approach to constructing limiting systems in the PSt-problem for nonautonomous systems was proposed by Hatvani [1983b], [1985b]. This approach is based on the study of the $\boldsymbol{y}$-behavior of the solutions of the original system by constructing the system $\dot{\boldsymbol{y}} = \boldsymbol{Y}_*(t, \boldsymbol{y})$ limiting to the original one relative to *each continuous function* $\boldsymbol{z} = \boldsymbol{z}(t) \to \infty$; we discussed this approach above for the case of autonomous systems. This approach also made it possible to obtain a number of conditions for asymptotic (including uniform) $\boldsymbol{y}$-stability based on the use of a single V-function with a sign-constant derivative. Here, it was possible to *eliminate* the requirement that solutions be $\boldsymbol{z}$-bounded.

Imposing certain additional constraints on the $\boldsymbol{Z}$-component $\boldsymbol{Z}(t, \boldsymbol{x})$ of the vector function $\boldsymbol{X} = \boldsymbol{X}(t, \boldsymbol{x})$ in right-hand side of the original nonautonomous system, Andreev [1991] introduced $\{t, \boldsymbol{z}\}$*-limiting systems* $\dot{\boldsymbol{x}} = \boldsymbol{X}^*(t, \boldsymbol{x})$ for them, that make it possible, in contrast to the case of Hatvani's limiting systems [1983b], [1985b], *to allow for the* $\boldsymbol{z}$*-properties of the solutions* of the original system as $\|\boldsymbol{z}\| \to \infty$.

Finally, Andreev [1991] derived general analytic continuity- and invariance-type properties for $\boldsymbol{y}$-positive limiting sets that make it possible to analyze the limiting $\boldsymbol{y}$-behavior of the solutions of the original system as a function of additional conditions imposed on its right-hand side.

0.4.6 Application of Differential Inequalities

The use of *differential inequalities* in the theory of stability goes back to the works by Corduneanu [1960], [1964].

Let the following differential inequality be satisfied in domain (0.3.2):

$$\dot{V} \le \omega(t, V(t, \boldsymbol{x})). \tag{0.4.7}$$

Here $\omega(t, v)$ is a function continuous for $t \ge 0$, $v \ge 0$ such that the right-hand side of the *comparison equation*

$$\dot{v} = \omega(t, v) \quad (\omega(t, v) \equiv 0) \tag{0.4.8}$$

satisfies conditions of existence and uniqueness of solutions at each point (t_0, v_0) from the domain of definition.

Theorem 0.4.10. (Corduneanu [1964]) *Suppose that for system* (0.3.1) *there exists a V-function satisfying the condition* $V \ge a(\|\boldsymbol{y}\|)$ *and differential inequality* (0.4.7) *in domain* (0.3.2).

Then

(1) *if the solution* $v = 0$ *of equation* (0.4.8) *is Lyapunov stable* (*asymptotically stable*), *then the unperturbed motion* $\boldsymbol{x} = \mathbf{0}$ *of system* (0.3.1) *is* $\boldsymbol{y}$*-stable* (*asymptotically* $\boldsymbol{y}$*-stable*);

(2) *if* $V \leq b(\|\boldsymbol{x}\|)$ *and the solution* $v = 0$ *of equation* (0.4.8) *is uniformly stable* (*uniformly asymptotically stable*), *then the unperturbed motion* $\boldsymbol{x} = \mathbf{0}$ *of system* (0.3.1) *is uniformly* $\boldsymbol{y}$*-stable (uniformly asymptotically* $\boldsymbol{y}$*-stable*).

Hatvani [1975b] proposed the generalization of Theorem 0.4.10 for the case when $\dot{V} \leq \omega(t, \boldsymbol{x}, V)$. These issues are highlighted in detail by Rumyantsev and Oziraner [1987].

0.4.7 Conditions for Global Asymptotic y-Stability

Rumyantsev and Oziraner [1987] demonstrated that *global* asymptotic $\boldsymbol{y}$-stability is guaranteed if we augment the conditions of Theorem 0.4.2 (which, in this case, are satisfied in the domain $t \geq 0$, $\|\boldsymbol{x}\| < \infty$) assuming

$$V(t, \boldsymbol{x}) \to \infty, \quad \|\boldsymbol{w}\| = (x_1^2 + \cdots + x_k^2)^{1/2} \to \infty. \tag{0.4.9}$$

Condition (0.4.9) might seem somewhat unexpected. Indeed, by analogy with the classic case, the condition

$$a(\|\boldsymbol{y}\|) \to \infty, \quad \|\boldsymbol{y}\| \to \infty \tag{0.4.10}$$

seems more natural.

It can be demonstrated (Vorotnikov [1993]), however, that the replacement of condition (0.4.10) with (0.4.9) does not guarantee global asymptotic $\boldsymbol{y}$-stability.

To this end, let us consider the system (Barbashin and Krasovskii [1952])

$$\dot{y}_1 = \frac{2(y_1 + z_1)}{(1 + z_1^2)^2}, \quad \dot{z}_1 = 2y_1 + \frac{2z_1}{(1 + z_1^2)^2}. \tag{0.4.11}$$

The analysis of the phase portrait of system (0.4.11) reveals that its zero solution is not globally asymptotically stable with respect to variable y_1. At the same time, the V-function

$$V = y_1^2 + \frac{z_1^2}{1 + z_1^2}$$

satisfies the conditions

$$a(\|\boldsymbol{y}\|) \leq V \leq b(\|\boldsymbol{x}\|), \quad \dot{V} \leq -c(\|\boldsymbol{x}\|),$$
$$a(\|\boldsymbol{y}\|) \to \infty, \quad \|\boldsymbol{y}\| \to \infty$$

over the entire (y_1, z_1)-plane.

However, the condition $V \to \infty$, $\|\boldsymbol{x}\| \to \infty$ fails to hold, since $V \to 1$ at $y_1 = 0$ as $|z_1| \to \infty$.

In contrast to Theorem 0.4.2, extending the conditions of theorems of type 0.4.4–0.4.6 (Risito [1970], Rumyantsev and Oziraner [1987]) to the problem of global asymptotic $\boldsymbol{y}$-stability, as in the classic case, requires (0.4.10).

0.4.8 Methods for Constructing V-Functions

The process of constructing V-functions inevitably raises the following question: Does a V-function with the corresponding properties always exist whenever a system exhibits one or another type of stability? The *problem of inverting* the MLF theorems arises, as first stated by Chetayev. As applied to the PSt-problem, these issues were considered by Zubov [1959], Halanay [1963], Corduneanu [1964], [1971], Oziraner [1971], [1986], and Savchenko and Ignat'ev [1989] and are presented in sufficient detail in the monograph by Rumyantsev and Oziraner [1987]. Here, as in the classic case, it has been possible to establish just the existence of V-functions for a number of MLF theorems. And although a positive answer to the question of existence of V-functions can promote the theoretical constructions that substantiate many problems of stability (in the linear approximation, in the presence of CAP, of parameters variations, etc.), the practically important problems of *constructively constructing* V-functions present major difficulties in the general theory of stability. The PSt-problem is no exception in this regard.

We distinguish four methods for constructing V-functions which, though not as general as desired and not always simple to use, nonetheless permit solving a number of applied problems within the scope of PSt-problems.

1. Chetayev's method of bundling first integrals [1946]. This method is rather well known and it is discussed in detail in the literature (Rouche et al. [1977]). It yields adequate results in analyzing the stability of *mechanical* systems. Similar matters are considered in the context of the PSt-problem by Rumyantsev [1957a], [1972b], [1987], Moiseev and Rumyantsev [1965], Rubanovskii and Rumyantsev [1979], Rumyantsev and Oziraner [1987], and Nabiullin [1990]. In definite cases, the PSt-analog of Pozharitskii's theorem (Rumyantsev and Oziraner [1987]) may be useful for constructing bundles of first integrals. As regards the question of using first integrals in the PSt-problem, we also note Risito [1971], which takes a somewhat different approach to the problem.

2. Zubov's method of constructing auxiliary systems [1959]. This method implies the availability of estimates (coarse as they might be) for uncontrolled $\boldsymbol{z}$-variables of system (0.3.1) which then can be used as a basis for constructing a certain auxiliary system of differential equations. Then the Lyapunov function for the auxiliary system, having been constructed, solves the PSt-problem for the original system.

3. The method of constructing different-type limiting systems and limiting V-functions. Of course, this method is not really a method of constructing V-functions, though the possibilities for finding "appropriate" V-function are extended within its scope. A similar approach to the PSt-problem has already been dealt with in Subsection 0.4.5.

4. The method of transformations of system (0.3.1) (Vorotnikov [1991a]). In this approach V-functions are constructed in the form of functions that are sign-definite with respect *not only* to $\boldsymbol{y}$-variables (relative to which stability is being investigated), but *also with respect to the auxiliary*

μ-variables that are some functions of the phase variables of the original system. The number and the specific form of such $\boldsymbol{\mu}$-variables depend on the PSt-problem under study. This method is described and developed as applied to the PSt-problem in Chapters 1 through 3.

0.4.9 Conditions for Asymptotic y-Stability with Two V-Functions

Matrosov's theorem [1962a] is fundamental to this approach; this theorem was later analyzed and generalized along various lines (see Rouche at al. [1977]). The application of this approach to the PSt-problem was investigated by Peiffer [1968], Fergola and Moauro [1970], Oziraner [1972c], Salvadori [1972], and Hatvani [1975a], [1984]. In the context of the PSt-problem Fergola and Moauro [1970] used the *one-parameter family of V-functions* introduced by Salvadori [1969]. The conditions for asymptotic $\boldsymbol{y}$-stability that go back to Marachkov's theorem are developed in Hatvani [1984].

0.4.10 Method of Lyapunov Vector Functions (MLVF) in PSt-Problem

Bellman [1962] and Matrosov [1962b] independently proposed that *several* Lyapunov functions satisfying a certain system of differential inequalities be simultaneously used in studying stability. Each of the Lyapunov functions in this set (*vector function*, following Bellman's terminology) is subject to less stringent requirements than those in theorems using a single Lyapunov function. The intensive development of this method resulted in a rather general *principle of comparison with the Lyapunov vector function* (Lakshmikantham et al. [1991]) being worked out. This principle has been successfully used in both theoretical and applied research on the dynamics of nonlinear systems, especially high-dimensional *complex systems.*

Matrosov [1965] applied the MLVF to the PSt-problem. The same problem was considered by Peiffer [1968], Peiffer and Rouche [1969], Shoichi et al. [1982], and Vorotnikov [1991a].

Theorem 0.4.11. (Matrosov [1969]) *Suppose that for system* (0.3.1) *in domain* (0.3.2) *it is possible to specify a vector function* $\boldsymbol{V}(t,\boldsymbol{x}) = [V_1(t,\boldsymbol{x}),\dots, V_k(t,\boldsymbol{x})]$ *satisfying the conditions*

(1) *functions* V_i *and* $\dot V_i$ *are continuous and* $V_i(t,\mathbf{0}) = \dot V_i(t,\mathbf{0}) = 0$;

(2) $\dot{\boldsymbol{V}}$ *satisfies the system of differential inequalities*

$$\dot{\boldsymbol{V}} \le \boldsymbol{f}(t,\boldsymbol{V}), \quad \boldsymbol{f}(t,\mathbf{0}) = \mathbf{0};$$

(3) *vector function* $\boldsymbol{f} = (f_1,\dots,f_k)$ *is defined and continuous in the domain* $t \ge 0$, $\|\boldsymbol{V}\| \le R$, *where* $R = \infty$ *or* $R > \sup[\|\boldsymbol{V}(t;\boldsymbol{x})\| : t \ge 0, \|\boldsymbol{y}\| \le H]$;

(4) *each function* $f_s(t,V_1,\dots,V_k)$ $(s = 1,\dots,k)$ *is nondecreasing with respect to* V_i $(i \ne s)$;

(5) *the condition*

$$\max V_i(t,\boldsymbol{x}) \ge a(\|\boldsymbol{y}\|) \quad (i=1,\dots,l)$$

is valid for some l $(1 \le l \le k)$; *and*

(6) *the solution* $\boldsymbol{\omega} = \mathbf{0}$ *of the comparison system* $\dot{\boldsymbol{\omega}} = \boldsymbol{f}(t,\boldsymbol{\omega})$ *is stable (asymptotically stable) with respect to* $\omega_1,\dots,\omega_l$.

Then the unperturbed motion $\boldsymbol{x} = \mathbf{0}$ *of system* (0.3.1) *is* $\boldsymbol{y}$*-stable (asymptotically* $\boldsymbol{y}$*-stable).*

We also note the investigations of the PSt-problem that involve

(*a*) *large-scale systems* on the basis of *Lyapunov matrix functions* (Martynyuk [1991a], [1991b]) and

(*b*) *complex perturbed interrelated systems* which split in the absence of perturbations (Strogaya [1990]).

0.4.11 Examples of Constructing Lyapunov V-Functions

To illustrate the use of certain of the MLF theorems discussed, let us consider some examples.

1. Let the system of equations of perturbed motion (0.3.1) have the form

$$\dot{y}_1 = -y_1 + z_1^2 z_2, \quad \dot{z}_1 = z_1 + y_1 z_1, \quad \dot{z}_2 = -3z_2 - 2y_1 z_2. \tag{0.4.12}$$

Among the roots of the characteristic equation of the system of linear approximation for equations (0.4.12), there is a root equal to 1. Therefore (Lyapunov [1892]), the unperturbed motion $y_1 = z_1 = z_2 = 0$ of system (0.4.12) is Lyapunov unstable.

As a Lyapunov function for system (0.4.12), let us take the nonlinear V-function

$$V = y_1^2 + 2y_1 z_1^2 z_2 + 2z_1^4 z_2^2. \tag{0.4.13}$$

Being a nonlinear function of variables y_1, z_1, and z_2, at the same time this function is a quadratic form in variables y_1, μ_1, where $\mu_1 = z_1^2 z_2$.

Since the following inequalities hold in the domain (0.3.2):

$$V \ge l_1 y_1^2 \quad (0 \le l_1 < 1/2),$$
$$\dot{V} = -2(y_1^2 + 2y_1 z_1^2 z_2 + 2z_1^4 z_2^2) \le -l_2 y_1^2 \quad (0 \le l_2 < 3/4),$$

then the unperturbed motion $y_1 = z_1 = z_2 = 0$ of system (0.4.12) is uniformly stable with respect to y_1 by Theorem 0.4.1.

In addition, V-function (0.4.13) satisfies the inequalities

$$V \le b(\|\boldsymbol{\xi}\|), \quad \dot{V} \le -c(\|\boldsymbol{\xi}\|), \quad \boldsymbol{\xi} = (y_1, z_1^2 z_2)$$

and, consequently, the unperturbed motion $y_1 = z_1 = z_2 = 0$ of system (0.4.12) is uniformly asymptotically stable with respect to y_1 by Corollary 0.4.1.

Let us note, by the way, that neither value of $k = 1, 2$ possible in this case permits the V-function to satisfy the conditions of Theorem 0.4.2.

The unperturbed motion $y_1 = z_1 = z_2 = 0$ of system (0.4.12) is uniformly stable not only with respect to y_1, but also with respect to $z_1^2 z_2$. This implies that for each $t_0 \ge 0$ there are numbers $\delta' > 0$ and $M > 0$ such that $|Y_1| \le M$, $Y_1 = -y_1 + z_1^2 z_2$ for all the solutions $\boldsymbol{x} = \boldsymbol{x}(t; t_0, \boldsymbol{x}_0)$ of system (0.4.12) for which $\|\boldsymbol{x}_0\| < \delta'$. Thus, V-function (0.4.13) satisfies the conditions of Theorem 0.4.3.

For V-function (0.4.13) in domain (0.3.2), the differential inequality

$$\dot{V} \leq \alpha V \quad (0 \leq \alpha \leq 2)$$

also holds, because the zero solution $v = 0$ of the comparison equation $\dot{v} = \alpha v$ is exponentially asymptotically stable. Therefore, the unperturbed motion $y_1 = z_1 = z_2 = 0$ of system (0.4.12) is uniformly stable and uniformly asymptotically stable with respect to y_1 by Theorem 0.4.10.

Finally, consider the vector function

$$\boldsymbol{V} = (V_1, V_2, V_3, V_4) = (y_1, -y_1, z_1^2 z_2, -z_1^2 z_2)$$

whose components V_i $(i = 1, \dots, 4)$ satisfy the differential inequalities

$$\dot{V}_1 \leq -\alpha_1 V_1 + V_3, \quad \dot{V}_3 \leq -\alpha_2 V_3,$$
$$\dot{V}_2 \leq -\alpha_1 V_2 + V_4, \quad \dot{V}_4 \leq -\alpha_2 V_4 \quad (0 \leq \alpha_{1,2} \leq 1).$$

Since the zero solution of the comparison system is Lyapunov asymptotically stable and $\max V_i \geq y_1^2$ $(i = 1, 2)$, then vector function $\boldsymbol{V}$ satisfies the conditions of Theorem 0.4.11.

We note that the equations of system (0.4.12) for $\dot{z}_i$ $(i = 1, 2)$ can be represented in the form

$$\dot{z}_1 = [1 + y_1(t; t_0, \boldsymbol{x}_0)]z_1, \quad \dot{z}_2 = -[3 + 2y_1(t; t_0, \boldsymbol{x}_0)]z_2.$$

Then, by virtue of stability with respect to y_1, taking into account that the component $y_1(t; t_0, \boldsymbol{x}_0)$ is defined for all $t \geq t_0$ for at least a sufficiently small value of $\|\boldsymbol{x}_0\|$, we conclude that the solutions of system (0.4.12) are $\boldsymbol{z}$-continuable.

2. Consider *Euler's dynamical equations*

$$\dot{y}_1 = \gamma_1 y_1 + \frac{1}{A}(B - C)y_3 z_1, \quad \dot{y}_2 = \gamma_2 y_2 + \frac{1}{B}(C - A)y_1 z_1,$$
$$\dot{z}_1 = \frac{1}{C}(A - B)y_1 y_2 \tag{0.4.14}$$

which describe the angular motion of a solid about its center of mass. The motion caused by initial perturbations occurs under the action of a linear dissipative moment

$$\boldsymbol{M} = (\gamma_1 y_1, \gamma_2 y_2, 0)$$

with incomplete dissipation. Here y_1, y_2, and z_1 are the projections of the instantaneous angular velocity of the solid onto its principal central axes of inertia; A, B, and C are the principal central moments of inertia of the solid.

As a V-function, we take the kinetic energy of the solid

$$V = \tfrac{1}{2}(Ay_1^2 + By_2^2 + Cz_1^2). \tag{0.4.15}$$

Since

$$\dot{V} = A\gamma_1 y_1^2 + B\gamma_2 y_2^2 \leq 0,$$

the equilibrium position of the solid is Lyapunov (nonasymptotically) stable. Hence, for each $t_0 \geq 0$ there are numbers $\delta' > 0$ and $M > 0$ such that $|Y_{1,2}| \leq M$ ($Y_{1,2}$ stands for expressions in the right-hand sides of the first two equations of system (0.4.14)) for all the solutions $\boldsymbol{x} = \boldsymbol{x}(t; t_0, \boldsymbol{x}_0)$ of system (0.4.14) for which $\|\boldsymbol{x}_0\| < \delta'$. Thus, V-function (0.4.15) satisfies the conditions of Theorem 0.4.3.

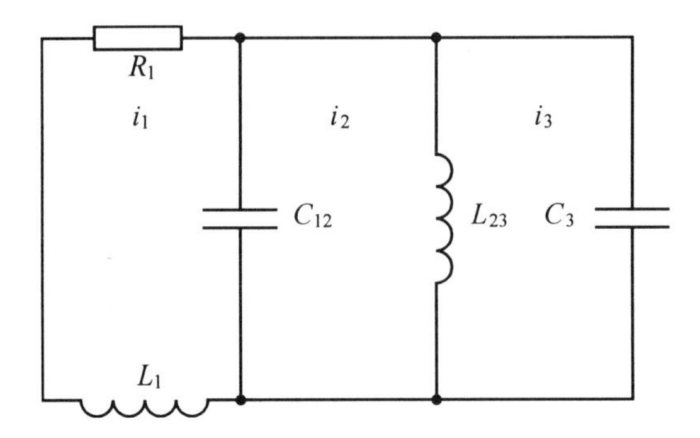

Figure 0.4.1. Scheme of an electric circuit.

If the conditions

$$V(\boldsymbol{x}) \geq a(\|\boldsymbol{x}\|), \quad \dot{V}(\boldsymbol{x}) \leq -c(\|\boldsymbol{x}\|)$$

of the Lyapunov theorem [1892] are satisfied in the problem of stability *with respect to all the variables*, the conditions of the Barbashin–Krasovskii theorem [1952] are satisfied automatically. In this connection, using system (0.4.14) as an example, let us show that, in the problem of stability *with respect to part of the variables*, the conditions

$$V(\boldsymbol{x}) \geq a(\|\boldsymbol{x}\|), \quad \dot{V}(\boldsymbol{x}) \leq -c(\|\boldsymbol{y}\|) \tag{0.4.16}$$

do not guarantee that the conditions of Theorems 0.4.4 and 0.4.6 are satisfied.

Indeed, the sets $M = \{\boldsymbol{x} : \dot{V}(\boldsymbol{x}) = 0\}$ and $M \cup \{\boldsymbol{x} : V(\boldsymbol{x}) > 0\}$ contain entire semitrajectories of system (0.4.14) of the form $y_1 = y_2 = 0$, $z_1 = z_1^* =$ const different from $y_1 = y_2 = z_1 = 0$. Therefore, V-function (0.4.15) does not satisfy all the conditions of Theorems 0.4.4 and 0.4.6.

At the same time, in this case the set $M_1 = \{\boldsymbol{x} : y_1 = y_2 = 0\}$ is an invariant set of system (0.4.14). The set $M \setminus M_1$ is empty and, consequently, contains no entire semitrajectories of this system. Therefore, V-function (0.4.15) satisfies all the conditions of Theorem 0.4.5.

However, there are examples of systems for which the conditions of none of Theorems 0.4.4–0.4.6 are satisfied, though conditions (0.4.16) hold. Such an example, as well as further modifications of Theorems 0.4.4–0.4.6, is cited in Chapter 3.

In the case $C < A, B$ or $C > A, B$ the *uniform* asymptotic stability with respect to y_1, y_2 can be established with the use of the V-function

$$V = A(A-C)y_1^2 + B(B-C)y_2^2,$$
$$\dot{V} = 2[A(A-C)\gamma_1 y_1^2 + B(B-C)\gamma_2 y_2^2],$$

which satisfy conditions of Theorem 0.4.2.

3. Consider an *electric circuit* presented in Figure 0.4.1.* The circuit contains inductors L_1, L_{23}, an ohmic resistor R_1, and capacitors C_{12}, C_3 (here L_1, L_{23}, R_1, C_{12}, and C_3 are constants).

---

*Gantmakher, F.R. [1966] *Lectures on Analytic Mechanics*, Moscow: Nauka (in Russian).

Using an electromechanical analogy, we can describe the evolution of a current i_s and a charge q_s ($i_s = \dot{q}_s$) in the circuit caused by the initial action of the electromotive force by the Lagrangian equations

$$\frac{d}{dt}\left(\frac{\partial T}{\partial \dot{q}_s}\right) - \frac{\partial T}{\partial q_s} = -\frac{\partial P}{\partial q_s} - \frac{\partial R}{\partial \dot{q}_s} \quad (s = 1, 2, 3), \tag{0.4.17}$$
$$T = {}^1\!/\!{}_2[L_1\dot{q}_1^2 + L_{23}(\dot{q}_2 - \dot{q}_3)^2],$$
$$R = {}^1\!/\!{}_2 R_1\dot{q}_1^2, \quad P = {}^1\!/\!{}_2[C_3^{-1}q_3^2 + C_{12}^{-1}(q_1 - q_2)^2],$$

where T and P are the kinetic and potential energies of the system, respectively, and R is Rayleigh's dissipative function.

Equations (0.4.17) reduce to the form

$$\begin{gathered} L_1\ddot{q}_1 = -R_1\dot{q}_1 - C_{12}^{-1}(q_1 - q_2), \\ L_{23}(\ddot{q}_2 - \ddot{q}_3) = C_{12}^{-1}(q_1 - q_2), \quad L_{23}(\ddot{q}_2 - \ddot{q}_3) = C_3^{-1}q_3. \end{gathered} \tag{0.4.18}$$

System (0.4.18) falls into the type of *linear singular* systems with constant coefficients unresolved for $\ddot{q}_s$ ($s = 1, 2, 3$) (alternative terms are *semistate* or *descriptor systems*). Such systems have been studied beginning with the work by Kronecker and Weierstrass.

Taking into account the equality

$$C_3(q_1 - q_2) = C_{12}q_3, \tag{0.4.19}$$

which is a consequence of the two last equations of system (0.4.18), we separate out the *regular* subsystem from system (0.4.18):

$$\begin{gathered} L_1\ddot{q}_1 = -R_1\dot{q}_1 - C_{12}^{-1}(q_1 - q_2), \\ C_3C_{12}^{-1}L_{23}\ddot{q}_1 - (L_{23} + C_3C_{12}^{-1})\ddot{q}_2 = C_{12}^{-1}(q_2 - q_1). \end{gathered} \tag{0.4.20}$$

As a result, instead of the singular system (0.4.18) we can deal with the *differential-algebraic* system comprising the regular system of differential equations (0.4.20) and the algebraic equation (0.4.19).

Let us divide the analysis of stability of the equilibrium position $q_s = \dot{q}_s$ ($s = 1, 2, 3$) of the circuit into six stages which constitute a sufficiently typical scheme for studying partial stability.

1st stage. As a V-function, we take the total energy of the system

$$V = T + P,$$
$$\begin{aligned} \dot{V} &= L_1\dot{q}_1\ddot{q}_1 + L_{23}(\dot{q}_2 - \dot{q}_3)(\ddot{q}_2 - \ddot{q}_3) + C_{12}^{-1}(q_1 - q_2)(\ddot{q}_1 - \ddot{q}_2) \\ &= -R_1\dot{q}_1^2 - C_{12}^{-1}\dot{q}_1(q_1 - q_2) + C_3^{-1}[q_3(\dot{q}_2 - \dot{q}_3) + q_3\dot{q}_3] + C_{12}^{-1}(q_1 - q_2)(\dot{q}_1 - \dot{q}_2) \\ &= -R_1\dot{q}_1^2 + C_3^{-1}\dot{q}_2q_3 - C_{12}^{-1}(q_1 - q_2)\dot{q}_2. \end{aligned}$$

Taking into account equality (0.4.19), we conclude that

$$\dot{V} = -R_1\dot{q}_1^2 \leq 0.$$

Since the V-function is positive-definite with respect to variables q_3, $\dot{q}_1$ and to variables $\mu_1 = q_1 - q_2$, $\mu_2 = \dot{q}_2 - \dot{q}_3$, the equilibrium position $q_s = \dot{q}_s = 0$ ($s = 1, 2, 3$) of the circuit is stable with respect to these variables by Theorem 0.4.1.

2nd stage. Analyzing the relationship obtained by differentiating both sides of equality (0.4.19) (with respect to t), we conclude that stability with respect to $\dot{q}_1$, μ_2 implies actually stability of the equilibrium position $q_s = \dot{q}_s = 0$ $(s = 1, 2, 3)$ of the circuit with respect to $\dot{q}_s$ $(s = 1, 2, 3)$.

3rd stage. Taking into account stability with respect to $\dot{q}_1$, μ_1, for any $t_0 \geq 0$, one can find numbers $\delta' > 0$ and $M > 0$ such that

$$\|Y(t, \boldsymbol{q}, \dot{\boldsymbol{q}})\| \leq M, \quad Y = L_1^{-1}[-R_1\dot{q}_1 - C_{12}^{-1}(q_1 - q_2)].$$

Hence, the equilibrium position $q_s = \dot{q}_s = 0$ $(s = 1, 2, 3)$ of the circuit is *asymptotically* stable with respect to $\dot{q}_1$ by Theorem 0.4.3.

4th stage. The structure of solutions of system (0.4.20) is such that asymptotic stability with respect to $\dot{q}_1$ is *exponential*. Therefore, the equilibrium position $q_s = \dot{q}_s = 0$ $(s = 1, 2, 3)$ of the circuit is not only asymptotically stable with respect to $\dot{q}_1$ but is also stable with respect to q_1.

5th stage. Since the equilibrium position $q_s = \dot{q}_s = 0$ $(s = 1, 2, 3)$ of the circuit is stable with respect to q_3, μ_1 (stage 1) and with respect to q_1 (stage 4), this position is stable with respect to q_s $(s = 1, 2, 3)$.

6th stage. Taking into account stability with respect to $\dot{q}_s$ $(s = 1, 2, 3)$ (stage 2) and exponential asymptotic stability with respect to $\dot{q}_1$ (stage 3), we conclude, finally, that the equilibrium position $q_s = \dot{q}_s = 0$ $(s = 1, 2, 3)$ of the circuit is
(a) Lyapunov stable (with respect to q_s, $\dot{q}_s$ $(s = 1, 2, 3)$) and
(b) exponentially asymptotically stable with respect to $\dot{q}_1$.

We should note that the equilibrium position of the circuit cannot be Lyapunov asymptotically stable. Indeed, any neighborhood of the equilibrium position $q_s = \dot{q}_s = 0$ $(s = 1, 2, 3)$ of the circuit contains nonzero equilibrium positions.

0.4.12 The MLF in the PSb-Problem

The fundamental Krasovskii theorem [1966] on optimal stabilization of motion (with respect to *all* the variables) is a modification of the classic Lyapunov theorem [1892] on asymptotic stability of motion. It was obtained employing Bellman's *dynamic programming* technique [1958]. Rumyantsev [1970], [1971b] extended this theorem to the problem of optimal stabilization with respect to *part* of the variables.

In what follows we use $\langle \cdot \rangle$ to denote the dot product of the corresponding vector functions.

Theorem 0.4.12. (Rumyantsev [1970], [1971b], [1993]) *Suppose that given system* (0.3.4) *it is possible to specify a function* $V^0 = V^0(t, \boldsymbol{x})$ *satisfying, in domain* (0.3.2), *the inequalities*

$$a(\|\boldsymbol{y}\|) \leq V^0(t, \boldsymbol{x}) \leq b(\|\boldsymbol{w}\|),$$
$$\|\boldsymbol{w}\| = (x_1^2 + \cdots + x_k^2)^{1/2}, \quad m \leq k \leq n$$

and a vector function $\boldsymbol{u} = \boldsymbol{u}^0(t, \boldsymbol{x}) \in U$ *such that*
(1) $W(t, \boldsymbol{x}) \equiv -\omega(t, \boldsymbol{x}, \boldsymbol{u}^0(t, \boldsymbol{x})) \leq -c(\|\boldsymbol{w}\|)$;

(2) $B[V^0, t, \boldsymbol{x}, \boldsymbol{u}^0(t, \boldsymbol{x})] = \dfrac{\partial V^0}{\partial t} + \langle \dfrac{\partial V^0}{\partial \boldsymbol{x}} \cdot \boldsymbol{X}(t, \boldsymbol{x}, \boldsymbol{u}^0) \rangle + \omega(t, \boldsymbol{x}, \boldsymbol{u}^0) \equiv 0$;

(3) $B[V^0, t, \boldsymbol{x}, \boldsymbol{u}] \geq 0$ *for any* $\boldsymbol{u} \in U$; *and*

(4) *for any* $\boldsymbol{u}^*(t, \boldsymbol{x}) \in U$ *that ensures asymptotic* $\boldsymbol{y}$*-stability of the unperturbed motion* $\boldsymbol{x} = \mathbf{0}$ *the following equality is true*:

$$\lim V^0(t, \boldsymbol{x}^*[t]) = 0, \quad t \to \infty.$$

Then vector function $\boldsymbol{u} = \boldsymbol{u}^0(t, \boldsymbol{x})$ *solves the problem of optimal* $\boldsymbol{y}$*-stabilization for system* (0.3.4). *In this case*

$$\int_{t_0}^{\infty} \omega(t, \boldsymbol{x}^0[t], \boldsymbol{u}^0[t])dt = \min \int_{t_0}^{\infty} \omega(t, \boldsymbol{x}^*[t], \boldsymbol{u}^*[t])dt = V^0(t_0, \boldsymbol{x}[t_0]).$$

We should note that when $k = m$, condition (4) is a consequence of the condition $V^0(t, \boldsymbol{x}) \leq b(\|\boldsymbol{w}\|)$. To warn one against inaccuracies, we also note that *in the case* $k \neq m$ condition (4) of Theorem 0.4.12 was initially omitted in a number of publications.

Thus far, no general constructive methods have been devised for constructing *optimal Lyapunov* V^0*-functions* that solve the PSb-problem. One possible approach to this problem was proposed by Rumyantsev [1970]. The approach suggests that the corresponding V-function for the uncontrolled part of system (0.3.4) should be taken as optimal. This results in a kind of "semireverse" choice of the performance functional I in which the integrand becomes completely defined (the structure is fixed in advance) only after solving the problem. This approach (including a number of examples drawn from mechanics) is presented in detail in the monograph by Rumyantsev and Oziraner [1987].

Another possible approach to solving the PSb-problem (Vorotnikov [1979a], [1982a], [1982b], [1990], [1991a], [1991b]) is discussed and developed in Chapter 4.

0.5 The Problem of Stability with Respect to Part of the Variables of Linear Systems, in Linear Approximation, and in Critical Cases

0.5.1 The PSt-Problem of Linear Systems

For linear systems, stability (with respect to either all or part of the variables) of one solution implies stability of any other, so we can speak simply of stability of the system.

Linear systems with constant coefficients (LSCC). Even in this case it is possible to find examples of systems that are stable (asymptotically stable) with respect to some variables, and simultaneously unstable (nonasymptotically stable) with respect to others. One such situation concerning a third-order LSCC is discussed in the monograph by Zubov [1959].

The transient process in the variable being examined for stability displayed high sensitivity to variation in the coefficients of the original system (though not all of them are equally involved). Peiffer [1968] also pointed out this inherent peculiarity of the PSt-problem—the dependence of the PSt-property *not only* on the roots of the characteristic equation of the LSCC, but also on the *form of its coefficients.*

By now the PSt-problem for LSCC has been analyzed in considerable detail (Vorotnikov [1991a]). Chapter 1 presents the material regarding the problem.

Linear systems with periodic coefficients (LSPC). In-depth analysis of the PSt-problem was performed for LSPC in which coefficients are *analytic* (Vorotnikov [1991a]). Chapter 1 presents details and advanced analysis of the problem.

Linear systems with continuous coefficients (LSConC). Three approaches have been proposed to handle the PSt-problem for this class of systems.

(1) *Zubov's approach* [1959] implies availability of *estimates* (*even though rough*) for the z-components of solutions of the original LSConC and reduces the PSt-problem to a problem of stability with respect to all the variables for an auxiliary linear system of the *same order.*

(2) *Corduneanu's approach* [1971] is connected with application of the MLF. In the context of this approach, *necessary and sufficient* conditions have been established for exponential asymptotic stability with respect to part of the variables.

Theorem 0.5.1. (Corduneanu [1971]) *A LSConC is exponentially asymptotically $\boldsymbol{y}$-stable if and only if there exists a V-function that satisfies the conditions*

$$\|\boldsymbol{y}\| \leq V(t, \boldsymbol{x}) \leq M\|\boldsymbol{x}\|, \quad \dot{V}(t, \boldsymbol{x}) \leq -\alpha V,$$
$$|V(t, \boldsymbol{x}') - V(t, \boldsymbol{x})| \leq M(\|\boldsymbol{y}' - \boldsymbol{y}\| + \|\boldsymbol{z}' - \boldsymbol{z}\|),$$

where M and α are positive constants.

(3) *The approach based on linear transformations of a LSConC* (Vorotnikov [1991a]) implies that part of the coefficients of the LSConC are sufficiently continuously differentiable functions. The approach is described and developed in Chapter 1.

0.5.2 The PSt-Problem in Linear Approximation

Let system (0.3.1) have the form

$$\dot{\boldsymbol{y}} = A\boldsymbol{y} + B\boldsymbol{z} + \boldsymbol{Y}(t, \boldsymbol{y}, \boldsymbol{z}), \quad \dot{\boldsymbol{z}} = C\boldsymbol{y} + D\boldsymbol{z} + \boldsymbol{Z}(t, \boldsymbol{y}, \boldsymbol{z}), \tag{0.5.1}$$

where A, B, C, and D are continuous matrix functions of appropriate dimensions for $t \in [t_0, \infty)$; $\boldsymbol{Y}$ and $\boldsymbol{Z}$ are nonlinear vector functions.

The first result in the PSt-problem in linear approximation was reported by Corduneanu [1971], who used a V-function satisfying the conditions of

Theorem 0.5.1, by virtue of the linear part of system (0.5.1), and a differential inequality obtained when deriving this V-function along the trajectories of the nonlinear system (0.5.1).

Then the issue of extending the Lyapunov theorem [1892] on stability in the linear approximation to the case of the PSt-problem was considered.

Theorem 0.5.2. (Oziraner [1973]) *Suppose that B is a zero matrix, the coefficients of matrices A, C, and D are constant, and in domain* (0.3.2) *the following condition is satisfied*:

$$||\boldsymbol{Y}(t, \boldsymbol{y}, \boldsymbol{z})|| \leq \alpha ||\boldsymbol{y}||, \tag{0.5.2}$$

where α is a sufficiently small positive constant.

Then, if all the eigenvalues of matrix A have negative real parts, the unperturbed motion $\boldsymbol{x} = \boldsymbol{0}$ of system (0.5.1) *is uniformly asymptotically stable* (*on the whole with respect to $\boldsymbol{z}_0$*).

Condition (0.5.2) is rather general. However its constructive verification is complicated. (An exclusion is the case when solutions of system (0.5.1) are known in advance to be uniformly $\boldsymbol{z}$-bounded). In such a situation further investigation has followed two lines:

(1) search for constructively verifiable conditions for $\boldsymbol{y}$-stability in the linear approximation for system (0.5.1) (Vorotnikov [1979a], [1979b], [1988a], [1991a]);

(2) consideration of narrower classes of systems (Oziraner [1975], Oziraner and Rumyantsev [1972], Vorotnikov [1988a], [1991a]).

These approaches are presented and developed in Chapter 2.

0.5.3 The PSt-Problem in Critical Cases

Following Lyapunov [1892], one can introduce the notion of *critical cases* in the PSt-problem (Prokop'ev [1975]).

Such cases occur if, under conditions of Theorem 0.5.2, there are no eigenvalues with positive real parts among the eigenvalues of matrix A, but there are eigenvalues with negative and zero real parts. Then, by an appropriate choice of vector function $\boldsymbol{Y}$ satisfying condition (0.5.2), one can ensure, if desired, either asymptotic $\boldsymbol{y}$-stability or $\boldsymbol{y}$-instability of the unperturbed motion $\boldsymbol{x} = \boldsymbol{0}$ of system (0.5.1).

The work by Prokop'ev [1975] studies one of critical cases in the PSt-problem, i.e., the case of *one zero root.* Other critical cases were considered by Shchennikov [1985] and Lizunova [1991].

0.5.4 The PSt-Problem for Some Specific Classes of Systems

A number of publications (see, for example, Cesari [1959], Hatvani and Terjeki [1985], Kertész [1991], Hatvani [1991]) consider the PSt-problem for specific classes of linear and nonlinear systems.

Thus, for example, Cesari [1959] cited a number of works of the 30s which consider the problem of attraction of the equilibrium position $x = \dot{x} = 0$ with respect to the coordinate x only (but not to the velocity $\dot{x}$) for the linear equation

$$\ddot{x} = \varphi(t)x.$$

Hatvani and Terjeki [1985] analyze the nonlinear system of the form

$$\frac{d}{dt}(p(t)\dot{x}) + g(t, x, \dot{x})\dot{x} + q(t)f(x) = 0, \quad xf(x) \geq 0,$$

and Kertész [1991] considers the nonlinear system

$$\ddot{x} + g(t, x, \dot{x})\dot{x} + q(t, x, \dot{x})x = f(t, x, \dot{x}).$$

These systems are significant for a number of applications, such as the dynamics of holonomic mechanical systems with one degree of freedom.

Also noteworthy is Avakumovic's* theorem of the 40s on x-attraction of the equilibrium position $x = \dot{x} = 0$ for the nonlinear equation

$$\ddot{x} = \varphi(t)x^{\lambda}, \quad \lambda > 1,$$

whose simpler proof was subsequently presented by Geluk in 1991.

Among still earlier results close to analysis of the PSt-problem, we should also note Esclangon's result**: if, for the equation

$$x^{(n)} + a_1(t)x^{(n-1)} + \cdots + a_0(t)x = 0, \tag{0.5.3}$$

whose coefficients are continuous and bounded functions of $t \in [t_0, \infty)$, the condition $|x(t)| \leq M_1 = \text{const} > 0$ is satisfied, then there exists a positive number M_2 such that $|x^{(k)}(t)| \leq M_2$ $(k = 1, \ldots, n-1)$. Bearing in mind that boundedness of solutions for equation (0.5.3) (with respect to all or to part of the variables) is equivalent to their stability (Bellman [1953]), we conclude that stability of the equilibrium position $x^{(n-1)} = \cdots = x = 0$ of equation (0.5.3) with respect to coordinate x implies stability with respect to derivatives $x^{(k)}$ $(k = 1, \ldots, n-1)$.

Esclangon's result has been extended to linear equations *with delays*, including *neutral-type* equations (Kolmanovskii and Nosov [1981]).

0.6 Special Features and Possibilities of the Problem of Stability with Respect to Part of the Variables

0.6.1 Preliminary Remarks

The special features of PSt-system performance deserve to be digested. The point is that the PSt-theory deals with rather delicate properties of the sys-

*Avakumovic, V.G. [1947] *Publ. Inst. Math.*, 1, 101–113.

Esclangon, E. [1917] Nouvelles Recherches sur les Fonctions Quasiperiodiques, *Ann. Obs. Bordeaux*, **16, 51–226.

tem. The necessary "persistence" of these properties depends on a larger number of factors than do the properties of "overall" stability. What is required is a deeper understanding of the nature of PSt-problems, of the laws governing the performance of PSt-systems, and of the mechanisms through which PSt-properties may be achieved (or lost). One needs also to recognize pitfalls along the path to practical application of alluring possibilities of PSt-problems.

The purpose of this section is to study these questions.

0.6.2 Special Features in the Performance of PSt-Systems

Now we formulate a few assumptions (Vorotnikov [1993], [1995a], [1995b]) that afford a deeper understanding of the special features of PSt-system performance.

1. *The predictability of structural changes as a precondition for normal PSt-system performance.* This assumption is motivated by the greater sensitivity of the PSt-property (as compared with stability with respect to all the variables) to changes in system structure. The implication is that the idea of "robustness" in the PSt-theory cannot be as general as it is in the theory of stability with respect to all the variables. This is natural, because PSt-theory is concerned with more "delicate" cases, in which "improved" or "better" stability is simply impossible. In addition, as was already noted, PSt-properties are sometimes not just desirable but absolutely necessary, whereas universal auxiliary function of the PSt-problem can be revealed in establishing various properties of the system regarding its "robustness."

For example, partial stabilization (a development of the partial stability problem as applied to controlled systems) plays an auxiliary role in the design of robust control rules for the angular motion of a solid (such as spacecraft), which is the subject of Chapter 4. Chapter 5 extends this approach to game-theoretical control problems under conditions of interference.

A better understanding of the problem may also be gained by clarifying the nature of the relationships among notions that determine whether PSt-properties are preserved. Among the latter are PSt-properties in the presence of constantly acting perturbations (CAP) and parametric perturbations.

2. *The PSt-problem in the presence of CAP is not generally equivalent to the PSt-problem of preserving stability even under small parametric perturbations.* This is even the case for linear autonomous systems. Though partially stable in the presence of CAP, such systems may lose their stability when even a slight "stir" is given to certain coefficients. This does not occur in the problem of stability with respect to all the variables.

Owing to this conclusion, any decision about whether to use the results of PSt-theory must be made at the design stage in each specific case. Here it is important to have constructively verifiable conditions for preserving PSt-properties at the researcher's disposal.

To this end, together with the original system (0.3.1), consider the "*perturbed system*"

$$\dot{\boldsymbol{x}} = \boldsymbol{X}(t, \boldsymbol{x}) + \boldsymbol{R}(t, \boldsymbol{x}), \quad \boldsymbol{R}(t, \boldsymbol{0}) \equiv \boldsymbol{0}. \tag{0.6.1}$$

System (0.6.1) is assumed to satisfy the same general requirements as system (0.3.1).

Theorem 0.6.1. (Corduneanu [1964]) *Suppose that for system* (0.3.1) *in domain* (0.3.2) *there is a V-function such that*

$$V(t, \boldsymbol{x}) \geq a(\|\boldsymbol{y}\|), \quad |V(t, \boldsymbol{x}_2) - V(t, \boldsymbol{x}_1)| \leq L\|\boldsymbol{x}_2 - \boldsymbol{x}_1\|,$$
$$\dot{V}(t, \boldsymbol{x}) \leq 0.$$

Then it is possible to specify a continuous monotone increasing function $d(r)$, $d(0) = 0$, *for* $r \in [0, H]$, *such that if conditions*

$$\|\boldsymbol{R}(t, \boldsymbol{x})\| \leq \varphi(t) d(\|\boldsymbol{y}\|), \quad \int_0^\infty \varphi(\tau) d\tau < +\infty$$

are satisfied, then the unperturbed motion $\boldsymbol{x} = \boldsymbol{0}$ *of system* (0.6.1) *is uniformly* $\boldsymbol{y}$*-stable.*

At the other end of the "fragility" scale for PSt-properties one has the following.

3. *A system that loses the PSt-property is nevertheless frequently "coarse" in the Andronov–Pontryagin sense.* In principle, the phase portrait of a "coarse" system is invariant under minor "stirs" of the parameters. Hence the loss of PSt-properties in such cases implies only a certain "*rotation*" of the phase portrait in the corresponding phase plane.

Thus, for example, the position $y_1 = z_1 = 0$ of the system

$$\dot{y}_1 = -y_1 + \varepsilon z_1, \quad \dot{z}_1 = z_1 \quad (\varepsilon = \text{const}) \tag{0.6.2}$$

is asymptotically y_1-stable for $\varepsilon = 0$ only. At the same time, for any ε, system (0.6.2) is a system of "general position" (Arnold [1990]) with a "saddle"-type phase portrait. When ε passes through the value $\varepsilon = 0$, one observes just the "rotation" of the phase portrait. However, the rotation is small, provided that ε is sufficiently small. When $\varepsilon = 0$, the separatrices of the "saddle" take the positions of the coordinate axes in the phase plane of system (0.6.2).

4. *The possibility of the invariance of PSt-properties in the presence of arbitrarily large CAP* in certain channels of system (0.3.1). This question is related to the general problem of invariance (for details, see Subsection 1.4.4).

Consider *Euler's dynamical equations*

$$A\dot{x}_1 = (B - C)x_2 x_3 + u_1 \quad (123, ABC) \tag{0.6.3}$$

which describe the angular motion of a solid driven by controlling torques u_i ($i = 1, 2, 3$). (Only one of the three equations of system (0.6.3) is written, the others are obtained by cyclic permutation of the indices $1 \to 2 \to 3$ and coefficients $A \to B \to C$.)

As has been shown in Example 0.4.1, in the case $u_i = \alpha_i x_i$ ($\alpha_i = \text{const} < 0$, $i = 1, 2$), $u_3 \equiv 0$, the equilibrium position $x_i = 0$ ($i = 1, 2, 3$) of the solid is asymptotically (x_1, x_2)-stable (property K) for any admissible values of A, B, and C. And then, if C is the greatest or smallest of A, B, and C, property K holds for any x_{30} (property K_1).

Property K_1 may be verified by using a Lyapunov function independent of x_3, namely, $V = A(A - C)y_1^2 + B(B - C)y_2^2$. Consequently, it is also preserved in the presence of CAP $\boldsymbol{R} = \boldsymbol{R}(t, \boldsymbol{x})$, $\boldsymbol{R}(t, \boldsymbol{0}) \not\equiv \boldsymbol{0}$ of the form $\boldsymbol{R} = (0, 0, R_3)$, i.e., in the presence of CAP in one channel of system (0.6.3). (Here $R_3(t, \boldsymbol{x})$ is any function such that the conditions for existence, uniqueness, and x_3-continuability of solutions of the "perturbed" system (0.6.3) hold in the domain $t \geq 0$, $|x_i| \leq H$ ($i = 1, 2$), $|x_3| < \infty$.) Thus, property K_1 is *invariant* in the presence of CAP of the type described.

Theoretically, cases are also possible in which a PSt-property is not only preserved but also becomes *asymptotic* when CAP are applied (see example in Section 4.1).

0.6.3 Mechanisms Giving Rise to PSt-Properties

The "degree of coarseness" of a PSt-system depends directly on the mechanism by which PSt-properties arise. Here we distinguish one aspect of this problem.

PSt-properties may be produced by different mechanisms even in linear systems. We shall dwell on this problem in Chapter 1.

0.6.4 Relationships between Conditions for y-Stability and z-Continuability of Solutions of System (0.3.1)

As has been noted, solutions of system (0.3.1) may not be $\boldsymbol{z}$-continuable even if they are $\boldsymbol{y}$-stable. For that reason $\boldsymbol{z}$-continuability and $\boldsymbol{y}$-stability are generally studied separately. In fact, $\boldsymbol{y}$-stability conditions in themselves may enable systems with $\boldsymbol{z}$-noncontinuable solutions to "pass" as admissible.

For example, consider the system

$$\dot{y}_1 = -y_1 + y_1^2 z_1, \quad \dot{z}_1 = z_1 - 2y_1 z_1^2. \tag{0.6.4}$$

The Lyapunov function $V = y_1^2 + (y_1 + y_1 z_1^2)^2 \geq y_1^2$, $\dot{V} = -4y_1^2 \leq 0$ satisfies the conditions of the Rumyantsev theorem on y_1-stability of the position $y_1 = z_1 = 0$. The conditions of Theorem 2.4.1 on asymptotic y_1-stability are also satisfied, since system (0.6.4) admits the construction of the auxiliary $\boldsymbol{\mu}$-system $\dot{y}_1 = -y_1 + \mu_1$, $\dot{\mu}_1 = -\mu_1$.

However, the solutions of system (0.6.4) (assuming, for simplicity, that $t_0 = 0$) have the form

$$y_1 = y_{10}(1 + y_{10}z_{10}t)e^{-t}, \quad z_1 = z_{10}(1 + y_{10}z_{10}t)^{-1}.$$

Consequently, they are not z_1-continuable. In an arbitrarily small neighborhood of the position $y_1 = z_1 = 0$, there are fixed values y_{10}, z_{10} ($y_{10}z_{10} < 0$) such that $z_1 = \infty$ for $t = -(y_{10}z_{10})^{-1} > 0$.

Thus, both the method of Lyapunov functions and the method of constructing auxiliary μ-systems being developed in this book in PSt-problems may enable systems with z-noncontinuable solutions to "pass" as admissible.

0.7 The Problem of Control with Respect to Part of the Variables in a Finite Time Interval

0.7.1 Preliminary Remarks

The problematics of PSt- and PSb-problems are closely related to the problem of *control with respect to part of the variables* in a finite time interval. Being rather natural for many controlled systems, this problem has long been studied in the literature. Apart from the term "control with respect to part of the variables" (Krasovskii [1968], Akulenko [1980], [1987], [1994a], Vorotnikov [1982b], [1991a], Mukharlyamov [1989], Kovalev [1993], [1994]), there are a number of widely used terms:

(1) "control *in output*" (Desoer and Vidyasagar [1975]);

(2) "control *with free endpoints*" (Pontryagin et al. [1961]);

(3) "control *in configuration space*" (Roitenberg [1987]); and

(4) "control *on manifolds*" (Zubov [1980], [1982]).

Besides, the control problems of the "*hard encounter*" type also relate to problems of control with respect to part of the variables (see, for example, Akulenko [1994b]). In such problems it is required to ensure a rendezvous of two controlled objects with arbitrary velocities at the time of encounter. In particular, these problems relate to military applications: hitting mobile and immobile targets. A new class of the problems of the "hard encounter" type (Vorotnikov [1997b], [1997d]) is distinguished in Sections 4.6 and 5.2. These are the problems of reorienting a spacecraft without damping its final angular velocity. As a result, a spacecraft just "passes" through a given angular position without making a stop in it.

We also note that the problems of control with respect to part of the variables (and the problems of partial stabilization) naturally arise when a controller falls into the class of *dynamic regulators*. In this case, the variables characterizing the state of the main object are taken as the controlled variables (or the variables being stabilized).

Two principal situations that provide motives stimulating the research of these problems can distinguished:

(1) when it is *sufficient* to solve control problems only with respect to part of the variables characterizing the state of a system and

(2) when controllability of a system with respect to all the variables is *not possible* at all (for example, because the system possesses some *first integrals*).

By now, the control theory has a long history. However, the problem of finding *conditions for controllability* of a system was given a mathematically rigorous formulation not so long ago (Kalman [1960]). Apparently, it was considered for a long time that, designed professionally, a control system cannot be uncontrollable.* However, the growth of complexity of systems being developed revised this point of view. So, an investigation of the problem of controllability became a vital necessity.

The literature on the subject presents a diversity of statements of the problem of controllability (see, for example, Krasovskii [1987]); a number of them concern the feasibility of control with respect to *part* of the variables (part of the components of the phase vector of a system).

0.7.2 Controllability of Linear Systems with Respect to Part of the Variables

Consider a linear controlled system with constant coefficients

$$\dot{\boldsymbol{x}} = A\boldsymbol{x} + B\boldsymbol{u}, \quad \boldsymbol{x} \in R^n, \quad \boldsymbol{u} \in R^r. \tag{0.7.1}$$

Theorem 0.7.1. (Kalman [1960]) *For system* (0.7.1) *to be completely controllable with respect to all the variables, it is necessary and sufficient that*

$$\operatorname{rank} K_n = n, \quad K_n = (B, AB, \ldots, A^{n-1}B).$$

Let us consider the *problem of complete controllability with respect to part of the variables*, whose strict formulation can be found in Section 1.6. To this end, we divide the variables constituting the phase vector of system (0.7.1) into two groups: $\boldsymbol{x}^{\mathsf{T}} = (\boldsymbol{y}^{\mathsf{T}}, \boldsymbol{z}^{\mathsf{T}})$, $\boldsymbol{y} \in R^m$, $\boldsymbol{z} \in R^p$, $m + p = n$.

Theorem 0.7.2. (Roitenberg [1987]) *For complete $\boldsymbol{y}$-controllability of system* (0.7.1) *it is necessary and sufficient that*

$$\operatorname{rank} CK_n = m,$$

where $C = (I_m, 0)$ is an $m \times n$ matrix.

Other (sufficient) conditions for complete $\boldsymbol{y}$-controllability were also obtained (Vorotnikov [1979a], [1982b], [1991a]), which are discussed in Section 1.6. These conditions rest on the possibility of constructing some auxiliary linear autonomous controlled system (through a special choice of control structure) with subsequent verification of the conditions of Theorem 0.7.1 for this system.

Among the possible tools for effectively solving the problems of *optimal* control with respect to part of the variables are, for example, *Pontryagin's maximum principle* (Pontryagin et al. [1961]) and *Krasovskii's method of extremum aiming* (Krasovskii [1970]); the latter applies to the case of control in the presence of uncontrollable interference.

---

*See, for example: Voronov A.A. [1979] *Stability. Controllability. Observability*, Moscow: Nauka (in Russian).

In particular, Pontryagin's maximum principle was employed to thoroughly investigate the time-optimal problem of the "hard encounter" of two mass points (Pontryagin et al. [1961], Akulenko [1994b]). Krasovskii's method enabled solving this problem stated as game-theoretic (Krasovskii [1970]).

0.7.3 Control of Nonlinear Systems with Respect to Part of the Variables

A number of methods are known for investigating the problems of control (including those of optimal control) with respect to part of the variables for nonlinear systems, among which are the following:

(1) *Pontryagin's maximum principle* for problems with free endpoints (Pontryagin et al. [1961]);

(2) *methods of the theory of games* in the case of control in the presence of uncontrollable interference (Krasovskii and Subbotin [1988]);

(3) *the asymptotic method* (Akulenko [1987], [1994a]);

(4) *the method of oriented manifolds* (Kovalev [1993], [1994]);

(5) the *method of nonlinear transformations of the variables* combined with a special choice of control structure (Vorotnikov [1994a]–[1994c], [1995c], [1997b]–[1997e]).

Method (5) is described in Sections 4.5, 4.6 and Chapter 5.

As appropriate examples of using methods (1), (3), and (4) in nonlinear applied control problems, we note the problem of controlling the motion of an aircraft (Akulenko [1980], [1987], [1994a]), a nonlinear version of the "hard encounter" problem (Akulenko [1994b]), and the problem of damping rotational motions of an asymmetric solid with a single flywheel (Kovalev [1993]).

Chapter 1

Linear Problems of Stability, Stabilization, and Control with Respect to Part of the Variables

In this chapter, the problem of stability (asymptotic stability) of motion with respect to a *given* part of the variables for linear systems with *constant* and *sufficiently differentiable periodic* coefficients is proved equivalent to a Lyapunov stability (asymptotic stability) problem for either an original or some auxiliary linear system of dimension less than that of the original one. The auxiliary system coefficients are constants or periodic (in the general case, *discontinuous*) functions according to whether the original system coefficients are constants or periodic functions.

An algorithm for constructing the above mentioned auxiliary system is presented along with algebraic conditions which provide a way of determining the auxiliary system dimension from the original system coefficients. Matrix relationships are obtained involving coefficients of both the original system and the equation used to determine characteristic numbers of the auxiliary system matrix. This allows stating the *necessary and sufficient* conditions concerning stability (asymptotic stability) with respect to a given part of the variables for the classes of linear systems under consideration and also permits deriving the rule for constructing Lyapunov functions.

The properties of Lyapunov functions constructed in this way are compared with those imposed on the auxiliary functions in well-known theorems of the method of Lyapunov functions. The peculiarities of constantly acting perturbations are analyzed which are not amenable to investigation in terms of stability with respect to all the variables.

A constructive procedure is described for constructing control rules in the problem of motion *stabilization* with respect to a given part of the variables for linear systems with constant coefficients. For the very class of controlled systems the sufficient condition for *complete controllability* with respect to part of the variables is obtained.

Possible extensions of the principal results to more general classes of linear, including controlled, systems are indicated.

1.1 Stability with Respect to Part of the Variables of Linear Systems with Constant Coefficients

1.1.1 Formulation of the Problem

Consider a real system of linear differential equations of perturbed motion with *constant* coefficients

$$\frac{dx_i}{dt} = \sum_{j=1}^{n} A_{ij} x_j \qquad (i = 1, \ldots, n),$$

$$\boldsymbol{x}^{\mathsf{T}} = (x_1, \ldots, x_n) = (y_1, \ldots, y_m, z_1, \ldots, z_p) = (\boldsymbol{y}^{\mathsf{T}}, \boldsymbol{z}^{\mathsf{T}}),$$
$$m > 0, \qquad p \geq 0, \qquad n = m + p.$$

In $\boldsymbol{y}$, $\boldsymbol{z}$ variables the system takes the form

$$\begin{aligned} \frac{dy_i}{dt} &= \sum_{k=1}^{m} a_{ik} y_k + \sum_{l=1}^{p} b_{il} z_l \qquad (i = 1, \ldots, m), \\ \frac{dz_j}{dt} &= \sum_{k=1}^{m} c_{jk} y_k + \sum_{l=1}^{p} d_{jl} z_l \qquad (j = 1, \ldots, p), \end{aligned} \tag{1.1.1}$$

where a_{ik}, b_{il}, c_{jk}, and d_{jl} are constants. We shall study the problem of stability (asymptotic stability) of the unperturbed motion $y_i = 0$, $z_j = 0$ $(i = 1, \ldots, m;\ j = 1, \ldots, p)$ of system (1.1.1) with respect to $y_1, \ldots, y_m$.

When analyzing the asymptotic stability with respect to all the variables, one can solve the problem with the use of *Hurwitz's algebraic criterion*. The criterion tests the roots of the characteristic equation* of system (1.1.1) for negativity of their real parts without explicitly evaluating the roots. Then it appears reasonable in studies of asymptotic stability with respect to part of the variables to find conditions that do not rely on evaluating the roots of the characteristic equation and the eigenvectors. Such conditions allow drawing the conclusion about asymptotic stability of the unperturbed motion of system (1.1.1) with respect to part of the variables based solely on the coefficients of the system. Besides, while solving the task just mentioned we

*From here on, for brevity, the characteristic equation $\det(A^* - \lambda I_n) = 0$ for the matrix $A^* = (A_{ij})$ $(i, j = 1, \ldots, n)$ of coefficients of system (1.1.1) is referred to just as the characteristic equation of this system.

shall try to elaborate an approach to analysis of stability with respect to part of the variables of linear systems with constant coefficients and also of linear systems with variable coefficients and nonlinear systems.

The idea of constructing some *auxiliary system* of differential equations is taken as the basis of such an approach. Proceeding from Lyapunov stability (asymptotic stability) analysis of an auxiliary system, we draw the conclusion about stability (asymptotic stability) of the unperturbed motion of the original system with respect to a given part of the variables.

1.1.2 An Auxiliary System of Equations

Consider vectors $\boldsymbol{b}_i = (b_{i1}, \dots, b_{ip})^{\mathsf{T}}$ $(i = 1, \dots, m)$ composed of the corresponding coefficients of system (1.1.1). Without loss of generality, suppose that the first m_1 of them $\boldsymbol{b}_1, \dots, \boldsymbol{b}_{m_1}$ $(m_1 \le m)$ are linearly independent. Then the vectors $\boldsymbol{b}_{m_1+1}, \dots, \boldsymbol{b}_m$ are linearly expressible in terms of the first ones. To construct an auxiliary system of equations, we introduce the new variables

$$\mu_i = \sum_{l=1}^{p} b_{il} z_l \qquad (i = 1, \dots, m_1). \tag{1.1.2}$$

The coefficients b_{il} $(i = 1, \dots, m_1;\ l = 1, \dots, p)$ in (1.1.2) form a set of linearly independent columns $\boldsymbol{b}_1, \dots, \boldsymbol{b}_{m_1}$.

With the new variables thus introduced, two cases are possible.

The first case. System (1.1.1) can be reduced to the form

$$\frac{dy_i}{dt} = \sum_{k=1}^{m} a_{ik} y_k + \sum_{l=1}^{m_1} \alpha_{il} \mu_l \qquad (i = 1, \dots, m),$$

$$\frac{d\mu_j}{dt} = \sum_{k=1}^{m} a_{jk}^* y_k + \sum_{l=1}^{p} b_{jl}^* z_l = \sum_{k=1}^{m} a_{jk}^* y_k + \sum_{l=1}^{m_1} \alpha_{jl}^* \mu_l \qquad (j = 1, \dots, m_1),$$

$$\frac{dz_s}{dt} = \sum_{k=1}^{m} c_{sk} y_k + \sum_{l=1}^{p} d_{sl} z_l \qquad (s = 1, \dots, p),$$

where α_{il}, α_{jl}^*, a_{jk}^*, and b_{jl}^* are constants. The behavior of variables $y_1, \dots, y_m$ of system (1.1.1), with respect to which the stability of the unperturbed motion is being studied, is completely determined by the system

$$\begin{aligned} \frac{dy_i}{dt} &= \sum_{k=1}^{m} a_{ik} y_k + \sum_{l=1}^{m_1} \alpha_{il} \mu_l \qquad (i = 1, \dots, m), \\ \frac{d\mu_j}{dt} &= \sum_{k=1}^{m} a_{jk}^* y_k + \sum_{l=1}^{m_1} \alpha_{jl}^* \mu_l \qquad (j = 1, \dots, m_1). \end{aligned} \tag{1.1.3}$$

In what follows, the system of type (1.1.3) is termed ***μ**-system* relative to the original system (1.1.1).

The second case. The variables defined in (1.1.2) do not reduce system (1.1.1) to the form (1.1.3). Suppose that only the first m_2 ($m_2 < m_1$) equalities are valid

$$\sum_{l=1}^{p} b_{jl}^{*} z_l = \sum_{v=1}^{p} b_{jv} \left(\sum_{l=1}^{p} d_{vl} z_l \right) = \sum_{l=1}^{m_1} \alpha_{jl}^{*} \mu_l, \tag{1.1.4}$$

whereas remaining $m_1 - m_2$ equalities (1.1.4) are impossible. In that case system (1.1.1) comprises the following equations:

$$\begin{aligned}
\frac{dy_i}{dt} &= \sum_{k=1}^{m} a_{ik} y_k + \sum_{l=1}^{m_1} \alpha_{il} \mu_l && (i = 1, \ldots, m), \\
\frac{d\mu_j}{dt} &= \sum_{k=1}^{m} a_{jk}^{*} y_k + \sum_{l=1}^{m_1} \alpha_{jl}^{*} \mu_l && (j = 1, \ldots, m_2), \\
\frac{d\mu_s}{dt} &= \sum_{k=1}^{m} a_{sk}^{*} y_k + \sum_{l=1}^{p} b_{sl}^{*} z_l && (s = m_2 + 1, \ldots, m_1), \\
\frac{dz_r}{dt} &= \sum_{k=1}^{m} c_{rk} y_k + \sum_{l=1}^{p} d_{rl} z_l && (r = 1, \ldots, p), \\
b_{sl}^{*} &= \sum_{v=1}^{p} b_{sv} d_{vl} && (s = m_2 + 1, \ldots, m_1).
\end{aligned} \tag{1.1.5}$$

Assuming that the first m_3 vectors of $\boldsymbol{b}_{m_2+1}^{*}, \ldots, \boldsymbol{b}_{m_1}^{*}$ ($m_3 \le m_1 - m_2$) are linearly independent and the remaining ones are linearly expressed via them, now we introduce the new variables once again:

$$\mu_{m_1+\gamma} = \sum_{l=1}^{p} b_{m_2+\gamma,l}^{*} z_l \qquad (\gamma = 1, \ldots, m_3).$$

Thereafter system (1.1.1) can be reduced either to system of type (1.1.3) or that of (1.1.5). In the first case the purpose of performing the transformations, i.e., constructing auxiliary $\boldsymbol{\mu}$-system, is attained. In the second case we should proceed to introduce new variables.

Now we show that the following lemma is valid.

Lemma 1.1.1. *An auxiliary $\boldsymbol{\mu}$-system for system* (1.1.1) *can always be constructed by introducing $R \le p$ sets of new variables of the form* (1.1.2) *(generally speaking, each set includes only one new variable). The dimension of $\boldsymbol{\mu}$-system does not exceed that of the original one.*

Proof. Suppose that by successively repeating the above described procedure of introducing new variables, we obtain a set of $p = n - m$ new linearly independent variables

$$\mu_j = \sum_{l=1}^{p} b_{jl}^{\vee} z_l \quad (j = 1, \ldots, p); \qquad b_{jl}^{\vee} = b_{jl} \quad (j = 1, \ldots, m_1), \text{ etc.}$$

Then, entering additional new variables loses its meaning, since each successive newly introduced variable

$$\mu_{p+r} = \sum_{l=1}^{p} b_{p+r,l}^{\vee} z_l \qquad (r = 1, 2, \dots)$$

can be represented as a linear combination of $\mu_1, \dots, \mu_p$. Indeed, $\det B^{\vee} \neq 0$, $B^{\vee} = (b_{jl}^{\vee})$ $(j, l = 1, \dots, p)$ and in this case the algebraic system of p equations

$$b_{11}^{\vee}\lambda_{1r} + \dots + b_{p1}^{\vee}\lambda_{pr} = b_{p+r,1}^{\vee}, \dots, b_{1p}^{\vee}\lambda_{1r} + \dots + b_{pp}^{\vee}\lambda_{pr} = b_{p+r,p}^{\vee}$$

for all $r = 1, 2, \dots$ has a nontrivial solution $\boldsymbol{\lambda}_1^0, \dots, \boldsymbol{\lambda}_p^0$. Consequently, $\mu_{p+r} = \lambda_{1r}^0\mu_1 + \dots + \lambda_{pr}^0\mu_p$ $(r = 1, 2, \dots)$.

Because the variables μ_{p+r} $(r = 1, 2, \dots)$ are linear combinations of the variables $\mu_1, \dots, \mu_p$, an auxiliary system will be closed relative to the variables $y_1, \dots, y_m, \mu_1, \dots, \mu_p$. Thus, the maximum dimension of $\boldsymbol{\mu}$-system is equal to $m + p = n$. The statement is proved. □

The concept of an auxiliary $\boldsymbol{\mu}$-system will be of primary importance in formulating the criterion for stability (asymptotic stability) with respect to part of the variables of the unperturbed motion of system (1.1.1).

Remark 1.1.1.

1. The equations and variables of $\boldsymbol{\mu}$-system split into subsets so that each equation of the corresponding subset except those of the last one can involve variables from all the previous and the single following subsets and does not include variables from the subsequent subsets.

2. A set of m_1 $(m_1 \leq m)$ linearly independent vectors out of $\boldsymbol{b}_1, \dots, \boldsymbol{b}_m$ may be chosen in a *nonunique* way. The same is true for the new variables $\mu_1, \dots, \mu_N$ $(N \leq p)$, required to form $\boldsymbol{\mu}$-system. However, as we show later, the dimension of $\boldsymbol{\mu}$-system and the roots of its characteristic equation, *do not depend* on ambiguity in the choice of the new variables.

1.1.3 Dimension of an Auxiliary System

We represent system (1.1.1) in the vector form

$$\dot{\boldsymbol{y}} = A\boldsymbol{y} + B\boldsymbol{z}, \qquad \dot{\boldsymbol{z}} = C\boldsymbol{y} + D\boldsymbol{z} \qquad (\dot{\boldsymbol{x}} = A^*\boldsymbol{x}), \tag{1.1.6}$$

where A^*, A, B, C, and D are constant matrices of appropriate sizes. We now find conditions which enable us to determine the dimension of an auxiliary $\boldsymbol{\mu}$-system solely from the coefficients of the original system (1.1.1).

The new variables μ_i $1 \leq i \leq p$, required to form an auxiliary $\boldsymbol{\mu}$-system, are chosen from the variables $\boldsymbol{\mu} = B\boldsymbol{z}$ (in the vector form) and also from the variables

$$\boldsymbol{\mu}^{(1)} = B\dot{\boldsymbol{z}} = BD\boldsymbol{z}, \qquad \boldsymbol{\mu}^{(2)} = B\ddot{\boldsymbol{z}} = BD^2\boldsymbol{z}, \dots,$$
$$\boldsymbol{\mu}^{(k)} = B\boldsymbol{z}^{(k)} = BD^k\boldsymbol{z} \qquad (1 \leq k \leq p - 1),$$

obtained by differentiating $\boldsymbol{\mu}$ k times by virtue of system (1.1.6). To determine a number of variables required to form an auxiliary $\boldsymbol{\mu}$-system (and, consequently, to determine the dimension of an auxiliary $\boldsymbol{\mu}$-system), consider the matrix

$$K_p = (B^\top, D^\top B^\top, \ldots, (D^\top)^{p-1} B^\top). \tag{1.1.7}$$

Lemma 1.1.2. *For a $\boldsymbol{\mu}$-system to be of dimension $m+h$, it is necessary and sufficient that* $\operatorname{rank} K_p = h$.

Proof. 1. First we note that if any column of matrix $(D^\top)^s B^\top$ is a linear combination of the columns of matrix K_s, then for any i $(i > s)$ an arbitrary column of matrix $(D^\top)^i B^\top$ is also a linear combination of the columns of matrix K_s.

Indeed, let $\boldsymbol{b}_i$ $(i = 1, \ldots, m)$ be the columns of matrix $B^\top$. Then $(D^\top)^s B^\top = ((D^\top)^s \boldsymbol{b}_1, \ldots, (D^\top)^s \boldsymbol{b}_m)$ and equalities

$$(D^\top)^s \boldsymbol{b}_j = \sum_{k=1}^m \lambda_{jk}^{(0)} \boldsymbol{b}_k + \sum_{k=1}^m \lambda_{jk}^{(1)} D^\top \boldsymbol{b}_k + \cdots + \sum_{k=1}^m \lambda_{jk}^{(s-1)} (D^\top)^{s-1} \boldsymbol{b}_k, \tag{1.1.8}$$

where $\lambda_{jk}^{(0)}, \ldots, \lambda_{jk}^{(s-1)}$ are constants, yield

$$\begin{aligned}(D^\top)^{s+1} \boldsymbol{b}_j &= D^\top((D^\top)^s \boldsymbol{b}_j) \\ &= \sum_{k=1}^m \lambda_{jk}^{(0)} D^\top \boldsymbol{b}_k + \cdots + \sum_{k=1}^m \lambda_{jk}^{(s-1)} (D^\top)^s \boldsymbol{b}_k.\end{aligned} \tag{1.1.9}$$

Taking into account that equality (1.1.8) holds for the last term of (1.1.9), we conclude that any column of matrix $(D^\top)^{s+1} B^\top$ is a linear combination of matrix K_s columns.

2. *Necessity.* Let the dimension of $\boldsymbol{\mu}$-system be equal to $m + h$. Using *reductio ad absurdum*, we assume that $\operatorname{rank} K_p = r \neq h$. Suppose that $r > h$. In this case there exists s $(s \leq p)$ such that matrix K_s has h linearly independent columns. Since the dimension of a $\boldsymbol{\mu}$-system is $m+h$ and the new variables required to form a $\boldsymbol{\mu}$-system are chosen from variables $\boldsymbol{\mu}^{(k)} = BD^k \boldsymbol{z}$ $(k = 1, \ldots, p-1)$, $\boldsymbol{\mu} = B\boldsymbol{z}$, then any column of matrix $(D^\top)^i B^\top$ $(i > s)$ is a linear combination of K_s columns. But this is impossible because in accordance with the statement in the first part of the proof, there are only h linearly independent columns in matrix K_p, which contradicts the assumption. Thus, $r \leq h$. If $r < h$, then the dimension of the $\boldsymbol{\mu}$-system does not reach $m + h$. This contradicts the condition of the lemma. Therefore, $r = h$.

3. *Sufficiency.* Let $\operatorname{rank} K_p = h$. We have already shown that the number of variables μ_j, which form the $\boldsymbol{\mu}$-with y_i $(i = 1, \ldots, m)$, system coincides with the number of linearly independent columns of matrix K_p. Consequently, the dimension of the $\boldsymbol{\mu}$-system equals $m + h$. The lemma is proved. □

Corollary 1.1.1. *For a $\boldsymbol{\mu}$-system have a dimension less than that of system* (1.1.1), *it is necessary and sufficient that* $\operatorname{rank} K_p < p$.

It is not necessary to construct the whole matrix K_p for calculating the dimension of a μ-system. It suffices to find a minimum number s for which $\operatorname{rank} K_{s-1} = \operatorname{rank} K_s$. Then the dimension of the μ-system according to Lemma 1.1.2 is equal to $m + \operatorname{rank} K_{s-1}$. This relationship makes it possible to reduce the computational burden while determining the dimension of an auxiliary μ-system. In particular, at the first step of introducing new variables a μ-system can be constructed if and only if $\operatorname{rank} B^{\mathsf{T}} = \operatorname{rank}(B^{\mathsf{T}}, D^{\mathsf{T}} B^{\mathsf{T}})$.

Remark 1.1.2.

1. The dimension of a μ-system depends on the coefficients of matrices B and D and does not depend on the coefficients of matrices A and C.

2. Matrices of type (1.1.7) are often dealt with in the *theory of control processes* (Kalman [1960b], Pontryagin et al. [1961]). To draw an analogy, along with (1.1.6) let us formally introduce the *controlled* system

$$\dot{\boldsymbol{\xi}} = D^{\mathsf{T}} \boldsymbol{\xi} + B^{\mathsf{T}} \boldsymbol{u}, \tag{1.1.10}$$

where $\boldsymbol{\xi}$ and $\boldsymbol{u}$ are the state and control vectors respectively. Then the rank of matrix (1.1.7) defines the *controllability properties* of system (1.1.10). However, the condition involving the rank of matrix (1.1.7) is to be interpreted rather as the condition for *observability* for the process $\dot{\boldsymbol{z}} = C\boldsymbol{y} + D\boldsymbol{z}$ in terms of $B\boldsymbol{z}$. This is quite in compliance with the essence of the phenomenon: if the processes in some part of the system are not observable in the remaining part, then they do not affect the stability of the latter. The relationship between the proposed method of investigating stability with respect to part of the variables and the *observability problem* was analyzed in detail (Cheremenskii [1987]).

1.1.4 Structure, Spectrum and Fundamental Matrix of an Auxiliary System

Let us assume that s is the minimum number for which $\operatorname{rank} K_{s-1} = \operatorname{rank} K_s$ holds. We define the matrices L_i $(i = 1, \ldots, 5)$ of the following type:

(*a*) The rows of $h \times p$ matrix L_1 are linearly independent columns of matrix K_{s-1}.

(*b*) The columns of $h \times h$ matrix L_2 are the first h columns of matrix L_1. (By appropriately reindexing variables $z_1, \ldots, z_p$ of system (1.1.1), if necessary, we can always obtain matrix L_1 whose *first* h columns are linearly independent.)

(*c*) The first h rows of $p \times h$ matrix L_3 are the rows of matrix L_2^{-1}, the inverse of matrix L_2. The remaining rows of matrix L_3 are zero row-vectors.

(*d*)

$$L_4 = \begin{pmatrix} I_m & 0 \\ 0 & L_1 \end{pmatrix}, \qquad L_5 = \begin{pmatrix} I_m & 0 \\ 0 & L_3 \end{pmatrix},$$

where I_m is the identity matrix of order m.

Thus the matrices L_i $(i = 1, \dots, 5)$ introduced enable us to establish the direct relationship between the coefficient values of the original system (1.1.1) and an auxiliary μ-system.

Lemma 1.1.3. 1. *An auxiliary μ-system for* (1.1.1) *comprises the equations*

$$\dot{\boldsymbol{\xi}} = L_4 A^* L_5 \boldsymbol{\xi}. \tag{1.1.11}$$

2. *The set of the characteristic equation roots of the μ-system is a subset of the set (in a particular case, coinciding with the whole set) of the characteristic equation roots of system* (1.1.1).

3. *The fundamental matrices $X(t)$, $G(t)$ of system* (1.1.1) *and the μ-system, respectively, satisfy the relationship*

$$G(t) = L_4 X(t), \qquad t \geq t_0.$$

Proof. 1. Let us show that equations (1.1.11) form a μ-system for (1.1.1). The passage from the original system (1.1.1) to a μ-system is equivalent to the linear change of variables $\boldsymbol{w} = L\boldsymbol{x}$ of the original system, where

$$L = \begin{pmatrix} I_m & 0 \\ 0 & L^* \end{pmatrix}, \qquad L^* = \begin{pmatrix} L_1 \\ L_* \end{pmatrix},$$

and an arbitrary $(n-m-h) \times p$ matrix L_* is such that matrix L is nonsingular. In that case the first $m+h$ equations of the system $\dot{\boldsymbol{w}} = LA^*L^{-1}\boldsymbol{w}$ will form the μ-system.

Let l^-_{kr} $(k, r = 1, \dots, n)$ be the elements of matrix L^{-1}, the inverse of L. Since the first h columns of L_1 are linearly independent, we can take zero columns as the first h columns of L_*. The elements of the inverse matrix are defined by the equations

$$l^-_{kr} = \frac{(-1)^{k+r} \det L_{rk}}{\det L} \qquad (k, r = 1, \dots, n),$$

where matrix L_{rk} is obtained from L by deleting its rth row and kth column.

Since $\det L_2 \neq 0$, L_* can be chosen as $L_* = (0, I_{p-h})$.

As a result, we arrive at the equalities

$$l^-_{kr} = \begin{cases} 1, & k = r, \\ 0, & k \neq r; \quad k, r = 1, \dots, m; \end{cases}$$

$$l^-_{kr} = 0 \quad \text{for} \quad \begin{cases} k = 1, \dots, m; & r = m+1, \dots, m+h; \\ k = m+1, \dots, m+h; & r = 1, \dots, m; \\ k = m+h+1, \dots, n; & r = 1, \dots, m+h; \end{cases}$$

$$l^-_{m+s, m+v} = \frac{(-1)^{s+v} \det L_{2vs}}{\det L} = \frac{(-1)^{s+v} \det L_{2vs}}{\det L_2} \qquad (v, s = 1, \dots, h),$$

where matrix L_{2vs} is obtained from L_2 by deleting its vth row and sth column. Thus, $L_5 = (l^-_{ij})$ $(i = 1, \dots, n;\ j = 1, \dots, m+h)$ and, consequently, the elements of matrix $L_4 A^* L_5$ are the elements of matrix $LA^* L^{-1}$ with indices $i, j = 1, \dots, m+h$. Hence, equations (1.1.11) form a $\boldsymbol{\mu}$-system for (1.1.1).

2. The first $m+h$ equations of the system $\dot{\boldsymbol{w}} = LA^* L^{-1} \boldsymbol{w}$ form an auxiliary $\boldsymbol{\mu}$-system for (1.1.1). Taking into account that the equations $\det(A^* - \lambda I_n) = 0$ and $\det(LA^* L^{-1} - \lambda I_n) = 0$ have the same roots, we infer that the set of the characteristic equation roots of the $\boldsymbol{\mu}$-system is a subset (in particular, the whole set) of the set of the characteristic equation roots of the original system (1.1.1).

3. The fundamental matrices $X(t)$ and $X^*(t)$ of solutions of system (1.1.1) and the system $\dot{\boldsymbol{w}} = LA^* L^{-1} \boldsymbol{w}$ obtained from it by nonsingular transformation of variables, respectively, satisfy the relationship $X^*(t) = LX(t)$, $t \geq t_0$. Since the first $m+h$ equations of the system $\dot{\boldsymbol{w}} = LA^* L^{-1} \boldsymbol{w}$ form the $\boldsymbol{\mu}$-system, the fundamental matrix $G(t)$ of solutions of the $\boldsymbol{\mu}$-system is specified by the relationship $\{G(t), 0\} = L_4 X(t)$, $t \geq t_0$. (The reason for writing the left-hand side in the form $\{G(t), 0\}$ is that matrices $G(t)$ and $L_4 X(t)$ are formally of different dimensions $(m+h) \times (m+h)$ and $(m+h) \times n$, respectively, though the last $n - m - h$ columns of $L_4 X(t)$ are actually zero.) In the equivalent form, $G(t) = L_4 X(t) L_5$, $t \geq t_0$. The proposition is proved. □

Remark 1.1.3.

1. In the case when all the roots λ_i $(i = 1, \dots, n)$ of the characteristic equation of system (1.1.1) are pairwise distinct (we take this case for simplicity and as the most illustrative), *fundamental matrices* $X(t)$, $G(t)$ of solutions of system (1.1.1) and the $\boldsymbol{\mu}$-system, respectively, can be represented in the form

$$X(t) = X_0 \left[e^{\lambda_1 t}, \dots, e^{\lambda_n t}\right]^{\mathsf{T}}$$
$$G(t) = G_0 \left[e^{\lambda_{i_1} t}, \dots, e^{\lambda_{i_{m+h}} t}\right]^{\mathsf{T}}.$$

Here $n \times n$ matrix X_0 and $(m+h) \times (m+h)$ matrix G_0 are constant; $\{\lambda_{i_j}\}$ is a subset Λ_μ of the set $\Lambda = \{\lambda_i\}$ $(i = 1, \dots, n)$ and consists of those $\{\lambda_i\}$ which are also the roots of the characteristic equation of the $\boldsymbol{\mu}$-system. Columns $\boldsymbol{X}_{0i}$ $(i = 1, \dots, n)$ of X_0 and $\boldsymbol{G}_{0j}$ $(j = 1, \dots, m+h)$ of G_0 are the *eigenvectors* of matrices A^* and $L_4 A^* L_5$ of system (1.1.1) and the $\boldsymbol{\mu}$-system, respectively (i.e., the eigenvectors of system (1.1.1) and the $\boldsymbol{\mu}$-system) which correspond to the *eigenvalues* λ_i $(i = 1, \dots, n)$ and λ_{i_j} $(j = 1, \dots, m+h)$.

Allowing for the structure of the fundamental matrices $X(t)$ and $G(t)$ indicated, from $G(t) = L_4 X(t) L_5$, $t \geq t_0$, we infer that the eigenvectors of system (1.1.1) and the $\boldsymbol{\mu}$-system satisfy the relationship $G_0 = L_4 X_0 L_5$ or, which is the same, $\{G_0, 0\} = L_4 X_0$. The eigenvectors $\boldsymbol{G}_{0j}$ of an auxiliary $\boldsymbol{\mu}$-system are obtained by projecting eigenvectors $\boldsymbol{X}_{0i_j}$, $\lambda_{i_j} \in \Lambda$ of the original system into a subspace of dimension $m+h$ (the transformation leaves the first m components unchanged)

$$\boldsymbol{X}_{0i_j} = \begin{pmatrix} \boldsymbol{X}_{0i_j y} \\ \boldsymbol{X}_{0i_j z} \end{pmatrix} \Rightarrow \boldsymbol{G}_{0j} = \begin{pmatrix} \boldsymbol{X}_{0i_j y} \\ L_1 \boldsymbol{X}_{0i_j z} \end{pmatrix} \qquad (j = 1, \dots, m+h),$$

where vectors $\boldsymbol{X}_{0i_j y}$ and $\boldsymbol{X}_{0i_j z}$ consist of the first m and the last p components of vector $\boldsymbol{X}_{0i_j}$, respectively.

2. Since the coefficients of system (1.1.1) and the μ-system are *real-valued* and, consequently, so are those of the characteristic equations of these systems, the characteristic equations together with each complex root (if any) have the root conjugate of it. Therefore, while constructing an auxiliary μ-system, we obtain the set of its characteristic equation roots Λ_μ which along with every complex root $\lambda_i \in \Lambda$ contains the root $\bar{\lambda}_i \in \Lambda$ conjugate of it. Thus, while the roots of the characteristic equation of system (1.1.1) are "passing" into the set Λ_μ, each of them enters the set inseparably with the root conjugate of it.

1.1.5 A Criterion for Stability with Respect to Part of the Variables

Let us show that the conditions for Lyapunov stability (asymptotic stability) of the zero solution of an auxiliary μ-system are necessary and sufficient for the unperturbed motion of the original system (1.1.1) to be stable (asymptotically stable) with respect to $y_1, \ldots, y_m$. This enables us to derive the criterion for asymptotic stability of the unperturbed motion of system (1.1.1) with respect to $y_1, \ldots, y_m$, which can be verified by applying Hurwitz's algebraic criterion directly to the coefficients of the original system (1.1.1).

Theorem 1.1.1. *For the unperturbed motion $\boldsymbol{x} = \mathbf{0}$ of system (1.1.1) to be asymptotically stable with respect to $y_1, \ldots, y_m$, it is necessary and sufficient that the zero solution of an auxiliary μ-system $\dot{\boldsymbol{\xi}} = L_4 A^* L_5 \boldsymbol{\xi}$ be Lyapunov asymptotically stable. This implies that all the roots of the equation*

$$\det(L_4 A^* L_5 - \lambda I_{m+h}) = 0 \qquad (1.1.12)$$

or, which is the same, of the equation

$$\det\begin{pmatrix} A - \lambda I_m & BL_3 \\ L_1 C & L_1 D L_3 - \lambda I_h \end{pmatrix} = 0$$

have negative real parts.

Proof. Sufficiency. The conditions for negativity of real parts of equation (1.1.12) roots are necessary and sufficient for Lyapunov asymptotic stability of the zero solution $\boldsymbol{\xi} = \mathbf{0}$ of an auxiliary μ-system. Hence, taking into account that $y_i = \xi_i$ $(i = 1, \ldots, m)$ hold for the first m components of $\boldsymbol{x}$ and $\boldsymbol{\xi}$ state vectors of system (1.1.1) and μ-system, respectively, we conclude that the conditions of Theorem 1.1.1 are sufficient for asymptotic stability of the unperturbed motion $\boldsymbol{x} = \mathbf{0}$ of the original system (1.1.1) with respect to $y_1, \ldots, y_m$.

Let us prove the *necessity* of the theorem conditions. First we note that asymptotic stability of variables $y_1, \ldots, y_m$ of system (1.1.1) is actually *exponential* asymptotic. Directly integrating the first m equations of system (1.1.1), we determine that, for this to be true, the following inequalities must be satisfied along the trajectories of system (1.1.1):

$$\left| \sum_{l=1}^{p} b_{il} z_l(t) \right| \le \alpha_i e^{-\beta_i (t - t_0)} \qquad (i = 1, \ldots, m),$$

where α_i and β_i are positive constants. Then, for asymptotic stability of the unperturbed motion of system (1.1.1) with respect to $y_1, \dots, y_m$, it is also necessary that variables (1.1.2) be constructed from the roots of the characteristic equation of system (1.1.1) with negative real parts. This completes the proof for the case when system (1.1.1) is reduced to (1.1.3). If system (1.1.1) is not reduced to the form (1.1.3) and, consequently, has the form (1.1.5), then we repeat the above arguments for system (1.1.5). Since constructing a $\boldsymbol{\mu}$-system is always possible, the statement is proved.

The equivalence of the equations appearing in the conditions of the theorem follows from the chain of equalities

$$\begin{aligned} L_4 A^* L_5 &= \begin{pmatrix} I_m & 0 \\ 0 & L_1 \end{pmatrix} \begin{pmatrix} A & B \\ C & D \end{pmatrix} \begin{pmatrix} I_m & 0 \\ 0 & L_3 \end{pmatrix} \\ &= \begin{pmatrix} A & B \\ L_1 C & L_1 D \end{pmatrix} \begin{pmatrix} I_m & 0 \\ 0 & L_3 \end{pmatrix} = \begin{pmatrix} A & B L_3 \\ L_1 C & L_1 D L_3 \end{pmatrix}. \end{aligned}$$

The theorem is proved. □

Theorem 1.1.2. *For the unperturbed motion* $\mathbf{x} = \mathbf{0}$ *of system* (1.1.1) *to be (nonasymptotically) stable with respect to* $y_1, \dots, y_m$, *it is necessary and sufficient that the zero solution of an auxiliary* $\boldsymbol{\mu}$*-system* $\dot{\boldsymbol{\xi}} = L_4 A^* L_5 \boldsymbol{\xi}$ *be Lyapunov stable.*

Proof. Necessity. Let us assume that the unperturbed motion of system (1.1.1) is (nonasymptotically) stable with respect to $y_1, \dots, y_m$. Allowing for the structure of solutions of linear system (1.1.1), we note that, along with the inequalities $|y_i(t)| < \varepsilon$ $(i = 1, \dots, m)$, the inequalities $|\dot{y}_i(t)| < \varepsilon_1$ $(i = 1, \dots, m)$ also hold, a number ε_1 being sufficiently small provided ε is sufficiently small. Then the form of the first m equations of system (1.1.1) yields the necessity of conditions

$$\left| \sum_{l=1}^{p} b_{il} z_l(t) \right| < \varepsilon_2 \qquad (i = 1, \dots, m),$$

where ε_2 is a sufficiently small positive constant. Thus, variables (1.1.2), and y_i $(i = 1, \dots, m)$, must satisfy the estimate $|\mu_j(t)| < \varepsilon_2$ $(j = 1, \dots, m_1)$. In the case when system (1.1.1) is reduced to the form (1.1.3), the inequalities $|y_i(t)| < \varepsilon$, $|\mu_j| < \varepsilon_2$ are possible, provided that the zero solution of the $\boldsymbol{\mu}$-system is Lyapunov (not necessarily asymptotically) stable. Therefore, for the case when system (1.1.1) is transformable to (1.1.3), the necessity is proved. Otherwise, we apply the above reasoning to system (1.1.5).

The *sufficiency* is obvious.

The theorem is proved. □

Remark 1.1.4. The set of the characteristic equation roots of a $\boldsymbol{\mu}$-system is a subset of the set of the characteristic equation roots of the original system (1.1.1). Thus, asymptotic stability of the unperturbed motion of system (1.1.1) with respect to $y_1, \dots, y_m$ requires that some roots with negative real parts be "in reserve." For example, if the dimension of the μ-system equals N, then, for asymptotic stability of the unperturbed motion of system (1.1.1) with respect to $y_1, \ldots, y_m$, it is necessary that at least N roots of the characteristic equation of this system have negative real parts.

Corollary 1.1.2. 1. *Let only m of the roots of the characteristic equation of system* (1.1.1) *have negative real parts (and the remaining roots have nonnegative real parts). For the unperturbed motion of system* (1.1.1) *to be asymptotically stable with respect to $y_1, \ldots, y_m$, it is necessary and sufficient that system* (1.1.1) *have the form*

$$\dot{y} = Ay, \qquad \dot{z} = Cy + Dz \tag{1.1.13}$$

and all the roots of the equation $\det(A - \lambda I_m) = 0$ *have negative real parts.*

2. *If the characteristic equation of system* (1.1.1) *has at least one root with a positive real part and* $\operatorname{rank} K_p = p$, *then unperturbed motion of system* (1.1.1) *is unstable with respect to $y_1, \ldots, y_m$.*

Proof. 1. The *sufficiency* is obvious.

To prove the *necessity*, we assume the opposite, i.e., that system (1.1.1) is not of the form (1.1.13). Then the dimension of the μ-system exceeds m. Since the roots of the μ-system belong to the set of the characteristic equation roots of the original system (1.1.1), the characteristic equation of the μ-system possesses at least one root with a nonnegative real part. From Theorem 1.1.1 it follows that the unperturbed motion of system (1.1.1) cannot be asymptotically stable with respect to $y_1, \ldots, y_m$, which contradicts the assumption made. System (1.1.1) has the form (1.1.13) and all the roots of the equation $\det(A - \lambda I_m) = 0$ have negative real parts.

2. Since $\operatorname{rank} K_p = p$, the dimension of the μ-system is equal to that of the original system. Thus, the characteristic equation of the μ-system has a root with a positive real part. By Theorem 1.1.2 we conclude that the unperturbed motion of system (1.1.1) is unstable with respect to $y_1, \ldots, y_m$. The proposition is proved. □

Corollary 1.1.3. *If* $\operatorname{rank} K_p = p$, *then, for the unperturbed motion of system* (1.1.1) *to be (asymptotically) stable with respect to $y_1, \ldots, y_m$, it is necessary and sufficient that the unperturbed motion of this system be Lyapunov (asymptotically) stable.*

Remark 1.1.5.

1. The first part of Corollary 1.1.2 was obtained earlier by a different method (Peiffer [1968]).

2. The proposed method for constructing μ-systems was extended to linear *discrete* systems (Il'yasov [1984]), linear *stochastic discrete* models (Phillis [1984]), and was also applied to investigating the problem of *polystability* of system (1.1.1) with respect to part of the variables (Martynyuk and Chernetskaya [1993a]).

Remark 1.1.6. While introducing new variables to construct an auxiliary μ-system, one can always take advantage of the process to get extra information on the properties of the solutions of the original system (1.1.1). By

Theorems 1.1.1 and 1.1.2, the problem of stability (asymptotic stability) of the unperturbed motion $\boldsymbol{y} = \mathbf{0}, \boldsymbol{z} = \mathbf{0}$ of system (1.1.1) with respect to $y_1, \ldots, y_m$ is reduced to the problem of *stability (asymptotic stability) of the invariant set* $\boldsymbol{y} = \mathbf{0}$, $L_1\boldsymbol{z} = \mathbf{0}$ of this system. This implies that the stability (asymptotic stability) of the unperturbed motion of system (1.1.1) with respect to $y_1, \ldots, y_m$ is actually possible only under the condition for stability (asymptotic stability) with respect to variables μ_i $(i = 1, \ldots, k \leq p)$. Since μ_i are linear combinations of variables $z_1, \ldots, z_p$, a certain type of "*group*" *stability* takes place.

In this regard the proposed method for constructing $\boldsymbol{\mu}$-systems supplements and elaborates the approach used by Rumyantsev (Moiseev and Rumyantsev [1965]) to investigate the stability of complex mechanical systems with respect to part of the variables. In the context of this approach, along with variables which are of interest in investigating the stability, *additional variables* (chosen depending on the features of a problem) are introduced to *integrally* characterize the state of the remaining variables of a system.

Example 1.1.1.

1. Let system (1.1.1) have the form

$$\dot{y}_1 = -y_1 + z_1 - 2z_2, \quad \dot{z}_1 = 4y_1 + z_1, \quad \dot{z}_2 = 2y_1 + z_1 - z_2. \tag{1.1.14}$$

The equations

$$\dot{y}_1 = -y_1 + \mu_1, \qquad \dot{\mu}_1 = -\mu_1 \qquad (\operatorname{rank} K_2 = 1)$$

compose the auxiliary $\boldsymbol{\mu}$-system which possesses the Lyapunov asymptotically stable zero solution. This means that the unperturbed motion $y_1 = z_1 = z_2 = 0$ of the original system (1.1.14) is asymptotically stable with respect to y_1 though it is Lyapunov unstable.

We note that for the unperturbed motion of system (1.1.14) to be asymptotically stable with respect to y_1, it should also be asymptotically stable with respect to the variable $\mu_1 = z_1 - 2z_2$.

2. Consider the equations of linear approximation of a *holonomic mechanical system with three degrees of freedom*

$$\ddot{y}_1 = ay_1 + b_1z_1 + b_2z_2, \qquad \ddot{z}_i = -b_iy_1 + dz_i \quad (i = 1, 2). \tag{1.1.15}$$

The matrices of coefficients of system (1.1.15)

$$A_1^* = \begin{pmatrix} a & 0 & 0 \\ 0 & d & 0 \\ 0 & 0 & d \end{pmatrix}, \qquad A_2^* = \begin{pmatrix} 0 & b_1 & b_2 \\ -b_1 & 0 & 0 \\ -b_2 & 0 & 0 \end{pmatrix}$$

determine linear *potential* and *nonconservative* forces, respectively.

Let us find the conditions for the equilibrium position $y_1 = \dot{y}_1 = z_1 = \dot{z}_1 = z_2 = \dot{z}_2 = 0$ of system (1.1.15) to be stable with respect to y_1, $\dot{y}_1$.

The equilibrium position of the auxiliary $\boldsymbol{\mu}$-system

$$\ddot{y}_1 = ay_1 + \mu_1, \quad \ddot{\mu}_1 = by_1 + d\mu_1, \quad b = -b_1^2 - b_2^2$$

is Lyapunov stable if both roots of the characteristic equation of the $\boldsymbol{\mu}$-system $\lambda^4 - (a + d)\lambda^2 + (ad - b) = 0$ solved for λ^2 are real and negative. For this to

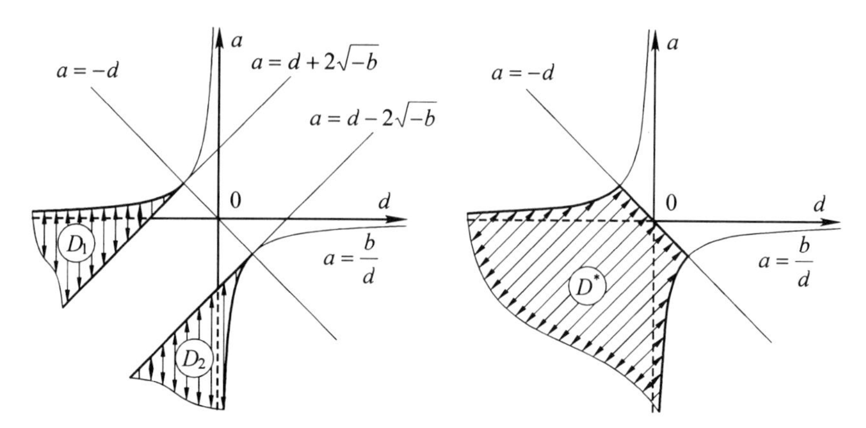

Figure 1.1.1. Region $D = D_1 \cup D_2$: $a + d < 0,\ ad - b > 0,$ $(a + d)^2 - 4(ad - b) > 0.$

Figure 1.1.2. Region D^*: $a + d < 0,$ $ad - b > 0.$

be true the coefficients and the discriminant of the characteristic equation must satisfy the conditions

$$a + d < 0, \quad ad - b > 0, \quad (a + d)^2 - 4(ad - b) > 0. \tag{1.1.16}$$

The region (1.1.16) shown in Figure 1.1.1 is the stability region of the equilibrium position of system (1.1.15) with respect to y_1, $\dot{y}_1$.

3. If the matrices A_1^* and A_2^* determine linear *dissipative*, *accelerating*, and *gyroscopic* forces, then instead of (1.1.15), we have the equations

$$\ddot{y}_1 = a\dot{y}_1 + b_1\dot{z}_1 + b_2\dot{z}_2, \qquad \ddot{z}_i = -b_i\dot{y}_1 + d\dot{z}_i \quad (i = 1, 2). \tag{1.1.17}$$

Let us find the conditions for the equilibrium position $y_1 = \dot{y}_1 = z_1 = \dot{z}_1 = z_2 = \dot{z}_2 = 0$ of system (1.1.17) to be stable with respect to y_1, $\dot{y}_1$ and asymptotically stable with respect to $\dot{y}_1$.

The equilibrium position of the auxiliary $\boldsymbol{\mu}$-system

$$\ddot{y}_1 = a\dot{y}_1 + \dot{\mu}_1, \quad \ddot{\mu}_1 = b\dot{y}_1 + d\dot{\mu}_1, \quad b = -b_1^2 - b_2^2$$

is Lyapunov stable and asymptotically stable with respect to $\dot{y}_1$, $\dot{\mu}_1$ provided that

$$a + d < 0, \qquad ad - b > 0. \tag{1.1.18}$$

Region (1.1.18) shown in Figure 1.1.2 is the region of stability with respect to y_1 and $\dot{y}_1$ and the region of asymptotic stability with respect to $\dot{y}_1$ of the equilibrium position of system (1.1.17).

Equations (1.1.17) describe, in particular, the *motion of an electron in a magnetostatic field*, the y_1-axis being perpendicular to the intensity vector $\boldsymbol{H}$ of the field. In this case (Tamm [1976])

$$a = d = 0, \qquad b_1 = -\frac{e}{mc}H_{z_1}, \qquad b_2 = -\frac{e}{mc}H_{z_2},$$
$$\boldsymbol{H} = (0, H_{z_1}, H_{z_2}),$$

where m and e are the mass and the charge of an electron, respectively, and c is the electromagnetic constant which is equal to the speed of light ($c = 3 \cdot 10^{10}$cm/s).

It follows from the general theory of linear gyroscopic systems that in the case under consideration the equilibrium position $y_1 = \dot{y}_1 = z_1 = \dot{z}_1 = z_2 = \dot{z}_2 = 0$ of system (1.1.17) is stable with respect to $\dot{y}_1$, $\dot{z}_1$, $\dot{z}_2$ but cannot be Lyapunov stable (with respect to y_1, $\dot{y}_1$, z_1, $\dot{z}_1$, z_2, $\dot{z}_2$).

Let us clarify the statement about instability. To this end we note that the equilibrium position of the $\boldsymbol{\mu}$-system, in the case in question, is Lyapunov stable. This implies that the equilibrium position of system (1.1.17) is stable with respect to $\dot{y}_1$, $\dot{z}_1$, $\dot{z}_2$ and also with respect to y_1. Moreover, stability with respect to μ_1 is only possible in the case of instability with respect to both z_1 and z_2.

Thus, in the case under consideration, the motion of an electron is stable with respect to y_1, $\dot{y}_1$, $\dot{z}_1$, $\dot{z}_2$ and unstable with respect to each variable z_1 and z_2.

1.1.6 The Possibility for Periodic Solutions being Asymptotically Stable with Respect to Part of the Variables

Theorem 1.1.3. *For system* (1.1.1) *to possess a periodic solution which is asymptotically stable with respect to* $y_1, \ldots, y_m$, *it is necessary and sufficient that*

(*a*) *one of the roots of the characteristic equation of system* (1.1.1) *is pure imaginary; and*

(*b*) *all the roots of equation* (1.1.12) *have negative real parts.*

Proof. Necessity. Let system (1.1.1) have a periodic solution which is asymptotically stable with respect to $y_1, \ldots, y_m$. The perturbed motion system constructed for this periodic solution is identical in form to (1.1.1). Thus, by Theorem 1.1.1 all the roots of equation (1.1.12) have negative real parts. The existence of a pure imaginary root of the characteristic equation of system (1.1.1) can be proved by contradiction. Indeed, otherwise none of the solutions of system (1.1.1) can be periodic. The necessity is proved.

Sufficiency. If condition (a) holds, then system (1.1.1) has a periodic solution. By Theorem 1.1.1, this solution is asymptotically stable with respect to $y_1, \ldots, y_m$, provided that condition (b) is true. The theorem is proved. □

Remark 1.1.7. Under conditions of Theorem 1.1.3, among the roots of the characteristic equation of system (1.1.1), there is a purely imaginary root. Thus, a periodic solution asymptotically stable with respect to $y_1, \ldots, y_m$ cannot be Lyapunov asymptotically stable. Moreover, it can even be Lyapunov unstable. This is the case when, among the roots of the characteristic equation of system (1.1.1) not in the set of the roots of the characteristic equation of a $\boldsymbol{\mu}$-type system, there is at least one root with a positive real part.

1.1.7 Application to the Problem of Constructing Lyapunov Functions

Let a Lyapunov function be obtained in solving the problem of stability (with respect to all or to part of the variables). This allows one to resolve the question of whether the system is (or is not) stable and also provides the means for investigating such problems as examining the quality of the transient processes, estimating the stability region, analyzing the influence of perturbations, etc.

Consider a $\boldsymbol{\mu}$-system of differential equations (1.1.11) for the original system (1.1.1). If the solution $\boldsymbol{\xi} = \mathbf{0}$ of system (1.1.11) is Lyapunov asymptotically stable, then there exists a *quadratic form* $v(\boldsymbol{\xi})$ satisfying one of the two following groups of conditions (Krasovskii [1959], Hahn [1967]):

(1) $c_1\|\boldsymbol{\xi}\|^2 \leq v(\boldsymbol{\xi}) \leq c_2\|\boldsymbol{\xi}\|^2$, $\dot{v}(\boldsymbol{\xi}) \leq -c_3\|\boldsymbol{\xi}\|^2$, where c_i $(i = 1,2,3)$ are arbitrary constants; and

(2) $v(\boldsymbol{\xi}) \geq c_1\|\boldsymbol{\xi}\|^2$, $\dot{v}(\boldsymbol{\xi}) \leq 0$, and the set $M = \{\boldsymbol{\xi} : \dot{v}(\boldsymbol{\xi}) = 0\}$ contains no entire semitrajectories of system (1.1.11), except for $\boldsymbol{\xi} = \mathbf{0}$.

It follows from Theorem 1.1.1 that combining the rules of constructing quadratic form $v(\boldsymbol{\xi})$ with those of constructing the $\boldsymbol{\mu}$-system yields the rules of constructing Lyapunov functions $v^*(\boldsymbol{x}) = v(\boldsymbol{\xi}(\boldsymbol{x}))$ which solve the problem of asymptotic stability of the unperturbed motion of system (1.1.1) with respect to $y_1, \ldots, y_m$. Let $v_1(\boldsymbol{\xi})$ and $v_2(\boldsymbol{\xi})$ be Lyapunov functions satisfying by virtue of $\boldsymbol{\mu}$-system (1.1.11), conditions (1) and (2), respectively. Then the rule of constructing function $v_1^*(\boldsymbol{x})$ for system (1.1.1), in particular, is reduced to solving Lyapunov's matrix equation

$$Q^{\mathsf{T}}G^* + G^*Q = -I_{m+h}, \qquad Q = L_4A^*L_5.$$

The sought for function $v_1(\boldsymbol{\xi}) = \boldsymbol{\xi}^{\mathsf{T}}G^*\boldsymbol{\xi}$ with $\boldsymbol{\xi}(\boldsymbol{x}) = L_4\boldsymbol{x}$ has the form $v_1^*(\boldsymbol{x}) = \boldsymbol{x}^{\mathsf{T}}L_4^{\mathsf{T}}G^*L_5\boldsymbol{x}$.

It might be interesting to compare the properties of functions $v_i^* = v_i(\boldsymbol{\xi}(\boldsymbol{x}))$ of variables $\boldsymbol{x} = (\boldsymbol{y}^{\mathsf{T}}, \boldsymbol{z}^{\mathsf{T}})$ obtained with the conditions imposed on Lyapunov functions in the well-known theorems on asymptotic stability of motion with respect to part of the variables. This will enable us to *relate* the *$\boldsymbol{\mu}$-system technique* and the *method of Lyapunov functions.*

Theorem 1.1.4. 1. *For function* $v_1^*(\boldsymbol{x})$ *to satisfy all the conditions of Theorem* 0.4.2, *it is necessary and sufficient that the coefficients of system* (1.1.1) *obey the relationships*

$$a) \quad b_{il} = 0, \quad d_{jl} = 0 \quad (i = 1, \ldots, m;\ j = 1, \ldots, k-m;\ l = k-m+1, \ldots, p); \qquad (1.1.19)$$

$$b) \qquad \operatorname{rank} K_p = k - m. \qquad (1.1.20)$$

2. *For function* $v_2^*(\boldsymbol{x})$ *to satisfy all the conditions of Theorem* 0.4.4, *it is necessary and sufficient that* $\operatorname{rank} K_p = p$.

3. (*a*) *Function* $v_1^*(\boldsymbol{x})$ *satisfies all the conditions of Theorem* 0.4.3; (*b*) *function* $v_2^*(\boldsymbol{x})$ *satisfies all the conditions of Theorem* 0.4.6; (*c*) *function* $v_3^*(\boldsymbol{x}) = \sqrt{v_1^*(\boldsymbol{x})}/\sqrt{c_1}$ *satisfies all the conditions of Theorem* 0.5.1.

4. *Vector function* $\boldsymbol{V}^\top = (\bar{\boldsymbol{V}}^\top, -\bar{\boldsymbol{V}}^\top) = (\bar{v}_1, \dots, \bar{v}_{m+h}, -\bar{v}_1, \dots, -\bar{v}_{m+h})$, *where* $\bar{v}_i = y_i$ $(i = 1, \dots, m)$, $\bar{v}_{m+j} = \boldsymbol{l}_j^\top \boldsymbol{z}$ $(j = 1, \dots, h)$, *with* $\boldsymbol{l}_j$ $(j = 1, \dots, h)$ *being linearly independent columns of matrix* K_p, *satisfies all the conditions of Theorem* 0.4.11.

Proof. 1. *Necessity.* Let $v_1^*(\boldsymbol{x})$ satisfy all the conditions of Theorem 0.4.2. Using *reductio ad absurdum*, let us assume that condition (1.1.19) does not hold. Then, it follows from the procedure of constructing the $\boldsymbol{\mu}$-system that vector $\boldsymbol{\xi}(\boldsymbol{x})$ has as a component at least one variable x_j of system (1.1.1) with index $k+1 \le j \le n$. Hence, the inequality

$$v_1^*(\boldsymbol{x}) \le b\left(\left(\sum_{i=1}^{k} x_i^2\right)^{\frac{1}{2}}\right) \tag{1.1.21}$$

is violated at $x_1 = \dots = x_k = 0$; $x_j \ne 0$. Consequently, our assumption fails, and condition (1.1.19) should hold.

Further, we assume that condition (1.1.20) does not hold. Then, by Lemma 1.1.2, the dimension of the $\boldsymbol{\mu}$-system for (1.1.1) and, therefore, that of vector $\boldsymbol{\xi}(\boldsymbol{x})$ are strictly less than k. This means that the equality $\boldsymbol{\xi}(\boldsymbol{x}) = \mathbf{0}$ is actually an algebraic system of equations

$$\begin{gathered} \xi_i = x_i = 0, \quad \xi_{m+j} = l_{m+j,1}x_{m+1} + \dots + l_{m+j,k-m}x_k = 0 \\ (i = 1, \dots, m; \quad 1 \le j < k-m). \end{gathered} \tag{1.1.22}$$

Since vectors $\boldsymbol{l}_{m+j} = (l_{m+j,1}, \dots, l_{m+j,k-m})$ are linearly independent, the number of equations of the above system is less than the number of unknowns $x_1, \dots, x_k$.

Let the dimension of vector $\boldsymbol{\xi}(\boldsymbol{x})$ be equal to $N < k$ $(N \ge m)$. Without loss of generality, also let the first h columns of matrix $L_1 = (l_{m+j,s})$ $(j = 1, \dots, N-m;\ s = 1, \dots, k-m)$ be linearly independent. Setting $x_{m+h+1} \ne 0, \dots, x_k \ne 0$ and solving system (1.1.22), $x_{m+1}^2 + \dots + x_{m+h}^2 \ne 0$. Hence, the equality $\boldsymbol{\xi}(\boldsymbol{x}) = \mathbf{0}$ is possible in the case

$$\begin{gathered} x_{m+h+1} \ne 0, \dots, x_k \ne 0, \quad x_{m+1}^2 + \dots + x_{m+h}^2 \ne 0, \\ x_i = 0 \quad (i = 1, \dots, m). \end{gathered} \tag{1.1.23}$$

Consequently, condition (1.1.23) results in the violation of the condition imposed on $\dot{v}_1^*$. Thus, the assumption that condition (1.1.20) does not hold is invalid. The necessity of conditions (1.1.19) and (1.1.20) is proved.

Sufficiency. Let conditions (1.1.19) and (1.1.20) hold. Then, by Lemma 1.1.2, the dimension of vector $\boldsymbol{\xi}(\boldsymbol{x})$ equals k. It follows from equality $\boldsymbol{\xi}(\boldsymbol{x}) = \mathbf{0}$ that $x_1^2 + \dots + x_k^2 = 0$. Thus, conditions $c_1\|\boldsymbol{\xi}\|^2 \le v_1(\boldsymbol{\xi}) \le c_2\|\boldsymbol{\xi}\|^2$, $\dot{v}_1(\boldsymbol{\xi}) \le -c_3\|\boldsymbol{\xi}\|^2$ yield

$$a(\|\boldsymbol{y}\|) \le v_1^*(\boldsymbol{x}) \le b\left(\left(\sum_{i=1}^{k} x_i^2\right)^{\frac{1}{2}}\right), \qquad \dot{v}_1^*(\boldsymbol{x}) \le -c\left(\left(\sum_{i=1}^{k} x_i^2\right)^{\frac{1}{2}}\right).$$

2. *Necessity.* Let $v_2^*(\boldsymbol{x})$ satisfy all the conditions of Theorem 0.4.4. Using *reductio ad absurdum*, let us assume that the condition $\operatorname{rank} K_p = p$ does not hold. But then the dimension of $\boldsymbol{\mu}$-system is less than that of the original system (1.1.1). As we indicated in the first part of the proof, in this case the equality $\boldsymbol{\xi}(\boldsymbol{x}) = \mathbf{0}$ is possible provided that $x_1^2 + \cdots + x_n^2 \neq 0$. Therefore, the set $M^* = \{\boldsymbol{x} : \dot{v}_2^*(\boldsymbol{x}) = 0\}$ contains entire semitrajectories of system (1.1.1) different from $\boldsymbol{x} = \mathbf{0}$. The necessity of the condition $\operatorname{rank} K_p = p$ is proved.

Sufficiency. If condition $\operatorname{rank} K_p = p$ holds, then the equality $\boldsymbol{\xi}(\boldsymbol{x}) = \mathbf{0}$ is equivalent to the equality $\boldsymbol{x} = \mathbf{0}$, and the set M^* contains no entire semitrajectories of system (1.1.1), except for $\boldsymbol{x} = \mathbf{0}$. Let us also take into account that for quadratic form $v_2(\boldsymbol{\xi})$ the condition $v_2(\boldsymbol{\xi}) \geq c_1\|\boldsymbol{\xi}\|^2$ yields $v_2^*(\boldsymbol{x}) \geq c_1\|\boldsymbol{y}\|^2$.

3. (*a*) It follows from the rule of constructing a $\boldsymbol{\mu}$-system stated in Subsection 1.1.2 that the expressions for the right-hand sides of the first equations of system (1.1.1) satisfy the equalities

$$Y_i(t, \boldsymbol{x}) = \sum_{k=1}^{m} a_{ik} y_k + \sum_{l=1}^{p} b_{il} z_l = \sum_{k=1}^{m} a_{ik} y_k + \sum_{l=1}^{m_1} \alpha_{il} \mu_l \qquad (i = 1, \ldots, m).$$

Since the zero solution of $\boldsymbol{\mu}$-system is Lyapunov asymptotically stable, for any $t_0 \geq 0$ there exist numbers $\delta' > 0$ and $M > 0$ such that $\|Y_i(t, \boldsymbol{x})\| \leq M$, $(t, \boldsymbol{x}) \in E(t_0, \delta')$. Moreover, the inequalities $v_1^*(\boldsymbol{x}) \geq a(\|\boldsymbol{y}\|)$ and $\dot{v}_1^*(\boldsymbol{x}) \leq -c(\|\boldsymbol{y}\|)$ hold for quadratic form $v_1^*(\boldsymbol{x})$. Consequently, function $v_1^*(\boldsymbol{x})$ satisfies all the conditions of Theorem 0.4.3.

(*b*) Since $v_2(\boldsymbol{\xi}) \geq c_1\|\boldsymbol{\xi}\|^2$, the set $M_2 = M_1 \cap M$, $M_1 = \{\boldsymbol{\xi} : v_2(\boldsymbol{\xi}) > 0\}$ coincides with the set M, except for $\boldsymbol{\xi} = \mathbf{0}$. Let us assume that the condition of Theorem 0.4.6 is not fulfilled and the set $M_2^* = M_1^* \cap M^*$, $M_1^* = \{\boldsymbol{x} : v_2^*(\boldsymbol{x}) > 0\}$ contains entire semitrajectories of system (1.1.1), different from $\boldsymbol{y} = \mathbf{0}$ and $\boldsymbol{z} = \mathbf{0}$. Components of vector $\boldsymbol{\xi}$ either coincide with those of vector $\boldsymbol{y}$ or are expressed linearly in terms of vector $\boldsymbol{z}$ components. We do not consider the case $\boldsymbol{\xi}(\boldsymbol{x}) = \mathbf{0}$ due to the condition $v_2^*(\boldsymbol{x}) = v_2(\boldsymbol{\xi}) > 0$. Thus, under the assumption, there exist entire semitrajectories of system (1.1.1), different from $\boldsymbol{\xi} = \mathbf{0}$, contained in M_2^*. Consequently, there also exist such semitrajectories in M_2, which contradicts the initial condition (2) for function $v_2(\boldsymbol{\xi})$. Thus, function $v_2^*(\boldsymbol{x})$ satisfies all the conditions imposed on Lyapunov functions in Theorem 0.4.6.

(*c*) The following estimates for function $v_1^*(\boldsymbol{x})$ are valid:

$$c_1\|\boldsymbol{\xi}(\boldsymbol{x})\|^2 \leq v_1^*(\boldsymbol{x}) \leq c_2\|\boldsymbol{\xi}(\boldsymbol{x})\|^2, \qquad \dot{v}_1^*(\boldsymbol{x}) \leq -c_3\|\boldsymbol{\xi}(\boldsymbol{x})\|.$$

Taking into account the equations $\xi_i = y_i$ $(i = 1, \ldots, m)$, $\xi_{m+j} = \boldsymbol{l}_j^{\mathsf{T}}\boldsymbol{z}$ $(j = 1, \ldots, h)$ (here $\boldsymbol{l}_j$ stands for linearly independent columns of matrix K_p), we derive the inequalities

$$v_3^*(\boldsymbol{x}) = \sqrt{\frac{v_1^*(\boldsymbol{x})}{c_1}} \geq \sqrt{\frac{c_1\|\boldsymbol{\xi}\|^2}{c_1}} = \|\boldsymbol{\xi}\| \geq \|\boldsymbol{y}\|,$$

$$v_3^*(\boldsymbol{x}) = \sqrt{\frac{v_1^*(\boldsymbol{x})}{c_1}} \leq \sqrt{\frac{c_2}{c_1}}\|\boldsymbol{\xi}\| \leq \sqrt{\frac{c_2}{c_1}}\big(\|\boldsymbol{y}\| + L\|\boldsymbol{z}\|\big) \leq M\big(\|\boldsymbol{y}\| + \|\boldsymbol{z}\|\big),$$

$$L = \sqrt{\|\boldsymbol{l}_1\|^2 + \cdots + \|\boldsymbol{l}_h\|^2}, \qquad M = \begin{cases} \sqrt{\dfrac{c_2}{c_1}}L, & L \geq 1, \\ \sqrt{\dfrac{c_2}{c_1}}, & L < 1, \end{cases}$$

$$\dot{v}_3^*(\boldsymbol{x}) = \frac{1}{2\sqrt{c_1}}\frac{\dot{v}_1^*(\boldsymbol{x})}{\sqrt{v_1^*(\boldsymbol{x})}} \leq -\frac{c_3}{2c_2\sqrt{c_1}}\frac{v_1^*(\boldsymbol{x})}{\sqrt{v_1^*(\boldsymbol{x})}} = -\frac{c_3}{2c_2}v_3^*(\boldsymbol{x}),$$

and

$$\big|v_3^*(\boldsymbol{x}') - v_3^*(\boldsymbol{x})\big| \leq M\big(\|\boldsymbol{y}' - \boldsymbol{y}\| + \|\boldsymbol{z}' - \boldsymbol{z}\|\big).$$

Hence, function $v_3^*(\boldsymbol{x})$ satisfies all the conditions of Theorem 0.5.1 for $\alpha = \dfrac{c_3}{2c_2}$.

4. We take

$$v^* = \max(\bar{v}_i, -\bar{v}_i) = \max|y_i| \qquad (i = 1, \ldots, m).$$

Since the equations

$$\dot{\bar{\boldsymbol{V}}} = (L_4 A^* L_5)\bar{\boldsymbol{V}}, \qquad (-\dot{\bar{\boldsymbol{V}}}) = (L_4 A^* L_5)(-\bar{\boldsymbol{V}})$$

constitute the comparison system for vector function $\boldsymbol{V}$ and $v^* \geq a(\|\boldsymbol{y}\|)$, vector function $\boldsymbol{V}$ satisfies all the conditions of Theorem 0.4.11. The theorem is proved. □

Remark 1.1.8.

1. The construction of functions $v_2^*(\boldsymbol{x})$ for system (1.1.1) is based on the equations of its $\boldsymbol{\mu}$-system. Therefore, it is admissible that (when rank $K_p < p$) among the roots of the characteristic equation of system (1.1.1) there are some roots with positive real parts. Consequently, among solutions $z_j(t)$ of system (1.1.1) beginning in a sufficiently small neighborhood of the origin of coordinates $\boldsymbol{y} = \boldsymbol{0}$, $\boldsymbol{z} = \boldsymbol{0}$, there are infinitely increasing solutions as $t \to \infty$. As a result, for linear system (1.1.1) under the conditions of Theorem 0.4.6, we can give up the assumption concerning $\boldsymbol{z}$-boundedness of solutions which begin in a sufficiently small neighborhood of the origin of coordinates $\boldsymbol{y} = \boldsymbol{0}$, $\boldsymbol{z} = \boldsymbol{0}$.

2. Using the method of constructing a $\boldsymbol{\mu}$-system, Oziraner [1986] demonstrated that in the case when the motion $\boldsymbol{x} = \boldsymbol{0}$ of system (1.1.1) is (nonasymptotically) stable with respect to $y_1, \ldots, y_m$, there exists a quadratic form $v(\boldsymbol{x})$ satisfying the conditions of Theorem 0.4.1.

Example 1.1.2. Let the equations of perturbed motion have the form

$$\dot{y}_1 = -y_1 + z_1 - 2z_2, \quad \dot{z}_1 = 4y_1 + z_1, \quad \dot{z}_2 = \frac{5}{2}y_1 + z_1 - z_2. \qquad (1.1.24)$$

Since the characteristic equation of system (1.1.24) possesses the root which equals one, the motion $y_1 = z_1 = z_2 = 0$ is Lyapunov unstable. The equations

$$\dot{y}_1 = -y_1 + \mu_1, \qquad \dot{\mu}_1 = -y_1 - \mu_1 \tag{1.1.25}$$

form the auxiliary $\boldsymbol{\mu}$-system.

The characteristic equation of system (1.1.25) has the roots $\lambda_{1,2} = -1 \pm i$ and the motion $y_1 = z_1 = z_2 = 0$ of system (1.1.24) is asymptotically stable with respect to y_1.

The functions

$$v_1^*(\boldsymbol{x}) = y_1^2 + (z_1 - 2z_2)^2, \quad \dot{v}_1^*(\boldsymbol{x}) = -2v_1^*,$$
$$v_2^*(\boldsymbol{x}) = y_1^2 - 2y_1(z_1 - 2z_2) + 3(z_1 - 2z_2)^2, \quad \dot{v}_2^*(\boldsymbol{x}) = -8(z_1 - 2z_2)^2$$

solve the problem of asymptotic stability of the unperturbed motion of the original system (1.1.24) with respect to y_1. These functions can be obtained by the change of variables $\mu_1 = z_1 - 2z_2$ from the functions

$$v_1(\boldsymbol{\xi}) = y_1^2 + \mu_1^2, \quad \dot{v}_1(\boldsymbol{\xi}) = -2v_1,$$
$$v_2(\boldsymbol{\xi}) = y_1^2 - 2y_1\mu_1 + 3\mu_1^2, \quad \dot{v}_2(\boldsymbol{\xi}) = -8\mu_1^2$$

which, in turn, solve the problem of Lyapunov asymptotic stability of the zero solution of the auxiliary $\boldsymbol{\mu}$-system.

We note the following:

(1) $\operatorname{rank} K_p = 1$.

(2) Function $v_1^*(\boldsymbol{x})$ does not satisfy the conditions of Rumyantsev's Theorem 0.4.2 for any $k = 1, 2$. At the same time, the conditions of Corollary 0.4.1 are met for $\boldsymbol{\xi} = (y_1, z_1 - 2z_2)$.

(3) Function $v_1^*(\boldsymbol{x})$ satisfies the differential inequality $\dot{v}_1^* \le -\alpha v_1^*$, $\alpha = 1$. As a consequence, the conditions of Corduneanu's Theorem 0.5.1 are satisfied.

(4) The set $M^* = \{\dot{v}_2^*(\boldsymbol{x}) = 0\}$ contains entire semitrajectories of system (1.1.24) of type $y_1 = 0, z_1 = 2z_2$, for instance, the entire semitrajectories $y_1 = 0$, $z_1 = \delta e^t$, $z_2 = {}^1\!/_2 \delta e^t$, where δ is an arbitrary number. This means that the conditions of Risito–Rumyantsev's Theorem 0.4.4 are not satisfied.

(5) The set $M^* \setminus M_1$, $M_1 = \{\boldsymbol{x} : \boldsymbol{y} = \boldsymbol{0}\}$ is empty and, consequently, contains no entire semitrajectories of system (1.1.24). At the same time, the set M_1 is not an invariant set of system (1.1.24). Thus, all the conditions of Risito's Theorem 0.4.5 fail to hold.

(6) The set $M_2^* = M_1^* \cap M^*$, $M_1^* = \{\boldsymbol{x} : v_2^*(\boldsymbol{x}) > 0\}$ is empty and, consequently, contains no entire semitrajectories of system (1.1.24). Furthermore, among solutions $z_i(t)$ $(i = 1, 2)$ of system (1.1.24), which begin in an arbitrarily small neighborhood of the point $y_1 = z_1 = z_2 = 0$, there are *unbounded* solutions as $t \to \infty$. Thus, all the conditions of Oziraner's Theorem 0.4.6 fail to hold.

(7) Vector function $\boldsymbol{V}^{\top} = (y_1, -y_1, z_1 - 2z_2, 2z_2 - z_1)$ satisfies the conditions of Matrosov's Theorem 0.4.11, $\boldsymbol{\mu}$-system (1.1.25) serving as a *comparison system*.

1.2 Stability with Respect to Part of the Variables of Linear Systems with Periodic Coefficients

1.2.1 Formulation of the Problem

Among linear equations with variable coefficients, the equations with *periodic* coefficients

$$\begin{aligned}\frac{dy_i}{dt} &= \sum_{k=1}^{m} a_{ik}(t)y_k + \sum_{l=1}^{p} b_{il}(t)z_l \qquad (i=1,\dots,m),\\ \frac{dz_j}{dt} &= \sum_{k=1}^{m} c_{jk}(t)y_k + \sum_{l=1}^{p} d_{jl}(t)z_l \qquad (j=1,\dots,p)\end{aligned} \tag{1.2.1}$$

are of particular importance in the theory of oscillations, engineering, and celestial mechanics. Here $a_{ik}(t), b_{il}(t), c_{jk}(t)$, and $d_{jl}(t)$ are continuous T-periodic functions of $t \in [0,\infty)$.

We consider the problem of stability (asymptotic stability) of the unperturbed motion $\boldsymbol{y} = \mathbf{0}, \boldsymbol{z} = \mathbf{0}$ of system (1.2.1) with respect to part of the variables, for definiteness, with respect to variables $y_1, \dots, y_m$.

When solving the problem, we first assume (Subsections 1.2.2–1.2.8) that the coefficients of system (1.2.1) are continuous and also *analytic* functions of $t \in [0,\infty)$. This assumption will allow all subsequent reasoning to avoid excessive details. In what follows, we shall show (Subsection 1.2.9) that the requirement of analyticity of the functions involved can be replaced with that of sufficient (no more than p times) *continuous differentiability*. In a number of cases, continuity of coefficients of system (1.2.1) may be sufficient.

1.2.2 Auxiliary Statements

Let us represent system (1.2.1) in the vector form

$$\dot{\boldsymbol{y}} = A(t)\boldsymbol{y} + B(t)\boldsymbol{z}, \quad \dot{\boldsymbol{z}} = C(t)\boldsymbol{y} + D(t)\boldsymbol{z} \quad \bigl(\dot{\boldsymbol{x}} = A^*(t)\boldsymbol{x}\bigr).$$

We introduce the matrices

$$G_1(t) = L_1(t), \qquad G_2(t) = \{L_1(t), L_2(t)\}, \dots,$$
$$G_j(t) = \{L_1(t), \dots, L_j(t)\} \quad (j = 3, \dots, p+1),$$

with the elements determined by the relationships

$$L_1(t) = B^{\mathsf{T}}(t), \dots, L_j(t) = \dot{L}_{j-1}(t) + D^{\mathsf{T}} L_{j-1}(t)$$
$$(j = 2, \dots, p+1).$$

Addition and multiplication of analytic functions and their differentiation, result in analytic functions. Therefore, all the elements of matrix functions $G_i(t)$ $(i = 1, \dots, p+1)$ will be analytic functions of $t \in [0,\infty)$.

We denote the set of points $t \in [0,T]$, except possibly their finite set M, by $t \in [0,T] \setminus M$.

Lemma 1.2.1. 1. *Each of the functions $F_i(t) = \operatorname{rank} G_i(t)$ $(i = 1, \dots, p+1)$ considered in the interval $[0,T] \setminus M$ maintains a constant value N_i $(1 \le N_i \le p$; $N_i = 0$, if all the elements of matrix $G_i(t)$ vanish identically over $[0,T])$. For all $t \in [0,T] \setminus M$, one and the same system of N_i column-vectors of matrix $G_i(t)$ $(i = 1, \dots, p+1)$ remains linearly independent.*

2. *There exists a constant number s (the minimal number s, $2 \le s \le p+1$, is implied) such that for all $t \in [0,T] \setminus M$ the following equality holds*

$$\operatorname{rank} G_{s-1}(t) = \operatorname{rank} G_s(t) = N \qquad (N = \text{const}, 1 \le N \le p). \tag{1.2.2}$$

Proof. 1. Consider the set $\Delta_i = \{F_{ij}(t)\}$ of all possible square matrices obtained from $G_i(t)$ by deleting columns and rows. Being the determinants of the matrices specified, the functions $\det F_{ij}(t)$ are analytic and, complying with the properties of analytic functions, can vanish only on a *finite* set M of values $t \in [0,T]$. Otherwise, they vanish identically for all $t \in [0,T]$. By definition, the function $F_i(t)$ $(i = 1, \dots, p+1)$ at each point $t \in [0,T]$ is equal to the maximum order $k_i(t)$ of a nonzero determinant $\det F_{ij}(t)$.

We take $k_i^+ = \max k_i(t), t \in [0,T]$. When all the elements of matrix $G_i(t)$ vanish identically on $[0,T]$, the statement of the lemma is obvious (in this case $F_i \equiv 0, t \in [0,T]$). So we assume $1 \le k_i^+ \le p$. This implies that the set Δ_i contains a square matrix $F_{ij}^*(t)$ of order k_i^+ such that $\det F_{ij}^*(t) \ne 0$ for at least one $t_i = t_i^* \in [0,T]$. But then, by the properties of analytic functions, $\det F_{ij}^*(t) \ne 0$ for all $t \in [0,T] \setminus M$. Therefore, $k_i(t) = k_i^+$ for all $t \in [0,T] \setminus M$. Setting $N_i = k_i^+$, we conclude that $F_i(t) = N_i$ $t \in [0,T] \setminus M$.

The above reasoning suggests that there exists $t_i = t_i^* \in [0,T]$ such that the system of N_i column-vectors of matrix $G_i(t)$ is linearly independent. The elements of these column-vectors form a square matrix of order N_i whose determinant is nonzero at $t_i = t_i^*$ (and, consequently, for all $t \in [0,T] \setminus M$). Thus, this system of N_i column-vectors is linearly independent for all $t \in [0,T] \setminus M$. This proves the first part of the lemma.

2. For all $t \in [0,T]$, there exists $s = s(t)$ such that equality (1.2.2) holds. Let $k = \max s(t)$, $t \in [0,T]$. We take into account that, for all $t \in [0,T]$, the value $s = s(t)$ is the minimal one for which (1.2.2) holds. Then, by the first part of the lemma, the number $N^+ = \operatorname{rank} G_k(t^+)$ (t^+ is the value of t at which $\max s(t)$ is attained) remains constant for all $t \in [0,T] \setminus M$. Therefore, the equality $\operatorname{rank} G_{k-1}(t) = \operatorname{rank} G_k(t)$ is true not only for $t = t_i$ but for all $t \in [0,T] \setminus M$. Setting $s = k$ completes the proof. □

Remark 1.2.1. The condition of Lemma (1.2.1) concerning the possible exception of a finite number of points from $[0,T]$ is essential.

The system $\dot{y}_1 = (\sin t) z_1 + z_2$, $\dot{z}_i = z_i$ $(i = 1, 2)$ clearly illustrates the situation. Indeed, $\operatorname{rank} G_2 = \operatorname{rank} G_3 = 2$ for all $t \in [0, 2\pi]$, except at $t = \pi/2, 3\pi/2$, whereas $\operatorname{rank} G_2 \ne \operatorname{rank} G_3$ at $t = \pi/2, 3\pi/2$.

1.2.3 Constructing an Auxiliary System

Let s be the minimal number such that for all $t \in [0,T] \setminus M$ equality (1.2.2) holds. By Lemma 1.2.1, in matrix $G_{s-1}(t)$ there exist N column-vectors linearly independent for $t \in [0,T] \setminus M$. Let us denote them by $\boldsymbol{g}_i(t) = [g_{i1}(t), \ldots, g_{ip}(t)]^{\top}$ $(i = 1, \ldots, N)$.

To construct an auxiliary linear system, we introduce the new variables

$$\mu_i = \sum_{l=1}^{p} g_{il}(t) z_l \qquad (i = 1, \ldots, N). \tag{1.2.3}$$

Because equality (1.2.2) is valid, we conclude, taking into account the structure of matrices G_i $(i = 1, \ldots, p+1)$, that introducing new variables (1.2.3) leads to an auxiliary linear $\boldsymbol{\mu}$-system

$$\dot{\boldsymbol{\xi}} = L(t)\boldsymbol{\xi}, \quad \boldsymbol{\xi} = (y_1, \ldots, y_m, \mu_1, \ldots, \mu_N)^{\top}. \tag{1.2.4}$$

Here $L(t)$ is a matrix of dimensions $(m+N) \times (m+N)$. The elements of this matrix are piecewise-continuous, analytic in the intervals of continuity, and T-periodic functions of $t \in [0, \infty)$. (Recall that a function is termed *piecewise-continuous* for $t \in [0, \infty)$, if it has only a finite number of discontinuity points in any finite interval.)

Let us dwell on the procedure of constructing a $\boldsymbol{\mu}$-system for system (1.2.1). By Lemma 1.2.1, among columns $\boldsymbol{b}_i(t) = [b_{i1}(t), \ldots, b_{ip}(t)]^{\top}$ $(i = 1, \ldots, m)$ of matrix $G_1(t)$ there exist m_1 $(m_1 \leq m)$ linearly independent vectors for all $t \in [0,T] \setminus M$. Without loss of generality, suppose that these are the first m_1 vectors $\boldsymbol{b}_i(t)$. To construct an auxiliary system, let us introduce the new variables

$$\mu_i = \sum_{l=1}^{p} b_{il}(t) z_l \qquad (i = 1, \ldots, m_1). \tag{1.2.5}$$

With the new variables thus introduced, two cases are possible.

The first case. System (1.2.1) can be transformed as follows:

$$\begin{gathered}
\frac{dy_i}{dt} = \sum_{k=1}^{m} a_{ik}(t) y_k + \sum_{l=1}^{m_1} \alpha_{il}(t) \mu_l, \\
\frac{d\mu_j}{dt} = \sum_{k=1}^{m} a^*_{jk}(t) y_k + \sum_{l=1}^{p} b^*_{jl}(t) z_l = \sum_{k=1}^{m} a^*_{jk}(t) y_k + \sum_{l=1}^{m_1} \alpha^*_{jl}(t) \mu_l, \\
\frac{dz_s}{dt} = \sum_{k=1}^{m} c_{sk}(t) y_k + \sum_{l=1}^{p} d_{sl}(t) z_l \\
(i = 1, \ldots, m; \ j = 1, \ldots, m_1, \ s = 1, \ldots, p).
\end{gathered} \tag{1.2.6}$$

Here $\alpha_{il}(t), \alpha^*_{jl}(t)$ are *piecewise-continuous*, analytic in the intervals of continuity, and T-periodic functions of $t \in [0, \infty)$. The discontinuities of functions $\alpha_{il}(t), \alpha^*_{jl}(t)$ are caused by linear independence of vectors $\boldsymbol{b}_i(t)$ in the interval $[0, T]$ being violated, generally speaking, at a finite number of points in this interval.

In this case the following linear system can be separated out from equations (1.2.1):

$$\begin{aligned} \frac{dy_i}{dt} &= \sum_{k=1}^{m} a_{ik}(t) y_k + \sum_{l=1}^{m_1} \alpha_{il}(t) \mu_l \qquad (i = 1, \ldots, m), \\ \frac{d\mu_j}{dt} &= \sum_{k=1}^{m} a^*_{jk}(t) y_k + \sum_{l=1}^{m_1} \alpha^*_{jl}(t) \mu_l \qquad (j = 1, \ldots, m_1). \end{aligned} \tag{1.2.7}$$

The system of type (1.2.7) is termed a ***μ**-system* relative to the original system (1.2.1).

Coefficients $\alpha_{il}(t), \alpha^*_{jl}(t)$ of an auxiliary $\boldsymbol{\mu}$-system *do not belong*, in general, to the *same* class of functions as the coefficients of the original system (1.2.1). This is the main peculiarity of the procedure of constructing an auxiliary $\boldsymbol{\mu}$-system compared to the case of linear systems with constant coefficients.

The second case. The variables defined in (1.2.5) do not reduce system (1.2.1) to the form (1.2.6). Suppose that only the first m_2 ($m_2 < m_1$) equalities

$$\sum_{l=1}^{p} b^*_{jl}(t) z_l = \sum_{l=1}^{m_1} \alpha^*_{jl}(t) \mu_l \qquad (j = 1, \ldots, m_1) \tag{1.2.8}$$

are valid, whereas remaining $m_1 - m_2$ equalities (1.1.4) are impossible. In this case we repeatedly introduce the new variables

$$\mu_{m_1+j} = \sum_{l=1}^{p} b^*_{m_2+j,l}(t) z_l \qquad (j = 1, \ldots, m_1 - m_2).$$

But now we select only those variables whose corresponding vectors $\boldsymbol{b}^*_j(t) = \left[b^*_{m_1+j,1}(t), \ldots, b^*_{m_1+j,p}(t)\right]^{\mathrm{T}}$ are linearly independent for all $t \in [0, T] \setminus M$.

Let us demonstrate that, successively repeating the indicated process of introducing new variables, for system (1.2.1) we can always construct a $\boldsymbol{\mu}$-system (1.2.4) whose dimension does not exceed that of the original system. Suppose while introducing new variables we get p variables $\mu_1, \ldots, \mu_p$ linearly independent for all $t \in [0, T] \setminus M$. Then entering additional new variables is meaningless. Indeed, each subsequent newly introduced variable can be expressed by a linear combination of variables $\mu_1, \ldots, \mu_p$ (with coefficients that are piecewise-continuous and analytic in the intervals of continuity). Hence, the maximum dimension of a $\boldsymbol{\mu}$-system equals n.

Lemma 1.2.2. *For system* (1.2.1) *an auxiliary linear* $\boldsymbol{\mu}$*-system can be constructed with coefficients that are piecewise-continuous, analytic in the intervals of continuity, and* T*-periodic functions. The dimension of such a system equals* $m + N$, $N = \operatorname{rank} G_{s-1}(t)$.

Proof. If $\operatorname{rank} G_{s-1}(t) = N$, then, by Lemma 1.2.1, for $t \in [0,T] \setminus M$, matrix $G_{s-1}(t)$ contains a system of N linearly independent column-vectors. Denote them by $\boldsymbol{g}_i(t) = \left[g_{i1}(t), \ldots, g_{ip}(t)\right]^{\mathrm{T}}$ $(i = 1, \ldots, N)$. New variables (1.2.3) correspond to these linearly independent column-vectors. Since relationship (1.2.2) is true, introducing additional new variables, while constructing the $\boldsymbol{\mu}$-system, is meaningless. In fact, each subsequent newly introduced variable will be a linear combination (with coefficients that are piecewise-continuous and analytic in the intervals of continuity) of variables introduced. Thus, the dimension of an auxiliary $\boldsymbol{\mu}$-system equals $m + \operatorname{rank} G_{s-1}(t)$. The statement is proved. □

The essential point in the proof of the possibility of constructing an auxiliary $\boldsymbol{\mu}$-system is the statement of Lemma 1.2.1 that for all $t \in [0,T] \setminus M$ *one and the same* system of column-vectors of matrix $G_{s-1}(t)$ can be taken as linearly independent. This means that for all $t \in [0, \infty)$ the dimension of an auxiliary $\boldsymbol{\mu}$-system will remain fixed, and one and the same set of variables y_i $(i = 1, \ldots, m)$, μ_j $(j = 1, \ldots, N)$ constitute the phase vector of the $\boldsymbol{\mu}$-system for all $t \in [0, \infty)$.

To implement the suggested procedure of constructing an auxiliary system for the case of nonanalytic coefficients of system (1.2.1), we should additionally require that an auxiliary system be formed by introducing one and the same set of new variables (Section 1.3).

Otherwise, the procedure of construction and the meaning of an auxiliary system (regarding the analysis of stability with respect to variables $y_1, \ldots, y_m$ of the original system (1.2.1)) should be revised in a certain way. To this end, in particular, the idea of using property (A), which will be considered in Subsection 1.2.4, might be fruitful.

The introduced concept of an auxiliary system will play a leading part, as well as in Section 1.1, in formulating the criteria for stability and asymptotic stability with respect to $y_1, \ldots, y_m$ of the unperturbed motion of system (1.2.1). Needless to say, a certain inconvenience will be caused by *discontinuous* behavior of coefficients of an auxiliary system. This will require treating the concepts of the solutions of an auxiliary system and their stability more precisely. It is essential that the class of discontinuous systems under investigation is not an arbitrary one but is obtained from the original system (1.2.1) with analytic coefficients by linear transformations of variables of this system. This will allow us to significantly simplify investigating the structure of the solutions of a discontinuous auxiliary system.

Before proceeding to such an analysis, we consider the cases separately when constructing an auxiliary system with analytic coefficients is possible.

1.2.4 Constructing an Auxiliary System with Analytic Coefficients

Definition 1.2.1. The coefficients of system (1.2.1) *possess property* (A) if, under condition (1.2.2), the first N columns $\boldsymbol{g}_i(t)$ of matrix $G_{s-1}(t)$ for all

$t \in [0, \infty)$ have the form:

$$\boldsymbol{g}_i = [g_{i1}(t), \ldots, g_{ip}(t)]^{\top} = g_i^*(t)[g_{i1}^*(t), \ldots, g_{ip}^*(t)]^{\top} \quad (i = 1, \ldots, N).$$

Here $g_j^*(t)$ and $g_{ij}^*(t)$ are analytic T-periodic functions of $t \in [0, \infty)$ such that

$$\operatorname{rank} G_{s-1}^*(t) = \operatorname{rank} G_s^*(t) = N, \qquad G_{s-1}^*(t) = \{g_{ij}^*(t)\}^{\top},$$
$$G_s^*(t) = \dot{G}_{s-1}^*(t) + D^{\top}(t) G_{s-1}^*(t) \quad (i = 1, \ldots, N) \quad \text{for all } t \in [0, \infty).$$

Now let us show that in this case an auxiliary linear system, with analytic T-periodic coefficients, can be constructed for system (1.2.1). To do this, first let us introduce, instead of (1.2.3), the new variables

$$\mu_i = \sum_{j=1}^{p} g_{ij}^*(t) z_j \qquad (i = 1, \ldots, N).$$

Since the equality $\operatorname{rank} G_{s-1}^* = \operatorname{rank} G_s^* = N$ holds for all $t \in [0, \infty)$, after introducing new variables, we get an auxiliary linear $\boldsymbol{\mu}$-system

$$\dot{\boldsymbol{\xi}} = L^*(t)\boldsymbol{\xi}, \qquad \boldsymbol{\xi} = (y_1, \ldots, y_m, \mu_1, \ldots, \mu_N)^{\top}, \tag{1.2.9}$$

where $L^*(t)$ is a matrix of dimensions $(m + N) \times (m + N)$ and the elements are analytic T-periodic functions of $t \in [0, \infty)$. For instance, let $m = 1$ (i.e., stability with respect to a single variable y_1 is analyzed). Then the coefficients of system (1.2.1) will possess property (A) if and only if at least one of two conditions holds:

(1) Functions $b_{1j}(t)$ $(j = 1, \ldots, p)$ do not have common zeros in $[0, T]$.

(2) There exist T-periodic functions $g(t)$ and $b_{1j}^*(t)$, analytic in the interval $t \in [0, \infty)$, such that $b_{1j}(t) = g(t) b_{1j}^*(t)$ $(j = 1, \ldots, p)$, for $t \in [0, \infty)$. Besides, functions $b_{1j}^*(t)$ $(j = 1, \ldots, p)$ do not have common zeros in $[0, T]$.

The specified requirements are realizable for a sufficiently wide class of linear systems with periodic coefficients. Thus, condition (A) does not appear to be too restrictive.

1.2.5 The Structure of an Auxiliary System

To directly relate the coefficients of the original system (1.2.1) to those of a corresponding $\boldsymbol{\mu}$-system, let us define the matrices of the following structure:

(1) $R_1(t)$ is an $N \times p$ matrix whose rows are linearly independent columns of either matrix $G_{s-1}(t)$ for $t \in [0, T] \setminus M$ or matrix G_{s-1}^* for $t \in [0, T]$.

(2) R_2 is an $N \times N$ matrix whose columns are the first h columns of matrix $R_1(t)$ either for $t \in [0, T] \setminus M$ or for $t \in [0, T]$. (By appropriately reindexing variables $z_1, \ldots, z_p$ of system (1.2.1), if necessary, we can always obtain matrix R_1 whose *first* h columns are linearly independent for $t \in [0, T]$.)

(3) $R_3(t)$ is a $p \times N$ matrix such that its first h rows are the rows of matrix R_2^{-1}. The remaining rows of matrix R_3 are identically zero row-vectors for $t \in [0,T]$. The elements of matrix $R_2^{-1}(t)$ and, consequently, those of $R_3(t)$, in general, may have the discontinuities at a finite set of points from $[0,T]$.

(4)
$$R_4 = \begin{pmatrix} I_m & 0 \\ 0 & R_1 \end{pmatrix}, \quad R_5 = \begin{pmatrix} I_m & 0 \\ 0 & R_3 \end{pmatrix}, \quad R_6 = \dot{R}_5.$$

Lemma 1.2.3. *The auxiliary linear μ-system* (1.2.4) (*in particular, μ-system* (1.2.9)) *comprises the equations*

$$\dot{\boldsymbol{\xi}} = R_4(t)\big[A^*(t)R_5(t) - R_6(t)\big]\boldsymbol{\xi}. \tag{1.2.10}$$

Proof. The passage from system (1.2.1) to system (1.2.4) is equivalent to the linear change of variables $\boldsymbol{w} = R(t)\boldsymbol{x}$. Here

$$R = \begin{pmatrix} I_m & 0 \\ 0 & R^* \end{pmatrix}, \qquad R^* = \begin{pmatrix} R_1 \\ R_* \end{pmatrix},$$

and R_* is an arbitrary $(p-N) \times p$ matrix, having as its elements analytic T-periodic functions, such that matrix $R(t)$ is nonsingular for $t \in [0,T] \setminus M$. In this notation, system (1.2.4) is made up of the the first $m+N$ equations of the system

$$\dot{\boldsymbol{w}} = R(t)\big[A^*(t)R^{-1}(t) - \dot{R}^{-1}(t)\big]\boldsymbol{w}.$$

Now we follow the scheme for analyzing the structure of matrices of type $R(t)A^*(t)R^{-1}(t)$ used in proving Lemma 1.1.3. It can be demonstrated that, for the first $m+N$ rows of this matrix, the following relationship holds for all $t \in [0,T]$:

$$R(t)A^*(t)R^{-1}(t) = R_4(t)A^*(t)R_5(t).$$

Furthermore, by direct validation, we ascertain that the first $m+N$ rows of the matrix $R(t)\dot{R}^{-1}(t)$, considered for all $t \in [0,T]$, satisfy the equality $R(t)\dot{R}^{-1}(t) = R_4(t)R_6(t)$. It follows from the relationships obtained that equations (1.2.10) are actually those of system (1.2.4). The statement is proved. □

Remark 1.2.2. System (1.2.10) can be rewritten in the equivalent form

$$\dot{\boldsymbol{\xi}} = \big[R_4(t)A^*(t) + R_6(t)\big]R_5(t)\boldsymbol{\xi},$$

provided $R_6 = \dot{R}_4$. Indeed, we can directly verify the relationship $R_4R_5 = I_n$, $t \in [0,T]$. Hence, $\dot{R}_4R_5 = -R_4\dot{R}_5$, $t \in [0,T]$. Allowing for this equality, for all $t \in [0,T]$,

$$(R_4A^* + \dot{R}_4)R_5 = R_4A^*R_5 - R_4\dot{R}_5 = R_4(A^*R_5 - \dot{R}_5),$$

which proves the statement.

1.2.6 Analyzing Solutions of an Auxiliary System

The conditions of a theorem for the existence and uniqueness of solutions hold in those intervals I_i where the coefficients of system (1.2.10) are continuous (and analytic) functions. Thus, in each interval I_i there exists a unique solution $\boldsymbol{\xi} = \boldsymbol{\xi}_{(1.2.10)}(t)$ of system (1.2.10) satisfying specified initial conditions in the interval and continuable to the entire interval I_i. Besides, these solutions are analytic functions of $t \in I_i$. Any possible sequence of such solutions (each sequence containing exactly one solution for each interval I_i) may be treated as a *solution* of system (1.2.10) for the infinite half-interval $[0, \infty)$. Indeed, these solutions will be defined for almost all $t \in [0, \infty)$ (to be precise, for all $t \in [0, \infty)$ with the exception of a denumerable set M^* of discontinuity points of coefficients of system (1.2.10) in $[0, \infty)$) and satisfy system (1.2.10) for $t \in [0, T] \setminus M$.

To find these solutions, it suffices to determine the fundamental matrix $G = G(t)$ of solutions of system (1.2.10) in the intervals I_i. Let us determine the structure of matrix G in the intervals I_i. By the *Floquet theorem* (Floquet [1883]), the fundamental matrix $X = X(t)$ of solutions of system (1.2.1) has the form

$$X(t) = \Phi(t)e^{tL}.$$

Here $\Phi(t)$ is an analytic and T-periodic $n \times n$ matrix-valued function such that $\det \Phi(t) \neq 0, t \in [0, \infty)$, and L is a constant matrix of the same order. Matrix-valued function e^{tL} is a fundamental matrix of solutions of the system $\dot{\boldsymbol{\eta}} = L\boldsymbol{\eta}$. The roots ρ_i $(i = 1, \ldots, n)$ of the equation $\det(L - \rho I_n)$ are termed *characteristic exponents* of system (1.2.1).

System (1.2.10) is obtained from (1.2.1) by a linear nonsingular, at $t \in I_i$, transformation of variables (whose coefficients are analytic T-periodic functions of $t \in [0, \infty)$). The dimension of system (1.2.10) equals $m + N$. Whence it follows that a fundamental matrix $G(t)$ of solutions of system (1.2.10) in the intervals I_i has the form

$$G(t) = R_4(t)X(t) = \Psi(t)e^{tK}.$$

Here $\Psi(t)$ is an analytic and T-periodic $(m+N) \times (m+N)$ matrix-valued function such that $\det \Phi(t) \neq 0, t \in I_i$ and K is a constant matrix of the same order. The roots ω_j $(j = 1, \ldots, m+N)$ of the equation $\det(K - \omega I_{m+N})$ will be called *characteristic exponents* of system (1.2.10). The set $\{\omega_1, \ldots, \omega_{m+N}\}$ is a subset of the set of characteristic exponents of system (1.2.1). In the sequel, we shall assume without loss of generality that the point $t = 0$ is not a point of discontinuity of coefficients of system (1.2.10). Then, taking into account the relationships

$$G(0) = \Psi(0), \qquad G(T) = \Psi(T)e^{TK}, \qquad \Psi(0) = \Psi(T),$$

we conclude that

$$K = \frac{\ln G(T) - \ln G(0)}{T}. \tag{1.2.11}$$

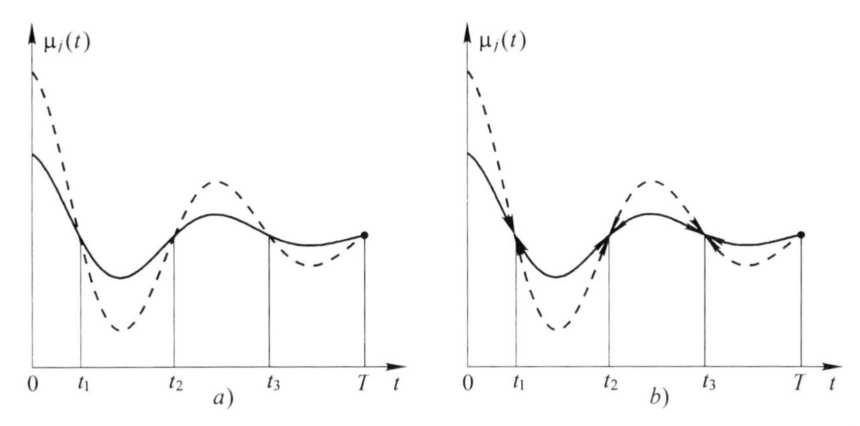

Figure 1.2.1. (a) Intersection of solutions $\mu_j(t)$ of system (1.2.1) at points $\{t_i\}$; (b) "Joint" of solutions $\mu_j(t)$ of system (1.2.10) at points $\{t_i\}$.

In analyzing the stability of the unperturbed motion $\boldsymbol{y} = \mathbf{0}$, $\boldsymbol{z} = \mathbf{0}$ of system (1.2.1) with respect to $y_1, \ldots, y_m$, the above interpretation of a solution of system (1.2.10) in the half-interval $[0, \infty)$ turns out to be *unnecessarily broad.* The way in which solutions *can be continued from one interval to another* is rather vague. To clarify the "joining" condition, we note that the essential purpose of constructing an auxiliary system (1.2.10) is to investigate the behavior of solutions

$$\mu_i(t) = \sum_{l=1}^{p} g_{il}(t) z_l(t) \qquad (i = 1, \ldots, N)$$

of system (1.2.1) determined by variables (1.2.3). This will have a significant impact on solving the problem of stability and asymptotic stability of the unperturbed motion of the original system (1.2.1) with respect to given variables $y_1, \ldots, y_m$. Therefore we are interested in those solutions $\boldsymbol{\xi} = \boldsymbol{\xi}_{(1.2.10)}(t)$ of system (1.2.10) which, for $t \in [0, \infty) \setminus M^*$, coincide with solutions $\boldsymbol{\xi}_{(1.2.1)}(t) = \left[y_1(t), \ldots, y_m(t), \mu_1(t), \ldots, \mu_N(t)\right]^{\mathsf{T}}$ of system (1.2.1).

Since system (1.2.10) in the intervals I_i is obtained from (1.2.1) by a linear nonsingular transformation of variables, the desired set of solutions of system (1.2.10) exists. Let us denote this set by $\mathfrak{M}$. Note that a solution $\boldsymbol{\xi}_{(1.2.10)}(t) \in \mathfrak{M}$ is *uniquely* defined for all $t_0 \in [0, \infty) \setminus M^*$ and $\boldsymbol{\xi}(t_0)$. The vectors $\boldsymbol{g}_i(t) = \left[g_{i1}(t), \ldots, g_{ip}(t)\right]^{\mathsf{T}}$ $(i = 1, \ldots, N)$ and, consequently, variables (1.2.3) are linearly dependent at points $\{t_i\}$ of discontinuity of the coefficients of system (1.2.10). This means that solutions $\mu_j(t)$ $(j = 1, \ldots, N)$ of system (1.2.1), corresponding to different initial conditions, *intersect* at points $\{t_i\}$ (Figure 1.2.1a). Therefore, points $\{t_i\}$ are also the *"joining" points* of components μ_j $(j = 1, \ldots, N)$ of solutions $\boldsymbol{\xi}_{(1.2.10)}(t) \in \mathfrak{M}$ (Figure 1.2.1b).

Thus, for auxiliary system (1.2.10) we can elucidate the mechanism of the discontinuities of its coefficients and also imagine the *phase portrait* of the system, i.e., the general layout of its integral curves. It follows from the above reasoning that the discontinuities of the coefficients of the auxiliary $\boldsymbol{\mu}$-system (1.2.10) result from vectors $\boldsymbol{g}_i(t)$, which correspond to new variables (1.2.3), required to form system (1.2.10), losing their linear independence at the finite set M of points of interval $[0, T]$. In this case, points $\{t_i\}$ are the "joining" points of components $\mu_j(t)$ $(j = 1, \dots, N)$ of solutions $\boldsymbol{\xi}_{(1.2.10)}(t) \in \mathfrak{M}$ of system (1.2.10), and the set of integral curves of the family of solutions from $\mathfrak{M}$ constitutes (generally, with an accuracy to the set of "joining" points, see Example 1.2.1) the entire possible set of integral curves of system (1.2.10).

Corresponding to points $\{t_i\}$ in the *phase plane* P of an auxiliary linear $\boldsymbol{\mu}$-system are also "*joining*" *points* of parts of phase trajectories. After solutions of the $\boldsymbol{\mu}$-system are analytically extended, the points of *intersection* of the phase trajectories correspond to points $\{t_i\}$. (A typical phase portrait of an asymptotically stable $\boldsymbol{\mu}$-system is presented in Figure 1.2.3.) We emphasize that, when considering a $\boldsymbol{\mu}$-system irrespective of the original system (1.2.1), it is impossible to unambiguously isolate phase trajectories with given initial conditions.

What matters in the sequel is not the explicit form of solutions $\boldsymbol{\xi}_{(1.2.10)}(t) \in \mathfrak{M}$ but only their general behavior for $t \in [0, \infty) \setminus M^*$ (boundedness and convergence to zero as $t \to \infty$). To this end, we define the concept of Lyapunov stability of the solution $\boldsymbol{\xi} = \mathbf{0}$ of system (1.2.10) more exactly. The solution $\boldsymbol{\xi} = \mathbf{0}$ of system (1.2.10) is understood in the sense that, for all $t \in [0, \infty) \setminus M^*$, it exists because the conditions of a theorem for existence and uniqueness are fulfilled and satisfies system (1.2.10) for all $t \in [0, \infty) \setminus M^*$.

1.2.7 The Stability of Solutions of an Auxiliary System

Definition 1.2.2. The solution $\boldsymbol{\xi} = \mathbf{0}$ of system (1.2.10) is said to be *Lyapunov stable* if, for any $\varepsilon > 0$, $t_0 > 0$ $(t_0 \in [0, \infty) \setminus M^*)$, there exists $\delta(\varepsilon) > 0$ such that, if $\|\boldsymbol{\xi}(t_0)\| < \delta$, then $\|\boldsymbol{\xi}_{(1.2.10)}(t; t_0, \boldsymbol{\xi}(t_0))\| < \varepsilon$, $\boldsymbol{\xi}_{(1.2.10)}(t) \in \mathfrak{M}$ for all $t \in [0, \infty) \setminus M^*$. In addition, if $\lim \|\boldsymbol{\xi}_{(1.2.10)}(t; t_0, \boldsymbol{\xi}(t_0))\| = 0$, as $t \to \infty$, then the solution $\boldsymbol{\xi} = \mathbf{0}$ is *Lyapunov asymptotically stable.*

Allowing for the structure of the fundamental matrix of solutions for system (1.2.10) and relationship (1.2.11), we conclude that the following lemma is valid.

Lemma 1.2.4. *The solution $\boldsymbol{\xi} = \mathbf{0}$ of system* (1.2.10) *is Lyapunov asymptotically stable if and only if all the roots of the equation*

$$\det \left[\frac{\ln G(T) - \ln G(0)}{T} - \omega I_{m+N} \right] = 0 \tag{1.2.12}$$

have negative real parts.

1.2.8 A Criterion for the Stability of the Unperturbed Motion of System (1.2.1) with Respect to Part of the Variables

Theorem 1.2.1. *For the unperturbed motion $\boldsymbol{y} = \boldsymbol{0}$, $\boldsymbol{z} = \boldsymbol{0}$ of system* (1.2.1) *to be asymptotically stable with respect to $y_1, \ldots, y_m$, it is necessary and sufficient that the solution $\boldsymbol{\xi} = \boldsymbol{0}$ of an auxiliary $\boldsymbol{\mu}$-system* (1.2.10) *be Lyapunov asymptotically stable. In this case all the roots of the equation* (1.2.12) *have negative real parts.*

Proof. The *sufficiency* can be proved in just the same way as for Theorem 1.1.1.

Necessity. The structure of the fundamental matrix of solutions for system (1.2.1) is determined by the Floquet theorem. This structure is such that, if unperturbed motion $\boldsymbol{y} = \boldsymbol{0}$, $\boldsymbol{z} = \boldsymbol{0}$ of system (1.2.1) is asymptotically stable with respect to $y_1, \ldots, y_m$, then the components $y_1(t), \ldots, y_m(t)$ of solutions of this system have the form

$$y_i(t) = c_{i1}\varphi_{i1}(t)e^{-\rho_1(t-t_0)} + \cdots + c_{in}\varphi_{in}(t)e^{-\rho_n(t-t_0)}, \qquad t \in [0, \infty)$$
$$(i = 1, \ldots, m).$$

Here $\varphi_{ij}(t)$ are T-periodic analytic functions, c_{ij} are constants depending on initial conditions, and $\rho_j > 0$ are the characteristic exponents of the system.

As a result, from the asymptotic stability of the unperturbed motion $\boldsymbol{y} = \boldsymbol{0}$, $\boldsymbol{z} = \boldsymbol{0}$ of system (1.2.1) with respect to $y_1, \ldots, y_m$, it follows that this motion is also asymptotically stable with respect to $\dot{y}_1, \ldots, \dot{y}_m$. Indeed, otherwise, at least one of the following relationships

$$\rho_j c_{ij}\varphi_{ij}(t)e^{-\rho_j(t-t_0)} \to 0, \quad c_{ij}\dot{\varphi}_{ij}(t)e^{-\rho_j(t-t_0)} \to 0$$

should be violated in expression for $\dot{y}_i(t)$.

Such a situation can occur only with corresponding functions φ_{ij} or $\dot{\varphi}_{ij}$ unbounded for $t \in [0, \infty)$. However, none of these functions, being analytic and T-periodic (T is a finite number), can be unbounded for $t \in [0, \infty)$.

Analytic and T-periodic coefficients $a_{ij}(t)$ of system (1.2.1) are also necessarily bounded for $t \in [0, \infty)$. Therefore, analyzing the structure of the first m equations of system (1.2.1), we infer that asymptotic stability of the unperturbed motion $\boldsymbol{y} = \boldsymbol{0}, \boldsymbol{z} = \boldsymbol{0}$ with respect to $y_1, \ldots, y_m$ necessitates that the inequalities

$$|\mu_i(t)| = \left|\sum_{l=1}^{p} b_{il}(t)z_l(t)\right| \leq \alpha_i e^{-\beta_i(t-t_0)} \qquad (i = 1, \ldots, m),$$

in which α_i and β_i are positive constants, are satisfied along the trajectories of system (1.2.1).

For the case when the system is reduced to (1.2.6), the necessity of conditions of the theorem is proved.

If system (1.2.1) is not reduced to the form (1.2.6) at the first step of introducing new variables and, consequently, has the form (1.2.8), then we follow the above reasoning for system (1.2.8). Since we always arrive at system (1.2.10) at some finite step of introducing new variables, the necessity of conditions of the theorem is also proved in the general case. Thus, the theorem is proved. □

Corollary 1.2.1. *For the unperturbed motion $\boldsymbol{y} = \mathbf{0}$, $\boldsymbol{z} = \mathbf{0}$ of system* (1.2.1) *to be (nonasymptotically) stable with respect to $y_1, \dots, y_m$, it is necessary and sufficient that the solution $\boldsymbol{\xi} = \mathbf{0}$ of auxiliary $\boldsymbol{\mu}$-system* (1.2.10) *be Lyapunov stable.*

Remark 1.2.3.

1. To evaluate the roots of equation (1.2.12), one can use numerical and approximate methods for finding fundamental matrices $G(0)$ and $G(T)$ of solutions of auxiliary $\boldsymbol{\mu}$-system (1.2.10). By this means the problem of investigating the stability with respect to part of the variables in the infinite half-interval $[0, \infty)$ amounts to that of numerically integrating over the periodicity interval $[0, T]$.

2. The class of $\boldsymbol{\mu}$-systems being considered in this section is *more general* than that of discontinuous systems with periodic coefficients which was studied by Yakubovich and Starzhinskii [1972].

3. The proposed approach to analyzing the stability of the unperturbed motion of system (1.2.1) with respect to part of the variables, based on constructing auxiliary $\boldsymbol{\mu}$-systems, was carried over to the problem of *polystability* with respect to part of the variables (Martynyuk and Chernetskaya [1993b]).

Example 1.2.1. Let system (1.2.1) have the form

$$\begin{gathered} \dot{y}_1 = -y_1 + (\sin 2t) z_1 + 2(\cos^2 t) z_2, \\ \dot{z}_1 = -(\cos t) y_1 - z_1 + z_2, \qquad \dot{z}_2 = (\sin t) y_1 - z_1 - z_2. \end{gathered} \tag{1.2.13}$$

In this case the auxiliary $\boldsymbol{\mu}$-system comprises the equations

$$\begin{gathered} \dot{y}_1 = -y_1 + \mu_1, \qquad \dot{\mu}_1 = -(1 + \tan t)\mu_1 \\ (\mu_1 = (\sin 2t) z_1 + 2(\cos^2 t) z_2). \end{gathered} \tag{1.2.14}$$

Among the coefficients of system (1.2.14) there is a *discontinuous* function $\tan t$. In the periodicity interval $[0, 2\pi]$ of the original system (1.2.13) coefficients, we note that discontinuities correspond to the points $\pi/2$ and $3\pi/2$, at which the coefficients $b_{11} = \sin 2t$ and $b_{12} = 2\cos^2 t$ of system (1.2.13) possess *common zeros.*

Integrating system (1.2.14), we find its solutions

$$\begin{aligned} y_1(t) &= \left\{ \left[y_1(t_0) - \tan t_0 \cdot \mu_1(t_0) \right] + \frac{\mu_1(t_0) \sin t}{\cos t_0} \right\} e^{-(t - t_0)} \\ &= \left\{ y_1(t_0) + 2(\sin t - \sin t_0) \left[\sin t_0 z_1(t_0) + \cos t_0 z_2(t_0) \right] \right\} e^{-(t - t_0)}, \\ \mu_1(t) &= \frac{\mu_1(t_0) \cos t e^{-(t - t_0)}}{\cos t_0} \\ &= 2 \left[\sin t_0 z_1(t_0) + \cos t_0 z_2(t_0) \right] \cos t e^{-(t - t_0)}, \qquad t \geq t_0 \geq 0. \end{aligned}$$

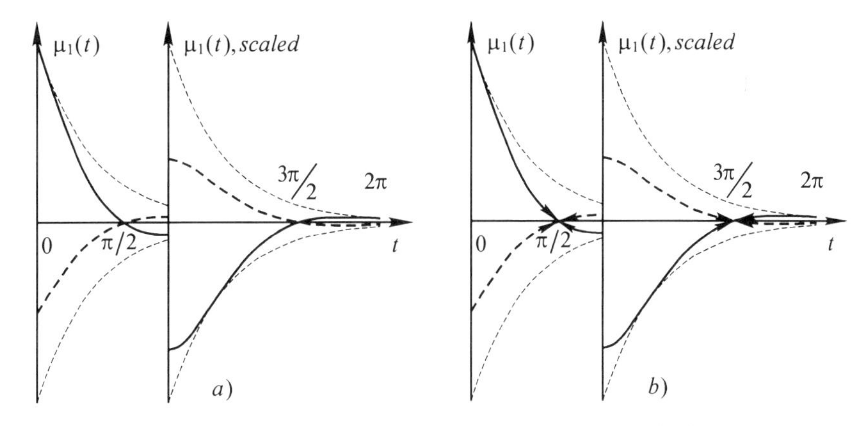

Figure 1.2.2. (a) Solutions $\mu_1(t)$ of systems (1.2.13) and (1.2.14); (b) Solutions $\mu_1(t)$ of class $\boldsymbol{\xi}_{(1.2.14)}(t) \in \mathfrak{M}$.

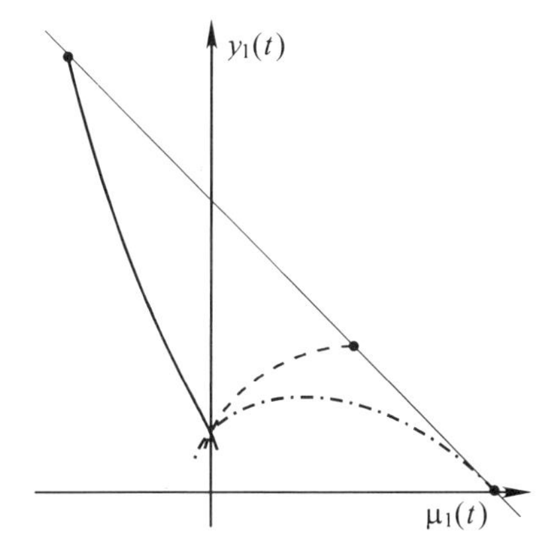

Figure 1.2.3. Phase portrait of system (1.2.14).

The solutions $y_1(t)$ and $\mu_1(t)$ of discontinuous system (1.2.14) are seen as *analytic* functions for all $t \in [0, \infty)$. (Figure 1.2.2a presents solutions $\mu_1(t)$ with different initial conditions in the periodicity interval $[0, 2\pi]$.) In this instance the caution of removing points $\{t_i\}$ from solutions $\boldsymbol{\xi}_{(1.2.14)}(t) \in \mathfrak{M}$ proved to be unnecessary (see Figure 1.2.2b).

However, from the standpoint of analyzing the stability problem studied it makes *no difference* whether solutions $\boldsymbol{\xi}_{(1.2.10)}(t) \in \mathfrak{M}$ are defined for all $t \in [0, \infty)$ or for all $t \in [0, \infty) \setminus M^*$. It is essential that for all $t \in [0, \infty) \setminus M^*$ they coincide with corresponding solutions of the original system (1.2.1) and with the use of analytic continuation can be extended to analytic solutions, for all $t \in [0, \infty)$, which coincide with corresponding solutions of the original system (1.2.1) for all $t \in [0, \infty)$.

Because the characteristic exponents $\omega_1 = \omega_2 = -1$ of system (1.2.14) have negative real parts, the unperturbed motion $y_1 = z_1 = z_2 = 0$ of system (1.2.13) is asymptotically stable with respect to y_1 by Theorem 1.2.1. Taking into account the equality

$$(\sin 2t, 2\cos^2 t) = 2\cos t(\sin t, \cos t),$$

where functions $\sin t$ and $\cos t$ do not have common zeros, we conclude that the coefficients of system (1.2.13) possess property (A). In this case system (1.2.9) is composed of the equations

$$\begin{gathered}\dot{y}_1 = -y_1 + 2(\cos t)\mu_2, \qquad \dot{\mu}_2 = -\mu_2 \\ (\mu_2 = (\sin t)z_1 + (\cos t)z_2).\end{gathered} \tag{1.2.15}$$

Systems (1.2.14) and (1.2.15) have the *same* characteristic exponents.

In conclusion we note that in the interval $[0, \pi/2]$, containing a single discontinuity point ($t_* = \pi/2$) of function $\tan t$, the relationships

$$\lim_{\substack{\varepsilon\to 0\\ \varepsilon>0}} \int_0^{\pi/2-\varepsilon} |\tan t| dt = \lim_{\substack{\varepsilon\to 0\\ \varepsilon>0}} \left(-\ln|\cos t| \Big|_0^{\pi/2-\varepsilon} \right) = \infty$$

hold true. Because of this, the coefficients of auxiliary μ-system (1.2.14) do not satisfy conditions of the work by Yakubovich and Starzhinskii [1972].

1.2.9 Weakening the Requirement of the Analyticity of Coefficients of System (1.2.1)

We distinguish two aspects of this problem.

(1) In the context of *constructing* an auxiliary μ-system (1.2.10), the analyticity of coefficients of system (1.2.1) is used for no reason except to ensure that the coefficients of a μ-system have a finite number of discontinuity points in the periodicity interval $[0, T]$.

Thus, the scheme for constructing μ-system can be expanded to include system (1.2.1) where coefficients b_{il} and d_{jl} are *sufficiently continuously differentiable* functions of $t \in [0, \infty)$. The coefficients and their derivatives should have a *finite* number of zeros in the interval $[0, T]$.

As to coefficients a_{ik} and c_{jk}, it suffices to demand their *continuity* for $t \in [0, \infty)$. This implies that, between discontinuity points of the coefficients of an auxiliary system, its solutions are continuously differentiable functions of t. Therefore, by smooth continuation, one can *uniquely* define solutions of an auxiliary system which are continuously differentiable functions for $t \in [0, \infty)$.

(2) However, if we additionally require that the problem of stability (asymptotic stability) of the unperturbed motion of system (1.2.1) with respect to $y_1, \ldots, y_m$ is *equivalent* to the problem of Lyapunov stability (asymptotic stability) of the zero solution of μ-system, then, to leave the proof scheme for the necessary conditions of Theorem 1.2.1 unchanged, not only coefficients b_{il} and d_{jl}, but also a_{ik} and c_{jk} are required to be (sufficiently) continuously differentiable for $t \in [0, \infty)$.

Let us dwell on the situation. First we assume that the construction of a μ-system is possible at the first step of introducing new variables of type (1.2.5). (For definiteness, we assume that all $\mu_1, \ldots, \mu_m$ are linearly independent for $t \in [0, T] \setminus M$.) In this case for the proof scheme of Theorem 1.2.1 to "work," it is sufficient for coefficients a_{ik}, c_{jk}, and d_{jl} to be continuous and for b_{il} to be continuously differentiable for $t \in [0, \infty)$.

Further, let the construction of the μ-system be possible at the second step of introducing the new variables specified. Once again assuming all $\mu_1, \ldots, \mu_m$ linearly independent for $t \in [0, T] \setminus M$, we conclude that successive variables $\mu_{m+1}, \ldots, \mu_{2m}$ are defined by the relationships

$$\ddot{y}_i = \sum_{k=1}^{m} \dot{a}_{ik}(t) y_k + \sum_{k=1}^{m} a_{ik}(t) \dot{y}_k + \dot{\mu}_{m+i} \qquad (1.2.16)$$
$$(i = 1, \ldots, m).$$

Just the possibility of writing relationships (1.2.16) requires the differentiability of coefficients a_{ik} and twofold differentiability of coefficients b_{il} for $t \in [0, \infty)$. Moreover, making the proof scheme of Theorem 1.2.1 "work" in the case under consideration necessitates that the unperturbed motion of system (1.2.1) be asymptotically stable with respect to $\ddot{y}_1, \ldots, \ddot{y}_m$. This implies that functions $\varphi_{ij}(t)$ involved into expressions for $y_1, \ldots, y_m$ are twice differentiable for $t \in [0, \infty)$. Twofold continuous differentiability of $\varphi_{ij}(t)$ is, in turn, guaranteed by the fact that all the coefficients of system (1.2.1) are continuously differentiable for $t \in [0, \infty)$.

Thus, for the proof scheme of Theorem 1.2.1 to "work" at the second step of introducing new variables, it is sufficient only that coefficients a_{ik}, c_{jk}, and d_{jl} be continuously differentiable and b_{il} be twice continuously differentiable for $t \in [0, \infty)$.

Summing up the above reasoning, we arrive at the following conclusion.

Conclusion. The requirement of analyticity for the coefficients of system (1.2.1) in Theorem 1.2.1 can be replaced by their *sufficient continuous differentiability* for $t \in [0, \infty)$. If s is a number of steps for introducing new variables to be used in constructing an auxiliary μ-system ($s \leq p$), then it is sufficient that coefficients a_{ik}, c_{jk}, and d_{jl} be continuously differentiable up to the $(s-1)$th order and b_{il} up to the sth order for $t \in [0, \infty)$. Besides, coefficients b_{il} and d_{jl}, along with their derivatives (up to the orders specified), should vanish only at a *finite* number of zeros in the periodicity interval $[0, T]$.

Example 1.2.2.

1. Let the following conditions hold true for system (1.2.1):

(1) $m = 1, p = 2$;

(2) a_{11}, c_{11}, c_{21}, d_{ij} $(i, j = 1, 2)$ are *continuous* functions of $t \in [0, \infty)$;

(3) b_{11}, b_{12} are *continuously differentiable* functions of $t \in [0, \infty)$ with a *finite* number of zeros in the periodicity interval $[0, T]$;

(4) the equality

$$\frac{\dot{b}_{11}}{b_{11}} + \frac{b_{12}}{b_{12}} d_{21} + d_{11} = \frac{\dot{b}_{12}}{b_{12}} + \frac{b_{11}}{b_{11}} d_{12} + d_{22} \overset{\text{def}}{=} d^*$$

is valid for $t \in [0, \infty)$.

The auxiliary μ-system takes the form

$$\dot{y}_1 = a_{11} y_1 + \mu_1, \quad \dot{\mu}_1 = c^* y_1 + d^* \mu_1 \qquad (1.2.17)$$
$$(c^* = b_{11} c_{11} + b_{12} c_{21}).$$

Thus, under conditions (1)–(4) the problem of stability (asymptotic stability) of the unperturbed motion of system (1.2.1) with respect to y_1 is *equivalent* to the problem of Lyapunov stability (asymptotic stability) of the zero solution of the auxiliary $\boldsymbol{\mu}$-system (1.2.17).

2. The second order system

$$\dot{x}_1 = x_2, \quad \dot{x}_2 = -c(t)x_1 \tag{1.2.18}$$

with a continuous T-periodic function $c(t)$ was considered by Hill [1886] in investigating *motion of the Moon* and, later on, by many other scientists in studying various oscillatory processes. The auxiliary $\boldsymbol{\mu}$-system for equations (1.2.18) is identical in form to (1.2.18). Therefore, as in the proof of Theorem 1.2.1, we infer that the zero solution $x_1 = x_2 = 0$ of system (1.2.18) cannot be simultaneously stable with respect to one of two variables and unstable with respect to the other.

1.2.10 Constructing Lyapunov Functions

For definiteness, let the coefficients of system (1.2.1) be analytic functions of $t \in [0, \infty)$.

1. First we suppose that the coefficients of system (1.2.1) also possess property (A) (see Subsection 1.2.4). Using transformation $\boldsymbol{\xi} = \Psi(t)\boldsymbol{\eta}$ in which $\Psi(t)e^{tK}$ is the fundamental matrix of solutions of system (1.2.9), we can reduce system (1.2.9) to a linear system with constant coefficients

$$\dot{\boldsymbol{\eta}} = K\boldsymbol{\eta}. \tag{1.2.19}$$

For system (1.2.19) there exists a quadratic form $v = v(\boldsymbol{\eta}) = \boldsymbol{\eta}^{\mathrm{T}}\Omega\boldsymbol{\eta}$ (Ω is a constant $(m+N)\times(m+N)$ matrix) which satisfies the conditions of the Lyapunov theorem on asymptotic stability:

$$c_1\|\boldsymbol{\eta}\|^2 \le v(\boldsymbol{\eta}) \le c_2\|\boldsymbol{\eta}\|^2, \qquad \dot{v}(\boldsymbol{\eta}) \le -c_3\|\boldsymbol{\eta}\|^2,$$
$$c_i = \text{const} > 0 \quad (i = 1, 2, 3),$$

where $\dot{v}(\boldsymbol{\eta})$ is the derivative of function $v = v(\boldsymbol{\eta})$ by virtue of system (1.2.19). Allowing for $\boldsymbol{\eta} = \Psi^{-1}(t)\boldsymbol{\xi}$ and $\boldsymbol{\xi} = R_4(t)\boldsymbol{x}$, we obtain the following expression for function $v = v(\boldsymbol{\eta}(\boldsymbol{\xi}(t, \boldsymbol{x}))) = v(t, \boldsymbol{y}, \boldsymbol{z})$:

$$v(t, \boldsymbol{y}, \boldsymbol{z}) = \boldsymbol{x}^{\mathrm{T}}\big[\Psi^{-1}(t)R_4(t)\big]^{\mathrm{T}}\Omega\big[\Psi^{-1}(t)R_4(t)\big]\boldsymbol{x}.$$

Since

$$v(t, \boldsymbol{y}, \boldsymbol{z}) \ge c_1\|\boldsymbol{\eta}\|^2, \qquad \boldsymbol{\xi} = \Psi(t)\boldsymbol{\eta}$$

and the equality $\boldsymbol{\eta} = \mathbf{0}$ for $t \in [0, \infty)$ holds if and only if $\boldsymbol{\xi} = \mathbf{0}$, quadratic form $v(t, \boldsymbol{y}, \boldsymbol{z})$ will be positive-definite relative to $\boldsymbol{\xi}$ for $t \in [0, \infty)$. Then,

$$v(t, \boldsymbol{y}, \boldsymbol{z}) \ge c_1^*\|\boldsymbol{\xi}\|^2, \qquad c_1^* = \text{const} > 0$$

and due to $\xi_i = y_i$ $(i = 1, \dots, m)$ the estimate

$$v(t, \boldsymbol{y}, \boldsymbol{z}) \ge c_1^*\|\boldsymbol{\xi}\|^2 \ge c_1^*\|\boldsymbol{y}\|^2$$

is valid for $t \in [0, \infty)$.

Consider the function $v^* = \sqrt{v(t,\boldsymbol{y},\boldsymbol{z})}/\sqrt{c_1^*}$. Being analytic T-periodic functions of $t \in [0,\infty)$, the elements of matrix $\Psi^{-1}(t)R_4(t)$ are bounded for $t \in [0,\infty)$. Consequently, $\|\Psi^{-1}(t)R_4(t)\| \le l = \text{const} < \infty, t \in [0,\infty)$. Therefore, for all $t \in [0,\infty)$, $\|\boldsymbol{x}\| < \infty$, the following estimates are true for function $v^*(t,\boldsymbol{y},\boldsymbol{z})$:

$$v^*(t,\boldsymbol{y},\boldsymbol{z}) \ge \frac{\sqrt{c_1^*}\,\|\boldsymbol{y}\|}{\sqrt{c_1^*}} = \|\boldsymbol{y}\|,$$

$$\begin{aligned} v^*(t,\boldsymbol{y},\boldsymbol{z}) &\le \sqrt{c_2}\,\|\boldsymbol{\eta}\| = \sqrt{c_2}\,\|\Psi^{-1}(t)R_4(t)\|\cdot\|\boldsymbol{x}\| \\ &\le \sqrt{c_2}\,\|\Psi^{-1}(t)R_4(t)\|\big(\|\boldsymbol{y}\|+\|\boldsymbol{z}\|\big) \le M\big(\|\boldsymbol{y}\|+\|\boldsymbol{z}\|\big) \\ &\big(M=\sqrt{c_2}\,l, \quad l=\max\|\Psi^{-1}(t)R_4(t)\|, \quad t\in[0,\infty)\big), \end{aligned}$$

$$\begin{aligned} \dot v^*(t,\boldsymbol{y},\boldsymbol{z}) &= \frac{1}{2\sqrt{c_1^*}}\frac{\dot v(t,\boldsymbol{y},\boldsymbol{z})}{\sqrt{v(t,\boldsymbol{y},\boldsymbol{z})}} \le -\frac{c_3}{2c_2\sqrt{c_1^*}}\frac{v(t,\boldsymbol{y},\boldsymbol{z})}{\sqrt{v(t,\boldsymbol{y},\boldsymbol{z})}} \\ &= -\alpha v^*(t,\boldsymbol{y},\boldsymbol{z}), \qquad \alpha=\frac{c_3}{2c_2}, \end{aligned}$$

$$\big|v^*(t,\boldsymbol{y},\boldsymbol{z}) - v^*(t,\boldsymbol{y}',\boldsymbol{z}')\big| \le M\big(\|\boldsymbol{y}-\boldsymbol{y}'\|+\|\boldsymbol{z}-\boldsymbol{z}'\|\big).$$

As a result, the function $v^*(t,\boldsymbol{y},\boldsymbol{z})$ satisfies all the conditions of Theorem 0.5.1.

2. Now we turn to the case when system (1.2.1), generally speaking, has a discontinuous $\boldsymbol{\mu}$-system (1.2.10).

Theorem 1.2.2. *The function*

$$v^*(t,\boldsymbol{x}) = c\sqrt{\boldsymbol{x}^{\mathsf T}\big[\Psi^{-1}(t)R_4(t)\big]^{\mathsf T}\Omega\big[\Psi^{-1}(t)R_4(t)\big]\boldsymbol{x}},$$

where Ω *is a constant* $(m+h)\times(m+h)$ *matrix which obeys the Lyapunov equation*

$$K^{\mathsf T}\Omega + \Omega K = -I_{m+h},$$

with a number $c>0$ *properly chosen, satisfies all the conditions of Theorem* 0.5.1.

Proof. First we show that a function $\Psi^{-1}(t)R_4(t)$ of $t\in[0,\infty)$ is analytic. We shall lean upon the equality $QX^{-1}Q^{-1} = (QXQ^{-1})^{-1}$ being valid for any two square matrices Q and X of the same order (Q is a nonsingular matrix). Setting $Q=R$ and $X=\Phi$,

$$R(t)\Phi^{-1}(t)R^{-1}(t) = \big[R(t)\Phi(t)R^{-1}(t)\big]^{-1}, \quad t\in[0,\infty).$$

Since $\Psi(t) = R_4(t)\Phi(t)R_5(t)$, $t\in[0,\infty)$, the above equation yields

$$\Psi^{-1}(t) = R_4(t)\Phi^{-1}(t)R_5(t), \quad t\in[0,\infty).$$

Taking into account that $R_5(t)R_4(t) = I_{m+p}$, $t\in[0,\infty)$, we arrive at the relationships

$$\begin{aligned} \Psi^{-1}(t)R_4(t) &= \big[R_4(t)\Phi(t)R_5(t)\big]^{-1}R_4(t) \\ &= R_4(t)\Phi^{-1}(t)R_5(t)R_4(t) = R_4(t)\Phi^{-1}(t). \end{aligned}$$

Matrix-valued functions $R_4(t)$ and $\Phi^{-1}(t)$ are analytic for $t \in [0, \infty)$. Hence, $\Psi^{-1}(t)R_4(t)$ is an *analytic* matrix-valued function of $t \in [0, \infty)$, whereas components of matrix-valued function $\Psi^{-1}(t)$, generally speaking, can have a finite number of discontinuities in $[0, T]$.

The rest of the proof follows that used in the case when the coefficients of system (1.2.1) possess property (A). □

Remark 1.2.4.

1. From the proof of Theorem 1.2.2, one can conclude that the *analyticity* of matrix-valued function $\Psi^{-1}(t)R_4(t)$ for all $t \in [0, \infty)$ and, consequently, the *analyticity* of the Lyapunov function $v^*(t, \boldsymbol{x})$ result from discontinuities of function $\Psi^{-1}(t)$ "being canceled" by those values of function $R_4(t)$ at which the linear independence of its rows is violated.

2. Function $v^*(t, \boldsymbol{x})$ in Theorem 1.2.2, generally, does not satisfy the conditions of Rumyantsev's Theorem 0.4.2 for any $k = 1, 2$. At the same time, this function obeys conditions of the type indicated in Corollary 0.4.1.

Example 1.2.3. Let us construct function v which solves the problem of asymptotic stability of the unperturbed motion of system (1.2.13) with respect to y_1. This can be arranged by constructing the corresponding Lyapunov function either for the auxiliary $\boldsymbol{\mu}$-system (1.2.14) or for (1.2.15).

1. First we implement the construction based on system (1.2.15). In this case the change of variables $\boldsymbol{\xi} = \Psi(t)\boldsymbol{\eta}$ with

$$\Psi(t) = \begin{pmatrix} 1 & 2\sin t \\ 0 & 1 \end{pmatrix}$$

transforms the auxiliary system (1.2.15) in the following way:

$$\dot{\boldsymbol{\eta}} = K\boldsymbol{\eta}, \qquad K = \begin{pmatrix} -1 & 0 \\ 0 & -1 \end{pmatrix}. \tag{1.2.20}$$

The Lyapunov function $v = \eta_1^2 + \eta_2^2$ for system (1.2.20), after reverting to the original variables

$$\boldsymbol{\eta} = \Psi^{-1}(t)\boldsymbol{\xi}, \qquad \boldsymbol{\xi} = R_4(t)\boldsymbol{x},$$

$$\Psi^{-1}(t) = \begin{pmatrix} 1 & -2\sin t \\ 0 & 1 \end{pmatrix}, \quad R_4(t) = \begin{pmatrix} 1 & 0 & 0 \\ 0 & \sin t & \cos t \end{pmatrix},$$

takes the form

$$\begin{aligned} v = v(t, \boldsymbol{y}, \boldsymbol{z}) &= \eta_1^2(t, \boldsymbol{y}, \boldsymbol{z}) + \eta_2^2(t, \boldsymbol{y}, \boldsymbol{z}) = (\xi_1 - 2\sin t\xi_2)^2 + \xi_2^2 \\ &= \left[y_1 - 2\sin t(z_1 \sin t + z_2 \cos t)\right]^2 + (z_1 \sin t + z_2 \cos t)^2. \end{aligned}$$

Since the inequalities $v \geq c_1^* y_1^2$, $0 < c_1^* < 1/5$, $\dot{v} = -v$ are valid for all $t \in [0, \infty)$, $\|\boldsymbol{x}\| < \infty$ (where the derivative of v is calculated by virtue of system (1.2.13)), the function $v^* = \sqrt{v}/\sqrt{c_1^*}$ for all $t \in [0, \infty)$, $\|\boldsymbol{x}\| < \infty$, satisfies the estimates $v^* \geq \|y_1\|$, $\dot{v}^* = -v^*/2$.

2. Now let us construct function v on the basis of discontinuous auxiliary system (1.2.14). Applying the change of variables $\boldsymbol{\xi} = \Psi^*(t)\boldsymbol{\eta}$ with

$$\Psi^*(t) = \begin{pmatrix} 1 & \sin t \\ 0 & \cos t \end{pmatrix},$$

we transform the auxiliary system (1.2.14) into (1.2.20).

The function $v = \eta_1^2 + \eta_2^2$, after reverting to the original variables

$$\boldsymbol{\eta} = \Psi^{*-1}(t)\boldsymbol{\xi}, \qquad \boldsymbol{\xi} = R_4^*(t)\boldsymbol{x},$$

$$\Psi^{*-1}(t) = \begin{pmatrix} 1 & -\tan t \\ 0 & 1/\cos t \end{pmatrix}, \quad R_4^*(t) = \begin{pmatrix} 1 & 0 & 0 \\ 0 & \sin 2t & 2\cos^2 t \end{pmatrix},$$

becomes

$$\begin{aligned} v = v(t, \boldsymbol{y}, \boldsymbol{z}) &= \eta_1^2(t, \boldsymbol{y}, \boldsymbol{z}) + \eta_2^2(t, \boldsymbol{y}, \boldsymbol{z}) = (\xi_1 - \xi_2 \tan t)^2 + (\xi_2/\cos t)^2 \\ &= \big[y_1 - 2\sin t(z_1 \sin t + z_2 \cos t)\big]^2 + 4(z_1 \sin t + z_2 \cos t)^2. \end{aligned}$$

Since the relationships $v \geq c_1^* y_1^2$, $0 < c_1^* < 1/2$, $\dot{v} = -v$ are valid for all $t \in [0, \infty)$, $\|\boldsymbol{x}\| < \infty$ (where the derivative of v is calculated by virtue of system (1.2.13)), the function $v^* = \sqrt{v}/\sqrt{c_1^*}$ for all $t \in [0, \infty)$, $\|\boldsymbol{x}\| < \infty$, satisfies the estimates $v^* \geq \|y_1\|$, $\dot{v}^* = -v^*/2$. We also note that V-function does not satisfy the conditions of Theorem 0.4.2 for any $k = 1, 2$.

1.3 Stability with Respect to Part of the Variables of Linear Systems with Continuous Sufficiently Differentiable Coefficients

1.3.1 Formulation of the Problem

Consider a system of differential equations of perturbed motion

$$\begin{aligned} \frac{dy_i}{dt} &= \sum_{k=1}^{m} a_{ik}(t) y_k + \sum_{l=1}^{p} b_{il}(t) z_l \qquad (i = 1, \ldots, m), \\ \frac{dz_j}{dt} &= \sum_{k=1}^{m} c_{jk}(t) y_k + \sum_{l=1}^{p} d_{jl}(t) z_l \qquad (j = 1, \ldots, p) \end{aligned} \tag{1.3.1}$$

In the matrix form the system can be written as

$$\dot{\boldsymbol{y}} = A(t)\boldsymbol{y} + B(t)\boldsymbol{z}, \quad \dot{\boldsymbol{z}} = C(t)\boldsymbol{y} + D(t)\boldsymbol{z},$$

where $A(t), B(t), C(t)$, and $D(t)$ are continuous matrix functions of appropriate dimensions. For all $t_0, \boldsymbol{y}_0$, and $\boldsymbol{z}_0$ there exists a unique solution of system (1.3.1), which is defined for all $t \geq 0$.

The analysis of linear system (1.3.1) is much more complicated than that of a system with coefficients which are constants or periodic functions. However, a number of problems in present-day engineering require that such systems should be analyzed.

The purpose of this section is to establish the least possible restrictive conditions that make the earlier proposed method of constructing $\boldsymbol{\mu}$-systems apply to the analysis of system (1.3.1). Conditions for stability with respect to part of the variables obtained in this way, in general, are only *sufficient*. The origin of the difficulties arising can be somewhat elucidated by investigating the mechanism of situations in which unperturbed motion of system (1.3.1) is stable with respect to one part of the variables and unstable with respect to another.

1.3.2 Auxiliary System and Conditions for Stability with Respect to Part of the Variables

When the coefficients of a linear system are constants or periodic, sufficiently differentiable functions of t, the problem of stability (asymptotic stability) with respect to part of the variables appeared to be reducible to the problem of stability (asymptotic stability) for some auxiliary linear system. For these two cases a constructive rule for constructing such an auxiliary system is obtained. For a more general class of linear systems with which this section deals, we shall just indicate the cases in which the technique mentioned is acceptable.

Suppose that $a_{ik}(t)$, $c_{jk}(t)$ are *continuous* and $b_{il}(t)$, $d_{jl}(t)$ are, respectively, p and $(p-1)$ *times continuously differentiable* functions of $t \in [0,\infty)$. Consider vectors $\boldsymbol{b}_i(t) = \left[b_{i1}(t),\dots,b_{ip}(t)\right]^{\mathsf{T}}$ $(i=1,\dots,m)$, assuming that the first m_1 of them $\boldsymbol{b}_1,\dots,\boldsymbol{b}_{m_1}$ are linearly independent for all $t \in [0,\infty)$. The remaining vectors $\boldsymbol{b}_j$ $(j=m_1+1,\dots,m)$ satisfy the equalities

$$\boldsymbol{b}_j(t) = \sum_{l=1}^{m_1} \alpha_{jl}(t)\boldsymbol{b}_l(t), \quad t \geq 0 \qquad (j=m_1+1,\dots,m),$$

where α_{jl} are continuous functions of $t \in [0,\infty)$. To transform system (1.3.1), let us introduce the new variables

$$\mu_i = \sum_{l=1}^{p} b_{il}(t) z_l \qquad (i=1,\dots,m_1). \tag{1.3.2}$$

Introducing the new variables in this way, we encounter two cases.

The first case. It is possible to make up a system

$$\begin{aligned} \frac{dy_i}{dt} &= \sum_{k=1}^{m} a_{ik}(t) y_k + \sum_{l=1}^{m_1} \alpha_{il}(t)\mu_l \qquad (i=1,\dots,m), \\ \frac{d\mu_j}{dt} &= \sum_{k=1}^{m} a^*_{jk}(t) y_k + \sum_{l=1}^{m_1} \alpha^*_{jl}(t)\mu_l \qquad (j=1,\dots,m_1) \end{aligned} \tag{1.3.3}$$

with coefficients continuous for $t \in [0,\infty)$. Following the terminology of Section 1.1, we shall call system (1.3.3) a $\boldsymbol{\mu}$*-system* for the original system (1.3.1).

The second case. Having introduced new variables (1.3.2), one fails to make up a $\boldsymbol{\mu}$-system. This implies that in the equations

$$\frac{dy_i}{dt} = \sum_{k=1}^{m} a_{ik}(t) y_k + \sum_{l=1}^{m_1} \alpha_{il}(t)\mu_l \qquad (i=1,\dots,m),$$

$$\frac{d\mu_j}{dt} = \sum_{k=1}^{m} a_{jk}^*(t)y_k + \sum_{l=1}^{p} b_{jl}^*(t)z_l \qquad (j = 1, \ldots, m_1),$$

$$\frac{dz_s}{dt} = \sum_{k=1}^{m} c_{sk}(t)y_k + \sum_{l=1}^{p} d_{sl}(t)z_l \qquad (s = 1, \ldots, p),$$

the equalities

$$\sum_{l=1}^{p} b_{jl}^*(t)z_l = \sum_{l=1}^{m_1} \alpha_{jl}^*(t)\mu_l, \quad t \geq 0 \qquad (j = 1, \ldots, m_1) \tag{1.3.4}$$

are impossible for at least one of the values of j. In this case we introduce the new variables once again

$$\mu_{m_1+j} = \sum_{l=1}^{p} b_{jl}^*(t)z_l \qquad (j = 1, \ldots, m_1),$$

selecting only those for which equalities (1.3.4) are violated.

Let us find conditions which, for the original system (1.3.1), allow constructing a $\boldsymbol{\mu}$-system with continuous coefficients of dimension less or equal to that of the original one. To do this, consider the matrix $G_{p+1} = (L_1, \ldots, L_{p+1})$ with elements defined by the relationships

$$L_1(t) = B^{\mathsf{T}}(t), \ldots, L_j(t) = \dot{L}_{j-1}(t) + D^{\mathsf{T}}(t)L_{j-1}(t) \qquad (j = 2, \ldots, p+1).$$

Theorem 1.3.1. *Let there exist a number $s \leq p+1$ such that*
(1) *matrix G_{s-1} contains $h \leq p$ linearly independent columns $\boldsymbol{l}_1^*, \ldots, \boldsymbol{l}_h^*$ for all $t \in [0, \infty)$; and*
(2) *all columns $\boldsymbol{l}_i$ of matrix G_s, except for $\boldsymbol{l}_1^*, \ldots, \boldsymbol{l}_h^*$, satisfy condition*

$$\boldsymbol{l}_i(t) = \sum_{r=1}^{h} \beta_{ir}(t)\boldsymbol{l}_r^*(t) \qquad (i = 1, \ldots, m \times s - h), \tag{1.3.5}$$

where $\beta_{ir}(t)$ are continuous functions of $t \in [0, \infty)$.

Then the dimension of the $\boldsymbol{\mu}$-system for (1.3.1) equals $m + h \leq n$. Conditions for Lyapunov stability of the zero solution of the $\boldsymbol{\mu}$-system are sufficient for the unperturbed motion of system (1.3.1) to be stable with respect to $y_1, \ldots, y_m$.

Proof. It follows from the above procedure for constructing an auxiliary $\boldsymbol{\mu}$-system that the new variables required to form an auxiliary system are chosen from the variables $\mu_i = \boldsymbol{l}_i^{\mathsf{T}}(t)\boldsymbol{z}$ $(i = 1, \ldots, m \times s)$. Here $\boldsymbol{l}_i$ stands for the columns of matrix G_{s-1}. By (1.3.5), all variables μ_i $(i = 1, \ldots, m \times s)$, except for $\mu_j^* = \boldsymbol{l}_j^{*\mathsf{T}}(t)\boldsymbol{z}$ $(j = 1, \ldots, h)$, satisfy the condition

$$\mu_i = \sum_{r=1}^{h} \beta_{ir}(t)\mu_r^* \qquad (i = 1, \ldots, m \times s - h).$$

Because functions $\beta_{ir}(t)$ in (1.3.5) are continuous, to construct an auxiliary system of equations which determines the evolution of variables y_i

$(i = 1, \dots, m)$ and μ_k $(k = 1, \dots, m \times s)$, it suffices to involve only variables y_i $(i = 1, \dots, m)$ and μ_j^* $(j = 1, \dots, h)$. Hence, the dimension of $\boldsymbol{\mu}$-system equals $m + h \leq n$. The transition from the original system (1.3.1) to a $\boldsymbol{\mu}$-system is carried out by a linear change of variables (with coefficients that are continuous functions of t). Variables y_i $(i = 1, \dots, m)$ are invariant under this transformation. Therefore, the Lyapunov stability of the zero solution of the $\boldsymbol{\mu}$-system leads to stability of the unperturbed motion $\boldsymbol{y} = \mathbf{0}$, $\boldsymbol{z} = \mathbf{0}$ of system (1.3.1) with respect to $y_1, \dots, y_m$. The theorem is proved. □

Corollary 1.3.1. *Let the following conditions hold:*

(1) *there exists a number $s \leq p + 1$ such that condition* (1) *of Theorem* 1.3.1 *is complied with;*

(2) *the problems of stability (asymptotic stability) of the zero solution of the $\boldsymbol{\mu}$-system in the Lyapunov sense and with respect to $y_1, \dots, y_m$ are equivalent.*

Then a necessary and sufficient condition for stability (asymptotic stability) of the unperturbed motion $\boldsymbol{y} = \mathbf{0}$, $\boldsymbol{z} = \mathbf{0}$ of system (1.3.1) *with respect to $y_1, \dots, y_m$ is that the zero solution of the $\boldsymbol{\mu}$-system be Lyapunov stable (asymptotically stable).*

Remark 1.3.1.

1. Condition (2) of Corollary 1.3.1 can be constructively verified in a number of cases. This is possible, for instance, when coefficients of the $\boldsymbol{\mu}$-system are constant or periodic.

2. With condition (2) omitted, Corollary 1.3.1 ceases to be true (in part of necessary conditions).

The system

$$\dot{y}_1 = z_1, \qquad \dot{z}_1 = a(t) y_1 \tag{1.3.6}$$

with continuous function $a(t)$ gives evidence to the fact. The auxiliary $\boldsymbol{\mu}$-system for (1.3.6) coincides with system (1.3.6). At the same time, conditions are known (Cesari [1959]) under which $y_1 \to 0, t \to \infty$ (i.e., solution $y_1 = z_1 = 0$ is asymptotically stable with respect to y_1 (Bellman [1953], Krivosheev and Lutsenko [1980])), whereas the Lyapunov asymptotic stability of this solution is impossible.

3. The problems of stability (asymptotic stability) of the unperturbed motion of system (1.3.1) with respect to $y_1, \dots, y_m$ and in the Lyapunov sense are not equivalent, in general, even if the dimension of system (1.3.1) *coincides* with that of its $\boldsymbol{\mu}$-system. This is what distinguishes system (1.3.1) from systems with constant or periodic coefficients.

Remark 1.3.2.

1. When constructing $\boldsymbol{\mu}$-systems for the original system (1.3.1), one can take advantage of the following occasional property of its coefficients. (This property is similar to property (A) introduced in Subsection 1.2.4.) Let the coefficients b_{il} of system (1.3.1), which determine new variables (1.3.2), satisfy conditions

$$b_{il}(t) = b_i(t) b_{il}^*(t), \quad t \in [0, \infty) \qquad (i = 1, \dots, m_1;\ l = 1, \dots, p),$$

where $b_i(t)$ are continuous for $t \in [0, \infty)$ (their differentiability is not assumed) and b_{il}^* have the same properties that we previously required for b_{il}. In this case, it is possible (and advisable) to introduce the new variables

$$\mu_i = \sum_{l=1}^{p} b_{il}^*(t) z_l \qquad (i = 1, \dots, m_1)$$

instead of (1.3.2), which allows extending the scope for constructing auxiliary μ-systems.

2. Suppose that coefficients $b_{il}(t)$ entering into the new μ-variables are uniformly *bounded* for all $t \in [0, \infty)$. Then the exponential asymptotic stability of the zero solution of μ-system implies the *exponential* asymptotic stability (in Corduneanu's sense [1971]) of the unperturbed motion $\mathbf{y} = \mathbf{0}$, $\mathbf{z} = \mathbf{0}$ of the original system (1.3.1) with respect to $y_1, \dots, y_m$.

Example 1.3.1.

1. Let system (1.3.1) have the form

$$\dot{y}_1 = ay_1 + be^{-2t} z_1, \quad \dot{z}_1 = ce^{2t} y_1 + dz_1 \quad (a, b, c, d = \text{const}). \tag{1.3.7}$$

The auxiliary μ-system comprises the equations

$$\dot{y}_1 = ay_1 + b\mu_1, \quad \dot{\mu}_1 = cy_1 + (d-2)\mu_1 \quad (\mu_1 = e^{-2t} z_1). \tag{1.3.8}$$

The zero solution of the μ-system (1.3.8) is Lyapunov (exponentially) asymptotically stable, provided that $a + d < 2$ and $a(d-2) - cb > 0$. Hence, the unperturbed motion $y_1 = z_1 = 0$ of system (1.3.7) is exponentially asymptotically stable with respect to y_1.

When $a > 0, d < 0$ and with the connection removed ($b = c = 0$), the subsystems $\dot{y}_1 = ay_1$ and $\dot{z}_1 = dz_1$ are Lyapunov unstable and asymptotically stable, respectively. Now we demonstrate that, entering the connection of the type specified ($b \neq 0, c \neq 0$), one can find, just in the case $a > 0, d < 0$, a domain in the parameters a, b, c, and d space for which the behavior of variables y_1 and z_1 of the entire system (1.3.7) changes. Namely, for $-2 < a + d < 2$ and $a(d-2) - cb > 0$, the unperturbed motion $y_1 = z_1 = 0$ of system (1.3.7) is asymptotically stable with respect to y_1 and unstable with respect to z_1.

The asymptotic stability with respect to y_1 has already been proved. To prove the instability with respect to z_1, construct the μ-system for examining the behavior of the variable, i.e., the system

$$\dot{z}_1 = dz_1 + \mu_2, \quad \dot{\mu}_2 = cbz_1 + (a+2)\mu_2 \quad (\mu_2 = ce^{2t} y_1). \tag{1.3.9}$$

According to the results of Section 1.1, the stability with respect to one of the variables for a linear system of two equations with constant coefficients (1.3.9) is only possible on condition of Lyapunov stability. Therefore, for the zero solution $z_1 = \mu_2 = 0$ of the auxiliary μ-system (1.3.9) to be stable with respect to z_1, it is necessary that $a + d \leq -2$.

Thus, for

$$a > 0, \quad d < 0, \quad -2 < a + d < 2,$$
$$a(d-2) - cb > 0,$$

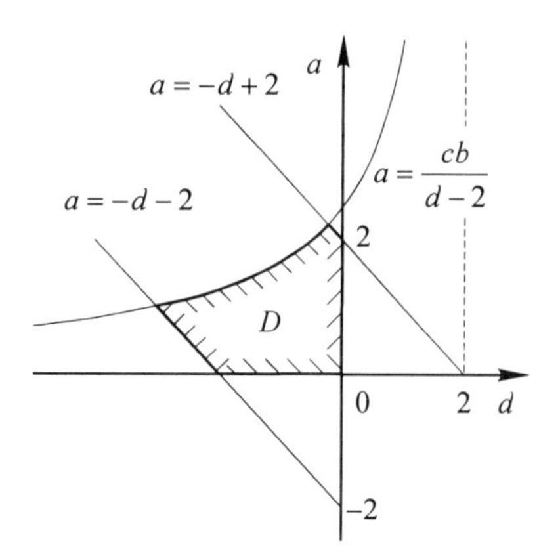

Figure 1.3.1. Region D: $a > 0$, $d < 0$, $-2 < a + d < 2$, $a(d-2) - cb > 0$. The case $cb < 0$.

the unperturbed motion $y_1 = z_1 = 0$ of system (1.3.7) is exponentially asymptotically stable with respect to y_1 and unstable with respect to z_1. The corresponding domain for the coefficients is shown in Figure 1.3.1.

2. Consider the *motion of a variable mass point* $m = m(t)$ in a vertical plane (ξ, η). We assume that the external force of the medium, defined as a function of the motion height $R = R(\eta)$, acts on the point tangentially. In this case the equations of linear approximation for the perturbed motion system take the form (Galiullin et al. [1971])

$$
\begin{aligned}
&\dot{y}_1 = (\sin\theta_0)z_1 + (v_0\cos\theta_0)z_2, \\
&\dot{z}_1 = -\left(\frac{R_{\eta 0}}{m}\right)y_1 - (g\cos\theta_0)z_2, \quad \dot{z}_2 = -\left(\frac{\dot{\theta}_0}{v_0}\right)z_1 + \left(\frac{g\sin\theta_0}{v_0}\right)z_2,
\end{aligned}
\tag{1.3.10}
$$

where y_1 is the perturbation of the height of the point, z_1 and z_2 are the perturbations of the magnitude and direction of the velocity of the point, respectively, $\xi_0 = \varphi(t)$, $\eta_0 = \psi(t)$ is the point's unperturbed motion, $v_0 = \left[\dot{\varphi}^2(t) + \dot{\psi}^2(t)\right]^{1/2}$, $\theta_0 = \arctan\left(\dfrac{\dot{\psi}}{\dot{\varphi}}\right)$, and $R_{\eta 0} = \partial R/\partial\eta$ along $\eta = \eta_0$.

Suppose that $\dot{\psi} = \text{const}$ in the unperturbed motion. This means that the point moves vertically with no acceleration and the condition $v_0\cos\theta_0 = \text{const}$ is valid. The auxiliary $\boldsymbol{\mu}$-system is composed of the equations

$$
\dot{y}_1 = \mu_1, \qquad \dot{\mu}_1 = -R_{\eta 0}\frac{\sin\theta_0}{m}y_1. \tag{1.3.11}
$$

The conditions for Lyapunov stability of the zero solution $y_1 = \mu_1 = 0$ of system (1.3.11) define the domain in which the unperturbed motion $y_1 = z_1 = z_2 = 0$ of system (1.3.10) is stable with respect to y_1.

The Lyapunov *asymptotic* stability of the zero solution of system (1.3.11) with continuous function $R_{\eta 0}\sin\theta_0/m$ is *impossible*. At the same time, as was already noted in Remark 1.3.1, there are a number of conditions which ensure asymptotical stability of the zero solution of system (1.3.11) with respect to y_1. These are conditions for asymptotic stability of the unperturbed motion of system (1.3.10) with respect to y_1.

The example under study shows that, when investigating the stability (asymptotic stability) of the unperturbed motion of system (1.3.1) with respect to part of the variables, it is *not always necessary* to consider the *Lyapunov* stability (asymptotic stability) problem for auxiliary $\boldsymbol{\mu}$-systems. In a number of cases, for the conditions for stability (asymptotic stability) of the original system (1.3.1) with respect to part of the variables to approach the *necessary* ones, it is also reasonable to consider the problem of stability (asymptotic stability) of auxiliary $\boldsymbol{\mu}$-systems with respect to part of the variables. In these cases constructing $\boldsymbol{\mu}$-system enables us to reduce the order of the system under investigation.

1.3.3 On the Mechanism of the Emergence of Stability with Respect to Part of the Variables

Let us analyze what leads to the situation in which, in spite of being Lyapunov unstable (or nonasymptotically stable), the unperturbed motion $\boldsymbol{y} = \mathbf{0}$, $\boldsymbol{z} = \mathbf{0}$ of system (1.3.1) is asymptotically stable with respect to $y_1, \ldots, y_m$. We distinguish a number of cases.

1. The situation under discussion results from the fact that those terms of linear combinations $\sum_{l=1}^{p} b_{il}(t) z_l$ $(i = 1, \ldots, m)$ that are associated with values of characteristic numbers undesirable for stability with respect to $y_1, \ldots, y_m$ mutually annihilate. In this case the problem of stability with respect to part of the variables is reduced to the Lyapunov stability problem for a linear system of lower dimension.

As an example, we refer to the solutions of system (1.1.14)

$$\begin{aligned}
y_1(t; t_0, \boldsymbol{x}_0) &= y_{10} e^{-t} + (z_{10} - 2z_{20}) t e^{-t}, \\
z_1(t; t_0, \boldsymbol{x}_0) &= (-2y_{10} - z_{10} + 2z_{20}) e^{-t} \\
&\quad - 2(z_{10} - 2z_{20}) t e^{-t} + 2(y_{10} + z_{10} - z_{20}) e^{t}, \\
z_2(t; t_0, \boldsymbol{x}_0) &= (-y_{10} - z_{10} + 2z_{20}) e^{-t} \\
&\quad - (z_{10} - 2z_{20}) t e^{-t} + (y_{10} + z_{10} - z_{20}) e^{t}.
\end{aligned}$$

The terms of the expressions for z_1 and z_2 containing e^t cancel out in the linear combination $z_1 - 2z_2$.

2. In linear combinations $\sum_{l=1}^{p} b_{il}(t) z_l$ $(i = 1, \ldots, m)$ cancellation of terms with undesirable values of characteristic numbers takes place, and also "compensation" for undesirable tendencies $b_{il}(t) \to \infty$, $t \to \infty$ occurs.

The system

$$\dot{y}_1 = -y_1 + e^t (z_1 - 2z_2), \quad \dot{z}_1 = 4y_1 + 2z_1, \quad \dot{z}_2 = 2y_1 + 3z_1 - 4z_2$$

illustrates the case: cancellation, in the linear combination $e^t(z_1 - 2z_2)$, of terms containing e^{2t} in the expressions for z_1 and z_2 is accompanied by undesirable asymptotic behavior $b_{11} = b_{12} = e^t \to \infty$, $t \to \infty$, being "compensated for."

3. Stability with respect to part of the variables stems from "compensation," in linear combinations $\sum_{l=1}^{p} b_{il}(t) z_l$ $(i = 1, \ldots, m)$, of undesirable ten-

dency $z_i(t) \to \infty$ $(i = 1, \dots, p)$ by functions $b_{il}(t)$ satisfying the condition $b_{il}(t) \to 0, t \to \infty$.

For simplicity, setting $a = -1, b = d = 1, c = 0$ in system (1.3.7), we see that undesirable tendency $z_1 = z_{10}e^t \to \infty, t \to \infty$ is "compensated" by the function $b_{11} = e^{-2t}$.

4. Another way in which the situation under study occurs is also possible. The structure of system (1.3.6), in particular, provides evidence for this statement. Indeed, the zero solution of the system $y_1 = z_1 = 0$, as already noted, can be asymptotically stable with respect to y_1 but not to z_1.

This situation is attributable to the fact that differentiating function $y_1(t)$, which satisfies relationship $y_1(t) \to 0, t \to \infty$, results in function $z_1 = \dot{y}_1$ losing tendency $z_1(t) \to 0, t \to \infty$ required for attraction. We note that this mechanism and the one indicated in (3) are "milder" than that based on canceling the terms with undesirable properties.

Apparently, any combination of the above mechanisms cannot be ruled out.

1.3.4 An Auxiliary Limiting System

Let coefficients of system (1.3.1), which are continuous and also *bounded* functions of $t \in [0, \infty)$ (unlike Subsection 1.3.2, this one does not suggest their differentiability), tend to numbers $a_{ik}^+, b_{il}^+, c_{jk}^+$, and d_{jl}^+ as $t \to \infty$. These numbers constitute matrices A^+, B^+, C^+, and D^+ of appropriate sizes.

Chetayev [1946] showed that if the unperturbed motion of a *limiting* system, for (1.3.1),

$$\dot{\boldsymbol{y}} = A^+\boldsymbol{y} + B^+\boldsymbol{z}, \qquad \dot{\boldsymbol{z}} = C^+\boldsymbol{y} + D^+\boldsymbol{z},$$

is Lyapunov asymptotically stable, then unperturbed motion of system (1.3.1) itself possesses the same property.

However, as the system

$$\dot{y}_1 = -y_1 + e^{-t}z_1, \qquad \dot{z}_1 = 2z_1 \tag{1.3.12}$$

demonstrates, this is not always the case for the problem of stability with respect to part of the variables. Indeed, the zero solution $y_1 = z_1 = 0$ of the limiting system, for (1.3.12),

$$\dot{y}_1 = -y_1, \qquad \dot{z}_1 = 2z_1$$

is asymptotically stable with respect to y_1. At the same time, for variable y_1 of system (1.3.12), which obeys the equation $\dot{y}_1 = -y_1 + z_{10}e^t$, the condition $|y_1(t)| \to \infty, t \to \infty, z_{10} \neq 0$ is fulfilled.

The conditions for asymptotic stability of the unperturbed motion of system (1.3.1) with respect to part of the variables can be derived by constructing limiting systems *not for system* (1.3.1) itself *but for its μ-system.*

Example 1.3.2. Let system (1.3.1) have the form

$$\begin{gathered} \dot{y}_1 = (e^{-t} - 1)y_1 + (e^{-t} + 1)(z_1 - 2z_2), \\ \dot{z}_1 = 2(\sin t)y_1 + z_1, \qquad \dot{z}_2 = (\sin t)y_1 + z_1 - z_2. \end{gathered} \tag{1.3.13}$$

The auxiliary $\boldsymbol{\mu}$-system is composed of the equations

$$\dot{y}_1 = (e^{-t} - 1)y_1 + (e^{-t} + 1)\mu_1, \qquad \dot{\mu}_1 = -\mu_1 \tag{1.3.14}$$
$$(\mu_1 = z_1 - 2z_2).$$

The roots of the characteristic equation of the system $\dot{y}_1 = -y_1 + \mu_1$, $\dot{\mu}_1 = -\mu_1$, which is a limiting system for (1.3.14), are real and negative. Hence, the unperturbed motion $y_1 = z_1 = z_2 = 0$ of system (1.3.13) is asymptotically stable with respect to y_1.

Note, that the coefficients of the original system (1.3.13) (unlike those of $\boldsymbol{\mu}$-system (1.3.14)) *do not allow* constructing a limiting system (for $t \to \infty$).

1.4 The Effect of Constantly Acting and Parametric Perturbations on Stability with Respect to Part of the Variables

1.4.1 General Features of the Problem

Let the unperturbed motion $\boldsymbol{y} = \boldsymbol{0}, \boldsymbol{z} = \boldsymbol{0}$ of system (1.1.6) be Lyapunov asymptotically stable. Then this motion is stable in the presence of constantly acting perturbations that are (integrally) small, and a small variation in the coefficients of system (1.1.6) preserves the asymptotic stability (Krasovskii [1959]).

Assuming that the unperturbed motion of system (1.1.6) is asymptotically stable with respect to $y_1, \ldots, y_m$, we shall prove the statements listed below.

(1) The unperturbed motion of system (1.1.6) is stable with respect to $y_1, \ldots,$ y_m in the presence of constantly acting perturbations. Some of the functions $R_i(t, \boldsymbol{x})$ constituting the perturbing vector function do not have to be (integrally) small. Given certain relationships between $R_i(t, \boldsymbol{x})$ ($R_i(t, \boldsymbol{x}) \not\equiv 0$), the *asymptotic* stability with respect to $y_1, \ldots, y_m$ is even possible.

(2) An arbitrarily small variation in the coefficients of system (1.1.6), generally speaking, can result in instability with respect to $y_1, \ldots, y_m$.

We note that a contradictory nature of statements (1) and (2) is a distinguishing feature of the problem of stability with respect to part of the variables.

1.4.2 Formulation of the Problem of Stability (Asymptotic Stability) with Respect to Part of the Variables in the Presence of Constantly Acting Perturbations Which are not Small

Consider, along with system (1.1.6), the "perturbed" system

$$\dot{\boldsymbol{y}} = A\boldsymbol{y} + B\boldsymbol{z} + \boldsymbol{R_y}(t, \boldsymbol{y}, \boldsymbol{z}), \qquad \dot{\boldsymbol{z}} = C\boldsymbol{y} + D\boldsymbol{z} + \boldsymbol{R_z}(t, \boldsymbol{y}, \boldsymbol{z}) \tag{1.4.1}$$

in which vector functions $\boldsymbol{R_y}$ and $\boldsymbol{R_z}$ describe constantly acting perturbations. (In general, $\boldsymbol{R_y}(t, \mathbf{0}, \mathbf{0}) \not\equiv \mathbf{0}$ and $\boldsymbol{R_z}(t, \mathbf{0}, \mathbf{0}) \not\equiv \mathbf{0}$.) We assume that functions $\boldsymbol{R_y}$ and $\boldsymbol{R_z}$ are such that the right-hand sides of system (1.4.1) satisfy conditions (a) and (b) imposed on system (0.3.1).

Let us split the components of vector $\boldsymbol{z}$ and vector function $\boldsymbol{R_z}$ into two groups and represent $\boldsymbol{z}$ and $\boldsymbol{R_z}$ in the form $\boldsymbol{z}^{\mathsf{T}} = (\boldsymbol{z}^+, \boldsymbol{z}^-)$ and $\boldsymbol{R}_{\boldsymbol{z}}^{\mathsf{T}} = (\boldsymbol{R}_{\boldsymbol{z}}^+, \boldsymbol{R}_{\boldsymbol{z}}^-)$. The purpose of such a subdivision is to establish conditions for portions of vector $\boldsymbol{z}_0$ and vector function $\boldsymbol{R_z}$ less restrictive than those commonly imposed on the entire vectors in the theory of stability of motion in the presence of constantly acting perturbations. This shows promise for analyzing limiting cases for which a dynamic system maintains the stability in the presence of constantly acting perturbations and also gives an insight into those peculiarities of the problem of stability with respect to part of the variables that are not present in investigating the stability with respect to all the variables.

Definition 1.4.1. The unperturbed motion $\boldsymbol{x} = \mathbf{0}$ of system (1.1.6) is said to be

(1) *stable with respect to* $y_1, \ldots, y_m$ *in the presence of constantly acting perturbations on the whole in* $\boldsymbol{z}_0^-, \boldsymbol{R}_{\boldsymbol{z}}^-$ if, for any $\varepsilon > 0$, there exist positive $\delta_i(\varepsilon)$ $(i = 1, 2)$ such that the inequalities

$$t \geq t_0, \quad ||\boldsymbol{y}(t; t_0, \boldsymbol{x}_0)|| < \varepsilon, \quad ||\boldsymbol{z}(t; t_0, \boldsymbol{x}_0)|| < +\infty \tag{1.4.2}$$

hold true for all the motions of system (1.4.1) that begin in the domain $||\boldsymbol{y}_0|| < \delta_1(\varepsilon)$, $||\boldsymbol{z}_0^+|| < \delta_1(\varepsilon)$, $||\boldsymbol{z}_0^-|| < +\infty$ for all $\boldsymbol{R_y}$ and $\boldsymbol{R_z}$ satisfying, in domain (1.4.2), the conditions

$$||\boldsymbol{R_y}|| < \delta_2(\varepsilon), \quad ||\boldsymbol{R}_{\boldsymbol{z}}^+|| < \delta_2(\varepsilon), \quad ||\boldsymbol{R}_{\boldsymbol{z}}^-|| < +\infty;$$

(2) *asymptotically stable with respect to* $y_1, \ldots, y_m$ *in the presence of constantly acting perturbations (on the whole in* $\boldsymbol{z}_0^-, \boldsymbol{R}_{\boldsymbol{z}}^-$) if this motion is stable in the sense of the above definition (1) for which

$$||\boldsymbol{z}_0^-|| < \delta_1(\varepsilon), \quad ||\boldsymbol{R}_{\boldsymbol{z}}^-|| < \delta_2(\varepsilon), \quad (||\boldsymbol{z}_0^-|| < \infty, \ ||\boldsymbol{R}_{\boldsymbol{z}}^-|| < \infty)$$

and, in addition,

$$\lim||\boldsymbol{y}(t; t_0, \boldsymbol{x}_0)|| = 0, \quad t \to \infty.$$

Remark 1.4.1.

1. In contrast to definitions of stability in the presence of constantly acting perturbations offered earlier (with respect to all as well as to part of the variables), Definition 1.4.1 allows *arbitrarily large values* for a portion of vector $\boldsymbol{z}_0$ and also of vector function $\boldsymbol{R_z}$.

2. The *asymptotic* stability of the unperturbed motion $\boldsymbol{x} = \mathbf{0}$ of system (1.1.6) with respect to *all* variables in the presence of constantly acting perturbations $R_i(t, \boldsymbol{x})$, $R_i(t, \mathbf{0}) \not\equiv 0$ $(i = 1, \ldots, n)$ is *impossible* even if $R_i(t, \boldsymbol{x})$

are small. Therefore, the definition of asymptotic stability in the presence of constantly acting perturbations, just introduced, has meaning only when the stability with respect to *some* of the components of a dynamic system phase vector is considered.

1.4.3 Conditions for Stability (Asymptotic Stability) with Respect to Part of the Variables in the Presence of Constantly Acting Perturbations Which are not Small

For class $\boldsymbol{R}$ under consideration, vector functions $\boldsymbol{R_y}$ and $\boldsymbol{R_z}$ involved in system (1.4.1) can always be represented as

$$\boldsymbol{R_y} = \boldsymbol{R}_{\boldsymbol{y}0} + \boldsymbol{R}^*_{\boldsymbol{y}}(t, \boldsymbol{y}, \boldsymbol{z}), \qquad \boldsymbol{R_z} = \boldsymbol{R}_{\boldsymbol{z}0} + \boldsymbol{R}^*_{\boldsymbol{z}}(t, \boldsymbol{y}, \boldsymbol{z}),$$
$$\boldsymbol{R}^*_{\boldsymbol{y}}(t, \boldsymbol{0}, \boldsymbol{0}) \equiv \boldsymbol{0}, \quad \boldsymbol{R}^*_{\boldsymbol{z}}(t, \boldsymbol{0}, \boldsymbol{0}) \equiv \boldsymbol{0}, \quad \boldsymbol{R}^\top = (\boldsymbol{R}^\top_{\boldsymbol{y}}, \boldsymbol{R}^\top_{\boldsymbol{z}}), \quad \boldsymbol{R}^{*\top} = (\boldsymbol{R}^{*\top}_{\boldsymbol{y}}, \boldsymbol{R}^{*\top}_{\boldsymbol{z}}),$$

where $\boldsymbol{R}_{\boldsymbol{y}0}$ and $\boldsymbol{R}_{\boldsymbol{z}0}$ are *constant* vectors of appropriate sizes.

Theorem 1.4.1. *Let all the roots of equation* (1.1.12) *have negative real parts.*

1. *If the columns of matrix L_1 with position numbers $i_1, \ldots, i_N$ are zeros, then the unperturbed motion $\boldsymbol{x} = \boldsymbol{0}$ of system* (1.1.6) *is stable with respect to $y_1, \ldots, y_m$ in the presence of constantly acting perturbations on the whole in $\boldsymbol{z}_0^-$ and $\boldsymbol{R}_{\boldsymbol{z}}^-$. In this case vector $\boldsymbol{z}^-$ and vector function $\boldsymbol{R}_{\boldsymbol{z}}^-$ are composed of variables z_s and functions R_{zs}, respectively, with indices $s = i_1, \ldots, i_N$.*

2. *Let the columns of matrix $B^\top$ be linearly independent and*

$$\boldsymbol{l}_j^\top \boldsymbol{R}_{\boldsymbol{z}0} = \sum_{k=1}^{m} \lambda_{jk} R_{y0k} + \sum_{r=m+1}^{k} \lambda_{jr} \boldsymbol{l}_r^\top \boldsymbol{R}_{\boldsymbol{z}0} \qquad (j = 1, \ldots, h), \tag{1.4.3}$$

where $\boldsymbol{l}_s$ $(s = 1, \ldots, h)$ are linearly independent columns of matrix K_{s-1} and λ_{ij} $(i, j = 1, \ldots, h)$ are the elements of matrix $L_1 D L_3$. Besides, in domain (0.3.2), *let the following condition be valid*

$$\|L_4 \boldsymbol{R}^*(t, \boldsymbol{y}, \boldsymbol{z})\| \le \alpha \|\boldsymbol{y}\|, \tag{1.4.4}$$

where α is a sufficiently small positive constant. Then the unperturbed motion of system (1.1.6) *is uniformly asymptotically stable with respect to $y_1, \ldots, y_m$ in the presence of small constantly acting perturbations that satisfy conditions* (1.4.3) *and* (1.4.4). *In addition, if the columns of matrix L_1 with position numbers $i_1, \ldots, i_N$ are zeros, then the uniform asymptotic stability of this motion with respect to $y_1, \ldots, y_m$ takes place on the whole in z_s and R_{zs} $(s = i_1, \ldots, i_N)$.*

Proof. 1. When transforming system (1.1.6) into a $\boldsymbol{\mu}$-system, from perturbed system (1.4.1), one can isolate the equations

$$\dot{\boldsymbol{\xi}} = L_4 A^* L_5 \boldsymbol{\xi} + L_4 \boldsymbol{R}. \tag{1.4.5}$$

Under conditions of the theorem, the zero solution of the $\boldsymbol{\mu}$-system for (1.1.6) is Lyapunov asymptotically stable. Hence, it is also stable with respect to all the variables in the presence of small constantly acting perturbations $L_4\boldsymbol{R}$.

Further, the columns of L_1 with numbers $i_1, \dots, i_N$ are zeros. This means that vector function $L_4\boldsymbol{R}$ contains no perturbations R_{zs} ($s = i_1, \dots, i_N$) and vector $\boldsymbol{\xi}$ does not include variables z_s with indices $s = i_1, \dots, i_N$. This fact admits relationships $\|\boldsymbol{z}_0^-\| < +\infty$, $\|\boldsymbol{R}_{\boldsymbol{z}}^-\| < +\infty$, vector $\boldsymbol{z}^-$ and vector function $\boldsymbol{R}_{\boldsymbol{z}}^-$ including variables z_s and functions R_{zs}, respectively, enumerated by $s = i_1, \dots, i_N$. Since the equalities $y_i = \xi_i$ ($i = 1, \dots, m$) hold in transforming system (1.4.1) into (1.4.5), the first part of the theorem is proved.

2. When (1.4.3) is true, on introducing the new variables

$$\mu_i = \boldsymbol{l}_i^{\mathsf{T}}\boldsymbol{z} + R_{y0i}, \qquad \mu_{m+j} = \boldsymbol{l}_{m+j}^{\mathsf{T}}\boldsymbol{z} + \boldsymbol{l}_{m+j}^{\mathsf{T}}\boldsymbol{R}_{\boldsymbol{z}0} \tag{1.4.6}$$
$$(i = 1, \dots, m;\ j = 1, \dots, h - m),$$

the system

$$\dot{\boldsymbol{y}} = A\boldsymbol{y} + B\boldsymbol{z} + \boldsymbol{R}_{\boldsymbol{y}0}, \qquad \dot{\boldsymbol{z}} = C\boldsymbol{y} + D\boldsymbol{z} + \boldsymbol{R}_{\boldsymbol{z}0} \tag{1.4.7}$$

can be rewritten as

$$\dot{\boldsymbol{\eta}} = L_4 A^* L_5 \boldsymbol{\eta}, \qquad \boldsymbol{\eta}^{\mathsf{T}} = (\boldsymbol{y}^{\mathsf{T}}, \boldsymbol{\mu}^{\mathsf{T}}). \tag{1.4.8}$$

Here $\boldsymbol{\mu}$ is an h-dimensional vector composed of variables (1.4.6). (Generally speaking, $h \geq m$, otherwise, instead of two groups of variables (1.4.6), just the first one of them is to be introduced.)

The transformation of system (1.4.7) into (1.4.8) is similar to that used in Section 1.1 to pass from system (1.1.6) (with which system (1.4.7) coincides if we take $\boldsymbol{R}_{\boldsymbol{y}0} = \boldsymbol{0}$ and $\boldsymbol{R}_{\boldsymbol{z}0} = \boldsymbol{0}$) to its $\boldsymbol{\mu}$-system. The eigenvalues of matrix $L_4 A^* L_5$ have negative real parts. Consequently, the solution $\boldsymbol{\eta} = \boldsymbol{0}$ of system (1.4.8) is Lyapunov asymptotically stable. Thus, for system (1.4.8), one can specify a quadratic form $v = v(\boldsymbol{\eta})$ satisfying conditions of the Lyapunov theorem on asymptotic stability. When conditions (1.4.4) are satisfied, the derivative of function v, by virtue of system

$$\dot{\boldsymbol{\eta}} = L_4 A^* L_5 \boldsymbol{\eta} + L_4 \boldsymbol{R}^*(t, \boldsymbol{y}, \boldsymbol{z}), \quad \dot{\boldsymbol{z}} = C\boldsymbol{y} + D\boldsymbol{z} + \boldsymbol{R}_{\boldsymbol{z}}^*(t, \boldsymbol{y}, \boldsymbol{z}), \tag{1.4.9}$$

will be (for a sufficiently small α) negative-definite relative to variables η_i ($i = 1, \dots, m+h$), and the solution $\boldsymbol{\eta} = \boldsymbol{0}, \boldsymbol{z} = \boldsymbol{0}$ of system (1.4.9) is uniformly asymptotically $\boldsymbol{\eta}$-stable by Corollary 0.4.1. Therefore, for any $\varepsilon > 0$, $t_0 \geq 0$, there exists $\lambda(\varepsilon) > 0$ such that $\|\boldsymbol{\eta}_0\| < \lambda$, $\|\boldsymbol{z}_0\| < \lambda$ yield $\|\boldsymbol{\eta}(t; t_0, \boldsymbol{\eta}_0, \boldsymbol{z}_0)\| < \varepsilon$ for all $t \geq t_0$ and, besides, $\lim\|\boldsymbol{\eta}(t; t_0, \boldsymbol{\eta}_0, \boldsymbol{z}_0)\| = 0$ as $t \to \infty$. Using λ, one can choose $\delta_i(\lambda) = \delta_i(\varepsilon) > 0$ ($i = 1, 2$) such that $\|\boldsymbol{x}_0\| < \delta_1$, $\|\boldsymbol{R}_{\boldsymbol{y}0}\| < \delta_2$, and $\|\boldsymbol{l}_j^{\mathsf{T}}\boldsymbol{R}_{\boldsymbol{z}0}\| < \delta_2$ ($j = 1, \dots, h$) yield $\|\boldsymbol{\eta}(t; t_0, \boldsymbol{x}_0)\| < \varepsilon$, $t \geq t_0$.

Provided $\|\boldsymbol{x}_0\| < \delta_1$, $\|\boldsymbol{R}_{\boldsymbol{y}0}\| < \delta_2$, and $\|\boldsymbol{l}_j^{\mathsf{T}}\boldsymbol{R}_{\boldsymbol{z}0}\| < \delta_2$ ($j = 1, \dots, h$), along trajectories of the system

$$\dot{\boldsymbol{y}} = A\boldsymbol{y} + B\boldsymbol{z} + \boldsymbol{R}_{\boldsymbol{y}0} + \boldsymbol{R}_{\boldsymbol{y}}^*(t, \boldsymbol{y}, \boldsymbol{z}), \quad \dot{\boldsymbol{z}} = C\boldsymbol{y} + D\boldsymbol{z} + \boldsymbol{R}_{\boldsymbol{z}0} + \boldsymbol{R}_{\boldsymbol{z}}^*(t, \boldsymbol{y}, \boldsymbol{z}),$$

the inequality $\|\boldsymbol{y}(t; t_0, \boldsymbol{x}_0)\| < \varepsilon$ is valid for all $t \geq t_0$ and, besides, the relationship $\lim\|\boldsymbol{y}(t; t_0, \boldsymbol{x}_0)\| = 0$ holds uniformly with respect to t_0, $\boldsymbol{x}_0$ as $t \to \infty$. This means that the unperturbed motion of system (1.1.6) is uniformly asymptotically stable with respect to $y_1, \ldots, y_m$ in the presence of small constantly acting perturbations which satisfy conditions (1.4.3) and (1.4.4).

If columns of matrix L_1 enumerated by $i_1, \ldots, i_N$ are zeros, then, just as in the proof of the first part of the theorem, one can show that uniform asymptotic stability with respect to $y_1, \ldots, y_m$ takes place on the whole in z_s and R_{zs} $(s = i_1, \ldots, i_N)$. The theorem is proved. □

Corollary 1.4.1. *Let all the roots of equation* (1.1.12) *have negative real parts. Then the unperturbed motion* $\boldsymbol{x} = \boldsymbol{0}$ *of system* (1.1.6) *is stable with respect to* $y_1, \ldots, y_m$ *in the presence of constantly acting perturbations that are small at each point in time.*

Proof. If all the roots of equation (1.1.12) have negative real parts, then the zero solution of the $\boldsymbol{\mu}$-system for (1.1.6) is stable in the presence of constantly acting perturbations $L_4\boldsymbol{R}$ that are small at each point. The smallness of $L_4\boldsymbol{R}$ is guaranteed in the case when constantly acting perturbations $\boldsymbol{R}$ are small at each point. Therefore, the solution of the $\boldsymbol{\mu}$-system for (1.1.6) is stable with respect to all the variables, whereas the unperturbed motion $\boldsymbol{x} = \boldsymbol{0}$ of system (1.1.6) is stable with respect to $y_1, \ldots, y_m$ in the presence of constantly acting perturbations $\boldsymbol{R}$ that are small at each time. The statement is proved. □

Corollary 1.4.2. *Let all the roots of equation* (1.1.12) *have negative real parts, let the columns of matrix* B^{T} *be linearly independent, and let the identity* $\boldsymbol{R}^* \equiv \boldsymbol{0}$ *hold true. Then, for the unperturbed motion* $\boldsymbol{x} = \boldsymbol{0}$ *of system* (1.1.6) *to be asymptotically stable with respect to* $y_1, \ldots, y_m$ *in the presence of small constantly acting perturbations* $\boldsymbol{R}^{\mathsf{T}} = (\boldsymbol{R}_{\boldsymbol{y}0}^{\mathsf{T}}, \boldsymbol{R}_{\boldsymbol{z}0}^{\mathsf{T}})$, *it is necessary and sufficient that condition* (1.4.3) *be satisfied.*

Proof. The *sufficiency* directly follows from Theorem 1.4.1.

Let us prove the *necessity*. On the contrary, we assume that the unperturbed motion $\boldsymbol{x} = \boldsymbol{0}$ of system (1.1.6) is asymptotically stable with respect to $y_1, \ldots, y_m$ in the presence of small constantly acting perturbations $\boldsymbol{R}^{\mathsf{T}} = (\boldsymbol{R}_{\boldsymbol{y}0}^{\mathsf{T}}, \boldsymbol{R}_{\boldsymbol{z}0}^{\mathsf{T}})$, but condition (1.4.3) is not fulfilled. Having introduced new variables (1.4.6), we arrive at the system

$$\dot{\boldsymbol{\eta}} = L_4 A^* L_5 \boldsymbol{\eta} + \boldsymbol{c}, \tag{1.4.10}$$

where $\boldsymbol{c}$ is a constant $(m+h)$-dimensional vector with at least one nonzero component c_i $(i = m+1, \ldots, m+h)$. (Components $c_1, \ldots, c_m$ are zeros identically by the construction of auxiliary system (1.4.10) with the help of new variables (1.4.6).) But then at least one of the components $\eta_i(t)$ $(i = 1, \ldots, m+h)$ of the general solution of system (1.4.10) satisfies the condition $\lim \eta_i(t) = \eta_i^0 = \text{const} \neq 0$, $t \to \infty$, and the structure of matrix $L_4 A^* L_5$ is such that $\lim y_i(t) = \eta_i^0 = \text{const} \neq 0$, $t \to \infty$ for at least one $i = 1, \ldots, m$. This means that the assumption that condition (1.4.3) does not hold contradicts the fact that the unperturbed motion $\boldsymbol{x} = \boldsymbol{0}$ of system (1.1.6) is asymptotically stable with respect to $y_1, \dots, y_m$ in the presence of small constantly acting perturbations $\boldsymbol{R}_0$. The statement is proved. □

Remark 1.4.2.

1. The assumption of linear independence of columns of matrix B^{T}, which we have incorporated in the second part of Theorem 1.4.1 and Corollary 1.4.2, is not critical and serves to simplify conditions and proofs.

2. If, in Corollary 1.4.2, perturbations $\boldsymbol{R}_{\boldsymbol{y}0}$ and $\boldsymbol{R}_{\boldsymbol{z}0}$ are arbitrary in magnitude, then the unperturbed motion $\boldsymbol{x} = \boldsymbol{0}$ of system (1.1.6) possesses the property of ***y**-attraction* ($\lim\|\boldsymbol{y}(t; t_0, \boldsymbol{x}_0)\| = 0,\ t \to \infty$ for *all* $t_0, \boldsymbol{x}_0$) in the presence of constantly acting perturbations $\boldsymbol{R}_{\boldsymbol{y}0}$ and $\boldsymbol{R}_{\boldsymbol{z}0}$.

3. The problem above on $\boldsymbol{y}$-stability in the presence of constantly acting perturbations that are not small can be posed in the case when unperturbed motion of system (1.6.1) is asymptotically stable with respect to part of the variables and also in the Lyapunov sense.

Remark 1.4.3. Massera [1956] demonstrated that, if the unperturbed motion $\boldsymbol{x} = \boldsymbol{0}$ of a linear system with continuous coefficients is stable in the presence of constantly acting perturbations, then it is *uniformly asymptotically* stable. In the problem of stability with respect to part of the variables, this result can be violated by restrictions on the structure of constant perturbations. For example, (nonasymptotically) $\boldsymbol{y}$-stable system (1.1.6) remains $\boldsymbol{y}$-stable, in case (1.4.3), in the presence of small constantly acting perturbations $\boldsymbol{R}_0^{\mathsf{T}} = (\boldsymbol{R}_{\boldsymbol{y}0}^{\mathsf{T}}, \boldsymbol{R}_{\boldsymbol{z}0}^{\mathsf{T}})$, $\boldsymbol{R}^* \equiv \boldsymbol{0}$.

Example 1.4.1. Let the equations of perturbed motion (1.1.6) have the form

$$\begin{gathered} \dot{y}_1 = -y_1 + z_1 - 2z_3, \\ \dot{z}_1 = 4y_1 + z_1 + 2z_2, \quad \dot{z}_2 = 8y_1 + 2z_1 + 4z_2, \quad \dot{z}_3 = 2y_1 + z_1 + z_2 - z_3. \end{gathered} \tag{1.4.11}$$

In the case in question

$$B = \begin{pmatrix} 1 & 0 & -2 \end{pmatrix}, \quad D = \begin{pmatrix} 1 & 2 & 0 \\ 2 & 4 & 0 \\ 1 & 1 & -1 \end{pmatrix}, \quad \operatorname{rank} K_3 = 1,$$

$$K_1 = \begin{pmatrix} 1 \\ 0 \\ -2 \end{pmatrix}, \quad \boldsymbol{l}_1^{\mathsf{T}} = \begin{pmatrix} 1 & 0 & -2 \end{pmatrix}, \quad L_4 A^* L_5 = \begin{pmatrix} -1 & 1 \\ 0 & -1 \end{pmatrix}$$

and all the roots of equation (1.1.12) have negative real parts. Because the second row of matrix K_1 is zero, the unperturbed motion $y_1 = z_1 = z_2 = z_3 = 0$ of system (1.4.11) is stable with respect to y_1 in the presence of constantly acting perturbations on the whole in z_2 and R_{z_2}.

Condition (1.4.3) in this case is written as

$$R_{z_1 0} - 2R_{z_3 0} = -R_{y_1 0}. \tag{1.4.12}$$

With conditions (1.4.4) and (1.4.12) satisfied, the unperturbed motion of system (1.4.11) is asymptotically stable with respect to y_1 in the presence of constantly acting perturbations on the whole in z_2 and R_{z_2}.

In this particular case condition (1.4.12) means that small constant perturbations $R_{z_i 0}$ $(i = 1, 3)$ and $R_{y_1 0}$ are, for example, equal in magnitude.

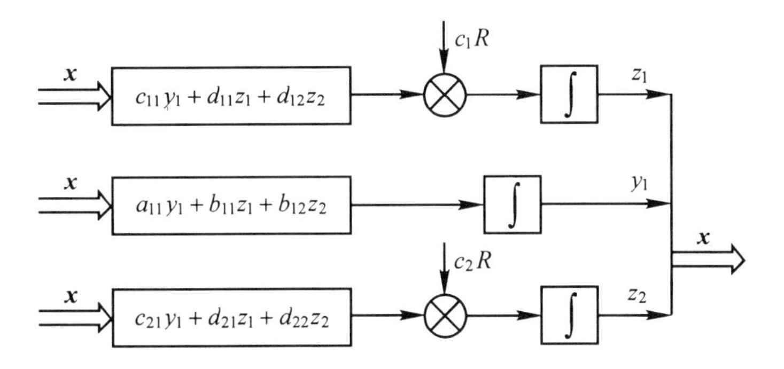

Figure 1.4.1. Invariance of y_1 relative to R in a system of order three ($m = 1, p = 2$).

Interestingly enough, if constantly acting perturbations are measured, the design of a control system providing asymptotic stability with respect to y_1 can be reduced not to counteracting disturbances but to maintaining the validity of equation (1.4.12). In so doing a substantial saving of control resources can be achieved. It is natural, however, that implementing such a control system in a particular situation can present certain difficulties.

1.4.4 Relation to the Invariance Theory

The condition of Theorem 1.4.1 concerning existence of zero rows in L_1 can be weakened under assumptions accepted in the *invariance theory* (Kukhtenko [1984])

$$\boldsymbol{R}_{\boldsymbol{y}} \equiv \boldsymbol{0}, \qquad \boldsymbol{R}_{\boldsymbol{z}} = \boldsymbol{c}R, \tag{1.4.13}$$

where R is a scalar perturbation function.

This condition means that a number of the channels through which perturbation R affects a system are connected to $\boldsymbol{z}$-output only.

Let all the roots of equation (1.1.12) have negative real parts and, in addition,

$$\boldsymbol{b}_i^{\mathsf{T}}\boldsymbol{c} = 0 \quad (i = 1, \ldots, m), \tag{1.4.14}$$

where $\boldsymbol{b}_i$ are columns of matrix B^{T}. Then the unperturbed motion of system (1.1.6) is asymptotically stable with respect to $y_1, \ldots, y_m$ on the whole in perturbation R. However, in this case, generally speaking, there is not even nonasymptotic stability with respect to $y_1, \ldots, y_m$ on the whole in any component of $\boldsymbol{z}_0$.

Remark 1.4.4.

1. The property discussed of stability in the presence of non-small constantly acting perturbations in the terminology of control theory, implies *invariance* of the property of stability with respect to part of the variables to external unmeasured (arbitrarily large) perturbations affecting some of the system channels. Conditions (1.4.13) and (1.4.14) of such invariance comply with the *two-channel principle* (Petrov [1960]) of invariant systems design (Figure 1.4.1). More extensive practical implementation of invariance conditions is associated with exploiting modern computers, which enable parametric identification of a system (i.e., determination of its parameters by measurements of its output) in real time.
2. The invariance indicated is possible for systems of the order three and higher.

1.4.5 The Condition for Parametric Stability

Along with (1.1.6) consider a system

$$\dot{\boldsymbol{y}} = (A + \varepsilon\bar{A})\boldsymbol{y} + (B + \varepsilon\bar{B})\boldsymbol{z}, \qquad \dot{\boldsymbol{z}} = (C + \varepsilon\bar{C})\boldsymbol{y} + (D + \varepsilon\bar{D})\boldsymbol{z} \tag{1.4.15}$$

in which $\bar{A}, \bar{B}, \bar{C}$, and $\bar{D}$ are constant matrices of appropriate sizes.

Theorem 1.4.2. *If the unperturbed motion $\boldsymbol{x} = \boldsymbol{0}$ of system* (1.1.6) *is asymptotically stable with respect to $y_1, \dots, y_m$ and the condition*

$$\operatorname{rank} K_p = \operatorname{rank}\{K_p, \bar{B}^{\mathsf{T}}, (L_1\bar{D})^{\mathsf{T}}\} \tag{1.4.16}$$

holds true, then, for sufficiently small ε, the unperturbed motion $\boldsymbol{x} = \boldsymbol{0}$ of system (1.4.15) *will possess the same property.*

Proof. Following the rule for constructing a $\boldsymbol{\mu}$-system for (1.1.6), in the case in question we obtain the equations

$$\dot{\boldsymbol{\xi}} = L_4 A^* L_5 \boldsymbol{\xi} + \varepsilon \begin{pmatrix} \bar{A} & \bar{B} \\ L_1\bar{C} & L_1\bar{D} \end{pmatrix} \boldsymbol{x}.$$

From (1.4.16) it follows that $\bar{B}\boldsymbol{z} = B^*\boldsymbol{\mu}$ and $L_1\bar{D}\boldsymbol{z} = D^*\boldsymbol{\mu}$, where B^* and D^* are constant $m \times h$ and $h \times h$ matrices, respectively, and $\boldsymbol{\mu}$ is the vector of auxiliary variables required to form a $\boldsymbol{\mu}$-system for (1.1.6). If (1.4.17) is valid, we arrive at the equations

$$\dot{\boldsymbol{\xi}} = (L_4 A^* L_5 + \varepsilon Q)\boldsymbol{\xi}, \tag{1.4.17}$$

where Q is a constant $(m + h) \times (m + h)$ matrix. If the unperturbed motion $\boldsymbol{x} = \boldsymbol{0}$ of system (1.1.6) is asymptotically stable with respect to $y_1, \dots, y_m$, then by Theorem 1.1.1 the solution $\boldsymbol{\xi} = \boldsymbol{0}$ of the auxiliary $\boldsymbol{\mu}$-system $\dot{\boldsymbol{\xi}} = L_4 A^* L_5 \boldsymbol{\xi}$ is Lyapunov asymptotically stable. Hence, for sufficiently small ε, the solution $\boldsymbol{\xi} = \boldsymbol{0}$ of system (1.4.17) is also Lyapunov asymptotically stable. Since system (1.4.17) is obtained from (1.4.15) by a linear change of variables and in this transformation $\xi_i = y_i$ $(i = 1, \dots, m)$, the unperturbed motion $\boldsymbol{x} = \boldsymbol{0}$ of system (1.4.15) is asymptotically stable with respect to $y_1, \dots, y_m$. The theorem is proved. □

Remark 1.4.5.

1. Condition (1.4.16) holds trivially for the zero matrices $\bar{B}$ and $\bar{D}$. This assumption means that coefficients of z_j $(j = 1, \dots, p)$ in system (1.1.6) are accurately specified, whereas coefficients of y_i $(i = 1, \dots, m)$ are known just up to an error ("*scatter*").

2. The proof scheme of Theorem 1.4.2 will not change under the condition that all the elements of matrices $\bar{A}$ and $\bar{C}$ are piecewise-continuous and bounded functions of $t \in [0, \infty)$. In this case the solution $\boldsymbol{\xi} = \boldsymbol{0}$ of an auxiliary system of type (1.4.17) will be (for sufficiently small ε) Lyapunov asymptotically stable.

3. In the general case, the asymptotic stability with respect to $y_1, \ldots, y_m$ *can be violated* by an *arbitrarily small* variation in the coefficients of system (1.1.1).

As an example, consider the equations

$$\dot{y}_1 = -y_1 + z_1 - 2z_2, \quad \dot{z}_1 = 4y_1 + (1+\Delta)z_1, \quad \dot{z}_2 = 2y_1 + z_1 - z_2, \qquad (1.4.18)$$

where Δ is a constant. Since rank $K_2 = -2\Delta \neq 0$ for an arbitrarily small $\Delta \neq 0$, the $\boldsymbol{\mu}$-system comprises all three equations of system (1.4.18). Having the root $\lambda = 1$ when $\Delta = 0$, the characteristic equation of system (1.4.18) also has a root with a positive real part for sufficiently small Δ. The unperturbed motion $\boldsymbol{x} = \mathbf{0}$ of system (1.4.18), therefore, is unstable with respect to y_1 for an arbitrarily small Δ. At the same time (see Example 1.1.1), for $\Delta = 0$, the unperturbed motion $\boldsymbol{x} = \mathbf{0}$ of system (1.4.18) is asymptotically stable with respect to y_1.

1.4.6 Conclusions

The analysis presented of perturbational influence shows that the property of asymptotic stability with respect to part of the variables for a linear system with constant coefficients is not devoid of *contradiction*. On the one hand, as it differs from Lyapunov asymptotic stability, this property allows greater flexibility, namely, the *invariance* relative to perturbations affecting some of the system channels and even *asymptotic* stability for a certain structure of perturbing vector function $\boldsymbol{R}(t, \boldsymbol{x})$ without violating the condition $\boldsymbol{R}(t, \mathbf{0}) \not\equiv \mathbf{0}$. On the other hand, though valid in the presence of small constantly acting perturbations, the property of asymptotic stability with respect to part of the variables *fails*, generally speaking, with an arbitrarily small *variation* in the coefficients of system (1.1.1).

Thus, compared to the property of Lyapunov asymptotic stability, the property of asymptotic stability with respect to part of the variables is *more sensitive* to a variation in the coefficients (b_{il} and d_{jl} but not a_{ik} and c_{jk}) of a linear system. This is quite reasonable, for the asymptotic stability with respect to part of the variables is a *more delicate* property of a system than the asymptotic stability in Lyapunov's sense.

1.5 Stabilization with Respect to Part of the Variables

1.5.1 Formulation of the Problem

Consider a *controlled object* whose dynamics are described by a system of linear differential equations with *constant* coefficients

$$\dot{\boldsymbol{x}} = A^*\boldsymbol{x} + B^*\boldsymbol{u},$$
$$\boldsymbol{x}^{\mathsf{T}} = (y_1, \ldots, y_m, z_1, \ldots, z_p) = (\boldsymbol{y}^{\mathsf{T}}, \boldsymbol{z}^{\mathsf{T}}), \quad m > 0, \quad p \geq 0, \quad n = m + p.$$

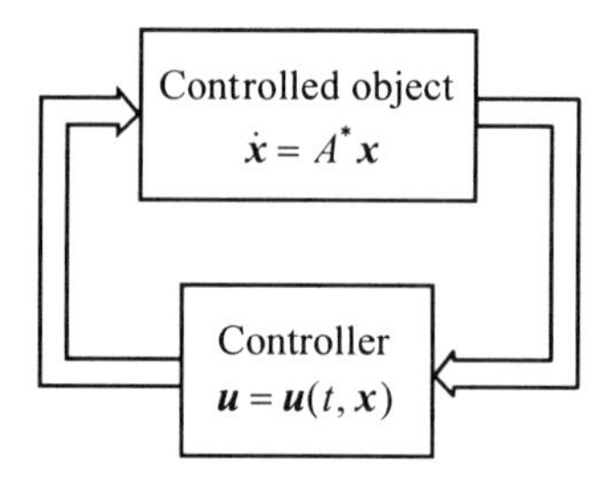

Figure 1.5.1. The structural model of system (1.5.1) for $\boldsymbol{u} = \boldsymbol{u}(t, \boldsymbol{x})$.

In $\boldsymbol{y}$, $\boldsymbol{z}$ variables the system takes the form

$$\dot{\boldsymbol{y}} = A\boldsymbol{y} + B\boldsymbol{z} + P\boldsymbol{u}, \qquad \dot{\boldsymbol{z}} = C\boldsymbol{y} + D\boldsymbol{z} + Q\boldsymbol{u}, \tag{1.5.1}$$

where A^*, B^*, A, B, C, and D are constant matrices of appropriate sizes, $\boldsymbol{x}$ is the vector characterizing the state of the system, and $\boldsymbol{u}^{\mathsf{T}} = (u_1, \dots, u_r)$ is the vector of control actions. Let $\mathfrak{U}$ be a class of *admissible* controls $\boldsymbol{u} = \boldsymbol{u}(t, \boldsymbol{x}), \boldsymbol{u}(t, \boldsymbol{0}) = \boldsymbol{0}$.

Problem 1.5.1. *Find $\boldsymbol{u} \in \mathfrak{U}$ that ensures the asymptotic stability of the unperturbed motion $\boldsymbol{x} = \boldsymbol{0}$ of system* (1.5.1) *with respect to $y_1, \dots, y_m$.*

This is a classical version of the problem of stabilization based on the *feedback control principle.* The structural model of the closed-loop control system is presented in Figure 1.5.1.

1.5.2 General Idea of Control Design

Now we show that the control vector solving Problem 1.5.1 can be constructed in the form

$$\boldsymbol{u} = \Gamma \boldsymbol{z} + \boldsymbol{u}^*, \tag{1.5.2}$$

where a constant matrix Γ is defined by a transformation that reduces system (1.5.1) to a particular form. Control $\boldsymbol{u}^*$ solves the problem of stabilization with respect to all the variables of the zero solution of some auxiliary linear system with constant coefficients of dimension not exceeding that of the original system (1.5.1).

To this end consider the system of equations

$$\dot{\boldsymbol{y}} = A\boldsymbol{y} + (B + P\Gamma)\boldsymbol{z}, \qquad \dot{\boldsymbol{z}} = C\boldsymbol{y} + (D + Q\Gamma)\boldsymbol{z} \tag{1.5.3}$$

and the matrix

$$K_p^0 = \{(B + P\Gamma)^{\mathsf{T}}, (D + Q\Gamma)^{\mathsf{T}}(B + P\Gamma)^{\mathsf{T}}, \dots, ((D + Q\Gamma)^{\mathsf{T}})^{p-1}(B + P\Gamma)^{\mathsf{T}}\}.$$

Let s be a minimum number such that $\operatorname{rank} K_s^0 = \operatorname{rank} K_{s-1}^0$, and, for definiteness, let $\operatorname{rank} K_{s-1}^0 = h$. Using the rules (a) through (d) from Subsection 1.1.4 (the only difference is that we manipulate K_p^0 instead of K_p), let us construct matrices L_i^0 $(i = 1, \ldots, 5)$.

Lemma 1.5.1. *If there exists a constant matrix* Γ *such that the control* $\boldsymbol{u}^* \in \mathfrak{U}$ *solves the problem of stabilization of the zero solution* $\boldsymbol{w} = \boldsymbol{0}$ *of the system*

$$\dot{\boldsymbol{w}} = L_4^0(A^* + B^*\Gamma)L_5^0\boldsymbol{w} + L_4^0B^*\boldsymbol{u}^* \tag{1.5.4}$$

with respect to all the variables, then control (1.5.2) *solves Problem* 1.5.1 *for system* (1.5.1).

Proof. Equations (1.5.4) are obtained from (1.5.1) as the result of transforming equations (1.5.3) into the $\boldsymbol{\mu}$-system. Hence $w_i = y_i$ $(i = 1, \ldots, m)$ and the Lyapunov asymptotic stability of the solution $\boldsymbol{w} = \boldsymbol{0}$ of system (1.5.4) closed by the control $\boldsymbol{u}^* \in \mathfrak{U}$ implies the asymptotic stability of the unperturbed motion $\boldsymbol{x} = \boldsymbol{0}$ of system (1.5.1), (1.5.2) with respect to $y_1, \ldots, y_m$. The lemma is proved. □

The representation of the sought for control $\boldsymbol{u} \in \mathfrak{U}$ in the form (1.5.2) allows one to reduce Problem 1.5.1 to the well-studied problem of stabilization with respect to all the variables for linear systems with constant coefficients. This fact *warrants* the chosen form of the desired control (1.5.2) and also renders its search *rational* based on this form.

Following the terminology of Section 1.1, system (1.5.4) will be called the *$\boldsymbol{\mu}$-system corresponding to matrix* Γ.

1.5.3 The Condition for Stabilizability with Respect to Part of the Variables

In constructing controls of the form (1.5.2), two points are of fundamental importance:

(1) validity of the converse of Lemma 1.5.1; and

(2) finding matrix Γ.

Let $\mathfrak{U}_1 = \{\boldsymbol{u} : \boldsymbol{u} = \Gamma_1\boldsymbol{y} + \Gamma_2\boldsymbol{z}\}$, where Γ_1 and Γ_2 are constant matrices of appropriate sizes.

Theorem 1.5.1.

1. *If* $\boldsymbol{u} \in \mathfrak{U}_1$, *then the converse of Lemma* 1.5.1 *is valid.*
2. *If there exists a constant matrix* Γ *such that*

$$\begin{gathered}\operatorname{rank} K_* = m + h,\\ K_* = L_4^0\big[B^*, (A^* + B^*\Gamma)B^*, \ldots, (A^* + B^*\Gamma)^{m+h-1}B^*\big],\end{gathered} \tag{1.5.5}$$

then Problem 1.5.1 *is solvable in the class* $\mathfrak{U}_1$.

Proof. 1. Let Problem 1.5.1 for system (1.5.1) be solvable, and let a control $\boldsymbol{u}^0 = \Gamma_{10}\boldsymbol{y} + \Gamma_{20}\boldsymbol{z}$ solve this problem. Let us show that $\boldsymbol{u}^0$ can be represented in the form $\boldsymbol{u}^0 = \Gamma\boldsymbol{z} + \boldsymbol{u}^*(\boldsymbol{y}, \boldsymbol{z})$, i.e., there exists a matrix Γ

such that the control $\boldsymbol{u}^*(\boldsymbol{y}, \boldsymbol{z}) \in \mathfrak{U}_1$ stabilizes the zero solution of the corresponding $\boldsymbol{\mu}$-system with respect to all the variables. We take $\Gamma = \Gamma_{20}$ and $\boldsymbol{u}^*(\boldsymbol{y}, \boldsymbol{z}) = \boldsymbol{u}^*(\boldsymbol{y}) = \Gamma_{10}\boldsymbol{y}$. Let us examine the control $\boldsymbol{u}^* = \Gamma_{10}\boldsymbol{y}$ for the desired property. To do this we consider system (1.5.1) for $\boldsymbol{u} = \boldsymbol{u}^0$

$$\dot{\boldsymbol{y}} = A\boldsymbol{y} + (B + P\Gamma_{20})\boldsymbol{z} + P\boldsymbol{u}^*(\boldsymbol{y}), \quad \dot{\boldsymbol{z}} = C\boldsymbol{y} + (D + Q\Gamma_{20})\boldsymbol{z} + Q\boldsymbol{u}^*(\boldsymbol{y}).$$

The control $\boldsymbol{u}^* = \boldsymbol{u}^*(\boldsymbol{y})$ does not depend on variables z_i $(i = 1, \ldots, p)$ and, by Lemma 1.1.2, cannot change the dimension of the $\boldsymbol{\mu}$-system corresponding to matrix Γ. Therefore, the zero solution of the $\boldsymbol{\mu}$-system is stabilized with respect to all the variables by the control $\boldsymbol{u}^* = \Gamma_{10}\boldsymbol{y}$.

2. We set

$$L^0 = \begin{pmatrix} I_m & 0 \\ 0 & L^{*0} \end{pmatrix}, \qquad L^{*0} = \begin{pmatrix} L_1^0 \\ L_*^0 \end{pmatrix},$$

where L_*^0 is an arbitrary $(p-h) \times p$ matrix that does not violate the nonsingularity of matrix L^0. Allowing for the fact that matrices L_4^0 and L_5^0 are composed of parts of coefficients of matrices L^0 and $(L^0)^{-1}$

$$\begin{aligned} (L^0)^{-1} &= (\bar{l}_{ij}^0)\ (i,j = 1, \ldots, n), \quad L_5^0 = (\bar{l}_{ij}^0)\ (i = 1, \ldots, n;\ j = 1, \ldots, m+h), \\ L^0 &= (l_{ij}^0)\ (i,j = 1, \ldots, n), \quad L_4^0 = (l_{ij}^0)\ (i = 1, \ldots, m+h;\ j = 1, \ldots, n), \end{aligned}$$

we infer that

$$\begin{aligned} L^0(A^* + B^*\Gamma)(L^0)^{-1} &= (l_{ij}^*) \qquad (i,j = 1, \ldots, n), \\ L_4^0(A^* + B^*\Gamma)L_5^0 &= (l_{ij}^*) \qquad (i,j = 1, \ldots, m+h). \end{aligned}$$

Hence, matrix $L_4^0(A^* + B^*\Gamma)L_5^0$ is composed of part of coefficients of matrix $L^0(A^* + B^*\Gamma)(L^0)^{-1}$. Using the equalities $(QRQ^{-1})^i = QR^iQ^{-1}$ $(i = 1, \ldots, n-1)$, which are valid for any quadratic matrices Q and R of order n (provided Q is nonsingular), and substituting L^0 for Q and $A^* + B^*\Gamma$ for R, we obtain the relationships

$$\left[L_4^0(A^* + B^*\Gamma)L_5^0\right]^i = L_4^0(A^* + B^*\Gamma)^i L_5^0 \qquad (i = 1, \ldots, m+h-1).$$

The equality $L_5^0 L_4^0 = I_{m+p}$ can also be validated. Consequently, the following equalities hold true:

$$\begin{aligned} &\left\{L_4^0 B^*, L_4^0(A^* + B^*\Gamma)B^*, \ldots, L_4^0(A^* + B^*\Gamma)^{m+h-1}B^*\right\} \\ &= \left\{L_4^0 B^*, L_4^0(A^* + B^*\Gamma)L_5^0 L_4^0 B^*, \ldots, L_4^0(A^* + B^*\Gamma)^{m+h-1}L_5^0 L_4^0 B^*\right\} \\ &= \left\{L_4^0 B^*, \left[L_4^0(A^* + B^*\Gamma)L_5^0\right]L_4^0 B^*, \ldots, \left[L_4^0(A^* + B^*\Gamma)L_5^0\right]^{m+h-1}L_4^0 B^*\right\}. \end{aligned}$$

Thus, from condition (1.5.5) it follows that the criterion for complete controllability (with respect to all the variables) is satisfied for system (1.5.4). This means that the problem of stabilization (with respect to all the variables) of the solution $\boldsymbol{w} = \boldsymbol{0}$ of system (1.5.4) is solvable in the class $\mathfrak{U}_1$. Relying on Lemma 1.5.1, we conclude that Problem 1.5.1 has a solution in the class $\mathfrak{U}_1$. The theorem is proved. □

1.5.4 The Condition for Stabilizability with Respect to Part of the Variables in the Case of an Uncontrollable μ-System

Let us assume that condition (1.5.5) of the second part of Theorem 1.5.1 does not hold, i.e., $\operatorname{rank} K_* = m^* < m + h$. As follows from the proof of Theorem 1.5.1, the auxiliary μ-system corresponding to matrix Γ is not completely controllable with respect to all the variables. In this case using a linear nonsingular transformation of variables $\boldsymbol{w}^* = R\boldsymbol{w}$, one can reduce system (1.5.4) to the form (Krasovskii [1968])

$$\dot{\boldsymbol{w}}^+ = A_1\boldsymbol{w}^+ + A_2\boldsymbol{w}^- + B_1\boldsymbol{u}^*, \qquad \dot{\boldsymbol{w}}^- = A_3\boldsymbol{w}^-,$$
$$\boldsymbol{w}^{*\top} = (\boldsymbol{w}^{+\top}, \boldsymbol{w}^{-\top}), \quad \boldsymbol{w}^+ \in R^{m^*}, \quad \boldsymbol{w}^- \in R^{m+h-m^*},$$

where A_i $(i = 1, 2, 3)$ and B_1 are constant matrices of appropriate sizes.

Corollary 1.5.1. *For Problem* 1.5.1 *to be solvable in the case when* $\operatorname{rank} K_* = m^* < m+h$, *it is necessary and sufficient that there exist a matrix* Γ *such that all the roots* λ_i $(i = 1, \ldots, m+h-m^*)$ *of the equation* $\det(A_3 - \lambda I_{m+h-m^*}) = 0$ *have negative real parts.*

Proof. The *sufficiency* is obvious.

Necessity. Let us assume that, for any matrix Γ among the roots λ_i $(i = 1, \ldots, m + h - m^*)$, there is at least one root with a nonnegative real part. Because the system $\dot{\boldsymbol{w}}^- = A_3\boldsymbol{w}^-$ is invariant relative to control $\boldsymbol{u}^*$, the solution $\boldsymbol{w} = \boldsymbol{0}$ of system (1.5.4) is not Lyapunov asymptotically stable for any $\boldsymbol{u} = \Gamma\boldsymbol{z} + \boldsymbol{u}^* \in \mathfrak{U}_1$. For any given values of coefficients of matrix Γ (and for any $\boldsymbol{u}^*$ chosen so that $\boldsymbol{u} = \Gamma\boldsymbol{z} + \boldsymbol{u}^* \in \mathfrak{U}_1$), system (1.5.4) is a μ-system for system (1.5.1), (1.5.2). By Theorem 1.1.1, the unperturbed motion $\boldsymbol{x} = \boldsymbol{0}$ of system (1.5.1), (1.5.2) is not asymptotically stable with respect to $y_1, \ldots, y_m$. The statement is proved. □

Remark 1.5.1.

1. Using the μ-system technique presented in Sections 1.2 and 1.3, the proposed approach to constructing control rules can be extended to the case of linear controlled systems with *variable* coefficients.

2. If a control $\boldsymbol{u} = \Gamma_1\boldsymbol{y} + \Gamma_2\boldsymbol{z} \in \mathfrak{U}_1$ solves Problem 1.5.1, then a control $\boldsymbol{u} = (\Gamma_1 + \varepsilon\bar{\Gamma}_1)\boldsymbol{y} + \Gamma_2\boldsymbol{z}$ will also solve Problem 1.5.1 for a sufficiently small number ε ($\bar{\Gamma}_1$ is a constant matrix of dimensions that Γ_1 has). This means that the control $\boldsymbol{u} = \Gamma_1\boldsymbol{y} + \Gamma_2\boldsymbol{z}$ is stable relative to small variations in coefficients of y_i $(i = 1, \ldots, m)$. At the same time, even small variations in coefficients of matrix Γ_2, generally speaking, can result in the unperturbed motion of a closed system losing the property of asymptotic $\boldsymbol{y}$-stability. With Theorem 1.4.2, more accurate tolerances for possible "*scatter*" in coefficients of matrix Γ_2 can be obtained.

Remark 1.5.2. The proposed approach to solving the problem of stabilization with respect to part of the variables was carried over to *discrete linear stochastic models* by Phillis [1984].

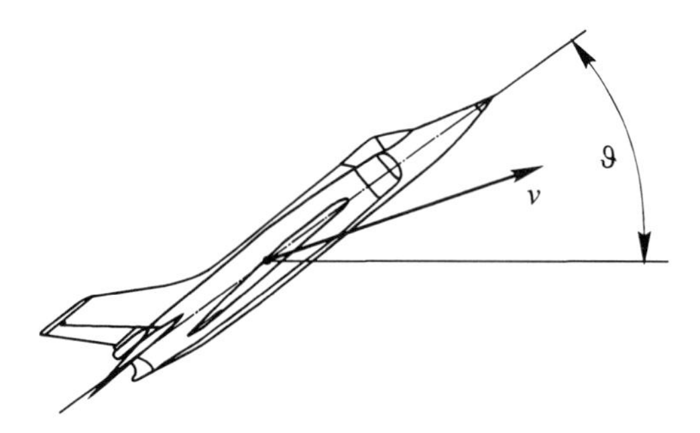

Figure 1.5.2. The pitch angle ϑ of an aircraft.

Example 1.5.1.

1. Consider the equations of *longitudinal motion of an aircraft* (Falb and Wolovich [1967])

$$\begin{gathered}\dot{y}_1 = y_2, \quad \dot{y}_2 = a_{22}y_2 + b_{21}z_1 + b_{22}z_2 + p_{21}u_1 + p_{22}u_2, \\ \dot{z}_1 = c_{11}y_1 + d_{11}z_1 + q_{11}u_1, \\ \dot{z}_2 = c_{22}y_2 + d_{22}z_2 + q_{22}u_2 + q_{23}u_3, \quad \dot{z}_3 = z_1, \quad \dot{z}_4 = z_2.\end{gathered} \tag{1.5.6}$$

Here y_1 and y_2 are perturbations of the pitch angle and its rate of change (Figure 1.5.1), respectively, and z_i $(i = 1, \dots, 4)$ are perturbations of horizontal and vertical components of the velocity $\boldsymbol{v}$ and coordinates of the center of mass of the aircraft; the coefficients of the system are aerodynamic constants.

Assuming $u_1 \equiv u_2 \equiv 0$, let us solve the problem of stabilization of the unperturbed motion of system (1.5.6) with respect to y_1, y_2 using control $u_3 \in \mathcal{U}_1$. Introduce the new variable $\mu_1 = a_{22}y_2 + b_{21}z_1 + b_{22}z_2$. As a result, system (1.5.6) takes the form

$$\begin{gathered}\dot{y}_1 = y_2, \quad \dot{y}_2 = \mu_1, \quad \dot{\mu}_1 = A_1y_1 + A_2y_2 + A_3z_1 + A_4z_2 + b_{22}q_{23}u_3, \\ \dot{z}_1 = c_{11}y_1 + d_{11}z_1, \quad \dot{z}_2 = c_{22}y_2 + d_{22}z_2 + q_{23}u_3, \quad \dot{z}_3 = z_1, \quad \dot{z}_4 = z_2, \\ A_1 = b_{21}c_{11}, \quad A_2 = a_{22}^2 + b_{22}c_{22}, \quad A_3 = b_{21}(a_{22} + d_{11}), \quad A_4 = b_{22}(a_{22} + d_{22}).\end{gathered}$$

Following (1.5.2), we construct control function u_3 as

$$\begin{gathered}u_3 = -\frac{A_3}{b_{22}q_{23}}z_1 - \frac{A_4}{b_{22}q_{23}}z_2 + u_3^* = \Gamma\boldsymbol{z} + u_3^*, \\ \Gamma = -\left(\frac{A_3}{b_{22}q_{23}}, \frac{A_4}{b_{22}q_{23}}\right).\end{gathered} \tag{1.5.7}$$

The auxiliary $\boldsymbol{\mu}$-system corresponding to matrix Γ comprises the equations

$$\dot{y}_1 = y_2, \quad \dot{y}_2 = \mu_1, \quad \dot{\mu}_1 = A_1y_1 + A_2y_2 + b_{22}q_{23}u_3^*. \tag{1.5.8}$$

If we choose $u_3^* = \dfrac{B_1y_1 + B_2y_2 + B_3\mu_1}{b_{22}q_{23}}$, the zero solution of system (1.5.8) will be Lyapunov asymptotically stable when $B_3 < 0$, $B_2 + A_2 < 0$, $B_3(B_2 + A_2) > -A_1 - B_1 > 0$. Consequently, by Lemma 1.5.1, control function (1.5.7) solves the problem of stabilization of the unperturbed motion of system (1.5.6) with respect to y_1, y_2.

Control rules (1.5.7) also ensure stabilization of the unperturbed motion of system (1.5.6) with respect to μ_1. In this case the closed system (1.5.6), (1.5.7) has the form

$$\dot{\boldsymbol{\xi}} = A_* \boldsymbol{\xi}, \qquad \boldsymbol{\xi} = (y_1, y_2, \mu_1, z_1, z_2, z_3, z_4),$$

$$A_* = \left(\begin{array}{ccc|cc|c} 0 & 1 & 0 & & & \\ 0 & 0 & 1 & & 0 & \\ A_1 + B_1 & A_2 + B_2 & B_3 & & & \\ \hline c_{11} & 0 & 0 & d_{11} & 0 & \\ b_{22}^{-1} B_1 & c_{22} + b_{22}^{-1} B_2 & b_{22}^{-1} B_3 & -b_{22}^{-1} A_3 & d_{22} - b_{22}^{-1} A_4 & 0 \\ \hline & & & 1 & 0 & \\ & 0 & & 0 & 1 & \end{array}\right).$$

Therefore, taking into account the equality $d_{22} - b_{22}^{-1} A_4 = -a_{22}$, we conclude that the control rules found also carry out stabilization with respect to variables z_1, z_2, provided at least one of the conditions $d_{11} < 0$ or $a_{22} > 0$ holds true. As this takes place, the unperturbed motion of system (1.5.6) is (nonasymptotically) Lyapunov stable (with respect to $y_1, y_2, z_1, z_2, z_3, z_4$).

2. The approach proposed can also be used in solving the problem of motion stabilization with respect to *all* the variables. The problem of stabilization with respect to part of the variables serves as an *intermediate link* in this research. Consider the problem of *stabilizing permanent rotations of a massive solid attached at a fixed point.* The linear approximation system for equations of perturbed motion has the form (for details, see Subsection 4.3.2)

$$\dot{x}_1 = \frac{B-C}{A}\omega(x_2\gamma + x_3\beta) + \frac{1}{A} mg(x_{3c}x_5 - x_{2c}x_6) + \frac{1}{A}u_1 \overset{\text{def}}{=} f_1(\boldsymbol{x}) + \frac{1}{A}u_1,$$
$$\dot{x}_4 = x_3\beta - x_2\gamma + x_5\omega\gamma - x_6\omega\beta \overset{\text{def}}{=} f_4(\boldsymbol{x}) \qquad (123, 456, ABC). \tag{1.5.9}$$

Only two of the six equations of system (1.5.9) are written. The remaining four equations can be obtained through the cyclic permutation of the indices $1 \to 2 \to 3$, $4 \to 5 \to 6$ and coefficients $A \to B \to C$. In equations (1.5.9), x_i $(i = 1, 2, 3)$ are perturbations of the scalar projections of the angular velocity vector of the body onto its principal axes of inertia; x_j $(j = 4, 5, 6)$ are perturbations of the scalar projections of the unit vector directed along a fixed vertical axis onto the principal axes; x_{ic} $(i = 1, 2, 3)$ are the coordinates of the center of mass of the solid in the principal axes; A, B, and C are the principal moments of inertia; α, β, and γ are constants characterizing the orientation of the permanent rotation axis; ω is the angular velocity of permanent rotation; g is the acceleration of gravity; m is the mass of the body; and u_i $(i = 1, 2, 3)$ are controlling moments realized by engines. The products mgx_{ic} $(i = 1, 2, 3)$ are assumed not too large, which warrants the structure of the control rules discussed below.

We shall use the designations $y_1 = x_2, y_2 = x_4, y_3 = x_6, z_1 = x_1, z_2 = x_3, z_3 = x_5$. Let us find control rules solving the problem of $\boldsymbol{y}$-stabilization of the unperturbed motion $y_i = z_i = 0, u_i = 0$ $(i = 1, 2, 3)$ of system (1.5.9).

We impose the following restrictions on the desired control rules:

$$\begin{aligned}
&f_3\beta - f_2\gamma + f_5\omega\gamma - f_6\omega\beta - \Gamma_1 y_2 - \Gamma_2 f_4 = \frac{u_2\gamma}{B} - \frac{u_3\beta}{C},\\
&f_1\gamma - f_3\alpha + f_6\omega\alpha - f_4\omega\gamma - \Gamma_3 y_3 - \Gamma_4 f_6 = \frac{u_3\alpha}{C} - \frac{u_1\gamma}{A},\\
&-f_2 + \Gamma_5 y_1 = \frac{u_2}{B}, \qquad \Gamma_i = \text{const} < 0 \quad (i = 1, \dots, 5).
\end{aligned} \tag{1.5.10}$$

Leaving out intermediate calculations, we conclude that variables y_i $(i = 1, 2, 3)$ of the closed system (1.5.9), (1.5.10) satisfy the equations

$$\begin{gathered}
\dot{y}_2 = \mu_1, \quad \dot{\mu}_1 = \Gamma_1 y_2 + \Gamma_2\mu_1, \quad \dot{y}_3 = \mu_2, \quad \dot{\mu}_2 = \Gamma_3 y_3 + \Gamma_4\mu_2,\\
\dot{y}_1 = \Gamma_5 y_1 \qquad (\mu_1 = f_4(\boldsymbol{y}, \boldsymbol{z}),\ \mu_2 = f_6(\boldsymbol{y}, \boldsymbol{z})).
\end{gathered} \tag{1.5.11}$$

Because the zero solution $y_1 = y_2 = y_3 = \mu_1 = \mu_2 = 0$ of system (1.5.11) is Lyapunov asymptotically stable, the unperturbed motion $y_i = z_i = 0$ $(i = 1, 2, 3)$ of system (1.5.9), (1.5.10) is asymptotically $\boldsymbol{y}$-stable.

From the equations of system (1.5.9) for $\dot{x}_4$ and $\dot{x}_6$, it follows that

$$z_1 = \frac{-\mu_2 + y_1\alpha - z_3\omega\alpha + y_2\omega\beta}{\beta}, \qquad z_2 = \frac{\mu_1 + y_1\gamma - z_3\omega\gamma + y_3\omega\beta}{\beta}. \tag{1.5.12}$$

By the geometric relationship $\alpha^2 + \beta^2 + \gamma^2 = 1$,

$$(y_2 + \alpha)^2 + (z_3 + \beta)^2 + (y_3 + \gamma)^2 = 1. \tag{1.5.13}$$

From (1.5.12) and (1.5.13) we conclude that $\boldsymbol{y}$-stabilization actually implies stabilization of the unperturbed motion $y_i = z_i = 0$ $(i = 1, 2, 3)$ of system (1.5.9), (1.5.10) with respect to *all* the variables.

Hence, control rules (1.5.10) solve the problem of stabilization for the unperturbed motion of system (1.5.9) with respect to all the variables.

3. Let us find conditions which permit stabilizing the unperturbed motion $y_i = z_i = 0$, $u_i = 0$ $(i = 1, 2, 3)$ of system (1.5.9) (with respect to all the variables) by using *only two* (*rather than three*, as previously) controlling torques u_i $(i = 1, 2, 3)$.

To this end, we set $u_2 \equiv 0$ and take u_1, u_3 in accordance with (1.5.10). We arrive at system (1.5.11) in which, however, the equation

$$\begin{gathered}
\dot{y}_1 = 2(C - A)B^{-1}\omega\alpha\gamma\beta^{-1} y_1 + f(t; t_0, \boldsymbol{x}_0),\\
|f| \le a_1 e^{-a_2(t - t_0)},
\end{gathered} \tag{1.5.14}$$

replaces $\dot{y}_1 = \Gamma_5 y_1$. Here a_1 and a_2 are positive constants.

Equation (1.5.14) is obtained by substituting expressions (1.5.12) and the solutions of the first two groups of equations of system (1.5.11) in the equation for $\dot{x}_2$ from (1.5.9).

Consequently, if the condition

$$(C - A)\omega\alpha\gamma\beta^{-1} < 0$$

is valid, the stabilization of the unperturbed motion $y_i = z_i = 0$, $u_i = 0$ $(i = 1, 2, 3)$ of system (1.5.9) is possible when *only two* controlling torques u_1, u_3 are used.

1.6 Control with Respect to Part of the Variables

1.6.1 Formulation of the Problem

In studies of controlled systems, just the possibility of transferring a system from one state to another is sometimes of interest. The *concept of controllability* of a system has its origin in the treatment of such a problem. A number of definitions for controllability find use in the control theory. They differ both by transferring conditions for a system and by restrictions imposed on control rules. The classification of controllability types based on these features is presented in A.A. Krasovskii [1987].

Discussed below is the problem of complete controllability of a linear autonomous control system in the *Kalman sense* with the difference that we pose the problem of transferring a system relative to *part* of its phase variables.

Problem 1.6.1. Let there be given a time interval $[t_0, t_1]$, *an initial state* $\boldsymbol{y}(t_0) = \boldsymbol{y}^0$, *and a finite state* $\boldsymbol{y}(t_1) = \boldsymbol{y}^1$ *of system* (1.5.1). *Find a control* $\boldsymbol{u}(t) \in \mathfrak{U}_2$, $t \in [t_0, t_1]$ ($\mathfrak{U}_2$ *is a class of functions measurable in the interval* $[t_0, t_1]$) *that transfers variables* $y_1, \ldots, y_m$ *of system* (1.5.1) *from a given initial state* $(t_0, \boldsymbol{y}^0)$ *to a specified finite state* $(t_1, \boldsymbol{y}^1)$. *The behavior of variables* $z_1, \ldots, z_p$ *in this process is of no importance for us.*

Let Problem 1.6.1 be solvable for any $\boldsymbol{y}^0$ and $\boldsymbol{y}^1$ in the interval $[t_0, t_1]$. Then system (1.5.1) is said to be *completely controllable with respect to* $y_1, \ldots, y_m$ *in the interval* $[t_0, t_1]$.

If

$$\operatorname{rank} K_n^* = m < n, \qquad K_n^* = \left(B^*, A^*B^*, \ldots, (A^*)^{n-1}B^*\right), \tag{1.6.1}$$

then, as noted in Subsection 1.5.4, by applying a linear transformation of a particular form, system (1.5.1) can be decomposed into two subsystems, controllable and uncontrollable, the dimension of the former equal to m.

But condition (1.6.1) alone does not guarantee complete controllability of system (1.5.1) with respect to $y_1, \ldots, y_m$.

This is the case for the system

$$\dot{y}_1 = y_1 + z_2, \qquad \dot{z}_1 = y_1 + z_1 + z_2 + u_1, \qquad \dot{z}_2 = z_2 \tag{1.6.2}$$

subject to condition (1.6.1) ($\operatorname{rank} K_3^* = 1$). However, this system is not controllable with respect to y_1, because the first and third equations of (1.6.2) form a system whose solutions do not depend upon control u_1.

Thus, the problem of complete controllability with respect to a *given* part of the variables requires separate analysis.

Now we show that conditions for complete controllability of system (1.5.1) with respect to $y_1, \ldots, y_m$ can be obtained

(*a*) by determining conditions for complete controllability with respect to all the variables of the $\boldsymbol{\mu}$-system corresponding to a properly chosen matrix Γ; and

(*b*) by complementing condition (1.6.1).

1.6.2 Conditions for Complete Controllability with Respect to Part of the Variables

Theorem 1.6.1.

1. *Let there exist a constant matrix* Γ *for which condition* (1.5.5) *is fulfilled. Then system* (1.5.1) *is completely controllable with respect to* $y_1, \dots, y_m$.

2. *If condition* (1.6.1) *holds true, then system* (1.5.1) *is completely controllable with respect to* m *of* n *variables* $x_1, \dots, x_n$. *The indices of the controllable variables coincide with the indices of* m *linearly independent rows of the matrix whose columns are linearly independent columns of matrix* K_m^*.

Proof. 1. By the second part of the proof of Theorem 1.5.1, the auxiliary $\boldsymbol{\mu}$-system (1.5.4) for the original system (1.5.1) which corresponds to matrix Γ is completely controllable (with respect to all the variables). This means that system (1.5.1) is completely controllable with respect to $y_1, \dots, y_m$.

2. The separation of system (1.5.1) into controllable and uncontrollable parts is performed by the linear change of variables $\boldsymbol{\xi} = G\boldsymbol{x}$ with matrix G containing m linearly independent columns of K_m^*. The remaining columns of matrix G are taken so as to provide its nonsingularity.

Let $i_1, \dots, i_m$ be the position numbers of linearly independent rows in linearly independent columns of matrix K_m^*. Place m linearly independent columns of matrix K_m^* into columns of matrix G with position numbers $i_1, \dots, i_m$ and set the elements g_{ij} $(i = i_1, \dots, i_m; j \neq i_1, \dots, i_m)$ of matrix G equal to zero, which can be done without violating the nonsingularity of G. From the equality $\boldsymbol{\xi} = G\boldsymbol{x}$ it follows that the variables x_j $(j = i_1, \dots, i_m)$ of system (1.5.1) are linear combinations of ξ_j $(j = i_1, \dots, i_m)$. Since the columns of matrix G with position numbers $i_1, \dots, i_m$ are linearly independent columns of K_m^* and, consequently, the variables ξ_j with indices $j = i_1, \dots, i_m$ will be completely controllable, the variables x_j $(j = i_1, \dots, i_m)$ of system (1.5.1) will possess the same property. The theorem is proved. □

Remark 1.6.1.

1. Comparing conditions (1) and (2) of Theorem 1.6.1, we note that for *higher order* system (1.5.1) condition (1) may appear to be much easier verifiable than condition (2). This is the case when the choice of the first term in control rule (1.5.2) allows reducing the order of an auxiliary $\boldsymbol{\mu}$-system corresponding to matrix Γ to a value which makes verifying conditions for complete controllability for $\boldsymbol{\mu}$-system easier than composing matrix K_m^* and analyzing its properties.

2. Linearly independent rows of the matrix whose columns are linearly independent columns of matrix K_m^* can be chosen in a *nonunique* way. (Ambiguity and unambiguity in choosing these rows are demonstrated in the second and third parts of Example 1.6.1, respectively.) Therefore, according to the statement in the second part of Theorem 1.6.1 in the case when condition (1.6.1) is satisfied, the choice of controllable variables (with its total number unchanged), as a rule, will also be ambiguous.

Example 1.6.1.

1. Let system (1.5.1) have the form

$$\begin{aligned} \dot{y}_1 &= a_{11}y_1 + b_{11}z_1 + b_{12}z_2 + p_1u_1, \\ \dot{y}_2 &= a_{22}y_2 + b_{21}z_1 + b_{22}z_2 + p_2u_1, \\ \dot{z}_1 &= c_{i1}y_1 + dz_i, \qquad p_i \neq 0 \quad (i = 1, 2). \end{aligned} \tag{1.6.3}$$

Taking

$$u_1 = \Gamma \boldsymbol{z} + u_1^*, \qquad \Gamma = -\frac{1}{p_2}(b_{21}, b_{22})$$

as the control rule, we obtain the auxiliary $\boldsymbol{\mu}$-system corresponding to matrix Γ

$$\begin{gathered} \dot{y}_1 = a_{11}y_1 + \mu_1 + p_1u_1^*, \quad \dot{y}_2 = a_{22}y_2 + p_2u_1^*, \quad \dot{\mu}_1 = \alpha y_1 + d\mu_1, \\ \mu_1 = \alpha_1 z_1 + \alpha_2 z_2, \qquad \alpha = \alpha_1 c_{11} + \alpha_2 c_{21}, \\ \alpha_1 = b_{11} - \frac{p_1 b_{21}}{p_2}, \qquad \alpha_2 = b_{12} - \frac{p_1 b_{22}}{p_2}. \end{gathered} \tag{1.6.4}$$

If

$$\alpha\big[a_{22}(a_{11} + d) + (a_{11}^2 + \alpha) - a_{11}(a_{11} + d) - a_{22}^2\big] \neq 0, \tag{1.6.5}$$

the auxiliary system (1.6.4) corresponding to matrix Γ is completely controllable with respect to all the variables. Hence, by the first part of Theorem 1.6.1 the original system (1.6.3) is completely controllable with respect to y_1 and y_2.

We note that condition (1.6.5) does not ensure complete controllability of the original system (1.6.3) with respect to all the variables, for instance, when $c_{11} = 0$ system (1.6.3) is known to be uncontrollable relative to z_1, though completely controllable with respect to y_1, y_2 (with condition (1.6.5) satisfied by $\alpha = \alpha_2 c_{12}$).

2. Let system (1.5.1) have the form

$$\dot{x}_1 = -x_1 + 3x_3, \qquad \dot{x}_2 = -x_2 + 3x_3, \qquad \dot{x}_3 = x_1 + x_2 + u_1. \tag{1.6.6}$$

In the case in question, $n = 3$, and

$$K_3^* = \begin{pmatrix} 0 & 3 & -3 \\ 0 & 3 & -3 \\ 1 & 0 & 6 \end{pmatrix}, \quad \operatorname{rank} K_3^* = m = 2 < n, \quad K_2^* = \begin{pmatrix} 0 & 3 \\ 0 & 3 \\ 1 & 0 \end{pmatrix}.$$

The matrix whose columns are linearly independent columns of K_2^* coincides with K_2^*. Since the maximum number of linearly independent rows in matrix K_2^* equals the rank of the matrix, the valid options for them are the first and third, the second and third but not the first two rows (numbered from the top). By the second part of Theorem 1.6.1, system (1.6.6) is completely controllable with respect to x_1, x_3 or x_2, x_3 but not to x_1, x_2.

The same result can be obtained by taking u_1 in the form $u_1 = -x_1 - x_2 + u_1^*$. The auxiliary $\boldsymbol{\mu}$-systems

$$\dot{x}_1 = -x_1 + 3x_3 \quad (\dot{x}_2 = -x_2 + 3x_3), \qquad \dot{x}_3 = u_1^*$$

are completely controllable with respect to $x_1(x_2), x_3$. By the first part of Theorem 1.6.1, system (1.6.6) is completely controllable with respect to $x_1(x_2), x_3$.

It is worth noting that, in the case under consideration, using the first part of Theorem 1.6.1 reduces computational burden.

3. For system (1.6.2), $n = 3$ and $\operatorname{rank} K_3^* = 1 < n$, $K_1^* = (0 \quad 1 \quad 0)^{\mathsf{T}}$. The matrix whose columns are linearly independent columns of K_1^* coincides with K_1^*. The only linearly independent row in matrix K_1^* is the middle one, for any other set of rows contains the zero row and, consequently, cannot be linearly independent. Therefore, system (1.6.2) is not completely controllable with respect to all or to any pair of the variables, though completely controllable relative to z_1.

1.6.3 Distinguishing Features of the Problems of Stabilization and Complete Controllability of Processes with Respect to Part of the Variables

Condition (1.5.5) for complete controllability with respect to part of the variables is *weaker* than the widely known Kalman criterion for complete controllability with respect to all the variables. From this viewpoint the notions of complete controllability and stabilizability with respect to part of the variables reflect *more delicate* properties of a system and allow revealing its *limiting capabilities*. However, one must be aware of the following circumstance: notions of stability, stabilizability, and complete controllability with respect to part of the variables are much *more sensitive* to changes in system structure than the same notions considered with respect to all the variables. This fact should be allowed for in designing particular systems of automatic control.

It stands to reason that the application of the methods developed for investigating stability, stabilization, and complete controllability of processes with respect to part of the variables is not confined to analyzing just these fine problems. A number of examples regarding design of *low-sensitivity* (*robust*) control systems are presented in Chapters 4 and 5. Further examples of this kind can be cited.

1.7 Overview of References

The solution of the problem of asymptotic stability with respect to all the variables for linear systems with constant coefficients was already given in the work by Routh and Hurwitz in the past century. However, this problem is still relevant for systems of *higher dimensions* where the effective use of Hurwitz's criterion is rather complicated.

As to the problem of *partial* asymptotic stability, it became the object of research in the 1930s (Armellini [1935], Sansone [1935], [1949], Tonelli [1935], Wiman [1936], Prodi [1950], etc.*). The investigations dealt with the prop-

*Armellini, G. [1935] Sopra un'Equazione Differenziale della Dinamica, *Rend. R. Acc. Naz. dei Lincei*, **21**, 111–116; Sansone, G. [1935] *Scritti Matematici Offerti a Luigi Berzolari*, Pavia, 385–403; Sansone, G. [1949] *Equazioni Differenziali nel Campo Reale*, Part 2, Seconda edizione, Bologna; Tonelli L. [1935] *Scritti Matematici Offerti a Luigi Berzolari*, Pavia, 404–405; Wiman, A. [1936] Ueber Eine Stabilitatsfrage in der Theorie der Linearen Differentialgleichungen, *Acta Math.*, **66**, 121–145; Prodi, G. [1950] Un'osservazione sugl'integrali dell'Equazione $y'' + A(x)y = 0$ nel caso $A(x) \to +\infty$ per $x \to \infty$, *Rend. Accad. Lincei.*, **8**, 462–464.

erty of x-attraction of the zero solution $x = \dot{x} = 0$ of the second-order linear differential equation $\ddot{x} = \varphi(t)x$. A bibliography of these works can be found in Cesari [1959]. The term "*partial stability*" seems to have appeared just at this stage of developing the general theory of stability. In addition, it was suggested that partial stability be called "*stability in Routh's sense*" and stability with respect to all the variables—"*stability in Dirichlet's sense*" (Sansone [1949]). However, one can say that only the term "partial stability" became conventional in the scientific literature. In this regard, we note that the term "stability with respect to part of the variables," which is used by V.V. Rumyantsev and his school, is apparently more accurate (though slightly verbose). This is probably the reason why both the terms mentioned are widely used (and are often manipulated as interchangeable by one and the same scientist).

Later, Zubov [1959], Matrosov [1965], and Corduneanu [1971] analyzed the problem on partial asymptotic stability for a general class of linear systems with coefficients continuous in t by using the method of Lyapunov functions and vector functions. Here, the result by Corduneanu [1971], which we cited in the Introduction (Theorem 0.5.1), is worth attention.

Chapter 1 reflects another approach to the problem of partial asymptotic stability for both autonomous (Vorotnikov and Prokop'ev [1978], Vorotnikov [1979a], [1983a]) and nonautonomous (Vorotnikov [1979a], [1988a]) systems of linear differential equations. This approach was summed up in the monograph by Vorotnikov [1991a].

The idea of this approach formed the basis for analyzing the problem of polystability with respect to part of the variables (Martynyuk [1993a], [1993b]) and for the problem of investigating partial stability by using quadratic forms (Oziraner [1986]). In addition, this approach was extended to other classes of linear systems:

(1) systems with retarded argument (Vorotnikov [1980a], [1991a]);
(2) stochastic systems (Vorotnikov [1983b], [1991a]);
(3) discrete systems (Il'yasov [1984]);
(4) discrete stochastic systems (Phillis [1984]).

These results are partially presented in Chapters 6 and 7.

Principal difficulties whose solution required essentially new ideas (Vorotnikov [1988a], [1991b]) stimulated the development of the method as applied to the linear systems with time-variable coefficients. These difficulties can be explained by the following factor. In this case we are forced to consider μ-systems with coefficients *discontinuous* in t (the discontinuity of the *second kind*) for nonautonomous systems with continuous coefficients and also analytic coefficients. Systems with such "strong" discontinuities are scarcely studied and the notions of solutions and stability of solutions of the auxiliary μ-system needed certain development.

The results of Chapter 1 allow the conclusion that is fundamental for understanding the problem of partial stability (Vorotnikov [1991a], [1993], [1995a], [1995b]): partial stability in the presence of constantly acting perturbations *is not equivalent* to partial stability in the presence of small parametric perturbations.

The idea of constructing control rules in the linear problem of partial stabilization in the form (Vorotnikov [1979a], [1980a], [1982a], [1982b])

$$\boldsymbol{u} = \Gamma \boldsymbol{z} + \boldsymbol{u}^*$$

also turned out to be rather fruitful in solving the nonlinear problems. Allowing for the nonlinear nature of the problem, it was suggested (Vorotnikov [1979a], [1982a]) that the controls be constructed in the form

$$\boldsymbol{u} = \boldsymbol{\varphi}(\boldsymbol{x}) + \boldsymbol{\psi}(\boldsymbol{x})\boldsymbol{u}^*.$$

This made it possible to constructively solve a number of problems of both partial stabilization and control and stabilization and control with respect to all the variables (Vorotnikov [1979a], [1982a], [1985], [1986a]–[1986c], [1988b], [1990], [1991a], [1991b], [1997b], [1997c]) and a number of game-theoretic problems of control in the presence of large-magnitude uncontrollable perturbations (Vorotnikov [1994a]–[1994c], [1995c], [1997b], [1997e]). The technique incorporates solving the problems of stabilization and control with respect to part of the variables as the first stage of solving the problems of stabilization and control with respect to all the variables.

Chapter 2

Nonlinear Problems of Stability with Respect to Part of the Variables in the First (Linear and Nonlinear) Approximation

Chapter 2 develops the method for investigating stability with respect to a given part of the variables of *nonlinear* systems. First we consider the problem of stability with respect to a given part of the variables in the *linear approximation*. To a much greater extent than in studies of stability with respect to all the variables, this problem depends on the structure of nonlinear terms. It is demonstrated that the class of nonlinear systems, for which the problem of stability with respect to part of the variables is solvable by using linear approximation, can be significantly extended. To this end, instead of the linear part of the original nonlinear system, we consider a linear approximating system constructed in a special way. The latter is "equivalent" to a certain nonlinear subsystem of the original system and can be obtained from it by special nonlinear transformations of variables. Then, from the analysis of stability of the linear system constructed with respect to all the variables, the conclusion is drawn about stability of the original nonlinear system with respect to a given part of the variables. Constructive procedures are described for constructing such linear approximating systems.

New theorems on stability with respect to part of the variables in the linear approximation are proved. The Lyapunov–Malkin theorem on stability in *critical cases*, frequently used in the theory of stability, is extended to cover a wider class of *nonlinear* systems. The same scheme is used in treating the problem of stability with respect to part of the variables in *nonlinear approximation*.

Taking the proposed investigative method as a basis, we analyze the problem of *absolute stability* for the Lur'e systems and the problem of damping of rotational motions of an asymmetric solid with a single "fixed" engine or a single flywheel.

2.1 Features of the Problem of Stability with Respect to Part of the Variables in a Linear Approximation

2.1.1 Formulation of the Problem

Let equations of perturbed motion (0.3.1) have the form

$$\dot{\boldsymbol{x}} = A^*\boldsymbol{x} + \boldsymbol{X}(t,\boldsymbol{x}),$$
$$\boldsymbol{x}^{\mathrm{T}} = (x_1,\ldots,x_n) = (y_1,\ldots,y_m,z_1,\ldots,z_p) = (\boldsymbol{y}^{\mathrm{T}},\boldsymbol{z}^{\mathrm{T}}), \tag{2.1.1}$$
$$m>0,\quad p\geq 0,\quad n=m+p.$$

In $\boldsymbol{y}$, $\boldsymbol{z}$ variables, system (2.1.1) can be written as

$$\frac{dy_i}{dt} = \sum_{k=1}^{m} a_{ik}y_k + \sum_{l=1}^{p} b_{il}z_l + Y_i(t,y_1,\ldots,y_m,z_1,\ldots,z_p),$$
$$\frac{dz_j}{dt} = \sum_{k=1}^{m} c_{jk}y_k + \sum_{l=1}^{p} d_{jl}z_l + Z_j(t,y_1,\ldots,y_m,z_1,\ldots,z_p) \tag{2.1.2}$$
$$(i=1,\ldots,m;\ j=1,\ldots,p),$$

where a_{ik}, b_{il}, c_{jk}, and d_{jl} are *constants*, and Y_i and Z_j are *nonlinear* perturbations. Functions Y_i $(i=1,\ldots,m)$ and Z_j $(j=1,\ldots,p)$ are assumed, regarding the variables in the aggregate, to be continuously differentiable in domain (0.3.2). This implies that these functions satisfy Lipschitz's condition relative to $\boldsymbol{x}$, which in turn ensures the uniqueness of solutions.

We shall study the problem of stability (asymptotic stability) of the unperturbed motion $\boldsymbol{y}=\boldsymbol{0}$, $\boldsymbol{z}=\boldsymbol{0}$ of nonlinear system (2.1.2) with respect to variables $y_1,\ldots,y_m$.

2.1.2 A Conventional Approach to the Study of the Problem

It is natural to begin investigating the problem of stability of the unperturbed motion $\boldsymbol{x}=\boldsymbol{0}$ of system (2.1.2) with respect to $y_1,\ldots,y_m$ by analyzing the linear part of the system, i.e., by analyzing the stability of system (1.1.1).

As outlined in Section 1.1, if the zero solution of the linear system (1.1.1) is asymptotically stable with respect to $y_1,\ldots,y_m$, there exists a function v

possessing certain properties. When applied to system (2.1.2) as an appropriate Lyapunov function, the function v enables one to derive conditions for the stability of unperturbed motion with respect to $y_1, \ldots, y_m$ in the form of inequalities met by nonlinear terms. This approach is typical in studies of motion stability in linear approximation. However, let us demonstrate why we cannot make substantial progress this way in investigating the problem of stability with respect to part of the variables in the linear approximation.

2.1.3 The Necessity of Constructing Linear Approximation Systems of a Special Type

Let $L_{\boldsymbol{x}}(L_{\boldsymbol{y}})$ be a class of vector functions $\boldsymbol{X}(\boldsymbol{x})$ in the domain $\|\boldsymbol{x}\| \leq H$ ($\|\boldsymbol{y}\| \leq H, \|\boldsymbol{z}\| < \infty$), where H is a sufficiently small positive constant, which can be expanded into convergent power series in integral powers of variables $x_1, \ldots, x_n$, The expansions begin with terms of degree two or higher. A constant matrix A^* will be termed $\boldsymbol{x}$-*Hurwitz* matrix if all the roots of its characteristic equation have negative real parts.

Lyapunov [1892] proved that, if A^* is an arbitrary $\boldsymbol{x}$-Hurwitz matrix, then for any vector function $\boldsymbol{X}(\boldsymbol{x}) \in L_{\boldsymbol{x}}$ the unperturbed motion $\boldsymbol{x} = \boldsymbol{0}$ of system (2.1.1) is asymptotically stable with respect to all the variables.

A different situation occurs in the problem of stability with respect to part of the variables. Matrix A^* will be called the $\boldsymbol{y}$-*Hurwitz* matrix if all the roots of the characteristic equation of matrix $L_4 A^* L_5$ have negative real parts. For such a matrix the zero solution $\boldsymbol{x} = \boldsymbol{0}$ of the system $\dot{\boldsymbol{x}} = A^* \boldsymbol{x}$ is asymptotically $\boldsymbol{y}$-stable. Let us demonstrate that, for $\boldsymbol{X}(\boldsymbol{x}) \in L_{\boldsymbol{y}}$, resolving the questions of $\boldsymbol{y}$-stability, asymptotic $\boldsymbol{y}$-stability and $\boldsymbol{y}$-instability of the unperturbed motion $\boldsymbol{x} = \boldsymbol{0}$ of system (2.1.1) can depend on a particular vector function $\boldsymbol{X}(\boldsymbol{x})$ and also on the values of coefficients constituting the $\boldsymbol{y}$-Hurwitz matrix A^*.

To be more precise, now we will show that, when the right-hand sides of system (2.1.2) are analytic, its unperturbed motion $\boldsymbol{y} = \boldsymbol{0}, \boldsymbol{z} = \boldsymbol{0}$ can be $\boldsymbol{y}$-unstable even if the zero solution of linear approximating system (1.1.1) is asymptotically $\boldsymbol{y}$-stable (*the first peculiarity*). Moreover (we call it *the second peculiarity*), with vector function $\boldsymbol{X}(\boldsymbol{x})$ fixed, the question of whether the unperturbed motion of system (2.1.2) is $\boldsymbol{y}$-stable, asymptotically $\boldsymbol{y}$-stable, or $\boldsymbol{y}$-unstable depends on the values of the coefficients constituting the $\boldsymbol{y}$-Hurwitz matrix A^*.

Consider the equations

$$\dot{y}_1 = a y_1 + z_1^2 z_2, \qquad \dot{z}_1 = b z_1 + y_1 z_1, \qquad \dot{z}_2 = c z_2 - 2 y_1 z_2, \tag{2.1.3}$$

where a, b and c are constants. After introducing the new variable $\mu_1 = z_1^2 z_2$, the system can be written as

$$\dot{y}_1 = a y_1 + \mu_1, \quad \dot{\mu}_1 = (2b + c)\mu_1, \quad \dot{z}_1 = b z_1 + y_1 z_1, \quad \dot{z}_2 = c z_2 - 2 y_1 z_2. \tag{2.1.4}$$

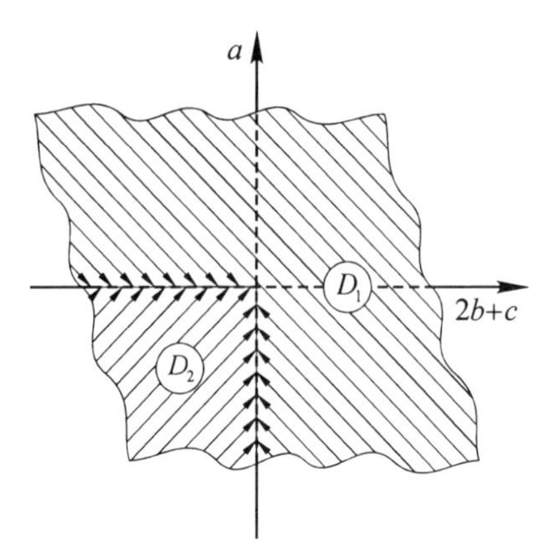

Figure 2.1.1. Regions of y_1-stability and y_1-instability of the unperturbed motion $y_1 = z_1 = z_2 = 0$ of system (2.1.3): D_1—the region of y_1-instability, D_2—the region of asymptotic y_1-stability, $D_3 = \{(a = 0) \wedge (2b + c) < 0\}$, and $D_4 = \{(a < 0) \wedge (2b + c) = 0\}$—the regions of y_1-stability.

From the first two equations of (2.1.4) it follows that variable y_1 of system (2.1.4) and, consequently, that of the original system (2.1.3) obey the equation

$$\dot{y}_1 = ay_1 + z_{10}^2 z_{20} e^{[(2b+c)(t-t_0)]}.$$

Therefore, the unperturbed motion $y_1 = z_1 = z_2 = 0$ of system (2.1.3) is
(a) asymptotically stable with respect to y_1 for $a < 0,\ 2b + c < 0$;
(b) stable with respect to y_1 for $a = 0,\ 2b + c < 0$ or $a < 0,\ 2b + c = 0$;
(c) unstable with respect to y_1 if at least one of two conditions $a > 0,\ 2b + c > 0$, or both $a = 0,\ 2b + c = 0$ hold true (Figure 2.1.1).

Taking

$$A^* = \begin{pmatrix} a & 0 & 0 \\ 0 & b & 0 \\ 0 & 0 & c \end{pmatrix}, \qquad \boldsymbol{X}(\boldsymbol{x}) = (z_1^2 z_2, y_1 z_1, -2y_1 z_2)^{\mathrm{T}}$$

confirms the propositions. Indeed, for $a < 0$, matrix A^* will be $\boldsymbol{y}$-Hurwitz, because the unperturbed motion $y_1 = z_1 = z_2 = 0$ is $\boldsymbol{y}$-unstable, provided $2b+c > 0$ (the first of the peculiarities noted). Then, $\boldsymbol{y}$-instability changes to asymptotic $\boldsymbol{y}$-stability as soon as the expression $2b + c$ reverses its sign (the second peculiarity).

The reasoning above indicates that the problem of stability of motion with respect to part of the variables in the linear approximation, to a much greater extent than in studies of stability with respect to all the variables, depends on the structure *of nonlinear* and *also of linear terms* of the original system. That is why, instead of linear part of the original nonlinear system, a specially constructed system of the linear approximation will be dealt with in the sequel. Having analyzed the Lyapunov stability of this system, we draw a conclusion about stability with respect to a given part of the variables of the original nonlinear system. This allows us to significantly expand the class of nonlinear systems for which the question about stability with respect to part of the variables is solved through linear approximation.

Applying this idea below, consider two versions of constructing systems of a linear approximation for (2.1.2). Taken together, they represent means that enable one to analyze the problem in a variety of situations. However, following the second version, we can obtain conditions for the $\boldsymbol{y}$-stability of unperturbed motion of the original system that are closer to the necessary conditions at the expense of increased computational effort.

2.2 A Method of Nonlinear Transformation of Variables in Investigating Stability with Respect to Part of the Variables in a Linear Approximation (1)

2.2.1 The Original System of Equations. Separating Special Forms of the Variables from Nonlinear Perturbations

Consider equations of perturbed motion (2.1.1) for $c_{jk} = 0$ $(j = 1, \dots, p;\ k = 1, \dots, m)$. In addition, let us assume that $b_{il} = 0$ $(i = 1, \dots, m;\ l = 1, \dots, p)$. This can be done without loss of generality because of results of Section 1.1 assuming that the linear part of system (2.1.1) has been properly transformed. As a result, system (2.1.1) takes the form

$$\begin{aligned} \frac{dy_i}{dt} &= \sum_{k=1}^{m} a_{ik} y_k + Y_i(t, y_1, \dots, y_m, z_1, \dots, z_p), \\ \frac{dz_j}{dt} &= \sum_{l=1}^{p} d_{jl} z_l + Z_j(t, y_1, \dots, y_m, z_1, \dots, z_p) \\ &\qquad (i = 1, \dots, m;\ j = 1, \dots, p). \end{aligned} \tag{2.2.1}$$

Let us write functions Y_i $(i = 1, \dots, m)$ as

$$\begin{aligned} Y_i(t, y_1, \dots, y_m, z_1, \dots, z_p) &= Y_i^0(z_1, \dots, z_p) \\ &+ \sum_{\gamma=1}^{N} Y_{i\gamma}^*(y_1, \dots, y_m) \bar{Y}_{i\gamma}^0(z_1, \dots, z_p) + Y_i^{**}(t, y_1, \dots, y_m, z_1, \dots, z_p) \\ &\qquad (i = 1, \dots, m), \end{aligned} \tag{2.2.2}$$

where functions Y_i^0 and $\bar{Y}_{i\gamma}^0$ are defined by the equalities

$$\begin{aligned} Y_i^0(z_1, \dots, z_p) &= U_2^{(i)}(z_1, \dots, z_p) + \dots + U_r^{(i)}(z_1, \dots, z_p), \\ \bar{Y}_{i\gamma}^0(z_1, \dots, z_p) &= \bar{U}_1^{(i\gamma)}(z_1, \dots, z_p) + \dots + \bar{U}_s^{(i\gamma)}(z_1, \dots, z_p), \end{aligned} \tag{2.2.3}$$

in which $U_v^{(i)}(z_1, \dots, z_p)$ and $\bar{U}_w^{(i\gamma)}(z_1, \dots, z_p)$ are *homogeneous forms* of variables $z_1, \dots, z_p$ of the order v $(v \le r)$ and w $(w \le s)$, respectively, and r and

s are *finite* numbers. Functions $Y^*_{i\gamma}(y_1, \ldots, y_m)$ $(i = 1, \ldots, m; \gamma = 1, \ldots, N)$ are assumed analytic in the domain $\|\boldsymbol{y}\| \leq H$ and $Y^*_{i\gamma}(0, \ldots, 0) \equiv 0$.

The essence of decomposition (2.2.2), (2.2.3) is that we separate out *forms of finite order* $Y_i^0, \bar{Y}^0_{i\gamma}$ from Y_i $(i = 1, \ldots, m)$ to use them as *additional variables* in constructing auxiliary *linear* system. In so doing the remaining expressions Y_i^{**} $(i = 1, \ldots, m)$ will be estimated (from the standpoint of the $\boldsymbol{y}$-stability problem) by choosing y_i and $Y_i^0, \bar{Y}^0_{i\gamma}$. Since decomposition (2.2.2), (2.2.3) for functions Y_i $(i = 1, \ldots, m)$ can be carried out in a *nonunique* way including one or another set of terms of the specified type into $Y_i^0, \bar{Y}^0_{i\gamma}$, the arbitrariness at our disposal should be used to *rationalize* the search for the most acceptable solution.

2.2.2 The Choice of a System of the First Approximation

As a system of the *first approximation* for system (2.2.1), let us take the equations

$$\begin{aligned} \frac{dy_i}{dt} &= \sum_{k=1}^{m} a_{ik} y_k + Y_i^0(z_1, \ldots, z_p) \\ &\quad + \sum_{\gamma=1}^{N} Y^*_{i\gamma}(y_1, \ldots, y_m) \bar{Y}^0_{i\gamma}(z_1, \ldots, z_p), \qquad (2.2.4) \\ \frac{dz_j}{dt} &= \sum_{l=1}^{p} d_{jl} z_l \qquad (i = 1, \ldots, m;\ j = 1, \ldots, p). \end{aligned}$$

and let us assume, without loss of generality, that $N = m$. Thus, *nonlinear subsystem* (2.2.4) but *not the linear part* of the original nonlinear system (2.2.1) is chosen as a system of the first approximation.

Let us demonstrate that, in investigating the stability of the unperturbed motion $\boldsymbol{x} = \boldsymbol{0}$ of system (2.2.1) with respect to $y_1, \ldots, y_m$, it is reasonable to introduce an *auxiliary linear system* which is the linear part of a specially constructed system obtained from (2.2.4) through nonlinear transformation of the variables. Based on the Lyapunov asymptotic stability analysis of this linear system, the conclusion is drawn whether or not the zero solution of system (2.2.4) and also (under additional conditions on Y_i^{**}, Z_j) the unperturbed motion of the original nonlinear system (2.2.1) are asymptotically $\boldsymbol{y}$-stable.

2.2.3 Constructing an Auxiliary System for the System of the First Approximation

Introduce the new variables

$$\begin{aligned} \mu_i^{(1)} &= Y_i^0(z_1, \ldots, z_p) = U_2^{(i)}(z_1, \ldots, z_p) + \cdots + U_r^{(i)}(z_1, \ldots, z_p), \\ \mu_{ij}^{(2)} &= \bar{Y}^0_{ij}(z_1, \ldots, z_p) = \bar{U}_1^{(ij)}(z_1, \ldots, z_p) + \cdots + \bar{U}_s^{(ij)}(z_1, \ldots, z_p), \qquad (2.2.5) \\ &\qquad (i, j = 1, \ldots, m). \end{aligned}$$

Since $U_v^{(i)}$ and $\bar{U}_w^{(ij)}$ are homogeneous forms of variables $z_1, \ldots, z_p$ of the orders v and w, respectively, they can be represented as

$$U_v^{(i)} = \sum_{\alpha_1+\cdots+\alpha_p=v} q_i^{\alpha_1\ldots\alpha_p} z_1^{\alpha_1} \ldots z_p^{\alpha_p},$$
$$\bar{U}_w^{(ij)} = \sum_{\alpha_1+\cdots+\alpha_p=w} \bar{q}_{ij}^{\alpha_1\ldots\alpha_p} z_1^{\alpha_1} \ldots z_p^{\alpha_p},$$

where $q_i^{\alpha_1\ldots\alpha_p}$ and $\bar{q}_{ij}^{\alpha_1\ldots\alpha_p}$ are constants.

Denote a number of distinct sets for which $\alpha_1 + \cdots + \alpha_p = v$ by N_v. (Two sets of numbers $\alpha_1, \ldots, \alpha_p$ and $\alpha_1', \ldots, \alpha_p'$ are supposed to be distinct if $\alpha_i \neq \alpha_i'$ at least for one i.)

Let us establish correspondence between form $U_v^{(i)}$ and vector $(q_{il}^1, \ldots, q_{il}^{N_v})$ composed of properly arranged coefficients $q_i^{\alpha_1\ldots\alpha_p}$ characterizing this form, i.e., let us associate vector $\boldsymbol{Q}_i^{(1)}$ with each new variable $\mu_i^{(1)}$:

$$\mu_i^{(1)} \to \boldsymbol{Q}_i^{(1)} = (q_{i2}^{11}, \ldots, q_{i2}^{1N_1}; \ldots; q_{ir}^{11}, \ldots, q_{ir}^{1N_r}).$$

Similarly, let us associate vector $\boldsymbol{Q}_{ij}^{(2)}$ with each new variable $\mu_{ij}^{(2)}$:

$$\mu_{ij}^{(2)} \to \boldsymbol{Q}_{ij}^{(2)} = (q_{ij1}^{21}, \ldots, q_{ij1}^{2N_1}; \ldots; q_{ijs}^{21}, \ldots, q_{ijs}^{2N_s}).$$

Suppose that vectors $\boldsymbol{Q}_i^{(1)}$ and $\boldsymbol{Q}_{ij}^{(2)}$ are linearly independent. Otherwise we take those which are linearly independent. In this introduction of new variables, two cases are possible.

The first case. System (2.2.4) can be reduced to the form

$$\begin{aligned}
\frac{dy_i}{dt} &= \sum_{k=1}^{m} a_{ik} y_k + \mu_i^{(1)} + \sum_{j=1}^{m} Y_{ij}^*(y_1, \ldots, y_m)\mu_{ij}^{(2)}, \\
\frac{d\mu_j^{(1)}}{dt} &= U_2^{(j)*}(z_1, \ldots, z_p) + \cdots + U_r^{(j)*}(z_1, \ldots, z_p) \\
&= \sum_{k=1}^{m} L_{jk}^{(1)} \mu_k^{(1)} + \sum_{k,\varepsilon=1}^{m} \bar{L}_{jk\varepsilon}^{(1)} \mu_{k\varepsilon}^{(2)}, \\
\frac{d\mu_{\gamma\theta}^{(2)}}{dt} &= \bar{U}_1^{(\gamma\theta)*}(z_1, \ldots, z_p) + \cdots + \bar{U}_s^{(\gamma\theta)*}(z_1, \ldots, z_p) \\
&= \sum_{k=1}^{m} L_{\gamma\theta k}^{(2)} \mu_k^{(1)} + \sum_{k,\varepsilon=1}^{m} \bar{L}_{\gamma\theta k\varepsilon}^{(2)} \mu_{k\varepsilon}^{(2)}, \\
\frac{dz_\sigma}{dt} &= \sum_{l=1}^{p} d_{\sigma l} z_l \qquad (i, j, \gamma, \theta = 1, \ldots, m;\ \sigma = 1, \ldots, p),
\end{aligned} \tag{2.2.6}$$

where $U_v^{(j)*}$ and $\bar{U}_w^{(\gamma\theta)*}$ are homogeneous forms of variables $z_1, \ldots, z_p$ of the order v $(v \leq r)$ and w $(w \leq s)$ and $L_{jk}^{(1)}$, $\bar{L}_{jk\varepsilon}^{(1)}$, $L_{\gamma\theta k}^{(2)}$, and $\bar{L}_{\gamma\theta k\varepsilon}^{(2)}$ are constants.

Because equations (2.2.6) result from a transformation of system (2.2.4) that does not change variables $y_1, \dots, y_m$, the system

$$
\begin{aligned}
\frac{dy_i}{dt} &= \sum_{k=1}^{m} a_{ik} y_k + \mu_i^{(1)} + \sum_{j=1}^{m} Y_{ij}^*(y_1, \dots, y_m)\mu_{ij}^{(2)}, \\
\frac{d\mu_j^{(1)}}{dt} &= \sum_{k=1}^{m} L_{jk}^{(1)} \mu_k^{(1)} + \sum_{k,\varepsilon=1}^{m} \bar{L}_{jk\varepsilon}^{(1)} \mu_{k\varepsilon}^{(2)}, \\
\frac{d\mu_{\gamma\theta}^{(2)}}{dt} &= \sum_{k=1}^{m} L_{\gamma\theta k}^{(2)} \mu_k^{(1)} + \sum_{k,\varepsilon=1}^{m} \bar{L}_{\gamma\theta k\varepsilon}^{(2)} \mu_{k\varepsilon}^{(2)} \\
&(i, j, \gamma, \theta = 1, \dots, m)
\end{aligned}
\tag{2.2.7}
$$

completely describes the behavior of variables $y_1, \dots, y_m$ of system (2.2.4). In the sequel such a system will be termed a ***μ*-*system relative to the system of the first approximation*** (2.2.4).

The second case. System (2.2.4) is not reduced to (2.2.7). Suppose that in the second and third groups of equations of system (2.2.6) only the first m_1 $(m_1 < m)$ and m_2 $(m_2 < m)$ equalities, respectively, hold which guarantee the formation of the auxiliary $\boldsymbol{\mu}$-system (2.2.7).

In this case system (2.2.4) looks like

$$
\begin{aligned}
\frac{dy_i}{dt} &= \sum_{k=1}^{m} a_{ik} y_k + \mu_i^{(1)} + \sum_{j=1}^{m} Y_{ij}^*(y_1, \dots, y_m)\mu_{ij}^{(2)} && (i = 1, \dots, m), \\
\frac{d\mu_j^{(1)}}{dt} &= \sum_{k=1}^{m} L_{jk}^{(1)} \mu_k^{(1)} + \sum_{k,\varepsilon=1}^{m} \bar{L}_{jk\varepsilon}^{(1)} \mu_{k\varepsilon}^{(2)} && (j = 1, \dots, m_1), \\
\frac{d\mu_{\gamma\theta}^{(2)}}{dt} &= \sum_{k=1}^{m} L_{\gamma\theta k}^{(2)} \mu_k^{(1)} + \sum_{k,\varepsilon=1}^{m} \bar{L}_{\gamma\theta k\varepsilon}^{(2)} \mu_{k\varepsilon}^{(2)} && (\gamma, \theta = 1, \dots, m_2), \\
\frac{d\mu_j^{(1)}}{dt} &= U_2^{(j)*}(z_1, \dots, z_p) + \dots + U_r^{(j)*}(z_1, \dots, z_p) && (j = m_1 + 1, \dots, m), \\
\frac{d\mu_{\gamma\theta}^{(2)}}{dt} &= U_1^{(\gamma\theta)*}(z_1, \dots, z_p) + \dots + U_s^{(\gamma\theta)*}(z_1, \dots, z_p) && (\gamma, \theta = m_2 + 1, \dots, m), \\
\frac{dz_\sigma}{dt} &= \sum_{l=1}^{p} d_{\sigma l} z_l && (\sigma = 1, \dots, p).
\end{aligned}
\tag{2.2.8}
$$

Once again we introduce the new variables

$$
\begin{aligned}
\bar{\mu}_j^{(1)} &= U_2^{(j)*}(z_1, \dots, z_p) + \dots + U_r^{(j)*}(z_1, \dots, z_p), \\
\bar{\mu}_{\gamma\theta}^{(2)} &= U_1^{(\gamma\theta)*}(z_1, \dots, z_p) + \dots + U_s^{(\gamma\theta)*}(z_1, \dots, z_p) \\
&(j = m_1 + 1, \dots, m;\ \gamma, \theta = m_2 + 1, \dots, m).
\end{aligned}
\tag{2.2.9}
$$

This permits reducing system (2.2.4) to a system of type (2.2.6) or (2.2.8). In the first case the aim of transforming system (2.2.4), i.e., constructing the auxiliary linear system (the linear part of the μ-system), is attained. In the second case it is necessary to continue introducing new variables. The question arises if it is possible to construct a *finite*-dimensional auxiliary μ-system at some final stage of introducing the new variables of the stated type. Let us demonstrate that the answer is affirmative.

2.2.4 Convergence of the Procedure for Transforming the System of the First Approximation

Lemma 2.2.1. *A finite-dimensional μ-system can always be constructed for system* (2.2.4).

Proof. Let us introduce a number of the new variables whose corresponding linearly independent, finite-dimensional vectors $\boldsymbol{Q}_i^{(1)}, \bar{\boldsymbol{Q}}_i^{(1)}, \ldots, \boldsymbol{Q}_{ij}^{(2)}, \bar{\boldsymbol{Q}}_{ij}^{(2)}, \ldots$ will compose columns of quadratic nonsingular matrices Q_1 and Q_2 of dimensions $N_1^* \times N_1^*$ and $N_2^* \times N_2^*$, respectively. Then each of the column vectors (denoted by $\hat{\boldsymbol{Q}}_j^{(1)}, \hat{\boldsymbol{Q}}_{kl}^{(2)}$) corresponding to the new variables subsequently introduced will be a linear combination of column vectors of Q_1 and Q_2. Indeed, the linear algebraic systems of equations

$$\hat{\boldsymbol{Q}}_j^{(1)} = Q_1 \boldsymbol{\lambda}_j, \qquad \hat{\boldsymbol{Q}}_{kl}^{(2)} = Q_2 \boldsymbol{\lambda}_{kl},$$
$$\boldsymbol{\lambda}_j = (\lambda_{j1}, \ldots, \lambda_{jN_1^*})^{\mathrm{T}}, \qquad \boldsymbol{\lambda}_{kl} = (\lambda_{kl1}, \ldots, \lambda_{klN_2^*})^{\mathrm{T}}$$

admit nonzero solutions $\boldsymbol{\lambda}_j^0$, $\boldsymbol{\lambda}_{kl}^0$ because matrices Q_1 and Q_2 are nonsingular.

Further introduction of new variables is thus meaningless, since each new variable subsequently introduced will be a linear combination of the preceding ones. Suppose that it is impossible to compose nonsingular quadratic matrices $Q^{(1)}$ and $Q^{(2)}$. This means that, at some finite stage of introducing new variables, each of the new variables subsequently introduced will also be a linear combination of the preceding ones. Since it is shown that a finite number of new variables is sufficient for constructing an auxiliary system of equations, then the μ-system will be finite-dimensional. The statement is proved. □

Remark 2.2.1.

1. The equations and variables of the finite-dimensional μ-system thus constructed are split into subsets so that each equation of the corresponding subset except the last can involve variables from the following subset and does not include variables from the subsequent subsets.

2. As distinct from linear systems, the dimension of a μ-system for nonlinear system (2.2.4) *may exceed* that of the system itself.

Lemma 2.2.2. *When* $\bar{Y}_{ij}^0(\boldsymbol{z}) \equiv 0$ $(i = 1, \ldots, m;\ j = 1, \ldots, N)$, *for the zero solution of nonlinear system* (2.2.4) *to be asymptotically stable with respect to* $y_1, \ldots, y_m$, *it is necessary and sufficient that the zero solution of linear μ-system (in the case in question μ-system is linear) be Lyapunov asymptotically stable.*

Proof. The *sufficiency* is obvious.

Let us prove the *necessity*. The asymptotic stability of the zero solution of system (2.2.4) (with respect to all and to part of the variables) is exponential, and asymptotic stability with respect to $y_1, \dots, y_m$ requires asymptotic stability with respect to variables $\mu_i = Y_i^0(\boldsymbol{z})$ $(i = 1, \dots, m)$. When system (2.2.4) is reduced to (2.2.7), the necessity is proved. If system (2.2.4) is not reduced to (2.2.7) and, consequently, has the form (2.2.8), we apply the reasoning above to system (2.2.8) once again. Since it is always possible to construct an auxiliary $\boldsymbol{\mu}$-system at some finite step of introducing new variables, the necessity of the statement is proved. □

Remark 2.2.2. When $\bar{Y}_{ij}^0(\boldsymbol{z}) \not\equiv 0$ (for at least one of the values i and j possible), the Lyapunov asymptotic stability of the zero solution of the linear part of $\boldsymbol{\mu}$-system is *not a necessary* condition for asymptotic stability of the zero solution of system (2.2.4) with respect to $y_1, \dots, y_m$ (as exemplified in Subsection 2.4.1). In the stated case the procedure for constructing an auxiliary system must be modified so that the conditions for asymptotic stability of the unperturbed motion of the original system (2.2.1) with respect to $y_1, \dots, y_m$ are obtained in to approach necessary conditions. This question is dealt with in Section 2.4.

2.2.5 Transformation of System (2.2.1)

Suppose, without loss of generality, that equations (2.2.4) are already reduced to $\boldsymbol{\mu}$-system (2.2.7) at the first step of introducing new variables (2.2.5). The original system (2.2.1) is transformed as follows:

$$
\begin{aligned}
\frac{dy_i}{dt} &= \sum_{k=1}^{m} a_{ik} y_k + \mu_i^{(1)} + \sum_{j=1}^{m} Y_{ij}^*(y_1, \dots, y_m)\mu_{ij}^{(2)} \\
&\qquad + Y_i^{**}(t, y_1, \dots, y_m, z_1, \dots, z_p), \\
\frac{d\mu_j^{(1)}}{dt} &= \sum_{k=1}^{m} L_{jk}^{(1)} \mu_k^{(1)} + \sum_{k,\varepsilon=1}^{m} \bar{L}_{jk\varepsilon}^{(1)} \mu_{k\varepsilon}^{(2)} + Z_j^{(1)}(t, y_1, \dots, y_m, z_1, \dots, z_p), \\
\frac{d\mu_{\gamma\theta}^{(2)}}{dt} &= \sum_{k=1}^{m} L_{\gamma\theta k}^{(2)} \mu_k^{(1)} + \sum_{k,\varepsilon=1}^{m} \bar{L}_{\gamma\theta k\varepsilon}^{(2)} \mu_{k\varepsilon}^{(2)} + Z_{\gamma\theta}^{(2)}(t, y_1, \dots, y_m, z_1, \dots, z_p), \quad (2.2.10) \\
\frac{dz_\sigma}{dt} &= \sum_{l=1}^{p} d_{\sigma l} z_l + Z_\sigma(t, y_1, \dots, y_m, z_1, \dots, z_p),
\end{aligned}
$$

$$
Z_j^{(1)} = \sum_{l=1}^{p} \frac{\partial \mu_j^{(1)}}{\partial z_l} Z_l, \qquad Z_{\gamma\theta}^{(2)} = \sum_{l=1}^{p} \frac{\partial \mu_{\gamma\theta}^{(2)}}{\partial z_l} Z_l
$$

$$
(i, j, \gamma, \theta = 1, \dots, m; \; \sigma = 1, \dots, p).
$$

In the general case, when constructing a $\boldsymbol{\mu}$-system for the system of the first approximation (2.2.4) is possible at some finite step of introducing new variables, system (2.2.1) is transformed into a system of type (2.2.10). The linear part of this system (excluding the last group of equations) forms a closed linear system relative to y_i $(i = 1, \ldots, m)$ and additional variables, i.e., an auxiliary system of the linear approximation for the original nonlinear system (2.2.1).

2.2.6 The Condition for Stability with Respect to Part of the Variables

The above replacement of nonlinear system (2.2.4) with the auxiliary $\boldsymbol{\mu}$-system allows one in studies of asymptotic stability of the unperturbed motion $\boldsymbol{x} = \mathbf{0}$ of system (2.2.1) with respect to $y_1, \ldots, y_m$ instead of the equations of the linear approximation

$$\frac{dy_i}{dt} = \sum_{k=1}^{m} a_{ik} y_k, \qquad \frac{dz_j}{dt} = \sum_{l=1}^{p} d_{jl} z_l$$

to consider the *specially constructed* system of the linear approximation, i.e., the *linear part of* a *$\boldsymbol{\mu}$-system.* Taking into account that in constructing $\boldsymbol{\mu}$-system for (2.2.4) the original system (2.2.1) is transformed into a system of type (2.2.10), let us term this linear system *$\boldsymbol{\mu}$-system of the linear approximation for the original nonlinear system* (2.2.1).

Thus, the proposed approach to the study of stability with respect to part of the variables is based on the idea of changing the concept of the linear approximation for the original nonlinear system and constructing the auxiliary system of the linear approximation.

This enables one to substantially expand the class of nonlinear systems for which the problem of stability with respect to part of the variables is solvable through the linear approximation.

Now we obtain the conditions for asymptotic stability of the unperturbed motion of system (2.2.1) with respect to $y_1, \ldots, y_m$ based on implementing the proposed approach to the study of stability in the linear approximation (see Scheme 2.2.1).

Further, without loss of generality, we assume that it is already possible to construct a $\boldsymbol{\mu}$-system of the linear approximation for (2.2.1) at the *first stage* of introducing the new variables of type (2.2.5). In this case the $\boldsymbol{\mu}$-system of the linear approximation for (2.2.1) is the linear part of system (2.2.7).

Let us introduce the designations:

(1) ξ_i $(i = 1, \ldots, m^2 + 2m)$ are components of vector $\boldsymbol{\xi}$ consisting of variables $y_i, \mu_j^{(1)}, \mu_{\gamma\theta}^{(2)}$ $(i, j, \gamma, \theta = 1, \ldots, m)$ which determine the state of the $\boldsymbol{\mu}$-system of the linear approximation;

(2) $Y_{*i}(t, \boldsymbol{y}, \boldsymbol{z})$ $(i = 1, \ldots, m^2 + 2m)$ signify components of vector function $\boldsymbol{Y}_*(t, \boldsymbol{y}, \boldsymbol{z})$ consisting of functions $Y_i^{**}, Z_j^{(1)}, Z_{\gamma\theta}^{(2)}$ $(i, j, \gamma, \theta = 1, \ldots, m)$ appearing in the right-hand sides of system (2.2.10).

Scheme 2.2.1. The scheme for analyzing asymptotic $\boldsymbol{y}$-stability of the unperturbed motion of system (2.2.1).

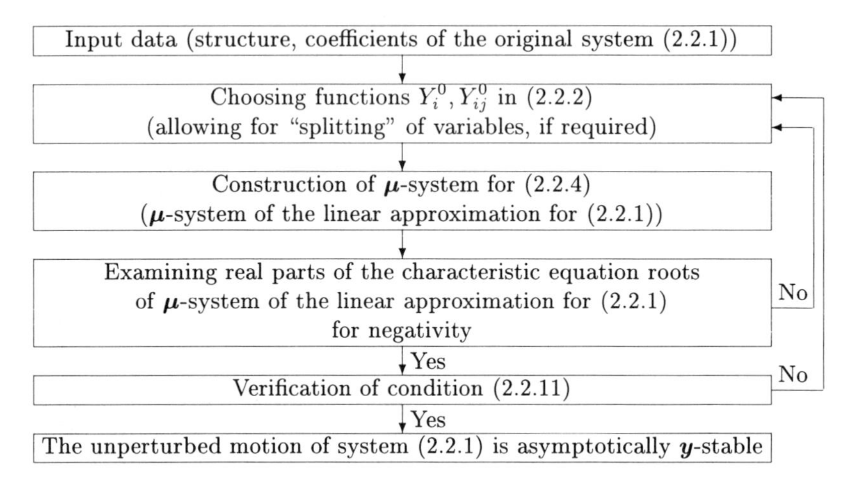

Suppose that the following condition is satisfied in the domain $t \geq 0$, $\|\boldsymbol{\xi}\| \leq H$, $\|\boldsymbol{z}\| < \infty$:

$$\big\|\boldsymbol{Y}_*(t, \boldsymbol{y}, \boldsymbol{z})\big\| \leq \alpha \|\boldsymbol{\xi}\|, \tag{2.2.11}$$

where α and H are sufficiently small positive constants.

Theorem 2.2.1. *Let all the roots of the characteristic equation of the linear part of $\boldsymbol{\mu}$-system* (2.2.7) (*$\boldsymbol{\mu}$-system of the linear approximation for* (2.2.1)) *have negative real parts. Then the unperturbed motion $\boldsymbol{x} = \boldsymbol{0}$ of system* (2.2.1) *will be uniformly asymptotically stable with respect to $y_1, \ldots, y_m$ if its nonlinear terms satisfy condition* (2.2.11).

Proof. Under the conditions of the theorem and designations introduced, for the linear part of $\boldsymbol{\mu}$-system (2.2.7), it is possible to specify function $V = V(\boldsymbol{\xi})$ which is subject to the conditions of the Lyapunov theorem on asymptotic stability

$$c_1 \|\boldsymbol{\xi}\|^2 \leq V(\boldsymbol{\xi}) \leq c_2 \|\boldsymbol{\xi}\|^2, \qquad \dot{V}_\mu(\boldsymbol{\xi}) \leq -c_3 \|\boldsymbol{\xi}\|^2, \tag{2.2.12}$$

where $\dot{V}_\mu$ is a derivative of function V by virtue of the linear part of $\boldsymbol{\mu}$-system and c_i $(i = 1, 2, 3)$ are positive constants. The expression for the derivative of this function V calculated along the trajectories of system (2.2.1) is

$$\dot{V}_{(2.2.1)} = \dot{V}_{(2.2.7)} + \sum_{i=1}^{m^2+2m} \frac{\partial V}{\partial \xi_i} Y_{*i}(t, \boldsymbol{y}, \boldsymbol{z})$$

$$= \dot{V}_\mu + \sum_{k=1}^{m} \frac{\partial V}{\partial \xi_k} \sum_{\gamma=1}^{N} Y_{k\gamma}^*(\boldsymbol{y}) \mu_{k\gamma} + \sum_{i=1}^{m^2+2m} \frac{\partial V}{\partial \xi_i} Y_{*i}(t, \boldsymbol{y}, \boldsymbol{z}).$$

Bearing in mind that $Y^*_{k\gamma}(\mathbf{0}) \equiv 0$ $(\gamma = 1, \dots, N)$ from (2.2.11) and (2.2.12), we obtain the estimate

$$\dot{V}_{(2.2.1)} \leq -c_3\|\boldsymbol{\xi}\|^2 + \beta\|\boldsymbol{\xi}\|^2,$$

which is valid in the domain $t \geq 0$, $\|\boldsymbol{\xi}\| \leq H$, $\|\boldsymbol{z}\| < \infty$ and in which β is a sufficiently small constant. Therefore, there is a number $c > 0$ such that $\dot{V}_{(2.2.1)} \leq -c\|\boldsymbol{\xi}\|^2$. Then, by virtue of system (2.2.1), function V satisfies the conditions of Corollary 0.4.1 on uniform asymptotic stability with respect to part of the variables, and the motion $\boldsymbol{x} = \mathbf{0}$ of system (2.2.1) is uniformly asymptotically $\boldsymbol{\xi}$-stable. The theorem is proved. □

Remark 2.2.3. The asymptotic stability of the unperturbed motion of system (2.2.1) with respect to variables $y_1, \dots, y_m$ can also be proved by introducing new variables of a type *simpler* than in (2.2.5) and (2.2.9). For instance, instead of (2.2.5) one can introduce the new variables μ_j $(j = 1, \dots, L)$:

$$\begin{aligned} Y^0_i(\boldsymbol{z}) &= \Phi_i(\mu_1, \dots, \mu_L), \\ \bar{Y}^0_{ij}(\boldsymbol{z}) &= \Phi_{ij}(\mu_1, \dots, \mu_L) \qquad (i, j = 1, \dots, m), \end{aligned} \tag{2.2.13}$$

where Φ_i and Φ_{ij} are certain forms of the new variables μ_j $(j = 1, \dots, L)$. We shall call such a technique the *method of "splitting" nonlinear terms* in constructing auxiliary systems.

In particular, instead of (2.2.5) the new variables of type (2.2.13) can be introduced as follows:

$$\mu^{(v)}_i = U^{(i)}_v(z_1, \dots, z_p), \qquad \mu^{(w)}_{ij} = \bar{U}^{(ij)}_w(z_1, \dots, z_p)$$
$$(i, j = 1, \dots, m;\ v = 2, \dots, r;\ w = 1, \dots, s).$$

In this case the auxiliary $\boldsymbol{\mu}$-system of the linear approximation for (2.2.1) has a simpler structure considered from the viewpoint of the Lyapunov stability analysis.

Example 2.2.1.

1. Let the equations of perturbed motion (2.2.1) have the form

$$\begin{aligned} \dot{y}_1 &= -y_1 + Y^*_1(y_1)z_1z_2 + z_2^2z_3 + Y^{**}_1(t, y_1, z_1, z_2, z_3), \\ \dot{z}_j &= \Sigma_j + Z_j(t, y_1, z_1, z_2, z_3) \qquad (j = 1, 2, 3) \\ &(\Sigma_1 = z_1, \quad \Sigma_2 = -2z_2, \quad \Sigma_3 = z_2 + 3z_3). \end{aligned} \tag{2.2.14}$$

In the domain (0.3.2), Y^{**}_1 and Z_j are analytic functions with continuous and bounded coefficients. Among the roots of the characteristic equation of the linear part of system (2.2.14), there are roots with positive real parts. Consequently, the unperturbed motion $y_1 = z_1 = z_2 = z_3 = 0$ of system (2.2.14) is *Lyapunov unstable.*

Let us examine this motion for the asymptotic stability with respect to y_1. As a system of the linear approximation for (2.2.14), we take the equations

$$\dot{y}_1 = -y_1 + Y^*_1(y_1)z_1z_2 + z_2^2z_3, \quad \dot{z}_1 = z_1, \quad \dot{z}_2 = -2z_2, \quad \dot{z}_3 = z_2 + 3z_3. \tag{2.2.15}$$

By introducing the new variables $\mu_1 = z_1 z_2$, $\mu_2 = z_2^2 z_3$, we construct the $\boldsymbol{\mu}$-system

$$\dot{y}_1 = -y_1 + Y_1^*(y_1)\mu_1 + \mu_2, \quad \dot{\mu}_1 = -\mu_1, \quad \dot{\mu}_2 = \mu_3, \quad \dot{\mu}_3 = -6\mu_2 - 7\mu_3 \qquad (2.2.16)$$

for system (2.2.15).

At the same time the original system (2.2.14) is transformed as follows:

$$\begin{aligned} \dot{y}_1 &= -y_1 + Y_1^*(y_1)\mu_1 + \mu_2 + Y_1^{**}(t, y_1, z_1, z_2, z_3), \\ \dot{\mu}_i &= \Sigma_i^* + Z_{3+i}(t, y_1, z_1, z_2, z_3), \\ \dot{z}_j &= \Sigma_i + Z_j(t, y_1, z_1, z_2, z_3) \qquad (i, j = 1, 2, 3) \end{aligned} \qquad (2.2.17)$$

$$(\Sigma_1^* = -\mu_1, \quad \Sigma_2^* = \mu_3, \quad \Sigma_3^* = -6\mu_2 - 7\mu_3, \quad Z_4 = z_1 Z_2 + z_2 Z_1, \quad Z_5 = 2z_2 z_3 Z_2 + z_2^2 Z_3, \quad Z_6 = -Z_3 + 3z_2^2 Z_2).$$

Suppose that, in the domain $t \geq 0$, $|y_1| \leq H$, $\|\boldsymbol{\mu}\| \leq H$, $\|\boldsymbol{z}\| < \infty$, the condition

$$|Y_1^{**}| + \sum_{j=1}^{3} |Z_{3+j}| \leq \alpha_0 |y_1| + \sum_{j=1}^{3} \alpha_j |\mu_j|,$$

holds true, where α_0 and α_j are sufficiently small positive constants. Taking into account that all the roots of the characteristic equation of the linear part of system (2.2.16) have negative real parts, we conclude that under the conditions specified the solution $y_1 = 0$, $\mu_i = z_i = 0$ $(i = 1, 2, 3)$ of system (2.2.17) is uniformly asymptotically stable with respect to y_1, μ_i $(i = 1, 2, 3)$. Consequently, the unperturbed motion $y_1 = z_1 = z_2 = z_3 = 0$ of the original system (2.2.14) is uniformly asymptotically stable with respect to variable y_1.

Notice that the conditions imposed on Y_1^{**}, Z_{3+i} $(i = 1, 2, 3)$ hold trivially if, for example, in the domain $t \geq 0$, $\|\boldsymbol{\xi}\| \leq H$ functions Y_1^{**}, Z_{3+i} can be expanded into convergent power series (with continuous bounded coefficients) relative to functions $y_1, \mu_1 = z_1 z_2, \mu_2 = z_2^2 z_3$ The expansions begin with terms of degree two or higher.

2. Let the equations of perturbed motion (2.2.1) have the form

$$\begin{aligned} \dot{y}_1 &= a y_1 + b z_1 z_2 + Y_1^{**}(t, y_1, z_1, z_2), \\ \dot{z}_i &= d_i z_i + Z_i(t, y_1, z_1, z_2) \qquad (i = 1, 2). \end{aligned} \qquad (2.2.18)$$

Here Y_1^{**} and Z_i $(i = 1, 2)$ are analytic functions in domain (0.3.2) with continuous and bounded coefficients, and a, b, d_1, and d_2 are constants. Introducing the new variable $\mu_1 = z_1 z_2$, we form the system

$$\begin{gathered} \dot{y}_1 = a y_1 + b\mu_1 + Y_1^{**}(t, y_1, z_1, z_2), \\ \dot{\mu}_1 = (d_1 + d_2)\mu_1 + Z_1^*(t, y_1, z_1, z_2), \qquad \dot{z}_i = d_i z_i + Z_i(t, y_1, z_1, z_2) \\ Z_1^* = z_1 Z_2 + z_2 Z_1 \qquad (i = 1, 2). \end{gathered}$$

Provided that

$$a < 0, \qquad d_1 + d_2 < 0, \qquad |Y_1^{**}| + |Z_1^*| \leq \alpha_1 |y_1| + \alpha_2 |z_1 z_2|,$$

where α_i $i = (1, 2)$ are sufficiently small positive constants (b stands for an arbitrary number), the unperturbed motion $y_1 = z_1 = z_2 = 0$ of system (2.2.18) is uniformly asymptotically stable with respect to y_1 by Theorem 2.2.1.

We note that the condition $d_1 + d_2 < 0$ can be valid when one of the constants d_1, d_2 is positive. Hence, the uniform asymptotic stability of the motion $y_1 = z_1 = z_2 = 0$ of system (2.2.18) with respect to y_1 is possible when this motion is *Lyapunov unstable.*

2.2.7 The Case of z-Boundedness of Solutions

Suppose that functions Y_i, Z_j appearing in the right-hand sides of system (2.2.1) do not explicitly depend on time. Let them also be analytic functions of variables $y_1, \dots, y_m$, $z_1, \dots, z_p$ in the domain $\|\boldsymbol{y}\| \leq H$, $\|\boldsymbol{z}\| \leq H_1 < \infty$. We distinguish three groups of terms contributing to functions Y_i:
(1) depending solely on $z_1, \dots, z_p$ and constituting functions $Y_i^0(\boldsymbol{z})$;
(2) linear in $y_1, \dots, y_m$ and constituting functions $y_j \bar{Y}_{ij}^0(\boldsymbol{z})$;
(3) the remaining expressions $Y_i^{**}(\boldsymbol{y}, \boldsymbol{z})$.
Then,

$$Y_i(\boldsymbol{y}, \boldsymbol{z}) = Y_i^0(\boldsymbol{z}) + \sum_{j=1}^{m} y_j \bar{Y}_{ij}^0(\boldsymbol{z}) + Y_i^{**}(\boldsymbol{y}, \boldsymbol{z})$$
$$(i = 1, \dots, m),$$

assuming that functions $Y_i^0, \bar{Y}_{ij}^0$ have the form (2.2.3), $N = m$.

The case of $\boldsymbol{z}$-boundedness of solutions of system (2.2.1) (originating from a sufficiently small neighborhood of the unperturbed motion $\boldsymbol{y} = \mathbf{0}$, $\boldsymbol{z} = \mathbf{0}$) is interesting because, after introducing the new variables $\mu_i = Y_i^0(\boldsymbol{z})$, $\mu_{ij} = \bar{Y}_{ij}^0(\boldsymbol{z})$ and constructing auxiliary system (2.2.7), functions $Y_i^{**}(\boldsymbol{y}, \boldsymbol{z})$ will admit estimates needed to draw conclusions about the $\boldsymbol{y}$-stability of the unperturbed motion of system (2.2.1).

Corollary 2.2.1. *Let all the roots of the characteristic equation of the linear part of the $\boldsymbol{\mu}$-system for* (2.2.4) *have negative real parts. Let nonlinear perturbations Z_j $(j = 1, \dots, p)$ in system* (2.2.1) *contain no terms depending solely on $z_1, \dots, z_p$ or terms linear in $y_1, \dots, y_m$. Then the unperturbed motion $\boldsymbol{x} = \mathbf{0}$ of system* (2.2.1) *is uniformly asymptotically stable with respect to $y_1, \dots, y_m$.*

Proof. The following estimates are valid in the case under consideration:

$$\dot{V}_{(2.2.1)} = \dot{V}_\mu + \sum_{k=1}^{m} \frac{\partial V}{\partial \xi_k} \left(\sum_{r=1}^{m} y_k \mu_{kr} \right) + \sum_{k=1}^{m} \frac{\partial V}{\partial \xi_k} Y_k^{**}(\boldsymbol{y}, \boldsymbol{z})$$
$$+ \sum_{j=m+1}^{m^2+2m} \sum_{l=1}^{p} \frac{\partial V}{\partial \xi_j} \frac{\partial \xi_j}{\partial z_l} Z_l(\boldsymbol{y}, \boldsymbol{z}).$$

Because solutions of system (2.2.1) are $\boldsymbol{z}$-bounded, the following inequalities are valid in the domain $\|\boldsymbol{\xi}\| \leq H$, $\|\boldsymbol{z}\| < H_1 < \infty$:

$$\left| \frac{\partial \xi_j}{\partial z_l} \right| \leq L, \qquad \left| Y_i^{**}(\boldsymbol{y}, \boldsymbol{z}) \right| \leq \sum_{k=1}^{m} \alpha_{ik} |y_k|, \qquad \left| Z_\gamma(\boldsymbol{y}, \boldsymbol{z}) \right| \leq \sum_{k=1}^{m} \beta_{\gamma k} |y_k|$$
$$(i = 1, \dots, m;\ j = 1, \dots, m^2 + 2m,\ \gamma = 1, \dots, p),$$

where α_{ik}, β_{jk} are sufficiently small positive constants.

Consequently,

$$\dot{V}_{(2.2.1)} \leq -c_3\|\boldsymbol{\xi}\|^2 + \sum_{k,r=1}^{m} \left|\frac{\partial V}{\partial \xi_k}\right| |y_k||\mu_{kr}|$$

$$+ \sum_{k,r=1}^{m} \alpha_{kr} \left|\frac{\partial V}{\partial \xi_k}\right| |y_r| + L \sum_{j=m+1}^{m^2+2m} \sum_{r=1}^{m} \sum_{l=1}^{p} \beta_{lr} \left|\frac{\partial V}{\partial \xi_j}\right| |y_r|.$$

This implies that there is $\bar{c} > 0$ such that $\dot{V}_{(2.2.1)} \leq -\bar{c}\|\boldsymbol{\xi}\|^2$. The remaining part of the proof follows the reasoning of Theorem 2.2.1. □

Remark 2.2.4.

1. Suppose that functions Y_i, Z_j appearing in the right-hand sides of system (2.2.1) do not depend on all variables $z_1, \ldots, z_p$ but only on part of them, for definiteness, on the first r $(r < p)$ variables $z_1, \ldots, z_r$. Then Corollary 2.2.1 will also be valid when solutions of system (2.2.1) beginning in a sufficiently small neighborhood of the origin of coordinates are not bounded in $z_{r+1}, \ldots, z_p$. Indeed, in that case the estimates derived while proving Corollary 2.2.1 hold good.

2. Assume that the unperturbed motion $\boldsymbol{x} = \boldsymbol{0}$ of system (2.2.1) is known to be (nonasymptotically) stable in the Lyapunov sense or with respect to $z_1, \ldots, z_p$. If all the roots of the characteristic equation of $\boldsymbol{\mu}$-system for the equations

$$\frac{dy_i}{dt} = \sum_{k=1}^{m} a_{ik} y_k + Y_i^0(z_1, \ldots, z_p), \qquad \frac{dz_j}{dt} = \sum_{l=1}^{p} d_{jl} z_l$$

$$(i = 1, \ldots, m;\ j = 1, \ldots, p)$$

have negative real parts and nonlinear perturbations $Z_j(\boldsymbol{y}, \boldsymbol{z})$ $(j = 1, \ldots, p)$ contain no terms that depend solely on $z_1, \ldots, z_p$, then the unperturbed motion $\boldsymbol{x} = \boldsymbol{0}$ of system (2.2.1) is asymptotically stable with respect to $y_1, \ldots, y_m$. To prove this, one can follow the scheme for the proof of Corollary 2.2.1.

3. Differing from the case of linear systems, the requirement of $\boldsymbol{z}$-boundedness of solutions of nonlinear system (2.2.1) is not equivalent to the requirement of $\boldsymbol{z}$-stability of its unperturbed motion. Therefore, when solutions of system (2.2.1) are $\boldsymbol{z}$-bounded the situation is also possible that when the unperturbed motion of system (2.2.1) is Lyapunov unstable, it is asymptotically $\boldsymbol{y}$-stable.

4. In the general case, it is important to know whether the zero solution of the linear part of $\boldsymbol{\mu}$-system is Lyapunov asymptotically stable. To estimate the remaining nonlinear terms of the original system we are also interested in obtaining variables $\mu_1, \ldots, \mu_N$ that form the $\boldsymbol{\mu}$-system.

2.2.8 An Application to the Problem of Stability in the Linear Nonautonomous Approximation

Let the equations of perturbed motion (0.3.1) have the form

$$\begin{aligned}\frac{dy_i}{dt} &= \sum_{k=1}^{m} a_{ik}(t)y_k + Y_i(t, y_1, \dots, y_m, z_1, \dots, z_p),\\ \frac{dz_j}{dt} &= \sum_{l=1}^{p} d_{jl}(t)z_l + Z_j(t, y_1, \dots, y_m, z_1, \dots, z_p)\\ &(i = 1, \dots, m;\ j = 1, \dots, p).\end{aligned} \tag{2.2.19}$$

Here $a_{ik}(t)$, $d_{il}(t)$ are continuous and bounded functions for $t \geq t_0$, and nonlinear perturbations Y_i, Z_j $(i = 1, \dots, m;\ j = 1, \dots, p)$ are expandable into convergent power series in integral powers of variables $y_1, \dots, y_m, z_1, \dots, z_p$ with real and bounded coefficients continuous with respect to t. The differentiability of the coefficients of system (2.2.19) will not be assumed.

Let us represent functions $Y_i(t, \boldsymbol{y}, \boldsymbol{z})$ $(i = 1, \dots, m)$ in system (2.2.19) as

$$Y_i(t, \boldsymbol{y}, \boldsymbol{z}) = \sum_{v=2}^{r} b_{iv}(t)U_v^{(i)}(\boldsymbol{z}) + Y_i^{**}(t, \boldsymbol{y}, \boldsymbol{z}),$$

$$U_v^{(i)}(\boldsymbol{z}) = A_i z_1^{\alpha_1} \dots z_p^{\alpha_p}, \quad A_i = \text{const}, \quad \alpha_1 + \dots + \alpha_p = v \leq r,$$

where $b_{iv}(t)$ are continuous and bounded functions for $t \geq t_0$ and r is a finite number, i.e., let us separate out finite-order forms of variables $z_1, \dots, z_p$ from $Y_i(t, \boldsymbol{y}, \boldsymbol{z})$ $(y = 1, \dots, m)$. As a system of the *first approximation* for (2.2.19), we take the equations

$$\begin{aligned}\frac{dy_i}{dt} &= \sum_{k=1}^{m} a_{ik}(t)y_k + \sum_{v=2}^{r} b_{iv}(t)U_v^{(i)}(\boldsymbol{z})\\ \frac{dz_j}{dt} &= \sum_{l=1}^{p} d_{jl}(t)z_l \qquad (i = 1, \dots, m;\ j = 1, \dots, p).\end{aligned} \tag{2.2.20}$$

To replace nonlinear system (2.2.20) with a linear system of equations, we introduce the new variables $\mu_{iv} = U_v^{(i)}(\boldsymbol{z})$ $(i = 1, \dots, m;\ v = 2, \dots, r)$. With the new variables thus introduced, two cases are possible.

The first case. It is possible to form the system

$$\frac{dy_i}{dt} = \sum_{k=1}^{m} a_{ik}(t)y_k + \sum_{v=2}^{r} b_{iv}(t)\mu_{iv}, \quad \frac{d\mu_{j\theta}}{dt} = \sum_{k=1}^{r} L_{j\theta k}(t)\mu_{k\theta} \tag{2.2.21}$$

$$(i, j = 1, \dots, m;\ \theta = 2, \dots, r),$$

where $L_{j\theta k}$ are functions expressed by corresponding linear combinations of coefficients in the second group of equations of system (2.2.20).

In this case the original system (2.2.19) is transformed as follows:

$$\begin{aligned}
\frac{dy_i}{dt} &= \sum_{k=1}^{m} a_{ik}(t)y_k + \sum_{v=2}^{r} b_{iv}(t)\mu_{iv} + Y_i^{**}(t,\boldsymbol{y},\boldsymbol{z}),\\
\frac{d\mu_{j\theta}}{dt} &= \sum_{k=1}^{m} L_{j\theta k}(t)\mu_{k\theta} + Z_{j\theta}^{*}(t,\boldsymbol{y},\boldsymbol{z}), \quad Z_{j\theta}^{*} = \sum_{l=1}^{p} \frac{\partial U_j^{(\theta)}}{\partial z_l} Z_l, \qquad (2.2.22)\\
\frac{dz_s}{dt} &= \sum_{l=1}^{p} d_{sl}(t)z_l + Z_s(t,\boldsymbol{y},\boldsymbol{z})\\
&(i,j = 1,\dots,m;\ \theta = 2,\dots,r;\ s = 1,\dots,p).
\end{aligned}$$

The linear system (2.2.21) will be called the *$\boldsymbol{\mu}$-system of the linear approximation for system* (2.2.19).

The second case. The formation of the linear system (2.2.21) by introducing the new variables μ_{iv} is impossible. However, it can be shown that, even in this case, system (2.2.19) can always be transformed into a system of type (2.2.22) at some *finite* step of introducing new variables.

Corollary 2.2.2. *Let the zero solution of the $\boldsymbol{\mu}$-system of the linear approximation for system* (2.2.19) *be Lyapunov exponentially asymptotically stable. Then the unperturbed motion $\boldsymbol{y} = \boldsymbol{0}$, $\boldsymbol{z} = \boldsymbol{0}$ of system* (2.2.19) *is uniformly asymptotically stable with respect to $y_1,\dots,y_m$ if nonlinear perturbations Y_i^{**}, Z_j of this system satisfy conditions of type* (2.2.11).

Proof. If the zero solution of linear $\boldsymbol{\mu}$-system (2.2.1) with continuous and *bounded* coefficients is exponentially asymptotically stable (with respect to all the variables), then there exists (Malkin [1966]) for this system a quadratic form $V(t,\boldsymbol{\xi})$ with continuous and *bounded* coefficients satisfying conditions of the Lyapunov theorem on asymptotic stability

$$\begin{aligned}
\frac{\partial V}{\partial t} &+ \sum_{i=1}^{m} \frac{\partial V}{\partial y_i}\left(\sum_{k=1}^{m} a_{ik}(t)y_k + \sum_{v=2}^{r} b_{iv}(t)\mu_{iv}\right)\\
&+ \sum_{j,\theta} \frac{\partial V}{\partial \mu_{j\theta}}\left(\sum_{k=1}^{m} L_{j\theta k}(t)\mu_{k\theta}\right) = -\|\boldsymbol{\xi}\|^2 \quad (j = 1,\dots,m;\ \theta = 1,\dots,r).
\end{aligned}$$

Now calculating the derivative of function $V(t,\boldsymbol{\xi})$ along the trajectories of system (2.2.19), we conclude that under conditions of type (2.2.11) this derivative will be negative-definite relative to variables y_i, μ_{iv} ($i = 1,\dots,m$; $v = 1,\dots,r$) in the domain $t \geq 0$, $\|\boldsymbol{\xi}\| \leq H$, $\|\boldsymbol{z}\| < \infty$. By Corollary 0.4.1, the unperturbed motion $\boldsymbol{y} = \boldsymbol{0}$, $\boldsymbol{z} = \boldsymbol{0}$ of system (2.2.19) is uniformly asymptotically stable with respect to $y_1,\dots,y_m$. The statement is proved. □

Example 2.2.2. Let the equations of perturbed motion (2.2.19) have the form

$$\begin{aligned}
\dot y_1 &= a(t)y_1 + b(t)z_1^2 z_2 + Y_1^{**}(t,y_1,z_1,z_2),\\
\dot z_i &= d_i(t)z_i + Z_i(t,y_1,z_1,z_2) \qquad (i = 1,2).
\end{aligned} \qquad (2.2.23)$$

Let us find conditions which ensure the asymptotic stability of the unperturbed motion $y_1 = z_1 = z_2 = 0$ of system (2.2.23) with respect to y_1. To this end, by introducing the new variable $\mu_1 = z_1^2 z_2$, we construct the system of the linear approximation

$$\begin{gathered} \dot{y}_1 = a(t)y_1 + b(t)\mu_1, \qquad \dot{\mu}_1 = \alpha(t)\mu_1 \\ \alpha(t) = 2d_1(t) + d_2(t) \end{gathered} \tag{2.2.24}$$

equivalent to the nonlinear subsystem

$$\dot{y}_1 = a(t)y_1 + b(t)z_1^2 z_2, \qquad \dot{z}_i = d_i(t)z_i \quad (i = 1, 2). \tag{2.2.25}$$

Suppose that the following condition is satisfied in the domain $t \geq 0$, $|y_1| \leq H$, $|\mu_1| \leq H$, $\|\boldsymbol{z}\| < \infty$:

$$\left|Y_1^{**}\right| + \left|Z_1^*\right| \leq \alpha_1|y_1| + \alpha_2|z_1^2 z_2|, \quad (Z_1^* = 2z_1 z_2 Z_1 + z_1^2 Z_2), \tag{2.2.26}$$

where α_i $i = (1, 2)$ are sufficiently small positive constants. If the solution $y_1 = \mu_1 = 0$ of system (2.2.24) is exponentially asymptotically stable with respect to both variables, then the unperturbed motion $y_1 = z_1 = z_2 = 0$ of system (2.2.23) is uniformly asymptotically stable with respect to y_1 by Corollary 2.2.2.

Let us consider separately the case when function $b(t)$ is *continuously differentiable* for all $t \geq t_0$ and, besides,

$$\dot{b}(t) + b(t)d(t) = \gamma b(t), \quad \gamma = \text{const}, \quad a(t) = \text{const} < 0.$$

In this case the $\boldsymbol{\mu}$-system for (2.2.23) comprises two linear equations with *constant* coefficients

$$\dot{y}_1 = ay_1 + \eta_1, \qquad \dot{\eta}_1 = \gamma\eta_1 \qquad (\eta_1 = b(t)z_1^2 z_2). \tag{2.2.27}$$

The solution $y_1 = \eta_1 = 0$ of system (2.2.27) is exponentially asymptotically stable with respect to both variables and condition (2.2.26) can be weakened to

$$\left|Y_1^{**}\right| + \left|Z_1^*\right| \leq \alpha_1|y_1| + \alpha_2|b(t)z_1^2 z_2|.$$

2.2.9 The Absolute Stability of Lur'e Systems

Let a controlled object in an automatic system be described by differential equations in the *Lur'e form* (Lur'e [1951]):

$$\begin{gathered} \dot{\boldsymbol{x}} = A^*\boldsymbol{x} + \boldsymbol{b}\varphi(\sigma), \qquad \sigma = \boldsymbol{\beta x}, \\ \boldsymbol{x}^{\mathsf{T}} = (x_1, \ldots, x_n) = (y_1, \ldots, y_m, z_1, \ldots, z_p) = (\boldsymbol{y}^{\mathsf{T}}, \boldsymbol{z}^{\mathsf{T}}), \\ m > 0, \quad p \geq 0, \quad n = m + p, \end{gathered} \tag{2.2.28}$$

where A^*, $\boldsymbol{b}$, and $\boldsymbol{\beta}$ are a constant matrix and vectors of dimensions $n \times n$, $n \times 1$, and $1 \times n$, respectively (see Figure 2.2.1). The characteristic $\varphi(\sigma)$ is *nonlinear* and its accurate determination presents some difficulties. We assume that

(1) function $\varphi(\sigma)$ is defined and continuous for all σ, $\varphi(0) = 0$;
(2) there exist constants $k_1 > 0$ and $k_2 > 0$ such that $k_1\sigma \leq \varphi(\sigma) \leq k_2\sigma$;

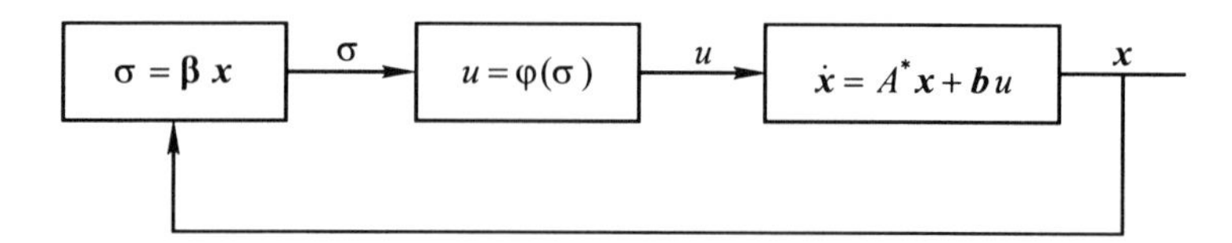

Figure 2.2.1. The structural model of a Lur'e system.

(3) $\int\limits_0^\infty \varphi(\sigma)d\sigma = \infty$.

Such characteristics will be called *admissible.*

Definition 2.2.1. System (2.2.28) is said to be *absolutely exponentially stable in the angle* $[k_1, k_2]$ *with respect to variables* $y_1, \dots, y_m$ (*absolutely exponentially* $\boldsymbol{y}$*-stable*) if there exist positive constants c_1 and c_2 such that, for any solution $\boldsymbol{x} = \boldsymbol{x}(t; t_0, \boldsymbol{x}_0)$ of system (2.2.28) and for *any admissible* characteristic $\varphi(\sigma)$, the condition

$$\|\boldsymbol{y}(t; t_0, \boldsymbol{x}_0)\| \le c_1 \|\boldsymbol{x}_0\| e^{-c_2(t-t_0)}, \quad t \ge t_0,$$

is fulfilled.

In $\boldsymbol{y}, \boldsymbol{z}$ variables, system (2.2.28) takes the form

$$\dot{\boldsymbol{y}} = A\boldsymbol{y} + B\boldsymbol{z} + \boldsymbol{b}^{(1)}\varphi(\sigma), \quad \dot{\boldsymbol{z}} = C\boldsymbol{y} + D\boldsymbol{z} + \boldsymbol{b}^{(2)}\varphi(\sigma), \quad \sigma = \boldsymbol{\beta}^{(1)}\boldsymbol{y} + \boldsymbol{\beta}^{(2)}\boldsymbol{z},$$

where A, B, C, and D are constant matrices and $\boldsymbol{b}^{(i)}$ and $\boldsymbol{\beta}^{(i)}$ $(i = 1, 2)$ are constant vectors of appropriate sizes. Consider the matrix

$$K_p^* = \big(B^*, D^*B^*, \dots, (D^*)^{p-1}B^*\big), \quad D^* = D^{\top}, \quad B^* = \big(B^{\top}, (\boldsymbol{\beta}^{(2)})^{\top}\big).$$

Let s be the minimal number such that $\operatorname{rank} K_{s-1}^* = \operatorname{rank} K_s^*$; assuming that $\operatorname{rank} K_{s-1}^* = h$ and using rules (a) through (d) from Subsection 1.1.4 (the only difference is that we manipulate K_{s-1}^* instead of K_{s-1}), let us construct matrices L_i^* $(i = 1, \dots, 5)$.

Let us also introduce the matrices

(1) L_6^* of dimensions $h \times (p-h)$ composed of those columns of matrix L_1^* not contained in L_2^*;

(2) L_7^* of dimensions $(p-h) \times p$ obtained from D by deleting its first h rows;

(3) L_8^*, L_9^* of dimensions $(p-h) \times h$ and $(p-h) \times (p-h)$, respectively, composed of the first h columns and the remaining columns of L_7^*, respectively;

(4) $L_{10}^* = L_9^* - L_8^*(L_2^*)^{-1}L_6^*$ of dimensions $(p-h) \times (p-h)$.

Theorem 2.2.2. 1. *If the system*

$$\dot{\boldsymbol{\xi}} = L_4^*A^*L_5^*\boldsymbol{\xi} + L_4^*\boldsymbol{b}\varphi(\sigma), \qquad \sigma = \boldsymbol{\beta}L_5^*\boldsymbol{\xi} \tag{2.2.29}$$

is absolutely exponentially stable in the angle $[k_1, k_2]$ *with respect to all the variables, then the original system* (2.2.28) *is absolutely exponentially* $\boldsymbol{y}$*-stable in the same angle.*

2. *If system* (2.2.29) *is absolutely exponentially stable in the angle* $[k_1, k_2]$ *with respect to all the variables and all the roots of the equation*

$$\det(L_{10}^* - \lambda I_{p-h}) = 0 \tag{2.2.30}$$

have negative real parts, then for sufficiently small ε *the system*

$$\dot{\boldsymbol{x}} = (A^* + \varepsilon \tilde{A}^*)\boldsymbol{x} + \boldsymbol{b}\varphi(\sigma), \qquad \sigma = (\boldsymbol{\beta} + \varepsilon \tilde{\boldsymbol{\beta}}^*)\boldsymbol{x} \tag{2.2.31}$$

($\tilde{A}^*$, $\tilde{\boldsymbol{\beta}}^*$ *are a constant matrix and a vector of dimensions* $n \times n$ *and* $1 \times n$, *respectively*) *will be absolutely exponentially stable in the angle* $[k_1, k_2]$.

(The additional matrix $\tilde{A}^*$ and vector $\tilde{\boldsymbol{\beta}}^*$ are introduced to extend the range of desired values of the coefficients of system (2.2.28).)

Proof. 1. The equalities $\xi_i = y_i$ $(i = 1, \dots, m)$ are valid in passing from the original system (2.2.28) to the $\boldsymbol{\mu}$-system (2.2.29). The conditions for absolute exponential stability with respect to all the variables in the angle $[k_1, k_2]$ for system (2.2.29), therefore, are sufficient for the original system (2.2.28) to be absolutely exponentially $\boldsymbol{y}$-stable in the angle $[k_1, k_2]$.

2. There is a linear nonsingular transformation $\boldsymbol{w} = L\boldsymbol{x}$ of the variables of system (2.2.28) that reduces this system to the form

$$\begin{aligned} &\dot{\boldsymbol{\xi}} = L_4^* A^* L_5^* \boldsymbol{\xi} + L_4^* \boldsymbol{b}\varphi(\sigma), \\ &\dot{\boldsymbol{\eta}} = C_1 \boldsymbol{\xi} + L_{10}^* \boldsymbol{\eta} + \boldsymbol{b}^* \varphi(\sigma), \qquad \sigma = \boldsymbol{\beta} L_5^* \boldsymbol{\xi}, \\ &\boldsymbol{\xi} \in R^{m+h}, \quad \boldsymbol{\eta} \in R^{n-m-h}, \quad \boldsymbol{w} = (\boldsymbol{\xi}, \boldsymbol{\eta}), \end{aligned} \tag{2.2.32}$$

where $C_1 = (C_2, L_8^*(L_2^*)^{-1})$ and $C_2, \boldsymbol{b}^*$ are, respectively, a matrix and vector of dimensions $(p-h) \times m$ and $(p-h) \times 1$ obtained from C and $\boldsymbol{b}^{(2)}$ by deleting their first h rows.

System (2.2.29) is absolutely exponentially stable. Allowing for the inequalities $k_1\sigma \le \varphi(\sigma) \le k_2\sigma$, the second group of equations (2.2.32) can be represented as

$$\begin{aligned} &\dot{\boldsymbol{\eta}} = L_{10}^* \boldsymbol{\eta} + \boldsymbol{f}(t), \\ &|\boldsymbol{f}(t)| = \left|C_1 \boldsymbol{\xi}(t) + \boldsymbol{b}^* \varphi(\sigma(t))\right| \le \alpha_1 \|\boldsymbol{\xi}_0\| e^{-\alpha_2(t-t_0)}, \end{aligned} \tag{2.2.33}$$

where α_i $(i = 1, 2)$ are positive constants.

Since all the roots of equation (2.2.30) have negative real parts, any solution of system (2.2.33) tends to zero exponentially for any admissible characteristic $\varphi(\sigma)$.

Consequently, system (2.2.32) is absolutely exponentially stable with respect to all the variables in the angle $[k_1, k_2]$. But in that case, the original system (2.2.28) and (for sufficiently small ε) system (2.2.31) possess the same property. The theorem is proved. □

Remark 2.2.5.

1. The decomposition of system (2.2.28) into two subsystems (2.2.29) and $\dot{\boldsymbol{\eta}} = L_{10}^* \boldsymbol{\eta}$ can be broken, generally speaking, even by small variations the

in coefficients of system (2.2.28). However, the conclusion about exponential stability drawn on the basis of this decomposition also remains valid for sufficiently small variations in the coefficients of the system.

2. Smets [1961] indicated that system (2.2.28) can describe the operation of a *nuclear reactor*. Vector $\boldsymbol{x}$ contains the variables characterizing the temperature of various components of the reactor's active zone .

Example 2.2.3. Consider the *critical case of two zero roots* when system (2.2.28) has the form

$$\dot{\boldsymbol{y}} = A\boldsymbol{y} + \boldsymbol{b}^{(1)}\varphi(\sigma),$$
$$\dot{z}_1 = b_1^{(2)}\varphi(\sigma), \quad \dot{z}_2 = b_2^{(2)}\varphi(\sigma), \quad \sigma = \boldsymbol{\beta}^{(1)}\boldsymbol{y} + \beta_1^{(2)}z_1 + \beta_2^{(2)}z_2. \tag{2.2.34}$$

Suppose that function $\varphi(\sigma)$ satisfies conditions (1) and (3) and condition (2) is replaced by $\sigma\varphi(\sigma) > 0$ $(\sigma \neq 0)$, commonly encountered in models of concrete control systems.

Since the equality $\sigma = 0$ is possible at points of type $\boldsymbol{y} = \boldsymbol{0}$, $z_i = z_i^*$ $(i = 1, 2)$, with z_i^* subject to the condition $\beta_1^{(2)}z_i^* + \beta_2^{(2)}z_2^*$, then system (2.2.34) cannot be absolutely stable with respect to all the variables. In this case the problem of stabilization of system (2.2.34) (with respect to all the variables) is said to be unsolvable by "*direct control.*"

Under these conditions consider the problem of absolute $\boldsymbol{y}$-stability of system (2.2.34) omitting, in general, the requirement that the convergence of the transient process should be exponential.

Let us introduce the new variable $\gamma\mu_1 = \beta_1^{(2)}z_1 + \beta_2^{(2)}z_2$ $(\gamma = \text{const} > 0)$. Then the auxiliary $\boldsymbol{\mu}$-system comprises the equations

$$\dot{\boldsymbol{y}} = A\boldsymbol{y} + \boldsymbol{b}^{(1)}\varphi(\sigma), \qquad \dot{\mu}_1 = b^*\varphi(\sigma),$$
$$\sigma = \boldsymbol{\beta}^{(1)}\boldsymbol{y} + \gamma\mu_1, \qquad b^* = \frac{\beta_1^{(2)}b_1^{(2)} + \beta_2^{(2)}b_2^{(2)}}{\gamma}. \tag{2.2.35}$$

The condition for absolute stability of system (2.2.35) with respect to all the variables (see, for example, Lefschetz [1965]) will be sufficient for the original system (2.2.34) to be absolutely $\boldsymbol{y}$-stable.

2.3 Damping (with Respect to Part of the Variables) of Angular Motions of an Asymmetric Solid

2.3.1 The System of Equations under Consideration

Consider *Euler's dynamical equations* describing the *angular motion of a solid (spacecraft) relative to the center of mass*

$$A\dot{x}_1 = (B - C)x_2x_3 + u_1,$$
$$B\dot{x}_2 = (C - A)x_1x_3 + u_2, \quad C\dot{x}_3 = (A - B)x_1x_2 + u_3. \tag{2.3.1}$$

Here A, B, and C are the principal central moments of inertia of the solid, x_i $(i = 1, 2, 3)$ are the projections of the instantaneous angular velocity vector

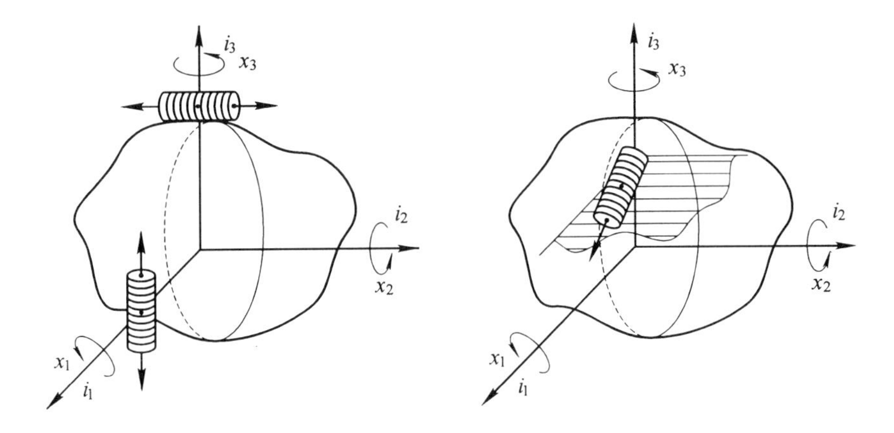

Figure 2.3.1. Positioning schemes of the engines.

of the body onto its principal central axes of inertia i_1, i_2, i_3, and u_i $(i = 1, 2, 3)$ are the controlling moments.

2.3.2 Some Known Results

The controlling moments

$$u_i = \alpha_i x_i \quad (\alpha_i = \text{const} < 0,\ i = 1, 2), \qquad u_3 \equiv 0 \tag{2.3.2}$$

ensure Lyapunov stability and asymptotic stability with respect to x_1, x_2 of the equilibrium position $x_1 = x_2 = x_3 = 0$ of the closed system (2.3.1), (2.3.2) for arbitrary $A \neq B \neq C$ satisfying the "*triangle inequality.*" The controlling moment

$$u_1 = \alpha_1 x_1, \qquad u_2 \equiv u_3 \equiv 0 \tag{2.3.3}$$

ensures (under the same assumptions about A, B, and C) Lyapunov stability and asymptotic stability with respect to x_1 of the equilibrium position $x_1 = x_2 = x_3 = 0$ of the closed system (2.3.1), (2.3.3).

Let us specify the positioning schemes of the engines realizing control rules (2.3.2) and (2.3.3). Rules (2.3.2) can be realized either with the help of two *fixed engines* I_1 and I_2 attached to the solid along axes i_3 and i_2, respectively, or with the help of a *vernier* (*swivel*) *engine* attached to the solid so that it can swivel in the plane perpendicular to the i_3-axis. Rule (2.3.3) can be realized with the help of a fixed engine I_2 (Figure 2.3.1). In both cases the thrust of the engines is assumed *continuously* variable.

2.3.3 Formulation of the Problem

Let us demonstrate that control action (2.3.3) ensures Lyapunov stability and asymptotic stability with respect to *two of three* variables x_i $(i = 1, 2, 3)$

of the equilibrium position $x_1 = x_2 = x_3 = 0$ of the closed system (2.3.1), (2.3.3). In other words, let us show that in the case $A \neq B \neq C$ it is possible to dampen angular rotations of a solid about its center of mass relative to two of three principal central axes of inertia using a *single* fixed engine I_1. (For comparison we note that, according to the results of Subsection 2.3.2, damping of rotations by engine I_1 is possible relative to only one of the axes of the ellipsoid of inertia.)

2.3.4 The Case of Small Initial Perturbations

Theorem 2.3.1. *If $B < A < C$ ($C < A < B$), then for any small initial perturbations the equilibrium position $x_1 = x_2 = x_3 = 0$ of the closed system (2.3.1), (2.3.3) is Lyapunov uniformly stable and uniformly asymptotically stable with respect to two of three variables x_i ($i = 1, 2, 3$). The domain of initial perturbations splits into three disjoint parts*

$$\text{(I)} \qquad \beta = \frac{C-A}{B}x_{30}^2 - \frac{A-B}{C}x_{20}^2 > 0 \qquad (\beta < 0), \tag{2.3.4}$$

$$\text{(II)} \qquad \beta < 0 \qquad (\beta > 0),$$

$$\text{(III)} \qquad \beta = 0$$

in which respective variables (x_1, x_2), (x_1, x_3), and (x_1, x_2, x_3) are asymptotically stable.

Proof. By introducing the new variable $\mu_1 = (B - C)x_2x_3/A$ we transform system (2.3.1), (2.3.3) as follows:

$$\begin{gathered}
\dot{x}_1 = \alpha_1^* x_1 + \mu_1, \qquad \dot{\mu}_1 = x_1 X(x_2, x_3), \\
\dot{x}_2 = \frac{C-A}{B}x_1x_3, \qquad \dot{x}_3 = \frac{A-B}{C}x_1x_2, \qquad \alpha_1^* = \frac{\alpha_1}{A}, \\
X(x_2, x_3) = \frac{(B-C)\left[C(C-A)x_3^2 + B(A-B)x_2^2\right]}{ABC}.
\end{gathered} \tag{2.3.5}$$

Let us examine the behavior of function $\Gamma(t) = X(x_2(t), x_3(t))$ along the trajectories of system (2.3.1), (2.3.3). To this end, we note that system (2.3.1), (2.3.3) possesses the *first integral*

$$V = \frac{C-A}{B}x_3^2 - \frac{A-B}{C}x_2^2 = \beta = \text{const}. \tag{2.3.6}$$

First assuming $\beta \neq 0$, let us show that when condition $C < A < B$ or $B < A < C$ holds along the trajectories of system (2.3.1), (2.3.3), the inequality $\Gamma(t) \leq -\gamma_0 = \text{const} < 0$ is valid. On the contrary, we assume that $\lim \Gamma(t) = 0$, $t \to \infty$ or there is a finite point in time $t = t_*$ such that $\Gamma(t_*) = 0$. Then, taking into account that $X(x_2, x_3)$ is a negative-definite function for all x_2, x_3, we arrive at $\lim x_i^2(t) = 0$, $t \to \infty$ ($i = 2, 3$), or $x_2(t_*) = x_3(t_*) = 0$. Consequently, equality (2.3.6) takes the form

$$0 = \lim V(t) = \beta = \text{const} \quad (t \to \infty)$$

or

$$0 = V(t_*) = \beta = \text{const},$$

which is impossible. The assumption made is thus invalid, and $\Gamma(t) \le -\gamma_0 < 0$ for all $t \ge t_0$.

Now the first two of equations (2.3.5) can be written as the $\boldsymbol{\mu}$-system

$$\dot{x}_1 = \alpha_1^* x_1 + \mu_1, \qquad \dot{\mu}_1 = \Gamma(t) x_1$$

or as a single equation

$$\ddot{x}_1 + p\dot{x}_1 + q(t)x_1 = 0, \quad p = -\alpha_1^*, \quad q(t) = -\Gamma(t) \ge \gamma_0. \tag{2.3.7}$$

Provided that

$$p > \sqrt{\Gamma_0} - \sqrt{\gamma_0}, \tag{2.3.8}$$

where $\Gamma_0 = \text{const} > 0$ is the upper bound of function $q(t)$, the solution $x_1 = \dot{x}_1 = 0$ of equation (2.3.7) is Lyapunov uniformly asymptotically stable (Merkin [1987]).

Let us verify whether condition (2.3.8) can be fulfilled in the case in question. Since the equilibrium position $x_1 = x_2 = x_3 = 0$ of system (2.3.1), (2.3.3) is (nonasymptotically) Lyapunov stable, in a sufficiently small neighborhood of the origin of coordinates, the estimate

$$\sqrt{\Gamma_0} - \sqrt{\gamma_0} < \varepsilon_1$$

is true, where ε_1 is a sufficiently small preassigned positive number. This implies that, when initial perturbations x_{i0} $(i = 1, 2, 3)$ (and $\varepsilon > 0$) are sufficiently small, inequality (2.3.8) holds for any fixed number p.

At the same time, solutions $x_1(t), \dot{x}_1(t)$ of equation (2.3.7) are solutions $x_1(t), \mu_1(t)$ of the original system (2.3.1), (2.3.3). This means that the equilibrium position $x_1 = x_2 = x_3 = 0$ of system (2.3.1), (2.3.3) possesses the following properties:

(a) for any ε, $t_0 \ge 0$, there exists $\delta(\varepsilon) > 0$ such that if $|x_{10}| < \delta$, $|x_{20}x_{30}| < \delta$, then

$$\big|x_1(t; t_0, \boldsymbol{x}_0)\big| < \varepsilon, \quad \big|x_2(t; t_0, \boldsymbol{x}_0)x_3(t; t_0, \boldsymbol{x}_0)\big| < \varepsilon \tag{2.3.9}$$

for all $t \ge t_0$;

(b) there exists $\Delta > 0$ such that

$$\lim x_1(t) = 0, \quad \lim\big|x_2(t)x_3(t)\big| = 0, \quad t \to \infty, \tag{2.3.10}$$

when $|x_{10}| < \Delta$, $|x_{20}x_{30}| < \Delta$.

Let us show that the relationships $|x_k(t)| < \varepsilon$, $\lim|x_k(t)| = 0$ $(t \to \infty)$, where $k = 2$ or $k = 3$ depending on the initial conditions, follow from (2.3.9), (2.3.10) for all $t \geq t_0$.

To determine the geometric boundary of the convergence domain of solutions of system (2.3.1), (2.3.3), we denote

$$x_2^2(t)x_3^2(t) = f(t) \tag{2.3.11}$$

and, omitting intermediate calculations, from (2.3.6), (2.3.11) find the following solutions:

$$\begin{aligned} x_2^2(t) &= -\frac{C\beta}{2(A-B)} + \sqrt{\frac{C^2\beta^2}{4(A-B)^2} + \frac{C(C-A)f(t)}{B(A-B)}}, \\ x_3^2(t) &= \frac{B\beta}{2(C-A)} + \sqrt{\frac{B^2\beta^2}{4(C-A)^2} + \frac{B(A-B)f(t)}{C(C-A)}} \end{aligned} \tag{2.3.12}$$

(solutions with a minus sign next to the radical are impossible). Now, with properties (2.3.9), (2.3.10) of functions $x_i(t)$ $(i = 1, 2)$ in mind, it is easy to verify that in the case $B < A < C$ $(C < A < B)$ the relationship $x_3^2(t) \to 0$ $(x_2^2(t) \to 0)$ is valid for $\beta < 0$ and $x_2^2(t) \to 0$ $(x_3^2(t) \to 0)$ is valid for $\beta > 0$. Thus, Theorem 2.3.1 is proved for $\beta \neq 0$.

Let us examine separately the case $\beta = 0$ when a finite time $t = t_*$ is possible such that $\Gamma(t_*) = 0$; the limiting relationship $\lim \Gamma(t) = 0$, $t \to \infty$, means the Lyapunov asymptotic stability of the equilibrium position $x_1 = x_2 = x_3 = 0$ of system (2.3.1), (2.3.3). Since the condition $\Gamma(t_*) = 0$ is equivalent to the condition $x_2(t_*) = x_3(t_*) = 0$ and, in addition, system (2.3.1), (2.3.3) has the solution $x_1 = x_1(t; t_0, \boldsymbol{x}_0)$, $x_2 \equiv x_3 \equiv 0$, then the conditions $x_1(t_*) \neq 0$, $x_2(t_*) = x_3(t_*) = 0$ determine the solution $x_1 = x_1(t; t_*, x_1(t_*))$, $x_2(t) \equiv x_3(t) \equiv 0$ for all $t \geq t_*$. This solution will be unique for the initial conditions specified because the conditions which guarantee the uniqueness of solutions are fulfilled for system (2.3.1), (2.3.3).

The equilibrium position $x_1 = x_2 = x_3 = 0$ of system (2.3.1), (2.3.3) is (nonasymptotically) Lyapunov stable, and function $x_1(t; t_*, x_1(t_*))$ obeys the differential equation $\dot{x}_1 = \alpha_1 x_1$ $(\alpha_1 = \text{const} < 0)$ for $t \geq t_*$. Therefore, the conditions $\beta = 0$, $\Gamma(t_*) = 0$ imply that the equilibrium position $x_1 = x_2 = x_3 = 0$ of system (2.3.1), (2.3.3) is Lyapunov asymptotically stable. In the case when $\beta = 0$, $\Gamma(t) \leq -\Gamma_0 = \text{const} < 0$, the same property of the equilibrium position of system (2.3.1), (2.3.3) follows from equalities (2.3.12). The theorem is proved. □

For the cases when $A < B < C$ $(A < C < B)$ or $C > B > A$ $(B > C > A)$, the following statement can be proved that is similar to the above one.

Corollary 2.3.1. *If $A < B < C$ $(A < C < B)$ or $C < B < A$ $(B < C < A)$, then for any small initial perturbations the equilibrium position $x_1 = x_2 = x_3 = 0$ of the closed system (2.3.1), (2.3.3) is Lyapunov uniformly stable and uniformly asymptotically stable with respect to two of three variables x_i*

$(i = 1, 2, 3)$. *The domain of initial perturbations splits into two parts in which respective variables* (x_1, x_2) *or* (x_1, x_3) *are asymptotically stable.*

Proof. We restrict the proof to the case when $A < B < C$ or $C < B < A$. Consider the set of initial perturbations x_{i0} $(i = 1, 2, 3)$ for which correspondent solutions $x_i(t; t_0, x_0)$ of system (2.3.1), (2.3.3) possess the property: there exists a finite (or nonfinite) point $t = t_*$ such that $x_1(t_*) = x_3(t_*) = 0$. Denote this set of initial perturbations by D_1 and the remaining set x_{i0} $(i = 1, 2, 3)$ by D_2.

Let us prove the asymptotic stability of solutions from D_1 with respect to x_1, x_3. By (2.3.6), equality $x_1(t_*) = x_3(t_*) = 0$ is equivalent to the equality $x_i(t_*) = 0$ $(i = 1, 2, 3)$ for $\beta = 0$, or equality $x_2(t_*) = x_2^* \neq 0$, $x_1(t_*) = x_3(t_*) = 0$ for $\beta \neq 0$. Since conditions which guarantee the uniqueness of solutions are fulfilled for system (2.3.1), (2.3.3), the identities $x_i(t) \equiv 0$ $(i = 1, 2, 3)$ $(\beta = 0)$ or $x_2(t) \equiv x_2^*$, $x_1(t) \equiv x_3(t) \equiv 0$ $(\beta \neq 0)$ will be true for all $t \geq t_*$. Because the equilibrium position of system (2.3.1), (2.3.3) is (nonasymptotically) Lyapunov stable, the asymptotic stability with respect to x_1, x_3 is proved.

Let us prove the asymptotic stability of solutions from D_2 with respect to x_1, x_2. To this end, introducing the new variable $\mu_2 = (C - A)x_1 x_3/B$, we transform system (2.3.1), (2.3.3) as follows:

$$\dot{x}_2 = \mu_2, \quad \dot{\mu}_2 = \alpha_1 \mu_2 + x_2 Y(x_1, x_3),$$

$$\dot{x}_1 = \alpha_1^* x_1 + \frac{(B - C)x_2 x_3}{A}, \quad \dot{x}_3 = \frac{(A - B)x_1 x_2}{C},$$

$$Y(x_1, x_3) = \frac{(C - A)\left[C(B - C)x_3^2 + A(A - B)x_1^2\right]}{ABC}.$$

In the case in question, the inequality $Y(x_1, x_3) = Q(t) \leq Q_0 = \text{const} < 0$ is fulfilled along the trajectories of this system, and its first two equations can be combined into a single equation:

$$\ddot{x}_2 + p\dot{x}_2 + q_1(t)x_2 = 0, \quad p = -\alpha_1^*, \quad q_1(t) \geq -Q_0 > 0.$$

The equation obtained is the same kind as (2.3.7), and therefore its solution $x_2 = \dot{x}_2 = 0$ is Lyapunov asymptotically stable. This means that the equilibrium position of system (2.3.1), (2.3.3) is asymptotically stable with respect to x_2, $(C - A)x_1 x_3/B$ and Lyapunov stable, so we conclude from the relationship $\dot{x}_1 = \alpha_1^* x_1 + (B - C)x_2 x_3/A$ that the equilibrium position of system (2.3.1), (2.3.3) is asymptotically stable with respect to x_1, x_2. The statement is proved. □

Remark 2.3.1. When $B < A < C$ or $C < A < B$, the question of whether initial perturbations belong to domains D_1 and D_2 is solved depending on the sign of the first integral β in (2.3.6). It follows from Corollary 2.3.1 that the equilibrium position $x_1 = x_2 = x_3 = 0$ of system (2.3.1), (2.3.3) is asymptotically stable with respect to function $Q^* = x_2 x_3$ for any sufficiently small initial perturbations x_{i0} $(i = 1, 2, 3)$. As the case with the proof of Theorem 2.3.1, it follows from (2.3.6), (2.3.11) that variables x_2, x_3 of system (2.3.1), (2.3.3) have the form (2.3.12). However, as easily verified, when $A < B < C$ $(A < C < B)$ or $C < B < A$ $(B < C < A)$, there are initial perturbations which belong simultaneously to domains D_1 and D_2 for both $\beta > 0$ and $\beta < 0$. Consequently, the question of geometrically characterizing domains D_1 and D_2 in Corollary 2.3.1 remains to be solved.

2.3.5 Arbitrary Initial Perturbations

Corollary 2.3.2. *If $B < A < C$ $(C < A < B)$, then the equilibrium position $x_1 = x_2 = x_3 = 0$ of the closed system (2.3.1), (2.3.3) is Lyapunov uniformly stable and uniformly asymptotically stable with respect to two of three variables x_i $(i = 1, 2, 3)$ in the domain*

$$\sqrt{Ax_{10}^2 + Bx_{20}^2 + Cx_{30}^2} < -\frac{\alpha_1^*}{\sqrt{l}},$$
$$l = \max\left\{\frac{(C-B)(C-A)}{ABC}, \frac{(C-B)(A-B)}{ABC}\right\}. \tag{2.3.13}$$

Domain (2.3.13) splits into three disjoint parts (2.3.4) in which respective variables (x_1, x_2), (x_1, x_3), and (x_1, x_2, x_3) are asymptotically stable.

Proof. If the inequality

$$-\alpha_1^* > \sqrt{\Gamma_0} - \sqrt{\gamma_0} \tag{2.3.14}$$

is valid for all t, x_i $(i = 1, 2, 3)$, then the solution $x_1 = \dot{x}_1 = 0$ of equation (2.3.7) is asymptotically stable with respect to both variables for any initial perturbations. Let us express inequality (2.3.14) in terms of initial conditions x_{i0} $(i = 1, 2, 3)$. To this end we note that function $V = Ax_1^2 + Bx_2^2 + Cx_3^2$ obeys the equation $\dot{V} = 2\alpha_1^* Ax_1^2$ by virtue of system (2.3.1), (2.3.3). Hence, the function $V(t) = Ax_1^2(t) + Bx_2^2(t) + Cx_3^2(t)$ is nonincreasing for all $t \geq t_0$. This means that $V(t) \leq V(t_0)$ and, consequently, function $\Gamma(t)$ admits the estimate

$$\begin{aligned} -\Gamma(t) &= \frac{(C-B)\big[C(C-A)x_3^2(t) + B(A-B)x_2^2(t)\big]}{ABC} \\ &\leq l\big[Cx_3^2(t) + Bx_2^2(t)\big] \leq lV(t) \leq lV(t_0) = \Gamma_0, \end{aligned} \tag{2.3.15}$$

for all $t \geq t_0$. Since inequality (2.3.15) will hold provided $(-\alpha_1^*) > \sqrt{\Gamma_0}$, then in the case when initial perturbations x_{i0} $i = (1, 2, 3)$ in system (2.3.1), (2.3.3) satisfy the condition

$$-\alpha_1^* > \sqrt{l}\sqrt{Ax_{10}^2 + Bx_{20}^2 + Cx_{30}^2},$$

the solution $x_1 = \dot{x}_1 = 0$ of equation (2.3.7) is asymptotically stable with respect to $x_1, \dot{x}_1$ and, consequently, the equilibrium position of system (2.3.1),

(2.3.3) is asymptotically stable with respect to $x_1, (B-C)x_2x_3/A$. The proof scheme further follows that for Theorem 2.3.1. The statement is proved. □

Remark 2.3.2.

1. By choosing a sufficiently large value of $|\alpha_1|$ in (2.3.3), one can force the fulfillment of the conditions of Corollary 2.3.2 for *any given* domain of initial perturbations.

2. The limiting value of the angular velocity of the solid (as $t \to \infty$) is determined by the equality $\omega^2 = \beta/E$, where $E = (C-A)/B$ or $E = (B-A)/C$, depending on about which axis of the ellipsoid of inertia the solid rotates. So, if the value β/E is sufficiently large, then when $B < A < C$ $(C < A < B)$, engine I_1 "*spins*" the solid about the major or minor (depending on the initial conditions) principal central axis of its ellipsoid of inertia. If the value β/E is small, engine I_1 practically suppresses rotations of the solid.

2.3.6 The Case of Nonlinear Feedback

Let us show that if function u_1 is chosen in the form of some nonlinear function of variables x_i $(i = 1,2,3)$ rather than in form (2.3.3), then the limiting relationship $\lim \boldsymbol{x}(t) = \mathbf{0}$, $t \to \infty$ is fulfilled with respect to *two specified variables* of x_i $(i = 1,2,3)$ for almost all initial perturbations from any bounded domain $\|\boldsymbol{x}_0\| \leq \Delta$, where Δ is a given number.

Theorem 2.3.2. *Let the control rules have the form*

$$u_1 = a_1x_1 + \frac{a_2x_2}{x_3} + \frac{a_3x_1^2x_2}{x_3} + a_4x_2x_3,$$
$$a_1 = A\Gamma_2, \quad a_2 = \frac{AB\Gamma_1}{C-A}, \quad a_3 = \frac{A(B-A)}{C}, \quad a_4 = C-B, \tag{2.3.16}$$
$$\Gamma_i = \text{const} < 0 \quad (i=1,2), \quad 4\Gamma_1 + \Gamma_2^2 > 0, \quad u_2 \equiv u_3 \equiv 0,$$

and let one of the following conditions hold

$$A > C > B, \quad B > C > A, \quad A > B > C, \quad C > B > A. \tag{2.3.17}$$

Then constants Γ_1, Γ_2 in rules (2.3.16) *can be chosen so that, for all finite x_{i0} $(i = 1,2,3)$, $x_{30} \neq 0$, solutions $x_i = x_i(t;t_0,\boldsymbol{x}_0)$ $(i = 1,2,3)$ of the closed system* (2.3.1), (2.3.16) *satisfy the relationships* $\lim x_j(t) = 0$, $t \to \infty$ $(j = 1,2)$. *In this case*

$$|u_1| \leq L = \text{const} > 0, \quad \lim|u_1| = 0 \quad (t \to \infty). \tag{2.3.18}$$

Proof. By direct calculations we ascertain that the following $\boldsymbol{\mu}$*-system*

$$\dot{x}_2 = \mu_1, \quad \dot{\mu}_1 = \Gamma_1x_2 + \Gamma_2\mu_1 \quad \left(\mu_1 = \frac{(C-A)x_1x_3}{B}\right) \tag{2.3.19}$$

can be singled out of the closed system (2.3.1), (2.3.16).

Under assumptions relative to Γ_1, Γ_2, the solutions of system (2.3.19) and, consequently, variables $x_2(t), \mu_1(t)$ of system (2.3.1), (2.3.16) satisfy the conditions

$$x_2(t) \to 0, \quad \mu_1(t) \to 0, \quad t \to \infty, \tag{2.3.20}$$

where zero is approached in (2.3.20) in accordance with an exponential law.

For system (2.3.1), (2.3.16) there is a first integral (2.3.6), and, consequently, the equality

$$x_3^2(t) = x_{30}^2 - \frac{(A-B)B}{(C-A)C}x_{20}^2 + \frac{(A-B)B}{(C-A)C}x_2^2(t) \tag{2.3.21}$$

holds for all $t \geq t_0$. Directly integrating system (2.3.19), we can verify that variables Γ_1, Γ_2 in (2.3.16) can be chosen so that, for all finite x_{i0} ($i = 1,2,3$), $x_{30} \neq 0$ and for all $t \geq t_0$, the relationships

$$\begin{gathered} \lim x_3^2(t) = \omega^2 = x_{30}^2 - \frac{(A-B)B}{(C-A)C}x_{20}^2, \quad t \to \infty, \\ x_3^2(t) \geq \varepsilon > 0, \end{gathered} \tag{2.3.22}$$

hold, where ε is a sufficiently small constant.

Provided (2.3.22) is fulfilled, it follows from the condition $\mu_1(t) \to 0$, $t \to \infty$, that $x_1(t) \to 0$, $t \to \infty$, for all x_{i0} ($i = 1,2,3$), $x_{30} \neq 0$, where zero is approached in accordance with an exponential law. In addition, if (2.3.22) holds, estimates (2.3.18) are valid. The theorem is proved. □

Remark 2.3.3.

1. Relations (2.3.20) and (2.3.22) imply that control rules (2.3.16) "*spin*" a solid about the i_3-axis with angular velocity $\omega^2 = x_{30}^2 - (A-B)Bx_{20}^2/(C-A)C$ that depends on x_{20}, x_{30} and does not depend on x_{10}.

2. Conditions (2.3.17) include the case $A > B > C$ when the "spinning" is about the *major* axis of the ellipsoid of inertia of the solid. This case is of interest in the ***passive stabilization of satellites***. The cases $C > B > A$ and $B > C > A$, $A > C > B$ mean "spinning" about the minor and medium axes of the ellipsoid of inertia of the solid, respectively.

3. Control rules (2.3.16) can be realized (when $x_{30}^2 \neq 0$) by using engine I_1 of low or limited (depending on the value of initial perturbations) capacity.

Remark 2.3.4. From the standpoint of the general principles of automatic control, the control problems in this section fall under the case of so-called *single-channel control.*

2.4 A Method of Nonlinear Transformation of the Variables in Investigating Stability with Respect to Part of the Variables in a Linear Approximation (2)

2.4.1 Preliminary Remarks

(1) The possibilities for forming an auxiliary linear system which arise when representing nonlinear perturbations Y_i $(i = 1, \dots, m)$ in form (2.2.2), (2.2.3) may be insufficient for solving the problem of stability of the unperturbed motion $\boldsymbol{x} = \mathbf{0}$ of system (2.1.1) with respect to $y_1, \dots, y_m$, using the approach described in Section 2.2.

Indeed, let the equations of perturbed motion (2.1.1) have the form

$$\begin{aligned} \dot{y}_1 &= -y_1 + y_2 z_1 z_2, & \dot{y}_2 &= -2y_2, \\ \dot{z}_1 &= 2z_1 - y_1 z_1, & \dot{z}_2 &= -z_2 + y_1 z_2. \end{aligned} \tag{2.4.1}$$

Following the recommendations of Section 2.2, let us introduce the new variable $\mu_1 = z_1 z_2$. So the $\boldsymbol{\mu}$-system comprises the equations

$$\dot{y}_1 = -y_1 + y_2 \mu_1, \qquad \dot{y}_2 = -2y_2, \qquad \dot{\mu}_1 = \mu_1. \tag{2.4.2}$$

Since there is a positive root among the roots of the characteristic equation of the linear part of $\boldsymbol{\mu}$-system (2.4.2), the results of Section 2.2 do not apply here.

But if the $\boldsymbol{\mu}$-system for the original system (2.4.1) is constructed by using the new variable $\mu_2 = y_2 z_1 z_2$, we arrive at the equations

$$\dot{y}_1 = -y_1 + \mu_2, \qquad \dot{\mu}_2 = -\mu_2, \qquad \dot{y}_2 = -2y_2 \tag{2.4.3}$$

whose correspondent characteristic equation has roots which are all real and negative.

The zero solution $y_1 = y_2 = \mu_2 = 0$ of the auxiliary $\boldsymbol{\mu}$-system (2.4.3) is Lyapunov asymptotically stable, and the unperturbed motion $y_1 = y_2 = z_1 = z_2 = 0$ of the original system (2.4.1) is asymptotically stable with respect to y_1, y_2. At the same time the zero solution of $\boldsymbol{\mu}$-system (2.4.2) is Lyapunov unstable.

(2) On the other hand, the extent to which we are close to the necessary conditions for stability with respect to part of the variables obtained by the proposed approach also depends on the way of constructing the auxiliary linear system.

Indeed, the domain

$$D_1 = \{a_1 < 0,\ a_2 < 0\} \cap D_1^*, \quad D_1^* = \{d_1 + d_2 < 0\}.$$

of the *sufficient* conditions for asymptotic stability with respect to y_1, y_2 of the unperturbed motion $y_1 = y_2 = z_1 = z_2 = 0$ of the system

$$\begin{aligned} \dot{y}_1 &= a_1 y_1 + y_2 z_1 z_2, & \dot{y}_2 &= a_2 y_2, \\ \dot{z}_1 &= d_1 z_1 - y_1 z_1, & \dot{z}_2 &= d_2 z_2 + y_1 z_2 \end{aligned}$$

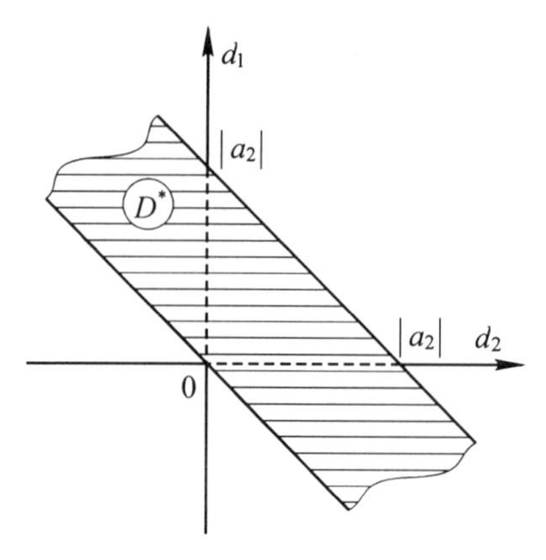

Figure 2.4.1. Region $D^* = D_2 \setminus D_1$.

(with coefficients a_i, d_i ($i = 1, 2$) constant) can be found by constructing the auxiliary $\boldsymbol{\mu}$-system

$$\dot{y}_1 = a_1 y_1 + y_2 \mu_1, \quad \dot{y}_2 = a_2 y_2, \quad \dot{\mu}_1 = (d_1 + d_2)\mu_1.$$

At the same time the domain

$$D_2 = \{a_1 < 0,\ a_2 < 0\} \cap D_2^*, \quad D_2^* = \{a_2 + d_1 + d_2 < 0\}$$

of the *necessary and sufficient* conditions for asymptotic stability with respect to y_1, y_2 is obtained by constructing the linear $\boldsymbol{\mu}$-system with the following structure:

$$\dot{y}_1 = a_1 y_1 + \mu_2, \quad \dot{y}_2 = a_2 y_2, \quad \dot{\mu}_2 = (a_2 + d_1 + d_2)\mu_2.$$

Domain D_2 is wider than domain D_1, and the size of domain $D^* = D_2 \setminus D_1$ depends on the value $|a_2|$ (Figure 2.4.1).

2.4.2 An Original System of Equations. Separating Special Forms of the Variables from Nonlinear Perturbations

Let us consider system (2.1.1) taking $b_{il} = 0$ ($i = 1, \ldots, m$; $l = 1, \ldots, p$). To extend the possibilities for constructing auxiliary linear systems stated in Section 2.2, we represent nonlinear perturbations Y_i ($i = 1, \ldots, m$) in system (2.1.1) in the form

$$Y_i(t, \boldsymbol{y}, \boldsymbol{z}) = Y_i^0(\boldsymbol{y}, \boldsymbol{z}) + Y_i^{**}(t, \boldsymbol{y}, \boldsymbol{z}) \qquad (i = 1, \ldots, m),$$

where

$$Y_i^0(\boldsymbol{y}, \boldsymbol{z}) = U_2^{(i)}(\boldsymbol{y}, \boldsymbol{z}) + \cdots + U_r^{(i)}(\boldsymbol{y}, \boldsymbol{z})$$

and functions $U_v^{(i)}(\boldsymbol{y}, \boldsymbol{z})$ are *homogeneous forms* of variables $y_1, \ldots, y_m, z_1, \ldots, z_p$ of order v, $v \le r$, with r a *finite* number.

The essence of the above representation is that we separate forms Y_i^0 of finite order in variables $\boldsymbol{y}, \boldsymbol{z}$ from Y_i $(i = 1, \ldots, m)$ to use them in constructing an auxiliary linear system. As in Section 2.2, the arbitrariness in the choice of Y_i^0 $(i = 1, \ldots, m)$ should be used, to rationalize the search for the most acceptable solution.

2.4.3 Constructing an Auxiliary System

To construct an auxiliary system of the linear approximation for equations (2.1.1), let us introduce the new variables

$$\mu_i = \sum_{v=2}^{r} U_v^{(i)}(\boldsymbol{y}, \boldsymbol{z}) \qquad (i = 1, \ldots, m) \tag{2.4.4}$$

which, without loss of generality, are assumed linear independent.

With the new variables thus introduced, two cases are possible.

The first case. System (2.1.1) is reduced to the form

$$\begin{aligned}
\frac{dy_i}{dt} &= \sum_{k=1}^{m} a_{ik} y_k + \mu_i + Y_i^{**}(t, \boldsymbol{y}, \boldsymbol{z}), \\
\frac{d\mu_j}{dt} &= \sum_{v=2}^{r} U_v^{(j)*}(\boldsymbol{y}, \boldsymbol{z}) + Z_j^*(t, \boldsymbol{y}, \boldsymbol{z}) = \sum_{k=1}^{m} e_{jk} \mu_k + Z_j^*(t, \boldsymbol{y}, \boldsymbol{z}), \\
\frac{dz_s}{dt} &= \sum_{k=1}^{m} c_{sk} y_k + \sum_{l=1}^{p} d_{sl} z_l + Z_s(t, \boldsymbol{y}, \boldsymbol{z}), \quad Z_j^* = \sum_{k=1}^{m} \frac{\partial Y_j^0}{\partial y_k} Y_k + \sum_{l=1}^{p} \frac{\partial Y_j^0}{\partial z_l} Z_l \\
&(i, j = 1, \ldots, m;\ s = 1, \ldots, p),
\end{aligned} \tag{2.4.5}$$

where e_{jk} are constants and $U_v^{(j)*}$ are homogeneous forms of variables $y_1, \ldots, y_m, z_1, \ldots, z_p$ of the order v, $v \le r$. In this case one can separate the linear system

$$\frac{dy_i}{dt} = \sum_{k=1}^{m} a_{ik} y_k + \mu_i, \quad \frac{d\mu_j}{dt} = \sum_{k=1}^{m} e_{jk} \mu_k \quad (i, j = 1, \ldots, m) \tag{2.4.6}$$

from system (2.4.5), where the former is a linear approximation for the first two groups of equations (2.4.5). Such a system will be termed the ***μ**-system of the linear approximation for the original system* (2.1.1).

The second case. After introducing the new variables (2.4.4), system (2.1.1) is not reduced to equations (2.4.5). Suppose that only the first m_1 equalities hold in the second group of equations of system (2.4.5) and the remaining $m - m_1$ such equalities are impossible. Then on introducing the new variables (2.4.4), system (2.1.1) is transformed as follows:

$$\frac{dy_i}{dt} = \sum_{k=1}^{m} a_{ik} y_k + \mu_i + Y_i^{**}(t, \boldsymbol{y}, \boldsymbol{z}),$$

$$\frac{d\mu_j}{dt} = \sum_{k=1}^{m} e_{jk} \mu_k + Z_j^{*}(t, \boldsymbol{y}, \boldsymbol{z}),$$

$$\frac{d\mu_\varepsilon}{dt} = \sum_{v=2}^{r} U_v^{(\varepsilon)*}(\boldsymbol{y}, \boldsymbol{z}) + Z_\varepsilon^{*}(t, \boldsymbol{y}, \boldsymbol{z}),$$

$$\frac{dz_s}{dt} = \sum_{k=1}^{m} a_{sk} y_k + \sum_{l=1}^{p} d_{sl} z_l + Z_s(t, \boldsymbol{y}, \boldsymbol{z})$$

$$(i = 1, \ldots, m;\ j = 1, \ldots, m_1;\ \varepsilon = m_1 + 1, \ldots, m;\ s = 1, \ldots, p).$$

In this case, once again let us introduce the new variables

$$\mu_\varepsilon = \sum_{v=2}^{r} U_v^{(\varepsilon)*}(\boldsymbol{y}, \boldsymbol{z}) \qquad (\varepsilon = m_1 + 1, \ldots, m).$$

Since the new variables of type (2.4.4) required to construct an auxiliary $\boldsymbol{\mu}$-system are homogeneous forms of finite order in variables $\boldsymbol{y}, \boldsymbol{z}$, they can be represented as

$$\mu_i = \sum_{v=2}^{r} \left(\sum_{\alpha_1 + \cdots + \alpha_n = v} q_i^{(\alpha_1 \ldots \alpha_n)} y_1^{\alpha_1} \ldots y_m^{\alpha_m} z_1^{\alpha_{m+1}} \ldots z_p^{\alpha_n} \right) \qquad (i = 1, \ldots, m),$$

where $q_i^{(\alpha_1 \ldots \alpha_n)}$ are constants.

We associate the vectors composed of properly arranged numbers $q_i^{(\alpha_1 \ldots \alpha_n)}$ characterizing the additional variables with these variables. We continue introducing the new variables until their associated linearly independent vectors constitute quadratic nonsingular matrices. Then further introducing new variables is meaningless because each subsequent newly introduced variable will be expressible as a linear combination of the previous ones. But this means that, when successively repeated, the indicated procedure of introducing new variables enables always transforming system (2.1.1) into a system of the type (2.4.5) from which the linear system can be separated to give the $\boldsymbol{\mu}$-system of the linear approximation for system (2.2.1).

2.4.4 The Condition for Stability with Respect to Part of the Variables

Theorem 2.4.1. *Let all the roots of the characteristic equation of $\boldsymbol{\mu}$-system (2.4.6) have negative real parts. Then the unperturbed motion $\boldsymbol{x} = \boldsymbol{0}$ of system (2.1.1) will be uniformly asymptotically stable with respect to $y_1, \ldots, y_m$, provided its nonlinear terms satisfy the following conditions in the domain $t \geq 0$,*

$\|\boldsymbol{y}\| \leq H,\ \|\boldsymbol{\mu}\| \leq H,\ \|\boldsymbol{z}\| < \infty$:

$$\left|Y_i^{**}(t,\boldsymbol{y},\boldsymbol{z})\right| + \left|Z_i^{*}(t,\boldsymbol{y},\boldsymbol{z})\right| \leq \sum_{k=1}^{m} \alpha_{ik}^{(1)} |y_k| + \sum_{k=1}^{m} \alpha_{ik}^{(2)} |\mu_k| \qquad (2.4.7)$$
$$(i = 1, \ldots, m),$$

where $\alpha_{ik}^{(s)}$ *and* H *are sufficiently small positive constants.*

Proof. The conditions of the theorem imply that, for $\boldsymbol{\mu}$-system (2.4.6), one can specify a Lyapunov function $V(\boldsymbol{\xi})$ of variables $\xi_i = y_i$, $\xi_{m+j} = \mu_j$ $(i, j = 1, \ldots, m)$ satisfying conditions (2.2.12). Then, differentiating this function along the trajectories of system (2.1.1)

$$\dot{V}_{(2.1.1)} = \dot{V}_\mu + \sum_{k=1}^{m} \frac{\partial V}{\partial y_k} Y_k^{**}(t,\boldsymbol{y},\boldsymbol{z}) + \sum_{k=1}^{m} \frac{\partial V}{\partial \mu_k} Z_k^{*}(t,\boldsymbol{y},\boldsymbol{z}),$$

from (2.2.12) and (2.4.7) we obtain the following estimate in the domain $t \geq 0$, $\|\boldsymbol{y}\| \leq H$, $\|\boldsymbol{\mu}\| \leq H$, $\|\boldsymbol{z}\| < \infty$:

$$V_{(2.1.1)} \leq -c\|\boldsymbol{\xi}\|^2 + \gamma\|\boldsymbol{\xi}\|^2 \leq -\bar{c}\|\boldsymbol{\xi}\|^2,$$

where γ and c are, respectively, a sufficiently small and some positive constants. By Corollary 0.4.1 we conclude that the unperturbed motion of system (2.1.1) is uniformly asymptotically stable with respect to $y_1, \ldots, y_m$. The theorem is proved. □

If estimates of type (2.4.7) cannot be established, one can employ the techniques given below to possibly obtain the desired ones.

(1) Each of the expressions

$$Y_i^0(\boldsymbol{y},\boldsymbol{z}) \qquad (i = 1, \ldots, m)$$

is represented as an aggregate of *several* new variables μ_{ij} (rather than a *single* variable μ_i) and variables $y_1, \ldots, y_m$. Like the basic procedure, this one always leads to a system, where the structure of (2.4.5) contains the linear part of type (2.4.6), and increases the possibility of establishing estimates of type (2.4.7).

(2) Separating one or another combination of nonlinear terms from Y_i^{**}, Z_j^{*} $(i, j = 1, \ldots, m)$ (and taking them as new variables), one can proceed to introduce new variables until an acceptable solution is obtained. Again, a system with the structure of (2.4.5) is constructed for which the possibility increases that estimates of type (2.4.7) will hold.

For example, by entering the new variable $\mu_1 = y_1^2 z_1$, the equations

$$\dot{y}_1 = -3y_1 + y_1^2 z_1, \qquad \dot{z}_1 = 2y_1 + z_1 \qquad (2.4.8)$$

are not reduced to a system closed relative to y_1, μ_1 (i.e., to the $\boldsymbol{\mu}$-system) because the equation $\dot{\mu}_1 = -5\mu_1 + 2y_1^3 + 2y_1^3 z_1^2$ includes the term $2y_1^3 z_1^2$ which cannot be (properly) expressed via y_1, μ_1. However, using the procedures (1) and (2), we can form the following auxiliary $\boldsymbol{\mu}$-systems:

1. $\dot{y}_1 = -3y_1 + y_1\mu_1,$
$\dot{\mu}_1 = -2\mu_1 + 2y_1^2 + \mu_1^2$
$(\mu_1 = y_1 z_1);$

2. $\dot{y}_1 = -3y_1 + \mu_1,$
$\dot{\mu}_1 = -5\mu_1 + 2y_1^3 + 2\mu_1\mu_2,$
$\dot{\mu}_2 = -2\mu_2 + 2y_1^2 + \mu_2^2$
$(\mu_1 = y_1^2 z_1, \quad \mu_2 = y_1 z_1).$

In either case the zero solution of the auxiliary system is uniformly Lyapunov asymptotically stable. Consequently, the unperturbed motion $y_1 = z_1 = 0$ of system (2.4.8) is uniformly asymptotically stable with respect to y_1, though *Lyapunov unstable*.

Example 2.4.1.

1. Let the equations of perturbed motion (2.1.1) with the form

$$\begin{aligned} \dot{y}_1 &= ay_1 + by_1^k(y_1 z_1)^r + Y_1^{**}(t, y_1, z_1), \\ \dot{z}_1 &= cy_1 + dz_1 + Z_1(t, y_1, z_1) \end{aligned} \tag{2.4.9}$$

include nonlinear perturbations which are assumed analytic functions with continuous bounded coefficients in domain (0.3.2). In these equations, a, b, c, and d are constants, and k and r are integral numbers such that $k \geq 2,\ r \geq 1,\ k > r$. The unperturbed motion $y_1 = z_1 = 0$ of system (2.4.9) is *Lyapunov unstable* for $d > 0$.

Consider the question of the asymptotic stability of this motion with respect to y_1. To this end, entering the new variable $\mu_1 = y_1 z_1$, let us form the system

$$\begin{aligned} \dot{y}_1 &= ay_1 + by_1^k \mu_1^r + Y_1^{**}(t, y_1, z_1), \\ \dot{\mu}_1 &= (a + d)\mu_1 + Z_1^*(t, y_1, z_1), \\ \dot{z}_1 &= cy_1 + dz_1 + Z_1(t, y_1, z_1), \\ Z_1^* &= cy_1^2 + by_1^{k-1}\mu_1^{r+1} + Z_1^{**}, \quad Z_1^{**} = y_1 Z_1 + z_1 Y_1^{**}. \end{aligned} \tag{2.4.10}$$

Let the following estimate be valid in the domain $t \geq 0,\ |y_1| \leq H,\ |\mu_1| \leq H,\ |z_1| < \infty$:

$$\left|Y_1^{**}(t, y_1, z_1)\right| + \left|Z_1^{**}(t, y_1, z_1)\right| \leq \alpha_1 |y_1| + \alpha_2 |y_1 z_1|,$$

where $\alpha_i\ (i = 1, 2)$ are sufficiently small positive constants. Then, provided $a < 0$, $a + d < 0$, the solution $y_1 = \mu_1 = z_1 = 0$ of system (2.4.10) is uniformly asymptotically stable with respect to y_1, μ_1 for all c and b. Consequently, the unperturbed motion $y_1 = z_1 = 0$ of system (2.4.9) is uniformly asymptotically stable with respect to y_1.

2. The system of differential equations

$$\begin{aligned} \dot{y}_1 &= \left(\frac{C - A}{B}\right) z_1 z_2 + u_1, \\ \dot{y}_2 &= -y_1 y_3 + z_2\sqrt{1 - y_2^2 - y_3^2}, \\ \dot{y}_3 &= y_1 y_2 - z_1\sqrt{1 - y_2^2 - y_3^2}, \\ \dot{z}_1 &= \left(\frac{B - C}{A}\right) y_1 z_2 + u_2, \quad \dot{z}_2 = \left(\frac{A - B}{C}\right) y_1 z_1 + u_3 \end{aligned} \tag{2.4.11}$$

describes the *angular motion of a spacecraft about its center of mass* in the neighborhood of the equilibrium position $y_1 = y_2 = y_3 = z_1 = z_2 = 0$ in which the spacecraft is given a uniaxial attitude in the inertial space. Here A, B, and C are the principal central moments of inertia of the spacecraft, y_1, z_1, and z_2 are the projections of the angular velocity vector of the spacecraft onto its principal central axes of inertia i_1, i_2, i_3, and u_i $(i = 1, 2, 3)$ are the controlling moments, realized by special engines.

Taking u_i as

$$u_j = \frac{f_i(\boldsymbol{y}, \boldsymbol{z}, \boldsymbol{u}^*)}{\sqrt{1 - y_2^2 - y_3^2}} \qquad (j = 2, 3),$$
$$u_1 = \left(\frac{A - C}{B}\right) z_1 z_2 + u_1^*,$$

forms f_j $(j = 2, 3)$ can be specified so as to separate the following auxiliary linear $\boldsymbol{\mu}$-system from system (2.4.11)

$$\dot{y}_1 = u_1^*, \qquad \ddot{y}_j = u_j^* \quad (j = 2, 3). \tag{2.4.12}$$

If auxiliary control rules u_i^* $(i = 1, 2, 3)$ are such that the zero solution $y_i = 0$ $(i = 1, 2, 3)$ and $\dot{y}_2 = \dot{y}_3 = 0$ of $\boldsymbol{\mu}$-system (2.4.12) is Lyapunov uniformly asymptotically stable, then $y_1 = y_2 = y_3 = z_1 = z_2 = 0$ of the original system (2.4.11) will be uniformly asymptotically stable with respect to y_i $(i = 1, 2, 3)$. Besides solving the equations of system (2.4.11) which involve $\dot{y}_1$ and $\dot{y}_2$ for z_1, z_2,

$$z_1 = \frac{-\dot{y}_3 + y_1 y_2}{\sqrt{1 - y_2^2 - y_3^2}}, \qquad z_2 = \frac{\dot{y}_2 + y_1 y_3}{\sqrt{1 - y_2^2 - y_3^2}},$$

we conclude that uniform asymptotic $\boldsymbol{y}$-stability actually means *Lyapunov* uniform asymptotic stability of the equilibrium position of the original system (2.4.11).

The scheme indicated for constructing $\boldsymbol{\mu}$-systems is employed in investigating a number of applied problems of stabilization and control considered in Chapters 4 and 5.

2.5 Stability with Respect to Part of the Variables in a Nonlinear Approximation

2.5.1 Formulation of the Problem

Let the linear part of system (2.1.2) be missing, i.e., let this system have the form

$$\frac{dy_i}{dt} = Y_i(t, \boldsymbol{y}, \boldsymbol{z}), \quad \frac{dz_j}{dt} = Z_j(t, \boldsymbol{y}, \boldsymbol{z}) \quad (i = 1, \ldots, m;\ j = 1, \ldots, p), \tag{2.5.1}$$

where Y_i and Z_j are nonlinear perturbations.

Suppose that, for system (2.5.1), there exist N continuously differentiable functions $\mu_i = \mu_i(\boldsymbol{x})$, $\mu_i(\boldsymbol{0}) \equiv 0$, continuous functions F_j and Φ_j

$(j = 1, \ldots, N)$, and m functions φ_i and ψ_i satisfying the condition $Y_i = \varphi_i + \psi_i$ such that

$$\begin{gathered}\frac{dy_i}{dt} = \varphi_i(\boldsymbol{y}, \boldsymbol{\mu}) + \psi_i(t, \boldsymbol{y}, \boldsymbol{\mu}, \boldsymbol{z}), \\ \frac{d\mu_j}{dt} = F_j(\boldsymbol{y}, \boldsymbol{\mu}) + \Phi_j(t, \boldsymbol{y}, \boldsymbol{\mu}, \boldsymbol{z}), \quad \frac{dz_s}{dt} = Z_s(t, \boldsymbol{y}, \boldsymbol{z}) \\ (i = 1, \ldots, m; \ j = 1, \ldots, N; \ s = 1, \ldots, p).\end{gathered} \tag{2.5.2}$$

Also suppose that, for the nonlinear subsystem

$$\frac{dy_i}{dt} = \varphi_i(\boldsymbol{y}, \boldsymbol{\mu}), \quad \frac{d\mu_j}{dt} = F_j(\boldsymbol{y}, \boldsymbol{\mu}) \quad (i = 1, \ldots, m; \ j = 1, \ldots, N), \tag{2.5.3}$$

the following conditions are satisfied.

(1) Functions φ_i and F_j are *homogeneous forms* of order $r \geq 2$ in variables $\boldsymbol{y}$, $\boldsymbol{\mu}$. This means that the following equalities are valid:

$$\begin{gathered}\varphi_i(c\boldsymbol{y}, c\boldsymbol{\mu}) = c^r \varphi_i(\boldsymbol{y}, \boldsymbol{\mu}), \quad F_j(c\boldsymbol{y}, c\boldsymbol{\mu}) = c^r F_j(\boldsymbol{y}, \boldsymbol{\mu}) \\ (i = 1, \ldots, m; \ j = 1, \ldots, p).\end{gathered}$$

(2) In the domain $t \geq 0$, $(\|\boldsymbol{y}\|^2 + \|\boldsymbol{\mu}\|^2)^{1/2} \leq H$, $\|\boldsymbol{z}\| < \infty$, the following estimate is true:

$$\sum_{i=1}^{m} |\psi_i(t, \boldsymbol{y}, \boldsymbol{\mu}, \boldsymbol{z})| + \sum_{j=1}^{N} |\Phi_j(t, \boldsymbol{y}, \boldsymbol{\mu}, \boldsymbol{z})| \leq \alpha(\|\boldsymbol{y}\|^2 + \|\boldsymbol{\mu}\|^2)^{r/2}, \tag{2.5.4}$$

where α and H are sufficiently small positive constants.

Under the assumptions made, system (2.5.3) will be called the auxiliary *$\boldsymbol{\mu}$-system of the nonlinear approximation of* rth *order* for the original nonlinear system (2.5.1).

Let us demonstrate that the problem of asymptotic stability of the unperturbed motion $\boldsymbol{x} = \boldsymbol{0}$ of system (2.5.1) with respect to $y_1, \ldots, y_m$ can be solved by analyzing the zero solution of the $\boldsymbol{\mu}$-system of nonlinear approximation (2.5.3) for Lyapunov asymptotic stability.

2.5.2 The Condition for Asymptotic Stability with Respect to Part of the Variables

Theorem 2.5.1. *Suppose that*

(*a*) *conditions* (1) *and* (2) *from Subsection* 2.5.1 *are satisfied*; *and*

(*b*) *the zero solution* $\boldsymbol{y} = \boldsymbol{0}$, $\boldsymbol{\mu} = \boldsymbol{0}$ *of system* (2.5.3) *is Lyapunov asymptotically stable.*

Then the unperturbed motion $\boldsymbol{x} = \boldsymbol{0}$ *of system* (2.5.1) *is asymptotically stable with respect to* $y_1, \ldots, y_m$.

Proof. If condition (1) from Subsection 2.5.1 is satisfied and the zero solution $\boldsymbol{y} = \boldsymbol{0}$, $\boldsymbol{\mu} = \boldsymbol{0}$ of system (2.5.3) is Lyapunov asymptotically stable, then there exist functions $V(\boldsymbol{y}, \boldsymbol{\mu})$ and $W(\boldsymbol{y}, \boldsymbol{\mu})$ such that (Zubov [1975])

(a) V and W are positive-definite in Lyapunov's sense (with respect to variables $\boldsymbol{y}, \boldsymbol{\mu}$);
(b) V and W are positively homogeneous of orders l and $l+r-1$, respectively;
(c) $\dot{V} = -W$ by virtue of system (2.5.3).

Differentiating the V-function by virtue of the original system (2.5.1), we arrive at the equality

$$\sum_{i=1}^{m} \frac{\partial V}{\partial y_i}(\varphi_i + \psi_i) + \sum_{j=1}^{N} \frac{\partial V}{\partial \mu_j}(F_j + \Phi_j) = W + W_1(t, \boldsymbol{y}, \boldsymbol{\mu}, \boldsymbol{z}).$$

If condition (2.5.4) holds true, the function $W+W_1$ will be positive-definite with respect to $\boldsymbol{y}$, $\boldsymbol{\mu}$ in Rumyantsev's sense. Therefore, the unperturbed motion $\boldsymbol{x} = \boldsymbol{0}$ of system (2.5.1) is asymptotically stable with respect to $y_1, \ldots, y_m$ by Corollary 0.4.1. The theorem is proved. □

Remark 2.5.1.

1. Conditions for asymptotic stability in the nonlinear approximation for the problem of stability with respect to *all* the variables were obtained in the early 50s by Malkin and Krasovskii (Krasovskii [1959]) and by Zubov [1975]; they were developed in the work by Aleksandrov [1996]. Theorem 2.5.1 is an extension of these conditions to the problem of *partial* stability.

2. The choice of μ_i-functions in constructing of systems of type (2.5.3) can be performed in a *nonunique* way. It is well to bear this circumstance in mind to expand the regions of asymptotic stability of the unperturbed motion $\boldsymbol{x} = \boldsymbol{0}$ of system (2.5.1) with respect to $y_1, \ldots, y_m$, obtained with this approach.

Example 2.5.1.

1. Let the equations of perturbed motion (2.5.1) have the form

$$\begin{aligned} \dot{y}_1 &= ay_1^3 + y_1(z_1 z_2)^2 + Y_1^{**}(t, y_1, z_1, z_2), \\ \dot{z}_i &= \Sigma_i + Z_i(t, y_1, z_1, z_2) \qquad (i = 1, 2), \\ (\Sigma_1 &= z_1^3 z_2^2 - y_1^2 z_1, \quad \Sigma_2 = (b-1) z_1^2 z_2^3), \end{aligned} \tag{2.5.5}$$

where a and b are constants. By introducing the new variable $\mu_1 = z_1 z_2$, we form the system

$$\dot{y}_1 = ay_1^3 + y_1\mu_1^2 + Y_1^{**}(t, y_1, z_1, z_2),$$
$$\dot{\mu}_1 = -y_1^2\mu_1 + b\mu_1^3 + Z_1^*(t, y_1, z_1, z_2), \quad \dot{z}_i = \Sigma_i + Z_i(t, y_1, z_1, z_2) \ (i = 1, 2).$$

The derivative $\dot{V}$ of the function $V = y_1^2 + \mu_1^2$ by virtue of the system

$$\dot{y}_1 = ay_1^3 + y_1\mu_1^2, \quad \dot{\mu}_1 = -y_1^2\mu_1 + b\mu_1^3 \tag{2.5.6}$$

satisfies the condition $\dot{V} = 2(ay_1^4 + b\mu_1^4)$. Therefore, if $a < 0$, $b < 0$, then the zero solution $y_1 = \mu_1 = 0$ of system (2.5.6) is asymptotically stable by the Lyapunov theorem.

If the condition

$$|Y_1^{**}| + |Z_1^*| \le \alpha\big[|y_1| + |z_1 z_2|\big]^3$$

holds true, system (2.5.6) is a μ-system of the nonlinear approximation of the third order for the original nonlinear system (2.5.5).

So, the unperturbed motion $y_1 = z_1 = z_2 = 0$ of system (2.5.5) is asymptotically stable with respect to y_1 by Theorem 2.5.1.

2. Consider the *motion of an artificial satellite in a circular orbit.* Suppose that the satellite is acted on only by gravitational forces that can be brought to a single resultant. In this case, the perturbed motion of the satellite center of mass is described by the equations (Merkin [1987])

$$\begin{gathered}
\dot{y}_1 = y_2, \quad \dot{y}_2 = (r_0 + y_1)[z_2^2 + (\omega + z_3)^2 \cos^2 z_1] - \frac{\varkappa}{(r_0 + y_1)^2}, \\
\dot{z}_1 = z_2, \quad \dot{z}_2 = -\frac{2y_2 z_2}{r_0 + y_1} - \frac{1}{2}(\omega + z_3)^2 \sin 2z_1, \\
\dot{z}_3 = -\frac{2y_2(\omega + z_3)}{r_0 + y_1} + 2z_2(\omega + z_3)\tan z_1 \quad (\varkappa = \omega^2 r_0^3),
\end{gathered} \tag{2.5.7}$$

where y_1 (y_2) is a perturbation of the magnitude (derivative of the magnitude) of the satellite radius vector r_0 in the orbit, z_1 (z_2) is a variable characterizing perturbation (velocity of perturbation) of the orbital plane, and z_3 is a perturbation of the magnitude ω of the orbital velocity.

By introducing the new variable

$$\mu_1 = (r_0 + y_1)[z_2^2 + (\omega + z_3)^2 \cos^2 z_1] - \frac{\varkappa}{(r_0 + y_1)^2},$$

we construct the auxiliary μ-system

$$\dot{y}_1 = y_2, \quad \dot{y}_2 = \mu_1, \quad \dot{\mu}_1 = -\frac{3y_2\mu_1}{r_0 + y_1} - \frac{\varkappa y_2}{(r_0 + y_1)^3}. \tag{2.5.8}$$

The dimension of the auxiliary μ-system (2.5.8) is less than that of the original system (2.5.7). When analyzing the zero solution $y_1 = y_2 = \mu_1 = 0$ of the auxiliary μ-system for Lyapunov stability, we encounter the *critical case of a single zero root and a pair of pure imaginary roots.* Such a case has been thoroughly studied in the general theory of stability (see, for example, Kamenkov [1971]).

It can be shown that the zero solution $y_1 = y_2 = \mu_1 = 0$ of the auxiliary μ-system is Lyapunov stable. Hence, the unperturbed motion $y_1 = y_2 = z_1 = z_2 = z_3 = 0$ of the original system (2.5.7) is stable with respect to y_1, y_2.

2.6 Stability with Respect to Part of the Variables in Lyapunov Critical Cases

2.6.1 Formulation of the Problem

Consider the system of differential equations of perturbed motion

$$\dot{\boldsymbol{y}} = A\boldsymbol{y} + B\boldsymbol{z} + \boldsymbol{Y}(t, \boldsymbol{y}, \boldsymbol{z}), \quad \dot{\boldsymbol{z}} = C\boldsymbol{y} + D\boldsymbol{z} + \boldsymbol{Z}(t, \boldsymbol{y}, \boldsymbol{z}), \tag{2.6.1}$$

where A, B, C, and D are constant matrices of appropriate sizes and $\boldsymbol{Y}$ and $\boldsymbol{Z}$ are nonlinear perturbations.

Suppose that the right-hand sides of system (2.6.1) are continuous in t and satisfy the Lipschitz condition with respect to $\boldsymbol{x}$ in the domain $t \geq 0$, $\|\boldsymbol{x}\| \leq H$. In addition, let nonlinear perturbations $\boldsymbol{Y}, \boldsymbol{Z}$ satisfy the conditions

$$\boldsymbol{Y}(t,\boldsymbol{0},\boldsymbol{0}) \equiv \boldsymbol{Y}(t,\boldsymbol{0},\boldsymbol{z}) \equiv \boldsymbol{0}, \quad \boldsymbol{Z}(t,\boldsymbol{0},\boldsymbol{0}) \equiv \boldsymbol{Z}(t,\boldsymbol{0},\boldsymbol{z}) \equiv \boldsymbol{0}, \tag{2.6.2}$$

$$\frac{\|\boldsymbol{Y}(t,\boldsymbol{y},\boldsymbol{z})\| + \|\boldsymbol{Z}(t,\boldsymbol{y},\boldsymbol{z})\|}{\|\boldsymbol{y}\|} \underset{(t\geq t_0)}{\rightrightarrows} 0, \quad \|\boldsymbol{y}\| + \|\boldsymbol{z}\| \to 0. \tag{2.6.3}$$

Also consider a more general (as regards nonlinear perturbations) system

$$\dot{\boldsymbol{y}} = A\boldsymbol{y} + B\boldsymbol{z} + \boldsymbol{Y}^0(\boldsymbol{z}) + \boldsymbol{Y}(t,\boldsymbol{y},\boldsymbol{z}), \quad \dot{\boldsymbol{z}} = C\boldsymbol{y} + D\boldsymbol{z} + \boldsymbol{Z}(t,\boldsymbol{y},\boldsymbol{z}), \tag{2.6.4}$$

where components $Y_i^0(\boldsymbol{z})$ $(i = 1,\ldots,m)$ of vector function $\boldsymbol{Y}^0(\boldsymbol{z})$ satisfy condition (2.2.3) and $\boldsymbol{Y}$ and $\boldsymbol{Z}$ satisfy conditions (2.6.2), (2.6.3).

Let us analyze whether the property of the Lyapunov stability and that of the asymptotic $\boldsymbol{y}$-stability of the zero solution of system (1.1.6) will be carried over to the unperturbed motion $\boldsymbol{y} = \boldsymbol{0}, \boldsymbol{z} = \boldsymbol{0}$ of nonlinear systems (2.6.1), (2.6.4).

2.6.2 Conditions for Simultaneous Lyapunov Stability and Asymptotic Stability with Respect to Part of the Variables

Theorem 2.6.1. *Let the solution $\boldsymbol{y} = \boldsymbol{0}$, $\boldsymbol{z} = \boldsymbol{0}$ of system (1.1.6) be exponentially asymptotically $\boldsymbol{y}$-stable and Lyapunov stable. Then this property is valid for*

(1) *the unperturbed motion $\boldsymbol{y} = \boldsymbol{0}$, $\boldsymbol{z} = \boldsymbol{0}$ of system (2.6.1);*

(2) *the unperturbed motion $\boldsymbol{y} = \boldsymbol{0}$, $\boldsymbol{z} = \boldsymbol{0}$ of system (2.6.4) if, in addition, all the elements of matrix B are zeros and all the roots of the characteristic equation of $\boldsymbol{\mu}$-system for equations*

$$\dot{\boldsymbol{y}} = A\boldsymbol{y} + \boldsymbol{Y}^0(\boldsymbol{z}), \qquad \dot{\boldsymbol{z}} = D\boldsymbol{z} \tag{2.6.5}$$

have negative real parts.

Proof. The first part of the theorem strengthens the familiar results (Malkin [1966], Oziraner [1975], Rumyantsev and Oziraner [1987]) which assume that matrices B, D or matrix B, respectively, are zeros. The second part of the theorem treats a wider (as compared to the results mentioned) class of nonlinear perturbations.

Let us demonstrate that if system (2.6.1) or (2.6.4) can be transformed by the procedures worked out of constructing an auxiliary $\boldsymbol{\mu}$-system, then the results of the above publications remain valid for the case under consideration.

1. Let us transform the linear part of system (2.6.1) using the change of variables $\boldsymbol{w} = L\boldsymbol{x}$ (L is defined in Subsection 1.1.4). We arrive at the equations

$$\dot{\boldsymbol{\xi}} = L_4 A^* L_5 \boldsymbol{\xi} + \boldsymbol{Y}^*(t,\boldsymbol{\xi},\boldsymbol{\eta}), \quad \dot{\boldsymbol{\eta}} = C_1\boldsymbol{\xi} + C_2\boldsymbol{\eta} + \boldsymbol{Z}^*(t,\boldsymbol{\xi},\boldsymbol{\eta}), \tag{2.6.6}$$

where $\boldsymbol{w} = (\boldsymbol{\xi}, \boldsymbol{\eta})$, C_1 and C_2 are constant matrices of appropriate sizes, and the solution $\boldsymbol{\xi} = \boldsymbol{0}, \boldsymbol{\eta} = \boldsymbol{0}$ of the linear part of system (2.6.6) is exponentially asymptotically $\boldsymbol{\xi}$-stable and Lyapunov stable. Components of vector functions $\boldsymbol{Y}^*, \boldsymbol{Z}^*$ are either those of vector functions $\boldsymbol{Y}, \boldsymbol{Z}$ or their linear combinations. Hence

$$\|\boldsymbol{Y}^*(t,\boldsymbol{\xi},\boldsymbol{\eta})\| \le \alpha_1 \|\boldsymbol{Y}(t,\boldsymbol{y},\boldsymbol{z})\|, \quad \|\boldsymbol{Z}^*(t,\boldsymbol{y},\boldsymbol{z})\| \le \alpha_2 \|\boldsymbol{Z}(t,\boldsymbol{y},\boldsymbol{z})\|,$$
$$\alpha_i = \text{const} > 0, \quad i = 1,2$$

and, consequently, the following estimate holds:

$$\frac{\|\boldsymbol{Y}(t,\boldsymbol{\xi},\boldsymbol{\eta})\| + \|\boldsymbol{Z}^*(t,\boldsymbol{\xi},\boldsymbol{\eta})\|}{\|\boldsymbol{\xi}\|} \le \alpha_3 \frac{\|\boldsymbol{Y}(t,\boldsymbol{y},\boldsymbol{z})\| + \|\boldsymbol{Z}(t,\boldsymbol{y},\boldsymbol{z})\|}{\|\boldsymbol{y}\|}, \tag{2.6.7}$$
$$\alpha_3 = \max(\alpha_1, \alpha_2).$$

Variables $\boldsymbol{\xi}, \boldsymbol{\eta}$ are uniquely expressible in terms of $\boldsymbol{y}$, $\boldsymbol{z}$, and the limiting relationship $\|\boldsymbol{\xi}\| + \|\boldsymbol{\eta}\| \to 0$ is valid if and only if $\|\boldsymbol{y}\| + \|\boldsymbol{z}\| \to 0$. This means that (2.6.3), (2.6.7) yield the condition

$$\frac{\|\boldsymbol{Y}^*(t,\boldsymbol{\xi},\boldsymbol{\eta})\| + \|\boldsymbol{Z}^*(t,\boldsymbol{\xi},\boldsymbol{\eta})\|}{\|\boldsymbol{\xi}\|} \rightrightarrows 0, \quad \|\boldsymbol{\xi}\| + \|\boldsymbol{\eta}\| \to 0. \tag{2.6.8}$$

Besides, by (2.6.2), the following identities are valid

$$\boldsymbol{Y}^*(t,\boldsymbol{0},\boldsymbol{\eta}) \equiv \boldsymbol{0}, \qquad \boldsymbol{Z}^*(t,\boldsymbol{0},\boldsymbol{\eta}) \equiv \boldsymbol{0}. \tag{2.6.9}$$

Consequently, the solution $\boldsymbol{\xi} = \boldsymbol{0}, \boldsymbol{\eta} = \boldsymbol{0}$ of system (2.6.6) is Lyapunov stable and exponentially asymptotically $\boldsymbol{\xi}$-stable, and the unperturbed motion of system (2.6.1) is Lyapunov stable and exponentially asymptotically $\boldsymbol{y}$-stable.

2. Using the procedure outlined in Section 2.2, one can form the equations

$$\dot{\boldsymbol{y}} = A\boldsymbol{y} + \boldsymbol{\mu}, \qquad \dot{\boldsymbol{\mu}} = D^*\boldsymbol{\mu} \tag{2.6.10}$$

for system (2.6.5). Here D^* is a constant matrix, and solution $\boldsymbol{y} = \boldsymbol{0}, \boldsymbol{\mu} = \boldsymbol{0}$ of system (2.6.10) is exponentially asymptotically stable (with respect to all the variables). With these equations, the original system (2.6.4) is transformed as follows:

$$\dot{\boldsymbol{y}} = A\boldsymbol{y} + \boldsymbol{\mu} + \boldsymbol{Y}(t,\boldsymbol{y},\boldsymbol{z}), \qquad \dot{\boldsymbol{\mu}} = D^*\boldsymbol{\mu} + \boldsymbol{Y}^*(t,\boldsymbol{y},\boldsymbol{z}),$$
$$\dot{\boldsymbol{z}} = C\boldsymbol{y} + D\boldsymbol{z} + \boldsymbol{Z}(t,\boldsymbol{y},\boldsymbol{z}), \tag{2.6.11}$$
$$\boldsymbol{Y}^*(t,\boldsymbol{y},\boldsymbol{z}) = \frac{\partial \boldsymbol{Y}^0(\boldsymbol{z})}{\partial \boldsymbol{z}}(C\boldsymbol{y} + \boldsymbol{Z}(t,\boldsymbol{y},\boldsymbol{z})), \quad \frac{\partial \boldsymbol{Y}^0}{\partial \boldsymbol{z}} = \left(\frac{\partial \boldsymbol{Y}^0}{\partial z_1}, \dots, \frac{\partial \boldsymbol{Y}^0}{\partial z_p}\right),$$

and the solution $\boldsymbol{y} = \boldsymbol{0}, \boldsymbol{\mu} = \boldsymbol{0}, \boldsymbol{z} = \boldsymbol{0}$ of the system of the linear approximation for (2.6.11) is Lyapunov stable and exponentially $(\boldsymbol{y}, \boldsymbol{\mu})$-stable. Since

$$\|\boldsymbol{Y}^*(t,\boldsymbol{y},\boldsymbol{z})\| \le \|C\| \left\|\frac{\partial \boldsymbol{Y}^0}{\partial \boldsymbol{z}}\right\| \|\boldsymbol{y}\| + \left\|\frac{\partial \boldsymbol{Y}^0}{\partial \boldsymbol{z}}\right\| \|\boldsymbol{Z}(t,\boldsymbol{y},\boldsymbol{z})\|,$$

where $\|C\|$ is a norm of matrix C, then, for $t \geq t_0$, we obtain the estimate

$$\frac{\|\boldsymbol{Y}(t,\boldsymbol{y},\boldsymbol{z})\| + \|\boldsymbol{Y}^*(t,\boldsymbol{y},\boldsymbol{z})\| + \|\boldsymbol{Z}(t,\boldsymbol{y},\boldsymbol{z})\|}{\|\boldsymbol{\xi}\|} \leq \frac{\|\boldsymbol{Y}(t,\boldsymbol{y},\boldsymbol{z})\| + \|\boldsymbol{Z}(t,\boldsymbol{y},\boldsymbol{z})\|}{\|\boldsymbol{y}\|}$$
$$+ \frac{\left\|\dfrac{\partial \boldsymbol{Y}^0}{\partial \boldsymbol{z}}\right\| \|\boldsymbol{Z}(t,\boldsymbol{y},\boldsymbol{z})\|}{\|\boldsymbol{y}\|} + \left\|\frac{\partial \boldsymbol{Y}^0}{\partial \boldsymbol{z}}\right\| \|C\|, \quad \boldsymbol{\xi} = (\boldsymbol{y}, \boldsymbol{\mu}). \quad (2.6.12)$$

As components $Y_i^0(\boldsymbol{z})$ $(i = 1,\dots,m)$ of vector function $\boldsymbol{Y}^0(\boldsymbol{z})$ satisfy condition (2.2.3), then, in the domain $t \geq 0$, $\|\boldsymbol{x}\| \leq H$, all the components of vector function $\partial \boldsymbol{Y}^0/\partial \boldsymbol{z}$ are bounded and $(\partial \boldsymbol{Y}^0/\partial \boldsymbol{z}) \rightrightarrows \boldsymbol{0}$ for $\|\boldsymbol{z}\| \to 0$. Thus, for $t \geq t_0$, conditions (2.6.12), (2.6.3) yield the relationship

$$\frac{\|\boldsymbol{Y}(t,\boldsymbol{y},\boldsymbol{z})\| + \|\boldsymbol{Y}^*(t,\boldsymbol{y},\boldsymbol{z})\| + \|\boldsymbol{Z}(t,\boldsymbol{y},\boldsymbol{z})\|}{\|\boldsymbol{\xi}\|} \rightrightarrows 0, \quad \|\boldsymbol{y}\| + \|\boldsymbol{z}\| \to 0.$$

Moreover, by (2.6.2), $\boldsymbol{Y}^*(t,\boldsymbol{0},\boldsymbol{z}) \equiv \boldsymbol{0}$, and therefore the unperturbed motion of system (2.6.4) is Lyapunov stable and exponentially asymptotically $\boldsymbol{y}$-stable. The theorem is proved. □

Remark 2.6.1.

1. In Theorem 2.6.1, the exponential asymptotic $\boldsymbol{y}$-stability of solution $\boldsymbol{y} = \boldsymbol{0}, \boldsymbol{z} = \boldsymbol{0}$ of systems (1.1.6) and (2.6.5) is important and also the form of variables constituting the $\boldsymbol{\mu}$-system. In this case conditions (2.6.2) in the first part of the theorem, in particular, can be replaced with *weaker* conditions (2.6.9).

2. With the results of Sections 1.2 and 1.3 and the results obtained by Rumyantsev and Oziraner [1987], Theorem 2.6.1 can be extended to the case when the elements of matrices A, B, C, and D *depend on* t.

Example 2.6.1.

1. Consider the system

$$\begin{aligned}
&\dot{y}_1 = -y_1 + z_1\xi^2 + Y_1(y_1, z_1, \dots, z_4),\\
&\dot{z}_i = \Sigma_i + Z_i(y_1, z_1, \dots, z_4) \quad (i = 1,\dots,4),\\
&(\Sigma_1 = -\xi, \quad \Sigma_2 = z_1 - z_2, \quad \Sigma_3 = -4z_1 + z_2 - z_3 - z_4,\\
&\qquad \Sigma_4 = 5z_1 + z_2 + 2z_3 + 2z_4, \quad \xi = 2z_1 + z_2 + z_3 + z_4),
\end{aligned} \quad (2.6.13)$$

with nonlinear perturbations Y_1, Z_i $(i = 1,\dots,4)$ satisfying conditions of type (2.6.2), (2.6.3) in the domain $\|\boldsymbol{x}\| \leq H$.

The characteristic equation for the system of the linear approximation for (2.6.13) has three roots with negative real parts and two zero roots with simple elementary divisors. So in studying the Lyapunov stability of the unperturbed motion $y_1 = z_1 = z_2 = z_3 = z_4 = 0$, we encounter the *critical case of two zero roots.*

System (2.6.13) has the structure of (2.6.4). In this case subsystem (2.6.5) comprises the equations

$$\dot{y}_1 = -y_1 + z_1\xi^2, \qquad \dot{z}_i = \Sigma_i \quad (i = 1,\dots,4). \quad (2.6.14)$$

The auxiliary $\boldsymbol{\mu}$-system for equations (2.6.14) has the form

$$\dot{y}_1 = -y_1 + \mu_1, \qquad \dot{\mu}_1 = \mu_2, \qquad \dot{\mu}_2 = -6\mu_1 - 5\mu_2. \tag{2.6.15}$$

Since all the roots of the characteristic equation of $\boldsymbol{\mu}$-system (2.6.15) have negative real parts, then system (2.6.13) satisfies all the conditions of the second part of Theorem 2.6.1. Therefore the unperturbed motion $y_1 = z_1 = z_2 = z_3 = z_4 = 0$ of system (2.6.13) is Lyapunov stable and exponentially asymptotically stable with respect to y_1.

2. The motion of a *holonomic mechanical system* with normal coordinates $x_1, \ldots, x_n$ under the action of linear gyroscopic, dissipative, and potential forces and of general nonlinear forces subject to constraints invariant in time is described by the equations

$$\begin{gathered} \ddot{x}_i = -a_i x_i - b_i \dot{x}_i + \sum_{j=1}^{n} g_{ij}\dot{x}_j + X_i(\boldsymbol{x}, \dot{\boldsymbol{x}}), \\ g_{ij} = -g_{ji}, \quad g_{ii} = 0, \quad (i, j = 1, \ldots, n). \end{gathered} \tag{2.6.16}$$

Here constants a_i, b_i, g_{ij} $(i, j = 1, \ldots, n)$ characterize the influence of linear potential, dissipative, and gyroscopic forces, respectively. Assuming that linear dissipative forces do not dissipate completely, for definiteness, we take

$$b_i > 0 \quad (i = 1, \ldots, m), \qquad b_j = 0 \quad (j = m+1, \ldots, n). \tag{2.6.17}$$

Because conditions (2.6.17) and $a_i > 0$ $(i = 1, \ldots, n)$ are fulfilled, the function $2V = \sum \dot{x}_i^2 + \sum a_i x_i^2$ (where i runs from 1 to n), by virtue of the *linear part* of system (2.6.16), satisfies the inequality

$$\dot{V} \leq -b^*(\dot{x}_1^2 + \cdots + \dot{x}_m^2), \quad b^* = \min(b_1, \ldots, b_m).$$

Consequently, the equilibrium position of this linear system is Lyapunov stable and exponentially asymptotically stable with respect to $\dot{x}_1, \ldots, \dot{x}_m$.

If nonlinear forces $X_i(\boldsymbol{x}, \dot{\boldsymbol{x}})$ $(i = 1, \ldots, n)$ vanish identically for $\dot{x}_1 = \cdots = \dot{x}_m = 0$, then the equilibrium position $x_i = \dot{x}_i = 0$ $(i = 1, \ldots, n)$ of the original nonlinear system (2.6.16) will also be Lyapunov stable and exponentially asymptotically stable with respect to $\dot{x}_1, \ldots, \dot{x}_m$ by Theorem 2.6.1.

Note that system (2.6.16) does not satisfy the conditions required in the work of Malkin [1966], Oziraner [1975], and Rumyantsev and Oziraner [1987], as in our case matrix B in (2.6.1) contains nonzero elements.

3. Consider an *asymmetric solid (spacecraft) fitted out with a symmetric flywheel whose rotational axis is fixedly attached along one of the principal central axes of inertia of the solid.* This mechanical system falls into the type of *gyrostats.* The angular motion of this system about its center of mass is described by the nonlinear system of differential equations

$$\begin{aligned} &(A - A_1)\dot{x}_1 = (B - C)x_2 x_3 - u_1, \quad && B\dot{x}_2 = (C - A)x_1 x_3 - A_1 x_3 \dot{\varphi}, \\ &C\dot{x}_3 = (A - B)x_1 x_2 + A_1 x_2 \dot{\varphi}, && A_1(\ddot{\varphi} + \dot{x}_1) = u_1. \end{aligned} \tag{2.6.18}$$

Here A, B, and C are the principal central moments of inertia of the gyrostat; x_i $(i = 1, 2, 3)$ are the projections of the angular velocity vector of the body onto its principal central axes of inertia i_1, i_2, i_3; A_1, φ are the axial moment of inertia and

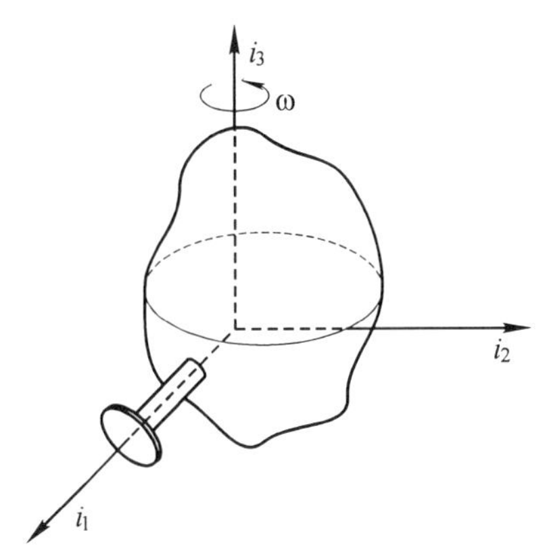

Figure 2.6.1. An asymmetric solid fitted out with a symmetric flywheel whose rotational axis is fixedly attached along the principal central axis of inertia i_1 of the solid.

the angular velocity of the proper rotation of the flywheel and u_1 is the controlling moment applied to the flywheel. Equations (2.6.18) correspond to the case when the flywheel rotation axis coincides with the i_1-axis.

Equations (2.6.18) admit the particular solution

$$x_1 = x_2 = 0, \quad x_3 = \omega = \text{const} > 0, \quad \dot{\varphi} = 0, \quad u_1 = 0, \tag{2.6.19}$$

corresponding to the permanent rotation ("spinning") of the solid with constant angular velocity ω about the i_3-axis. This implies that the flywheel is static (the control engine is turned off) and the angular momentum vector $\boldsymbol{K}$ of the gyrostat is aligned with the i_3-axis.

Suppose that mode (2.6.19) of the permanent rotation of the gyrostat is violated by some perturbations. To stabilize it let us introduce the new variables $y_j = x_j$ $(j = 1, 2)$, $y_3 = \dot{\varphi}$, $z_1 = x_3 - \omega$ and compose the system of equations of perturbed motion in deviations from (2.6.19)

$$\begin{aligned}(A - A_1)\dot{y}_1 &= (B - C)y_2(z_1 + \omega) - u_1,\\ B\dot{y}_2 &= \big[(C - A)y_1 - A_1 y_3\big](z_1 + \omega),\\ (A - A_1)\dot{y}_3 &= (C - B)y_2(z_1 + \omega) + AA_1^{-1}u_1,\\ C\dot{z}_1 &= \big[(A - B)y_1 + A_1 y_3\big]y_2.\end{aligned} \tag{2.6.20}$$

Solution (2.6.19) is the unperturbed motion $y_1 = y_2 = y_3 = z_1 = 0$ of system (2.6.20).

Consider the problem of stabilizing the unperturbed motion of system (2.6.20) with respect to variables y_1, y_2, y_3 by the control u_1. The stabilization with respect to y_1, y_2 will mean damping of small precessional and nutational oscillations of the angular momentum vector $\boldsymbol{K}$ of the gyrostat with respect to axes attached to the solid. Such a problem is of interest in spacecraft dynamics where it is important to "spin" a spacecraft about one (as a rule, *major*) of the principal central axes of inertia.

Let us describe the mechanical essence of stabilization with respect to y_1, y_2, y_3. To do so, we note that, by using the flywheel, the total angular momentum of

the gyrostat *cannot be changed* but it *can be "redistributed."* So in the process of stabilization with respect to y_1, y_2, two cases are possible:
(1) the flywheel "*takes up*" the perturbations of the angular momentum of the gyrostat caused by precessional and nutational oscillations;
(2) the flywheel just "*transfers*" (without "taking up") these perturbations into "extra rotation" of the gyrostat about the axis of rotation i_3.

The stabilization of the unperturbed motion $y_1 = y_2 = y_3 = z_1 = 0$ of system (2.6.20) with respect to variables y_1, y_2, y_3 will correspond to case (2). In this case the stabilization with respect to variable y_3 acts as if it determined the technical means for solving the problem of stabilization with respect to y_1, y_2.

Now we turn to solving the indicated problem of stabilizing the unperturbed motion of system (2.6.20) with respect to y_1, y_2, y_3.

Let us obtain the solution in two stages.

1st stage. Let us solve the problem of stabilizing the zero solution with respect to y_1, y_2, y_3 for the equations of the *linear approximation* for system (2.6.20). This can be done by control rule u_1 which is linear in these variables, since the linear subsystem

$$\begin{gathered}(A - A_1)\dot{y}_1 = (B - C)\omega y_2 - u_1, \quad B\dot{y}_2 = \big[(C - A)y_1 - A_1 y_3\big]\omega, \\ (A - A_1)\dot{y}_3 = (C - B)\omega y_2 + AA_1^{-1}u_1,\end{gathered} \tag{2.6.21}$$

describing the behavior of variables y_1, y_2, y_3 of the linear part of system (2.6.20), is completely controllable. (The Kalman criterion for complete controllability of system (2.6.21) holds for all the cases except for $B = C$ which we exclude from consideration because of asymmetry of the gyrostat.)

2nd stage. Let us show that control rule u_1, constructed for the equations of the linear approximation, actually guarantees that the unperturbed motion $y_1 = y_2 = y_3 = z_1 = 0$ of system (2.6.20) is (nonasymptotically) Lyapunov stable and asymptotically stable with respect to y_1, y_2, y_3. Indeed, the right-hand sides of the equations of system (2.6.20) vanish at $y_1 = y_2 = y_3 = 0$. Therefore the indicated property takes place by the first part of Theorem 2.6.1.

The linear (in variables y_1, y_2, y_3) control rule u_1 thus constructed solves the nonlinear problem of stabilizing the unperturbed motion $y_1 = y_2 = y_3 = z_1 = 0$ of system (2.6.20) with respect to y_1, y_2, y_3. In this case the flywheel just "transfers" (without "taking up") the specified perturbations into "extra rotation" of the gyrostat about the axis of rotation i_3.

Technically, the stabilization proceeds as follows. While the gyrostat is in the steady rotation, the flywheel is static (control engine is turned off). When nutational and precessional oscillations of vector $\boldsymbol{K}$ occur, controlling moment u_1 (formed according to the feedback principle as a linear function of variables y_1, y_2, y_3) is applied to the flywheel and makes the flywheel rotate. The flywheel in turn, dampens the precessional and nutational oscillations of vector $\boldsymbol{K}$ also passing to the state of rest.

2.7 Overview of References

Fundamental results concerning stability with respect to all the variables in the linear approximation were obtained by Lyapunov [1892], the founder of the modern theory of stability.

Corduneanu [1971] and Oziraner [1973] were the first to analyze the *partial* asymptotic stability of nonlinear systems in the linear approximation based on the method of Lyapunov functions.

A number of issues concerning the problems of partial stability and instability in the linear approximation and in critical cases are presented in the work by Lutsenko and Stadnikova [1973], Prokop'ev [1975], Vu Tuan [1980], Shchennikov [1985], Rumyantsev and Oziraner [1987], and Lizunova [1991].*

The matter of Chapter 2 reflects another approach to this problem which was proposed in the late 70s (Vorotnikov [1979a], [1979b], [1986a], [1988a], [1991a], [1993]). This approach rests on nonlinear transformations of the variables of an original nonlinear system of equations. Early studies (Vorotnikov [1979a], [1979b]) require z-boundedness of solutions of the system under investigation. Later, this requirement was removed.

We note that the proposed procedure for constructing auxiliary systems does not imply that nonlinear terms of a system are subdivided into "resonant" and "nonresonant." This essentially distinguishes the proposed approach from the procedure of *normalization* of nonlinear systems, which goes back to the work of Henri Poincaré.

Below are some endnotes relevant to the problems Chapter 2 deals with.

The problem of partial absolute stability, considered in Section 2.2, was posed in the work by Vorotnikov [1979b], [1991a]. In relation to this problem, it is pertinent to note the work by Wada et al. [1995] which studies a more general problem of *parametric absolute stability in output* for the Lur'e systems.

Section 2.3 discusses the problem of damping the rotational motions of an asymmetric solid with a single engine fixed relative to the solid's axes (Vorotnikov [1986a], [1991a]). Stated differently (the case of a "skew-fixed" engine**), this problem was analyzed using alternative methods in the work by Agrachev and Sarychev [1983], Akulenko [1987], [1994a], and Kovalev [1991].*** Akulenko [1987], [1994a] demonstrated the possibility of completely damping the rotational motions of an asymmetric solid *just by two impulses* of constant intensity. Also noteworthy is the result presented in Subsection 2.3.6 (Vorotnikov [1991a]). It considerably extends the possibilities of one-channel control proper in damping the rotational motions of an asymmetric solid (recall that in this case control appears just in one of Euler's dynamical equations).

The stability of the center of mass of an Earth's artificial satellite in a circular orbit was investigated in a variety of ways, among which were ap-

*See also Shchennikov V.N. [1985] Solving the Problem of Stability in a Critical Case, in *Comparison Methods and Lyapunov's Methods*, 8–17, Saransk: Izd. Mord. Gos. Univ. (in Russian).

**In this case, the controlling moment have nonzero projections onto all three principal central axes of the solid's ellipsoid of inertia.

***Agrachev, A.A. and Sarychev, A.V. [1983] Controlling the Rotational Motion of an Asymmetric Solid, *Problemy Upravlen. Teor. Inform.*, **12**, 5, 385–387; Kovalev, A.M. [1991] Oriented Manifolds and Controllability of Dynamic Systems, *Prikl. Mat. Mekh.*, **55**, 4, 639–646.

proaches employed the Routh–Lyapunov theorem (Rumyantsev [1967]) and the method of Lyapunov functions (Merkin [1987]).

Example 2.6.1 exhibits a phenomenon of fundamental importance in the dynamics of spacecraft with a single flywheel: when damping the oscillations of the rotational axis of a spacecraft, the flywheel "lateral" to this axis can *only act as a "redistributor"* of the angular momentum of the system; at the end of a damping session the flywheel itself comes to rest without "taking up" the perturbations. This fact deserves further investigation. So, there arises a question of determining the admissible domain of initial perturbations of a system. The answer to this question depends on whether the domain of attraction for the theorems of the Lyapunov–Malkin type (Theorem 2.6.1) can be estimated. Apparently, this problem have not been touched upon in the literature.

Chapter 3

"Essentially" Nonlinear Problems of Stability with Respect to Part of the Variables

Chapter 3 deals mainly with *"essentially" nonlinear problems* of stability with respect to part of the variables. To solve them we employ the *method of Lyapunov V-functions* which is being developed here along the lines of the approach proposed in this book, namely, the basic idea of introducing new variables suggested in the first two chapters is used here to construct not the auxiliary μ-systems but the auxiliary V-functions of an appropriate structure. As a result, we manage to advance in the following directions.

(1) The concept of V-function sign-definite with respect to part of the variables is extended, which allows the investigation to use V-functions not satisfying the conditions of the fundamental Rumyantsev theorem on stability with respect to part of the variables and to state new conditions for stability involving two Lyapunov vector functions.

(2) New conditions of asymptotic stability with respect to part of the variables are obtained which complement the familiar theorems of the Barbashin–Krasovskii type. Such conditions are applicable in those cases when the conditions of Theorems 0.4.4–0.4.6 fail to hold.

The problem of stability with respect to part of the variables under large initial perturbations of "uncontrollable" variables is posed, and conditions for solvability based on the Lyapunov functions method are given.

Also further developed is the method proposed in Chapters 1 and 2. Differential inequalities which extend the constructibility of auxiliary μ-systems are considered and the instability with respect to part of the variables is also analyzed.

In the conclusion of the chapter the notion of stability with respect to part of the variables is broadened, and stability with respect to a specified number of the variables (their composition not being fixed) is considered. Conditions for stability of this kind are obtained for systems with a known first integral.

3.1 Using a New Class of Lyapunov Functions

3.1.1 Formulation of the Problem

Consider the problem of stability of the unperturbed motion $\boldsymbol{x} = \mathbf{0}$ of system (0.3.1) with respect to part of the variables. The motives and "aftereffects" of the designation of admissible boundaries within which the "uncontrollable" variables are allowed to vary in the given problem are analyzed. Though barely highlighted in the literature, these factors substantially influence research productivity. Now we demonstrate that the imposition of additional constraints on the "uncontrollable" variables while examining a system for stability with respect to part of the variables makes it possible (and desirable) to use Lyapunov functions with new properties ("intermediate" relative to the conventional ones). We shall also show how to define the notion of sign-definiteness of Lyapunov functions with respect to part of the variables to avoid such "intermediate" properties. The main partial stability theorem will be extended to handle the newly arising conditions.

3.1.2 Motives for Designating Admissible Boundaries of Variation for "Uncontrollable" Variables

When studying the $\boldsymbol{y}$-stability of the unperturbed motion $\boldsymbol{x} = \mathbf{0}$ of system (0.3.1), in principle, one need not monitor the behavior of $\boldsymbol{z}$-variables (provided certain general conditions are observed). In the coupled system (0.3.1), however, they exert an important influence on the "main" $\boldsymbol{y}$-variables.

Let us distinguish the factors that determine the admissibility of the "uncontrollable" $\boldsymbol{z}$-variables. Since such a classification in the theory of stability with respect to part of the variables has not yet been conducted, we do not claim that this issue is given a comprehensive description here.

(1) *Allowance for the "worst" case scenario (general conditions being the same) in the variation of "uncontrollable" variables.* This entails the assumption $||\boldsymbol{z}|| < \infty$ and, consequently, the study of $\boldsymbol{y}$-stability of the unperturbed motion $\boldsymbol{x} = \mathbf{0}$ of system (0.3.1) in domain (0.3.2).

Such considerations may prove overcautious. Indeed, one does not use inequalities $|z_j| \leq H$ that are valid (or admissible) for certain $\boldsymbol{z}$-components or relationships like $|f_i(t, \boldsymbol{x})| \leq H$. Such relationships may considerably facilitate examining the system for $\boldsymbol{y}$-stability. In a sense, allowance for the "worst" case scenario is comparable with the game-theoretic approach.

(2) *Allowance for specification of requirements imposed on the "uncontrollable" variables.* An alternative to the "worst" case scenario. This approach has various meanings.

Rationalizing the formulation of the problem of stability with respect to part of the variables. This requires "subjecting" the system to certain general estimates (possibly including integral estimates) for the "uncontrollable" variables. This significantly simplifies the solution. An example is the study of the stability of the motion of bodies containing cavities filled with liquid (Moiseev and Rumyantsev [1965]).

"Built-in" possibilities for facilitating the solution of problems of stability with respect to part of the variables. Put differently, the use of additional relationships linking the components of the phase vector of system (0.3.1). The validity of such relationships must somehow be confirmed when solving the problem. This approach provides the basis, for example, for the method of solving problems of stability with respect to part of the variables by constructing auxiliary systems being developed in this book.

(3) *Allowance for availability of estimates (even though rough) for "uncontrollable" variables.* In such cases the problem of the stability of system (0.3.1) with respect to part of the variables can be reduced to a problem of stability with respect to all the variables in an auxiliary system of differential equations of the *same* dimensions (Zubov [1959]).

3.1.3 Extensions of the Main Partial Stability Theorem

These extensions imply that the notion of sign-definiteness with respect to part of the variables has been defined in some specific manner for the Lyapunov V-function used in Theorem 0.4.1. This means that the possible set of variables with respect to which the V-function is sign-definite has been *enlarged* (by including *the phase variables* of system (0.3.1) and also *certain functions of those variables*). It is also assumed that the requirements imposed on the "uncontrollable" variables have been clearly specified.

To formulate further results, let us introduce the following functions: (1) $a(r)$ continuous and monotone increasing for $r \in [0, H]$; (2) a scalar $V(t, \boldsymbol{x})$ and a vector function $\boldsymbol{\mu}(t, \boldsymbol{x})$ both continuously differentiable in domain (0.3.2). It is assumed that $a(0) = V(t, \mathbf{0}) \equiv 0$, $\boldsymbol{\mu}(t, \mathbf{0}) \equiv \mathbf{0}$. Let $\dot{V}$ denote the derivative of V in virtue of system (0.3.1).

Theorem 3.1.1. *Suppose that, given system* (0.3.1), *one can find a scalar function* $V(t, \boldsymbol{x})$ *and a vector function* $\boldsymbol{\mu}(t, \boldsymbol{x})$ *such that, in the domain*

$$t \geq 0, \quad \|\boldsymbol{y}\| + \|\boldsymbol{\mu}(t, \boldsymbol{x})\| \leq H, \quad \|\boldsymbol{z}\| < \infty \tag{3.1.1}$$

the following conditions are satisfied

$$V(t, \boldsymbol{x}) \geq a\big(\|\boldsymbol{y}\| + \|\boldsymbol{\mu}(t, \boldsymbol{x})\|\big) \tag{3.1.2}$$

$$\dot{V} \leq 0. \tag{3.1.3}$$

Then the unperturbed motion $\boldsymbol{x} = \mathbf{0}$ *of system* (0.3.1) *is* $\boldsymbol{y}$*-stable.*

Proof. It follows from the continuity of functions V, a and from the conditions $a(0) = 0$, $V(t, \mathbf{0}) \equiv 0$ that for any $\varepsilon > 0$, $t_0 \geq 0$, $\varepsilon \in (0, H)$, one can find a number $\delta(\varepsilon, t_0) > 0$ such that if $\|\boldsymbol{x}_0\| < \delta$, then $V(t_0, \boldsymbol{x}_0) \leq a(\varepsilon)$. By conditions (3.1.2), (3.1.3), if $\boldsymbol{x} = \boldsymbol{x}(t; t_0, \boldsymbol{x}_0)$ is a solution such that $\|\boldsymbol{x}_0\| < \delta$, then for all $t \geq t_0$,

$$\begin{aligned} a\big(\|\boldsymbol{y}(t; t_0, \boldsymbol{x}_0)\| + \|\boldsymbol{\mu}(t, \boldsymbol{x}(t; t_0, \boldsymbol{x}_0))\|\big) &\leq V(t, \boldsymbol{x}(t; t_0, \boldsymbol{x}_0)) \\ &\leq V(t_0, \boldsymbol{x}_0) \leq a(\varepsilon). \end{aligned}$$

Taking into account the properties of function $a(r)$, we get $\|\boldsymbol{y}(t; t_0, \boldsymbol{x}_0)\| + \|\boldsymbol{\mu}(t, \boldsymbol{x}(t; t_0, \boldsymbol{x}_0))\| < \varepsilon$, $t \geq t_0$. Consequently, $\|\boldsymbol{y}(t; t_0, \boldsymbol{x}_0)\| < \varepsilon$, $t \geq t_0$, which proves the theorem. □

Remark 3.1.1.

1. If $\boldsymbol{\mu} = \mathbf{0}$, we have Rumyantsev's partial stability theorem (Rumyantsev [1957a]), i.e., Theorem 0.4.1. When $\boldsymbol{\mu} \neq \mathbf{0}$, the conditions of Theorem 3.1.1 are more general than those of Rumyantsev's theorem, though inequalities (0.3.2) (which represent the "worst" case as regards the variation of the "uncontrollable" $\boldsymbol{z}$-variables) are replaced by the more restrictive inequalities (3.1.1). Indeed, even in the case when $\dim(\boldsymbol{y}) = \dim(\boldsymbol{z}) = 1$, function $V(y_1, z_1)$ satisfying condition (3.1.2) *may not be sign-definite either in Lyapunov's sense or with respect to* y_1 (in the sense of Rumyantsev [1957a]).

An example is the function

$$V = \frac{y_1^2(1 + z_1^2)}{1 + y_1^4 z_1^4}. \tag{3.1.4}$$

Indeed, although condition (3.1.2) holds for $\mu_1 = y_1 z_1$ and sufficiently small H in domain (3.1.1), one has $\lim V = 0$ for any fixed y_1 and $z_1 \to \infty$. Hence the function is not sign-definite in Lyapunov's sense ($V = 0$ for $y_1 = 0$ and any z_1) or even in the sense that $V(t, \boldsymbol{x}) \geq a(\|\boldsymbol{y}\|)$ in domain (0.3.2), i.e., the condition for a function to be $\boldsymbol{y}$-sign-definite in the sense of Rumyantsev [1957a].

That is why in investigating the $\boldsymbol{y}$-stability of the position $\boldsymbol{x} = \mathbf{0}$ of system (0.3.1), it is reasonable to choose V-functions not only among functions sign-definite relative to $\boldsymbol{y}$ (or even relative to a larger number of the phase variables). It is also advisable to consider V-functions that are sign-definite relative to both $\boldsymbol{y}$ and (simultaneously) certain functions $\boldsymbol{\mu} = \boldsymbol{\mu}(t, \boldsymbol{x})$. The properties of such V-functions, even in the case $\dim(\boldsymbol{y}) = \dim(\boldsymbol{z}) = 1$, may be "intermediate" between sign-definiteness with respect to $\boldsymbol{y}$ and in Lyapunov's sense.

Note that, for each $0 < c < 1/2$, the curves defining the level surfaces $V = c$ of the V-function (3.1.4) split into two disjoint classes (see Figure 3.1.1). The first class consists of the *open* curves encircling the axis $y_1 = 0$, which are typical of classic y_1 sign-definite V-functions (in the sense of Rumyantsev [1957a]). The special feature here is the existence of the second, additional class of curves. It also consists of the open curves that recede from the position $y_1 = z_1 = 0$, as c is decreased, and asymptotically approach the axes $y_1 = 0$ and $z_1 = 0$, as $z_1 \to \infty$ and $y_1 \to \infty$, respectively.

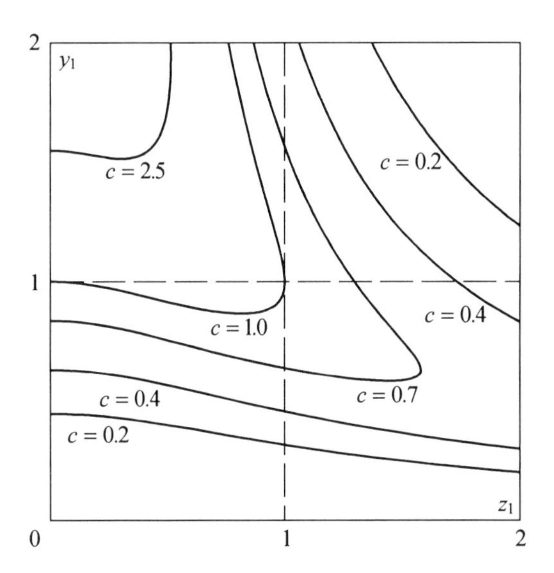

Figure 3.1.1. The level surfaces $V = c$ ($c = 0.2; 0.4; 0.7; 1.0; 2.5$) of V-function (3.1.4). As $z_1 \to \infty$, all the curves asymptotically approach the Oz_1-axis ($y_1 \to 0$) and asymptotically approach the Oy_1-axis ($z_1 \to 0$), as $y_1 \to \infty$. In the region of small c ($c < 0.5$), there are two branches in the quadrant $y_1 \geq 0$, $z_1 \geq 0$ and, when $c = 0$ and $c > 0.5$, there is one branch of surface $V = c$.

2. If V depends (explicitly) on t, then even when $\dim(\boldsymbol{y}) = \dim(\boldsymbol{z}) = 1$, one may have "intermediate" properties not only as $z_1 \to \infty$.

For example, consider the function

$$V = \frac{y_1^2(1 + e^{2t} z_1^2)}{1 + e^{4t} y_1^4 z_1^4} \tag{3.1.5}$$

that is not sign-definite in Lyapunov's sense.

For $\mu_1 = e^t y_1 z_1$ and sufficiently small H, condition (3.1.2) holds in domain (3.1.1). At the same time, V-function (3.1.5) is not y_1-sign-definite in the sense of Rumyantsev [1957a], because $V \to 0$ as $z_1 \to \infty$ (or $t \to \infty$) for any fixed t, y_1 (or y_1, z_1, respectively).

3. Inequality (3.1.2) means that the V-function is sign-definite relative to $\boldsymbol{y}$ and the components of the $\boldsymbol{\mu}$-function, i.e., $(\boldsymbol{y}, \boldsymbol{\mu})$-sign-definite in the sense of Rumyantsev [1957a]. However, in this case it is not a priori obvious how to choose a suitable $\boldsymbol{\mu}$-function; it must be defined when solving the problem. In this sense the $\boldsymbol{\mu}$-function plays the part of a second (vector-valued) Lyapunov function (together with the first, scalar V-function).

4. In the case

$$V(t, \boldsymbol{x}) \equiv V^*(t, \boldsymbol{y}, \boldsymbol{\mu}(t, \boldsymbol{x})), \tag{3.1.6}$$

the verification of condition (3.1.2) reduces to verifying that V^* is sign-definite in Lyapunov's sense. In particular, if V^* is a quadratic form in the variables $\boldsymbol{y}$ and $\boldsymbol{\mu}$, one can use the generalized Hurwitz criterion. Incidentally, if the

V-function possesses the structure of (3.1.6), the possibilities for solving the problem posed in Rumyantsev and Oziraner [1987]—analysis of the $\boldsymbol{y}$-stability of the unperturbed motion $\boldsymbol{x} = \mathbf{0}$ of system (0.3.1) in terms of *quadratic forms*—are improved considerably. The structure represented by (3.1.6) enables using "*essentially nonlinear*" V-functions in solving that problem.

Example 3.1.1. Consider the *motion of a unit mass point in a constant gravitational field*, constrained to move on the surface

$$x_3 = f(x_1, x_2), \quad f = \frac{x_1^2(1+x_2^2)}{1+x_1^4x_2^4} \tag{3.1.7}$$

in three-dimensional $Ox_1x_2x_3$-space, with x_3-axis pointing vertically upward. The kinetic T and potential P energies of the point are expressed by

$$T = \frac{1}{2}\left\{\dot{x}_1^2 + \dot{x}_2^2 + \left[\frac{\partial f}{\partial x_1}\dot{x}_1 + \frac{\partial f}{\partial x_2}\dot{x}_2\right]^2\right\},$$
$$P = gf(x_1, x_2), \quad g = \text{const} > 0.$$

Setting $\boldsymbol{y} = (x_1, \dot{x}_1, \dot{x}_2)$, $z = x_2$ and introducing auxiliary functions $V = T + P$ (*the total energy of the point*), $\mu_1 = x_1x_2$, we obtain the relationships

$$V \geq \frac{1}{2}(\dot{x}_1^2 + \dot{x}_2^2) + g\frac{x_1^2 + \mu_1^2}{1+\mu_1^4}, \quad \dot{V} \equiv 0. \tag{3.1.8}$$

Consequently, conditions (3.1.2) and (3.1.3) hold in domain (3.1.1) (for sufficiently small H). As a result, the equilibrium position of the point

$$x_i = \dot{x}_i = 0 \quad (i = 1, 2, 3) \tag{3.1.9}$$

is $\boldsymbol{y}$-stable by virtue of Theorem 3.1.1.

At the same time, the V-function is not sign-definite whether relative to $\boldsymbol{y}$ in the sense of Rumyantsev [1957a] ($V \to 0$ for $\dot{x}_1 = \dot{x}_2 = 0$, $|x_2| \to \infty$ and any fixed x_1) or in Lyapunov's sense ($V = 0$ for $x_1 = \dot{x}_1 = \dot{x}_2 = 0$ and any x_2).

By (3.1.8), position (3.1.9) is also stable with respect to μ_1. Summing up, in view of (3.1.7), we conclude that position (3.1.9) is stable with respect to $x_1, x_3, \dot{x}_1, \dot{x}_2$.

Theorem 3.1.2. *Suppose that, given system* (0.3.1), *one can find two scalar functions* $V(t, \boldsymbol{x})$, $U(t, \boldsymbol{x})$ *and a vector function* $\boldsymbol{\mu}(t, \boldsymbol{x})$ *such that, besides* (3.1.3), *the following conditions also hold in domain* (3.1.1):

$$V(t, \boldsymbol{x}) \geq a(\|\boldsymbol{y}\|) \tag{3.1.10}$$
$$U(t, \boldsymbol{x}) \geq b(\|\boldsymbol{\mu}(t, \boldsymbol{x})\|), \quad \dot{U} \leq 0. \tag{3.1.11}$$

Then the unperturbed motion $\boldsymbol{x} = \mathbf{0}$ *of system* (0.3.1) *is* $\boldsymbol{y}$-*stable.*

Proof. If conditions (3.1.3) and (3.1.10) hold in domain (3.1.1) for any $\varepsilon > 0$, $t_0 > 0$, $\varepsilon \in (0, H)$, one can find a number $\delta_1(\varepsilon, t_0) > 0$ such that, if $\boldsymbol{x} = \boldsymbol{x}(t; t_0, \boldsymbol{x}_0)$ is a solution, $\|\boldsymbol{x}_0\| < \delta_1$, satisfying the inequality $\|\boldsymbol{y}(t; t_0, \boldsymbol{x}_0)\| + \|\boldsymbol{\mu}(t, \boldsymbol{x}(t; t_0, \boldsymbol{x}_0))\| \leq H$ for all $t \geq t_0$, then $\|\boldsymbol{y}(t; t_0, \boldsymbol{x}_0)\| < \varepsilon$, $t \geq t_0$.

On the other hand, if conditions (3.1.11) hold in domain (3.1.1), then, for any $t_0 \geq 0$, one can find a number $\delta_2(t_0, \varepsilon) > 0$ such that, if $\|\boldsymbol{x}_0\| < \delta_2$, then $\|\boldsymbol{\mu}(t, \boldsymbol{x}(t; t_0, \boldsymbol{x}_0))\| \leq H - \varepsilon$, $t \geq t_0$.

Taking $\delta = \min\{\delta_1, \delta_2\}$, we conclude as a result that, for any $\varepsilon > 0$, $t_0 \geq 0$, $\varepsilon \in (0, H)$, and every solution $\boldsymbol{x} = \boldsymbol{x}(t; t_0, \boldsymbol{x}_0)$ such that $\|\boldsymbol{x}_0\| < \delta$, the inequality $\|\boldsymbol{y}(t; t_0, \boldsymbol{x}_0)\| < \varepsilon$ is valid for $t \geq t_0$. The theorem is proved. □

Remark 3.1.2.

1. A V-function satisfying inequality (3.1.2) in domain (3.1.1) also satisfies inequality (3.1.10). Consequently, it is $\boldsymbol{y}$-sign-definite in domain (3.1.1) (but not in domain (0.3.2), as in Rumyantsev [1957a]). However, the validity of conditions (3.1.3) and (3.1.10) in domain (3.1.1) does not guarantee that the inequality $\|\boldsymbol{y}(t; t_0, \boldsymbol{x}_0)\| + \|\boldsymbol{\mu}(t, \boldsymbol{x}(t; t_0, \boldsymbol{x}_0))\| \leq H$, $t \geq t_0$ holds true along all solutions of system (0.3.1) for sufficiently small $\|\boldsymbol{x}_0\|$. (There is no guarantee for the verification of constraints (3.1.1) imposed here on the "uncontrollable" variables.) Therefore, a V-function that satisfies conditions (3.1.3) and (3.1.10) in domain (3.1.1) does not guarantee $\boldsymbol{y}$-stability of the position $\boldsymbol{x} = \mathbf{0}$ of system (0.3.1). But if condition (3.1.3) holds, a $(\boldsymbol{y}, \boldsymbol{\mu})$-sign-definite V-function does guarantee $\boldsymbol{y}$-stability. In that case the constraints imposed on the "uncontrollable" variables are really observed.
2. A V-function that is $\boldsymbol{y}$-sign-definite in (3.1.1) guarantees $\boldsymbol{y}$-stability if it can be proved that $\|\boldsymbol{\mu}(t, \boldsymbol{x})\| \leq H$, $t \geq t_0$ along the corresponding solutions of system (0.3.1). For example, this can be done using one more Lyapunov function. This approach (conditions (3.1.11)) is implemented in Theorem 3.1.2.

3.1.4 Specification of the Notion of Sign-Definiteness of a V-Function with Respect to Part of the Variables

One can avoid the situations in which the above mentioned "intermediate" properties of V-functions come about. To this end, the verification of whether V is (or is not) $\boldsymbol{y}$-sign-definite should not be performed in domain (0.3.2) but over the *set* $M = \{\boldsymbol{x} : \boldsymbol{x}(t; t_0, \boldsymbol{x}_0)\}$ of solutions of system (0.3.1) with sufficiently small $\|\boldsymbol{x}_0\|$. Such a verification (if possible) will suffice to establish $\boldsymbol{y}$-stability.

Theorem 3.1.3. *Suppose that, given system* (0.3.1), *one can find a function* $V(t, \boldsymbol{x})$ *such that, besides* (3.1.3), *the following condition also holds in the set* M:

$$V\big(t, \boldsymbol{x}(t; t_0, \boldsymbol{x}_0)\big) \geq a\big(\|\boldsymbol{y}(t; t_0, \boldsymbol{x}_0)\|\big). \tag{3.1.12}$$

Then the unperturbed motion $\boldsymbol{x} = \mathbf{0}$ *of system* (0.3.1) *is* $\boldsymbol{y}$*-stable.*

Remark 3.1.3.

1. The conditions of Theorem 3.1.3 differ from those adopted in Rumyantsev [1957a] in that $\boldsymbol{y}$-sign-definiteness of V-function is not verified in domain (0.3.2) but over the set M. A similar result (unrelated to the questions under discussion here) was established in Halanay [1963].
2. If condition (3.1.3) is satisfied, V-functions (3.1.4) and (3.1.5) will be y_1-positive-definite in the sense of (3.1.12). Indeed, by virtue of (3.1.3), the inequality

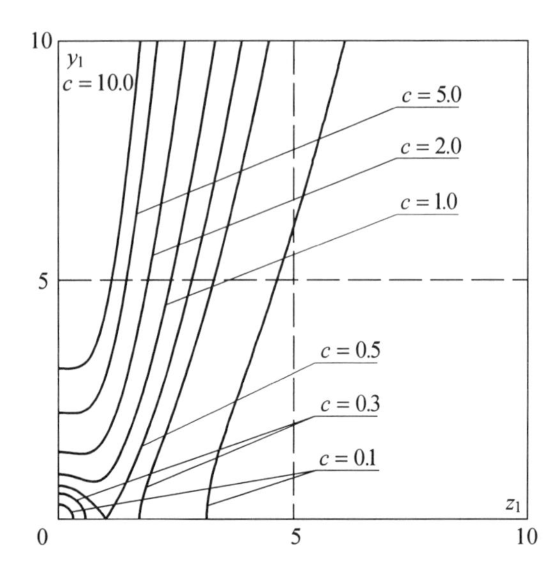

Figure 3.1.2. The level surfaces $V = c$ ($c = 0.1; 0.3; 0.5; 1.0; 2.0; 5.0; 10.0$) of V-function (3.1.13). In the region of small c ($c < 0.5$), there are two branches in the quadrant $y_1 \geq 0$, $z_1 \geq 0$ and, when $c = 0$ and $c > 0.5$, there is one branch of surface $V = c$.

$|\mu_1(t; \boldsymbol{x}(t; t_0, \boldsymbol{x}_0))| \leq H$ holds in the set M. Hence, cases that lead to "intermediate" properties are eliminated. Similarly, the V-function in Example 3.1.1 will be $\boldsymbol{y}$-positive-definite in the sense of (3.1.12).

3. The function

$$V = \frac{y_1^2 + z_1^2}{1 + z_1^4} \tag{3.1.13}$$

(see Rumyantsev and Oziraner [1987]) is sign-definite in Lyapunov's sense but not relative to y_1 (in the sense of Rumyantsev [1957a]). But if $\dot{V} \leq 0$, function (3.1.13) is y_1-sign-definite in the sense of (3.1.12).

The level surfaces $V = c$ of V-function (3.1.13) have not been analyzed, but we note that, for each $0 < c < 1/2$, the curves defining them split into two disjoint classes (see Figure 3.1.2). The first comprises the *closed* curves encircling the position $y_1 = z_1 = 0$, which are typical of classic Lyapunov V-functions. The special feature is the existence of the second, additional class; it consists of the *open* curves that recede from the position $y_1 = z_1 = 0$ as c is decreased.

3.1.5 Conditions for Stability with Respect to Part of the Variables Using Two Lyapunov Vector Functions

Theorem 3.1.4. *Suppose that, given system* (0.3.1), *along with a main function* $\boldsymbol{V}(t, \boldsymbol{x}) = [V_1(t, \boldsymbol{x}), \ldots, V_k(t, \boldsymbol{x})]$, *one can find a second vector function* $\boldsymbol{\mu}(t, \boldsymbol{x})$ *such that the following conditions are satisfied in domain* (3.1.1):

(1) $\boldsymbol{V}$ *is continuously differentiable,* $\boldsymbol{\mu}$ *is continuous and, besides,*

$$\boldsymbol{V}(t, \boldsymbol{0}) \equiv \dot{\boldsymbol{V}}(t, \boldsymbol{0}) \equiv \boldsymbol{0}, \quad \boldsymbol{\mu}(t, \boldsymbol{0}) \equiv \boldsymbol{0};$$

(2) *there is a number* l $(1 \leq l \leq k)$ *such that the following inequalities are valid*

$$V_i(t,\boldsymbol{x}) \geq 0 \ \ (i=1,\dots,l), \quad \sum_{i=1}^{l} V_i(t,\boldsymbol{x}) \geq a\big(\|\boldsymbol{y}\| + \|\boldsymbol{\mu}(t,\boldsymbol{x})\|\big);$$

(3) $\dot{\boldsymbol{V}} \leq \boldsymbol{F}(t,\boldsymbol{V})$, $\boldsymbol{F}(t,\boldsymbol{0}) \equiv \boldsymbol{0}$, *where vector function* $\boldsymbol{F}$ *is defined and continuous in the domain* $t \geq 0$, $\|\boldsymbol{V}\| < R$ *with* $R=\infty$ *or* $R > \sup\big[\|\boldsymbol{V}(t,\boldsymbol{x})\| : t \geq 0,\ \|\boldsymbol{y}\| + \|\boldsymbol{\mu}(t,\boldsymbol{x})\| \leq H\big]$;

(4) *each of the functions* F_s $(s=1,\dots,k)$ *is a component of vector function* $\boldsymbol{F}$ *and is nondecreasing in* $V_1,\dots,V_{s-1},V_{s+1},\dots,V_k$;

(5) *the zero solution* $\boldsymbol{\omega}=\boldsymbol{0}$ *of the system* $\dot{\boldsymbol{\omega}}=\boldsymbol{F}(t,\boldsymbol{\omega})$ *is stable (asymptotically stable) with respect to* $\omega_1,\dots,\omega_l$ *on condition that* $\omega_{10} \geq 0,\dots,\omega_{l0} \geq 0$.

Then the unperturbed motion $\boldsymbol{x}=\boldsymbol{0}$ *of system* (0.3.1) *is* $\boldsymbol{y}$*-stable (asymptotically* $\boldsymbol{y}$*-stable).*

Remark 3.1.4.

1. If $\boldsymbol{\mu}=\boldsymbol{0}$, we have a theorem of the Matrosov type (Matrosov [1965], see also Peiffer and Rouche [1969]). If $\boldsymbol{\mu} \neq \boldsymbol{0}$, the situations are possible when the expression $\sum_{i=1}^{l} V_i(t,\boldsymbol{x})$ may not be $\boldsymbol{y}$-sign-definite in the sense of Rumyantsev [1957a], even in the case of $\dim(\boldsymbol{y})=\dim(\boldsymbol{z})=1$. Besides, the vector differential inequality $\dot{\boldsymbol{V}} \leq \boldsymbol{F}(t,\boldsymbol{V})$, valid in domain (3.1.1), may not hold (or may be hard to verify) in domain (0.3.2).

2. If $k=1$, then Theorem 3.1.4 elaborates on the Corduneanu theorem (Corduneanu [1964]) on $\boldsymbol{y}$-stability (asymptotic $\boldsymbol{y}$-stability).

3. Other conditions for partial stability with two Lyapunov vector functions are given in Section 3.4.

Example 3.1.2. Let system (0.3.1) have the form

$$\dot{y}_1 = -{}^1\!/_2 y_1(1+z_1^2), \quad \dot{z}_1 = {}^1\!/_2 z_1(1+z_1^2). \tag{3.1.14}$$

We take V-function (3.1.4) and $\mu_1 = y_1 z_1$ as the auxiliary Lyapunov functions. Because relationships (3.1.2) together with $\dot{V}=-V$ are valid in domain (3.1.1), the unperturbed motion $y_1=z_1=0$ of system (3.1.14) is asymptotically stable with respect to y_1 by Theorem 3.1.4.

3.1.6 Extension of Marachkov's Theorem

This subsection presents system (0.3.1) in the form

$$\dot{\boldsymbol{y}} = \boldsymbol{Y}(t,\boldsymbol{y},\boldsymbol{z}), \quad \dot{\boldsymbol{z}} = \boldsymbol{Z}(t,\boldsymbol{y},\boldsymbol{z}). \tag{3.1.15}$$

Theorem 3.1.5. *Suppose that, given system* (3.1.15), *one can find a scalar function* $V(t,\boldsymbol{x})$ *and a vector function* $\boldsymbol{\mu}(t,\boldsymbol{x})$ *such that the following conditions are satisfied in domain* (3.1.1):

(1) $V(t,\boldsymbol{x}) \geq a\big(\|\boldsymbol{y}\| + \|\boldsymbol{\mu}(t,\boldsymbol{x})\|\big)$;

(2) $\dot{V}(t,\boldsymbol{x}) \leq -c(\|\boldsymbol{y}\|)$;

(3) *there is a number* $M > 0$ *such that* $\|\boldsymbol{Y}(t, \boldsymbol{x})\| \leq M$.

Then the unperturbed motion $\boldsymbol{x} = \mathbf{0}$ *of system* (3.1.15) *is asymptotically* $\boldsymbol{y}$*-stable.*

Proof. If conditions (1) and (2) are satisfied, the unperturbed motion $\boldsymbol{x} = \mathbf{0}$ of system (3.1.15) is $\boldsymbol{y}$-stable by Theorem 3.1.1. This means that, for any $t_0 \geq 0$, there is $\Delta(t_0)$, $0 < \Delta < \delta$, such that $\|\boldsymbol{y}(t; t_0, \boldsymbol{x}_0)\| + \|\boldsymbol{\mu}(t, \boldsymbol{x}(t; t_0, \boldsymbol{x}_0))\| < H$, $t \geq t_0$ for $\|\boldsymbol{x}_0\| < \Delta$.

Let us demonstrate that, for $\|\boldsymbol{x}_0\| < \Delta$, also,

$$\lim \|\boldsymbol{y}(t; t_0, \boldsymbol{x}_0)\| = 0, \quad t \to \infty.$$

On the contrary, let us assume that there exists a number $l > 0$, a value $\boldsymbol{x}_*$ with $\|\boldsymbol{x}_*\| < \Delta$, and a sequence $t_k \to \infty$ such that

$$\|\boldsymbol{y}(t_k; t_0, \boldsymbol{x}_*)\| \geq l, \quad k = 1, 2, 3, \ldots \tag{3.1.16}$$

For definiteness, suppose that $t_k - t_{k-1} \geq \alpha > 0$, $k = 1, 2, 3, \ldots$, which can always be attained by passing to the corresponding subsequence. Further analysis will make a specific value of α apparent. Taking into account that, in a neighborhood of point $t = t_k$, we have equalities

$$\boldsymbol{y}(t; t_0, \boldsymbol{x}_*) = \boldsymbol{y}(t_k; t_0, \boldsymbol{x}_*) + \int_{t_k}^{t} \boldsymbol{Y}(\tau; \boldsymbol{x}(\tau; t_0, \boldsymbol{x}_*)) d\tau,$$

by (3.1.16) we get

$$\|\boldsymbol{y}(t_k; t_0, \boldsymbol{x}_*)\| \geq l - M(t - t_k). \tag{3.1.17}$$

Taking $\beta = \tfrac{1}{2} M^{-1} l$ and specifying α so that $\alpha > 2\beta$, from (3.1.17), we derive the inequalites

$$\tfrac{1}{2} l \leq \|\boldsymbol{y}(t; t_0, \boldsymbol{x}_*)\| \leq H, \quad t \in T_k = [t_k - \beta, t_k + \beta], \tag{3.1.18}$$

which are valid for $\|\boldsymbol{x}_*\| < \Delta$ and for all $k = 1, 2, 3, \ldots$.

Then, by (3.1.18), for $\|\boldsymbol{x}_*\| < \Delta$ along the solutions $\boldsymbol{x} = \boldsymbol{x}(t; t_0, \boldsymbol{x}_*)$,

$$\dot{V}(t, \boldsymbol{x}) \leq -c(\|\boldsymbol{y}\|) \leq -c(\tfrac{1}{2} l), \quad t \in T_k.$$

Therefore,

$$\begin{aligned} 0 \leq V(t_k + \beta, \boldsymbol{x}(t_k + \beta; t_0, \boldsymbol{x}_0)) &\leq V(t_0, \boldsymbol{x}_0) + \sum_{i=1}^{k} \int_{t_i - \beta}^{t_i + \beta} \dot{V}(\tau; \boldsymbol{x}(\tau; t_0, \boldsymbol{x}_*)) d\tau \\ &\leq V(t_0, \boldsymbol{x}_0) - 2k\beta c(\tfrac{1}{2} l), \end{aligned}$$

which is impossible for large k.

This means that the assumption (3.1.16) is wrong. The theorem is proved. □

Remark 3.1.5.

1. When $\boldsymbol{\mu} = \mathbf{0}$, Theorem 3.1.5 becomes the Peiffer–Rouche theorem (Peiffer and Rouche [1969]), which extends the classical Marachkov result [1940] to the case of partial asymptotic stability.

2. When $\boldsymbol{\mu} \neq \mathbf{0}$, essential peculiarities may appear even in the case $\dim(\boldsymbol{y}) = \dim(\boldsymbol{z}) = 1$. Not only V but $\dot{V}$ can be non-sign-definite either with respect to $\boldsymbol{y}$ (in Rumyantsev's sense [1957a]) or in Lyapunov's sense. Moreover, the condition $\|\boldsymbol{Y}(t, \boldsymbol{x})\| \leq M$ is verified in domain (3.1.1) rather than in domain (0.3.2), which extends the usage of the theorem.

Example 3.1.3. Let system (3.1.15) have the form

$$
\begin{aligned}
\dot{y}_1 &= -\tfrac{1}{2} y_1 (1 + e^{2t} z_1^4 z_2^4)^{-1} + z_1^2 z_2^2, \\
\dot{z}_1 &= \tfrac{1}{2} z_1 + e^{-t} y_1 z_1 + z_1 (1 + e^{2t} z_1^4 z_2^4)^{-1} + z_1^3, \\
\dot{z}_2 &= -z_2 - 2e^{-t} y_1 z_2 - \tfrac{3}{2} z_2 (1 + e^{2t} z_1^4 z_2^4)^{-1} - z_1^2 z_2.
\end{aligned}
\tag{3.1.19}
$$

The position $y_1 = z_1 = z_2 = 0$ of system (3.1.19) is Lyapunov unstable (with respect to all the variables). Let us consider the question of whether it is asymptotically stable with respect to y_1.

We introduce the following Lyapunov functions

$$
V = y_1^2 + e^t z_1^2 z_2^2, \quad \mu_1 = e^{\frac{1}{2}t} z_1 z_2.
$$

Since the following relationships hold in domain (3.1.1):

$$
\begin{gathered}
V = y_1^2 + \mu_1^2, \quad \dot{V} = -(y_1^2 + \mu_1^2)(1 + \mu_1^4)^{-1} \leq l_1 y_1^2 \quad (l_1 = \text{const} > 0), \\
|Y| = |-\tfrac{1}{2} y_1 (1 + \mu_1^4)^{-1} + e^{-t} \mu_1^2| \leq M,
\end{gathered}
$$

the conditions of Theorem 3.1.5 are satisfied. Therefore, the position $y_1 = z_1 = z_2 = 0$ of system (3.1.19) is asymptotically stable with respect to y_1.

In this case we note that the derivative $\dot{V}$ is not sign-definite with respect to y_1 (in Rumyantsev's sense [1957a]) or with respect to any part of or all the variables y_1, z_1, z_2.

In addition, the V-function does not admit an infinitely small upper bound with respect to all or to any part of variables y_1, z_1, z_2.

3.2 Developing Theorems of the Barbashin–Krasovskii Type

3.2.1 Formulation of the Problem

In this section the problem of asymptotic $\boldsymbol{y}$-stability of the equilibrium position $\boldsymbol{x}^{\mathsf{T}} = (\boldsymbol{y}^{\mathsf{T}}, \boldsymbol{z}^{\mathsf{T}}) = \mathbf{0}$ is considered for a system of ordinary differential equations

$$
\dot{\boldsymbol{x}} = \boldsymbol{X}(\boldsymbol{x}), \quad \boldsymbol{X}(\mathbf{0}) = \mathbf{0},
\tag{3.2.1}
$$

whose right-hand sides *do not depend explicitly on time.* In this case system (3.2.1) is constructed as a system of perturbed motion of the process to be examined for stability.

To solve the problem posed, we shall employ the Lyapunov functions method appropriately modified. In what follows using this method, we derive conditions for asymptotic $\boldsymbol{y}$-stability which elaborate on the well-known theorems of the *Barbashin–Krasovskii type* (Risito [1970], Rumyantsev [1971a], Oziraner [1973])—Theorems 0.4.4–0.4.6. These conditions are applicable when conditions of Theorems 0.4.4–0.4.6 are not satisfied.

To be more specific, in the conditions suggested below, as distinguished from those of Theorems 0.4.4–0.4.6, the sets $M = \{\dot{V} = 0\}$ (Theorem 0.4.4) and $\{V > 0\} \cap M$ (Theorem 0.4.6) can contain entire semitrajectories of system (3.2.1) and, besides, the set $\{\boldsymbol{y} = \boldsymbol{0}\}$ (Theorem 0.4.5) need not be an invariant set of system (3.2.1).

Suppose that, in the domain $\|\boldsymbol{y}\| \leq H$, $\|\boldsymbol{z}\| \leq H_1 < \infty$, condition ($a$) of Section 0.3 is satisfied for system (3.2.1). The requirement of $\boldsymbol{z}$-continuability of solutions of system (3.2.1) will be a consequence of the assumptions which we adopt below.

Further we consider the following functions: (1) a scalar function $a(r)$ continuous and monotone increasing for $r \in [0, H]$, $a(0) = 0$; (2) a scalar function $V(\boldsymbol{x})$, $V(\boldsymbol{0}) = 0$, and a vector function $\boldsymbol{\mu}(\boldsymbol{x})$, $\boldsymbol{\mu}(\boldsymbol{0}) = \boldsymbol{0}$. The last two are assumed to be continuously differentiable and continuous, respectively, in domains of variation of variables $\boldsymbol{y}$, $\boldsymbol{z}$ which we shall consider below.

3.2.2 Conditions for Asymptotic Stability with Respect to Part of the Variables

Theorem 3.2.1. *Suppose that*

(1) *solutions of system* (3.2.1) *originating from a sufficiently small neighborhood of the unperturbed motion* $\boldsymbol{x} = \boldsymbol{0}$ *are* $\boldsymbol{z}$*-bounded*;

(2) *there exist functions* $V(\boldsymbol{x})$ *and* $\boldsymbol{\mu}(\boldsymbol{x})$ *such that, in the domain* $\|\boldsymbol{y}\| \leq H$, $\|\boldsymbol{z}\| \leq H_1 < \infty$ (*where* H *is a sufficiently small positive constant*), *the following conditions are satisfied*:

   (2a) $V(\boldsymbol{x}) \geq a(\|\boldsymbol{y}\|)$, $\dot{V}(\boldsymbol{x}) \leq 0$;

   (2b) $\boldsymbol{y} = \boldsymbol{0}$, $\boldsymbol{\mu}(\boldsymbol{x}) = \boldsymbol{0}$ *is an invariant set of system* (3.2.1);

   (2c) *the set* $\{\dot{V}(\boldsymbol{x}) = 0\} \setminus \{\boldsymbol{y} = \boldsymbol{0},\ \boldsymbol{\mu}(\boldsymbol{x}) = \boldsymbol{0}\}$ *does not contain entire semitrajectories of system* (3.2.1) *for* $t \in [0, \infty)$.

Then the unperturbed motion $\boldsymbol{x} = \boldsymbol{0}$ *of system* (3.2.1) *is asymptotically* $\boldsymbol{y}$*-stable.*

Proof. If conditions (2a) of the theorem are satisfied, the unperturbed motion $\boldsymbol{x} = \boldsymbol{0}$ of system (3.2.1) is $\boldsymbol{y}$-stable. This means that, given any $\varepsilon > 0$, there is $\delta(\varepsilon) > 0$ such that, if $\|\boldsymbol{x}_0\| < \delta$, then $\|\boldsymbol{y}(t; 0, \boldsymbol{x}_0)\| < \varepsilon$ for all $t \geq 0$.* To prove the asymptotic $\boldsymbol{y}$-stability, let us show that, for all $\|\boldsymbol{x}_0\| < \delta$, the relationship $\lim\|\boldsymbol{y}(t; 0, \boldsymbol{x}_0)\| = 0$ holds, as $t \to \infty$.

On the contrary, suppose that

$$\lim\|\boldsymbol{y}(t; 0, \boldsymbol{x}_0)\| = \alpha = \text{const} > 0, \quad t \to \infty. \tag{3.2.2}$$

*Since system (3.2.1) is *autonomous*, we use the designation $\boldsymbol{x}(t; \boldsymbol{x}_0) = \boldsymbol{x}(t; 0, \boldsymbol{x}_0)$.

By virtue of condition (1) of the theorem, there exists a sequence $t_k \to \infty$ such that $\boldsymbol{x}(t;0,\boldsymbol{x}_0) \to \boldsymbol{x}_* \in \Gamma^+$, where Γ^+ is a nonempty *set of ω-limiting points* (Nemytskii and Stepanov [1949], Krasovskii [1959]) of the solution $\boldsymbol{x}(t;0,\boldsymbol{x}_0)$. In this case $\boldsymbol{y}_* \neq \mathbf{0}$ (and, consequently, $\boldsymbol{\xi}_* = (\boldsymbol{y}_*, \boldsymbol{\mu}(\boldsymbol{x}_*)) \neq \mathbf{0}$) by (3.2.2).

It is known that $\boldsymbol{x}(t;0,\boldsymbol{x}_*) \in \Gamma^+$ for all $t \geq 0$ and $\Gamma^+ \in M = \{\dot{V} = 0\}$. Since $\boldsymbol{\xi}_* \neq \mathbf{0}$, then by condition $(2b)$ of the theorem, we conclude that $\boldsymbol{\xi}(t;0,\boldsymbol{x}_*) \neq \mathbf{0}$ for all $t \geq 0$. (Here $\boldsymbol{\xi}(t;0,\boldsymbol{x}_*) = \{\boldsymbol{y}(t;0,\boldsymbol{x}_*), \boldsymbol{\mu}(\boldsymbol{x}(t;0,\boldsymbol{x}_*))\}$.) But then $\boldsymbol{x}(t;0,\boldsymbol{x}_*) \neq \mathbf{0}$ for all $t \geq 0$ as well, since otherwise not only $\boldsymbol{y}(t;0,\boldsymbol{x}_*) = 0$, $t \geq 0$ but also $\boldsymbol{\mu}(\boldsymbol{x}(t;0,\boldsymbol{x}_*)) = \mathbf{0}$, $t \geq 0$, by virtue of $\boldsymbol{\mu}(\mathbf{0}) = \mathbf{0}$, which leads to the relationship $\boldsymbol{\xi}(t;0,\boldsymbol{x}_*) = \mathbf{0}$ for all $t \geq 0$. Therefore, $\boldsymbol{x}(t;0,\boldsymbol{x}_*) \in M \setminus \{\boldsymbol{y} = \mathbf{0},\ \boldsymbol{\mu}(\boldsymbol{x}) = \mathbf{0}\}$ for any $t \geq 0$, which is impossible by condition $(2c)$ of the theorem.

This means that the assumption of (3.2.2) is erroneous and, consequently, $\lim \|\boldsymbol{y}(t;0,\boldsymbol{x}_0)\| = 0$, as $t \to \infty$, which proves the theorem. ☐

Remark 3.2.1.

1. If $\boldsymbol{\mu} = \mathbf{0}$, we have the Risito theorem (Risito [1970])—Theorem 0.4.5. If $\boldsymbol{\mu} \neq \mathbf{0}$, the conditions of Theorem 3.2.1 are more general than those of Theorem 0.4.5.

It should be stressed that the conditions of Theorem 0.4.5 and of the extended version stated may be satisfied in those cases when the sets $M = \{\dot{V} = 0\}$ (Rumyantsev [1971a]) and $\{V > 0\} \cap M$ (Oziraner [1973]) contain entire semitrajectories of system (3.2.1) for $t \in [0, \infty)$. Hence, the conditions of Theorem 3.2.1 are applicable when conditions of Theorems 0.4.4–0.4.6 are not satisfied.

2. If condition (1) of Theorem 3.2.1 is satisfied, the requirement of $\boldsymbol{z}$-continuability of solutions of system (3.2.1), which belongs to the basic assumptions accepted in the theory of stability with respect to part of the variables, is automatically met.

3. Condition $(2c)$ of Theorem 3.2.1 implies that the set $M = \{\dot{V} = 0\}$ can contain those entire semitrajectories of system (3.2.1) whose existence does not contradict the asymptotic $\boldsymbol{y}$-stability of the unperturbed motion $\boldsymbol{x} = \mathbf{0}$ of this system. As a result, Theorem 3.2.1 is a natural outgrowth of the Risito theorem (Risito [1970]) because it replaces the invariant set $\boldsymbol{y} = \mathbf{0}$ with $\boldsymbol{y} = \mathbf{0},\ \boldsymbol{\mu}(\boldsymbol{x}) = \mathbf{0}$.

4. Hatvani [1983a] proposed another (more general) variant of modifying the Risito theorem.

Now let the right-hand sides of system (3.2.1) be continuous and satisfy the conditions of uniqueness of solutions over the entire domain $\|\boldsymbol{x}\| < \infty$.

Theorem 3.2.2. *Suppose that*

(1) *each solution $\boldsymbol{x}(t;0,\boldsymbol{x}_0)$ of system (3.2.1) is $\boldsymbol{z}$-bounded;*

(2) *conditions $(2a)$–$(2c)$ of Theorem 3.2.1 hold in the domain $\|\boldsymbol{x}\| \leq H_1 < \infty$ and, besides,*

$$\lim a\big(\|\boldsymbol{y}\|\big) = \infty, \quad \|\boldsymbol{y}\| \to \infty.$$

Then the unperturbed motion $\boldsymbol{x} = \mathbf{0}$ *of system* (3.2.1) *is globally asymptotically* $\boldsymbol{y}$*-stable.*

Example 3.2.1. Consider *Euler's dynamical equations*

$$\begin{gathered} A\dot{y}_1 = \alpha_1 y_1 + (B - C)z_1 z_2, \\ B\dot{z}_1 = (C - A)y_1 z_2, \quad C\dot{z}_2 = (A - B)y_1 z_1 \end{gathered} \tag{3.2.3}$$

describing the angular motion of a solid under the action of a linear dissipative force with incomplete dissipation ($\alpha_1 = \text{const} < 0$).

Though Lyapunov stable, the equilibrium position of the solid cannot be asymptotically stable with respect to all the variables. We shall study the problem of asymptotic stability of the equilibrium position with respect to y_1. We take the *kinetic energy* of the solid as a V-function:

$$V = \frac{1}{2}(Ay_1^2 + Bz_1^2 + Cz_2^2), \quad \dot{V} = \alpha_1 y_1^2. \tag{3.2.4}$$

The sets $M = \{\dot{V} = 0\}$ (Rumyantsev [1971a]) and $M \cap \{V > 0\}$ (Oziraner [1973]) contain the entire trajectories $y_1 = z_1 = 0$, $z_2 = \text{const} \neq 0$ of system (3.2.3). Besides, the set $M_1 = \{y_1 = 0\}$ (Risito [1970]) is not an invariant set of system (3.2.3), though the set $M \setminus M_1$ (Risito [1970]) is empty and, consequently, contains no entire semitrajectories of this system for $t \in [0, \infty)$. As a result, the conditions of Theorems 0.4.4–0.4.6 fail to hold.

Let us demonstrate, however, that V-function (3.2.4) satisfies all the conditions of Theorem 3.2.1. To do so, we take the expression $\mu_1 = z_1 z_2$ as the second auxiliary function (in the case in question, it is a *scalar* function). The set $M_1^* = \{y_1 = \mu_1 = 0\}$ is an invariant set of system (3.2.3). If $B \neq C$, the set $M \setminus M_1^*$ contains no entire trajectories of this system. Indeed, the set $M \setminus M_1^*$ can contain only entire semitrajectories of the kind

$$y_1 = 0, \quad z_i = f_i(t), \; f_i > 0, \; (i = 1, 2), \quad t \geq 0. \tag{3.2.5}$$

(Otherwise, one can find $t = t^* > 0$ such that $y_1(t^*) = \mu_1(\boldsymbol{x}(t^*)) = 0$, which is impossible.) But the semitrajectories of type (3.2.5) cannot be entire semitrajectories of system (3.2.3), when $B \neq C$, by virtue of the structure of the first equation of the system.

In addition, being Lyapunov stable, all the solutions of system (3.2.3) originating from a sufficiently small neighborhood of the position $y_1 = z_1 = z_2 = 0$ are bounded. As a result, functions V and μ_1 satisfy all the conditions of Theorem 3.2.1 when $B \neq C$.

It can be shown that if $B \neq C$, the conditions of Theorem 3.2.2 also hold.

3.3 Partial Stability in the Presence of Large Initial Perturbations

3.3.1 Formulation of the Problem

The fact that stability in the small can be observed in one group of the variables while a different group is subjected to arbitrary initial perturbations has drawn the attention of Rouche and Peiffer [1967], Fergola and Moauro

[1970]. They demonstrated that, if the Lyapunov function V satisfies all the conditions of Theorem 0.4.1 and, besides, $V \leq b(\|\boldsymbol{y}\|)$, then the unperturbed motion $\boldsymbol{x} = \mathbf{0}$ of system (0.3.1) possesses the following property (*$\boldsymbol{y}$-stability on the whole with respect to $\boldsymbol{z}_0$*): given any $\varepsilon > 0$, $t_0 \geq 0$, one can find $\delta(\varepsilon) > 0$ such that if $\|\boldsymbol{y}_0\| < \delta$, $\|\boldsymbol{z}_0\| < \infty$, then $\|\boldsymbol{y}(t; t_0, \boldsymbol{x}_0)\| < \varepsilon$ for all $t_0 \geq 0$.

However, the desire to have this property implemented in a particular system conflicts with conditions that ensure it. Therefore, it seems reasonable to modify the definition above to weaken the conditions that make it feasible. The gist of the matter, as it appears from the standpoint of applications, is the same. To this end, we note that, in solving applied problems, the requirement $\|\boldsymbol{z}_0\| < \infty$ can be replaced by the condition $\|\boldsymbol{z}_0\| < \Delta$, where Δ is a *given* positive number.

Throughout this section it is assumed that the right-hand sides of system (0.3.1) satisfy condition (a) of Section 0.3. In addition, we suppose that solutions of system (0.3.1) beginning in the domain $\|\boldsymbol{y}\| \leq H_1 < H$, $\|\boldsymbol{z}\| \leq H_2 < \infty$ at $t = t_0$ are $\boldsymbol{z}$-continuable.

Definition 3.3.1. Let Δ be a given positive number. The unperturbed motion $\boldsymbol{x} = \mathbf{0}$ of system (0.3.1) is said to be

(1) *$\boldsymbol{y}$-stable for large $\boldsymbol{z}_0$* if, given any $\varepsilon > 0$, $t_0 \geq 0$, one can find $\delta(\varepsilon, t_0) > 0$ such that from $\|\boldsymbol{y}_0\| < \delta$, $\|\boldsymbol{z}_0\| < \Delta$ it follows that $\|\boldsymbol{y}(t; t_0, \boldsymbol{x}_0)\| < \varepsilon$ for all $t \geq t_0$;

(2) *uniformly $\boldsymbol{y}$-stable for large $\boldsymbol{z}_0$* if in definition (1) above a number δ does not depend on t_0;

(3) *asymptotically $\boldsymbol{y}$-stable for large $\boldsymbol{z}_0$* if it is $\boldsymbol{y}$-stable for large $\boldsymbol{z}_0$ and, besides, for any $t_0 \geq 0$, there is $\delta_1(t_0) > 0$ such that condition (0.3.3) is satisfied for all t_0, $\boldsymbol{x}_0$ from the domain $t_0 \geq 0$, $\|\boldsymbol{y}_0\| < \delta_1$, $\|\boldsymbol{z}_0\| < \Delta$.

As will be apparent, it is this kind of stability that is really established on numerous occasions, using the investigative technique devised.

Remark 3.3.1.

1. The requirement of $\boldsymbol{z}$-continuability of solutions of system (0.3.1) that originate at $t = t_0$ from the domain $\|\boldsymbol{y}\| \leq H_1 < H$, $\|\boldsymbol{z}\| \leq H_2 < \infty$ (and not just from a sufficiently small neighborhood of the unperturbed motion $\boldsymbol{x} = \mathbf{0}$) stems from the specificity of Definition 3.3.1. Indeed, in the context of Definition 3.3.1, the analysis embraces solutions of system (0.3.1) subjected to initial deviations from the unperturbed motion $\boldsymbol{x} = \mathbf{0}$ that are not small with respect to $\|\boldsymbol{z}_0\|$.

2. Definition 3.3.1 can be modified to treat the case of *exponential* asymptotic $\boldsymbol{y}$-stability for large $\boldsymbol{z}_0$.

3.3.2 Using the Method of Lyapunov Functions

Let us clarify conditions which are to be imposed on a Lyapunov function to use it in establishing stability in the sense of Definition 3.3.1.

Theorem 3.3.1. 1. *If it is possible to find a Lyapunov function $V(t, \boldsymbol{x})$ for system* (0.3.1) *satisfying all the conditions of Theorem* 0.4.1 *and, besides, $V(t, \mathbf{0}, \boldsymbol{z}) \equiv 0$, then the unperturbed motion $\boldsymbol{x} = \mathbf{0}$ of system* (0.3.1) *is $\boldsymbol{y}$-stable for large $\boldsymbol{z}_0$.*

2. *If all the conditions of Theorem* 0.4.1 *are satisfied and, in addition,*

$$V(t, \mathbf{0}, \boldsymbol{z}) \equiv 0, \qquad V(t, \boldsymbol{y}, \boldsymbol{z}) \leq b(\|\boldsymbol{x}\|), \tag{3.3.1}$$

then the unperturbed motion $\boldsymbol{x} = \mathbf{0}$ *of system* (0.3.1) *is uniformly* $\boldsymbol{y}$*-stable for large* $\boldsymbol{z}_0$.

3. *If one can find a vector function* $\boldsymbol{V} = (V_1, \ldots, V_k)$ *for system* (0.3.1), *satisfying conditions* (1)–(6) *of Theorem* 0.4.11 *and, in addition,*

$$V_i(t, \mathbf{0}, \boldsymbol{z}) \equiv 0 \qquad (i = 1, \ldots, k), \tag{3.3.2}$$

then the unperturbed motion $\boldsymbol{x} = \mathbf{0}$ *of system* (0.3.1) *is asymptotically* $\boldsymbol{y}$*-stable for large* $\boldsymbol{z}_0$.

Proof. 1. By virtue of $V(t, \mathbf{0}, \mathbf{0}) \equiv V(t, \mathbf{0}, \boldsymbol{z}) \equiv 0$, for any $\varepsilon > 0$, $t_0 \geq 0$ one can find $\delta(\varepsilon, t_0) > 0$ such that from $\|\boldsymbol{y}_0\| < \delta$, $\|\boldsymbol{z}_0\| < \Delta$ it follows that $V(t_0, \boldsymbol{x}_0) < a(\varepsilon)$. By the inequality $\dot{V} \leq 0$, for a solution that satisfies $\|\boldsymbol{y}_0\| < \delta$, $\|\boldsymbol{z}_0\| < \Delta$,

$$a(\|\boldsymbol{y}(t; t_0, \boldsymbol{x}_0)\|) \leq V(t, \boldsymbol{x}(t; t_0, \boldsymbol{x}_0)) \leq V(t_0, \boldsymbol{x}_0) < a(\varepsilon), \quad t \geq t_0. \tag{3.3.3}$$

Taking into account the properties of function $a(r)$, from (3.3.3) we derive $\|\boldsymbol{y}(t; t_0, \boldsymbol{x}_0)\| < \varepsilon$, $t \geq t_0$.

2. The condition $V(t, \boldsymbol{x}) \leq b(\|\boldsymbol{x}\|)$ implies that $V(t, \boldsymbol{x})$ can be reduced to an arbitrarily small value only by decreasing $\|\boldsymbol{x}\|$. Hence, for $t \geq 0$, $\|\boldsymbol{x}\| \leq N < +\infty$, the function $V(t, \boldsymbol{x})$ will be bounded ($|V(t, \boldsymbol{x})| \leq L < +\infty$) and, when number N is sufficiently small, number L will also be sufficiently small. Consequently, by virtue of $V(t, \mathbf{0}, \mathbf{0}) \equiv V(t, \mathbf{0}, \boldsymbol{z}) \equiv 0$, for any $\varepsilon > 0$, $t_0 \geq 0$, one can find $\delta(\varepsilon) > 0$ such that from $\|\boldsymbol{y}_0\| < \delta$, $\|\boldsymbol{z}_0\| < \Delta$ it follows that $V(t_0, \boldsymbol{x}_0) < a(\varepsilon)$. By the inequality $\dot{V} \leq 0$, for a solution of system (0.3.1) that satisfies $\|\boldsymbol{y}_0\| < \delta$, $\|\boldsymbol{z}_0\| < \Delta$, we have inequalities (3.3.3), whence it follows that $\|\boldsymbol{y}(t; t_0, \boldsymbol{x}_0)\| < \varepsilon$, $t \geq t_0$.

3. In view of conditions (2)–(4) of Theorem 0.4.11 and by the theorem of Wazewski [1950], the following relationship holds for all $t \geq t_0$:

$$a(\|\boldsymbol{y}\|) \leq \max_{i=1,\ldots,l} V_i(t, \boldsymbol{x}) \leq \max_{i=1,\ldots,l} \omega_i^+(t; t_0, \boldsymbol{V}(t_0, \boldsymbol{x}_0)). \tag{3.3.4}$$

Here $\boldsymbol{\omega}^+(t; t_0, \boldsymbol{\omega}_0) = \boldsymbol{\omega}^+(t; t_0, \boldsymbol{V}(t_0, \boldsymbol{x}_0))$ is the upper solution of system $\dot{\boldsymbol{\omega}} = \boldsymbol{f}(t, \boldsymbol{\omega})$, satisfying the initial condition $\boldsymbol{\omega}^+(t; t_0, \boldsymbol{\omega}_0) = \boldsymbol{\omega}_0$. The solution $\boldsymbol{\omega} = \mathbf{0}$ of the comparison system is asymptotically stable with respect to $\omega_1, \ldots, \omega_l$. Thus, for any $\varepsilon > 0$, $t_0 \geq 0$, one can find $\eta(\varepsilon, t_0) > 0$ such that from $\|\boldsymbol{\omega}_0\| < \eta$ it follows that $\max \omega_i(t; t_0, \boldsymbol{\omega}_0) < \varepsilon$ $(i = 1, \ldots, l)$ for all $t \geq t_0$.

By condition (1) of Theorem 0.4.11 and condition (3.3.2), there exists $\delta^*(\eta, t_0) = \delta(\varepsilon, t_0) > 0$ such that

$$\|\boldsymbol{w}_0\| = \|\boldsymbol{V}(t_0, \boldsymbol{x}_0)\| < \eta(\varepsilon, t_0) \text{ for } \|\boldsymbol{y}_0\| < \delta \quad \|\boldsymbol{z}_0\| < \Delta.$$

But then (3.1.5) implies that $\|\boldsymbol{y}(t; t_0, \boldsymbol{x}_0)\| < \varepsilon$ for all $t \geq t_0$.

For any $t_0 \geq 0$, there is $\delta_1^*(t_0) \geq 0$ such that the condition

$$\lim[\max \omega_i(t; t_0, \boldsymbol{\omega}_0)] = 0, \quad t \to \infty$$

holds for all $\|\boldsymbol{\omega}_0\| < \delta_1^*$. Therefore, by (3.3.2), for any $t_0 \geq 0$, there exists $\delta_1(t_0) > 0$ such that condition (0.3.3) is valid for $\|\boldsymbol{y}_0\| < \delta_1$, $\|\boldsymbol{z}_0\| < \Delta$. The theorem is proved. □

Remark 3.3.2.

1. Conditions (3.3.1) are *weaker* than the condition $V(t, \boldsymbol{x}) \leq b(\|\boldsymbol{y}\|)$. At the same time, from the practical point of view, the $\boldsymbol{y}$-stability for large $\boldsymbol{z}_0$ is equivalent to the $\boldsymbol{y}$-stability on the whole with respect to $\boldsymbol{z}_0$.

2. The problem of synthesizing a system *autonomous to within ε in a group of coordinates* is one of the problems of the automatic control theory that were posed by Letov [1981]. If the desired control is chosen so that the closed system (0.3.1) satisfies the conditions of Theorem 3.1.1, then the system will be autonomous, in particular (for $\delta = 0$, $\Delta = R$), to within ε with respect to variables $y_1, \ldots, y_m$.

Example 3.3.1.

1. Consider the *Lagrangian equations of a holonomic mechanical system under the action of potential and gyroscopic forces subject to constraints invariant in time*

$$\frac{d}{dt}\left(\frac{\partial T}{\partial \dot{q}_i}\right) - \frac{\partial T}{\partial q_i} = \sum_{j=1}^{n} g_{ij}\dot{q}_j - \frac{\partial P}{\partial q_i} \qquad (i = 1, \ldots, n). \tag{3.3.5}$$

Equations (3.3.5) are obtained from (3.2.3) by setting $Q_i = 0$ $(i = 1, \ldots, n)$ and admit the first integral $V = T + P$.

Let us denote $\boldsymbol{y} = (q_1, \ldots, q_m)$, $\boldsymbol{z} = (q_{m+1}, \ldots, q_n)$ and suppose that function $P(\boldsymbol{y}, \boldsymbol{z})$ is $\boldsymbol{y}$-positive-definite and $P(\boldsymbol{0}, \boldsymbol{z}) = 0$.

Let us also assume that T is positive-definite with respect to $\dot{q}_i$ $(i = 1, \ldots, n)$. Taking into account that $T(\boldsymbol{0}, \boldsymbol{q}) = 0$, we conclude that the function $V = V(\boldsymbol{y}, \dot{\boldsymbol{y}}, \boldsymbol{z}, \dot{\boldsymbol{z}})$ is $(\boldsymbol{y}, \dot{\boldsymbol{y}}, \dot{\boldsymbol{z}})$-positive-definite and $V(\boldsymbol{0}, \boldsymbol{0}, \boldsymbol{z}, \boldsymbol{0}) = 0$.

Hence, under the assumptions made, the equilibrium position $\boldsymbol{q} = \dot{\boldsymbol{q}} = \boldsymbol{0}$ of system (3.3.5) is uniformly $(\boldsymbol{y}, \dot{\boldsymbol{y}}, \dot{\boldsymbol{z}})$-stable *for large $\boldsymbol{z}_0$* by the second part of Theorem 3.3.1.

2. Consider the *motion of a unit mass point under the action of a constant gravitational force field* on the surface $x_1 = 0.5x_2^2(1 + x_3^2)$ in $Ox_1x_2x_3$-space, with the Ox_1-axis pointing vertically upward. This motion was analyzed in the work of Peiffer and Rouche [1969] and Hatvani [1983a]. In this case the kinetic T and potential P energies have the form

$$T = 0.5\{\dot{x}_3^2 + \dot{x}_2^2 + x_2^2[\dot{x}_2(1 + x_3^2) + \dot{x}_3 x_3 x_2]^2\},$$
$$P = 0.5gx_2^2(1 + x_3^2), \qquad g = \text{const} > 0.$$

Taking $\boldsymbol{y} = (x_2, \dot{x}_2, \dot{x}_3)$, $z = x_3$, we conclude that the function $V = T + P$ satisfies all the conditions of the second part of Theorem 3.3.1. Consequently, the equilibrium position $x_i = \dot{x}_i = 0$ $(i = 1, 2, 3)$ of the point is uniformly $\boldsymbol{y}$-stable *for large $\boldsymbol{z}_0$*. This means that if, at the initial moment, the deviation of the point from the equilibrium position along Ox_1- and Ox_2-axes is sufficiently small, that along the Ox_3-axis is prescribed to be arbitrary, and the velocity of the point is sufficiently small, then the deviation of the point along the Ox_2-axis and the projections of its velocity onto the Ox_2- and Ox_3-axes will remain sufficiently small.

We note that the condition $V \leq b(\|\boldsymbol{y}\|)$ for function V fails to hold.

3. In Example 3.1.1 the equilibrium position (3.1.9) of the point is stable with respect to $x_1, x_3, \dot{x}_1, \dot{x}_2$ *for large* x_{20}. Indeed, $V = 0$ once $x_1 = \dot{x}_1 = \dot{x}_2 = 0$ and $x_2 \neq 0$.

3.3.3 Relationship to Procedures for Constructing $\boldsymbol{\mu}$-Systems

We shall show just one possible usage of procedures for constructing $\boldsymbol{\mu}$-systems to analyze $\boldsymbol{y}$-stability in the sense of Definition 3.3.1.

Theorem 3.3.2. *Let all the conditions of Theorem* 2.4.1 *be satisfied. In addition, if functions* $\mu_j = \mu_j(\boldsymbol{y}, \boldsymbol{z})$ *in system* (2.4.5) *satisfy the conditions*

$$\mu_j(\boldsymbol{0}, \boldsymbol{z}) = 0 \qquad (j = 1, \ldots, m), \tag{3.3.6}$$

then the unperturbed motion $\boldsymbol{x} = \boldsymbol{0}$ *of system* (2.1.1) *is asymptotically* $\boldsymbol{y}$*-stable for large* $\boldsymbol{z}_0$.

Proof. Under the conditions of Theorem 2.4.1, the unperturbed motion $\boldsymbol{x} = \boldsymbol{0}$ of system (2.1.1) is asymptotically stable with respect to $y_1, \ldots, y_m$. This means that, for any ε, $t_0 \geq 0$, one can find $\eta_1(\varepsilon, t_0) > 0$ such that from $\|\boldsymbol{y}_0\| < \eta_1$, $\|\boldsymbol{\mu}_0\| < \eta_1$, it follows that $\|\boldsymbol{y}(t; t_0, \boldsymbol{x}_0)\| < \varepsilon$ for all $t_0 \geq 0$. Besides, for any $t_0 \geq 0$, there exists $\eta_2(t_0) > 0$ such that the condition $\lim\|\boldsymbol{y}(t; t_0, \boldsymbol{x}_0)\| = 0$, $t \to \infty$, is true for all the solutions that meet the condition $\|\boldsymbol{y}_0\| < \eta_2$, $\|\boldsymbol{\mu}_0\| < \eta_2$. Since the function $V = V(\boldsymbol{y}, \boldsymbol{\mu})$, which is employed to prove $\boldsymbol{y}$-stability in Theorem 2.4.1, does not explicitly depend on $z_1, \ldots, z_p$, the condition $\|\boldsymbol{z}_0\| < \eta_i$ $(i = 1, 2)$ is not implied (only if it does not follow from the conditions $\|\boldsymbol{y}_0\| < \eta_i$, $\|\boldsymbol{\mu}_0\| < \eta_i$).

If condition (3.3.6) is satisfied, the inequalities $\|\boldsymbol{y}_0\| < \eta_i$, $\|\boldsymbol{\mu}_0\| < \eta_i$ $(i = 1, 2)$ do not necessitate the condition $\|\boldsymbol{z}_0\| < \eta_i$ $(i = 1, 2)$. At the same time, there exist numbers $\delta_1(\varepsilon, t_0) > 0$, $\delta_2(t_0) > 0$ such that the inequalities $\|\boldsymbol{\mu}_0\| < \eta_i$ $(i = 1, 2)$ hold when $\|\boldsymbol{y}_0\| < \delta_i$, $\|\boldsymbol{z}_0\| < \Delta$, where Δ is a prescribed number. But then the unperturbed motion $\boldsymbol{x} = \boldsymbol{0}$ of system (2.1.1) is asymptotically $\boldsymbol{y}$-stable for large $\boldsymbol{z}_0$ in the sense of Definition 3.3.1. The theorem is proved. □

Corollary 3.3.1. *Let all the conditions of Theorem* 2.2.1 *be satisfied. In addition, if* $\boldsymbol{z}^{\mathsf{T}} = (\boldsymbol{z}^+, \boldsymbol{z}^-)$ *and functions* $\mu_j^{(1)} = \mu_j^{(1)}(\boldsymbol{z}^+, \boldsymbol{z}^-)$, $\mu_{\gamma\theta}^{(2)} = \mu_{\gamma\theta}^{(2)}(\boldsymbol{z}^+, \boldsymbol{z}^-)$ *in system* (2.2.7) *satisfy the conditions*

$$\mu_j^{(1)}(\boldsymbol{0}, \boldsymbol{z}^-) \equiv 0, \qquad \mu_{\gamma\theta}^{(2)}(\boldsymbol{0}, \boldsymbol{z}^-) \equiv 0 \qquad (j, \gamma, \theta = 1, \ldots, m),$$

then the unperturbed motion $\boldsymbol{x} = \boldsymbol{0}$ *of system* (2.2.1) *is asymptotically* $\boldsymbol{y}$*-stable for large* $\boldsymbol{z}_0^-$.

Example 3.3.2.

1. In Example 2.4.1 the new variable $\mu_1 = y_1 z_1$, through which the auxiliary system (2.4.10) is constructed, satisfies the equality $\mu_1(0, z_1) = 0$. Therefore, if conditions $a < 0$, $a + d < 0$ are valid, the unperturbed motion $y_1 = z_1 = 0$ of system (2.4.9) is asymptotically stable with respect to y_1 *for large* z_{10}.

2. Consider the problem of damping angular motions of a solid by one fixed engine, analyzed in Section 2.3.

The new variable $\mu_1 = A^{-1}(B - C)x_2 x_3$, through which the auxiliary $\boldsymbol{\mu}$-system (equivalent to equation (2.3.7)) is constructed, satisfies the equalities $\mu_1(0, x_3) = \mu_1(x_2, 0) = 0$.

Therefore, the equilibrium position $x_1 = x_2 = x_3 = 0$ of system (2.3.1), (2.3.3) is asymptotically stable with respect to x_1 *for large* x_{20} *or* x_{30}.

3.4 Using Differential Inequalities

3.4.1 Preliminary Remarks

The use of *differential inequalities* in the procedures for constructing auxiliary systems developed in Chapter 2 looks promising; evidently, differential inequalities make it *easier* to form an appropriate system. However, such an approach will require the *two-sided estimation* of the variables of the original system which are to be studied. This circumstance and an additional requirement imposed on the right-hand sides of the resulting auxiliary system (which also stems from involving the inequalities), restrict the scope for investigating applied problems with the approach outlined. However, the situation is typical of most methods for studying the stability of nonlinear systems. Turning to particular results and examples, we begin with the statement of general conditions that will allow one to use the proposed construction procedures in combination with differential inequalities. These conditions rest on the *principle of comparison with Lyapunov vector function*, which is further developed here.

3.4.2 Two-Sided Differential Inequalities and Conditions for Stability with Respect to Part of the Variables

Theorem 3.4.1. 1. *Let there exist in domain* (0.3.2) *two continuously differentiable vector functions* $\boldsymbol{V} = (V_1, \ldots, V_k)$ *and* $\boldsymbol{W} = (W_1, \ldots, W_r)$ *such that*

$$V_i = W_i = y_i \quad (i = 1, \ldots, m), \quad V_j = \mu_j(t, \boldsymbol{x}), \quad W_s = \bar{\mu}_s(t, \boldsymbol{x}),$$
$$\mu_j(t, \mathbf{0}) \equiv 0, \quad \bar{\mu}_s(t, \mathbf{0}) \equiv 0 \quad (j = m+1, \ldots, k;\ s = m+1, \ldots, r).$$

In addition, let these functions satisfy the following conditions:

(1) *by virtue of system* (0.3.1), *derivatives* $\dot{\boldsymbol{V}}$, $\dot{\boldsymbol{W}}$ *obey the inequalities*

$$\dot{V}_\mu \leq \varphi_\mu(t, V_1, \ldots, V_k) \qquad (\mu = 1, \ldots, k), \tag{3.4.1}$$

$$\dot{W}_\theta \geq f_\theta(t, W_1, \ldots, W_r) \qquad (\theta = 1, \ldots, r), \tag{3.4.2}$$

where vector functions $\boldsymbol{\varphi}(t, \boldsymbol{V}) = (\varphi_1, \ldots, \varphi_k)$ *and* $\boldsymbol{f}(t, \boldsymbol{W}) = (f_1, \ldots, f_r)$, $\boldsymbol{\varphi}(t, \mathbf{0}) \equiv \mathbf{0}$, $\boldsymbol{f}(t, \mathbf{0}) \equiv \mathbf{0}$, *are defined and continuous in the domain* D : $t \geq 0$, $\|\boldsymbol{V}^*\| \leq R$ *with* $R = \infty$ *or*

$$R > \sup\big[\|\boldsymbol{V}^*(t, \boldsymbol{x})\| : t \geq 0, \ \|\boldsymbol{y}\| \leq H\big], \quad \boldsymbol{V}^* = (\boldsymbol{V}, \boldsymbol{W});$$

(2) *each of the functions is nondecreasing in* $V_i, i \neq \mu$ ($W_j, j \neq \theta$);

(3) *the solution* $\boldsymbol{\omega} = \mathbf{0}$ ($\boldsymbol{u} = \mathbf{0}$) *of the comparison system* $\dot{\boldsymbol{\omega}} = \boldsymbol{\varphi}(t, \boldsymbol{\omega})$ ($\dot{\boldsymbol{u}} = \boldsymbol{f}(t, \boldsymbol{u})$) *is asymptotically stable with respect to* $\omega_1, \ldots, \omega_m$ ($u_1, \ldots, u_m$).

Then the unperturbed motion $\boldsymbol{x} = \mathbf{0}$ *of system* (0.3.1) *is asymptotically stable with respect to* $y_1, \ldots, y_m$.

2. *In the case* $m = 1$, *the construction of a system of differential inequalities* (3.4.1) *can be performed in the domain*

$$\boldsymbol{\Lambda}_1 : t \geq 0, \ \ 0 \leq y_1 \leq H, \ \ \|\boldsymbol{z}\| < \infty,$$

and the construction of a system (3.4.2) *in the domain*

$$\boldsymbol{\Lambda}_2 : t \geq 0, \ \ -H \leq y_1 < 0, \ \ \|\boldsymbol{z}\| < \infty.$$

Proof. For simplicity, we assume that, in any closed subdomain of D bounded with respect to $\boldsymbol{V}$ ($\boldsymbol{W}$), functions $\boldsymbol{\varphi}$ ($\boldsymbol{f}$) satisfy the Lipschitz condition relative to $\boldsymbol{V}$ ($\boldsymbol{W}$), i.e., the conditions of the uniqueness of solutions are fulfilled for the right-hand sides of the comparison systems.

1. Condition (3.4.2) is equivalent to the condition

$$-\dot{W}_\theta \leq -f_\theta(t, W_1, \ldots, W_r) = f_\theta^\wedge(t, -W_1, \ldots, -W_r) \quad (\theta = 1, \ldots, r).$$

Because functions $f_\theta(t, \boldsymbol{W})$ are nondecreasing in W_j ($j \neq \theta$),

$$f_\theta(t, \boldsymbol{W}) = \sum_{j=1}^{r} f_{\theta j}^*(t, \boldsymbol{W}) W_j$$

and $f_{\theta j}^*(t, \boldsymbol{W}) \geq 0$ for $j \neq \theta$. Hence,

$$f_\theta^\wedge(t, -\boldsymbol{W}) = \sum_{j=1}^{r} f_{\theta j}^*(t, \boldsymbol{W})(-W_j)$$

and, consequently functions $f_\theta^\wedge$ ($\theta = 1, \ldots, r$) do not decrease in $-W_j$ $j \neq \theta$.

The system of equations $\dot{\boldsymbol{\xi}} = \boldsymbol{f}^\wedge(t, \boldsymbol{\xi})$ is obtained by multiplying both sides of the vector equality $\dot{\boldsymbol{u}} = \boldsymbol{f}(t, \boldsymbol{u})$ by (-1) and designating $\boldsymbol{\xi} = -\boldsymbol{u}$. Therefore, from the asymptotic stability of the solution $\boldsymbol{u} = \mathbf{0}$ of the system

$\dot{\boldsymbol{u}} = \boldsymbol{f}(t, \boldsymbol{u})$ with respect to $u_1, \ldots, u_m$, it follows that the solution $\boldsymbol{\xi} = \boldsymbol{0}$ of the system $\dot{\boldsymbol{\xi}} = \boldsymbol{f}^{\wedge}(t, \boldsymbol{\xi})$ is asymptotically stable with respect to $\xi_1, \ldots, \xi_m$.

Let us construct vector function $\bar{\boldsymbol{V}} = (V_1, \ldots, V_k, -W_1, \ldots, -W_r)$ and set $\bar{V}^* = \max(V_s, -W_s) = \max|y_s|$, $(s = 1, \ldots, m)$. Taking into account that vector function $\bar{\boldsymbol{V}}$ satisfies the system of differential inequalities

$$\dot{V}_\mu \le \varphi_\mu(t, \boldsymbol{V}), \quad -\dot{W}_\theta \le f_\theta^{\wedge}(t, -\boldsymbol{W}) \quad (\mu = 1, \ldots, k;\ \theta = 1, \ldots, r)$$

and that $\bar{V}^* \ge a(\|\boldsymbol{y}\|)$, we conclude that vector function $\bar{\boldsymbol{V}}$ satisfies the conditions of Theorem 0.4.11. By this theorem, the unperturbed motion $\boldsymbol{x} = \boldsymbol{0}$ of system (0.3.1) is asymptotically stable with respect to $y_1, \ldots, y_m$.

2. By the theorem of Wazewski [1950], the inequalities

$$\begin{aligned} \boldsymbol{V}(t; t_0, \boldsymbol{V}_0) &\le \boldsymbol{\omega}(t; t_0, \boldsymbol{\omega}_0), & \boldsymbol{V}_0 &\le \boldsymbol{\omega}_0, & V_{10} &\ge 0, \\ -\boldsymbol{W}(t; t_0, \boldsymbol{W}_0) &\le -\boldsymbol{u}(t; t_0, \boldsymbol{u}_0), & -\boldsymbol{W}_0 &\le -\boldsymbol{u}_0, & W_{10} &< 0 \end{aligned}$$

are valid for all $t \ge t_0$ at which $V_1(t) \ge 0$ and $W_1(t) < 0$, respectively.

Hence,

$$\begin{aligned} y_1(t; t_0, \boldsymbol{x}_0) &\le \omega_1(t; t_0, \boldsymbol{\omega}_0), & y_{10} &\ge 0, & (y_1(t) &\ge 0), \\ -y_1(t; t_0, \boldsymbol{x}_0) &\le -u_1(t; t_0, \boldsymbol{u}_0), & y_{10} &< 0 & (y_1(t) &< 0). \end{aligned}$$

As a result, by virtue of condition (3) of the theorem, for any $\varepsilon > 0$, $t_0 \ge 0$, one can find $\delta(\varepsilon, t_0) > 0$ such that from $\|\boldsymbol{x}_0\| < \delta$ (in both cases $y_{10} \ge 0$ and $y_{10} < 0$), it follows that $y_1(t; t_0, \boldsymbol{x}_0) < \varepsilon$ for all $t \ge t_0$ at which $y_1(t) \ge 0$ and, besides, $\lim|y_1(t; t_0, \boldsymbol{x}_0)| = 0$, $t \to \infty$.

In the same way, we can prove that $y_1(t; t_0, \boldsymbol{x}_0) > -\varepsilon$ for all $t \ge t_0$ at which $y_1(t) < 0$ and $\lim|y_1(t; t_0, \boldsymbol{x}_0)| = 0$, $t \to \infty$. The theorem is proved. □

Remark 3.4.1.

1. Theorem 3.4.1 proposes to obtain the inequalities $|y_i(t; t_0, \boldsymbol{x}_0)| < \varepsilon$, $t \ge t_0$ $(i = 1, \ldots, m)$ required to draw the conclusion on the stability with respect to $y_1, \ldots, y_m$ in *two stages*: first we prove the inequalities $y_i(t; t_0, \boldsymbol{x}_0) < \varepsilon$ and then the inequalities $y_i(t; t_0, \boldsymbol{x}_0) > -\varepsilon$, $t \ge t_0$ $(i = 1, \ldots, m)$. So, it is suggested that in each of the indicated cases *its own* system of differential inequalities be constructed, i.e., systems (3.4.1) and (3.4.2), respectively.

2. If $m = 1$, then, as was shown in the second part of Theorem 3.4.1, the differential inequalities (3.4.1) and (3.4.2) can be constructed not in domain (0.3.2), but in domains $\boldsymbol{\Lambda}_1$ and $\boldsymbol{\Lambda}_2$, respectively. This extends the potentialities of the proposed approach.

3.4.3 The Relationship to Procedures for Constructing μ-Systems

We shall now dwell on one possible application of differential inequalities to the procedures for constructing auxiliary systems developed earlier. Consider the system of equations (2.1.1)

$$\begin{aligned} \frac{dy_i}{dt} &= \sum_{k=1}^{m} a_{ik} y_k + Y_i^0(\boldsymbol{y}, \boldsymbol{z}) + Y_i^{**}(t, \boldsymbol{y}, \boldsymbol{z}), \\ \frac{dz_j}{dt} &= \sum_{k=1}^{m} c_{jk} y_k + \sum_{l=1}^{p} d_{jl} z_l + Z_j(t, \boldsymbol{y}, \boldsymbol{z}) \\ &(i = 1, \ldots, m;\ j = 1, \ldots, p), \end{aligned} \tag{3.4.3}$$

under the assumptions made in Subsection 2.4.2.

In addition, suppose that the equalities

$$Y_i^{(1)}(\boldsymbol{y}, \boldsymbol{z}) \leq Y_i^0(\boldsymbol{y}, \boldsymbol{z}) \leq Y_i^{(2)}(\boldsymbol{y}, \boldsymbol{z}) \quad (i = 1, \ldots, m) \tag{3.4.4}$$

hold in domain (0.3.2), where $Y_i^{(r)}$ $(i = 1, \ldots, m;\ r = 1, 2)$ have the the same type of forms as Y_i^0.

Let us consider the systems of equations

$$\begin{aligned} \frac{dy_i}{dt} &= \sum_{k=1}^{m} a_{ik} y_k + Y_i^{(r)}(\boldsymbol{y}, \boldsymbol{z}) + Y_i^{**}(t, \boldsymbol{y}, \boldsymbol{z}), \\ \frac{dz_j}{dt} &= \sum_{k=1}^{m} c_{jk} y_k + \sum_{l=1}^{p} d_{jl} z_l + Z_j(t, \boldsymbol{y}, \boldsymbol{z}) \\ &(i = 1, \ldots, m;\ j = 1, \ldots, p) \end{aligned} \tag{3.4.5}$$

corresponding to the values $r = 1, 2$. (These systems differ only in expressions for $Y_i^{(r)}(\boldsymbol{y}, \boldsymbol{z})$.)

To construct auxiliary systems, in line with Subsection 2.4.3, we introduce the new variables $\mu_i^{(r)} = Y_i^{(r)}(\boldsymbol{y}, \boldsymbol{z})$ $(i = 1, \ldots, m;\ r = 1, 2)$ and suppose that at the first stage of introducing new variables one can already form (for $r = 1$ and $r = 2$, respectively) the systems

$$\begin{aligned} \frac{dy_i}{dt} &= \sum_{k=1}^{m} a_{ik} y_k + \mu_i^{(r)} + Y_i^{**}(t, \boldsymbol{y}, \boldsymbol{z}), \\ \frac{d\mu_j^{(r)}}{dt} &= \sum_{k=1}^{m} e_{jk}^{(r)} \mu_k^{(r)} + Z_j^{(r)*}(t, \boldsymbol{y}, \boldsymbol{z}), \\ \frac{dz_s}{dt} &= \sum_{k=1}^{m} c_{sk} y_k + \sum_{l=1}^{p} d_{sl} z_l + Z_s(t, \boldsymbol{y}, \boldsymbol{z}), \\ Z_j^{(r)*} &= \sum_{k=1}^{m} \frac{\partial Y_j^{(r)}}{\partial y_k} (Y_k^{(r)} + Y_k^{**}) + \sum_{l=1}^{p} \frac{\partial Y_j^{(r)}}{\partial z_l} Z_l \\ &(i, j = 1, \ldots, m;\ s = 1, \ldots, p;\ r = 1, 2). \end{aligned} \tag{3.4.6}$$

In Subsection 2.4.3 the linear parts of the first two groups of equations of this system were termed auxiliary $\boldsymbol{\mu}$-systems for the original systems (3.4.5). In Subsection 2.4.4 it was shown that, in the domains $t \geq 0$, $|y_i| \leq H$, $|\mu_i^{(r)}| \leq H$, $|z_j| < \infty$ $(i = 1, \ldots, m;\ j = 1, \ldots, p;\ r = 1, 2)$, if the conditions

$$\left|Y_i^{**}(t, \boldsymbol{y}, \boldsymbol{z})\right| + \left|Z_i^{(r)*}(t, \boldsymbol{y}, \boldsymbol{z})\right| \leq \sum_{k=1}^{m} \alpha_{ik} |y_k| + \sum_{k=1}^{m} \bar{\alpha}_{ik}^{(r)} |\mu_k^{(r)}| \qquad (3.4.7)$$

$$(i = 1, \ldots, m;\ r = 1, 2)$$

are satisfied, the solutions $y_i = 0$, $z_j = 0$ $(i = 1, \ldots, m;\ j = 1, \ldots, p)$ of systems (3.4.5) are asymptotically stable with respect to $y_1, \ldots, y_m$.

To weaken conditions (3.4.7) we assume that the following inequalities are valid in the domain $t \geq 0$, $|y_i| \leq H$, $|\mu_i^{(r)}| \leq H$, $|z_j| < \infty$ $(i = 1, \ldots, m;\ j = 1, \ldots, p;\ r = 1, 2)$:

$$\begin{gathered} Y_i^{**}(t, \boldsymbol{y}, \boldsymbol{z}) \underset{(\geq)}{\leq} \Phi_i^{(r)}(t, \boldsymbol{y}, \boldsymbol{\mu}^{(r)}), \\ Z_j^{(r)*}(t, \boldsymbol{y}, \boldsymbol{z}) \underset{(\geq)}{\leq} \Psi_j^{(r)}(t, \boldsymbol{y}, \boldsymbol{\mu}^{(r)}), \\ \boldsymbol{\mu}^{(r)} = (\mu_1^{(r)}, \ldots, \mu_m^{(r)}), \quad r = 1, 2 \quad (i, j = 1, \ldots, m), \end{gathered} \qquad (3.4.8)$$

where functions $\Phi_i^{(r)}$ and $\Psi_j^{(r)}$ are defined and continuous in the specified domains. In inequalities (3.4.8) and further in Subsection 3.4.3, the sign "$\leq$" corresponds to $r = 1$ and the sign "$\geq$" corresponds to $r = 2$.

From equations (3.4.3) one can segregate the systems of inequalities

$$\begin{gathered} \frac{dy_i}{dt} \underset{(\geq)}{\leq} \sum_{k=1}^{m} a_{ik} y_k + \mu_i^{(r)} + \Phi_i^{(r)}(t, \boldsymbol{y}, \boldsymbol{\mu}^{(r)}), \\ \frac{d\mu_j^{(r)}}{dt} \underset{(\geq)}{\leq} \sum_{k=1}^{m} e_{jk}^{(r)} \mu_k^{(r)} + \Psi_j^{(r)}(t, \boldsymbol{y}, \boldsymbol{\mu}^{(r)}) \\ (i, j = 1, \ldots, m;\ r = 1, 2) \end{gathered} \qquad (3.4.9)$$

which we call *$\boldsymbol{\mu}$-systems of inequalities* for (3.4.3) in accordance with the terminology adopted. Replacing (3.4.7) with weaker conditions (3.4.8) allows us to strengthen the conditions of asymptotic stability with respect to $y_1, \ldots, y_m$ established in Theorem 2.4.1.

Corollary 3.4.1. *Let $a_{ik} \geq 0$ $(i, k = 1, \ldots, m;\ i \neq k)$, $e_{jk}^{(r)} \geq 0$, $(j, k = 1, \ldots, m;\ j \neq k;\ r = 1, 2)$. Also let each of the functions $\Phi_i^{(r)}$, $\Psi_j^{(r)}$ $(i, j = 1, \ldots, m;\ r = 1, 2)$ be nondecreasing in y_k, $\mu_s^{(r)}$ $(k, s = 1, \ldots, m)$ for $k \neq i$ and $s \neq j$, respectively. If the zero solutions of the comparison systems for (3.4.9) are Lyapunov asymptotically stable, then the unperturbed motion $\boldsymbol{x} = \boldsymbol{0}$ of system (3.4.3) is asymptotically stable with respect to $y_1, \ldots, y_m$.*

Proof. Let us compose vector functions $\boldsymbol{W}^{(r)} = (y_1, \ldots, y_m, \mu_1^{(r)}, \ldots, \mu_m^{(r)})$, $r = 1, 2$. By (3.4.4), (3.4.8), in the domain $t \geq 0$, $|y_i| \leq H$, $|\mu_i^{(r)}| \leq H$,

$|z_j| < \infty$ $(i = 1, \dots, m;\ j = 1, \dots, p;\ r = 1, 2)$, estimates (3.4.9) are valid for system (3.4.3).

Therefore, in the specified domain, vector functions $\boldsymbol{W}^{(r)}$ satisfy the inequalities

$$\dot W_i^{(r)} \underset{(\geq)}{\leq} \sum_{k=1}^{m} a_{ik} W_k^{(r)} + W_{m+i}^{(r)} + \Phi_i^{(r)}(t, \boldsymbol{W}^{(r)}),$$

$$\dot W_{m+j}^{(r)} \underset{(\geq)}{\leq} \sum_{l=1}^{m} e_{jl}^{(r)} W_{m+l}^{(r)} + \Psi_j^{(r)}(t, \boldsymbol{W}^{(r)})$$

$$(i, j = 1, \dots, m;\ r = 1, 2).$$

This means that vector functions $\boldsymbol{W}^{(r)}$ satisfy the conditions of Theorem 3.4.1 and, consequently, the unperturbed motion $\boldsymbol{x} = \boldsymbol{0}$ of system (3.4.3) is asymptotically stable with respect to $y_1, \dots, y_m$. The statement is proved. □

Remark 3.4.2.

1. When constructing auxiliary systems of inequalities for (3.4.5) instead of $\mu_i^{(r)} = Y_i^{(r)}(\boldsymbol{y}, \boldsymbol{z})$, one can use new variables obtained by "*splitting*" $Y_i^{(r)}$ as indicated in Section 2.2 (see Remark 2.2.3).

2. The procedure for constructing $\boldsymbol{\mu}$-systems of inequalities can also be extended to the case when the coefficients of the linear part and the expressions Y_i^0 *depend on* t. (For details see Subsection 2.2.8.)

Example 3.4.1.

1. Let the equations of perturbed motion (3.4.3) have the form

$$\begin{gathered} \dot y_1 = -y_1 + y_1^2 z_1^2 - z_1^2 z_2^2, \\ \dot z_1 = 0.5 z_1 - y_1 z_1^3 + z_1 f(t, y_1, z_1, z_2), \quad \dot z_2 = -z_2 + y_1 z_1 z_2, \end{gathered} \tag{3.4.10}$$

where f is a function continuous in domain (0.3.2) such that $f \leq 0$. Bearing in mind that the estimates

$$-z_1^2 z_2^2 \leq Y_1(y_1, z_1, z_2) = y_1^2 z_1^2 - z_1^2 z_2^2 \leq y_1^2 z_1^2$$

are valid for all y_1, z_1, z_2, let us introduce the new variables $\mu_1^{(1)} = y_1^2 z_1^2$, $\mu_2^{(2)} = -z_1^2 z_2^2$ and construct $\boldsymbol{\mu}$-systems of inequalities

$$\begin{aligned} \dot y_1 &\leq -y_1 + \mu_1^{(1)}, & \dot\mu_1^{(1)} &= -\mu_1^{(1)} + 2 y_1^2 z_1^2 f \leq -\mu_1^{(1)}, \\ \dot y_1 &\geq -y_1 + \mu_1^{(2)}, & \dot\mu_1^{(2)} &= -\mu_1^{(2)} - 2 z_1^2 z_2^2 f \geq -\mu_1^{(2)}. \end{aligned} \tag{3.4.11}$$

Since the zero solutions of the comparison systems $\dot y_1 = -y_1 + \mu_1^{(r)}$, $\dot\mu_1^{(r)} = -\mu_1^{(r)}$ $(r = 1, 2)$ for the differential inequalities obtained are asymptotically stable, the unperturbed motion $y_1 = z_1 = z_2 = 0$ of system (3.4.10) is asymptotically stable with respect to y_1 by Corollary 3.4.1.

For comparison with the results obtained in Chapter 2, we note that introducing the new variable $\mu_1 = y_1^2 z_1^2 - z_1^2 z_2^2$ leads to the system

$$\begin{gathered} \dot y_1 = -y_1 + \mu_1, \quad \dot\mu_1 = -\mu_1 + \mu_1 f, \\ \dot z_1 = 0.5 z_1 - y_1 z_1^3 + z_1 f, \quad \dot z_2 = -z_2 + y_1 z_1 z_2. \end{gathered} \tag{3.4.12}$$

By Theorem 2.3.1, the unperturbed motion $y_1 = z_1 = z_2 = 0$ of system (3.4.10) is asymptotically stable with respect to y_1 if the following condition is valid in the domain $t \geq 0$, $|y_1| \leq H$, $|\mu_1| \leq H$, $\|\boldsymbol{z}\| < \infty$:

$$|\mu_1 f(t, y_1, z_1, z_2)| \leq \alpha_1 |y_1| + \alpha_2 |\mu_1|, \tag{3.4.13}$$

where α_1, α_2 are sufficiently small positive constants. Condition (3.4.13) holds (in the domain specified) for $|f| \leq \alpha$, with α a sufficiently small positive constant. At the same time, using $\boldsymbol{\mu}$-systems of differential inequalities (3.4.11), one can handle the case of *unbounded* function f.

2. Consider *Euler's dynamical equations*

$$\begin{gathered} A\dot{y}_1 = \alpha_1 y_1 + (B - C) z_1 z_2, \\ B\dot{z}_1 = \alpha_2 z_1 + (C - A) y_1 z_2, \quad C\dot{z}_2 = \alpha_3 z_2 + (A - B) y_1 z_1 \end{gathered} \tag{3.4.14}$$

describing rotational motions of a solid about its center of mass under the action of dissipative and accelerating forces. Here constant coefficients α_i $(i = 1, 2, 3)$ characterize the total effect of these forces.

Suppose that A is the *medium* moment of inertia of the solid. Transforming equations (3.4.14) to $\boldsymbol{\mu}$-system of inequalities, we obtain the following estimates:

(a)

$$\begin{gathered} \dot{y}_1 = \alpha_1^* y_1 + \mu_1, \\ \dot{\mu}_1 = (\alpha_2^* + \alpha_3^*)\mu_1 + \frac{B - C}{A} y_1 \left[\frac{C - A}{B} z_2^2 + \frac{A - B}{C} z_1^2 \right] \leq (\alpha_2^* + \alpha_3^*)\mu_1 \end{gathered} \tag{3.4.15}$$

in the domain $0 \leq y_1 \leq H$, $-\infty < z_i < +\infty$ $(i = 1, 2)$;

(b)

$$\dot{y}_1 = \alpha_1^* y_1 + \mu_1, \quad \dot{\mu}_1 \geq (\alpha_2^* + \alpha_3^*)\mu_1 \tag{3.4.16}$$

in the domain $-H \leq y_1 < 0$, $-\infty < z_i < +\infty$ $(i = 1, 2)$.

In relationships (3.4.15), (3.4.16) we set

$$\alpha_1 = \alpha_1^* A \qquad (123, ABC).$$

Since the solution $\xi_1 = \xi_2 = 0$ of the system

$$\dot{\xi}_1 = \alpha_1^* \xi_1 + \xi_2, \qquad \dot{\xi}_2 = (\alpha_2^* + \alpha_3^*)\xi_2 \tag{3.4.17}$$

is exponentially asymptotically stable (with respect to both variables) for $\alpha_1^* < 0$, $\alpha_2^* + \alpha_3^* < 0$, the equilibrium position $y_1 = z_1 = z_2 = 0$ of system (3.4.14) is exponentially asymptotically stable with respect to y_1 by Corollary 3.4.1.

We note that $\boldsymbol{\mu}$-systems of inequalities (3.4.15), (3.4.16) can be constructed for any $H > 0$. In addition, the zero solution of system (3.4.17) is globally exponentially asymptotically stable. Therefore, in the case when $\alpha_1^* < 0$, $\alpha_2^* + \alpha_3^* < 0$ the equilibrium position of system (3.4.14) is *globally* exponentially asymptotically stable with respect to y_1.

By virtue of identities $\mu_1(0, z_2) \equiv \mu_1(z_1, 0) \equiv 0$, the equilibrium position of system (3.4.14) is also exponentially asymptotically stable with respect to y_1 *for large* z_{10} *or* z_{20}. It should be stressed that this property (in view of particularity of Definition 3.3.1) is not a consequence of global exponential asymptotic stability with respect to y_1.

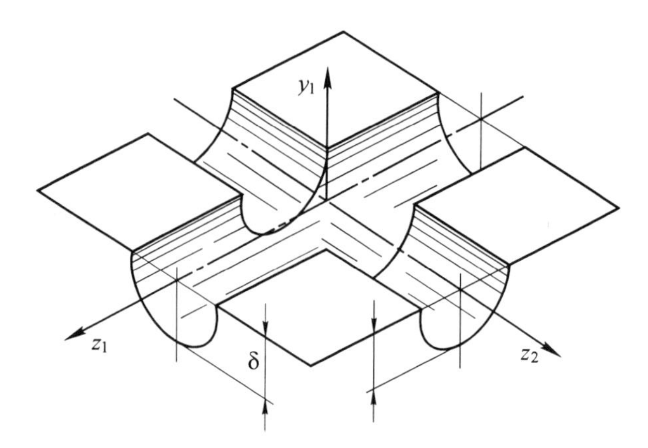

Figure 3.4.1. $\delta(\varepsilon)$-channels characterizing the properties of the solutions of system (3.2.13).

3. Consider a *system of Lorenz differential equations*

$$\begin{gathered} \dot{y}_1 = -\beta y_1 + z_1 z_2, \\ \dot{z}_1 = s(z_2 - z_1), \quad \dot{z}_2 = r z_1 - z_2 - z_1 y_1, \end{gathered} \tag{3.4.18}$$

where constants r, β, and s are strictly positive. Certain *models of convective turbulence* and *quantum optical generator* can be described in terms of equations (3.4.18) (Lorenz [1963]).

The equilibrium position $y_1 = z_1 = z_2 = 0$ of system (3.4.18) is Lyapunov asymptotically stable for $0 < r < 1$.

In addition, we show that, for all admissible r, β, s and for any $\varepsilon > 0$ (not necessarily sufficiently small only), an arbitrary solution of system (3.4.18), which originated from within $\delta(\varepsilon)$ channels of a prescribed fixed length (see Figure 3.4.1), will not cross the boundary defined by the plane $y_1 = -\varepsilon$ for all $t \geq t_0$.

This property can be interpreted in the terms introduced in Subsection 3.3.1 as a *one-sided stability with respect to y_1 for large z_{10} or z_{20}*.

To prove this fact, we note that the $\boldsymbol{\mu}$-system of inequalities

$$\begin{gathered} \dot{y}_1 = -\beta y_1 + \mu_1 \quad (\mu_1 = z_1 z_2), \\ \dot{\mu}_1 = -(s+1)\mu_1 + s z_2^2 + r z_1^2 - z_1^2 y_1 \geq -(s+1)\mu_1 \end{gathered}$$

can be constructed for variable y_1 of system (3.4.18) in the domain $-H \leq y_1 \leq 0$, $\|\boldsymbol{z}\| < \infty$.

By Corollary 3.4.1, we conclude that the equilibrium position $y_1 = z_1 = z_2 = 0$ of system (3.4.18) has the following property: for any $\varepsilon > 0$, $t_0 \geq 0$, there exists $\delta(\varepsilon) > 0$ such that, from $|y_{10}| < \delta$, $|z_{10} z_{20}| < \delta$, it follows that $y_1(t; t_0, y_{10}, z_{10}, z_{20}) > -\varepsilon$, $t \geq t_0$. Besides, $\lim y_1(t) = 0$, $t \to \infty$, where zero is approached by the exponential law. Since the inequality $|z_{10} z_{20}| < \delta$ is possible for $|z_{10}| < \delta_1$, $|z_{20}| < \Delta$ or for $|z_{10}| < \Delta$, $|z_{20}| < \delta_1$, where δ_1 is a sufficiently small number and Δ is a prescribed number, then the one-sided stability of the equilibrium position of system (3.4.18) with respect to y_1 for large z_{10} or z_{20} is observed.

3.5 Instability with Respect to Part of the Variables

3.5.1 Formulation of the Problem

The interest in the *instability* problem is explained by the fact that, having found the instability, one can spend fruitless efforts proving the stability. In line with the subject of this book, we consider the problem of instability *with respect to part* of the variables.

Let us make the corresponding notion of instability more specific.

Definition 3.5.1. The unperturbed motion $\boldsymbol{x} = \mathbf{0}$ of system (0.3.1) is said to be *unstable with respect to* $y_1, \ldots, y_m$ if this motion is not stable with respect to $y_1, \ldots, y_m$, i.e., there exist numbers $\varepsilon > 0$, $t_0 \geq 0$, such that for any $\delta(\varepsilon, t_0) > 0$, no matter how small, system (0.3.1) has a solution $\boldsymbol{x} = \boldsymbol{x}(t; t_0, \boldsymbol{x}_0)$ with $\|\boldsymbol{x}_0\| < \delta$ that satisfies the inequality $\|\boldsymbol{y}(t; t_0, \boldsymbol{x}_0\|) \geq \varepsilon$ at some time.

3.5.2 Auxiliary Definitions and Proposition

Definition 3.5.2. The unperturbed motion $\boldsymbol{x} = \mathbf{0}$ of system (0.3.1) is said to be *upper (lower) unstable with respect to* $y_1, \ldots, y_m$ if there exist numbers $\varepsilon > 0$, $t_0 \geq 0$ such that, for any $\delta(\varepsilon, t_0) > 0$, no matter how small, system (0.3.1) has a solution $\boldsymbol{x} = \boldsymbol{x}(t; t_0, \boldsymbol{x}_0)$ with $\|\boldsymbol{x}_0\| < \delta$ that satisfies the inequalities $\|y_i(t; t_0, \boldsymbol{x}_0\|) \geq \varepsilon$ ($\|y_i(t; t_0, \boldsymbol{x}_0\|) \leq -\varepsilon$) at some time, for at least one $i = 1, \ldots, m$.

Definition 3.5.3. Let $U_1(t, \boldsymbol{x})$ be a scalar continuous function. The set of points $(t, \boldsymbol{x}^{\mathsf{T}}) = (t, \boldsymbol{y}^{\mathsf{T}}, \boldsymbol{z}^{\mathsf{T}})$ contained in the domain

$$\begin{gathered} t \geq 0, \quad 0 \leq y_l \leq H, \quad \|\boldsymbol{x}^*\| < \infty, \\ \boldsymbol{x}^* = (y_1, \ldots, y_{l-1}, y_{l+1}, \ldots, y_m, z_1, \ldots, z_p) \end{gathered} \tag{3.5.1}$$

such that the inequality $U_1(t, \boldsymbol{x}) > 0$ is satisfied for at least one $l = 1, \ldots, m$, is termed *domain* $U_1(l)$.

Corollary 3.5.1. *Given system* (0.3.1), *suppose that one can find a function* $U_1(t, \boldsymbol{x})$ *such that the following conditions are satisfied*:

(1) *for any* $t \geq 0$ *and an arbitrary small* $\|\boldsymbol{x}\|$ *there exists domain* $U_1(l)$;

(2) *function* U_1 *is bounded in domain* $U_1(l)$;

(3) *for any* $\varepsilon > 0$ *there is* $\delta(\varepsilon) > 0$ *such that, for all* $(t, \boldsymbol{x})$ *that belong to domain* (3.5.1) *and that satisfy* $U_1(t, \boldsymbol{x}) \geq \varepsilon$, *the inequality* $\dot{U}_1(t, \boldsymbol{x}) \geq \delta$ *is valid, where* $\dot{U}_1$ *is a derivative of* U_1 *by virtue of system* (0.3.1);

(4) *the surface* $U_1(t, \boldsymbol{x}) = 0$ *contains no points for which* $y_l < 0$.

Then the unperturbed motion $\boldsymbol{x} = \mathbf{0}$ *of system* (0.3.1) *is upper unstable with respect to* $y_1, \ldots, y_m$.

Proof. Conditions (1)–(3) are identical to those of the Chetayev theorem on instability (Chetayev [1946]) and the Chetayev theorem amended by Rumyantsev [1971a] to treat instability with respect to part of the variables. However, these theorems deal with the domains $t \geq 0$, $\|\boldsymbol{x}\| \leq H$ and (0.3.2), respectively, whereas Corollary 3.5.1 concerns domain (3.5.1). Condition (4) is a new one. Conditions (1)–(3) ensure that the corresponding solutions of system (0.3.1) leave domain (3.5.1) in *finite* time. Because solutions of system (0.3.1) are $\boldsymbol{z}$-continuable and in view of (4), these solutions reach the surface $y_1 = H$. The proposition is proved. □

Example 3.5.1. Let the equations of perturbed motion have the form

$$\dot{y}_1 = y_1^3 + 2y_1 z_1, \qquad \dot{z}_1 = y_1. \tag{3.5.2}$$

Now we demonstrate that the unperturbed motion $y_1 = z_1 = 0$ is upper unstable with respect to y_1. To this end, we consider the function $V = y_1 - z_1^2$, with domain $V > 0$ bounded by the parabola $y_1 = z_1^2$. The derivative $\dot{V} = y_1^3$ is positive for all $y_1 > 0$ and any z_1. Therefore, conditions (1)–(3) are satisfied in the domain $V > 0$. Since the boundary of the domain $V > 0$ contains no points of type $y_1 < 0$, the unperturbed motion $y_1 = z_1 = 0$ of system (3.5.2) is upper unstable with respect to y_1 by Corollary 3.5.1.

3.5.3 Conditions for Instability with Respect to Part of the Variables

Theorem 3.5.1. *In domain* (0.3.2) *let there exist, for at least one* l, $1 \leq l \leq m$, *one of two continuously differentiable vector functions* $\boldsymbol{V} = (V_1, \ldots, V_k)$, $\boldsymbol{W} = (W_1, \ldots, W_r)$ *in which*

$$V_1 = W_1 = y_l, \quad V_j = \mu_j(t, \boldsymbol{x}),$$
$$W_s = \bar{\mu}_s(t, \boldsymbol{x}), \quad j(s) = 2, \ldots, k(r)$$

and such that the following conditions are satisfied:

(1) $V_j(t, \boldsymbol{0}) \equiv 0$, $W_s(t, \boldsymbol{0}) \equiv 0$ $(j \neq 1,\ s \neq 1)$;

(2) *the derivative* $\dot{\boldsymbol{V}}$ $(\dot{\boldsymbol{W}})$, *by virtue of system* (0.3.1), *in the domain*

$$t \geq 0, \quad -H \leq y_l < 0 \ \ (0 \leq y_l \leq H), \quad \|\boldsymbol{x}^*\| < \infty$$

satisfies inequality (3.4.1) (*inequality* (3.4.2)), *and vector function* $\boldsymbol{\varphi}(t, \boldsymbol{V})$ $(\boldsymbol{f}(t, \boldsymbol{W}))$ *is defined and continuous in the domain* D:

$$t \geq 0,\ \|\boldsymbol{V}\| < R,\ R = \infty \text{ or } R > \sup\big[\|\boldsymbol{V}(t, \boldsymbol{x})\| : t \geq 0,\ -H \leq y_l \leq 0\big]$$
$$(t \geq 0,\ \|\boldsymbol{W}\| < R^*,\ R^* = \infty \text{ or } R^* > \sup\big[\|\boldsymbol{W}(t, \boldsymbol{x})\| : t \geq 0,\ 0 < y_l < H\big])$$

and satisfies the Lipschitz condition relative to $\boldsymbol{V}$ $(\boldsymbol{W})$ *in any closed subdomain* G *of* D *bounded with respect to* $\boldsymbol{V}$ $(\boldsymbol{W})$;

(3) *condition* (2) *of Theorem* 3.4.1;

(4) *the solution* $\boldsymbol{\omega} = \boldsymbol{0}$ $(\boldsymbol{u} = \boldsymbol{0})$ *of the system* $\dot{\boldsymbol{\omega}} = \boldsymbol{\varphi}(t, \boldsymbol{\omega})$ $(\dot{\boldsymbol{u}} = \boldsymbol{f}(t, \boldsymbol{u}))$ *is lower (upper) unstable with respect to* ω_1 (u_1).

Then the unperturbed motion $\boldsymbol{x} = \boldsymbol{0}$ *of system* (0.3.1) *is unstable with respect to* $y_1, \ldots, y_m$.

Proof. Because the solution $\boldsymbol{\omega} = \mathbf{0}$ ($\boldsymbol{u} = \mathbf{0}$) of the system $\dot{\boldsymbol{\omega}} = \boldsymbol{\varphi}(t, \boldsymbol{\omega})$ ($\dot{\boldsymbol{u}} = \boldsymbol{f}(t, \boldsymbol{u})$) is lower (upper) unstable with respect to ω_1 (u_1) and in view of condition (1), for any $\delta(\varepsilon, t_0) > 0$, one can find numbers $\varepsilon > 0$, $t_0 \geq 0$, such that there exist solutions $\boldsymbol{\omega}(t; t_0, \boldsymbol{\omega}_0) = \boldsymbol{\omega}(t; t_0, \boldsymbol{V}(t_0; t_0, \boldsymbol{x}_0))$ (solutions $\boldsymbol{\omega}(t; t_0, \boldsymbol{\omega}_0) = \boldsymbol{u}(t; t_0, \boldsymbol{W}(t_0; t_0, \boldsymbol{x}_0))$, respectively) with initial conditions $\|\boldsymbol{x}_0\| < \delta$, $y_{l0} \leq 0$ ($y_{l0} \geq 0$) that satisfy the inequality $\omega_1(t; t_0, \boldsymbol{\omega}_0) \leq -\varepsilon$ ($u_1(t; t_0, \boldsymbol{u}_0) \geq \varepsilon$) at some moment. By condition (2), for all $t \geq t_0$, the inequality $y_l(t; t_0, \boldsymbol{x}_0) \leq \omega_1(t; t_0, \boldsymbol{\omega}_0)$, $\boldsymbol{\omega}_0 = \boldsymbol{V}(t_0, \boldsymbol{x}_0)$ (inequality $y_l(t; t_0, \boldsymbol{x}_0) \geq u_1(t; t_0, \boldsymbol{u}_0)$, $\boldsymbol{u}_0 = \boldsymbol{W}(t_0, \boldsymbol{x}_0)$, respectively) is valid. Consequently, the relationships $y_l(t; t_0, \boldsymbol{x}_0) \leq -\varepsilon$ ($y_l(t; t_0, \boldsymbol{x}_0) \geq \varepsilon$) are satisfied at some moment. The theorem is proved. □

Remark 3.5.1.

1. If the Lipschitz condition for the systems $\dot{\boldsymbol{\omega}} = \boldsymbol{\varphi}(t, \boldsymbol{\omega})$ and $\dot{\boldsymbol{u}} = \boldsymbol{f}(t, \boldsymbol{u})$ fails to hold, there is no guarantee as to the uniqueness of their solutions. In this case the notion of upper (lower) instability with respect to part of the variables should be refined in terms of *lower (upper) solutions.*

2. Theorem 3.5.1 elaborates on the theorems instability based on the Lyapunov vector functions method (Lakshmikantham et al. [1991]).

Example 3.5.2. Consider the issue of instability of *permanent rotations of a solid in the Euler–Poinsot case.* To this end, we write Euler's dynamical equations

$$\begin{gathered} A\dot{x}_1 = (B - C)x_2x_3, \\ B\dot{x}_2 = (C - A)x_1x_3, \quad C\dot{x}_3 = (A - B)x_1x_2. \end{gathered} \tag{3.5.3}$$

Three particular solutions of equations (3.5.3) of interest are

$$x_1 = p = \text{const} > 0, \quad x_2 = x_3 = 0 \qquad (123). \tag{3.5.4}$$

These solutions correspond to permanent rotations of a solid about the principal central axes of its ellipsoid of inertia.

We shall confine ourselves solely to the particular solution presented because two others can be written in the form (3.5.4) by a different notation. To analyze the instability, we introduce the new variables $y_1 = x_1 - p$, $z_i = x_i$ ($i = 1, 2$) and compose the system of perturbed motion in deviations from (3.5.4):

$$\begin{gathered} A\dot{y}_1 = (B - C)z_1z_2, \quad B\dot{z}_1 = (C - A)z_2(y_1 + p), \\ C\dot{z}_2 = (A - B)z_1(y_1 + p). \end{gathered} \tag{3.5.5}$$

Suppose that A is the *medium* moment of inertia of the solid. In the domain $y_1 < 0$, $y_1 + p > 0$, $\|\boldsymbol{z}\| < \infty$, we obtain the following $\boldsymbol{\mu}$-system of inequalities for (3.5.5):

$$\dot{y}_1 = \mu_1, \quad \dot{\mu}_1 = \frac{(y_1 + p)(B - C)\big[C(C - A)z_2^2 + B(A - B)z_1^2\big]}{ABC} \leq 0.$$

The solution $\omega_1 = \omega_2 = 0$ of the system $\dot{\omega}_1 = \omega_2$, $\dot{\omega}_2 = 0$ is lower unstable with respect to ω_1. Therefore, by Theorem 3.5.1 we conclude that the permanent rotation $y_1 = z_1 = z_2 = 0$ is unstable with respect to y_1. In other words, the permanent rotation about the medium axis of the ellipsoid of inertia is unstable relative to the quantity which is the projection of the instantaneous angular velocity vector of the body onto the axis of rotation.

Chetayev [1946] proved the instability of the permanent rotation of a body about the medium axis of the ellipsoid of inertia using the function $V = z_1 z_2$. This result also gives evidence of the instability of the permanent rotation with respect to z_1 and z_2, i.e., the variables characterizing the deviation of the axis of the instantaneous angular velocity vector of the body from its initial position in coordinate axes attached to the body.

3.5.4 Developing the Krasovskii Theorem

Let system (0.3.1) be *autonomous*, i.e., its right-hand sides do not explicitly depend on t.

Theorem 3.5.2. *Suppose that*

(1) *condition* (1) *of Theorem* 3.2.1 *is satisfied*;

(2) *there exist* $V(\boldsymbol{x})$ *and* $\boldsymbol{\mu}(\boldsymbol{x})$ *such that the following conditions hold in the domain* $\|\boldsymbol{y}\| \leq H$, $\|\boldsymbol{z}\| \leq H_1 < +\infty$:

  (*a*) *in any neighborhood of the position* $\boldsymbol{x} = \mathbf{0}$ *there are points* $\boldsymbol{x}$ *for which* $V(\boldsymbol{x}) < 0$;

  (*b*) $V(\mathbf{0}, \boldsymbol{z}) \geq 0$ *for any* $\boldsymbol{z}$;

  (*c*) $\boldsymbol{y} = \mathbf{0}$, $\boldsymbol{\mu}(\boldsymbol{x}) = \mathbf{0}$ *is an invariant set of system* (0.3.1);

  (*d*) $\dot{V}(\boldsymbol{x}) \leq 0$; *and*

  (*e*) *the set*

$$(\{\dot{V}(\boldsymbol{x}) = 0\} \setminus \{\boldsymbol{y} = \mathbf{0}, \boldsymbol{\mu}(\boldsymbol{x}) = \mathbf{0}\}) \cap \{V(\boldsymbol{x}) < 0\}$$

  contains no entire trajectories of system (0.3.1).

Then the position $\boldsymbol{x} = \mathbf{0}$ *of system* (0.3.1) *is* $\boldsymbol{y}$*-unstable.*

Proof. On the contrary, suppose that the position $\boldsymbol{x} = \mathbf{0}$ of system (0.3.1) is $\boldsymbol{y}$-stable. Let us choose $\boldsymbol{x}_0$ so that the conditions $V(\boldsymbol{x}_0) < 0$, $\|\boldsymbol{y}(t; 0, \boldsymbol{x}_0)\| < H$ are satisfied for $t \geq 0$. Then, by conditions (2a), (2d),

$$V(\boldsymbol{x}(t; 0.\boldsymbol{x}_0)) \leq V(\boldsymbol{x}_0) < 0. \tag{3.5.6}$$

By condition (1) there exists a sequence $t_k \to \infty$ such that $\boldsymbol{x}(t_k; 0, \boldsymbol{x}_0) \to \boldsymbol{x}_* \in \Gamma^+$ (Γ^+ is a nonempty set of ω-limiting points). If $\lim \|\boldsymbol{y}(t; 0, \boldsymbol{x}_0)\| = 0$, $t \to \infty$, then $\boldsymbol{y}_* = \mathbf{0}$. Hence, passing to the limit in inequalities (3.5.6), we obtain $0 \leq V(\mathbf{0}, \boldsymbol{z}_*) \leq V(\boldsymbol{x}_0) < 0$, which is impossible in view of condition (2b). Consequently, $\|\boldsymbol{y}(t_k; 0, \boldsymbol{x}_0)\| \geq \eta > 0$ for some sequence $t_k \to \infty$, and one can take $\boldsymbol{\xi}_* = (\boldsymbol{y}_*, \boldsymbol{\mu}(\boldsymbol{x}_*)) \neq \mathbf{0}$. Since $\boldsymbol{\xi}_* \neq \mathbf{0}$, then, bearing in mind condition (2c), we conclude that $\boldsymbol{\xi}(t; 0, \boldsymbol{x}_*) \neq \mathbf{0}$ for all $t \geq 0$. (Here $\boldsymbol{\xi}(t; 0, \boldsymbol{x}_*) = \{\boldsymbol{y}(t; 0, \boldsymbol{x}_*), \boldsymbol{\mu}(\boldsymbol{x}(t; 0, \boldsymbol{x}_*))\}$.) Then, by virtue of the invariance of Γ^+ and the property $\Gamma^+ \in M$ for any $t \geq 0$,

$$\boldsymbol{x}(t_k; 0, \boldsymbol{x}_0) \in (M \setminus \{\boldsymbol{y} = \mathbf{0}, \boldsymbol{\mu}(\boldsymbol{x}) = \mathbf{0}\}). \tag{3.5.7}$$

From (3.5.6) it also follows that

$$\boldsymbol{x}(t_k; 0, \boldsymbol{x}_0) \in \{V(\boldsymbol{x}) < 0\} \tag{3.5.8}$$

for all $t \geq 0$. But a simultaneous fulfillment of conditions (3.5.7) and (3.5.8) for all $t \geq 0$ contradicts condition ($2e$). This means that the assumption on $\boldsymbol{y}$-stability of position $\boldsymbol{x} = \mathbf{0}$ of system (0.3.1) is wrong. The theorem is proved. □

Remark 3.5.2.

1. Theorem 3.5.2 is the development of the Krasovskii theorem (Krasovskii [1959] on Lyapunov instability. When $\boldsymbol{\mu} = \mathbf{0}$, it turns into the Risito theorem on $\boldsymbol{y}$-instability (Risito [1970]) (amended by Oziraner [1973]). When $\boldsymbol{\mu} \neq \mathbf{0}$, the conditions of Theorem 3.5.2 are more general than in the work mentioned.
2. The reader can find other conditions for instability with respect to part of the variables and the corresponding bibliography in the monograph by Rumyantsev and Oziraner [1987].

3.6 Stability with Respect to a Specified Number of the Variables

3.6.1 Formulation of the Problem

Now we extend the notion of stability with respect to part of the variables and consider the stability with respect to a *specified* number of variables (specified *quantitatively*), the set of which *changes* depending on the initial conditions.

To do so, we assume that the set of variables with respect to which the stability is examined is not specified in advance, and for any sufficiently small initial perturbations it is necessary only to guarantee the stability with respect to a specified number of the variables (specified quantitatively). Which of the variables will turn out to be stable does not matter, and it is assumed that different variables can be stable depending on the initial conditions (which necessarily embrace the whole sufficiently small or even larger domain of the initial perturbations). To be more precise, let us introduce the following definition.

Definition 3.6.1. The unperturbed motion $\boldsymbol{x} = \mathbf{0}$ of system (0.3.1) is said to be *stable with respect to a specified number m of the variables* if for any ε, $t_0 \geq 0$ one can find numbers $\delta(\varepsilon, t_0) > 0$, $L > 0$ (where L is assumed to be independent of t_0, ε) such that the domain $\|\boldsymbol{x}_0\| < \delta$ is divided into L parts D_j

$$D_1 \cup D_2 \cup \cdots \cup D_L = \{\boldsymbol{x}_0 : \|\boldsymbol{x}_0\| < \delta\} = D, \quad D_1 \cap D_2 \cap \cdots \cap D_L = \{0\}$$

and from $\|\boldsymbol{x}_0\| < \delta$ it follows that $\|\tilde{\boldsymbol{y}}_j(t; t_0, \boldsymbol{x}_0)\| < \varepsilon$, $(j = 1, \ldots, L)$ for all $t \geq t_0$ if $\boldsymbol{x}_0 \in D_j$; here $\tilde{\boldsymbol{y}}_j = (x_{j1}, \ldots, x_{jm})$ are various sets of m variables from $x_1, \ldots, x_n$. In addition, if the conditions $\lim\|\tilde{\boldsymbol{y}}_j(t; t_0, \boldsymbol{x}_0)\| = 0$, $t \to \infty$, hold, then the unperturbed motion $\boldsymbol{x} = \mathbf{0}$ of system (0.3.1) is *asymptotically stable with respect to a specified number m of the variables.*

Remark 3.6.1.

1. We assume that the right-hand sides of system (0.3.1) satisfy the conditions imposed on them in the domain

$$t \geq 0, \quad \|\tilde{\boldsymbol{y}}_j\| \leq H, \quad \|\tilde{\boldsymbol{z}}_j\| < \infty, \quad \tilde{\boldsymbol{x}}_j = (\tilde{\boldsymbol{y}}_j, \tilde{\boldsymbol{z}}_j), \quad \|\boldsymbol{x}\| = \|\tilde{\boldsymbol{x}}_j\| \\ (j = 1, \ldots, L). \tag{3.6.1}$$

2. In the sense of the definition $L \geq 2$, when $L = 1$, it becomes Definition 0.3.1, if $D_1 = D$, or that of *conditional* $\boldsymbol{y}$*-stability*, if $D_1 \neq D$. (We note that the term "*conditional stability*" was introduced by Lyapunov [1892].) When $L \geq 2$, Definition 3.6.1 is broader than that of the conditional stability with respect to part of the variables, for the *whole* sufficiently small domain of initial perturbations is encompassed by this definition. Number L can equal the maximum possible number of different sets $\tilde{\boldsymbol{y}}_j$, though this is not necessarily the case.

3. Definition 3.6.1 permits various modifications with allowance for various interpretations of the concept of $\boldsymbol{y}$-stability. In particular, we can refer to *uniform* or *exponential* stability with respect to a specified number of variables, etc.

Remark 3.6.2. We also distinguish the case when the unperturbed motion of a system is stable with respect to a (quantitatively) specified number of variables in the whole domain Ω of variation of the parameters of the system, but the particular composition of the variables depends on the parameters of the system belonging to some or other part of domain Ω. We shall call this stability the *parametric stability with respect to a* (quantitatively) *specified number of variables.*

3.6.2 Using the Method of Lyapunov Functions

Theorem 3.6.1. *Let $W(\boldsymbol{x})$, $W(\boldsymbol{0}) = 0$ be the first integral of system* (0.3.1), *i.e., $\dot{W}(\boldsymbol{x}) = 0$ by virtue of system* (0.3.1).

1. *If there exist two functions $V_j(t, \boldsymbol{x})$ $(j = 1, 2)$ such that in the domain*

$$\{t \geq 0, \quad \|\tilde{\boldsymbol{y}}_j\| \leq H, \quad \|\tilde{\boldsymbol{z}}_j\| < \infty\} \cap \{\boldsymbol{x} : (-1)^j W(\boldsymbol{x}) \leq 0\} \tag{3.6.2}$$

the following conditions hold:

$$V_j(t, \boldsymbol{x}) \geq a_j(\|\tilde{\boldsymbol{y}}_j\|), \quad \dot{V}_j(t, \boldsymbol{x}) \leq 0, \tag{3.6.3}$$

then the unperturbed motion $\boldsymbol{x} = \boldsymbol{0}$ is stable with respect to a specified number m of the variables. In this case the domain of initial perturbations $\|\boldsymbol{x}_0\| < \delta$ is divided into three parts $W(\boldsymbol{x}_0) > 0$, $W(\boldsymbol{x}_0) < 0$, and $W(\boldsymbol{x}_0) = 0$; in each part the variables that constitute the vectors $(\tilde{\boldsymbol{y}}_1)$, $(\tilde{\boldsymbol{y}}_2)$, and $(\tilde{\boldsymbol{y}}_1, \tilde{\boldsymbol{y}}_2)$, respectively, are stable.

2. *In addition, if the conditions $V_j \downarrow 0$ $(j = 1, 2)$ hold (the notation $V_j \downarrow 0$ implies that V_j approaches zero, monotonically decreasing), then the stability with respect to the specified number of variables is an asymptotic type.*

Proof. Because conditions (3.6.3) are valid in domain (3.6.2), it follows that functions V_j $(j = 1, 2)$ are $\tilde{\boldsymbol{y}}_j$-positive-definite and nonincreasing in domain (3.6.1) only under the additional conditions $(-1)^j W(\boldsymbol{x}) < 0$. This means that function $W(\boldsymbol{x})$ explicitly enters into the expression for functions V_j (additively or multiplicatively). In other words, functions V_j can be written in one of the following ways:

(1) $V_j(t, \boldsymbol{x}) = V_j^*(t, \boldsymbol{x}) + (-1)^j W(\boldsymbol{x})$ and, by virtue of $\dot{W} = 0$, the inequalities $V_j^* \geq a_j(\|\tilde{\boldsymbol{y}}_j\|)$, $\dot{V}_j^* = \dot{V}_j \leq 0$ hold in domain (3.6.1);

(2) $V_j(t, \boldsymbol{x}) = V_j(t, \boldsymbol{x}, W(\boldsymbol{x}))$, $V_j(t, \mathbf{0}, W(\boldsymbol{x})) \equiv 0$ and, by virtue of $\dot{W}(\boldsymbol{x}) = 0$, $V_j(t, \boldsymbol{x}) = V_j(t, \boldsymbol{x}, W(\boldsymbol{x}_0))$, $V_j(t, \mathbf{0}, W(\boldsymbol{x}_0)) \equiv 0$.

In case (1), by relationships $V_j(t, \mathbf{0}) \equiv 0$, $W(\mathbf{0}) = 0$, for any ε, $t_0 \geq 0$, one can find $\delta(\varepsilon, t_0) > 0$ such that from

$$\|\tilde{\boldsymbol{y}}_{j0}\| < \delta, \quad \|\tilde{\boldsymbol{z}}_{j0}\| < \delta, \quad (-1)^j W(\boldsymbol{x}_0) \leq 0 \quad (j = 1, 2) \tag{3.6.4}$$

it follows that $V_j(t_0, \boldsymbol{x}_0) < a(\varepsilon)$. By virtue of $\dot{V}_j \leq 0$, for a solution $\tilde{\boldsymbol{x}}_j = (\tilde{\boldsymbol{y}}_j, \tilde{\boldsymbol{z}}_j) = \tilde{\boldsymbol{x}}_j(t; t_0, \tilde{\boldsymbol{x}}_{j0})$ with the initial conditions from domain (3.6.4), the estimates

$$a_j(\|\tilde{\boldsymbol{y}}_j(t; t_0, \tilde{\boldsymbol{x}}_{j0})\|) \leq V_j(t; \tilde{\boldsymbol{x}}_j(t; t_0, \tilde{\boldsymbol{x}}_{j0})) \leq V(t_0, \boldsymbol{x}_0) \leq a(\varepsilon) \tag{3.6.5}$$

are valid for all $t \geq t_0$.

Taking into account the properties of function $a(r)$, from (3.6.5) we deduce that $\|\tilde{\boldsymbol{y}}_j(t; t_0, \tilde{\boldsymbol{x}}_{j0})\| < \varepsilon$, $t \geq t_0$.

In case (2), by relationships $W(\mathbf{0}) \equiv 0$, $V_j(t, \mathbf{0}, W(\mathbf{0})) \equiv 0$, for any ε, $t_0 \geq 0$, one can also find $\delta(\varepsilon, t_0) > 0$ such that, for all $t \geq t_0$, estimates (3.6.5) are true for solutions $\tilde{\boldsymbol{x}}_j = \tilde{\boldsymbol{x}}_j(t; t_0, \tilde{\boldsymbol{x}}_{j0})$ with the initial conditions from domain (3.6.4).

Because the conditions $V_j \downarrow 0$ are satisfied, relationship (3.6.5) implies that $a_j(\|\tilde{\boldsymbol{y}}_j(t)\|) \downarrow 0$ and, consequently,

$$\lim \|\tilde{\boldsymbol{y}}_j(t; t_0, \tilde{\boldsymbol{x}}_{j0})\| = 0, \quad t \to \infty.$$

The theorem is proved. □

Remark 3.6.3.

1. The conditions $V_j \downarrow 0$ can be verified in the same way as in the well-known theorems on stability with respect to part of the variables. For example, following Rumyantsev [1957a], [1970], one can replace the conditions $V_j \downarrow 0$ with $V_j \leq b(\|\tilde{\boldsymbol{y}}_j\|)$, $\dot{V}_j \leq -c(\|\tilde{\boldsymbol{y}}_j\|)$ in domain (3.6.2).

2. Theorem 3.6.1 can be extended to the study of *other* types of stability with respect to a given part of the variables

Example 3.6.1.

1. Consider *Lancaster's equations*

$$\dot{x}_1 = -k_1 x_1 x_2, \quad \dot{x}_2 = k_2 x_1 x_2 \quad (k_i \text{ are const}, \ i = 1, 2) \tag{3.6.6}$$

which describe a "*military standoff.*" A direct verification readily shows that the equilibrium position $x_1 = x_2 = 0$ of system (3.6.6) is Lyapunov stable and asymptotically stable with respect to one of the variables, namely, when $k_1 > 0$, $k_2 > 0$, the domain of initial perturbations is divided into three parts $\gamma = k_2 x_{10} + k_1 x_{20} > 0$, $\gamma < 0$, and $\gamma = 0$, each one corresponding to the asymptotic stability relative to (x_1), (x_2), and (x_1, x_2), respectively.

2. Let us construct the Lyapunov functions for the problem of damping rotational motions of an asymmetric solid with respect to two (of three) axes of its ellipsoid of inertia, which was discussed in Section 2.3.

To do so, we use the functions

$$V = x_1^2 + \frac{1}{q(t)}\mu_1^2 \qquad (V = x_1^2 + \mu_1^2), \tag{3.6.7}$$

which under conditions (2.3.8) are the Lyapunov functions for the $\boldsymbol{\mu}$-system equivalent to equation (2.3.7) (Lim and Kazda [1964], Merkin [1987]). The condition $V \downarrow 0$ is satisfied along the trajectories of the $\boldsymbol{\mu}$-system.

Let us write expressions (3.6.7) in the original variables x_i, $(i = 1, 2, 3)$ (by substituting $\mu_1 = (B - C)A^{-1}x_2x_3)$. The first integral (2.3.6) taken into account, the functions

$$V_1 = x_1^2 + \frac{(B-C)^2}{A^2 q(t)} \frac{B\beta}{(C-A)} x_2^2 + \frac{(B-C)^2(A-B)B}{A^2(C-A)Cq(t)} x_2^4,$$
$$V_2 = x_2^2 + \frac{(B-C)^2}{A^2 q(t)} \frac{C\beta}{(B-A)} x_3^2 + \frac{(B-C)^2(C-A)C}{A^2(A-B)Bq(t)} x_3^4$$

are obtained for the first of functions (3.6.7).

In the domain

$$\{t \geq 0,\ \|\tilde{\boldsymbol{y}}_j^+\| \leq H,\ \|\tilde{\boldsymbol{z}}_j^+\| < \infty\} \cap \{(-1)^j W < 0\},$$
$$W = \begin{cases} \beta & (B < A < C), \\ -\beta & (C < A < B), \end{cases} \qquad \beta = \frac{C-A}{B} x_{30}^2 - \frac{A-B}{C} x_{20}^2,$$
$$\tilde{\boldsymbol{y}}_1^+ = (x_1, x_2), \quad \tilde{\boldsymbol{y}}_2^+ = (x_2, x_3), \quad j = 1, 2,$$

functions V_1, V_2, W satisfy the conditions of Theorem 3.6.1.

3. Consider the *system of the Lotka–Volterra equations* (Volterra [1931])

$$\begin{gathered} \dot{x}_1 = (k_1 + \beta_1^{-1} a_{13} x_3) x_1, \quad \dot{x}_2 = (k_2 + \beta_2^{-1} a_{23} x_3) x_2, \\ \dot{x}_3 = (k_3 - \beta_3^{-1}(a_{13} x_1 + a_{23} x_2)) x_3, \end{gathered} \tag{3.6.8}$$

describing the dynamics of a tripartite system in which two species feed on a third. Here $x_i(t) \geq 0$ is the size of the population of the ith species, k_i is the difference between the birth and death rates of the ith species, under the assumption that there is no intervention, and the parameters β_i are positive and reflect the fact that the reproduction of one predator is connected with the death of more than one victim; a_{ij} characterize the influence of the interaction between the different species on the population growth rate: $k_1 < 0$, $k_2 < 0$, $k_3 > 0$, $a_{13} > 0$, $a_{23} > 0$.

Equations (3.6.8) possess three and only three equilibrium positions (except for singular case $\dfrac{k_1\beta_1}{a_{13}} = \dfrac{k_2\beta_2}{a_{23}}$)

(a) $x_i = 0$ $(i = 1, 2, 3)$;

(b) $x_1 = 0$, $x_2 = \dfrac{k_3\beta_3}{a_{23}} = n_1 > 0$, $x_3 = -\dfrac{k_2\beta_2}{a_{23}} = -n_2 > 0$;

(c) $x_1 = \dfrac{k_3\beta_3}{a_{13}} = n_3 > 0$, $x_2 = 0$, $x_3 = -\dfrac{k_1\beta_1}{a_{13}} = -n_4 > 0$.

Now we show that if

$$k_1\beta_1 - a_{13}n_2 < 0 \quad (k_2\beta_2 - a_{23}n_4 < 0), \tag{3.6.9}$$

then position (b) (position (c)) is asymptotically stable with respect to x_1 (x_2) for any initial perturbations belonging to the domain $\Psi_1(\Psi_2) = \{x_1 \geq 0 \ (x_1 > 0),\ x_2 > 0 \ (x_2 \geq 0),\ x_3 > 0\}$. For this purpose we introduce two Lyapunov functions (Rouche et al. [1977])

$$W_1 = \beta_1 x_1 + \beta_2 n_1 \left(\frac{x_2}{n_1} - \ln\frac{x_2}{n_1}\right) + \beta_3 n_2 \left(\frac{x_3}{n_2} - \ln\frac{x_3}{n_2}\right),$$

$$\left(W_2 = \beta_2 x_2 + \beta_1 n_3 \left(\frac{x_1}{n_3} - \ln\frac{x_1}{n_3}\right) + \beta_3 n_4 \left(\frac{x_3}{n_4} - \ln\frac{x_3}{n_4}\right)\right).$$

If (3.6.9) is true, function W_1^* (W_2^*) equal to the difference between function W_1 (W_2) and its value in position (b) (position (c)) satisfies the following conditions in domain Ψ_1 (Ψ_2)

$$W_i^* \geq a(\|\boldsymbol{x}\|), \quad W_i^* \to \infty \quad (\|\boldsymbol{x}\| \to \infty), \quad \dot{W}_i^* \leq 0.$$

Therefore, the solutions of system (3.6.8) originating from domain Ψ_1 (Ψ_2) are *bounded* (Yoshizawa [1966]). This means that, for any t_0, λ there is $M(t_0, \lambda)$ such that the right-hand sides X_i $(i = 1, 2, 3)$ of system (3.6.8) satisfy the condition

$$\left(\sum_{i=1}^{3} X_i^2\right) \leq M$$

for all $x_i = x_i(t; t_0, \boldsymbol{x}_0)$, $\|\boldsymbol{x}_0\| < \lambda$, $t \geq t_0$.

But then function W_1^* (W_2^*) also satisfies the conditions of the theorem on *global* asymptotic stability with respect to part of the variables (Peiffer and Rouche [1969]). In accordance with this theorem the position (b) (position (c)) is asymptotically stable with respect to x_1 (x_2) for arbitrary initial perturbations from domain Ψ_1 (Ψ_2).

Taking into account that inequalities (3.6.9) are mutually exclusive, we conclude that position (a), which is unstable with respect to all the variables, at the same time is (parametrically) asymptotically stable *with respect to one of the variables* for any initial perturbations from the domain $\Psi_1 \cap \Psi_2$. In other words, for any solution of system (3.6.8) beginning in the region $x_i > 0$ $(i = 1, 2, 3)$, one of the variables x_i asymptotically tends to zero. This statement is the exact formulation (Rouche et al. [1977]) of the statement known as the *Lotka–Volterra ecological principle of extinction.*

3.7 Overview of References

Motives for designating admissible boundaries of variation of "uncontrollable" variables in problems on partial stability were formulated quite recently (Vorotnikov [1995a], [1995b]).

Inequality (3.1.2), as applied to the analysis of $\boldsymbol{y}$-stability of the unperturbed motion $\boldsymbol{x} = \mathbf{0}$ of system (0.3.1), was first introduced (Vorotnikov [1979a]) to facilitate constructing Lyapunov functions. However, the understanding of the fact that this condition leads to a *new class* of Lyapunov functions came later (Vorotnikov [1993], [1995a], [1995b]). This occurred, to some extent, because of the example of V-function (3.1.13) which appeared in Rumyantsev and Oziraner [1987].

Theorems 3.1.1–3.1.3, 3.2.1, and 3.2.2 were obtained in the work by Vorotnikov [1995a], [1995b], [1997a] (the original version [1979a]). Theorems 3.1.4 and 3.1.5 are formulated for the first time.

The matter of Sections 3.3 and 3.4 relates to the idea of using differential inequalities, which goes back to the work by Chaplygin [1919] and Wazewski [1950] and to the method of Lyapunov vector functions.

The notion of stability with respect to a specified number of the variables (Vorotnikov [1986a]) turns out to be fruitful in such fields as biocenose dynamics and spacecraft dynamics. In particular, the material of Section 2.3. gives an important example of this type of stability.

Investigation of partial stability for Lotka–Volterra systems can be found in the monograph by Rouche et al. [1977]. Analysis of system (3.6.8) somewhat supplements this research.

Chapter 4

Nonlinear Problems of Stabilization and Control with Respect to Part of the Variables

In this chapter the approach proposed in Chapters 1–3 is further developed as applied to the class of nonlinear *controlled systems.*

Three groups of problems are considered:

(1) Problems of partial stabilization of controlled systems or partial stabilization of uncontrolled system by means of additional forces.

(2) Problems of optimal (in the sense of minimizing some functional of integral type) stabilization of controlled systems with respect to all the variables; the corresponding problem of partial stabilization is solved at the first (preliminary) stage.

The specified problems are referred to the classic type of stabilization problems considered in an *infinite* time interval.

(3) Control problems with respect to part of the variables (partial controllability) in a *finite* time interval; "geometric" constraints on the control functions are taken into account.

Although these problems differ significantly (as separately discussed) an approach can be suggested to handle them on a unified basis. Such an approach hinges on the method of constructing auxiliary linear μ-systems proposed in Chapters 1–3 which is appropriately modified here to embrace the case of controlled systems, namely, in control problems it is suggested that auxiliary linear controlled μ-systems be constructed which extend the special choice of structure to the controls themselves, substantially enhancing constructibility of these linear μ-systems.

As a result, we obtain the constructively verifiable conditions for solvability of the above groups of problems which can be applied to a sufficiently general class of nonlinear controlled systems.

The effectiveness of the proposed approach is evidenced by a series of constructively solved problems on control of the angular motion of a solid which have an independent theoretic and applied meaning: stabilization and partial stabilization of permanent rotations (of equilibrium positions) of a solid; stabilization and partial stabilization of orbital relative equilibrium positions of a satellite; partial stabilization of stationary modes of a mechanical system "solid + flywheels" (gimballed gyroscope) by specially arranged rotation of the (gyroscope) flywheels; reorientation of a solid and a system of solids.

Also briefly outlined are the two following groups of issues:

(1) some extra possibilities arising in solving the problems of stabilization or control with respect to part of the variables which are not encountered when solving the corresponding problems of stabilization and control with respect to all the variables;

(2) the possibility of using control rules in the problems of partial stabilization (in an infinite time interval), formed according to the feedback principle as a fractional function of the phase variables and operating in the mode "uncertainty of type 0/0" (as $t \to \infty$).

4.1 Stabilization with Respect to Part of the Variables

4.1.1 Preliminary Remarks

The approach to solving the problem of partial stabilization based on the Lyapunov functions method (Rumyantsev [1970], Rumyantsev and Oziraner [1987]) is just one of the possible ways for analyzing it. The present section is aimed at outlining another approach to this problem which develops the method of constructing auxiliary μ-systems worked out in Chapters 1–3. Thus we extend the class of admissible control rules and indicate some key features of the problem of partial stabilization not encountered when solving the problem of stabilization with respect to all the variables.

4.1.2 Formulation of the Problem

$$\begin{gathered} \dot{\boldsymbol{x}} = \boldsymbol{X}(t, \boldsymbol{x}, \boldsymbol{u}), \qquad \boldsymbol{X}(t, \boldsymbol{0}, \boldsymbol{0}) \equiv \boldsymbol{0} \\ \boldsymbol{x}^{\mathsf{T}} = (y_1, \ldots, y_m, z_1, \ldots, z_p) = (\boldsymbol{y}^{\mathsf{T}}, \boldsymbol{z}^{\mathsf{T}}), \quad \boldsymbol{u}^{\mathsf{T}} = (u_1, \ldots, u_r), \end{gathered} \tag{4.1.1}$$

where $\boldsymbol{x}$ is the vector of phase variables and $\boldsymbol{u}$ is the vector of controls. The right-hand sides of system (4.1.1) are defined and continuous in domain (0.3.2) for $\|\boldsymbol{u}\| < \infty$.

The class $\mathfrak{U}^* = \{\boldsymbol{u}(t, \boldsymbol{x})\}$ of *admissible controls* includes vector functions $\boldsymbol{u} = \boldsymbol{u}(t, \boldsymbol{x})$ defined in domain (0.3.2). It is also assumed that, for any $\boldsymbol{u} \in \mathfrak{U}^*$,

(1) common conditions imposed on the right-hand sides of system (0.3.1) in domain (0.3.2) hold true;

(2) there exists the equilibrium position $\boldsymbol{x} = \boldsymbol{0}$ of system (4.1.1).

Condition (2) is stipulated for the reason that the identity $\boldsymbol{u}(t, \boldsymbol{0}) \equiv \boldsymbol{0}$ *might not be valid* for controls $\boldsymbol{u} \in \mathfrak{U}^*$. (Otherwise condition (2) is a consequence of condition $\boldsymbol{X}(t, \boldsymbol{0}, \boldsymbol{0}) \equiv \boldsymbol{0}$.) We note that to satisfy condition (2) when $\boldsymbol{u}(t, \boldsymbol{0}) \not\equiv \boldsymbol{0}$, it is necessary that $\boldsymbol{X}(t, \boldsymbol{0}, \boldsymbol{u}) \equiv \boldsymbol{0}$. In this case the structure of system (4.1.1) is as follows

$$\dot{\boldsymbol{x}} = \boldsymbol{X}^*(t, \boldsymbol{x}) + \boldsymbol{R}(t, \boldsymbol{x}, \boldsymbol{u}), \tag{4.1.2}$$

and $\boldsymbol{X}^*(t, \boldsymbol{0}) \equiv \boldsymbol{R}(t, \boldsymbol{0}, \boldsymbol{0}) \equiv \boldsymbol{R}(t, \boldsymbol{0}, \boldsymbol{u}) \equiv \boldsymbol{0}$.

Problem 4.1.1 (*on $\boldsymbol{y}$-stabilization of the equilibrium position $\boldsymbol{x} = \boldsymbol{0}$ of system* (4.1.1) *in the class $\boldsymbol{u} \in \mathfrak{U}^*$*). *Find a set of control rules $\boldsymbol{u}(t, \boldsymbol{x}) \in \mathfrak{U}^*$ such that the equilibrium position $\boldsymbol{x} = \boldsymbol{0}$ of system* (4.1.1) *is asymptotically $\boldsymbol{y}$-stable.*

4.1.3 Discussion of the Formulation of Problem 4.1.1

1. If the condition $\boldsymbol{u}(t, \boldsymbol{0}) \equiv \boldsymbol{0}$ is required for controls $\boldsymbol{u} \in \mathfrak{U}^*$, then system (4.1.1) can be treated as a *system of perturbed motion* for some *control system* under study. Such a system of perturbed motion is constructed (each time anew) for each particular program motion and for program control corresponding to it in the system under study, where vectors $\boldsymbol{x}, \boldsymbol{u}$ characterize the variations from the program motion and program control, respectively. (Regarding this issue, see Section 0.3.)

In this case Problem 4.1.1 corresponds to the classic version of the problem of *stabilizing program motions of controlled systems* (Krasovskii [1966]); the only difference is that we consider the stabilization with respect to a *given part* of the variables.

2. If the condition $\boldsymbol{u}(t, \boldsymbol{0}) \equiv \boldsymbol{0}$ is not obligatory for controls $\boldsymbol{u} \in \mathfrak{U}^*$, then system (4.1.1) is represented in the form (4.1.2). The expression $\boldsymbol{R}(t, \boldsymbol{x}, \boldsymbol{u})$ in (1.4.2) can be treated as a vector of *additional actions* (involving controlled parameters) applied to the system

$$\dot{\boldsymbol{x}} = \boldsymbol{X}^*(t, \boldsymbol{x}). \tag{4.1.3}$$

In this case Problem 4.4.1 is reduced to the choice of controls $\boldsymbol{u} \in \mathfrak{U}^*$ so as to render, by vector $\boldsymbol{R}$ of additional actions, the equilibrium position (or the unperturbed motion) $\boldsymbol{x} = \boldsymbol{0}$ of system (4.1.3) asymptotically $\boldsymbol{y}$-stable. Such a problem (and the problem of stabilization of program motions) is frequently encountered in applications.

When solving this problem for the case $\boldsymbol{u}(t, \boldsymbol{0}) \not\equiv \boldsymbol{0}$, the requirement

$$\left|\boldsymbol{u}(t; t_0, \boldsymbol{x}_0)\right| \to 0, \quad t \to 0, \tag{4.1.4}$$

is natural for all solutions $\boldsymbol{x}(t; t_0, \boldsymbol{x}_0)$ of system (4.1.1) starting in the domain of initial perturbations.

3. In contrast to familiar work on partial stabilization (see Rumyantsev [1970], Rumyantsev and Oziraner [1987]), the *continuity* of control functions $\boldsymbol{u} \in \mathcal{U}^*$ in domain (0.3.2) *is not assumed*, generally speaking, in Problem 4.1.1. Such a possible property of controls does not contradict the condition $\boldsymbol{u} \in \mathcal{U}^*$ in the case $\boldsymbol{X}(t, \boldsymbol{0}, \boldsymbol{u}) \equiv \boldsymbol{0}$. (This means that *intentional* introduction of such discontinuous control functions does not yield systems with *variable structure*.) The class of admissible controls being extended increases the solvability of partial stabilization problems. Apparently, the question of whether the controls $\boldsymbol{u} \in \mathcal{U}^*$ are implementable requires a separate study.

Consider, for example, the stabilization of the unstable equilibrium position $y_1 = z_1 = 0$ (for $Q_1 = 0$) of the system

$$\dot{y}_1 = ay_1 + Q_1, \quad \dot{z}_1 = bz_1 \quad (a, b = \text{const} > 0) \tag{4.1.5}$$

with respect to y_1 through additional force $Q_1 = z_1 u_1$. The force can be interpreted as a linear force that depends on z_1 and includes a controlled parameter.

The control

$$u_1 = \frac{cy_1}{z_1}, \quad c = \text{const} < 0 \tag{4.1.6}$$

belongs to class $\mathcal{U}^*$. For $a + c < 0$, the equilibrium position $y_1 = z_1 = 0$ of system (4.1.5), (4.1.6) is asymptotically stable with respect to variable y_1.

Though discontinuous in domain (0.3.2) ($|u_1| = \infty$ for $y_1 \neq 0$, $z_1 = 0$), control function (4.1.6) is *continuous* on the set M of solutions of system (4.1.5), (4.1.6) for $z_{10} \neq 0$. Moreover, relationship (4.1.4) is valid when $|a + c| > b$.

At the same time, for $y_{10} \neq 0$, $z_{10} = 0$, control function (4.1.6) is also discontinuous on set M. To avoid this disadvantage, one should set $u_1 = 0$ at the first stage of the control process, thereby "taking away" the unstable system (4.1.5) quickly enough from the undesirable value $z_{10} = 0$.

4. In the subsequent sections of the book (Section 4.5, Chapter 5) we shall *intentionally* introduce discontinuous controls leading (as distinct from discontinuous controls $\boldsymbol{u} \in \mathcal{U}^*$) to systems with *variable structure*. But unlike Problem 4.1.1, we shall discuss the control in a *finite* time interval.

5. When $\boldsymbol{u}(t, \boldsymbol{0}) \not\equiv \boldsymbol{0}$, $\boldsymbol{R}(t, \boldsymbol{0}, \boldsymbol{u}) \not\equiv \boldsymbol{0}$ system (4.1.2) does not have the position $\boldsymbol{x} = \boldsymbol{0}$. This implies that the stabilization of the position $\boldsymbol{x} = \boldsymbol{0}$ of system (4.1.3) with respect to *all* the variables (by means of additional actions $\boldsymbol{R}$) is *impossible*. However, in this case the stabilization *with respect to part* of the variables *can be carried out*.

For instance, the controlled system

$$\begin{gathered} \dot{y}_1 = ay_1 + z_1 - z_2 - y_1 u_1, \\ \dot{z}_1 = dy_1 + bz_1 + u_1, \quad \dot{z}_2 = bz_2 + u_1 \end{gathered} \tag{4.1.7}$$

does not have the position $y_1 = z_1 = z_2 = 0$ for $u_1 = c = \text{const}$. Therefore, solutions of system (4.1.7) originating from the neighborhood of the position $y_1 = z_1 = z_2 = 0$ cannot (simultaneously with respect to all the variables) asymptotically approach zero.

At the same time, the $\boldsymbol{\mu}$-system for (4.1.7), when $u_1 = c$, comprises the equations

$$\dot{y}_1 = (a-c)y_1 + \mu_1, \qquad \dot{\mu}_1 = dy_1 + b\mu_1; \tag{4.1.8}$$

the variable $y_1(t)$ of system (4.1.8) satisfies the relationship $\lim|y_1(t)| = 0$, $t \to \infty$, for $a - c + b < 0$, $(a-c)b - d > 0$. So all (but not only those originating from a sufficiently small neighborhood of the position $y_1 = z_1 = z_2 = 0$) solutions $\boldsymbol{x}(t) = \boldsymbol{x}(t; t_0, \boldsymbol{x}_0)$ of system (4.1.7) satisfy the condition $\lim|y_1(t; t_0, \boldsymbol{x}_0)| = 0, t \to \infty$; zero is approached by an exponential law.

We note that the zero (at $u_1 = 0$) equilibrium position $y_1 = z_1 = z_2 = 0$ of system (4.1.7) is unstable with respect to y_1, when $a+b > 0$. Thus, for $a-c+b > 0$, $(a-c)b-d > 0$, the control rule $u_1 = c$ has a *stabilizing effect* relative to variable y_1.

6. The stabilizing effect indicated for system (4.1.7) is not devoid of interest from the standpoint of the problem of *partial stability in the presence of constantly acting perturbations* $\boldsymbol{R}(t, \boldsymbol{x})$, $\boldsymbol{R}(t, \boldsymbol{0}) \not\equiv \boldsymbol{0}$, namely, superposition of perturbations (*intentional* inclusive) may change an equilibrium position of a system unstable with respect to some of the variables to one globally exponentially asymptotically stable with respect to these variables. In the problem of stability with respect to all the variables, such a situation is impossible, when $\boldsymbol{R}(t, \boldsymbol{0}) \not\equiv \boldsymbol{0}$.

To ascertain that this is true, let us modify the example of system (4.1.7). Indeed, the equilibrium position $y_1 = z_1 = z_2 = 0$ of the linear autonomous system

$$\dot{y}_1 = ay_1 + z_1 - z_2, \quad \dot{z}_1 = dy_1 + bz_1, \quad \dot{z}_2 = bz_2$$

is unstable with respect to y_1 for $a + b > 0$.

However, when superposing a vector function of the form $\boldsymbol{R} = (-cy_1, c, c)^{\mathsf{T}}$, $\boldsymbol{R} \not\equiv \boldsymbol{0}$ one can choose constant c so that the equilibrium position becomes globally exponentially asymptotically stable with respect to y_1.

4.1.4 The Principles of Constructing Control Rules

Theorem 4.1.1. 1. *Let* $m = r = 1$ *and let the right-hand sides of system* (4.1.1) *in domain* (0.3.2) *satisfy the conditions*

$$\begin{array}{lll} X_i(t, \boldsymbol{x}, \boldsymbol{u}) \equiv X_i^*(t, \boldsymbol{x}) + X_i^{**}(t, \boldsymbol{x})u_1 & & (i = 2, \ldots, n), \\ X_1^*(t, \boldsymbol{x}) \equiv X_1^*(\boldsymbol{x}), & X_1^{**}(t, \boldsymbol{x}) \equiv 0, & X_1^*(\boldsymbol{0}) = 0, \end{array} \tag{4.1.9}$$

for any u_1, *and*

$$\begin{gathered} u_1 = \frac{u_1^0 - \displaystyle\sum_{i=1}^{n} \frac{\partial X_1^*(t, \boldsymbol{x})}{\partial x_i} X_i^*(t, \boldsymbol{x})}{f(t, \boldsymbol{x})} \in \mathfrak{U}^*, \\ f(t, \boldsymbol{x}) = \sum_{i=2}^{n} \frac{\partial X_1^*(\boldsymbol{x})}{\partial x_i} X_i^{**}(t, \boldsymbol{x}), \\ u_1^0 = \Gamma_1 x_1 + \Gamma_2 X_1^*(\boldsymbol{x}), \quad \Gamma_i = \text{const} < 0 \quad (i = 1, 2). \end{gathered} \tag{4.1.10}$$

Then, for $y_1 = x_1$, $z_{i-1} = x_i$ $(i = 2, \ldots, n)$, *the set of control rules* (4.1.10) *solves Problem* 4.1.1 *for system* (4.1.1).

2. *Let there exist a control* $\boldsymbol{u} \in \mathfrak{U}^*$ *and* N *continuously differentiable functions* $\mu_i = \mu_i(t, \boldsymbol{x})$, $\mu_i(t, \boldsymbol{0}) \equiv 0$ *such that from the closed system* (4.1.1) *one can segregate the* $\boldsymbol{\mu}$*-system*

$$\dot{\boldsymbol{y}} = \boldsymbol{Y}_1(t, \boldsymbol{y}, \boldsymbol{\mu}), \qquad \dot{\boldsymbol{\mu}} = \boldsymbol{Y}_2(t, \boldsymbol{y}, \boldsymbol{\mu}),$$
$$\boldsymbol{\mu}^{\mathsf{T}} = (\mu_1, \ldots, \mu_N), \quad \boldsymbol{Y}_1(t, \boldsymbol{0}, \boldsymbol{0}) \equiv \boldsymbol{0}, \quad \boldsymbol{Y}_2(t, \boldsymbol{0}, \boldsymbol{0}) \equiv \boldsymbol{0} \tag{4.1.11}$$

(*in the domain* $D: t \geq 0$, $\|\boldsymbol{y}\| \leq H$, $\|\boldsymbol{\mu}\| \leq H$, *the right-hand sides of system* (4.1.11) *are continuous and satisfy conditions for uniqueness of solutions*).

If the solution $\boldsymbol{y} = \boldsymbol{0}$, $\boldsymbol{\mu} = \boldsymbol{0}$ *of system* (4.1.11) *is Lyapunov asymptotically stable, then the control* $\boldsymbol{u} \in \mathfrak{U}^*$ *solves Problem* 4.1.1 *of stabilizing system* (4.1.1) *with respect to variables* $y_1, \ldots, y_m$.

Proof. 1. Consider the closed system (4.1.1), (4.1.9), and (4.1.10). Introducing the new variable $\mu_1 = X_1^*(\boldsymbol{x})$, by a direct check we ascertain that the variables and equations of the transformed system split into two groups. One of these two groups of equations forms the *linear* $\boldsymbol{\mu}$*-system*

$$\dot{x}_1 = \mu_1, \qquad \dot{\mu}_1 = \Gamma_1 x_1 + \Gamma_2 \mu_1. \tag{4.1.12}$$

Since any solution $x_1(t)$, $\mu_1(t)$ of system (4.1.12) is also a solution of the original nonlinear system, then system (4.1.1), (4.1.9), and (4.1.10) possesses a family of solutions

$$x_1 = 0, \qquad \mu_1 = X_1^*(x_1, \ldots, x_n) = 0. \tag{4.1.13}$$

By virtue of equality $X_1^*(\boldsymbol{0}) = 0$ in a sufficiently small neighborhood of position $x_i = 0$ $(i = 1, \ldots, n)$, the inequalities $|x_1| < \delta$, $|X_1^*| < \delta$ are valid, where δ is a sufficiently small positive number. Therefore, from the exponential asymptotic stability of solution (4.1.13), it follows that all the solutions of the original system (4.1.1), (4.1.9), and (4.1.10) beginning in a neighborhood of the position $x_i = 0$ $(i = 1, \ldots, n)$ possess the property

$$|x_1(t)| \leq M\left(|x_{10}| + |X_1^*(\boldsymbol{x}_0)|\right)e^{-\Gamma(t-t_0)},$$
$$t \geq t_0, \quad \Gamma, M = \text{const} > 0.$$

So, the set of control rules (4.1.10) solves Problem 4.1.1 for system (4.1.1).

2. System (4.1.11) is obtained from system (4.1.1) (closed by a control $\boldsymbol{u} \in \mathfrak{U}^*$) by nonlinear transformation of variables. Therefore, being solutions of system (4.1.11), vector functions $\boldsymbol{y}(t)$, $\boldsymbol{\mu}(t)$ will also be solutions of the original system (4.1.1). This implies that system (4.1.1) possesses a family of solutions $\boldsymbol{y} = \boldsymbol{0}$, $\boldsymbol{\mu}(t, \boldsymbol{x}) = \boldsymbol{0}$ for $\boldsymbol{u} \in \mathfrak{U}^*$, and the inequalities $\|\boldsymbol{y}\| < \delta$, $\|\boldsymbol{\mu}\| < \delta$ (δ is a sufficiently small positive number) are valid in a sufficiently small neighborhood of the unperturbed motion $\boldsymbol{x} = \boldsymbol{0}$ by virtue of $\boldsymbol{\mu}(t, \boldsymbol{0}) \equiv \boldsymbol{0}$.

As a result, from the Lyapunov asymptotic stability of the solution $\boldsymbol{y} = \boldsymbol{0}$, $\boldsymbol{\mu} = \boldsymbol{0}$ of system (4.1.11), we conclude that, for $\boldsymbol{u} \in \mathfrak{U}^*$, all the solutions of system (4.1.1) originating from a sufficiently small neighborhood of the unperturbed motion $\boldsymbol{x} = \mathbf{0}$ possess the property $\|\boldsymbol{y}(t; t_0, \boldsymbol{x}_0)\| < \varepsilon$ $(t \geq t_0)$, $\lim\|\boldsymbol{y}(t; t_0, \boldsymbol{x}_0)\| = 0$ $(t \to \infty)$. The theorem is proved. □

Remark 4.1.1.

1. In constructing control rules of the form (4.1.10), it might be reasonable to consider u_1 and the right-hand sides of the closed system (4.1.1), (4.1.9), (4.1.10) not in the phase space of the variables $x_1, \ldots, x_n$ but in some *auxiliary* $\boldsymbol{\mu}$*-space* Ω. Together with $x_1, \ldots, x_n$, space Ω contains as elements some of their functions such that in Ω the structures of u_1 and the right-hand sides of the closed system (4.1.1), (4.1.9), and (4.1.10) are more convenient for investigation.

2. The statement in the second part of Theorem 4.1.1 allows us to reduce the study of the problem of $\boldsymbol{y}$-stabilization to that of Lyapunov stability of the zero solution of auxiliary $\boldsymbol{\mu}$-system (4.1.11).

3. In their structure control rules (4.1.1) are *fractional functions* of phase variables.

Example 4.1.1.

1. Consider *Euler's dynamical equations*

$$A\dot{x}_1 = (B - C)x_2x_3 + u_1,$$
$$B\dot{x}_2 = (C - A)x_1x_3 + u_2, \quad C\dot{x}_3 = (A - B)x_1x_2 + u_3. \tag{4.1.14}$$

In the absence of the controlling moments ($u_i = 0$, $i = 1, 2, 3$), equations (4.1.14) admit a particular solution

$$x_1 = \omega = \text{const}, \quad x_2 = x_3 = 0 \qquad (123) \tag{4.1.15}$$

corresponding to the *permanent rotation* of a solid about one of the principal central axes of the ellipsoid of inertia.

Let us consider the problem of stabilizing the specified permanent rotation of a solid by using one of the controlling moments, namely, controlling moment u_2 ($u_1 = u_3 = 0$). To this end, introducing the new variables $y_j = x_j$ ($j = 1, 2$), $z_1 = x_3 - \omega$, we form the system of equations of perturbed motion in variations from state (4.1.15)

$$A\dot{y}_1 = (B - C)y_2(z_1 + \omega), \quad B\dot{y}_2 = (C - A)y_1(z_1 + \omega) + u_2,$$
$$C\dot{z}_1 = (A - B)y_1y_2. \tag{4.1.16}$$

Using the rule from Theorem 4.1.1, let us choose a set of *single-channel* control rules

$$u_2 = B\left[\Gamma_1 \frac{y_1}{z_1 + \omega} + \Gamma_2 y_2 + \frac{A - C}{B} y_1(z_1 + \omega) + \frac{(B - A)y_1y_2^2}{C(z_1 + \omega)}\right], \tag{4.1.17}$$

where

$$\Gamma_2 < 0, \quad \Gamma_1(B - C) < 0, \quad \Gamma_j = \text{const} \quad (j = 1, 2). \tag{4.1.18}$$

Following the above recommendations, let us consider the closed system (4.1.16)–(4.1.18) not in the phase space of the variables y_1, y_2, z_1 but in $\boldsymbol{\mu}$-space Ω of the variables

$$y_1, y_2, z_1, \mu_1 = A^{-1}(B - C)y_2(z_1 + \omega), \; \mu_2 = y_1(z_1 + \omega)^{-1}.$$

Omitting the intermediate calculations, we obtain the following $\boldsymbol{\mu}$-system in Ω:

$$\begin{gathered}\dot{y}_1 = \mu_1, \quad \dot{\mu}_1 = \frac{B-C}{A}\Gamma_1 y_1 + \Gamma_2 \mu_1, \\ \dot{y}_2 = \Gamma_1\mu_1 + \Gamma_2 y_2 - \frac{A-B}{C} y_2^2 \mu_2, \quad \dot{\mu}_2 = \frac{B-C}{A} y_2 - \frac{A-B}{C} y_2 \mu_2^2, \\ \dot{z}_1 = \frac{A-B}{C} y_1 y_2. \end{gathered} \tag{4.1.19}$$

It can be verified that, if conditions (4.1.18) are fulfilled, all the roots of the characteristic equation of the *linear part* of the first two groups of equations of system (4.1.19) have negative real parts. As a result, by the Lyapunov theorem on asymptotic stability in the linear approximation, the solution $y_1 = y_2 = \mu_1 = \mu_2 = z_1 = 0$ of system (4.1.19) is asymptotically stable on the whole with respect to y_1, μ_1 and asymptotically stable "in the small" with respect to y_1, y_2, μ_1, μ_2. The use of the cited Lyapunov theorem is well-founded, if

(1) initial perturbations y_{10}, y_{20}, z_{10} are sufficiently small;

(2) the angular velocity ω of rotation of the solid is, on the contrary, not too small in magnitude.

Under the specified conditions not only the values y_{10}, y_{20}, z_{10} but also $\mu_{10} = A^{-1}(B-C)y_{20}(z_{10}+\omega)$ and $\mu_{20} = y_{10}(z_{10}+\omega)^{-1}$ are sufficiently small.

We emphasize that conditions (1) and (2) ensure not only asymptotic stability of the solution $y_1 = y_2 = \mu_1 = \mu_2 = z_1 = 0$ of system (4.1.19) with respect to y_1, y_2, μ_1, μ_2 but also (by Theorem 2.6.1) its Lyapunov stability.

Thus, the set of control rules (4.1.17), (4.1.18) guarantees the Lyapunov stability and asymptotic stability of the unperturbed motion $y_1 = y_2 = z_1 = 0$ of system (4.1.16)–(4.1.18) with respect to variables y_1, y_2.

In $\boldsymbol{\mu}$-space Ω control u_2 has the form

$$u_2 = B\left[\Gamma_1\mu_2 + \Gamma_2 y_2 + \frac{A-C}{B} y_1(z_1+\omega) + \frac{B-A}{C} y_2^2 \mu_2\right].$$

Consequently, if conditions (1) and (2) are fulfilled, the estimate $|u_2| \le \varepsilon$ is valid, where ε is a sufficiently small positive number. In addition, $\lim|u_2| = 0,\ t \to 0$.

Pragmatically, the stabilization performed with respect to y_1, y_2 means that the control rules (4.1.17), (4.1.18) *restore* (*asymptotically*) the solid to the initial *mode of "spin"* about one of the principal central axes of inertia when the solid is disturbed by small perturbations. Let us note that the considered problem is of interest in dynamics of spacecraft where it is important to have the mode of "spin" of a spacecraft about the major axis of its ellipsoid of inertia.

2. When solving specific control problems, apparently, one should not omit a "singular case" from consideration, when control rule u_1 of the form (4.1.10) operates in the mode "*uncertainty of type* 0/0" for sufficiently large t. Requirement of type (4.1.4) might also be fulfilled.

Because there is no possibility of a comprehensive analysis of this problem, let us present some results of computer simulation for the problem of controlling the angular motion of a solid by a *single* homogeneous symmetric flywheel with a fixed (along one of the principal central axes of inertia of the body) rotational axis. The angular motion of the mechanical system under consideration (*gyrostat*) is described by a system of ordinary differential equations

$$\begin{gathered}(A-J_1)\dot{x}_1=(B-C)x_2x_3-u_1, \quad B\dot{x}_2=(C-A)x_1x_3-J_1x_3\dot{\varphi}_1, \\ C\dot{x}_3=(A-B)x_1x_2+J_1x_2\dot{\varphi}_1, \quad J_1(\dot{x}_1+\ddot{\varphi}_1)=u_1.\end{gathered} \tag{4.1.20}$$

Here x_i $(i=1,2,3)$ are the projections of the angular velocity vector of the body onto its principal central axes of inertia $Oxyz$, A, B, and C are the moments of inertia of the gyrostat with respect to the coordinate axes $Oxyz$, $J_1, \dot{\varphi}_1$ are the axial moment of inertia and the relative angular velocity of the flywheel rotation, and u_1 is the controlling moment applied to the flywheel and realized by a special engine. Equations (4.1.20) describe the dynamics of spacecraft of the type "*Dual Spinners*" when

$$A-J_1<C<B,$$

where the flywheel's rotational axis is fixed along the major axis of the system's ellipsoid of inertia.

Consider the control rule

$$\begin{gathered}u_1=\frac{1}{B}\Big\{(C-B)(A-B-J_1)x_2x_3+\frac{C(A-J_1)}{x_2}\big[u_1^*(\boldsymbol{x})-f_1(\boldsymbol{x})\big]\Big\}, \\ f_1(\boldsymbol{x})=x_3(CB)^{-1}\big[(C-A)x_1-J_1\dot{\varphi}_1\big]f_2(\boldsymbol{x}), \\ u_1^*(\boldsymbol{x})=\Gamma_1x_3+\Gamma_2C^{-1}x_2f_2(\boldsymbol{x}), \quad f_2(\boldsymbol{x})=(A-B)x_1+J_1\dot{\varphi}_1, \\ \Gamma_i=\text{const}<0 \quad (i=1,2).\end{gathered} \tag{4.1.21}$$

In the closed system (4.1.20), (4.1.21), one can distinguish the linear $\boldsymbol{\mu}$-system

$$\dot{x}_3=\mu_1, \quad \dot{\mu}_1=\Gamma_1x_3+\Gamma_2\mu_1 \quad \big(\mu_1=C^{-1}x_2f_2(\boldsymbol{x})\big).$$

This means that, in the course of control, the relationships

$$x_3(t)\to 0, \quad \mu_1(t)\to 0, \quad t\to\infty,$$

are valid; this allows for one of two possibilities:

(1) $x_i(t)\to 0, \quad t\to\infty \quad (i=2,3)$;

(2) $x_3(t)\to 0, \quad f_2(\boldsymbol{x})\to 0, \quad t\to\infty$.

Case (1) ensures partial (with respect to the variables x_2, x_3) damping of rotations of the main body of the gyrostat; as time goes by, the gyrostat enters the *mode* of "*spinning*" about the *major* axis of its ellipsoid of inertia.

In case (2) the gyrostat is brought to the position

$$x_3=f_2(\boldsymbol{x})=0, \tag{4.1.22}$$

where $u_1=0$ by (4.1.21).

It is known (Guelman [1989]) that the equilibrium position (4.1.22) is a *stable* equilibrium position of the gyrostat for $u_1=0$; having brought the gyrostat to this position by subsequently applying the controlling moment u_1 which is *constant* in magnitude, one can *completely* damp rotations of the gyrostat's main body.

In case (1) the expression (4.1.21) for u_1 contains the term

$$\frac{x_3}{x_2}\frac{C(A-J_1)}{B}\Big\{\Gamma_1-\big[(C-A)x_1-J_1\dot{\varphi}_1\big]f_2(\boldsymbol{x})\Big\},$$

which is a *fractional* function of the phase variables and thus will operate in the mode "*uncertainty of type* 0/0" *for* $t\to\infty$.

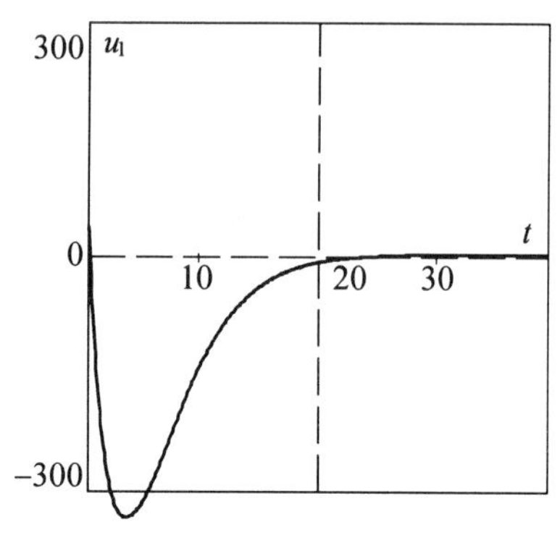

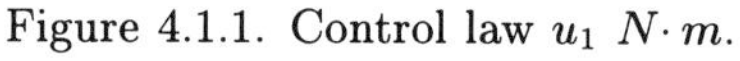
Figure 4.1.1. Control law u_1 $N\cdot m$.

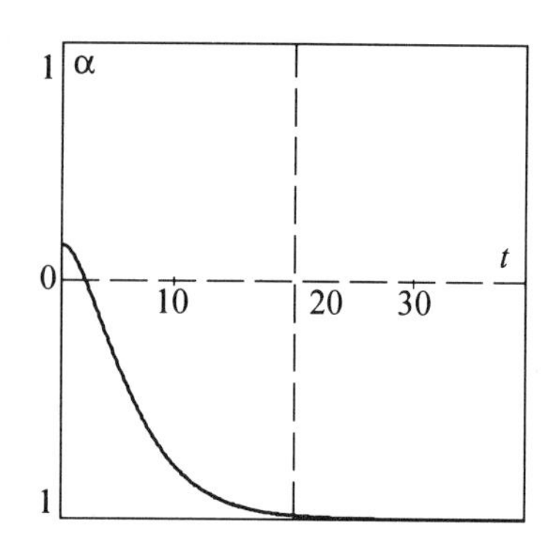

Figure 4.1.2. Function α rad.

Computer simulation demonstrates that in many cases such a mode of operation of u_1 might be quite acceptable.

For example, consider the case when $A = 559$, $B = 982$, $C = 900$, $J_1 = 400$ $kg{\cdot}m^2$; $x_1(t_0) = 1.0$, $x_2(t_0) = 3.0$ $rad{\cdot}s^{-1}$, $x_3(t_0) = \dot{\varphi}_1(t_0) = 0$; $\Gamma_1 = -0.05$ s^{-2}, $\Gamma_2 = -0.5$ s^{-1}. In this case we have partial (with respect to x_2, x_3) damping of the rotations of the gyrostat's main body. Figure 4.1.1 presents the behavior of u_1 (as a function of time t) for this case. Figure 4.1.2 shows the transient process relative to the variable

$$\alpha = \cos\theta = (Ax_1 + J_1\dot{\varphi}_1)K^{-1},$$
$$K^2 = (Ax_1 + J_1\dot{\varphi}_1)^2 + (Bx_2)^2 + (Cx_3)^2$$

characterizing the angle θ of variation of the angular momentum vector $\boldsymbol{K}$ of the gyrostat from the major axis of its ellipsoid of inertia (axis of "spinning").

It is apparent that, by using control u_1, the partial damping of rotation takes just about 20 seconds; u_1 in this interval is the *impulse of variable intensity* with maximum value $|u_1| = 332$ $N\cdot m$.

4.2 The Auxiliary Function of the Partial Stabilization Problem in Studying Problems of Stabilization with Respect to All the Variables and Polystabilization

4.2.1 Preliminary Remarks

One of the important motives for investigating the problem of partial stabilization is the *auxiliary function* of this problem in studying stabilization with respect to *all* or to a *given* (*major*) *part* of the variables. Such an approach to the problem may turn out to be effective; in this section it is applied for a sufficiently general class of nonlinear controlled systems. The proposed approach is based on the method of constructing *auxiliary linear* *controlled μ-systems*; such systems are obtained from the original nonlinear controlled system through a special choice of the structure of controls (in accordance with the *feedback principle*) and nonlinear transformations of variables.

Within the framework of this approach, solving the problem of stabilization with respect to *all the variables* is carried out in *two stages.*

In the *first stage* (Subsections 4.2.3–4.2.8), we state and solve the problem of finding a set of control rules ensuring Lyapunov asymptotic stability of the unperturbed motion of the system in question; the preassigned domain S is the domain of attraction of the unperturbed motion. At this stage the problem of *partial* stabilization is an *intermediate link* of the study.

The *second stage* (Subsection 4.2.9) shows how the control rules satisfying additional performance criteria or, more specifically, *optimal* in the sense of a minimum of some integral-type functional can be found in the chosen set.

Then (Section 4.2.11) the problem of *polystabilization* is considered in a similar way. In this case the Lyapunov (nonasymptotic) stability and asymptotic stability with respect to a given part of the variables are guaranteed for unperturbed motion.

4.2.2 The Class of Systems under Consideration

Let the perturbed motion of a control object be described by a nonlinear system of ordinary differential equations (in vector form)

$$\dot{\boldsymbol{\xi}} = \boldsymbol{X}^{(1)}(t, \boldsymbol{\xi}, \boldsymbol{\eta}, \boldsymbol{u}), \qquad \dot{\boldsymbol{\eta}} = \boldsymbol{X}^{(2)}(t, \boldsymbol{\xi}, \boldsymbol{\eta})$$
$$(\boldsymbol{\xi}, \boldsymbol{X}^{(1)} \in R^l; \quad \boldsymbol{\eta}, \boldsymbol{X}^{(2)} \in R^{n-l}; \quad \boldsymbol{u} \in R^r), \tag{4.2.1}$$

where $\boldsymbol{x} = (\xi_1, \dots, \xi_l, \eta_1, \dots, \eta_{n-l}) = (\boldsymbol{\xi}^\top, \boldsymbol{\eta}^\top)$ is the vector of the phase variables and $\boldsymbol{u}$ is the vector of controls.

The right-hand sides of systems (4.2.1) satisfy the conditions:

(1) vector functions $\boldsymbol{X}^{(1)}, \boldsymbol{X}^{(2)}$ are defined and continuous together with their partial derivatives with respect to $t, \boldsymbol{x}, \boldsymbol{u}$ in the domain

$$t \geq 0, \quad \|\boldsymbol{x}\| \leq H, \quad \|\boldsymbol{u}\| < \infty; \tag{4.2.2}$$

(2) $\boldsymbol{X}^{(1)}(t, \boldsymbol{0}, \boldsymbol{0}) \equiv \boldsymbol{0}, \quad \boldsymbol{X}^{(2)}(t, \boldsymbol{0}, \boldsymbol{0}) \equiv \boldsymbol{0}.$

Besides, let us assume that the phase variables of system (4.2.1) are related as follows:

$$G_s(t, \boldsymbol{x}) = 0, \quad G_s(t, \boldsymbol{0}) \equiv 0 \quad (s = 1, \dots, n_1 \geq n - 2r); \tag{4.2.3}$$

functions G_s $(s = 1, \dots, n_1)$ are continuous together with their partial derivatives with respect to $t, \boldsymbol{x}$ in the domain $t \geq 0$, $\|\boldsymbol{x}\| \leq H$.

Structure (4.2.1), (4.2.3) of the equations of perturbed motion is characteristic, for example, of the *problems of controlling the angular velocity of a solid (spacecraft)*. In this case the first group of equations (4.2.1) includes *dynamical equations,* and the second group (containing no controls) embraces *kinematic equations.* Relationships (4.2.3) are typical of kinematic equations whose variables are the components of *vectors*, *matrices of directional cosines*, or *Euler's quaternion* components (*Rodrigues–Hamilton variables*).

If $2r \geq n$ (i.e., the doubled dimension of vector $\boldsymbol{u}$ is not less than the dimension of system (4.2.1)), then the technique considered below will also be valid for system (4.2.1) whose phase variables are not bound by additional relationships (4.2.3). (See Subsection 4.2.12 regarding this issue.)

4.2.3 Formulation of the Problem of Stabilization with Respect to All the Variables

We shall choose controls in the class $\mathcal{U}_* = \{\boldsymbol{u}(t, \boldsymbol{x})\}$ of vector functions $\boldsymbol{u} = \boldsymbol{u}(t, \boldsymbol{x})$, $\boldsymbol{u}(t, \boldsymbol{0}) \equiv \boldsymbol{0}$ that are continuous in the domain $t \geq 0$, $||\boldsymbol{x}|| \leq H$. It is assumed that, for any $\boldsymbol{u} \in \mathcal{U}_*$, the conditions for existence and uniqueness of all the solutions of system (4.2.1) considered below are satisfied.

Problem 4.2.1 (*on stabilization of the unperturbed motion $\boldsymbol{x} = \boldsymbol{0}$ of system* (4.2.1) *with respect to all the variables*). *Find the set M of control rules $\{\boldsymbol{u}(t, \boldsymbol{x}) \in \mathcal{U}_*\}$ such that*

(*a*) *the unperturbed motion $\boldsymbol{x} = \boldsymbol{0}$ of system* (4.2.1) *is Lyapunov asymptotically stable*;

(*b*) *the given domain $S = \{\boldsymbol{x}_0 : ||\boldsymbol{x}_0|| \in S\}$ of initial perturbations $\boldsymbol{x}_0$ is the domain of attraction of the unperturbed motion $\boldsymbol{x} = \boldsymbol{0}$, which means that* $\lim ||\boldsymbol{x}(t; t_0, \boldsymbol{x}_0)|| = 0$, $t \to \infty$ *for $\boldsymbol{x}_0 \in S$.*

4.2.4 A Scheme Proposed for Solving Problem 4.2.1

(1) First we solve the problem of stabilizing the unperturbed motion $\boldsymbol{x} = \boldsymbol{0}$ of system (4.2.1) with respect to some initially selected *part* of the variables. We managed to reduce this problem to the problem of stabilizing (with respect to *all* the phase variables) the zero solution of the auxiliary linear controlled $\boldsymbol{\mu}$-system.

(2) Then we impose additional conditions under which the *partial* stabilization of the unperturbed motion of system (4.2.1) is actually the stabilization of this motion with respect to *all* the variables as well. The additional conditions also ensure that the preassigned domain S is the domain of attraction of the unperturbed motion of system (4.2.1).

We note that in this scheme for solving Problem 4.2.1 two types of stabilization (with respect to all and to part of the variables) appeared to be *closely interrelated* and also *mutually complementary* in the investigation. Indeed, for stabilizing the unperturbed motion $\boldsymbol{x} = \boldsymbol{0}$ of system (4.2.1) *with respect to all the variables*, it suffices (under certain conditions) to *partially* stabilize this motion (with respect to some initially selected part of the variables). This *partial* stabilization problem is, in turn, reduced to the problem of stabilization *with respect to all the variables* for the specially constructed auxiliary *linear controlled $\boldsymbol{\mu}$-system* (which is essentially simpler than the original one).

4.2.5 The Construction of an Auxiliary Linear Controlled System

Suppose that $n - l \geq r$, and consider the functions

$$\Phi_j(t, \boldsymbol{x}, \boldsymbol{u}) = \frac{\partial X_j^{(2)}}{\partial t} + \sum_{i=1}^{l} \frac{\partial X_j^{(2)}}{\partial \xi_i} X_i^{(1)} + \sum_{s=1}^{n-l} \frac{\partial X_j^{(2)}}{\partial \eta_s} X_s^{(2)} \tag{4.2.4}$$
$$(j = 1, \ldots, n - l),$$

where $X_i^{(1)}$, $X_j^{(2)}$, ξ_i, and η_j $(i = 1, \ldots, l; j = 1, \ldots, n - l)$ are components of vector functions $\boldsymbol{X}^{(1)}$, $\boldsymbol{X}^{(2)}$ and vectors $\boldsymbol{\xi}$, $\boldsymbol{\eta}$, respectively.

Introduce the *Jacobi matrix* $F(t, \boldsymbol{x}, \boldsymbol{u}) = (\partial \Phi_j / \partial u_k)$ of functions Φ_j $(j = 1, \ldots, n - l)$ with respect to components u_k $(k = 1, \ldots, r)$ of vector $\boldsymbol{u}$. We assume that, in the domain

$$t \geq 0, \quad \|\boldsymbol{x}\| \leq h_1, \quad \|\boldsymbol{u}\| \leq h_2 \quad (h_1 \leq H), \tag{4.2.5}$$

the condition

$$\operatorname{rank} F(t, \boldsymbol{x}, \boldsymbol{u}) = r \tag{4.2.6}$$

is satisfied. Also suppose that, for all $t, \boldsymbol{x}, \boldsymbol{u}$ from domain (4.2.5), the *same* rows of matrix F are linearly independent. For definiteness, let these be the *first* r rows.

By the *theory of implicit functions*, in the case of (4.2.6) the system of equations

$$\Phi_j(t, \boldsymbol{x}, \boldsymbol{u}) = u_j^* \qquad (j = 1, \ldots, n - l) \tag{4.2.7}$$

has the solution

$$\boldsymbol{u} = \boldsymbol{f}(t, \boldsymbol{x}, \boldsymbol{u}^*), \qquad \boldsymbol{u}^* = (u_1^*, \ldots, u_r^*) \tag{4.2.8}$$

in domain (4.2.5). Under assumptions concerning the right-hand sides of system (4.2.1), vector function $\boldsymbol{f}(t, \boldsymbol{x}, \boldsymbol{u}^*)$ is a *single-valued continuous* function (regarding the variables in the aggregate) in domain (4.2.5). Besides, $\boldsymbol{f}(t, \boldsymbol{0}, \boldsymbol{0}) \equiv \boldsymbol{0}$.

We shall interpret u_k^* $(k = 1, \ldots, r)$ as "*auxiliary controls.*" As a result, under condition (4.2.6) one can segregate the *auxiliary linear controlled* $\boldsymbol{\mu}$*-system*

$$\dot{\eta}_k = \mu_k, \quad \dot{\mu}_k = u_k^* \qquad (k = 1, \ldots, r) \tag{4.2.9}$$

from the closed nonlinear system (4.2.1), (4.2.8).

Further we propose to construct the solution of the original nonlinear Problem 4.2.1 based on solving the corresponding stabilization problem for the linear controlled $\boldsymbol{\mu}$-system (4.2.9). As a result, construction (4.2.8) can be regarded as a *general structural form of the controls* in Problem 4.2.1. The parameters of this form, the auxiliary controls u_k^*, are determined by solving the corresponding linear problems of stabilization.

4.2.6 Extending the Constructibility of an Auxiliary Linear Controlled System

We can attain this aim by considering the equalities

$$\begin{aligned} X_i^{(1)}(t,\boldsymbol{x},\boldsymbol{u}) &= u_i^* \qquad (i=1,\ldots,q<r<l), \\ \Phi_j(t,\boldsymbol{x},\boldsymbol{u}) &= u_{q+j}^* \qquad (j=1,\ldots,n-l) \end{aligned} \tag{4.2.10}$$

instead of (4.2.7) and replacing conditions $n-l \geq r$, $n_1 \geq n-2r$ with $n-l \geq r-q$, $n_1 \geq n-2r+q$. The condition $n-l \geq r-q$ is weaker than $n-l \geq r$ and means that, for fixed r, the number of equations containing no controls explicitly may be greater than in the case of $n-l \geq r$.

The condition $n_1 \geq n-2r+q$ is stronger than $n_1 \geq n-2r$ and means a greater number of additional relationships. Thus the weakening of the condition $n-l \geq r$ occurs at the expense of strengthening the condition $n_1 \geq n-2r$.

In applied problems we note that the condition $n_1 \geq n-2r$ may be satisfied, for instance, in stabilizing a satellite in an orbit (see Section 4.3, $n=12$, $n=r=3$), where the number of additional relationships (4.3.23) equals $n-2r$ or not satisfied, for instance, in stabilizing permanent rotations of a solid fixed at one point (see Section 4.3, $n=6$, $n=r=3$), where the number of additional relationships (4.3.5) does not equal $n-2r$. The specified modification of the conditions $n-l \geq r$, $n_1 \geq n-2r$ with the purpose of extending the constructibility of the auxiliary linear system of type (4.2.9), therefore, is caused by the desire to broaden the possibilities of the proposed approach in its theoretical aspect and also by the needs of applications.

Let us introduce the Jacobi matrix $Q(t,\boldsymbol{x},\boldsymbol{u})$ of functions $X_i^{(1)}$ $(i=1,\ldots,q)$, Φ_j $(j=1,\ldots,n-l)$ with respect to the components of vector $\boldsymbol{u}$. Suppose that condition

$$\operatorname{rank} Q(t,\boldsymbol{x},\boldsymbol{u}) = r \tag{4.2.11}$$

is satisfied in domain (4.2.5). Also suppose that, for all $t,\boldsymbol{x},\boldsymbol{u}$ from domain (4.2.5), the same rows of matrix Q are linearly independent. For definiteness, let these be the *first* r rows.

In the case of (4.2.11) the system of equations (4.2.10) admits the single-valued solution

$$\boldsymbol{u} = \boldsymbol{f}^*(t,\boldsymbol{x},\boldsymbol{u}^*). \tag{4.2.12}$$

In domain (4.2.5) vector function $\boldsymbol{f}^*(t,\boldsymbol{x},\boldsymbol{u}^*)$ is continuous with respect to the variables in the aggregate. Besides, $\boldsymbol{f}^*(t,\boldsymbol{0},\boldsymbol{0}) \equiv \boldsymbol{0}$.

In the closed system (4.2.1), (4.2.12) one can distinguish the *linear controlled $\boldsymbol{\mu}$-system*

$$\dot{\xi}_i = u_i^* \qquad (i=1,\ldots,q<r<l), \tag{4.2.13}$$

$$\dot{\eta}_j = \mu_j, \quad \dot{\mu}_j = u_{q+j}^* \qquad (j=1,\ldots,r-q). \tag{4.2.14}$$

4.2.7 Solving Problem 4.2.1 Based on a Linear Controlled μ-System (4.2.9)

Along with the matrices F, Q introduced, let us also introduce the Jacobi matrix $\Psi(t, \boldsymbol{x})$ of functions $X_k^{(2)}$ $(k = 1, \ldots, r)$, G_s $(s = 1, \ldots, n_1)$ with respect to variables $\xi_1, \ldots, \xi_l, \eta_{r+1}, \ldots, \eta_{n-l}$.

In the domain

$$t \geq 0, \quad \|\boldsymbol{x}\| \leq h_3 \leq H, \tag{4.2.15}$$

let the condition

$$\operatorname{rank} \Psi(t, \boldsymbol{x}) = n - r \tag{4.2.16}$$

be satisfied. Also suppose that, for all $t, \boldsymbol{x}$ from domain (4.2.15), the *same* rows of matrix Ψ are linearly independent. For definiteness, let these be the *first* $n - r$ rows.

We note that the conditions $n - l \geq r$, $n_1 \geq n - 2r$ introduced earlier imply the relationship $r + n_1 \geq n - r$ required to fulfil condition (4.2.16).

By the theory of implicit functions, the equalities $\mu_k = X_k^{(2)}$ $(k = 1, \ldots, r)$, $G_s = 0$ $(s = 1, \ldots, n_1)$ in the case of (4.2.16) admit the single-valued solution (in the vector form)

$$\begin{gathered} \boldsymbol{z} = \boldsymbol{\varphi}(t, \boldsymbol{y}), \qquad \boldsymbol{\varphi}(t, \mathbf{0}) \equiv \mathbf{0}, \\ \boldsymbol{y} = (\eta_1, \ldots, \eta_r, \mu_1, \ldots, \mu_r), \quad \boldsymbol{z} = (\xi_1, \ldots, \xi_l, \eta_{r+1}, \ldots, \eta_{n-l}). \end{gathered} \tag{4.2.17}$$

Vector function $\boldsymbol{\varphi}(t, \boldsymbol{y})$ is continuous in domain (4.2.15).

Consider the class

$$\mathfrak{U}_1 = \left\{u_k^* = \Gamma_k^{(1)} \eta_k + \Gamma_k^{(2)} \mu_k;\ \Gamma_k^{(1)}, \Gamma_k^{(2)} = \text{const} < 0,\ k = 1, \ldots, r\right\}$$

of the auxiliary linear control rules u_k^* $(i = 1, \ldots, r)$ solving the problem of stabilizing the zero solution of system (4.2.9) (or system (4.2.13), (4.2.14), if $\Gamma_i^{(2)} = 0$, $\eta_i = \xi_i$, $i = 1, \ldots, q$) with respect to all the variables.

Since control rules (4.2.8), (4.2.12) are functions of $\boldsymbol{x}$ for $\boldsymbol{u}^* \in \mathfrak{U}_1$, then the inequality $\|\boldsymbol{u}\| \leq h_2$ (involved in (4.2.5)) defines some domain $\|\boldsymbol{x}\| \leq h_4$ of variation of $\boldsymbol{x}$.

Introduce the designation $h^* = \min(h_1, h_3, h_4)$. It it obvious that $h^* \leq H$.

Theorem 4.2.1. *Let the following conditions be satisfied*:

(1) *condition* (4.2.6) *in domain* (4.2.5);

(2) *condition* (4.2.16) *in domain* (4.2.15);

(3) $\{\boldsymbol{x}_0 : \|\boldsymbol{x}(t; t_0, \boldsymbol{x}_0)\| \leq h^*\} \in S$;

(4) $\|\boldsymbol{\varphi}(t, \boldsymbol{y})\| \to 0$ *as* $\|\boldsymbol{y}\| \to 0$.

Then the set of control rules (4.2.8) *with* $\boldsymbol{u}^* \in \mathfrak{U}_1$ *guarantees that Problem* 4.2.1 *is solved.*

Proof. Equations (4.2.9) are obtained by nonlinear transformation of the variables of system (4.2.1), (4.2.8). So from the Lyapunov asymptotic stability of the solution $\eta_k = \mu_k = 0$ $(k = 1, \ldots, r)$ of system (4.2.9) for $\boldsymbol{u}^* \in \mathfrak{U}_1$, it follows that the unperturbed motion $\boldsymbol{x} = \boldsymbol{0}$ of system (4.2.1), (4.2.8) is asymptotically stable with respect to the variables $\eta_k, X_k^{(2)}(\boldsymbol{x})$ $(k = 1, \ldots, r)$ (also for $\boldsymbol{u}^* \in \mathfrak{U}_1$).

Equalities (4.2.17) relate those phase variables of system (4.2.1), which are not included in vector $\boldsymbol{y}$, to variables $\eta_k, X_k^{(2)}(\boldsymbol{x})$. By virtue of $\boldsymbol{\varphi}(t, \boldsymbol{0}) \equiv \boldsymbol{0}$ and condition (4), this relationship means that the $\boldsymbol{y}$-stabilization of the unperturbed motion $\boldsymbol{x} = \boldsymbol{0}$ of system (4.2.1), (4.2.8) (for $\boldsymbol{u}^* \in \mathfrak{U}_1$) is, in fact, the stabilization of this motion with respect to all the phase variables.

By condition (3) the given domain S is that of attraction of the unperturbed motion $\boldsymbol{x} = \boldsymbol{0}$. The theorem is proved. □

Remark 4.2.1.

1. Using the approach proposed, we can find solutions of system (4.2.1), (4.2.8) *explicitly*. That is why the constructive check of condition (3) of Theorem 4.2.1 presents no crucial difficulties.
2. Under conditions of Theorem 4.2.1 the problem of stabilizing the unperturbed motion of *nonlinear* system (4.2.1) with respect to all the variables is reduced to the problem of stabilizing the zero solution of the *linear* controlled $\boldsymbol{\mu}$-system (4.2.9) (also with respect to all the variables).

4.2.8 Solving Problem 4.2.1 Based on a Linear Controlled $\boldsymbol{\mu}$-System (4.2.13), (4.2.14)

Introduce the Jacobi matrix $P(t, \boldsymbol{x})$ of functions $X_j^{(2)}$ $(j = 1, \ldots, r - q)$, G_s $(s = 1, \ldots, n_1)$ with respect to variables $\xi_{q+1}, \ldots, \xi_l, \eta_{r-q+1}, \ldots, \eta_{n-l}$. Also suppose that $r - q + n_1 \geq n - r$.

In domain (4.2.15) let the condition

$$\operatorname{rank} P(t, \boldsymbol{x}) = n - r \tag{4.2.18}$$

be satisfied.

In this case the relationship of type (4.2.17) takes the form

$$\boldsymbol{z}^* = \boldsymbol{\varphi}^*(t, \boldsymbol{y}^*), \qquad \boldsymbol{\varphi}^*(t, \boldsymbol{0}) \equiv \boldsymbol{0}, \tag{4.2.19}$$

where vector $\boldsymbol{y}^*$ consists of the variables defining the state of the linear controlled $\boldsymbol{\mu}$-system (4.2.13), (4.2.14) and vector $\boldsymbol{z}^*$ is composed of those phase variables of system (4.2.1), which are not included in $\boldsymbol{y}^*$.

Corollary 4.2.1. *Let the following conditions be satisfied:*
(1) *condition* (4.2.11) *in domain* (4.2.5);
(2) *condition* (4.2.18) *in domain* (4.2.15);
(3) *condition* (3) *of Theorem* 4.2.1;
(4) $\|\boldsymbol{\varphi}^*(t, \boldsymbol{y}^*)\| \to 0$, *as* $\|\boldsymbol{y}^*\| \to 0$.

Then the set of control rules (4.2.12) *with* $\boldsymbol{u}^* \in \mathfrak{U}_1$ *guarantees that Problem* 4.2.1 *is solved.*

4.2.9 Choosing Optimal Control Rules

From the set M of stabilizing control rules (4.2.8) or (4.2.12) previously found, let us choose the control rules satisfying the additional performance criteria, optimal in the sense of a minimum of some integral-type functional

$$I = \int_{t_0}^{\infty} R(t, \boldsymbol{x}, \boldsymbol{u})dt. \tag{4.2.20}$$

As a *generalized performance criterion of the control*, functional (4.2.20) should be designated in accordance with the practical requirements to the systems being designed (i.e., with the physical essence of the problem). But such requirements are often contradictory and are not always susceptible to formalization. Moreover, the control rules obtained may turn out to be excessively complicated and, as a consequence, difficult to implement.

As a result, solving the problem of optimal stabilization is often an *iterative process*: after obtaining the first solution, the correction is entered into the functional being minimized, and the solution is obtained again. The procedure is repeated until an acceptable result is achieved. The less time one iteration takes, the sooner the desired result is obtained.

Using the method proposed in Subsection 4.2.4, Problem 4.2.1 is solved based on the linear controlled $\boldsymbol{\mu}$-systems (4.2.9) or (4.2.13), (4.2.14). In this case the search for optimal control rules in Problem 4.2.1 can be carried out only in the context of the general structural form of (4.2.8) or (4.2.12). In solving this optimization problem, it is natural to take the integrand of functional (4.2.20) as a quadratic form of the phase variables of $\boldsymbol{\mu}$-system and auxiliary controls u_k^* $(k = 1, \ldots, r)$. For example, in the case of $\boldsymbol{\mu}$-system (4.2.9), we set

$$R(t, \boldsymbol{x}, \boldsymbol{u}) = \sum_{k=1}^{r} \{a_k \eta_k^2 + b_k \mu_k^2(t, \boldsymbol{x}) + c_k \left[u_k^*(t, \boldsymbol{x}, \boldsymbol{u})\right]^2\}, \tag{4.2.21}$$

where a_k, b_k, and c_k are positive constants.

As a result, for the linear controlled $\boldsymbol{\mu}$-system, we arrive at the classic *linear-quadratic problem* of optimal stabilization (Kalman [1960a], Letov [1960]). Its solution has the form

$$u_k^* = -\sqrt{\frac{a_k}{c_k}}\eta_k - \sqrt{\frac{b_k + 2\sqrt{a_k c_k}}{c_k}}\mu_k \qquad (k = 1, \ldots, r). \tag{4.2.22}$$

By selecting (iteratively) the corresponding values of a_k, b_k, and c_k, one can choose the control rules satisfying additional performance criteria from set (4.2.8).

At the same time, in variables $\boldsymbol{x}, \boldsymbol{u}$ characterizing the original nonlinear problem, functional (4.2.20), (4.2.21) is *essentially nonlinear* and falls into the type of *semidefinite* functionals (the values of constants a_k, b_k, and c_k are not preassigned).

We note that semidefinite functionals are widely used in applications. (See, for example, the work of A.A. Krasovskii et al. [1977], Rumyantsev and Oziraner [1987], A.A. Krasovskii [1992], and bibliographies to them.) To simplify the solution technique, even semidefinite functionals are often used that do not possess the classic Kalman–Letov structure, i.e., their integrand $R(t, \boldsymbol{x}, \boldsymbol{u})$ is either only sign-constant (non-sign-definite with respect to $\boldsymbol{x}, \boldsymbol{u}$) or even non-sign-constant. It is often difficult to interpret such functionals from the standpoint of physics.

In this connection let us show that under sufficiently general assumptions concerning the right-hand sides of system (4.2.1), the semidefinite performance functional (4.2.20), (4.2.21) used possesses a structure characteristic of the Kalman–Letov functionals: its integrand $R(t, \boldsymbol{x}, \boldsymbol{u})$ is *sign-definite with respect to* $\boldsymbol{x}, \boldsymbol{u}$. In a number of cases important for applications (see examples in Section 4.3), this functional has a clear mechanical meaning.

Theorem 4.2.2. *Let the conditions of Theorem* 4.2.1 *be satisfied. Then control rules* (4.2.8) *with* u_k^* *having the form* (4.2.22) *guarantee that Problem* 4.2.1 *is solved and minimized functional* (4.2.20), (4.2.21) *has the form*

$$I = \int_{t_0}^{\infty} \left(\sum_{k=1}^{r} \left\{ a_k \eta_k^2 + b_k \left[X_k^{(2)}(t, \boldsymbol{x}) \right]^2 + c_k \left[\Phi_k(t, \boldsymbol{x}, \boldsymbol{u}) \right]^2 \right\} \right) dt. \qquad (4.2.23)$$

If functions $X_k^{(2)}, \Phi_k$ $(k = 1, \ldots, r)$ *do not explicitly depend on* t, *then the integrand in* (4.2.23) *is a function that is positive-definite with respect to variables* $\boldsymbol{x}, \boldsymbol{u}$ *in the domain* $\|\boldsymbol{x}\| \leq h^*$, $\|\boldsymbol{u}\| \leq h_2$.

Proof. System (4.2.9), (4.2.22) splits into r independent equations, which leads to minimizing the functional $I = \sum_{k=1}^{r} I_k$ on the trajectories of this system. The integrand in I_k has the form

$$a_k \eta_k^2 + b_k \mu_k^2(t, \boldsymbol{x}) + c_k \left[u_k^*(t, \boldsymbol{x}, \boldsymbol{u}) \right]^2.$$

By equalities (4.2.7) and $\mu_k = X_k^{(2)}(t, \boldsymbol{x})$ and because of the additivity property of the integral, functional I has the form (4.2.23).

This means that functional (4.2.23) is minimized on the trajectories of system (4.2.1), (4.2.8), and (4.2.22) originating from domain S of attraction of the unperturbed motion.

Let us show that, if functions $X_k^{(2)}, \Phi_k$ $(k = 1, \ldots, r)$ do not explicitly depend on t, then the integrand $R(\boldsymbol{x}, \boldsymbol{u})$ in (4.2.23) is a positive-definite function with respect to $\boldsymbol{x}, \boldsymbol{u}$ in the domain $\|\boldsymbol{x}\| \leq h^*$, $\|\boldsymbol{u}\| \leq h_2$. Since $R(\boldsymbol{x}, \boldsymbol{u}) \geq 0$, it suffices to prove that $R = 0$ if and only if $\boldsymbol{x} = \boldsymbol{0}$, $\boldsymbol{u} = \boldsymbol{0}$.

Let $\boldsymbol{x} = \boldsymbol{0}$, $\boldsymbol{u} = \boldsymbol{0}$. Taking into account that $X_k^{(2)}(\boldsymbol{0}) = 0$, $\Phi_k(\boldsymbol{0}, \boldsymbol{0}) = 0$, $(k = 1, \ldots, r)$, we conclude that $R(\boldsymbol{0}, \boldsymbol{0}) = 0$.

On the contrary, let $R(\boldsymbol{x}, \boldsymbol{u}) = 0$. Since $a_k > 0$, $b_k > 0$, $c_k > 0$ $(k = 1, \ldots, r)$, then the following equalities:

$$\eta_k = 0, \tag{4.2.24}$$

$$X_k^{(2)} = 0, \tag{4.2.25}$$

$$\Phi_k = 0 \tag{4.2.26}$$

are satisfied $(k = 1, \ldots, r)$.

Because dependency (4.2.17) is valid in domain (4.2.15), $\boldsymbol{z} = \mathbf{0}$ by (4.2.24), (4.2.25). Then, allowing for (4.2.24), we obtain the equality $\boldsymbol{x} = \mathbf{0}$.

In domain (4.2.5), relationship (4.2.8) is also valid, which together with $\boldsymbol{x} = \mathbf{0}$ and (4.2.26) yield the equality $\boldsymbol{u} = \mathbf{0}$.

Thus, when (4.2.24)–(4.2.26) are fulfilled, we arrive at $\boldsymbol{x} = \mathbf{0}, \boldsymbol{u} = \mathbf{0}$. The theorem is proved. □

4.2.10 Discussion of the Proposed Approach to Solving the Problem of Stabilization

1. Solving Problem 4.2.1 based on the proposed method of constructing linear controlled $\boldsymbol{\mu}$-systems implies that the following issues are solved:

(*a*) computing the ranks of the functional matrices F and Ψ (or Q and P);

(*b*) solving the functional equations (4.2.7) (or (4.2.10), respectively) to find the set of control rules (4.2.8) (or (4.2.12), respectively);

(*c*) deriving dependencies (4.2.17) (or (4.2.19), respectively), which relate the $\boldsymbol{y}$-variables being stabilized when solving the auxiliary linear problem with the remaining $\boldsymbol{z}$-variables of system (4.2.1);

(*d*) choosing the functional to be optimized in solving the auxiliary linear problem of stabilization and selecting control rules satisfying given performance criteria from the set (4.2.8) (or (4.2.12), respectively).

2. Semidefinite functional (4.2.23), whose variables a_k, b_k, c_k are adjusted while getting a solution, is a *generalized performance criterion for the control* in Problem 4.2.1.

3. Control rules obtained on the basis of the proposed approach are *robust* (*stable*) relative to errors and perturbations. Modification of the approach for the case of high-level interference is considered in Chapter 5.

4. Within the framework of the proposed approach, solutions of the closed systems (4.2.1), (4.2.8) and (4.2.1), (4.2.12) can be *explicitly* found. Thus, control rules (4.2.8), (4.2.12) constructed in accordance with the feedback principle can generally be used as *program* (i.e., depending on time t and initial perturbations $\boldsymbol{x}_0 \in S$) "*quieters*" of the system. This possible "double use" of the control rules obtained widens the fields for applying the approach.

4.2.11 Application to the Problem of Polystabilization

Let the perturbed motion of the controlled object be described by a nonlinear system of differential equations (in vector form)

$$\begin{gathered}\dot{\boldsymbol{w}} = \boldsymbol{X}^{(0)}(t, \boldsymbol{w}, \boldsymbol{x}), \quad \dot{\boldsymbol{\xi}} = \boldsymbol{X}^{(1)}(t, \boldsymbol{w}, \boldsymbol{x}, \boldsymbol{u}), \quad \dot{\boldsymbol{\eta}} = \boldsymbol{X}^{(2)}(t, \boldsymbol{x}), \\ \boldsymbol{w}, \boldsymbol{X}^{(0)} \in R^m; \; \boldsymbol{\xi}, \boldsymbol{X}^{(1)} \in R^l; \; \boldsymbol{\eta}, \boldsymbol{X}^{(2)} \in R^{n-l}; \; \boldsymbol{u} \in R^r; \; \boldsymbol{x} = (\boldsymbol{\xi}^{\mathsf{T}}, \boldsymbol{\eta}^{\mathsf{T}}),\end{gathered} \tag{4.2.27}$$

where $\boldsymbol{w}, \boldsymbol{x}$ are the vectors of phase variables and $\boldsymbol{u}$ is the vector of controls.

The right-hand sides of system (4.2.27) satisfy the following conditions:

(1) vector functions $\boldsymbol{X}^{(0)}, \boldsymbol{X}^{(1)}, \boldsymbol{X}^{(2)}$ are defined and continuous together with their partial derivatives with respect to $t, \boldsymbol{x}, \boldsymbol{u}$ in the domain

$$t \geq 0, \quad \|\boldsymbol{w}\| + \|\boldsymbol{x}\| \leq H, \quad \|\boldsymbol{u}\| \leq \infty;$$

(2) $\boldsymbol{X}^{(0)}(t, \mathbf{0}, \mathbf{0}) \equiv \mathbf{0}, \; \boldsymbol{X}^{(1)}(t, \mathbf{0}, \mathbf{0}, \mathbf{0}) \equiv \mathbf{0}, \; \boldsymbol{X}^{(2)}(t, \mathbf{0}) \equiv \mathbf{0}$.

In addition, suppose that the phase $\boldsymbol{x}$-variables of system (4.2.27) are related by (4.2.3).

For example, structure (4.2.27), (4.2.3) of equations of perturbed motion is characteristic of two groups of problems:

(1) problems of stabilizing relative equilibrium positions of an artificial satellite with perturbations of its orbit taken into account;

(2) problems of stabilizing equilibrium positions of a solid (spacecraft) by rotating masses (flywheels, gyroscope).

In these problems the first group of equations of system (4.2.27) comprises the equations describing the perturbed motion of the center of mass of the satellite (or the motion of masses attached to the solid) whereas the second and third groups comprise, respectively, dynamical and kinematic equations describing the angular motion and attitude of the satellite (the solid containing the masses).

We note that the first group of problems does not require that the unperturbed motion $\boldsymbol{w} = \boldsymbol{x} = \mathbf{0}$ of system (4.2.27) be asymptotically stabilized with respect to all the phase variables, because the satellite can function properly without the asymptotic stability with respect to the variables characterizing the perturbation of the orbit and it is reasonable to attain just (nonasymptotic) stability with respect to these variables.

In contrast, when stabilizing an equilibrium position of a solid by a gimballed gyroscope, as a rule, the (nonasymptotic) stability of the gyroscope's angular motion with respect to the solid is necessary whereas the asymptotic stability is merely unattainable (for details, see Subsection 4.3.6).

These (and a number of other) situations justify the statement of the polystabilization problem given below.

Problem 4.2.2 (on polystabilization of the unperturbed motion $\boldsymbol{w} = \boldsymbol{x} = \mathbf{0}$ *of system (4.2.27)). Find the set* $M = \{\boldsymbol{u}(t, \boldsymbol{x}, \boldsymbol{w})\} \in \mathfrak{U}_*$ *of control rules such that*

(a) the unperturbed motion $\boldsymbol{w} = \mathbf{0}$, $\boldsymbol{x} = \mathbf{0}$ *of system (4.2.27) is Lyapunov stable and asymptotically* $\boldsymbol{x}$*-stable;*

(b) *the given domain* $S = \{\boldsymbol{x}_0 : \|\boldsymbol{x}_0\| \in S\}$ *of initial perturbations of* $\boldsymbol{x}_0$ *is the domain of* $\boldsymbol{x}$*-attraction of the unperturbed motion* $\boldsymbol{w} = \boldsymbol{0}$, $\boldsymbol{x} = \boldsymbol{0}$, *i.e.,* $\lim\|\boldsymbol{x}(t; t_0, \boldsymbol{x}_0, \boldsymbol{w}_0)\| = 0$, $t \to \infty$, *for* $\boldsymbol{x}_0 \in S$ *and a sufficiently small value of* $\|\boldsymbol{w}_0\|$.

The solution technique of Problem 4.2.2 can be briefly described as follows.

(1) First we solve the problem of *partially* stabilizing the unperturbed motion $\boldsymbol{w} = \boldsymbol{0}$, $\boldsymbol{x} = \boldsymbol{0}$, i.e., the problem of stabilization with respect to variables $\eta_1, \ldots, \eta_r$ is part of vector $\boldsymbol{\eta}$ coordinates. If, in the domain

$$t \geq 0, \quad \|\boldsymbol{w}\| + \|\boldsymbol{x}\| \leq h_1, \quad \|\boldsymbol{u}\| \leq h_2 \quad (h_1 \leq H), \tag{4.2.28}$$

the condition

$$\operatorname{rank} F(t, \boldsymbol{w}, \boldsymbol{x}) = r \tag{4.2.29}$$

is satisfied (F is the Jacobi matrix of functions Φ_j of type (4.2.4) with respect to u_k $(k = 1, \ldots, r)$), then this problem of partial stabilization is reduced to one of stabilizing the zero solution $\boldsymbol{y} = \boldsymbol{0}$ of the linear controlled $\boldsymbol{\mu}$-system (4.2.9) (with respect to *all* the variables).

Thus one can find the set of control rules

$$\boldsymbol{u} = \boldsymbol{f}_*(t, \boldsymbol{w}, \boldsymbol{x}), \qquad \boldsymbol{f}_*(t, \boldsymbol{0}, \boldsymbol{0}) \equiv \boldsymbol{0} \tag{4.2.30}$$

that solve the problem of partial stabilization in question.

(2) Then, we impose the additional conditions (conditions (2)–(4) of Theorem 4.2.1) under which the performed partial stabilization of the unperturbed motion $\boldsymbol{w} = \boldsymbol{0}, \boldsymbol{x} = \boldsymbol{0}$ (with respect to $\eta_1, \ldots, \eta_r$) is actually the stabilization of this motion with respect to a *greater number* of the variables (i.e., $\boldsymbol{x}$-stabilization); the given domain S is the domain of $\boldsymbol{x}$-attraction of the unperturbed motion $\boldsymbol{w} = \boldsymbol{0}, \boldsymbol{x} = \boldsymbol{0}$.

(3) Further we distinguish the "*truncated*" system

$$\dot{\boldsymbol{w}} = \boldsymbol{X}^*(t, \boldsymbol{w}), \qquad \boldsymbol{X}^*(t, \boldsymbol{w}) \stackrel{\text{def}}{=} \boldsymbol{X}^{(0)}(t, \boldsymbol{w}, \boldsymbol{0}) \tag{4.2.31}$$

derived from the first group of equations of system (4.2.27) by setting $\boldsymbol{x} = \boldsymbol{0}$. (Note, that $\boldsymbol{X}^*(t, \boldsymbol{0}) \equiv \boldsymbol{0}$.) Analyzing the Lyapunov stability of the zero solution $\boldsymbol{w} = \boldsymbol{0}$ of the "truncated" system (4.2.31), we show that the unperturbed motion $\boldsymbol{w} = \boldsymbol{0}, \boldsymbol{x} = \boldsymbol{0}$ of the closed system (4.2.27), (4.2.30) is exponentially asymptotically $\boldsymbol{x}$-stable and also Lyapunov stable. In so doing, we apply the *Lyapunov–Malkin "reduction principle"* (see Dykhman [1950]) to the system composed of the equations (4.2.9) (with $\boldsymbol{u}^* \in \mathfrak{U}_1$) and $\dot{\boldsymbol{w}} = \boldsymbol{X}^{(0)}(t, \boldsymbol{w}, \boldsymbol{x})$, where $\boldsymbol{x} = (\eta_1, \ldots, \eta_r, \boldsymbol{\varphi}(t, \boldsymbol{y}))$ and $\boldsymbol{\varphi}(t, \boldsymbol{y})$ is defined in accordance with (4.2.17).

Similar to the case of Problem 4.2.1, solving Problem 4.2.2 shows that the issues of stabilization with respect to part of, major part of, or all the variables and those of polystabilization are closely interrelated and also complement each other in the investigation.

Let us sum up the proposed scheme for solving Problem 4.2.2 in the form of the following theorem, condition (4) of which substantiates the applicability of the "reduction principle" (in the Dykhman form).

Theorem 4.2.3. *Let the following conditions be satisfied:*

(1) *condition* (4.2.29) *in domain* (4.2.28);

(2) *conditions* (2)–(4) *of Theorem* 4.2.1;

(3) *the zero solution* $\boldsymbol{w} = \boldsymbol{0}$ *of the "truncated" system* (4.2.31) *is Lyapunov stable*;

(4) *vector function* $\boldsymbol{X}^{(0)}(t, \boldsymbol{w}, \boldsymbol{y}, \boldsymbol{\varphi}(t, \boldsymbol{y}))$ *in the domain* $t \geq 0$, $\|\boldsymbol{w}\| + \|\boldsymbol{y}\| \leq H$ *is an analytic function whose coefficients are continuous and bounded functions of* t.

Then the set of control rules (4.2.30), *in which* $\boldsymbol{u}^* \in \mathfrak{U}_1$, *guarantees that Problem* 4.2.2 *is solved.*

4.2.12 The Case of $n_1 = 0$ for System (4.2.1), (4.2.3)

Here we will use the relationships obtained earlier that ground the solvability of Problem 4.2.1 based on linear controlled $\boldsymbol{\mu}$-systems (4.2.9) or (4.2.13), (4.2.14). Thus, to solve Problem 4.2.1 based on $\boldsymbol{\mu}$-system (4.2.9) for $n_1 = 0$, the conditions $n - l \geq r$ and $2r \geq n$ should be met. The fact that these conditions are satisfied simultaneously (with allowance for the natural assumption that $r \leq l$) yields the equality

$$n = 2l = 2r. \tag{4.2.32}$$

Now, to ensure that Problem 4.2.1 is solved based on the linear controlled $\boldsymbol{\mu}$-system (4.2.13), (4.2.14), one should require that $n - l \geq r - q$ and $r - q \geq n - r$. In this case, for $r \leq l$, we obtain

$$n + q = 2l = 2r. \tag{4.2.33}$$

Condition (4.2.32) (or the respective condition (4.2.33)) guarantees that Problem 4.2.1 is solved based on the linear controlled $\boldsymbol{\mu}$-system (4.2.9) (or the respective $\boldsymbol{\mu}$-system (4.2.13), (4.2.14)) in the case of $n_1 = 0$, when the phase variables of system (4.2.1) do not obey the additional relationships (4.2.3).

However, we note that the structure (4.2.1), (4.2.3) (the case of $n_1 \neq 0$) of the system of perturbed motion is quite *natural* for many applied problems, for instance, the problems of controlling the angular motion of a solid. This circumstance justifies investigating the problem of stabilizing the controlled system possessing the structure of (4.2.1), (4.2.3) and also makes it reasonable.

Besides, in the case of $n_1 = 0$, one can often find some *first integrals* for system (4.2.1). This also results in the system of type (4.2.1), (4.2.3).

4.3 Stabilization and Partial Stabilization of Permanent Rotations (of Equilibrium Positions) of a Solid and a Satellite in Orbit

4.3.1 Problems to be Considered

Now we apply the approach to investigating optimal stabilization suggested in Section 4.2 for solving the following problems:

(1) *stabilization and partial stabilization of permanent rotations (of equilibrium positions) of a solid attached at a fixed point* (rotating about its center of mass);

(2) *stabilization (partial stabilization) of a relative equilibrium position of a satellite* in a circular (geostationary) orbit;

(3) *partial stabilization of stationary modes of the mechanical system of "a solid + flywheels" (gimballed gyroscope)* by specially arranged rotation of the flywheels (gyroscope). In this case the variables to be stabilized are those characterizing the state (the angular velocity and the attitude) of the solid or part of them.

We solve the above problems in the *two stages* indicated in Section 4.2: (1) finding the set of stabilizing control rules; (2) selecting control rules belonging to the set found which satisfy additional performance criteria and are optimal in some sense.

In the *first* of the stages outlined, we select some part of the variables characterizing the state of the system of perturbed motion under study and carry out the stabilization with respect to these variables. Then we show that the partial stabilization performed is the stabilization with respect to all or to a given (major) part of the variables. To substantiate the polystabilization, we also use the *Lyapunov–Malkin "reduction principle."*

In the *second* stage of solving the problem, the fact that the proposed approach allows finding the solutions of the closed control system in the *explicit form* essentially facilitates the choice of optimal control rules. As a result, one can use a simple *iterative* search of the control rules to provide the desired quality of the transient in the closed system.

4.3.2 Stabilization of Permanent Rotations of a Massive Solid

Consider *Euler's dynamical* and *Poisson's kinematic equations*, which, taken together, describe the angular motion and the attitude of a massive solid attached at a fixed point O:

$$\begin{aligned} A\dot{x}_1 &= (B-C)x_2x_3 + mg(x_{3c}q_2 - x_{2c}q_3) + u_1, \\ B\dot{x}_2 &= (C-A)x_1x_3 + mg(x_{1c}q_3 - x_{3c}q_1) + u_2, \\ C\dot{x}_3 &= (A-B)x_1x_2 + mg(x_{2c}q_1 - x_{1c}q_2) + u_3, \end{aligned} \tag{4.3.1}$$

$$\dot{q}_1 = x_3 q_2 - x_2 q_3,$$
$$\dot{q}_2 = x_1 q_3 - x_3 q_1,$$
$$\dot{q}_3 = x_2 q_1 - x_1 q_2.$$

Here A, B, and C are the principal moments of inertia of the solid, x_i ($i = 1, 2, 3$) are the projections of the angular velocity vector of the body onto its principal axes of inertia; q_i ($i = 1, 2, 3$) are the projections of the unit vector directed along a fixed vertical axis onto the principal axes, x_{ic} ($i = 1, 2, 3$) are the coordinates of the center of mass of the solid in the principal axes; u_i ($i = 1, 2, 3$) are the controlling moments; and mg is the weight of the body. The products mgx_{ic} ($i = 1, 2, 3$) are assumed not too large, which justifies the structure of the control rules considered below.

When $u_i = 0$ ($i = 1, 2, 3$), the massive solid admits an infinite set of *permanent rotations* about the fixed vertical axis passing through the attachment point. The following solutions correspond to these rotations:

$$x_1 = \omega\alpha,\ x_2 = \omega\beta,\ x_3 = \omega\gamma,\ q_1 = \alpha,\ q_2 = \beta,\ q_3 = \gamma, \tag{4.3.2}$$
$$\alpha^2 + \beta^2 + \gamma^2 = 1 \tag{4.3.3}$$

of system (4.3.1) in which ω is a constant angular velocity.

The specified class of motions of a solid was independently found by Mlodzeevsky and Staude in 1894. The necessary conditions for their stability were stated by Grammel in 1920. The rigorous sufficient conditions for permanent rotations of a massive solid were obtained by Rumyantsev [1956], [1957b].

We note that a massive solid with a single fixed point is a *conservative* mechanical system. Therefore, the *asymptotic* stability of permanent rotations is impossible, and the problem of stabilizing these rotations by controlling moments is of interest.

Consider one of the body axes which, pointing vertically upward, can serve as an axis of permanent rotation with a constant angular velocity ω. (The set of such admissible axes was characterized by Staude [1894]; see also Rumyantsev [1956].) As a result, we have one of the possible modes (4.3.2), (4.3.3) of permanent rotation of the solid.

Introducing the new variables

$$\xi_1 = x_1 - \omega\alpha, \qquad \xi_2 = x_2 - \omega\beta, \qquad \xi_3 = x_3 - \omega\gamma,$$
$$\xi_4 = q_1 - \alpha, \qquad \xi_5 = q_2 - \beta, \qquad \xi_6 = q_3 - \gamma,$$

let us compose the system of perturbed motion in variations from the mode of permanent rotation under study:

$$A\dot{\xi}_1 = (B - C)\big[\xi_2\xi_3 + \omega(\xi_2\gamma + \xi_3\beta)\big] + mg(x_{3c}\xi_5 - x_{2c}\xi_6) + u_1 \overset{\text{def}}{=} Af_1(\boldsymbol{\xi}) + u_1,$$
$$B\dot{\xi}_2 = (C - A)\big[\xi_1\xi_3 + \omega(\xi_1\gamma + \xi_3\alpha)\big] + mg(x_{1c}\xi_6 - x_{3c}\xi_4) + u_2 \overset{\text{def}}{=} Bf_2(\boldsymbol{\xi}) + u_2,$$
$$C\dot{\xi}_3 = (A - B)\big[\xi_1\xi_2 + \omega(\xi_1\beta + \xi_2\alpha)\big] + mg(x_{2c}\xi_4 - x_{1c}\xi_5) + u_3 \overset{\text{def}}{=} Cf_3(\boldsymbol{\xi}) + u_3,$$

$$\begin{aligned}
\dot{\xi}_4 &= \xi_3\beta - \xi_2\gamma + \xi_5(\xi_3 + \omega\gamma) - \xi_6(\xi_2 + \omega\beta) \stackrel{\text{def}}{=} f_4(\boldsymbol{\xi}),\\
\dot{\xi}_5 &= \xi_1\gamma - \xi_3\alpha + \xi_6(\xi_1 + \omega\alpha) - \xi_4(\xi_3 + \omega\gamma) \stackrel{\text{def}}{=} f_5(\boldsymbol{\xi}),\\
\dot{\xi}_6 &= \xi_2\alpha - \xi_1\beta - \xi_5(\xi_1 + \omega\alpha) + \xi_4(\xi_2 + \omega\beta) \stackrel{\text{def}}{=} f_6(\boldsymbol{\xi}).
\end{aligned} \tag{4.3.4}$$

From geometric relationship (4.3.3), we derive the equality

$$G_1 = (\xi_4 + \alpha)^2 + (\xi_5 + \beta)^2 + (\xi_6 + \gamma)^2 - 1 = 0, \tag{4.3.5}$$

valid for variables ξ_j $(j = 4, 5, 6)$ of system (4.3.4).

With the notation for the variables changed to $\eta_i = \xi_{3+i}$ $(i = 1, 2, 3)$, the structure of system (4.3.4), (4.3.5) is identical to that of (4.2.1), (4.2.3). By setting $q = 1$, $X_1^{(1)} = f_2 + B^{-1}u_2$, $X_1^{(2)} = f_4$, and $X_2^{(2)} = f_6$, one can check that, if $\xi_5 + \beta \neq 0$, conditions (1) and (2) of Corollary 4.2.1 hold for system (4.3.4), (4.3.5).

Indeed, in this case matrices Q and P have the form

$$Q = \begin{pmatrix} 0 & B^{-1} & 0 \\ 0 & (\xi_6 + \gamma)B^{-1} & (\xi_5 + \beta)C^{-1} \\ (\xi_5 + \beta)A^{-1} & (\xi_4 + \alpha)B^{-1} & 0 \\ & \cdots & \end{pmatrix},$$

$$P = \begin{pmatrix} 0 & \xi_5 + \beta & \xi_3 + \omega\gamma \\ -(\xi_5 + \beta) & 0 & -(\xi_1 + \omega\alpha) \\ 0 & 0 & 2(\xi_5 + \beta) \end{pmatrix}.$$

Consequently, when $\xi_5 + \beta \neq 0$, the condition $\operatorname{rank} Q = \operatorname{rank} P = 3$ is valid. Since $n = 6$, $r = l = 3$, $q = 1$ for system (4.3.4), (4.3.5), the conditions of Corollary 4.2.1 hold if $\xi_5 + \beta \neq 0$.

Theorem 4.3.1. *If $\beta \neq 0$, then the set of control rules*

$$\begin{aligned}
u_1 &= \frac{A\Phi_1(\boldsymbol{\xi})}{\beta + \xi_5} + \frac{A(\alpha + \xi_4)u_2}{B(\beta + \xi_5)} - \frac{Au_1^*}{\beta + \xi_5}, \qquad u_2 = B(-f_2 + u_2^*),\\
u_3 &= \frac{C\Phi_2(\boldsymbol{\xi})}{\beta + \xi_5} + \frac{C(\gamma + \xi_6)u_2}{B(\beta + \xi_5)} + \frac{Cu_3^*}{\beta + \xi_5},
\end{aligned} \tag{4.3.6}$$

where

$$\begin{gathered}
\Phi_1(\boldsymbol{\xi}) = f_2(\alpha + \xi_4) - f_1(\beta + \xi_5) - f_5(\xi_1 + \omega\alpha) + f_4(\xi_2 + \omega\beta),\\
\Phi_2(\boldsymbol{\xi}) = -f_3(\beta + \xi_5) - f_5(\xi_3 + \omega\gamma) + f_2(\gamma + \xi_6) + f_6(\xi_2 + \omega\beta),\\
u_1^* = \Gamma_1\xi_6 + \Gamma_2 f_6, \quad u_2^* = \Gamma_3\xi_2, \quad u_3^* = \Gamma_4\xi_4 + \Gamma_5 f_4,\\
\Gamma_k = \text{const} < 0 \quad (k = 1, \ldots, 5),
\end{gathered}$$

is the set of stabilizing control rules which ensure exponential asymptotic stability of the unperturbed motion $\xi_j = 0$ $(j = 1, \ldots, 6)$ of system (4.3.4) with respect to ξ_j $(j = 1, \ldots, 6)$.

Proof. From technical calculations omitted here, it follows that one can segregate the *linear $\boldsymbol{\mu}$-system* ($\mu_1 = f_4$, $\mu_2 = f_6$)

$$\begin{gathered} \dot{\xi}_4 = \mu_1, \quad \dot{\mu}_1 = \Gamma_4\xi_4 + \Gamma_5\mu_1, \\ \dot{\xi}_6 = \mu_2, \quad \dot{\mu}_2 = \Gamma_1\xi_6 + \Gamma_2\mu_2, \quad \dot{\xi}_2 = \Gamma_3\xi_2 \end{gathered} \tag{4.3.7}$$

from the closed system (4.3.4)–(4.3.6). Since $\Gamma_k = \text{const} < 0$ ($k = 1,\ldots,5$), the solution $\xi_2 = \xi_4 = \xi_6 = \mu_1 = \mu_2 = 0$ of system (4.3.7) is exponentially asymptotically stable (with respect to all the variables). Consequently, the unperturbed motion $\xi_j = 0$ ($j = 1,\ldots,6$) of the closed system (4.3.4)–(4.3.6) is exponentially asymptotically stable with respect to ξ_2, ξ_4, ξ_6.

Solving equation (4.3.5) for ξ_5, we arrive at the following alternative:

$$\xi_5(t) = -\beta + \sqrt{\beta^2 - \xi_4^2(t) - \xi_6^2(t) - 2\alpha\xi_4(t) - 2\gamma\xi_6(t)}, \tag{4.3.8}$$

$$\xi_5(t) = -\beta - \sqrt{\beta^2 - \xi_4^2(t) - \xi_6^2(t) - 2\alpha\xi_4(t) - 2\gamma\xi_6(t)}. \tag{4.3.9}$$

Let $\beta > 0$. Since β is some fixed number, at the initial moment $t = t_0$, solution (4.3.9) will not belong to a sufficiently small neighborhood of the unperturbed motion $\xi_j = 0$ ($j = 1,\ldots,6$). Therefore, it should be eliminated from consideration when studying the Lyapunov stability of the unperturbed motion $\xi_j = 0$ ($j = 1,\ldots,6$) of system (4.3.4)–(4.3.6). The remaining relationship (4.3.8) shows that, when the initial perturbations $\xi_{j0} = 0$ ($j = 1,\ldots,6$) are sufficiently small, the unperturbed motion $\xi_j = 0$ ($j = 1,\ldots,6$) of the closed system (4.3.4)–(4.3.6) will be exponentially asymptotically stable with respect to ξ_2, ξ_4, ξ_6 and also with respect to ξ_5. The case $\beta < 0$ leads to a similar result with the difference that solutions (4.3.8) should be excluded from consideration.

From the equations of system (4.3.4) for $\dot{\xi}_4$ and $\dot{\xi}_6$, one can derive the equalities

$$\begin{aligned} \xi_1(t) &= \frac{-\mu_2 + \xi_2\alpha - \xi_5\omega\alpha + \xi_4(\xi_2 + \omega\beta)}{\beta + \xi_5}, \\ \xi_3(t) &= \frac{\mu_1 + \xi_2\gamma - \xi_5\omega\gamma + \xi_6(\xi_2 + \omega\beta)}{\beta + \xi_5}. \end{aligned} \tag{4.3.10}$$

By virtue of the exponential asymptotic stability of the unperturbed motion $\xi_j = 0$ ($j = 1,\ldots,6$) of system (4.3.4)–(4.3.6) with respect to ξ_2, ξ_4, ξ_5, ξ_6, μ_1, and μ_2, for any fixed number β, one can find numbers $\Delta(\beta) > 0$ and $\delta(\beta) > 0$ such that the inequality $\|\boldsymbol{\xi}_0\| < \delta(\beta)$ yields the estimate $|\beta + \xi_5| > \Delta$. This means that, if $\|\boldsymbol{\xi}_0\| < \delta(\beta)$, the components ξ_1, ξ_3 of the solution $\boldsymbol{\xi}(t; t_0, \boldsymbol{\xi}_0)$ of system (4.3.4)–(4.3.6) satisfy the inequalities

$$\begin{aligned} |\xi_1(t)| &\le \frac{1}{\Delta}|-\mu_2 + \xi_2\alpha - \xi_5\omega\alpha + \xi_4(\xi_2 + \omega\beta)|, \\ |\xi_3(t)| &\le \frac{1}{\Delta}|\mu_1 + \xi_2\gamma - \xi_5\omega\gamma + \xi_6(\xi_2 + \omega\beta)|. \end{aligned} \tag{4.3.11}$$

From estimates (4.3.11), we draw the conclusion that the unperturbed motion $\xi_j = 0$ $(j = 1, \dots, 6)$ of system (4.3.4)–(4.3.6) is exponentially asymptotically stable with respect to all the variables. The theorem is proved. □

Remark 4.3.1.

1. The condition $\beta \neq 0$ of Theorem 4.3.1 is not crucial. Indeed, $\alpha^2 + \beta^2 + \gamma^2 = 1$, and at least one of the numbers α, β, γ is distinct from zero. For example, if $\alpha \neq 0$, the stabilizing control rules can be obtained from (4.3.6) by a permutation of the indices and coefficients. Without loss of generality, in Theorem 4.3.1 we can thus assume that $\beta^2 > 1/3$; this inequality is important for estimating control rules (4.3.6) and the domain of attraction of the unperturbed motion $\xi_j = 0$ $(j = 1, \dots, 6)$ of system (4.3.4)–(4.3.6).
2. It should be stressed that condition (1) of Theorem 4.2.1 fails to hold for system (4.3.4)–(4.3.6) no matter how its variables are renamed.

Let us find the *domain of attraction* of the unperturbed motion $\xi_j = 0$ $(j = 1, \dots, 6)$ of system (4.3.4)–(4.3.6), i.e., the set S of $\xi_{j0} = \xi_j(t_0)$ such that $\left|\xi_j(t; t_0, \boldsymbol{\xi}_0)\right| \to 0$, $t \to \infty$. For definiteness, we assume that $\Gamma_{v+1}^2 + 4\Gamma_v > 0$ $(v = 1, 4)$.

Corollary 4.3.1. *The domain* $S = S_1 \cap S_2$ *of initial perturbations* $\boldsymbol{\xi}_0 \in S$ *such that*

$$S_1 = \left\{\boldsymbol{\xi}_0 : K_1^2 + K_2^2 + 2|\alpha| K_1 + 2|\gamma| K_2 \le r^2 < \beta^2\right\}, \quad S_2 = \left\{\xi_{50}^2 \le r^2 < \beta^2\right\},$$

$$\begin{aligned} K_1 &= \frac{2|\mu_{10}|}{\sqrt{\Gamma_5^2 + 4\Gamma_4}} + \frac{1}{2}|\xi_{40}|\left(|l_1 - 1| + |l_1 + 1|\right), \\ K_2 &= \frac{2|\mu_{20}|}{\sqrt{\Gamma_2^2 + 4\Gamma_1}} + \frac{1}{2}|\xi_{60}|\left(|l_2 - 1| + |l_2 + 1|\right), \\ l_1 &= \frac{\Gamma_5}{\sqrt{\Gamma_5^2 + 4\Gamma_4}}, \qquad l_2 = \frac{\Gamma_2}{\sqrt{\Gamma_2^2 + 4\Gamma_1}}, \\ \mu_{10} &= f_4(\boldsymbol{\xi}_0), \quad \mu_{20} = f_6(\boldsymbol{\xi}_0), \quad \boldsymbol{\xi}_0 = \boldsymbol{\xi}(t_0) \end{aligned} \tag{4.3.12}$$

is the domain of attraction of the unperturbed motion $\xi_j = 0$ $(j = 1, \dots, 6)$ *of the closed system* (4.3.4)–(4.3.6).

Proof. Directly integrating the equations

$$\dot{\xi}_4 = \mu_1, \qquad \dot{\mu}_1 = \Gamma_4 \xi_4 + \Gamma_5 \mu_1$$

entering into system (4.2.7), we conclude that, for all $t \ge t_0$, solution $\xi_4(t)$ of system (4.3.7) has the form

$$\begin{gathered} \xi_4(t) = C_1 e^{\lambda_1 (t - t_0)} + C_2 e^{\lambda_2 (t - t_0)}, \\ C_1 = \frac{\mu_{10} - \lambda_2 \xi_{40}}{\lambda_1 - \lambda_2}, \quad C_2 = \frac{\mu_{10} - \lambda_1 \xi_{40}}{\lambda_2 - \lambda_1}, \\ \lambda_1 = 1/2\left(\Gamma_5 + \sqrt{\Gamma_5^2 + 4\Gamma_4}\right), \quad \lambda_2 = 1/2\left(\Gamma_5 - \sqrt{\Gamma_5^2 + 4\Gamma_4}\right). \end{gathered}$$

Taking into account the inequalities

$$|C_1| + |C_2| \le \frac{2|\mu_{10}| + |\xi_{40}|(|\lambda_2| + |\lambda_1|)}{\sqrt{\Gamma_5^2 + 4\Gamma_4}} \le K_1,$$

we arrive at $|\xi_4(t)| \le K_1, t \ge t_0$.

Similarly, $|\xi_6(t)| \le K_2, t \ge t_0$.

Therefore, for all $t \ge t_0$, the following relationship is valid:

$$|f(t)| \le K_1^2 + K_2^2 + 2|\alpha| K_1 + 2|\gamma| K_2,$$
$$f = \xi_4^2 + \xi_6^2 + 2\alpha\xi_4 + 2\gamma\xi_6.$$

Bearing in mind the inequalities

$$|-\beta \underset{(-)}{+} \sqrt{\beta^2 + f}| \le \sqrt{|f|} \quad (\beta \underset{(<)}{>} 0),$$
$$|f| \le r^2 < \beta^2, \quad t \ge t_0,$$

from (4.3.8), (4.3.9), we deduce that $|\xi_5(t)| \le |r| < |\beta|, t \ge t_0$.

As a result, allowing for equalities (4.3.10), we conclude that the estimates $|\xi_j(t; t_0, \boldsymbol{\xi}_0)| \to 0, t \to \infty$ $(j = 1, \ldots, 6)$ hold for all the solutions $\xi_j(t; t_0, \boldsymbol{\xi}_0), \boldsymbol{\xi}_0 \in S$ of system (4.3.4)–(4.3.6). The statement is proved. □

Remark 4.3.2.

1. Following the technique of Subsection 4.2.9, one can choose constants Γ_k $(k = 1, \ldots, 5)$ in (4.3.6) to minimize a functional I of type (4.2.23) on the trajectories of system (4.3.4)–(4.3.6) that originate from domain S. In the case in question, the integrand of I has the form

$$R(\boldsymbol{\xi}, \boldsymbol{u}) = a_1\xi_4^2 + a_2\xi_6^2 + b_1\mu_1^2(\boldsymbol{\xi}) + b_2\mu_2^2(\boldsymbol{\xi}) + \sum_{i=1}^{3} c_i \left[u_i^*(\boldsymbol{\xi}, \boldsymbol{u})\right]^2, \tag{4.3.13}$$

where a_i, c_i $(i = 1, 2, 3), b_1$, and b_2 are positive constants.

By employing the proof scheme for Theorem 4.2.2, one can show that R is a *positive-definite* function of the variables constituting vectors $\boldsymbol{\xi}, \boldsymbol{u}$.

2. The "optimal" control rules u_i $(i = 1, 2, 3)$ thus obtained are *robust* (*stable*) relative to errors and perturbations.

3. In Example 1.5.1 we found the conditions under which stabilizing the unperturbed motion $\xi_j = 0$ $(j = 1, \ldots, 6)$ of system (4.3.4) in the *linear approximation* can be achieved by *only two* of three controlling torques u_i $(i = 1, 2, 3)$. In this connection we note that the stabilization by *three* controlling torques presented here has the following points in its favor:

(*a*) no restrictions are imposed on the parameters of system (4.3.4) in the course of stabilization;

(*b*) the solutions of the closed system (4.3.4)–(4.3.6) are obtained in the explicit form, which makes it comparatively easy to find the domain of attraction of the unperturbed motion $\xi_j = 0$ $(j = 1, \ldots, 6)$ and to choose the control rules satisfying additional performance criteria from the set specified by (4.3.6).

Example 4.3.1. Let us analyze the structure of controls (4.3.6) and integrand (4.3.13) for stabilizing an *equilibrium position* of a solid with $\alpha = \gamma = 0$, $\beta = 1$, $\omega = 0$, $x_{ic} = 0$ $(i = 1, 2, 3)$. Such a problem arises, for example, when *stabilizing the orientation of spacecraft in interplanetary missions.*

In the case under consideration

$$\mu_1 = \xi_3 + \ldots, \quad \mu_2 = -\xi_1 + \ldots, \quad Au_1^* = -u_1 + \ldots, \quad Bu_2^* = u_2 + \ldots, \quad Cu_3^* = -u_3 + \ldots,$$

where ellipsis dots stand for the terms of *higher than the first* order in variables ξ_j $(j = 1, \ldots, 6)$, u_i $(i = 1, 2, 3)$.

Hence, expression (4.3.13) can be written as

$$\begin{aligned} R(\boldsymbol{\xi}, \boldsymbol{u}) &= a_1\xi_4^2 + a_2\xi_6^2 + a_3\xi_2^2 + b_1\xi_3^2 + b_2\xi_1^2 \\ &\quad + (c_1/A^2)u_1^2 + (c_2/B^2)u_2^2 + (c_3/C^2)u_3^2 + \Lambda(\boldsymbol{\xi}, \boldsymbol{u}), \end{aligned}$$

where function Λ represents the terms of order *three and higher* in ξ_j $(j = 1, \ldots, 6)$, u_i $(i = 1, 2, 3)$. Taking $b_1 = C$, $b_2 = A$, $b_3 = B$, we conclude that the quadratic part of function $R(\boldsymbol{\xi}, \boldsymbol{u})$ is the sum of the following expressions:

(1) the kinetic energy of the body $A\xi_1^2 + B\xi_2^2 + C\xi_3^2$;

(2) the quadratic form $a_1\xi_4^2 + a_2\xi_6^2$ characterizing the deviation of the body from the equilibrium position;

(3) the quadratic form $(c_1/A^2)u_1^2 + (c_2/B^2)u_2^2 + (c_3/C^2)u_3^2$ characterizing the resources expended in generating the controls.

Let us also dwell on the structure of controls (4.3.6) for the case under consideration. Taking into account the expansion

$$\frac{1}{\xi_5 + 1} = 1 - \xi_5 + \xi_5^2 + \ldots,$$

one can show that the control rules (4.3.6) minimizing functional (4.3.13) have the form

$$u_1 = -A\left(-\sqrt{\frac{a_2}{c_1}}\xi_6 + \sqrt{\frac{b_2 + 2\sqrt{a_2c_1}}{c_1}}\xi_1\right) + \ldots, \qquad u_2 = B\left(-\sqrt{\frac{a_3}{c_2}}\xi_2\right) + \ldots,$$

$$u_3 = C\left(\sqrt{\frac{a_1}{c_3}}\xi_4 + \sqrt{\frac{b_1 + 2\sqrt{a_1c_3}}{c_3}}\xi_3\right) + \ldots,$$

where ellipsis dots stand for the terms of order two and higher in ξ_j $(j = 1, \ldots, 6)$.

The structure of the optimal control rules obtained is such that, in the *first approximation*, the independent variables ξ_4, ξ_6 enter none of the control actions u_i $(i = 1, 2, 3)$ simultaneously, and the same is true for variables ξ_i $(i = 1, 2, 3)$. In the first approximation, the optimal control is performed through the three channels *independently*, namely, in the first approximation, variables ξ_1, ξ_6 are *invariant* relative to controls u_2, u_3, variables ξ_3, ξ_4 to u_1, u_2, and ξ_2 to u_1, u_3, respectively.

4.3.3 Partial Stabilization of Permanent Rotations (of Equilibrium Positions) of a Solid

Let us discuss certain possibilities of *partially* stabilizing permanent rotations of a massive solid by *two* controlling moments.

1. *The case when the ellipsoid of inertia of a solid relative to a fixed point O is a spheroid.* Let $A = C$, and let the center of mass be arbitrarily placed within the solid. The particular case of $x_{1c} = x_{3c} = 0$ is known as the *Lagrangian gyroscope.*

For the case in question, let us consider stabilizing admissible permanent rotations with a finite angular velocity ω with respect to *part* of the variables, namely, with respect to variables ξ_s $(s = 4, 5, 6)$ determining the vertical deviation of that body axis which is an admissible axis of permanent rotations.

Corollary 4.3.2. *Let $u_2 \equiv 0$, and let u_1, u_3 have the form* (4.3.6). *Then in the case of $A = C, \beta \neq 0$, the unperturbed motion $\xi_j = 0$ $(j = 1, \ldots, 6)$ possesses the following properties:*

(1) *it is Lyapunov stable and exponentially asymptotically stable with respect to ξ_s $(s = 4, 5, 6)$;*

(2) *for any $\boldsymbol{\xi}_0 \in S$ (where S is defined by relationships* (4.3.12)) *the following limiting (as $t \to \infty$) equalities are valid:*

$$\begin{gathered} \xi_1 = \alpha\beta^{-1}\xi_{20}, \quad \xi_2 = \xi_{20}, \quad \xi_3 = \gamma\beta^{-1}\xi_{20}, \\ \xi_s = 0 \quad (s = 4, 5, 6). \end{gathered} \tag{4.3.14}$$

Proof. In the given case, from the closed nonlinear system (4.3.4)–(4.3.6), one can segregate the $\boldsymbol{\mu}$-system embracing the first two groups of equations (4.3.7). As a result, the unperturbed motion $\xi_j = 0$ $(j = 1, \ldots, 6)$ of system (4.3.4)–(4.3.6) turns out to be exponentially asymptotically stable with respect to $\xi_4, \xi_6, \mu_1, \mu_2$.

The exponential asymptotic stability of the unperturbed motion $\xi_j = 0$ $(j = 1, \ldots, 6)$ with respect to ξ_5 can be proved in the same manner as for Theorem 4.3.1. The domain S defined by relationships (4.3.12) is the domain of attraction of the unperturbed motion $\xi_j = 0$ $(j = 1, \ldots, 6)$ with respect to variables ξ_s $(s = 4, 5, 6)$, μ_1, μ_2.

Then, substituting the solutions of the $\boldsymbol{\mu}$-system in the equation for $\dot{\xi}_2$ from (4.3.4), we obtain the relationship

$$\dot{\xi}_2 = f(t; t_0, \boldsymbol{\xi}_0), \tag{4.3.15}$$

with $|f| \leq a_1 e^{-a_2(t-t_0)}$ (a_1, a_2 are positive constants) for any $\boldsymbol{\xi}_0 \in S$.

As a result, $\xi_2(t) \to \xi_{20}$ $(t \to \infty)$ for any $\boldsymbol{\xi}_0 \in S$ and by (4.3.10), we conclude that the unperturbed motion $\xi_j = 0$ $(j = 1, \ldots, 6)$ is exponentially asymptotically stable with respect to ξ_s $(s = 4, 5, 6)$ and also Lyapunov stable. Besides, by virtue of the structure of equalities (4.3.10), the required limiting relationships (4.3.14) are valid. The statement is proved. □

Remark 4.3.3.

1. In view of limiting relationships (4.3.14), in the course of the partial stabilization described, as time passes, the solid exponentially asymptotically approaches the state

$$q_1 = \alpha, \quad x_1 = \alpha(\omega + \omega^*) \ (123, \alpha\beta\gamma), \quad \omega^* = \beta^{-1}\xi_{20}. \tag{4.3.16}$$

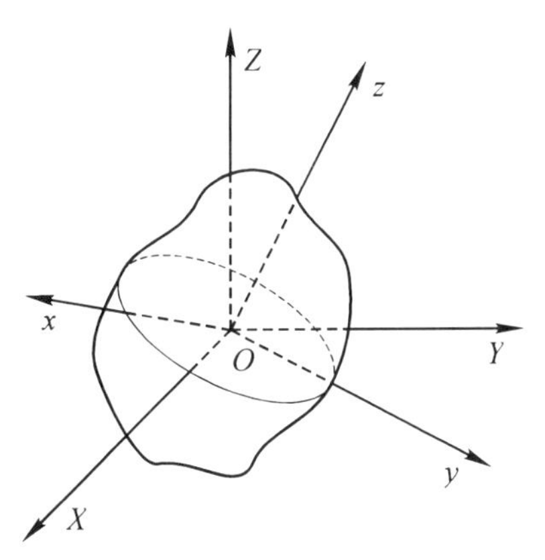

Figure 4.3.1. Translationally moving $OXYZ$-axes; $Oxyz$ are the principal central axes of inertia of a solid. The directions of Oz- and OZ-axes coincide in the unperturbed motion.

2. The cases of $B = C, \alpha \neq 0$ and $A = B, \gamma \neq 0$ can also be handled by following the proposed scheme.

3. When investigating the problem of stabilization with respect to ξ_s ($s = 4,5,6$) by two controlling torques for *arbitrary* values of A, B, and C, we deal with the nonlinear *Riccati equation*

$$\dot{\xi}_2 = b_1(t)\xi_2^2 + b_2(t)\xi_2 + b_3(t) \tag{4.3.17}$$

instead of relationship (4.3.15).

So, some inferences about performing the partial stabilization in the general case can be made only with allowance for qualitative conclusions concerning solutions of this equation. Such a problem requires further investigation.

2. ***The case when a solid with arbitrary values of*** A, B, ***and*** C ***is in translational motion (or in equilibrium) relative to an inertial frame of reference.*** We take this motion as the unperturbed one. To study the deviation of the body from the specified unperturbed motion, we introduce the axes $OXYZ$ in translational motion. The origin of the axes coincides with the center of mass O of the solid, and the OZ-axis points vertically upward (see Figure 4.3.1). If the Oz- and OZ-axes have the same sense of direction in the unperturbed motion of the solid, the perturbed motion is described by system (4.3.4) with

$$\omega = 0, \quad x_{ic} = 0 \ (i = 1,2,3), \quad \alpha = \gamma = 0, \quad \beta = 1 \tag{4.3.18}$$

(see also Example 4.3.1).

Corollary 4.3.3. *Let* $u_2 \equiv 0$, *and let* u_1, u_3 *have the form* (4.3.6). *Then in the case of* (4.3.18), *the unperturbed motion* $\xi_j = 0$ $(j = 1,\dots,6)$ *of system* (4.3.4)–(4.3.6) *is Lyapunov stable and exponentially asymptotically stable with respect to* ξ_1, ξ_3, ξ_s $(s = 4,5,6)$.

Proof. In the case in question, from the closed system (4.3.4)–(4.3.6), one can segregate the equations

$$\dot{\xi}_1 = \Gamma_4\xi_1 - \Gamma_3\xi_6 + G_1(\boldsymbol{\xi}), \quad \dot{\xi}_2 = G_2(\boldsymbol{\xi}), \quad \dot{\xi}_3 = \Gamma_2\xi_3 + \Gamma_1\xi_4 + G_3(\boldsymbol{\xi}),$$
$$\dot{\xi}_4 = \xi_3 + G_4(\boldsymbol{\xi}), \quad \dot{\xi}_6 = -\xi_1 + G_6(\boldsymbol{\xi}), \tag{4.3.19}$$
$$\xi_5 = -1 - \sqrt{1 - \xi_4^2 - \xi_6^2} = {}^1\!/_2(\xi_4^2 + \xi_6^2 + \ldots),$$

where functions G_w $(w = 1, \ldots, 4, 6)$ represent the terms of order two and higher in ξ_w $(w = 1, \ldots, 4, 6)$. (Variable ξ_5 is expressed in terms of ξ_4, ξ_6 in G_w.)

The zero solution $\xi_w = 0$ $(w = 1, \ldots, 4, 6)$ of the system of linear approximation for system (4.3.19) is Lyapunov stable and exponentially asymptotically stable with respect to $\xi_1, \xi_3, \xi_4, \xi_6$. Functions G_w $(w = 1, \ldots, 4, 6)$ vanish identically for $\xi_1 = \xi_3 = \xi_4 = \xi_6 = 0$.

Therefore, in this case, the unperturbed motion $\xi_j = 0$ $(j = 1, \ldots, 6)$ of system (4.3.4)–(4.3.6) is Lyapunov stable and exponentially asymptotically stable with respect to $\xi_1, \xi_3, \xi_4, \xi_5, \xi_6$ by Theorem 2.6.1. The statement is proved. □

Remark 4.3.4.

1. In the course of the partial stabilization performed, the translational motion of the body Oz-axis is maintained rather than that of the entire body.

2. Corollary 4.3.3 guarantees the partial stabilization "in the small." (We emphasize that, in contrast to the stabilization with respect to all the variables, investigating just the system of linear approximation is insufficient to this end.) If initial perturbations are not sufficiently small, one can employ the proof scheme of Corollary 4.3.2 to establish partial stabilizability; in that case the analysis of the Riccati equation of type (4.3.17) should be carried out with $b_3(t) \equiv 0$.

4.3.4 Stabilization of a Relative Equilibrium Position of a Satellite in a Circular Orbit

Consider an *artificial satellite* (a solid) orbiting the Earth in a *circular orbit* under the action of the central Newtonian force field. To identify the position of the satellite in the orbit, we introduce the "*orbital*" coordinate system $Ox'y'z'$ whose center is located in the center of mass of the satellite (point O). The Oz'-axis is directed along the current radius vector of the orbit. The Oy'- and Oz'-axes are, respectively, parallel to the normal to the orbital plane and the transversal. We also use the system $Oxyz$ whose axes are directed along the principal central axes of inertia of the satellite (see Figure 4.3.2).

The position of the satellite relative to the orbital coordinate system is specified by the directional cosines presented in Table 4.3.1. $\alpha_i, \beta_i, \gamma_i$ $(i = 1, 2, 3)$ are uniquely expressible in terms of three independent angles: θ, be-

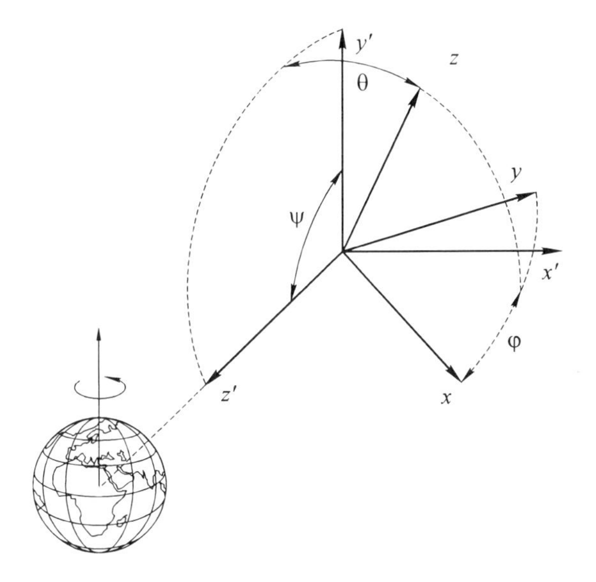

Figure 4.3.2. The coordinate axes characterizing the position of an artificial satellite of the Earth.

tween the Oz-axis of the satellite and its projection onto the (y', z')-plane; ψ, between this projection and the Oz'-axis; φ, the rotation about the Oz-axis.

Table 4.3.1. Table of Directional Cosines

O	x'	y'	z'
x	α_1	α_2	α_3
y	β_1	β_2	β_3
z	γ_1	γ_2	γ_3

The system of equations describing the angular motion of the satellite in a circular orbit comprises twelve nonlinear differential equations (Beletskii [1965])

$$\begin{aligned} A\dot{x}_1 &= (B-C)(x_2+\omega\beta_2)(x_3+\omega\beta_3) \\ &\quad - A\omega(\beta_2 x_3 - \beta_3 x_2) + \frac{3\varkappa(C-B)\gamma_2\gamma_3}{R^3} + u_1, \\ \dot{\alpha}_1 &= \alpha_2(x_3+\omega\beta_3) - \alpha_3(x_2+\omega\beta_2) - \omega\gamma_1, \\ \dot{\beta}_1 &= \beta_2(x_3+\omega\beta_3) - \beta_3(x_2+\omega\beta_2), \qquad (123, ABC) \\ \dot{\gamma}_1 &= \gamma_2(x_3+\omega\beta_3) - \gamma_3(x_2+\omega\beta_2) + \omega\alpha_1. \end{aligned} \tag{4.3.20}$$

Here A, B, and C are the principal moments of inertia of the satellite with respect to axes $Oxyz$, x_1, x_2, x_3 are the projections of its angular velocity vector onto the very axes; $\varkappa$ is the gravitational constant; $\omega^2 = \varkappa R^{-3}$ is the orbital velocity (R is the radius of the orbit); and u_i ($i = 1,2,3$) are the controlling moments.

The position of the satellite to be stabilized is specified by the equalities

$$\begin{gathered} p = q = r = 0, \\ \alpha_1 = \beta_2 = \gamma_3 = 1, \\ \alpha_2 = \alpha_3 = \beta_1 = \beta_3 = \gamma_1 = \gamma_2 = 0 \end{gathered} \tag{4.3.21}$$

or, which is the same, $p = q = r = \varphi = \psi = \theta = 0$. This position (a particular solution of system (4.3.20)) corresponds to the *equilibrium of the satellite in the orbital coordinate system*, i.e., the case when the center of mass of the satellite moves along a plane circular orbit and the satellite rotates about the Oy'-axis perpendicular to the plane of the orbit (at the same angular velocity). In this motion (relative orbital equilibrium), one and the same side of the satellite faces the Earth.

The conditions for stability of the relative equilibrium of a satellite in a circular orbit were first stated in the course of celestial mechanics by Tisserand [1891] on the basis of approximate analysis of motion equations and qualitative reasoning. The rigorous proof of these conditions was conducted by Beletskii [1965] by using the method of Lyapunov functions and by Rumyantsev [1967] based on the Routh–Lyapunov theorem.

Because the mechanical system under consideration is conservative, the *asymptotic* stability of relative equilibrium of the satellite in the orbit is impossible. Therefore, in a number of cases, a control system is required which enables stabilizing the relative equilibrium position of the satellite sufficiently quickly and accurately when the satellite deviates from this position. The use of the systems of "*passive*" stabilization (which employ "natural" factors, such as gravitational and magnetic moments, moments of light pressure forces, etc.) is not always possible for this purpose because they have a limited operating range and small controlling moments are involved. As a rule, we should have a system of "*active*" stabilization realized via special engines and other power devices (flywheels, gyroscopes). Such systems are capable of generating sufficiently large controlling moments, though this requires considerable consumption of energy or working medium to implement the control.

Based on the approach proposed in Section 4.2, we further construct control rules u_i ($i = 1,2,3$) which can be used to actively stabilize the relative equilibrium position of a satellite (4.3.21). We emphasize that we consider the so-called *restricted formulation* of the problem (without taking into account the perturbations of the satellite orbit) which is well justified in applications.

To solve the problem of stabilization indicated, let us introduce the new variables

$$\xi_1 = x_1,\ \xi_2 = x_2,\ \xi_3 = x_3,\ \xi_4 = \alpha_1 - 1,\ \xi_5 = \alpha_2,\ \xi_6 = \alpha_3,\ \xi_7 = \beta_1,$$
$$\xi_8 = \beta_2 - 1,\ \xi_9 = \beta_3,\ \xi_{10} = \gamma_1,\ \xi_{11} = \gamma_2,\ \xi_{12} = \gamma_3 - 1,$$

and compose the system of equations of perturbed motion in variations from the position (4.3.21)

$$\begin{aligned}
A\dot{\xi}_1 &= \omega\big[(B-C-A)\xi_3(\xi_8+1) + (B-C+A)\xi_2\xi_9\big] \\
&\quad + (B-C)\big\{\omega^2\big[\xi_9(\xi_8+1) - 3\xi_{11}(\xi_{12}+1)\big] + \xi_2\xi_3\big\} + u_1 \overset{\text{def}}{=} R_1(\boldsymbol{\xi}) + u_1, \\
B\dot{\xi}_2 &= \omega\big[(C-A-B)\xi_1\xi_9 + (C-A+B)\xi_3\xi_7\big] \\
&\quad + (C-A)\big\{\omega^2\big[\xi_7\xi_9 - 3\xi_{10}(\xi_{12}+1)\big] + \xi_1\xi_3\big\} + u_2 \overset{\text{def}}{=} R_2(\boldsymbol{\xi}) + u_2, \\
C\dot{\xi}_3 &= \omega\big[(A-B-C)\xi_2\xi_7 + (A-B+C)\xi_1(\xi_8+1)\big] \\
&\quad + (A-B)\big\{\omega^2\big[\xi_7(\xi_8+1) - 3\xi_{10}\xi_{11}\big] + \xi_1\xi_2\big\} + u_3 \overset{\text{def}}{=} R_3(\boldsymbol{\xi}) + u_3, \\
\dot{\xi}_4 &= \xi_3\xi_5 - \xi_2\xi_6 + \omega\big[-\xi_{10} + \xi_5\xi_9 - \xi_6(\xi_8+1)\big] \overset{\text{def}}{=} R_4(\boldsymbol{\xi}), \\
\dot{\xi}_5 &= \xi_1\xi_6 - \xi_3(\xi_4+1) + \omega\big[-\xi_{11} + \xi_6\xi_7 - \xi_9(\xi_4+1)\big] \overset{\text{def}}{=} R_5(\boldsymbol{\xi}), \qquad (4.3.22) \\
\dot{\xi}_6 &= -\xi_1\xi_5 + \xi_2(\xi_4+1) + \omega\big[\xi_4 + \xi_8 - \xi_{12} + \xi_4\xi_8 - \xi_5\xi_7\big] \overset{\text{def}}{=} R_6(\boldsymbol{\xi}), \\
\dot{\xi}_7 &= \xi_3(\xi_8+1) - \xi_2\xi_9 \overset{\text{def}}{=} R_7(\boldsymbol{\xi}), \qquad \dot{\xi}_8 = \xi_1\xi_9 - \xi_3\xi_7 \overset{\text{def}}{=} R_8(\boldsymbol{\xi}), \\
\dot{\xi}_9 &= -\xi_1(\xi_8+1) + \xi_2\xi_7 \overset{\text{def}}{=} R_9(\boldsymbol{\xi}), \\
\dot{\xi}_{10} &= \xi_3\xi_{11} - \xi_2(\xi_{12}+1) + \omega\big[\xi_4 - \xi_8 - \xi_{12} + \xi_9\xi_{11} - \xi_8\xi_{12}\big] \overset{\text{def}}{=} R_{10}(\boldsymbol{\xi}), \\
\dot{\xi}_{11} &= \xi_1(\xi_{12}+1) - \xi_3\xi_{10} + \omega\big[\xi_5 - \xi_9\xi_{10} + \xi_7(\xi_{12}+1)\big] \overset{\text{def}}{=} R_{11}(\boldsymbol{\xi}), \\
\dot{\xi}_{12} &= \xi_2\xi_{10} - \xi_1\xi_{11} + \omega\big[\xi_6 - \xi_7\xi_{11} + \xi_{10}(\xi_8+1)\big] \overset{\text{def}}{=} R_{12}(\boldsymbol{\xi}).
\end{aligned}$$

From geometric relationships among the directional cosines, we obtain the following equalities:

$$\begin{aligned}
G_1 &= (\xi_4+1)^2 + \xi_5^2 + \xi_6^2 - 1 = 0, \\
G_2 &= \xi_7^2 + (\xi_8+1)^2 + \xi_9^2 - 1 = 0, \\
G_3 &= \xi_{10}^2 + \xi_{11}^2 + (\xi_{12}+1)^2 - 1 = 0, \qquad (4.3.23) \\
G_4 &= (\xi_4+1)\xi_7 + (\xi_8+1)\xi_5 + \xi_6\xi_9 = 0, \\
G_5 &= (\xi_4+1)\xi_{10} + \xi_5\xi_{11} + \xi_6(\xi_{12}+1) = 0, \\
G_6 &= \xi_7\xi_{10} + (\xi_8+1)\xi_{11} + \xi_9(\xi_{12}+1) = 0.
\end{aligned}$$

After renaming the variables $\eta_k = \xi_{3+k}$ $(k = 1, \ldots, 9)$, system (4.3.22), (4.3.23) has the structure of (4.2.1), (4.2.3). Taking $q = 0, X_1^{(2)} = R_7, X_2^{(2)} = R_9, X_3^{(2)} = R_{10}$, one can readily verify that, when $\xi_8 + 1 \neq 0$, system (4.3.22), (4.3.23) satisfies conditions (1) and (2) of Theorem 4.2.1.

Theorem 4.3.2. *The set of control rules*

$$u_1 = \frac{A\Psi_1(\boldsymbol{\xi})}{1+\xi_8} + \frac{A\xi_7 u_2}{B(1+\xi_8)} - \frac{Au_3^*}{1+\xi_8}, \quad u_2 = \frac{B(1+\xi_8)\big[\Psi_2^*(\boldsymbol{\xi}) - u_2^*\big]}{(1+\xi_8)(1+\xi_{12}) - \xi_9\xi_{11}},$$
$$u_3 = \frac{C\Psi_3(\boldsymbol{\xi})}{1+\xi_8} + \frac{C\xi_9 u_2}{B(1+\xi_8)} - \frac{Cu_1^*}{1+\xi_8}, \tag{4.3.24}$$

where

$$\Psi_1 = -\frac{(1+\xi_8)R_1}{A} + \frac{\xi_7 R_2}{B} + \xi_2 R_7 - \xi_1 R_8, \quad \Psi_2^* = -\Psi_2 + \frac{\xi_{11}(\Psi_3 - u_1^*)}{1+\xi_8},$$
$$\Psi_2 = \frac{(1+\xi_{12})R_2}{B} - \omega(R_4 - R_8 - R_{12}) - (\omega\xi_9 + \xi_3)R_{11} - \omega\xi_{11}R_9$$
$$- \frac{\xi_{11}R_3}{C} - \frac{\xi_{11}\Psi_3}{1+\xi_8} + \xi_2 R_{12} + \omega(\xi_{12}R_9 + \xi_9 R_{12}),$$
$$\Psi_3 = -\frac{(1+\xi_8)R_3}{C} - \xi_3 R_8 + \frac{\xi_9 R_2}{B} + \xi_2 R_9,$$
$$u_1^* = L_1\xi_7 + L_2 R_7, \quad u_2^* = L_3\xi_9 + L_4 R_9, \quad u_3^* = L_5\xi_{10} + L_6 R_{10},$$
$$L_j = \text{const} < 0 \quad (j = 1, \ldots, 6),$$

is the set of stabilizing control rules ensuring the exponential asymptotic stability of the unperturbed motion of system (4.3.22)–(4.3.24) *with respect to* ξ_l $(l = 1, \ldots, 12)$.

Proof. Omitting the technical calculations, we conclude that, from the closed system (4.3.22)–(4.3.24), one can segregate the *linear* $\boldsymbol{\mu}$*-system* ($\mu_1 = R_7$, $\mu_2 = R_9$, $\mu_3 = R_{10}$)

$$\dot{\xi}_7 = \mu_1, \quad \dot{\mu}_1 = L_1\xi_7 + L_2\mu_1;$$
$$\dot{\xi}_9 = \mu_3, \quad \dot{\mu}_3 = L_5\xi_9 + L_6\mu_3; \quad \dot{\xi}_{10} = \mu_2, \quad \dot{\mu}_2 = L_3\xi_{10} + L_4\mu_2. \tag{4.3.25}$$

Since $L_j = \text{const} < 0$ $(j = 1, \ldots, 6)$, the solution $\xi_7 = \xi_9 = \xi_{10} = \mu_1 = \mu_2 = \mu_3 = 0$ of system (4.3.25) is exponentially asymptotically stable. Therefore, the unperturbed motion $\xi_l = 0$ $(l = 1, \ldots, 12)$ of system (4.3.22)–(4.3.24) is exponentially asymptotically stable with respect to ξ_7, ξ_9, ξ_{10}.

Solving the system for G_2, G_3, G_6, we get the relationships

$$\xi_8(t) = -1 \pm \sqrt{1 - \xi_7^2(t) - \xi_9^2(t)}, \tag{4.3.26}$$

$$\xi_{11}(t) = -\frac{\xi_7(t)\xi_{10}(t) + \xi_9(t)\big[1 + \xi_{12}(t)\big]}{1 + \xi_8(t)}, \tag{4.3.27}$$

$$\big[1 + \xi_{12}(t)\big]^2 + \frac{\big\{-\xi_7(t)\xi_{10}(t) - \xi_9(t)\big[1 + \xi_{12}(t)\big]\big\}^2}{\big[1 + \xi_8(t)\big]^2} = 1 - \xi_{10}^2(t). \tag{4.3.28}$$

Solving equation (4.3.28) as a quadratic for $1 + \xi_{12}$ and omitting the solutions that do not belong to a sufficiently small neighborhood of the unperturbed motion $\xi_l = 0$ $(l = 1, \ldots, 12)$, from (4.3.26), (4.3.28), we choose the following relationships for further consideration:

$$\xi_8(t) = -1 + \sqrt{1 - \xi_7^2(t) - \xi_9^2(t)}, \tag{4.3.29}$$

$$\xi_{12}(t) = \frac{-1 + \xi_7^2(t) - \xi_7(t)\xi_9(t)\xi_{10}(t) + \sqrt{1 + \bar{R}_1(t)}}{1 - \xi_7^2(t)}, \tag{4.3.30}$$

$$\bar{R}_1 = 2\xi_8 + \xi_8^2 - \xi_{10}^2 - 2\xi_8\xi_{10}^2 - \xi_8^2\xi_{10}^2 - \xi_7^2 - 2\xi_8\xi_7^2 - \xi_8^2\xi_7^2 + 2\xi_8\xi_{10}^2\xi_7^2 + \xi_8^2\xi_{10}^2\xi_7^2 + \xi_9^2\xi_{10}^2\xi_7^2.$$

Following the same technique, from the equations for G_1, G_4, G_5, we obtain

$$\xi_4(t) = -1 + \sqrt{1 - \xi_5^2(t) - \xi_6^2(t)}, \tag{4.3.31}$$

$$\xi_5(t) = \frac{\bar{R}_3(t)}{\bar{R}^*(t)}, \qquad \xi_6(t) = \frac{\bar{R}_4(t)}{\bar{R}^*(t)}, \tag{4.3.32}$$

$$\bar{R}^* = \sqrt{\bar{R}_2^2 + \bar{R}_3^2 + \bar{R}_4^2}, \qquad \bar{R}_2 = 1 + \xi_8\xi_{12} + \xi_8 + \xi_{12} - \xi_9\xi_{11},$$
$$\bar{R}_3 = -\xi_7(1 + \xi_{12}) + \xi_9\xi_{10}, \qquad \bar{R}_4 = -(1 + \xi_8)\xi_{10} - \xi_7\xi_{11}.$$

The equations for R_7, R_9, R_{10} allow us to obtain the equalities

$$\xi_3(t) = \frac{-\xi_9(t)\left[\dot{\xi}_{10}(t) - \omega\bar{R}_5(t) + \dot{\xi}_7(t)\right]\left[1 + \xi_{12}(t)\right]}{\bar{R}_2(t)},$$
$$\xi_2(t) = \frac{\dot{\xi}_7(t)\xi_{11} - \left[1 + \xi_8(t)\right]\left[\dot{\xi}_{10}(t) - \omega\bar{R}_5(t)\right]}{\bar{R}_2(t)}, \tag{4.3.33}$$
$$\xi_1(t) = \frac{-\dot{\xi}_9(t) + \xi_2(t)\xi_7(t)}{1 + \xi_8(t)}, \quad \bar{R}_5 = \xi_4 - \xi_8 - \xi_{12} + \xi_9(\xi_{11} - \xi_{12}).$$

By virtue of the exponential asymptotic stability of the unperturbed motion $\xi_l = 0$ ($l = 1, \ldots, 12$) of system (4.3.22)–(4.3.24) with respect to ξ_7, ξ_9, ξ_{10}, one can find numbers Δ, δ such that $|\xi_{l0}| < \delta$ ($l = 1, \ldots, 12$) yield the estimate $|1 - \xi_7^2(t)| > \Delta$. Thus, by (4.3.30),

$$|\xi_{12}(t)| \leq \frac{1}{\Delta}\left|-1 + \xi_7^2(t) - \xi_7(t)\xi_9(t)\xi_{10}(t) + \sqrt{1 + \bar{R}_1(t)}\right|. \tag{4.3.34}$$

From (4.3.29), (4.3.34), it follows that the unperturbed motion $\xi_l = 0$ ($l = 1, \ldots, 12$) of system (4.3.22)–(4.3.24) is exponentially asymptotically stable with respect to ξ_7, ξ_9, ξ_{10} and also with respect to ξ_8, ξ_{12}. This means that one can find Δ', δ' such that $|\xi_{l0}| < \delta'$ ($l = 1, \ldots, 12$) yield

$$|1 + \xi_8(t)| > \Delta', \quad |\bar{R}_2(t)| > \Delta', \quad |\bar{R}^*(t)| > \Delta',$$

and from (4.3.27), (4.3.32), we conclude that

$$|\xi_{11}(t)| \leq \frac{1}{\Delta'}\left|-\xi_7(t)\xi_{10}(t) - \xi_9(t)\left[1 + \xi_{12}(t)\right]\right|,$$
$$|\xi_5(t)| \leq \frac{1}{\Delta'}|\bar{R}_3(t)|, \qquad |\xi_6(t)| \leq \frac{1}{\Delta'}|\bar{R}_4(t)|.$$

Therefore, the unperturbed motion $\xi_l = 0$ $(l = 1, \ldots, 12)$ is also exponentially asymptotically stable with respect to ξ_5, ξ_6, ξ_{11}.

Finally, on the basis of (4.3.31), (4.3.33) we draw the conclusion that the unperturbed motion $\xi_l = 0$ $(l = 1, \ldots, 12)$ of system (4.3.22)–(4.3.24) is exponentially asymptotically stable with respect to all the variables. The Theorem is proved. □

Remark 4.3.5.

1. The explicit form of solutions of system (4.3.22)–(4.3.24) allows one to estimate the attraction domain of the unperturbed motion $\xi_l = 0$ $(l = 1, \ldots, 12)$ of the system. Besides, following the scheme of Remark 4.3.2, it is possible to iteratively search control rules satisfying the additional performance criteria.

2. By the technique of Subsection 4.2.9, one can choose the constants L_j $(j = 1, \ldots, 6)$ appearing in u_i^* $(i = 1, 2, 3)$ so that the functional of type (4.2.23) is minimized on the trajectories of system (4.3.22)–(4.3.24). The integrand

$$R(\boldsymbol{\xi}, \boldsymbol{u}) = a_1\xi_7^2 + a_2\xi_9^2 + a_3\xi_{10}^2 + \sum_{i=1}^{3} b_i\mu_i^2(\boldsymbol{\xi}) + \sum_{i=1}^{3} c_i\big[u_i^*(\boldsymbol{\xi}, \boldsymbol{u})\big]^2 \qquad (4.3.35)$$

is a *positive-definite* function in variables ξ_l $(l = 1, \ldots, 12)$, u_i $(i = 1, 2, 3)$, whose coefficients a_i, b_i, c_i $(i = 1, 2, 3)$ can be chosen in a rational way, i.e., based on expedient requirements for the quality of the system transient.

To analyze the structure of the function $R(\boldsymbol{\xi}, \boldsymbol{u})$ in more detail, we bring in the relationships

$$\begin{aligned}
&\alpha_1 = \cos\theta\cos\varphi, \quad \alpha_2 = -\cos\theta\sin\varphi, \quad \alpha_3 = \sin\theta,\\
&\beta_1 = \cos\psi\sin\varphi + \sin\psi\sin\theta\cos\varphi, \quad \beta_2 = \cos\psi\cos\varphi - \sin\psi\sin\theta\sin\varphi,\\
&\beta_3 = -\sin\psi\cos\theta, \quad \gamma_1 = \sin\psi\sin\varphi - \cos\psi\sin\theta\cos\varphi,\\
&\gamma_2 = \sin\psi\sin\varphi + \cos\psi\sin\theta\sin\varphi, \quad \gamma_3 = \cos\psi\cos\theta.
\end{aligned}$$

As a result, we get the following relationships:

$$\begin{aligned}
&\xi_4 = 0 + \ldots, && \xi_5 = -\varphi + \ldots, && \xi_6 = \theta + \ldots, && \xi_7 = \varphi + \ldots,\\
&\xi_8 = 0 + \ldots, && \xi_9 = -\psi + \ldots, && \xi_{10} = -\theta + \ldots, && \xi_{11} = \psi + \ldots,\\
&\xi_{12} = 0 + \ldots, && Cu_1^* = u_3 + \ldots, && Au_2^* = -u_1 + \ldots, && Bu_3^* = -u_2 + \ldots,\\
&\mu_1 = x_3 + \ldots, && \mu_2 = -x_2 + \ldots, && \mu_3 = -x_1 + \ldots, &&
\end{aligned} \qquad (4.3.36)$$

where the ellipsis dots stand for the terms of higher than the first order in $x_i, \varphi, \psi, \theta, \omega, u_i$ $(i = 1, 2, 3)$.

Bearing in mind (4.3.36), one can write expression (4.3.35) as

$$\begin{aligned}
R(\boldsymbol{\xi}, \boldsymbol{u}) = {} & a_1\varphi^2 + a_2\psi^2 + a_3\theta^2 + b_1x_3^2 + b_2x_2^2 + b_3x_1^2\\
& + \left(c_2/A^2\right)u_1^2 + \left(c_3/B^2\right)u_2^2 + \left(c_1/C^2\right)u_3^2 + \Lambda(\boldsymbol{\xi}, \boldsymbol{u}, \omega),
\end{aligned}$$

where function Λ includes the terms not lower than the *third* order in $x_i, \varphi, \psi, \theta, \omega, u_i$ $(i = 1, 2, 3)$.

Setting $b_1 = C, b_2 = B, b_3 = A$, we deduce that the integrand in the functional to be minimized is the sum of four expressions:

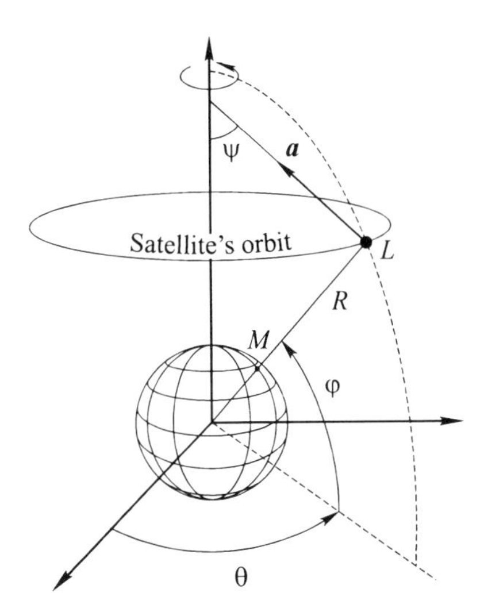

Figure 4.3.3. A geostationary satellite L "hovering" over an arbitrary point M of the Earth's surface. The center of mass of the satellite moves along a circular non-Keplerian orbit lying in a plane parallel to the equatorial plane with angular velocity equal to that of the Earth's rotation.

(1) the kinetic energy of the satellite $Ax_1^2 + Bx_2^2 + Cx_3^2$;
(2) the quadratic form $a_1\varphi^2 + a_2\psi^2 + a_3\theta^2$ characterizing the deviation of the satellite from the equilibrium position;
(3) the quadratic form $(c_2/A^2)u_1^2 + (c_3/B^2)u_2^2 + (c_1/C^2)u_3^2$ characterizing the resources expended in generating the controls;
(4) the function Λ including the terms of order three and higher in $x_i, \varphi, \psi, \theta, \omega, u_i$ $(i = 1, 2, 3)$.

4.3.5 Partial Stabilization of a Satellite in a Geostationary Orbit

Various scientific work has discussed the ways of implementing the idea of Arthur Clarke, a science-fiction writer, to design *satellites* which can "*hover*" *over an arbitrary point on the Earth's surface* for a sufficiently long period of time (*geostationary satellites*). To this end, as indicated in Nita [1973], it suffices that a reactive acceleration $\boldsymbol{a}$, constant in magnitude, be applied to the satellite. The acceleration vector makes an angle ψ with the Earth's rotational axis (see Figure 4.3.3).

The rotational-translational motion of a satellite in a geostationary orbit can be described by the equations (Veretennikov [1984]), in which the sums are taken over i from 1 to 3:

$$\ddot{R} - R(\dot{\theta}^2 \cos^2\varphi + \dot{\varphi}^2) = a\sum \gamma_i\sigma_i + \frac{1}{m}\frac{\partial U}{\partial R},$$

$$R^2\ddot{\varphi} + 2R\dot{R}\dot{\varphi} + \frac{1}{2}R^2\dot{\theta}^2 \sin 2\varphi = Ra\sum \beta_i\sigma_i,$$

$$\begin{aligned}
\frac{d(R^2\dot{\theta}\cos^2\varphi)}{dt} &= Ra\cos\varphi\sum\alpha_i\sigma_i, \\
A\dot{x}_1 = (B-C)x_2x_3 &+ \frac{3\varkappa(B-C)\gamma_2\gamma_3}{R^3} + u_1, \qquad (4.3.37)\\
\dot{\alpha}_1 &= r\alpha_2 - q\alpha_3 + \dot{\theta}(\beta_1\sin\varphi - \gamma_1\cos\varphi), \\
\dot{\beta}_1 &= r\beta_2 - q\beta_3 - \dot{\theta}\alpha_1\sin\varphi - \dot{\varphi}\gamma_1, \qquad (123, ABC)\\
\dot{\gamma}_1 &= r\gamma_2 - q\gamma_3 - \dot{\theta}\alpha_1\cos\varphi - \dot{\varphi}\beta_1.
\end{aligned}$$

Here R, φ, θ, x_i $(i = 1,2,3)$ are the spherical coordinates of the center of mass of the satellite and the projection of its angular velocity vector onto the principal central axes $Oxyz$; A, B, C are the principal central moments of inertia of the satellite; $\sigma_i, \alpha_i, \beta_i, \gamma_i$ $(i = 1,2,3)$ are the respective directional cosines of vector $\boldsymbol{a}$ and the axes of system $Oxyz$ with respect to the orbital system of coordinates (σ_i $(i = 1,2,3)$ are constants because vector $\boldsymbol{a}$ of reactive thrust is rigidly fixed to the satellite frame); U is the potential of the gravitational forces; $\varkappa$ is the gravitational constant; and m is the mass of the satellite.

The control is implemented by three pairs of fixed engines which generate the controlling moments u_i $(i = 1,2,3)$ with respect to the axes of system $Oxyz$.

In the absence of controls (the engines are off), equations (4.3.37) admit the particular solution (Veretennikov [1984])

$$\begin{gathered}
R = R_0 = \text{const}, \quad \varphi = \varphi_0 = \text{const}, \quad \dot{\theta} = \omega_0 = \text{const}, \\
\alpha_1 = 1, \quad \alpha_2 = \alpha_3 = \beta_1 = \gamma_1 = \sigma_1 = 0, \quad \beta_2 = \gamma_3 = \cos\lambda, \\
\gamma_2 = -\beta_3 = -\sin\lambda, \quad x_1 = 0, \quad x_2 = \omega_0\cos(\varphi_0 + \lambda), \\
x_3 = \omega_0\sin(\varphi_0+\lambda), \quad \tan\psi = (\rho - 1)\cot\varphi_0, \\
a = \frac{\varkappa\sin\varphi_0}{R_0^2\cos\psi}, \quad \rho = \frac{\omega_0^2R_0^3}{\varkappa}, \quad \tan 2\lambda = \frac{-\rho\sin 2\varphi_0}{3+\rho\cos 2\varphi_0}
\end{gathered} \qquad (4.3.38)$$

which corresponds to the *relative equilibrium position* of the satellite in the circular non-Keplerian orbit of radius R_0 at an arbitrary latitude φ_0, where one of the satellite's principal central axes of inertia is always aligned with the velocity vector of the center of mass and two others are located in the meridian plane making an angle λ with the radius vector and the normal to it, respectively. This angle is uniquely determined by the inclination of the thrust vector with respect to the Earth's rotational axis and latitude φ_0. This orientation can be performed naturally without employing a special control system. To do so, one just needs to properly orient the satellite initially when it is separated from the last stage of the booster.

Nevertheless, at the moment the satellite is separated, perturbations arise (which can also be caused by other reasons), and it is necessary to choose the controlling actions u_i $(i = 1,2,3)$ to ensure the asymptotic stability of the satellite's motion in the orbit.

To solve the problem being considered of stabilizing the satellite's motion, let us compose the equations of perturbed motion

$$\dot{\xi}_k = X_k(\xi_1, \dots, \xi_{18}, u_1, u_2, u_3) \qquad (k = 1, \dots, 18) \tag{4.3.39}$$

in deviations

$$\xi_1 = R - R_0, \quad \xi_2 = \dot{R}, \quad \xi_3 = \varphi - \varphi_0, \quad \xi_4 = \dot{\varphi}, \quad \xi_5 = \theta - \omega_0 t,$$
$$\xi_6 = \dot{\theta} - \omega_0, \quad \xi_7 = x_1, \quad \xi_8 = x_2 - \omega_0 \cos(\varphi_0 + \lambda), \quad \xi_9 = x_3 - \omega_0 \sin(\varphi_0 + \lambda),$$
$$\xi_{10} = \alpha_1 - 1, \quad \xi_{11} = \alpha_2, \quad \xi_{12} = \alpha_3, \quad \xi_{13} = \beta_1, \quad \xi_{14} = \beta_2 - \cos\lambda,$$
$$\xi_{15} = \beta_3 - \sin\lambda, \quad \xi_{16} = \gamma_1, \quad \xi_{17} = \gamma_2 + \sin\lambda, \quad \xi_{18} = \gamma_3 - \cos\alpha$$

from the relative equilibrium position (4.3.38).

To specify the structure of control rules u_i $(i = 1, 2, 3)$ solving the problem of partial stabilization (with respect to variables $\xi_{10}, \dots, \xi_{18}$) of the unperturbed motion $\xi_k = 0$ $(k = 1, \dots, 18)$ of system (4.3.39), it suffices to know the structure of just a few of X_k functions, i.e., the functions

$$X_{11} = \xi_7\xi_{12} - \xi_9(\xi_{10} + 1)X_{11}^*(\boldsymbol{\xi}), \quad X_{12} = -\xi_7\xi_{11} + \xi_8(\xi_{10} + 1) + X_{12}^*(\boldsymbol{\xi}),$$
$$X_{17} = -\big[\xi_9 + \omega_0 \sin(\varphi_0 + \lambda)\big]\xi_{16} + \xi_7(\xi_{18} + \cos\lambda) + X_{17}^*(\boldsymbol{\xi}),$$
$$X_{11}^*(\mathbf{0}) = X_{12}^*(\mathbf{0}) = X_{17}^*(\mathbf{0}) = 0.$$

Let us consider the set of control rules

$$u_1 = \frac{A(\xi_{10} + 1)F_1}{(\cos\lambda + \xi_{18})(\xi_{10} + 1)\xi_{12}\xi_{16}},$$
$$u_2 = \frac{B(AF_2 + \xi_{11}u_1)}{A(\xi_{10} + 1)}, \qquad u_3 = \frac{C(AF_3 + \xi_{12}u_1)}{A(\xi_{10} + 1)}, \tag{4.3.40}$$

$$F_1 = u_3^* - (\cos\lambda + \xi_{18})X_7 - \xi_7 X_{18} + \xi_9 X_{16} + \xi_{16} X_9 - \dot{X}_{17}^* - \frac{\xi_{16}F_3}{\xi_{10} + 1},$$
$$F_2 = u_2^* - \xi_8 X_{10} - (\xi_{10} + 1)X_8 + \xi_{11}X_7 + \xi_7 X_{11} - \dot{X}_{12}^*,$$
$$F_3 = u_1^* - \xi_7 X_{12} - \xi_{12}X_7 + \xi_9 X_{10} + (\xi_{10} + 1)X_9 - \dot{X}_{11}^*,$$
$$u_1^* = \Gamma_1\xi_{11} + \Gamma_2 X_{11}, \quad u_2^* = \Gamma_3\xi_{12} + \Gamma_4 X_{12}, \quad u_3^* = \Gamma_5\xi_{17} + \Gamma_6 X_{17},$$
$$\Gamma_j = \text{const} < 0 \quad (j = 1, \dots, 6),$$

where $\dot{X}_s^*$ $(s = 11, 12, 17)$ are the derivatives of function X_s^* by virtue of system (4.3.33).

Theorem 4.3.3. *If the control rules u_i $(i = 1, 2, 3)$ have the form* (4.3.40), *then the unperturbed motion $\xi_j = 0$ $(j = 1, \dots, 18)$ of system* (4.3.39), (4.3.40) *is asymptotically stable with respect to variables $\xi_{10}, \dots, \xi_{18}$.*

Proof. Omitting technical calculations, we infer that variables ξ_s $(s = 11, 12, 17)$ of system (4.3.39), (4.3.40) satisfy the *$\boldsymbol{\mu}$-system of differential equations*

$$\dot{\xi}_{11} = \mu_1, \qquad \dot{\mu}_1 = \Gamma_1\xi_{11} + \Gamma_2\mu_1, \qquad (\mu_1 = X_{11}),$$
$$\dot{\xi}_{12} = \mu_2, \qquad \dot{\mu}_2 = \Gamma_3\xi_{12} + \Gamma_4\mu_2, \qquad (\mu_2 = X_{12}),$$
$$\dot{\xi}_{17} = \mu_3, \qquad \dot{\mu}_3 = \Gamma_5\xi_{17} + \Gamma_6\mu_3, \qquad (\mu_3 = X_{17}).$$

Since $\Gamma_j = \text{const} < 0$ $(j = 1, \ldots, 6)$, the unperturbed motion $\xi_k = 0$ $(k = 1, \ldots, 18)$ of system (4.3.39), (4.3.40) is exponentially asymptotically stable with respect to ξ_s $(s = 11, 12, 17)$.

Variables ξ_v $(v = 10, \ldots, 18; v \neq 11, 12, 17)$ can be explicitly expressed in terms of $\xi_{11}, \xi_{17}, \xi_{18}$ (see the corresponding relationships in Subsection 4.3.4). Hence the unperturbed motion $\xi_j = 0$ $(j = 1, \ldots, 18)$ of system (4.3.39), (4.3.40) is asymptotically stable with respect to ξ_{11}, ξ_{17}, ξ_{18} and also with respect to ξ_v $(v = 10, \ldots, 18)$. The theorem is proved. □

4.3.6 Partial Stabilization of Stationary Modes of a Mechanical System of "Solid + Flywheels" (Gimballed Gyroscope)

We confine the consideration to those stationary modes in which the main body of the system is in an equilibrium position. The problems of stabilizing such stationary modes are typical for partial stabilization. Indeed, in the absence of external forces, the system under consideration is isolated and subject only to *internal* controlling moments, which are generated by the engines and applied to the flywheels (gyroscope). The law of conservation of total angular momentum is valid with respect to the center of mass of the system. This means that the total angular momentum of the system is invariable no matter which internal moments are applied to the flywheels (gyroscope). As a result, *it is impossible* to ensure the stabilization of the stationary modes of the system under study with respect to *all* of its phase variables only by internal controlling moments.

Though not changing the total angular momentum of the system, the specially chosen internal controlling moments can *redistribute* it. This enables us to *partially* stabilize the stationary modes of the system under study, i.e., to stabilize them with respect to the variables which determine the velocities and also (as a more detailed analysis shows) the attitude of the main body. The specifics of such an (incomplete) stabilization is that the flywheels (gyroscope) "take up" the perturbations of the angular momentum of the system.

The problems of partial stabilization outlined are of significant interest for applications. Though these are the problems of *partial* stabilization in the mechanical systems considered, they are also the problems of *full* (with respect to all the variables) stabilization of an equilibrium position of the main body of such systems.

Technically, this partial stabilization is performed as follows. While the main body of the system is in the equilibrium position, the masses (flywheels, gyroscope) attached to it either rest or rotate uniformly with respect to the body. When perturbations appear, the main body departs from the equilibrium position thereby acquiring some angular velocity and changing the attitude. In this case the engines are turned on, which form the specially constructed (in solving the corresponding mathematical problem) rules of altering the controlling moments that make the masses rotate appropriately. As time passes, this results in the body returning to the initial equilibrium position, the engines are turned off, and the flywheels (gyroscope) still rotate by inertia, "taking up" the perturbations of the angular momentum of the system.

In a number of cases, we note that just partially stabilizing the equilibrium position of the main body of the system may be sufficient. In these cases (as was demonstrated in Example 2.6.1), the rotating masses attached to the body may sometimes *just redistribute* (without "taking up") the perturbations of the angular momentum of the system among separate groups of the variables determining the state (angular velocity and attitude) of the main body.

The necessity of only partially (incompletely) stabilizing the equilibrium position of the main body of the system can stem from an *emergency*: for example, the body has three symmetric homogeneous flywheels whose rotational axes are fixedly aligned with its principal central axes of inertia, because the controlling engine of at least one of the flywheels is out of order.

Further, to analyze the issues touched on, we consider three problems of this type which are solved based on the approach proposed in this book:

(1) the problem of *uniaxial* stabilization of an equilibrium position of a solid by *three* symmetric homogeneous flywheels whose rotational axes are fixedly aligned with the principal central axes of inertia of the solid;

(2) the problem of *uniaxial* partial stabilization of an equilibrium position of a solid by *two* symmetric homogeneous flywheels attached to the solid in the same way. In this case, as distinct from Example 2.6.1, the flywheels redistribute the perturbations of the angular momentum of the system and also "take up" its respective part. This situation is believed to be more typical than that of Example 2.6.1.

(3) the problem of *triaxial* stabilization of an equilibrium position of a solid by a gimballed gyroscope. We also ensure (nonasymptotic) stability with respect to all the variables of the corresponding stationary mode of the system which is necessary from the standpoint of proper operation and also to ensure the realizability of the proposed control scheme.

1. ***The problem of full uniaxial stabilization of a solid by three flywheels.*** Let us consider a mechanical system: a solid and three homogeneous symmetric flywheels whose axes are aligned with the principal central axes of inertia of the solid. The flywheels are brought into rotation by special engines. This system falls into the type of *gyrostats*, namely, the systems whose mass distribution is invariable in motion. In the absence of external forces, the angular motion of the solid under study with respect to the center of mass is described by the equations

$$\begin{gathered}(A-J_1)\dot{x}_1+(C-B)x_2x_3-J_2x_3\dot{\varphi}_2+J_3x_2\dot{\varphi}_3=-v_1\\ J_1(\ddot{\varphi}_1+\dot{x}_1)=v_1, \qquad (ABC,123)\end{gathered} \tag{4.3.41}$$

Here x_i $(i=1,2,3)$ are the projections of the instantaneous angular velocity vector of the body onto its principal central axes of inertia $Oxyz$; A, B, and C are the moments of inertia of the system with respect to the $Oxyz$-axes; $J_i, \dot{\varphi}_i$ $(i = 1, 2, 3)$ are the axial moments of inertia and the relative angular velocities of the flywheels; and v_i $(i = 1, 2, 3)$ are the controlling moments applied to the flywheels.

We shall consider equations (4.3.42) together with Poisson's kinematic equations that determine the uniaxial attitude of a solid

$$\dot{q}_1 = x_3 q_2 - x_2 q_3, \quad \dot{q}_2 = x_1 q_3 - x_3 q_1, \quad \dot{q}_3 = x_2 q_1 - x_1 q_2, \tag{4.3.42}$$

where q_i $(i = 1, 2, 3)$ are the projections of the unit vector directed along a fixed vertical axis.

For $v_i = 0$ $(i = 1, 2, 3)$, equations (4.3.41), (4.3.42) admit the particular solution

$$x_i = 0, \quad \dot{\varphi}_i = 0, \quad v_i = 0 \quad (i = 1, 2, 3), \quad q_1 = q_3 = 0, \quad q_2 = 1, \tag{4.3.43}$$

which is an equilibrium position of the system (both the solid and the flywheels). In this position one of the principal central axes of inertia of the solid (Oy-axis) is aligned with the fixed vertical axis. The controlling engines are off.

Under the action of initial perturbations let the solid depart from the equilibrium position (4.3.43) acquiring values x_i, q_i $(i = 1, 2, 3)$ different from those specified in (4.3.43); we still suppose that the flywheels' relative angular velocities are initially zero.

To study the deviations of the perturbed solutions from the equilibrium position (4.3.43), we introduce the new variables

$$\xi_i = x_i, \quad \xi_4 = q_1, \quad \xi_5 = q_2 - 1, \quad \xi_6 = q_3, \quad \xi_{6+i} = \dot{\varphi}_i,$$
$$u_i = -v_i \quad (i = 1, 2, 3)$$

and compose the system of equations of perturbed motion

$$\begin{aligned} (A - J_1)\dot{\xi}_1 &= (B - C)\xi_2\xi_3 + J_2\xi_3\xi_8 - J_3\xi_2\xi_9 + u_1 \\ &\stackrel{\text{def}}{=} (A - J_1) f_1^*(\boldsymbol{\xi}) + u_1 \quad (ABC, 123, 789), \\ \dot{\xi}_{3+i} &= f_{3+i}^*(\boldsymbol{\xi}), \quad J_i(\dot{\xi}_i + \dot{\xi}_{6+i}) = -u_i \quad (i = 1, 2, 3), \end{aligned} \tag{4.3.44}$$

where f_{3+i}^* are obtained from f_{3+i} $(i = 1, 2, 3)$ in (4.3.4) taking $\omega = \alpha = \gamma = 0$ and $\beta = 1$.

Let us consider the problem of partially stabilizing the equilibrium position $\xi_k = 0$ $(k = 1, \dots, 9)$ of system (4.3.44), i.e., stabilization with respect to variables ξ_j $(j = 1, \dots, 6)$ that determine the perturbed state of the main body. This is the problem of full uniaxial stabilization of the equilibrium position of the body by three flywheels.

Theorem 4.3.4. *The set of control rules* (4.3.6), *where one should take* $f_j^*(\boldsymbol{\xi})$ $(j = 1, \dots, 6)$ *instead of* $f_j(\boldsymbol{\xi})$, *is the set of stabilizing control rules that ensure the exponential asymptotic stability of the unperturbed motion* $\xi_k = 0$ $(k = 1, \dots, 9)$ *of system* (4.3.44) *with respect to* ξ_j $(j = 1, \dots, 6)$.

The proof follows the proof scheme for Theorem 4.3.1.

2. ***The problem of partial uniaxial stabilization of a solid by two flywheels.*** As distinct from the the problem studied earlier, let us consider *partial* uniaxial stabilization of the equilibrium position of the main body of the system, i.e., the problem of stabilization with respect to variables ξ_1, ξ_3, ξ_4, ξ_5, ξ_6 by *two* (and *not three*) controlling moments u_1, u_3 $(u_2 \equiv 0)$.

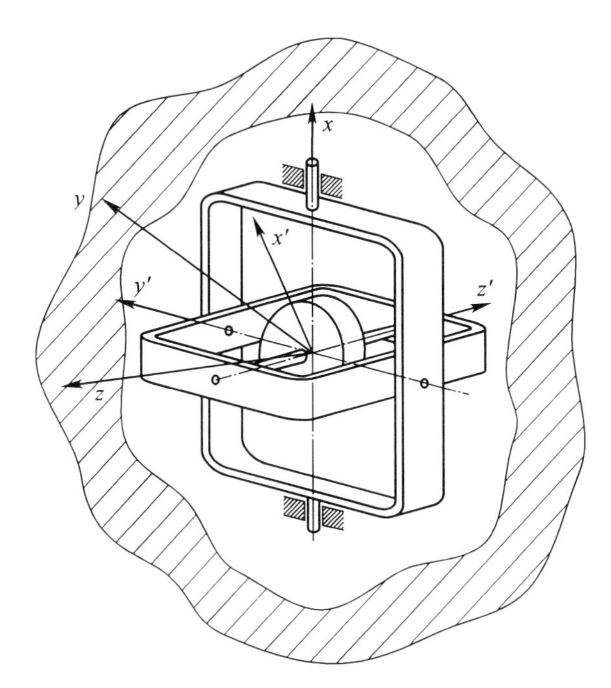

Figure 4.3.4. A free solid with a gimballed gyroscope on board.

Corollary 4.3.4. *If* $u_2 \equiv 0$ *and* u_1, u_3 *have the form* (4.3.6), *where one should take* $f_j^*(\boldsymbol{\xi})$ $(j = 1, \ldots, 6)$ *instead of* $f_j(\boldsymbol{\xi})$, *the equilibrium position* $\xi_k = 0$ $(k = 1, \ldots, 9)$ *of system* (4.3.44) *is Lyapunov stable and exponentially asymptotically stable with respect to variables* $\xi_1, \xi_3, \xi_4, \xi_5, \xi_6$.

From the practical point of view, we note that the problems considered of stabilizing the equilibrium position of a solid with respect to all the variables and with respect to part of the variables bear close similarity: in both the cases the process of stabilization returns the solid to the position in which one of its principal central axes of inertia is aligned with the fixed vertical axis, the only difference being that, in the case of partial stabilization, the solid's rotation about the vertical axis is not dampened (which, let us note, may even be desirable from the standpoint of making the solid "spin" about this axis in the future). At the same time, fewer flywheels are required to perform partial stabilization.

3. ***The problem of triaxial stabilization of a solid by a gimballed gyroscope***. We consider a free solid with principal central axes of inertia $Oxyz$ which has a homogeneous symmetric *balanced* gimballed gyroscope installed so that the axis of its external gimbal is directed along Ox and its fixed point coincides with the center of mass of the solid (see Figure 4.3.4). The gyroscope is regulated by *three* engines generating torques v_i $(i = 1, 2, 3)$ about the axes of the external and internal gimbals and the gyroscope axis.

The equations of the angular motion of the given mechanical system have the form (Krementulo [1977])

$$A\dot{x}_1 = (B - C)x_2x_3 + u_1 \qquad (ABC, 123),$$

$$u_1 = -v_1, \quad u_2 = -v_1 \sin\alpha \tan\beta - v_2 \cos\alpha + v_3 \sin\alpha \sec\beta,$$

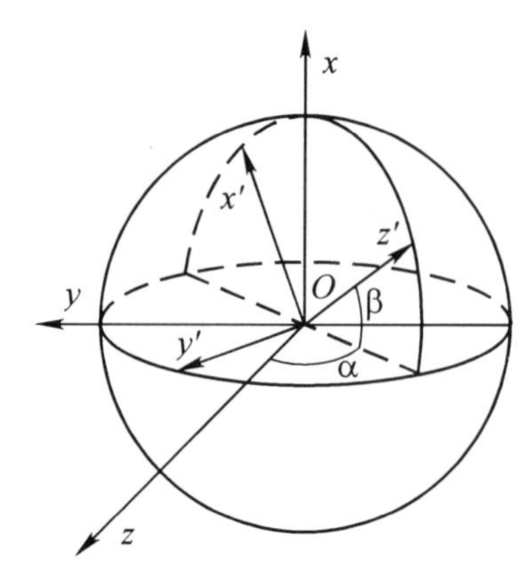

Figure 4.3.5. Krylov's angles α and β determining the position of the gyroscope's Oz'-axis with respect to a solid.

$$\begin{aligned}
u_3 &= v_1 \cos\alpha \tan\beta - v_2 \sin\alpha - v_3 \cos\alpha \sec\beta, \\
J_1\dot{\alpha} &= -(A + J_1)x_1 - (B + J_1)(x_3 \cos\alpha - x_2 \sin\alpha)\tan\beta \\
&\quad + K_1 + (K_2 \sin\alpha - K_3 \cos\alpha)\tan\beta, \\
J_1\dot{\beta} &= -(B + J_1)x_2 \cos\alpha - (C + J_1)x_3 \sin\alpha + K_2 \cos\alpha + K_3 \sin\alpha, \\
\dot{\varepsilon}_{i1} &= x_3\varepsilon_{i2} - x_2\varepsilon_{i3} \quad (i = 1,2,3) \quad (123), \\
K_i &= \sum_{k=1}^{3} h_k \varepsilon_{ki}, \quad \dot{h}_k = 0 \quad (i,k = 1,2,3).
\end{aligned} \tag{4.3.45}$$

In system (4.3.45), x_i $(i = 1,2,3)$ are the projections of the instantaneous angular velocity vector of the solid onto the $Oxyz$-axes; A, B, C are the moments of inertia of the solid with respect to the same axes; J_1 is the equatorial moment of inertia of the gyroscope; α, β are Krylov's angles determining the position of the gyroscope's axis relative to the solid (see Table 4.3.2 and Figure 4.3.5, where $Ox'y'z'$ is the system of coordinates rigidly fixed to the internal gimbal (housing): Oy' is the axis of the internal gimbal and Oz' is the rotor axis); ε_{ij} $(i,j = 1,2,3)$ are the directional cosines of the angles between the $Oxyz$-axes and the axes of the inertial system $OXYZ$ (see Table 4.3.3) related by six additional dependencies

$$\sum_{i=1}^{3} \varepsilon_{\delta i}\varepsilon_{\rho i} = \begin{cases} 1, & \delta = \rho \\ 0, & \delta \neq \rho \end{cases} \qquad (\delta, \rho = 1,2,3);$$

and K_i, h_i $(i = 1,2,3)$ are the projections of the angular momentum vector of the system onto the $Oxyz$-axes and $OXYZ$-axes, respectively.

Equations (4.3.45) were derived under two assumptions:

(1) $\cos\beta \neq 0$, i.e., the external and internal gimbals lie in different planes;

(2) the mass of the gimbal is neglected.

We note that, because of condition (2), the gyroscope affects the body only in the presence of controlling moments v_i $(i = 1,2,3)$. Otherwise $u_i = 0$ $(i = 1,2,3)$ and, consequently, the gyroscope in no way affects the solid.

Table 4.3.2. Krylov's Angles α, β

O	x'	y'	z'
x	$\cos\beta$	0	$\sin\beta$
y	$\sin\alpha\sin\beta$	$\cos\alpha$	$-\sin\alpha\cos\beta$
z	$-\cos\alpha\sin\beta$	$\sin\alpha$	$\cos\alpha\cos\beta$

Table 4.3.3. Table of Directional Cosines

O	x	y	z
X	ε_{11}	ε_{12}	ε_{13}
Y	ε_{21}	ε_{22}	ε_{23}
Z	ε_{31}	ε_{32}	ε_{33}

Equations (4.3.45) admit the particular solution

$$x_i = 0, \quad \varepsilon_{ij} = \begin{cases} 1, & i = j \\ 0, & i \neq j \end{cases}, \quad K_i = h_i^0, \quad v_i = 0 \tag{4.3.46}$$
$$\alpha = \alpha_0, \quad \beta = \beta_0, \quad \dot\gamma = \dot\gamma_0 \quad (\alpha_0, \beta_0, \dot\gamma_0, h_i^0 \text{ are constants}) \quad (i, j = 1, 2, 3)$$

corresponding to the equilibrium position of the solid in the inertial system of coordinates; $\dot\gamma_0$ is the velocity of the gyroscope's proper rotation in unperturbed motion.

Taking into account the expressions

$$\begin{aligned} K_1 &= (J_1\cos^2\beta + J_2\sin^2\beta)\dot\alpha + J_2\dot\gamma\sin\beta + (A + J_1\cos^2\beta + J_2\sin^2\beta)x_1 \\ &\quad + (J_1 - J_2)(x_2\sin\alpha - x_3\cos\alpha)\sin\beta\cos\beta, \\ K_2 &= (J_1 - J_2)\dot\alpha\sin\alpha\sin\beta\cos\beta + J_1\dot\beta\cos\alpha - J_2\dot\gamma\sin\alpha\sin\beta \\ &\quad + (J_1 - J_2)x_1\sin\alpha\sin\beta\cos\beta + \left[B + J_1 + (J_2 - J_1)\sin^2\alpha\cos^2\beta\right]x_2 \\ &\quad + (J_1 - J_2)x_3\sin\alpha\cos\alpha\cos^2\beta, \\ K_3 &= (J_2 - J_1)\dot\alpha\cos\alpha\sin\beta\cos\beta + J_1\dot\beta\sin\alpha + J_2\dot\gamma\cos\alpha\cos\beta \\ &\quad + (J_2 - J_1)x_1\cos\alpha\sin\beta\cos\beta + \left[C + J_1 + (J_2 - J_1)\cos^2\beta\cos^2\alpha\right]x_3 \\ &\quad + (J_1 - J_2)x_2\sin\alpha\cos\alpha\cos^2\beta, \end{aligned}$$

where J_2 is the gyroscope's axial moment of inertia, one can show that

$$\sin\alpha_0 = -\frac{h_2^0}{\sqrt{(h_2^0)^2 + (h_3^0)^2}}, \quad \cos\beta_0 = \frac{h_1^0}{\sqrt{(h_1^0)^2 + (h_2^0)^2 + (h_3^0)^2}}. \tag{4.3.47}$$

Further we restrict consideration to the case of $\alpha_0 = \beta_0 = 0$ (as the most common one in engineering), when, in an equilibrium position of the solid, the planes of the internal and external gimbals are at right angles to one another, and the axes of the external and internal gimbals and the rotor are directed along the respective principal central axes of inertia of the solid. In this case by (4.3.47), $h_1^0 = h_2^0 = 0$, $h_3^0 = J_2\dot{\gamma}_0$.

Let us consider the problem of stabilizing the equilibrium position (4.3.46) through controls v_i $(i = 1, 2, 3)$ generated by the engines. To this end, we introduce the new variables

$$\begin{gathered}\xi_i = x_i, \quad \xi_4 = \varepsilon_{11} - 1, \quad \xi_5 = \varepsilon_{12}, \quad \xi_6 = \varepsilon_{13}, \quad \xi_7 = \varepsilon_{21}, \\ \xi_8 = \varepsilon_{22} - 1, \quad \xi_9 = \varepsilon_{23}, \quad \xi_{10} = \varepsilon_{31}, \quad \xi_{11} = \varepsilon_{32}, \quad \xi_{12} = \varepsilon_{33} - 1, \\ \xi_{13} = \alpha, \quad \xi_{14} = \beta, \quad \xi_{14+i} = h_i - h_i^0 \quad (i = 1, 2, 3), \quad h_1^0 = h_2^0 = 0\end{gathered}$$

and compose the system of equations of perturbed motion

$$\begin{gathered}A\dot{\xi}_1 = (B - C)\xi_2\xi_3 + u_1 \stackrel{\text{def}}{=} R_1^*(\boldsymbol{\xi}) + u_1 \qquad (ABC, 123), \\ u_1 = -v_1, \quad u_2 = -v_1 \sin\xi_{13} \tan\xi_{14} - v_2 \cos\xi_{13} + v_3 \sin\xi_{13} \sec\xi_{14}, \\ u_3 = v_1 \cos\xi_{13} \tan\xi_{14} - v_2 \sin\xi_{13} - v_3 \cos\xi_{13} \sec\xi_{14}, \\ \dot{\xi}_{3+j} = R_{3+j}^*(\boldsymbol{\xi}) \quad (j = 1, \ldots, 9), \\ J_1\dot{\xi}_{13} = -(A + J_1)\xi_1 - (B + J_1)(\xi_2 \sin\xi_{13} - \xi_3 \cos\xi_{13}) \tan\xi_{14} + \xi_{15}(\xi_4 + 1) \\ + \xi_7\xi_{16} + (h_3^0 + \xi_{17})\xi_{10} + (\Lambda_1 \sin\xi_{13} - \Lambda_2 \cos\xi_{13}) \tan\xi_{14}, \\ J_1\dot{\xi}_{14} = -(B + J_1)\xi_2 \cos\xi_{13} - (A + J_1)\xi_3 \sin\xi_{13} + \Lambda_1 \cos\xi_{13} + \Lambda_2 \sin\xi_{13}, \\ \Lambda_1 = \xi_{16}(1 + \xi_8) + \xi_5\xi_{15} + (h_3^0 + \xi_{17})\xi_{11}, \\ \Lambda_2 = \xi_6\xi_{15} + \xi_9\xi_{16} + (h_3^0 + \xi_{17})(1 + \xi_{12}), \\ \dot{\xi}_{14+i} = 0 \qquad (i = 1, 2, 3),\end{gathered} \tag{4.3.48}$$

to which we should also add equalities (4.3.23).

Let us specify control rules u_i $(i = 1, 2, 3)$ in the form (4.3.24) but we take R_l^* $(l = 1, \ldots, 12)$ instead of R_l. Now we can demonstrate by analogy with the proof of Theorem 4.3.2 that the unperturbed motion $\xi_s = 0$ $(s = 1, \ldots, 17)$ of system (4.3.48) is asymptotically stable with respect to ξ_l $(l = 1, \ldots, 12)$.

To implement the specified rules u_i $(i = 1, 2, 3)$, the controlling engines should generate the torques

$$\begin{gathered}v_1 = -u_1, \quad v_2 = -u_2 \cos\xi_{13} - u_3 \sin\xi_{13}, \\ v_3 = -u_1 \sin\xi_{14} + u_2 \sin\xi_{13} \cos\xi_{14} + u_3 \cos\xi_{13} \cos\xi_{14}.\end{gathered} \tag{4.3.49}$$

Together with system (4.3.48), consider also the "*truncated*" *system*

$$\begin{aligned}J_1\dot{\xi}_{13} &= \xi_{15} + \left[\xi_{16} \sin\xi_{13} - (h_3^0 + \xi_{17}) \cos\xi_{13}\right] \tan\xi_{14}, \\ J_1\dot{\xi}_{14} &= \xi_{16} \cos\xi_{13} + (h_3^0 + \xi_{17}) \sin\xi_{13}, \\ \dot{\xi}_{14+i} &= 0 \qquad (i = 1, 2, 3),\end{aligned} \tag{4.3.50}$$

the first two equations of which are obtained from system (4.3.48), (4.3.24) (by replacing R_l with R_l^* in (4.3.24)) for $\xi_l = 0$ $(l = 1, \ldots, 12)$. System (4.3.50) describes the perturbed motion of an uncontrolled gyroscope on an immobile base.

The "rigidity" property of the axis of the gyroscope on an immobile base is well known. It can be formulated as follows: if perturbations of the angular momentum of the system are small (when the initial values of ξ_{14+i} $(i = 1, 2, 3)$ are sufficiently small), the zero solution $\xi_{13} = \xi_{14} = \xi_{14+i} = 0$ $(i = 1, 2, 3)$ is stable with respect to ξ_{13}, ξ_{14} for any finite (not necessarily large) value of $\dot{\gamma}_0$. But if the initial values of variables ξ_{14+i} $(i = 1, 2, 3)$ are arbitrary and finite, the stability with respect to ξ_{13}, ξ_{14} will be observed only for a sufficiently large value of $\dot{\gamma}_0$.

On the basis of the Lyapunov–Malkin "*reduction principle*" (in the Dykhman form [1950]), the similar conclusion on stability with respect to ξ_{13}, ξ_{14} is also valid for the unperturbed motion $\xi_s = 0$ $(s = 1, \dots, 17)$ of system (4.3.48), (4.3.24) (when R_l^* is substituted for R_l $(l = 1, \dots, 12)$ in (4.3.24)). (We note that, when applying the "reduction principle," we use ξ_{13}, ξ_{14} as the critical variables.)

Let us sum up the above reasoning in the following theorem.

Theorem 4.3.5. *Control rules* (4.3.49), *in which the set of control rules* u_i $(i = 1, 2, 3)$ *has the form* (4.3.24) (*when* R_l^* *is substituted for* R_l $(l = 1, \dots, 12)$), *for an appropriate value of* $\dot{\gamma}_0$ *is the set of stabilizing control rules ensuring the stability of the unperturbed motion* $\xi_s = 0$ $(s = 1, \dots, 17)$ *of system* (4.3.48) *with respect to* ξ_{13}, ξ_{14} *and its exponential asymptotic stability with respect to variables* ξ_l $(l = 1, \dots, 12)$.

Remark 4.3.6.

1. The stability of the unperturbed motion $\xi_s = 0$ $(s = 1, \dots, 17)$ of system (4.3.48) with respect to ξ_{13}, ξ_{14} justifies the initial assumption that $\cos\beta \neq 0$. Therefore, the stability with respect to ξ_{13}, ξ_{14} is the necessary condition for implementing the proposed control scheme. The specified property is provided by the fact that, in the equilibrium position of the solid, the gyroscope should have a nonzero (and even sufficiently large) angular velocity $\dot{\gamma}_0$ of its proper rotation. We note that this is not necessary when stabilizing the equilibrium position of the solid by flywheels, in which case the flywheels in the equilibrium position may even be at rest with respect to the solid.

2. If the mass of the gimbal is neglected (as was assumed earlier), then the proposed approach is applicable to stabilizing an arbitrary position $\alpha = \alpha_0$, $\beta = \beta_0$ $(\cos\beta_0 \neq 0)$, not only the position $\alpha_0 = \beta_0 = 0$. Otherwise (when the gimbal's mass is taken into account), we have the stability of the axis of the gyroscope on an immobile base only in the case $\beta_0 = 0$. If $\beta_0 \neq 0$ $(\cos\beta_0 \neq 0)$, then we have the stability of the gyroscope axis only *with respect to part of the variables*: the stability with respect to β and the instability with respect to α. This phenomenon, which was first discovered by Magnus [1955] followed by Plymale and Goodstein [1955], was termed the "*drift of the gyroscope's axis*" (see Figure 4.3.6 showing the evolution of the vertex of the gyroscope's axis).

Let us point out that the phenomenon of the "*drift of the gyroscope's axis*" does not contradict the principal property of the balanced gyroscope in accordance with which the axis of a rapidly rotating free gyroscope maintains its position in inertial space. This is explained by the fact that the specified principal property is valid for a single gyroscope whereas the gimballed gyroscope is a system of three solids (gyroscope, housing, and external gimbal) interacting in a complicated manner.

3. What is essential in the proposed scheme of gyroscopic stabilization is that the angular momentum of the gyroscope, not necessarily large in magnitude, *varies*. Another conventional scheme of gyroscopic stabilization uses a large *constant* angular momentum of the gyroscope, in which case the system of equations under study is much simplified.

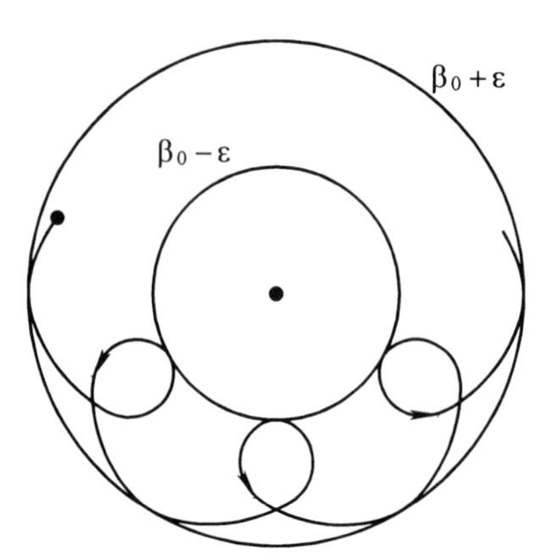

Figure 4.3.6. Evolution of the vertex of the axis of a gyroscope on an immobile base.

4. Following the technique of Subsection 4.2.4, one can choose the controls u_i^* ($i = 1, 2, 3$) in the set of stabilizing control rules u_i ($i = 1, 2, 3$) to minimize functional (4.2.20) with integrand $R(\boldsymbol{\xi}, \boldsymbol{u})$ of the kind (4.3.35).

For a more detailed analysis of the functional structure, we transform from variables ε_{ij} to three independent *Krylov's angles* (Lur'e [1961]): φ (*yaw*), ψ (*trim*), and θ (*tilt*), taking the OX_2- and Ox_3-axes as fundamental. Then

$$\xi_5 = \beta_{12} = -\varphi + \dots, \quad \xi_6 = \beta_{13} = \psi + \dots, \quad \xi_7 = \beta_{21} = \varphi + \dots,$$
$$\xi_9 = \beta_{23} = -\theta + \dots, \quad \xi_{10} = \beta_{31} = -\psi + \dots, \quad \xi_{11} = \beta_{32} = \theta + \dots,$$
$$\mu_1 = q_3 + \dots, \quad \mu_2 = -q_2 + \dots, \quad \mu_3 = -q_1 + \dots,$$
$$A_3 u_1^* = -u_3 + \dots, \quad A_1 u_2^* = u_1 + \dots, \quad A_2 u_3^* = u_2 + \dots,$$

where ellipsis dots stand for the terms *higher than the first* order in $\xi_i, \varphi, \psi, \theta, \xi_{13}$, ξ_{14}, u_i ($i = 1, 2, 3$). Therefore, in this case the main (*quadratic*) part R_2 of function $R(\boldsymbol{\xi}, \boldsymbol{u})$ has the form

$$\begin{aligned} R_2(\boldsymbol{\xi}, \boldsymbol{u}) &= a_1\varphi^2 + a_2\theta^2 + a_3\psi^2 + b_1\xi_3^2 + b_2\xi_2^2 + b_3\xi_1^2 \\ &\quad + (c_2/A^2)u_1^2 + (c_3/B^2)u_2^2 + (c_1/C^2)u_3^2. \end{aligned}$$

Setting $b_1 = C$, $b_2 = B$, $b_3 = A$, we deduce that R_2 is the sum of the following expressions:
(1) the kinetic energy of the solid $A\xi_1^2 + B\xi_2^2 + C\xi_3^2$;
(2) the quadratic form $a_1\varphi^2 + a_2\theta^2 + a_3\psi^2$ characterizing the deviation of the solid from the equilibrium position $\varphi = \theta = \psi = 0$;
(3) the quadratic form $(c_2/A^2)u_1^2 + (c_3/B^2)u_2^2 + (c_1/C^2)u_3^2$ characterizing the resources expended in generating the controls.

4.3.7 Conclusions

1. From the viewpoint of this book's objectives, we note that the examples considered in this Section vividly demonstrate the situations giving rise to problems of partial stabilization. Summing up, we list the cases where such situations are encountered:

(*a*) In the *preliminary stages* of investigating the problems of stabilization with respect to all or to a given (major) part of the variables and the problems of polystabilization.

(*b*) When *reducing the number* of controlling actions, which results in the *impossible* stabilization with respect to all the variables. This situation can be treated in another way: partial stabilization is *sufficient* for the object to *operate properly* and *fewer* (as compared to the stabilization with respect to all the variables) controlling actions are required.

(*c*) When stabilizing equilibrium positions of a solid by rotating masses (or some other *specially arranged* relative motion).

(*d*) When solving the problems of stabilization with respect to all the variables presents purely *mathematical difficulties.*

(*e*) When in the course of stabilization with respect to all the variables only the quality of the transient is important for some given part of the variables.

The list above can be extended by the situations arising when the partial stabilization *subject to direct restrictions* on controls is carried out (see Section 4.5 regarding this issue).

2. The examples cited also indicate that, in systems synthesis, the problems of stabilization with respect to all and to part of the variables are *closely interrelated* and also *complement each other.*

4.4 Reorientation of an Asymmetric Solid and Coordinated Control of a System of Solids (Manipulator Model)

4.4.1 Preliminary Remarks

In this section the method of constructing auxiliary controlled μ-systems is further developed as applied to the problems of *reorienting an asymmetric solid* and *coordinating a system of solids.* The main peculiarity here lies in considering the *global behavior* of the systems under study: reorienting a solid (or coordinating a system of solids) into a given position from an *arbitrary* initial state. Similar possibilities have been stated in general (see Remark 4.3.2) earlier. A more detailed and specific analysis of these possibilitiés is given below applied to the following problems:

(1) *uniaxial reorientation* of an asymmetric solid;

(2) *triaxial reorientation* of an asymmetric solid;

(3) *coordination of a system of solids* having a "tree" structure.

The problems indicated can be treated as those of *stabilization "on the whole"* (*with respect to all the variables*) of given equilibrium positions of a solid or a system of solids. In the *first stage* of the study the corresponding problem of *partial stabilization* (with respect to some initially chosen part of the variables) is solved. The auxiliary function of the problem of partial stabilization thus is demonstrated in studying the global behavior of the class of mechanical systems important from the viewpoint of applications.

4.4.2 Uniaxial Reorientation of an Asymmetric Solid

Consider *Euler's dynamical equations*

$$A\dot{x}_1 = (B - C)x_2x_3 + u_1,$$
$$B\dot{x}_2 = (C - A)x_1x_3 + u_2, \quad C\dot{x}_3 = (A - B)x_1x_2 + u_3. \tag{4.4.1}$$

describing the angular motion of a solid with respect to its center of mass O under the influence of controlling moments u_i $(i = 1, 2, 3)$.

Let in the inertial space $OXYZ$ there be given an orientation of a unit vector $\boldsymbol{s}$; its projections onto principal central axes of inertia of the solid $Oxyz$ are s_i $(i = 1, 2, 3)$. In spacecraft attitude control, the values s_i $(i = 1, 2, 3)$ are often understood as directions toward some physical reference points, for example, the Sun, planets, and other objects. Since vector $\boldsymbol{s}$ is in rotational motion with respect to the system $Oxyz$ with angular velocity $-\boldsymbol{\omega}$, then $\dot{\boldsymbol{s}} = -\boldsymbol{\omega}\times\boldsymbol{s}$, where $\boldsymbol{\omega}$ is the angular velocity vector of the solid. The projections to the axes of system $Oxyz$ are given by the Poisson equations

$$\dot{s}_1 = s_2x_3 - s_3x_2, \quad \dot{s}_2 = s_3x_1 - s_1x_3, \quad \dot{s}_3 = s_1x_2 - s_2x_1, \tag{4.4.2}$$
$$s_1^2 + s_2^2 + s_3^2 = 1. \tag{4.4.3}$$

The control rules u_i $(i = 1, 2, 3)$ must be chosen so that a vector $\boldsymbol{r}$ rigidly attached to the solid is aligned with vector $\boldsymbol{s}$ and the angular velocity of the body is suppressed. For simplicity, without loss of generality, vector $\boldsymbol{r}$ is assumed to be directed along one of the principal central axes of inertia of the body, let it be the Oy-axis.

We further consider two versions of this problem for which

(1) $\boldsymbol{r} \to \pm\boldsymbol{s}, \quad \boldsymbol{\omega} \to \boldsymbol{0}$;

(2) $\boldsymbol{r} \to \boldsymbol{s}, \quad \boldsymbol{\omega} \to \boldsymbol{0}$.

Case (2) is more important, of course, but case (1) is also of interest in applications. Besides, the technique for solving the problem in case (1) serves as the basis for solving that in case (2).

1. ***Uniaxial reorientation of type*** $\boldsymbol{r} \to \pm\boldsymbol{s}$, $\boldsymbol{\omega} \to \boldsymbol{0}$. Consider the set of control rules

$$u_1 = \frac{A}{s_2}f_1^0(\boldsymbol{x}, \boldsymbol{s}) + \frac{As_1u_2}{Bs_2} - \frac{Au_1^*}{s_2}, \quad u_2 = B\big[-\psi_2(\boldsymbol{x}) + u_2^*\big],$$
$$u_3 = \frac{C}{s_2}f_2^0(\boldsymbol{x}, \boldsymbol{s}) + \frac{Cs_3u_2}{Bs_2} + \frac{Cu_3^*}{s_2}, \tag{4.4.4}$$

where

$$f_1^0 = \psi_2s_1 - \psi_1s_2 + \varphi_1x_2 - \varphi_2x_1,$$
$$f_2^0 = \psi_2s_3 - \psi_3s_2 + \varphi_3x_2 - \varphi_2x_3,$$
$$\varphi_1 = s_2x_3 - s_3x_2 \quad (123), \qquad A\psi_1 = (B - C)x_2x_3 \quad (ABC, 123).$$

Parameters u_i^* in (4.4.4) are the controllable ones, which will be specified in what follows based on the requirements for control quality.

The structure of controls (4.4.4) is such that from the closed system (4.4.1)–(4.4.4) one can segregate the auxiliary *linear controlled* $\boldsymbol{\mu}$*-system*

$$\begin{gathered}\dot{s}_1 = \mu_1, \quad \dot{\mu}_1 = u_3^*; \quad \dot{x}_2 = u_2^*; \quad \dot{s}_3 = \mu_2, \quad \dot{\mu}_2 = u_1^* \\ (\mu_1 = s_2 x_3 - s_3 x_2, \quad \mu_2 = s_1 x_2 - s_2 x_1).\end{gathered} \tag{4.4.5}$$

Theorem 4.4.1. 1. *If $s_2^2(t_0) \geq \varepsilon^2 \neq 0$, then, for any ε, the constants $\Gamma_1, \ldots, \Gamma_5$ in the expressions*

$$\begin{gathered}u_1^* = \Gamma_1 s_3 + \Gamma_2 \mu_2, \quad u_2^* = \Gamma_3 s_2, \quad u_3^* = \Gamma_4 s_1 + \Gamma_5 \mu_1, \\ \Gamma_k = \text{const} < 0 \ \ (k = 1, \ldots, 5), \quad 4\Gamma_v + \Gamma_{v+1}^2 > 0 \ \ (v = 1, 3)\end{gathered} \tag{4.4.6}$$

can be chosen so that, under the action of the controlling moments (4.4.4), (4.4.6) for any motion of the solid, its angular velocity is exponentially asymptotically dampened and vector $\boldsymbol{r}$ exponentially asymptotically approaches the direction collinear with that of vector $\boldsymbol{s}$. In the case of $s_2(t_0) \geq \varepsilon > 0$, vector $\boldsymbol{r}$ exponentially asymptotically approaches the direction coinciding with that of vector $\boldsymbol{s}$ and, in the case of $s_2(t_0) \leq -\varepsilon < 0$, it approaches the direction opposite that of vector $\boldsymbol{s}$.

2. *The equilibrium positions*

$$\boldsymbol{\omega} = \mathbf{0}, \quad \boldsymbol{r} = \pm \boldsymbol{s} \quad (s_1 = s_3 = 0, \ s_2 = \pm 1), \tag{4.4.7}$$

in which the solid is oriented, are exponentially asymptotically stable with respect to x_i, s_i $(i = 1, 2, 3)$.

Proof. Variables $s_1, s_3, \mu_1, \mu_2, x_2$ of the closed system (4.4.1)–(4.4.4), (4.4.6) are determined in solving the auxiliary $\boldsymbol{\mu}$-system (4.4.5), (4.4.6). Thus, for all the motions of the solid the following relationships are valid:

$$x_2(t) \to 0, \quad \mu_w(t) \to 0 \ \ (w = 1, 2), \quad s_v(t) \to 0 \ \ (v = 1, 3), \quad t \to \infty; \tag{4.4.8}$$

zero is approached in (4.4.8) by an exponential law.

From equations (4.4.2), we derive the equalities

$$x_1 = \frac{-\mu_2 + x_2 s_1}{s_2}, \quad x_3 = \frac{\mu_1 + x_2 s_3}{s_2}. \tag{4.4.9}$$

By (4.4.9) we conclude that, if for some α the condition

$$s_2^2(t) \geq \alpha^2 = \text{const} \neq 0 \tag{4.4.10}$$

holds for all $t \geq t_0$, then variables x_1, x_3 also approach zero by an exponential law.

By virtue of (4.4.3), for all $t \geq t_0$, condition (4.4.10) is equivalent to

$$s_1^2(t) + s_3^2(t) \leq 1 - \alpha^2. \tag{4.4.11}$$

Directly integrating $\boldsymbol{\mu}$-system (4.4.5), (4.4.6), we deduce that in the case of $s_2^2(t_0) \geq \varepsilon^2 \neq 0$ for any fixed initial values of variables x_i, s_i $(i = 1, 2, 3)$, one can choose constants α and Γ_k $(k = 1, \ldots, 5)$ so that, for any fixed ε, condition (4.4.11) will hold for all $t \geq t_0$.

Because condition (4.4.10) is well-founded, along with (4.4.8) for all the motions of the solid (for which $s_2^2(t_0) \geq \varepsilon^2 \neq 0$),

$$x_v(t) \to 0, \ \ t \to \infty \quad (v = 1, 3). \tag{4.4.12}$$

Solving equation (4.4.3) for s_2, for all $t \geq t_0$, we obtain the equalities

$$s_2(t) = \pm f(t), \quad f = \sqrt{1 - s_1^2 - s_3^2}. \tag{4.4.13}$$

Bearing in mind condition (4.4.11), from (4.4.13), we deduce that

$$s_2(t) = \begin{cases} f(t), & s_2(t_0) \geq \varepsilon > 0 \\ -f(t), & s_2(t_0) \leq -\varepsilon < 0. \end{cases} \tag{4.4.14}$$

Relationships (4.4.14) can be substantiated *by contradiction*. For example, let $s_2(t) = f(t)$ for $s_2(t_0) \leq -\varepsilon < 0$. Then $s_2(t) \to 1$, $t \to \infty$ by (4.4.8). As a result, one can specify a closed interval of time $[t_0, t_1] \in [t_0, \infty)$, at the end points of which the continuous function $s_2(t)$ assumes values different in sign. Then by the Bolzano–Cauchy theorem, well-known in analysis, there exists a moment $t^* \in [t_0, t_1]$ such that $s_2(t^*) = 0$. By (4.4.3), $s_1^2(t^*) + s_3^2(t^*) = 1$, which contradicts condition (4.4.11).

Allowing for relationships (4.4.8), (4.4.12), and (4.4.14), we conclude that the *first part* of proposition of Theorem 4.4.1 is proved.

The *second part* of proposition of Theorem 4.4.1 can be proved by following the scheme suggested in Subsection 4.3.2. The theorem is proved. □

Remark 4.4.1.

1. The case of small ε in condition $s_2^2(t_0) = \varepsilon^2$ means that, initially, vector $\boldsymbol{r}$ practically makes an angle of $\pi/2$ with the desired direction of orientation. This case and that, when the initial angular velocity of the solid is sufficiently large, characterize the limiting capabilities of control rules (4.4.4), (4.4.6).

There are two ways of avoiding the use of the proposed control rules in "limiting modes."

(1) By preliminary damping the angular velocity of the solid. This problem is solved solely based on Euler's dynamical equations and is not considered here.

(2) By successively using two constructions of control rules of type (4.4.4), (4.4.6) none of which enters the "limiting mode" of operation (see Theorem 4.4.2).

2. Constructions of control rules of type (4.4.4), (4.4.6) possess the property of *robustness* (*stability*) with respect to small errors and perturbations.

3. By using the proposed approach, the solutions of the closed system (4.4.1)–(4.4.4), (4.4.6) are obtained in *explicit form*. This means that control rules (4.4.4), (4.4.6), designed in accordance with the *feedback principle*, can also be represented as control *programs* $u_i = u_i(t)$ $(i = 1, 2, 3)$ which are continuous functions of t for given initial values $x_i(t_0)$, $s_i(t_0)$.

2. *Uniaxial reorientation of type* $\boldsymbol{r} \to \boldsymbol{s}$, $\boldsymbol{\omega} \to \boldsymbol{0}$. Along with the set of control rules (4.4.4), also consider the following two sets of control rules

$$(1) \qquad u_1 = \frac{A}{s_3} f_1^{\vee}(\boldsymbol{x}, \boldsymbol{s}) + \frac{A s_1 u_3}{C s_3} + \frac{A u_1^*}{s_3},$$
$$u_2 = \frac{B}{s_3} f_2^{\vee}(\boldsymbol{x}, \boldsymbol{s}) + \frac{B s_2 u_3}{C s_3} - \frac{B u_2^*}{s_3}, \quad u_3 = C\big[-\psi_3(\boldsymbol{x}) + u_3^*\big], \tag{4.4.15}$$

where

$$f_1^{\vee} = -\psi_1 s_3 + \psi_3 s_1 + \varphi_1 x_3 - \varphi_3 x_1,$$
$$f_2^{\vee} = -\psi_2 s_3 + \psi_3 s_2 + \varphi_2 x_3 - \varphi_3 x_2;$$

and

$$(2) \quad u_1 = A\big[-\psi_1(\boldsymbol{x}) + u_1^*\big], \quad u_2 = \frac{B}{s_1} f_1^{\wedge}(\boldsymbol{x}, \boldsymbol{s}) + \frac{B s_2 u_1}{A s_1} + \frac{B u_2^*}{s_1},$$
$$u_3 = \frac{C}{s_1} f_2^{\wedge}(\boldsymbol{x}, \boldsymbol{s}) + \frac{C s_3 u_1}{A s_1} - \frac{C u_3^*}{s_1}, \tag{4.4.16}$$

where

$$f_1^{\wedge} = \psi_1 s_2 - \psi_2 s_1 + \varphi_2 x_1 - \varphi_1 x_2,$$
$$f_2^{\wedge} = \psi_3 s_1 - \psi_1 s_3 + \varphi_1 x_3 - \varphi_3 x_1.$$

From the closed system (4.4.1)–(4.4.3), (4.4.15) one can segregate the *linear controlled* $\boldsymbol{\mu}$*-system*

$$\dot{s}_1 = \mu_1, \ \dot{\mu}_1 = u_2^*; \quad \dot{s}_2 = \mu_3, \ \dot{\mu}_3 = u_1^*; \quad \dot{x}_3 = u_3^*$$

and, from the closed system (4.4.1)–(4.4.3), (4.4.16), the *linear controlled* $\boldsymbol{\mu}$*-system*

$$\dot{s}_2 = \mu_3, \ \dot{\mu}_3 = u_3^*; \quad \dot{s}_3 = \mu_2, \ \dot{\mu}_2 = u_2^*; \quad \dot{x}_1 = u_1^*.$$

Theorem 4.4.2. *Let $s_2^2(t_0) \geq \varepsilon^2 \neq 0$. For any ε, control rules u_i $(i = 1, 2, 3)$ can be chosen so that vector $\boldsymbol{r}$ exponentially asymptotically approaches the direction coinciding with that of vector $\boldsymbol{s}$. In the case of $s_2(t_0) \geq \varepsilon > 0$, these are control rules of type* (4.4.4), (4.4.6). *In the case of $s_2(t_0) \leq -\varepsilon < 0$, it is necessary to successively use two constructions of control rules:*

(1) *first, control rules* (4.4.15) *or* (4.4.16) *to bring vector $\boldsymbol{r}$ to the position $s_2(t_0) \geq \varepsilon > 0$ (in a finite time);*

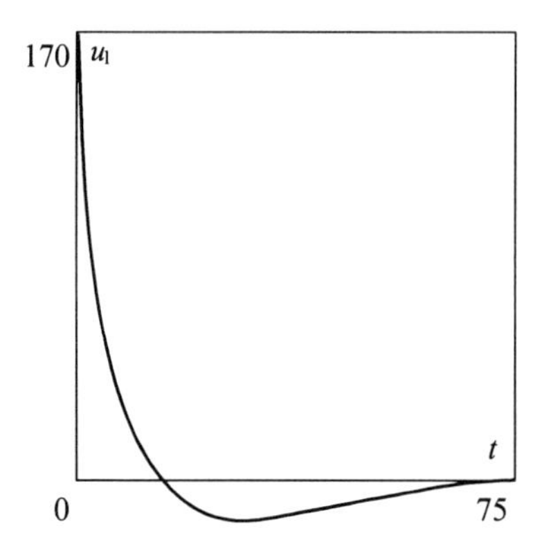

Figure 4.4.1. Control law u_1.

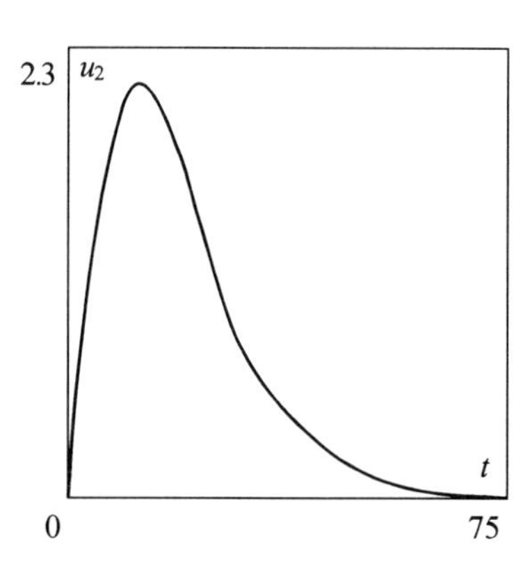

Figure 4.4.2. Control law u_2.

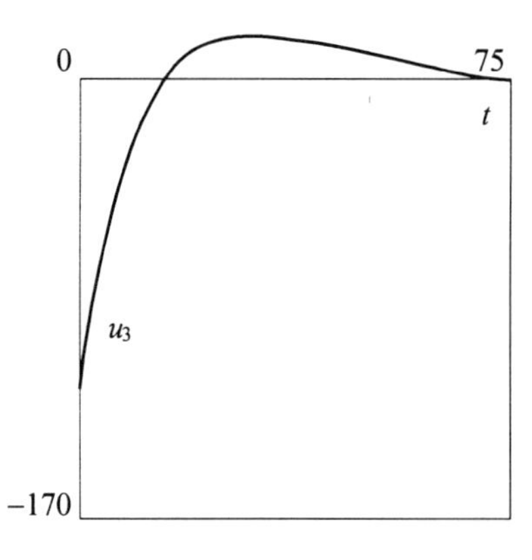

Figure 4.4.3. Control law u_3.

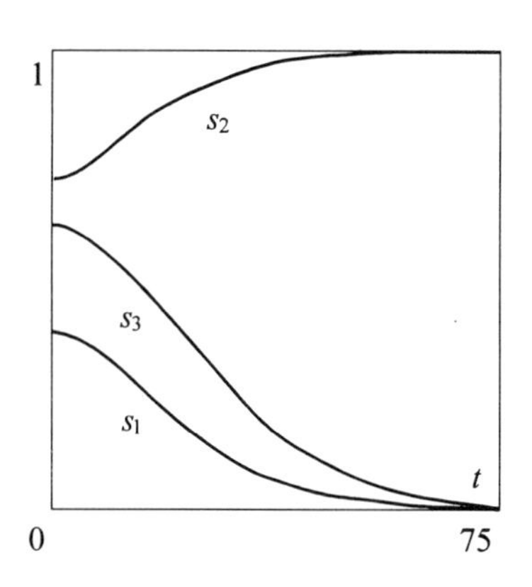

Figure 4.4.4. Functions s_i.

(2) *then, control rules* (4.4.4), (4.4.6).

The proof scheme follows that for Theorem 4.4.1.

Example 4.4.1. Suppose that, initially, a solid (spacecraft) with $A = 4 \cdot 10^4$, $B = 8 \cdot 10^4$, and $C = 5 \cdot 10^4$ $kg{\cdot}m^2$ is in the equilibrium position $x_i(t_0) = 0$ and $s_1(t_0) = 0.4$, $s_3(t_0) = 0.6$ *rad*.

Consider the problem of reorienting the solid from the specified equilibrium position into the final equilibrium position $x_i(t_0) = 0$, $s_1 = s_3 = 0$ $(s_2 = 1)$. Let us take the original constraints on the controls in the form $|u_i| \leq 170$ $N{\cdot}m$.

The computations show that, by constructing control rules (4.4.4)–(4.4.6), these constraints will be satisfied, for example, for $\Gamma_1 = \Gamma_3 = -4.907 \cdot 10^{-3}$ s^{-2}, $\Gamma_2 = \Gamma_4 = -1.410 \cdot 10^{-1}$ s^{-1}.

Functions u_i, s_i (with t as an argument) are shown in Figures 4.4.1–4.4.4. We point out that u_1 alone attains the boundary value, whereas $|u_2| \leq 2.3$ and $|u_3| \leq 141.7$ $N{\cdot}m$. It is apparent that control u_2 performs just the *corrective* function, i.e., compensates for one of the components of the *vector of the solid's gyroscopic moment.*

To assess the performance of the control rules u_i used, we note that a tenfold decrease in $s_j(t_0)$ $(j = 1, 3)$ is achieved in 56 s, a hundredfold in 96 s, and a thousandfold in 133 s.

4.4.3 The Triaxial Reorientation of a Solid

In the inertial space $OXYZ$ let the orientation of three mutually perpendicular unit vectors $\boldsymbol{s}^{(i)}$ $(i = 1, 2, 3)$ be given; their projections onto principal central axes of inertia of the solid $Oxyz$ are s_{ji} $(j = 1, 2, 3)$.

The control rules u_i $(i = 1, 2, 3)$ must be chosen so that three mutually perpendicular vectors $\boldsymbol{r}_i$ $(i = 1, 2, 3)$ rigidly attached to the solid are respectively aligned with vectors $\boldsymbol{s}^{(i)}$ and the angular velocity of the body is suppressed. For simplicity, without loss of generality, vectors $\boldsymbol{r}_i$ are assumed to be directed along the body's principal central axes of inertia.

The solution of this problem essentially depends on the form of kinematic equations determining the solid's orientation. For example, one can use three types of kinematic equations involving the following quantities as their variables:
(1) the components of the *directional cosines matrix*;
(2) the components of *Euler's quaternion* (the *Rodrigues–Hamilton variables*);
(3) *Euler's angles.*

When choosing a specific type of equations, we shall take the following circumstance into account. As a rule, the components of the angular velocity of a body are measured by rate gyroscopes. In this case the variables determining the attitude of the body can be found by integrating the kinematic equations on a real-time computer. In this respect the *algebraic* form of kinematic equations of type (1) and (2) is preferable compared with the *trigonometric* form of equations of type (3). Thus we proceed with solving the problem of triaxial reorientation based on Euler's dynamical equations and kinematic equations of type (1) and (2).

1. *The use of kinematic equations of type* (1). These equations have the form

$$\dot{\boldsymbol{s}}^{(i)} = -\boldsymbol{\omega} \times \boldsymbol{s}^{(i)} \qquad (i = 1, 2, 3)$$

or, in the coordinate representation,

$$\dot{s}_{1i} = s_{2i}x_3 - s_{3i}x_2 \quad (123) \quad (i = 1, 2, 3). \tag{4.4.17}$$

The variables appearing in nine equations (4.4.17) are related by six dependencies

$$\begin{gathered}\sum_{j=1}^{3} s_{ij}^2 = 1 \qquad (i = 1, 2, 3), \\ \sum_{i=1}^{3} s_{i1}s_{i2} = \sum_{i=1}^{3} s_{i1}s_{i3} = \sum_{i=1}^{3} s_{i2}s_{i3} = 0.\end{gathered} \tag{4.4.18}$$

Consider the set of control rules

$$u_1 = \frac{A}{s_{22}}\Phi_1(\boldsymbol{x}, \boldsymbol{s}) + \frac{As_{12}u_2}{Bs_{22}} - \frac{Au_2^*}{s_{22}}, \quad u_2 = -\frac{Bs_{22}\left[\Phi_2(\boldsymbol{x}, \boldsymbol{s}) + u_3^*\right]}{s_{33}s_{22} - s_{23}s_{32}},$$
$$u_3 = \frac{C}{s_{22}}\Phi_3(\boldsymbol{x}, \boldsymbol{s}) + \frac{Cs_{32}u_2}{Bs_{22}} + \frac{Cu_3^*}{s_{22}}, \tag{4.4.19}$$

where

$$\Phi_1 = s_{12}\psi_2 - s_{22}\psi_1 + \varphi_{12}x_2 - \varphi_{22}x_1,$$
$$\Phi_2 = s_{33}\psi_2 - s_{23}\psi_3 + \varphi_{33}x_2 - \varphi_{23}x_3 - \frac{s_{23}\Phi_3}{s_{22}},$$
$$\Phi_3 = s_{32}\psi_2 - s_{22}\psi_3 + \varphi_{32}x_2 - \varphi_{22}x_3,$$
$$\varphi_{1i} = s_{2i}x_3 - s_{3i}x_2 \ (123), (i = 1, 2, 3); \quad A\psi_1 = (B - C)x_2x_3 \ (ABC, 123).$$

u_i^* $(i = 1, 2, 3)$ in (4.4.19) are the controllable parameters, which will be specified in what follows.

Theorem 4.4.3. 1. *If* $s_{22}^2(t_0) \geq \varepsilon^2 > 0$, $s_{12}^2(t_0) + s_{13}^2(t_0) < 1 - \varepsilon^2$, *then for any* ε *the constants* L_l $(l = 1, \ldots, 6)$ *in the expressions*

$$u_1^* = L_1s_{12} + L_2\varphi_{12}, \quad u_2^* = L_3s_{32} + L_4\varphi_{32}, \quad u_3^* = L_5s_{13} + L_6\varphi_{13},$$
$$L_l = \text{const} < 0 \ \ (l = 1, \ldots, 6), \quad 4L_f + L_{f+1}^2 > 0 \ \ (f = 1, 3, 5) \tag{4.4.20}$$

can be chosen so that, under the action of the controlling moments (4.4.19), (4.4.20), *for any motion of the solid, its angular velocity is exponentially asymptotically dampened and vectors* $\boldsymbol{r}_i$ $(i = 1, 2, 3)$ *characterizing the directions of the principal central axes of inertia* Ox, Oy, Oz *of the solid exponentially asymptotically approach directions collinear with those of vectors* $\boldsymbol{s}^{(1)}$, $\boldsymbol{s}^{(2)}$, *and* $\boldsymbol{s}^{(3)}$, *respectively.*

2. *The equilibrium positions*

$$\boldsymbol{\omega} = \boldsymbol{0}, \quad s_{ij} = \begin{cases} \pm 1, & i = j, \\ 0, & i \neq j \end{cases} \quad (\boldsymbol{r}_i = \pm \boldsymbol{s}^{(i)}), \tag{4.4.21}$$

in which the solid is oriented, are exponentially asymptotically stable with respect to x_i, s_{ij} $(i, j = 1, 2, 3)$.

Proof. It can be verified that, from the closed system (4.4.1), (4.4.17)–(4.4.20), one can segregate the auxiliary *linear* $\boldsymbol{\mu}$*-system*

$$\dot{s}_{12} = \mu_1, \quad \dot{\mu}_1 = L_1s_{12} + L_2\mu_1;$$
$$\dot{s}_{32} = \mu_3, \quad \dot{\mu}_3 = L_5s_{32} + L_6\mu_3; \quad \dot{s}_{13} = \mu_2, \quad \dot{\mu}_2 = L_3s_{13} + L_4\mu_2 \tag{4.4.22}$$
$$(\mu_1 = \varphi_{12}, \quad \mu_2 = \varphi_{32}, \quad \mu_3 = \varphi_{13}).$$

Therefore, for all the motions of the solid, variables s_{12}, s_{13}, s_{32}, μ_i $(i = 1, 2, 3)$ of the closed system (4.4.1), (4.4.17)–(4.4.20) satisfy the relationships

$$s_{12}(t) \to 0, \ \ s_{13}(t) \to 0, \ \ s_{32}(t) \to 0,$$
$$\mu_i(t) \to 0 \ \ (i = 1, 2, 3), \quad t \to \infty, \tag{4.4.23}$$

where zero is approached in (4.4.23) by an exponential law.

Again, as in Subsection 4.3.4, using relationships (4.4.18) and equations for $\dot{s}_{12}$, $\dot{s}_{13}$, $\dot{s}_{32}$ from (4.4.17), one can express variables s_{ij}, x_i $(i, j = 1, 2, 3)$ (except for s_{12}, s_{13}, s_{32}) in terms of s_{12}, s_{13}, s_{32}, μ_i $(i = 1, 2, 3)$. The corresponding expressions will be called expressions of type (A). These expressions will be nonsingular for all $t \geq t_0$ if, for some positive number α, the inequalities

$$\begin{gathered} F_i \geq \alpha \quad (i = 1, 2, 3), \\ F_1 = \left|1 - s_{12}^2\right|, \quad F_2 = \left|s_{22}\right|, \quad F_3 = \left|s_{22}s_{33} - s_{23}s_{32}\right|. \end{gathered} \tag{4.4.24}$$

are valid for all $t \geq t_0$.

In the case of (4.4.24), bearing in mind (4.4.23), by expressions of type (A), we conclude that the relationships

$$x_i(t) \to 0, \quad s_{ij}(t) \to \begin{cases} \pm 1, & i = j \\ 0, & i \neq j \end{cases} \quad (i, j = 1, 2, 3)$$

are valid that guarantee fulfilling the first part of the theorem conditions.

Let us establish the validity of inequalities (4.4.24).

Directly integrating $\boldsymbol{\mu}$-system (4.4.22), we conclude that, under conditions of the theorem for any fixed initial values of variables x_i, s_{ij} $(i, j = 1, 2, 3)$, one can choose constants α and L_l $(l = 1, \ldots, 6)$ so that for all $t \geq t_0$ the inequalities $F_1 \geq \alpha, F_2 \geq \alpha$ will be satisfied for any fixed ε.

Now we turn to the inequality $F_3 \geq \alpha$. First we note that if $F_1 \geq \alpha, F_2 \geq \alpha$ are valid, the expressions of type (A) yield $F_3 = 1$ as $t \to \infty$. Therefore, the assumption (of the opposite) that the inequality $F_3 \geq \alpha$ fails to hold implies that there exists a *finite* time $t = t^*$ at which $F_3(t^*) = 0$.

Following this assumption, we set $t = t^*$ and solve the system

$$s_{22}s_{33} - s_{23}s_{32} = 0, \quad s_{13}^2 + s_{23}^2 + s_{33}^2 = 1$$

for s_{23}, s_{33}. Obtaining the solutions

$$s_{33}^2 = \frac{s_{32}^2(1 - s_{13}^2)}{1 - s_{12}^2}, \quad s_{23}^2 = \frac{(1 - s_{13}^2)s_{22}^2}{1 - s_{12}^2},$$

then we substitute in the equality $s_{23}s_{22} = -s_{12}s_{13} - s_{32}s_{33}$ that follows from the expressions of type (A) when $F_2 \geq \alpha$. As a result, for $t = t^*$,

$$\left(\pm\sqrt{\frac{1 - s_{13}^2}{1 - s_{12}^2}\, s_{22}^2}\right) s_{22} = -s_{12}s_{13} - s_{32}\left(\pm\sqrt{\frac{1 - s_{13}^2}{1 - s_{12}^2}\, s_{32}^2}\right).$$

After transformations,

$$\left(\pm\sqrt{\frac{1 - s_{13}^2}{1 - s_{12}^2}}\right)(s_{22}^2 + s_{32}^2) = -s_{12}s_{13}, \quad s_{22}^2 + s_{32}^2 = 1 - s_{12}^2,$$

and because $a\sqrt{a^2} = a^2$, we obtain the equality

$$\sqrt{(1 - s_{13}^2)(1 - s_{12}^2)} = \pm s_{12} s_{13}. \tag{4.4.25}$$

Squaring both sides of equality (4.4.25), we arrive at the relationship

$$1 - s_{13}^2(t^*) - s_{12}^2(t^*) = 0$$

which contradicts the inequality $F_1 \geq \alpha$, $t \geq t_0$.

This means that the assumption we have made that the inequality $F_3 \geq \alpha$, $t \geq t_0$ does not hold is wrong. Inequalities (4.4.24) are justified.

The proof scheme of the second part of the theorem follows that for Theorem 4.3.2. The theorem is proved. □

Let us introduce logical symbols $\wedge$ and $\vee$ denoting "and" and "or," respectively.

Corollary 4.4.1. *In each of the following cases,*

$$\begin{aligned}
&(1)\ \ (s_{22}^2(t_0) \geq \varepsilon^2 \neq 0) \vee (s_{31}^2(t_0) + s_{32}^2(t_0) < 1 - \varepsilon^2),\\
&(2)\ \ (s_{11}^2(t_0) \geq \varepsilon^2 \neq 0) \wedge (s_{31}^2(t_0) + s_{32}^2(t_0) < 1 - \varepsilon^2\\
&\qquad\qquad\qquad \vee\, s_{21}^2(t_0) + s_{23}^2(t_0) < 1 - \varepsilon^2),\\
&(3)\ \ (s_{33}^2(t_0) \geq \varepsilon^2 \neq 0) \wedge (s_{21}^2(t_0) + s_{23}^2(t_0) < 1 - \varepsilon^2\\
&\qquad\qquad\qquad \vee\, s_{12}^2(t_0) + s_{13}^2(t_0) < 1 - \varepsilon^2),
\end{aligned}$$

control rules solving the problem of triaxially orienting a solid are derived from (4.4.19), (4.4.20) *by permutations of indices and coefficients. For any motion of the solid, its angular velocity is exponentially asymptotically dampened and vectors $\boldsymbol{r}_i$ $(i = 1, 2, 3)$ characterizing the directions of the principal central axes of inertia Ox, Oy, Oz of the solid exponentially asymptotically approach directions collinear with those of vectors $\boldsymbol{s}^{(1)}$, $\boldsymbol{s}^{(2)}$, and $\boldsymbol{s}^{(3)}$, respectively. The equilibrium positions* (4.4.21) *are exponentially asymptotically stable with respect to x_i, s_{ij} $(i, j = 1, 2, 3)$.*

Remark 4.4.2.

1. In applications the triaxial reorientation is preferable at which $\boldsymbol{r}_i \to \boldsymbol{s}^{(i)}$ rather than $\boldsymbol{r}_i \to \pm\boldsymbol{s}^{(i)}$. To this end (as in the case with uniaxial reorientation), generally speaking, it is not sufficient to use just one construction of control rules of type (4.4.19), (4.4.20).

2. A more demonstrative solution of the triaxial reorientation problem can be obtained by using kinematic equations in the Rodrigues–Hamilton variables (equations of type (2)).

2. *The use of kinematic equations of type* (2). According to the classic Euler theorem, a solid can be transferred from one position to another by a turn about *one* axis, which is called *Euler's axis.* Euler's quaternion components (Rodrigues–Hamilton variables) are defined by the relationships

$$\eta_i = e_i \sin \tfrac{1}{2}\phi \ \ (i = 1, 2, 3), \quad \eta_4 = \cos \tfrac{1}{2}\phi,$$

where ϕ is the angle of rotation about Euler's axis and $\boldsymbol{e} = (e_1, e_2, e_3)$ is a unit directional vector of this axis.

Kinematic equations in variables η_i $(i = 1, 2, 3), \eta_4$ have the form (Lur'e [1961])

$$\begin{gathered} 2\dot{\eta}_1 = \eta_4 x_1 + x_3\eta_2 - x_2\eta_3, \quad 2\dot{\eta}_2 = \eta_4 x_2 + x_1\eta_3 - x_3\eta_1, \\ 2\dot{\eta}_3 = \eta_4 x_3 + x_2\eta_1 - x_1\eta_2, \quad 2\dot{\eta}_4 = -\sum_{i=1}^{3} x_i\eta_i, \end{gathered} \tag{4.4.26}$$

where variables η_i $(i = 1, 2, 3)$, η_4 are also related as

$$\sum_{i=1}^{3} \eta_i^2 + \eta_4^2 = 1. \tag{4.4.27}$$

Let there be given two right-handed Cartesian coordinate systems whose origin is in the center of mass of a solid: $OXYZ$ is an inertial system and $Oxyz$ is a system rigidly attached to the body; the axes of the latter system coincide with the principal central axes of inertia of the solid. To bring system $Oxyz$ into coincidence with system $OXYZ$ implies that the relationships

$$\eta_i = 0 \ \ (i = 1, 2, 3), \quad \eta_4 = \pm 1 \tag{4.4.28}$$

are satisfied in the final position.

Since the sign "$\pm$" in (4.4.28) does not alter the orientation of the solid relative to the inertial coordinate system, we shall further confine ourselves solely to the case

$$\eta_i = 0 \ \ (i = 1, 2, 3), \quad \eta_4 = 1. \tag{4.4.29}$$

One can always obtain the case of (4.2.29) by appropriately choosing the inertial frame of reference.

The convenience of the kinematic equations (4.4.26), (4.4.27) is that, apart from the algebraic form, the solution technique for the triaxial reorientation problem is close in form to that of the uniaxial reorientation problem studied in detail in Subsection 4.4.2.

Indeed, one can use nonlinear linearizing controls of the type

$$u_i = \frac{1}{\eta_4} f_i(\boldsymbol{x}, \boldsymbol{\eta}, \boldsymbol{u}^*) \quad (i = 1, 2, 3) \tag{4.4.30}$$

by which the *linear controlled $\boldsymbol{\mu}$-system*

$$\dot{\eta}_i = \mu_i, \quad \dot{\mu}_i = u_i^* \quad (i = 1, 2, 3) \tag{4.4.31}$$

s separated from system (4.4.1), (4.4.26).

We shall often use the approach indicated in the sequel, i.e., when solving the problems of triaxial time-suboptimal reorientation (Sections 4.5 and 4.6) and game-theoretic problems of triaxial reorientation of a solid (Sections 5.2, 5.4, and 5.5).

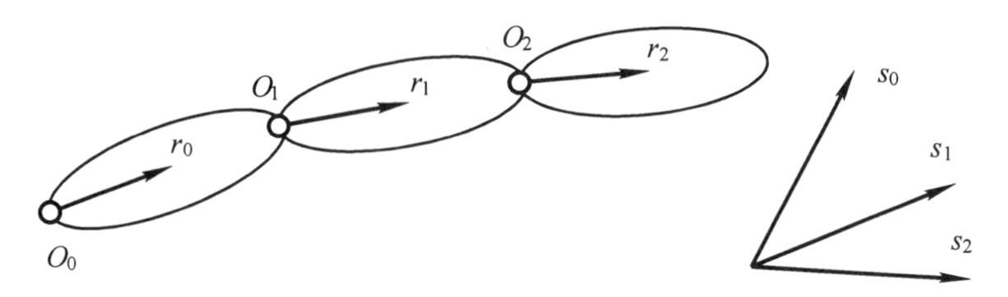

Figure 4.4.5. A chain of solids T_j $(j = 0, 1, 2)$.

Remark 4.4.3.

1. Besides Lur'e [1961], equations (4.4.26), (4.4.27) were first (independently) obtained in the work of Robinson [1958], Margulies [1963], Martensen [1963], Harding [1964].*

2. At present kinematic equations (4.4.26), (4.4.27) are favored when describing spacecraft angular motion; for example, they were employed in the *Space Shuttle* and *Galileo* projects.

4.4.4 Coordinated Control of a System of Solids (Manipulator Model)

Let T_0 be some solid to which a *chain of solids* T_i $(i = 1, \ldots, n)$ is attached that are pointwise hinged so that solid T_i can freely rotate about point O_i immobile both in solid T_{i-1} and in T_i (see Figure 4.4.5).

The motion equations of the mechanical system under study have the form (Zubov [1975])

$$m_j(\dot{\boldsymbol{\nu}}_j + \boldsymbol{\omega}_j \times \boldsymbol{\nu}_j + \dot{\boldsymbol{\omega}}_j \times \boldsymbol{\rho}_j + \boldsymbol{\omega}_j \times \boldsymbol{\omega}_j \times \boldsymbol{\rho}_j) = \boldsymbol{F}_j, \tag{4.4.32}$$

$$A_j \dot{x}_{j1} = (B_j - C_j) x_{j2} x_{j3} - Q_j + u_j \quad (A_j B_j C_j, 123), \tag{4.4.33}$$

$$(j = 0, 1, \ldots, n).$$

Here m_j, A_j, B_j, C_j $(j = 0, 1, \ldots, n)$ are the mass and the principal moments of inertia of solid T_j; x_{j1}, x_{j2}, x_{j3} are the projections of the angular velocity vector $\boldsymbol{\omega}_j$ of solid T_j onto its principal axes of inertia $O_j x_j y_j z_j$; $\boldsymbol{\rho}_j$ is the radius vector of the center of mass of solid T_j with respect to point O_j; $\boldsymbol{\nu}_j$ is the vector of absolute velocity of point O_j; $\boldsymbol{F}_j$ is the resultant of the forces exerted at point O_j; Q_j are the projections of vector $-m_j \boldsymbol{\rho}_j \times (\dot{\boldsymbol{\nu}}_j + \boldsymbol{\omega}_j \times \boldsymbol{\nu}_j)$ onto the axes of system $O_j x_j y_j z_j$; and u_j are the controlling moments generated by actuators.

---

*Robinson, A.S. [1958] On the Use of Quaternions in Simulation of Rigid Body Motion, *Wright Air Development Center, Wright–Patterson AFB, WADC Tech. Rept.* 58–17. Margulies, G. [1963] On Real Four-Parameter Representations of Satellite Attitude Motions, *Ford-Philco, Palo Alto, CA, Rept.* 52. Martensen, R.E. [1963] On Systems for Automatic Control of the Rotation of Rigid Body, *Electronics Research Laboratory, University of California, Berkeley, Calif., Rept.* 63–23, Nov. 27. Harding, C.F. [1964] Solution to Euler's Gyrodynamics. I., *J. Appl. Mech.*, 325–328.

Let there be given a system of unit vectors $\boldsymbol{s}_j$ $(j = 0, 1, \dots, n)$ in inertial space pointing toward some reference objects; projections of the vectors onto the axes of system $O_j x_j y_j z_j$ are s_{j1}, s_{j2}, s_{j3}. Vector $\boldsymbol{s}_j$ will rotate about solid T_j with angular velocity $(-\boldsymbol{\omega}_j)$ and, consequently,

$$\dot{s}_{j1} = s_{j2}x_{j3} - s_{j3}x_{j2} \quad (123) \qquad (j = 0, 1, \dots, n). \tag{4.4.34}$$

Suppose that a unit vector $\boldsymbol{r}_j$ $(j = 0, 1, \dots, n)$ is rigidly attached to each of the solids T_j $(j = 0, 1, \dots, n)$.

Our task is to choose the controlling moments u_j $(j = 0, 1, \dots, n)$ so that vectors $\boldsymbol{r}_j$ are oriented in the directions of vectors $\boldsymbol{s}_j$ and the angular velocities of the solids are dampened. For simplicity, let us assume that vector $\boldsymbol{r}_j$ is aligned with one of the principal axes of inertia of solid T_j, i.e., the $O_j y_j$-axis.

Theorem 4.4.4. *If $s_{j2}(t_0) \geq \varepsilon_j > 0$ $(j = 0, 1, \dots, n)$, then for any ε_j the constants $\Gamma_{j1}, \dots, \Gamma_{j5}$ can be chosen so that under the action of the controlling moments*

$$u_{j1} = \frac{A_j}{s_{j2}} f_{j1} + \frac{A_j s_{j1} u_{j2}}{B_j s_{j2}}, \qquad u_{j2} = B_j(\Gamma_{j5}x_{j2} - \psi_{j2}),$$
$$u_{j3} = \frac{C_j}{s_{j2}} f_{j2} + \frac{C_j s_{j3} u_{j2}}{B_j s_{j2}},$$
$$\begin{aligned} f_{j1} &= -\Gamma_{j3}s_{j3} - \Gamma_{j4}\varphi_{j3} + \psi_{j2}s_{j1} - \psi_{j1}s_{j2} + \varphi_{j1}x_{j2} - \varphi_{j2}x_{j1}, \\ f_{j2} &= \Gamma_{j1}s_{j3} - \Gamma_{j2}\varphi_{j1} + \psi_{j2}s_{j3} - \psi_{j3}s_{j2} + \varphi_{j3}x_{j2} - \varphi_{j2}x_{j3}, \end{aligned} \tag{4.4.35}$$
$$\varphi_{j1} = s_{j2}x_{j3} - s_{j3}x_{j2} \quad (123),$$
$$\psi_{j1} = \frac{1}{A_j}\left[(B_j - C_j)x_{j2}x_{j3} - Q_j\right] \quad (A_j B_j C_j, 123),$$
$$\Gamma_{jk} = \text{const} < 0 \ \ (k = 1, \dots, 5), \quad 4\Gamma_{jv} + \Gamma_{j,v+1}^2 > 0 \ \ (v = 1, 3)$$
$$(j = 0, 1, \dots, n)$$

for any motion of the mechanical system under study, the angular velocities of the solids are exponentially asymptotically dampened and vectors $\boldsymbol{r}_j$ $(j = 0, 1, \dots, n)$ exponentially asymptotically approach directions coinciding with that of vectors $\boldsymbol{s}_j$.

The equilibrium position

$$\boldsymbol{\omega}_j = \boldsymbol{0}, \quad \boldsymbol{r}_j = \boldsymbol{s}_j \ \ (s_{j1} = s_{j3} = 0, \ \ s_{j2} = 1) \ \ (j = 0, 1, \dots, n), \tag{4.4.36}$$

at which the system is oriented, is exponentially asymptotically stable with respect to x_{jl}, s_{jl} $(j = 0, 1, \dots, n;\ l = 1, 2, 3)$.

Proof. Variables s_{j1}, s_{j3}, x_{j2} $(j = 0, 1, \dots, n)$ of the closed system (4.4.33)–(4.4.35) form the *$\boldsymbol{\mu}$-system*

$$\dot{s}_{j1} = \mu_{j1}, \quad \dot{\mu}_{j1} = \Gamma_{j1}s_{j1} + \Gamma_{j2}\mu_{j1} \quad (\mu_{j1} = s_{j2}x_{j3} - s_{j3}x_{j2}),$$

$$\dot{s}_{j3} = \mu_{j2}, \quad \dot{\mu}_{j2} = \Gamma_{j3}s_{j3} + \Gamma_{j4}\mu_{j2} \quad (\mu_{j2} = s_{j1}x_{j2} - s_{j2}x_{j1}),$$
$$\dot{x}_{j2} = \Gamma_{j5}x_{j2} \quad (j = 0, 1, \ldots, n).$$

Therefore, the following conditions hold for any motions of the system:

$$\begin{gathered} s_{j1}(t) \to 0, \quad s_{j3}(t) \to 0, \quad x_{j2}(t) \to 0, \\ \mu_{j1}(t) \to 0, \quad \mu_{j2}(t) \to 0, \quad (t \to \infty), \quad (j = 0, 1, \ldots, n). \end{gathered} \tag{4.4.37}$$

Equations (4.4.34) yield the relationships

$$x_{j1} = \frac{-\mu_{j2} + x_{j2}s_{j1}}{s_{j2}}, \quad x_{j3} = \frac{\mu_{j1} + x_{j2}s_{j3}}{s_{j2}},$$
$$(j = 0, 1, \ldots, n),$$

and we conclude (by analogy with the proof of Theorem 4.4.1) that for any ε_j and any finite initial values $x_{jl}(t_0)$ $(l = 1, 2, 3)$, constants $\Gamma_{j1}, \ldots, \Gamma_{j4}$ can be chosen so that $s_{j2}(t) \geq \alpha > 0$ for all $t \geq t_0$.

Hence, along with (4.4.37), the following relationships will be valid

$$x_{jv}(t) \to 0, \quad s_{j2}(t) \to 1 \quad (t \to \infty) \quad (j = 0, 1, \ldots, n; \ v = 1, 3).$$

Thus we obtain $\boldsymbol{\omega}_j \to \mathbf{0}$, $\boldsymbol{r}_j \to \boldsymbol{s}_j$ $(j = 0, 1, \ldots, n)$ and the equilibrium position (4.4.36) is exponentially asymptotically stable with respect to x_{jl}, s_{jl} $(j = 0, 1, \ldots, n;\ l = 1, 2, 3)$. The theorem is proved. □

Remark 4.4.4.

1. If $-1 < s_{j2}(t_0) < -\varepsilon_j < 0$ $(j = 0, 1, \ldots, n)$, then to fulfill the relationship $\boldsymbol{r}_j \to \boldsymbol{s}_j$ $(j = 0, 1, \ldots, n)$, one can successively use two constructions of control rules, namely, (4.4.4) and the construction of type (4.4.15) or (4.4.16) derived from (4.4.4) by the corresponding permutation of indices and coefficients.

2. By appropriately choosing constants $\Gamma_{j1}, \ldots, \Gamma_{j5}$ $(j = 0, 1, \ldots, n)$ in (4.4.35) the desirable quality of the transient can be achieved in the course of reorienting the system.

3. To thoroughly describe the motion of the system, it is necessary to specify the law of motion for point O_0, since the motion of this point, together with $\boldsymbol{\omega}_i$, prescribes the motion of points O_i $(i = 1, \ldots, n)$. By appropriately choosing the resultant vector $\boldsymbol{F}_0$ in equations (4.4.32), one can obtain the motion of point O_0 in accordance with the given program.

4. The problem considered of controlling a system of solids is of interest from the standpoint of applied research in the attitude control of a *composite orbital space station* and *manipulator* control. However, a number of additional technical issues arise here, for example, allowance for the actuator dynamics.

4.5 Finite-Time Control with Respect to Part of the Variables with Constraints on Controls

4.5.1 Preliminary Remarks

Growing needs of practice call for further developing the problem of partial stabilization. The analysis of realizability of partial stabilization is required to account for a *given* domain of initial perturbations and also *direct constraints* imposed on the control rules. The damping time (that of "quieting" a system) should be as short as possible. It is just this outlined statement of the problem that makes the expedience of partial stabilization (as compared to stabilization with respect to all the variables) most conspicuous. Indeed, damping of perturbations with respect to part of the variables can be achieved in significantly less time than that with respect to all the variables, "geometric" constraints being the same. Besides, energy consumed for implementing the controls can be reduced appreciably. Again, the controls may turn out to be simpler in structure and easier to implement.

Such a situation is observed, for example, even in the case of "*double integrator*" $\ddot{x} = u$, the simplest controlled object (here x is the phase variable and u is the control). In particular, this object can describe the controlled motion of a unit mass point along some fixed direction in a smooth horizontal plane. Dampening the deviation of the point from the equilibrium position $x = \dot{x} = 0$ is required in a minimum time subject to the constraint $|u| \leq \alpha$ (where $\alpha > 0$ is a given constant). We consider two versions of the problem:

(1) Damping of deviation of the point only with respect to coordinate x. In this case it is necessary to ensure fulfilling the equality $x = 0$ (velocity $\dot{x}$ can be arbitrary) at a terminal time.

(2) Damping of deviation of the point with respect to both coordinate x and velocity $\dot{x}$. In this case it is necessary to ensure fulfilling the equality $x = \dot{x} = 0$ at a terminal time.

These problems are called "*hard*" and "*soft*" *encounter*, respectively, with position $x = \dot{x} = 0$. From the standpoint of mechanics, it is apparent that the control in problem (1) must be *constant* in time and directed toward the position $x = 0$. In problem (2) the control is *piecewise-constant* and works in the mode of "*acceleration–deceleration*" of the point. (The strict solution of these problems can be obtained by Pontryagin's maximum principle (Pontryagin et al. [1961]).) Though "geometric" constraint on u is one and the same in either problem, the objective of problem (1) is attained in less time and at a lower expenditure of energy than that of problem (2). (For example, if $\dot{x}(t_0) = 0$, then the time necessary for arriving at the final state in problem (1) differs from that in problem (2) by the factor $\sqrt{2}$.) In addition, the control law in problem (1) is simpler in structure than that in problem (2).

Since the position $x = \dot{x} = 0$ is the equilibrium position for $u \equiv 0$, one can consider problem (1) a problem of partial time-optimal stabilization of this position with respect to x.

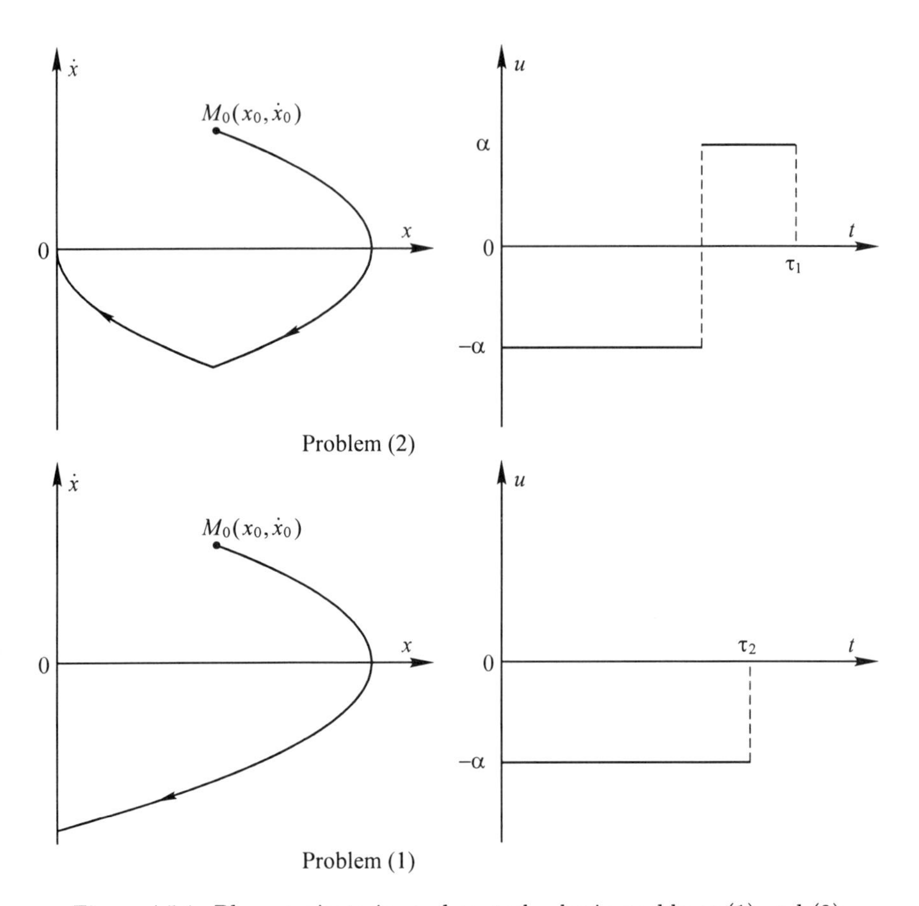

Figure 4.5.1. Phase trajectories and control rules in problems (1) and (2).

At the same time, even the above example reveals that the "transfer" of the partial stabilization problem from an infinite to a *finite* time interval, strictly speaking, can cause the classical treatment of the term "stabilization" to lose its meaning. Indeed, a controlled object often *just "passes"* through a final state without "making a stop" in it.

Taking this reasoning into account, we shall use the term *control with respect to part of the variables* for the problem which extends the problem of partial stabilization to the case of a *finite* time interval.

An insistent demand for developing such a problem is also a consequence of investigating systems with interference in the case when the interference level is not small. This problem is dealt with in Chapter 5.

In what follows, the problem of controllability with respect to part of the variables in a finite time interval subject to direct "geometric" (nonintegral) constraints on the control functions is solved for a class of nonlinear systems (in the absence of interference). The sufficient conditions for *nonlocal* solvability of this problem are obtained. The technique for estimating the control time is presented.

Construction of the controls is reduced to solving corresponding auxiliary *linear* problems of optimal control with respect to part of the variables. Some additional conditions make it possible to verify prescribed constraints on the controls along the trajectories of auxiliary linear systems.

As a result, the method for solving the problems of stability and stabilization based on constructing auxiliary linear $\boldsymbol{\mu}$-systems, elaborated in this book, is further developed as applied to the problems of control with respect to part of the variables in a finite time interval.

In the context of the proposed approach, the solution of the nonlinear problem of reorienting (without damping the angular velocity) an asymmetric solid is given in Section 4.6. In the initial state, the solid is at rest (or in a state close to it). The value of the angular velocity in the final state is of no importance. Possible applications of such a problem in spacecraft dynamics are indicated.

4.5.2 Formulation of the Problem

Let a controlled object motion be described by a system of nonlinear ordinary differential equations

$$\dot{\xi}_i = X_i^{(0)}(\boldsymbol{\xi}, \boldsymbol{\eta}) + \sum_{k=1}^{r} X_{ik}^{(1)}(\boldsymbol{\xi}, \boldsymbol{\eta}) u_k, \quad \dot{\eta}_j = X_j^{(2)}(\boldsymbol{\xi}, \boldsymbol{\eta}) \tag{4.5.1}$$
$$(i = 1, \ldots, l;\ j = 1, \ldots, n - l).$$

Here $\boldsymbol{x} = (\boldsymbol{\xi}, \boldsymbol{\eta})$ is the vector of phase variables ξ_i, η_j; vector $\boldsymbol{u}$ is composed of controls u_k. Functions $X_i^{(0)}, X_{ik}^{(1)}, X_j^{(2)}$ are defined and continuous ($X_j^{(2)}$ together with its partial derivatives with respect to ξ_i, η_j) in the domain

$$\boldsymbol{\Lambda}: \ ||\boldsymbol{x}|| \leq H = \text{const} > 0.$$

Unlike functions $X_i^{(0)}, X_j^{(2)}$ that vanish at $\boldsymbol{x} = \boldsymbol{0}$, functions $X_{ik}^{(1)}$ must not be zero at $\boldsymbol{x} = \boldsymbol{0}$. The converse will contradict conditions introduced below.

Controls $\boldsymbol{u} \in \mathcal{K}$ are chosen in class $\mathcal{K}$ of vector functions $\boldsymbol{u} = \boldsymbol{u}(\boldsymbol{x}, \boldsymbol{x}_0)$ (where $\boldsymbol{x}_0$ is the initial value of vector $\boldsymbol{x}$) *continuous* with respect to $\boldsymbol{x}$ and *piecewise-continuous* with respect to $\boldsymbol{x}_0$ (in an admissible domain of variation of $\boldsymbol{x}, \boldsymbol{x}_0$). Besides, controls $\boldsymbol{u} \in \mathcal{K}$ satisfy given "*geometric*" (nonintegral) constraints

$$|u_k| \leq \alpha_k = \text{const} > 0 \quad (k = 1, \ldots, r). \tag{4.5.2}$$

Let $\boldsymbol{y}$ be the vector of those phase variables of system (4.5.1) which are of interest in the course of controlling the system. Suppose that $\dim \boldsymbol{y} = \dim \boldsymbol{u} = r < n$, where $\boldsymbol{y}$ is composed of the *first* q ($q < r \leq m$) variables ξ_i and the *first* $r - q$ variables η_j. In the context of the specified requirements, number q can be arbitrary.

Problem 4.5.1. *Find control rules* $\boldsymbol{u} \in \mathcal{K}$ *transferring system* (4.5.1) *from a given domain* S *of initial perturbations* $\boldsymbol{x}_0 = \boldsymbol{x}(t_0) \in S$ *to the position* $\boldsymbol{y}_1 = \boldsymbol{y}(t_1) = 0$ *in a finite time. (The values of vector* $\boldsymbol{x}$ *components not entered into* $\boldsymbol{y}$ *can be arbitrary at* $t = t_1$.)

Remark 4.5.1.

1. Because the position $\boldsymbol{x} = \boldsymbol{0}$ is an equilibrium position of system (4.5.1) for $\boldsymbol{u} = \boldsymbol{0}$, Problem 4.5.1 can be interpreted as a version of the problem of *partial stabilization* (*$\boldsymbol{y}$-stabilization*) of this position. However, compared with the classic formulation of the problem of partial stabilization Problem 4.5.1 has the following peculiarities:

(*a*) this is a problem of partial stabilization in a *finite* time interval (in contrast to that in an *infinite* one) and, besides, *rigorous* damping of perturbations from domain S (with respect to given variables) is guaranteed;

(*b*) "*geometric*" constraints (4.5.2), instead of *integral* constraints are imposed on the control functions;

(*c*) domain S of initial perturbations is assumed to be *given*.

We note that though controls $\boldsymbol{u} \in \mathcal{K}$ do not solve Problem 4.5.1 in the form of *synthesis*, they extend the solvability of the problem. Moreover, the proposed approach to solving Problem 4.5.1 does not rule out constructing the control rules in the form close to that of synthesis (see Subsection 4.5.9).

2. Once Problem 4.5.1 has been solved, after the controls are turned off ($\boldsymbol{u} \equiv \boldsymbol{0}$), three cases are possible.

(1) The position $\boldsymbol{y} = \boldsymbol{0}$ is an *invariant set* of system (4.5.1). After being taken to position $\boldsymbol{y} = \boldsymbol{0}$, the system remains in it from that time on. In this case the position $\boldsymbol{y} = \boldsymbol{0}$ is also called a "*partial equilibrium position*" or "*balanced motion*" of the system and, in principle, Problem 4.5.1 can be treated as a problem of stabilization.

(2) The position $\boldsymbol{y} = \boldsymbol{0}$ is not an invariant set. The system just "*passes through* this position without "making a stop" in it. In this case the term "stabilization" loses any meaning whatever. But it should be stressed that Problem 4.5.1 is a specific version of the stabilization problem, i.e., *partial* (with respect *not to all* the variables) stabilization of the "*hard*" *encounter type* in a *finite* interval of time. This problem differs drastically from the classic problem of stabilization. Such dynamics may turn out to be sufficient to control the system. One of the possible situations, as applied to reorienting a spacecraft, is considered in Section 4.6.

(3) It is not the position $\boldsymbol{y} = \boldsymbol{0}$ but $\boldsymbol{y} = \boldsymbol{0}$, $\boldsymbol{\mu}(\boldsymbol{x}) = \boldsymbol{0}$ that is an invariant set. In this case the stabilization with respect to $\boldsymbol{\mu}$ is "forced upon" us in addition to $\boldsymbol{y}$-stabilization. How to choose vector function $\boldsymbol{\mu}$ is not a priori obvious but depends on the features of the problem being studied. As for investigating the partial stability (see Sections 3.1 and 3.2), the proposed approach may turn out to be effective in solving the original problem of $\boldsymbol{y}$-stabilization (in both specified above and classic sense).

4.5.3 An Auxiliary Linear System

Let us introduce the designations

$$X_i^{(1)}(\boldsymbol{x}, \boldsymbol{u}) \stackrel{\text{def}}{=} X_i^{(0)}(\boldsymbol{x}) + \sum_{k=1}^{r} X_{ik}^{(1)}(\boldsymbol{x}) u_k \quad (i = 1, \ldots, q) \tag{4.5.3}$$

for the functions defining the right-hand sides of the first q equations of system (4.5.1). Bearing in mind expressions (4.5.3), we consider functions (4.2.10) and introduce Jacobi matrix $F(\boldsymbol{x}) = (\partial\Phi_s/\partial u_k)$ $(s, k = 1, \ldots, r)$.

We note that in this case (as differed from Section 4.2) matrix F is a *quadratic* matrix of dimensions $r \times r$ whose components do not depend on $\boldsymbol{u}$.

Suppose that in the domain

$$\boldsymbol{\Lambda}_1 : \ \|\boldsymbol{x}\| \leq H_1 = \text{const} > 0 \quad (\boldsymbol{\Lambda}_1 \subseteq \boldsymbol{\Lambda}) \tag{4.5.4}$$

the following condition is valid:

$$\operatorname{rank} F(\boldsymbol{x}) = r. \tag{4.5.5}$$

In the case of (4.5.5), in domain (4.5.4) the system

$$\Phi_k(\boldsymbol{x}, \boldsymbol{u}) = u_k^* \quad (k = 1, \ldots, r)$$

(where u_k^* are the auxiliary controls defined below and constituting vector $\boldsymbol{u}^*$) has the solution

$$\boldsymbol{u} = \boldsymbol{f}(\boldsymbol{x}, \boldsymbol{u}^*) \tag{4.5.6}$$

such that $\boldsymbol{f}(\boldsymbol{0}, \boldsymbol{0}) = \boldsymbol{0}$. Components f_k of vector function $\boldsymbol{f} \in R_r$ are continuous with respect to $\boldsymbol{x}, \boldsymbol{u}^*$ for $\boldsymbol{x} \in \boldsymbol{\Lambda}_1$ and all $\boldsymbol{u}^*$.

As a result, under the assumptions made from the closed nonlinear system (4.5.1), (4.5.6), one can segregate *linear controlled $\boldsymbol{\mu}$-system*

$$\dot{\xi}_i = u_i^* \qquad (i = 1, \ldots, q), \tag{4.5.7}$$

$$\dot{\eta}_j = \mu_j, \quad \dot{\mu}_j = u_{q+j}^* \qquad (j = 1, \ldots, r - q). \tag{4.5.8}$$

In this system u_k^* $(k = 1, \ldots, r)$ are treated as the controls.

Based on solving the corresponding problems of optimal control *with respect to part of the variables* for the auxiliary linear system (4.5.7), we further construct the solution of the original nonlinear Problem 4.5.1. So construction (4.5.6) can be regarded as a *general structural form* of the control functions in Problem 4.5.1. The parameters of this form are the auxiliary controls u_k^* defined by solving the corresponding linear control problems.

4.5.4 An Auxiliary Linear Control Problem

Now we solve the problem of the time-optimal control of system (4.5.7), (4.5.8) *with respect to part of the variables*, i.e., the problem of transferring the system to the position

$$\xi_i = 0, \quad \eta_j = 0 \quad (i = 1, \dots, q;\ j = 1, \dots, r - q) \tag{4.5.9}$$

in a minimum time.

The control is realized through u_k^*, where the terminal values of μ_j are arbitrary. Therefore, by the essence of Problem 4.5.1, control rules u_k^* must *simultaneously* take variables ξ_i, η_j $(i = 1, \dots, q;\ j = 1, \dots, r - q)$ to position (4.5.9).

In solving this problem we should first specify the maximum possible magnitudes (levels) of u_k^*. Following (4.5.2) we shall take the corresponding constraints as

$$|u_k^*| \le \alpha_k^* = \text{const} > 0 \quad (k = 1, \dots, r). \tag{4.5.10}$$

The procedure of choosing constants α_k^* must take into account constraints (4.5.2) on the "original" control rules u_k $(k = 1, \dots, r)$.

With α_k^* $(k = 1, \dots, r)$ fixed, the solution of the above problem of time-optimal control of system (4.5.7), (4.5.8) with respect to part of the variables (when all $\xi_{i0} \ne 0, \eta_{j0} \ne 0$) is given by the control rules (Pontryagin et al. [1961])

$$u_i^* = -\alpha_i^* \operatorname{sgn} \xi_i \qquad (i = 1, \dots, q), \tag{4.5.11}$$

$$u_{q+j}^* = -\alpha_{q+j}^* \operatorname{sgn} \eta_j \qquad (j = 1, \dots, r - q). \tag{4.5.12}$$

The relationships

$$\begin{gathered} \tau = \left(\alpha_i^*\right)^{-1} |\xi_{i0}|, \quad \xi_{i0} \ne 0, \\ \tau = \left(\alpha_{q+j}^*\right)^{-1} \left[\mu_{j0} \operatorname{sgn} \eta_{j0} + \left(\mu_{j0}^2 + 2|\eta_{j0}|\alpha_{q+j}^*\right)^{1/2}\right], \quad \eta_{j0} \ne 0, \end{gathered} \tag{4.5.13}$$

define the minimum time τ (*one and the same* for all variables ξ_i, η_j) to arrive at position (4.5.9) for the auxiliary linear problem.

The trajectories of linear system (4.5.7), (4.5.8), (4.5.11), and (4.5.12) are described by

$$\xi_i = \xi_{i0} - \left(\alpha_i^* \operatorname{sgn} \xi_{i0}\right)t \quad (i = 1, \dots, q), \tag{4.5.14}$$

$$\begin{gathered} \mu_j = \mu_{j0} - \left(\alpha_{q+j}^* \operatorname{sgn} \eta_{j0}\right)t, \\ \eta_j = \eta_{j0} + \mu_{j0}t - \frac{1}{2}\left(\alpha_{q+j}^* \operatorname{sgn} \eta_{j0}\right)t^2 \quad (j = 1, \dots, r - q). \end{gathered} \tag{4.5.15}$$

If *all* ξ_{i0}, η_{j0} are equal to zero, the control aim is attained. If *part* of ξ_{i0}, η_{j0} are equal to zero, then for the corresponding values of indices i, j, it is necessary that $u_i^* = 0, u_{q+j}^* = -\left(\alpha_{q+j}^* \operatorname{sgn} \mu_{j0}\right)$ for $\mu_{j0} \ne 0$ and $u_{q+j}^* = 0$ for $\mu_{j0} = 0$. In this case $\xi_i \equiv 0, \eta_j \equiv 0$ ($\mu_{j0} = 0$) for the specified values of i, j over the entire control interval $t \in [t_0, t_0 + \tau]$, τ being determined from expressions (4.5.13) for nonzero values of ξ_{i0}, η_{j0}. And if $\mu_{j0} \neq 0$, then for the corresponding values of j in (4.5.13), one should set $\tau = 2(\alpha^*_{q+j})^{-1}\mu_{j0}(\operatorname{sgn}\mu_{j0})$.

The procedure for assigning levels α^*_k is considered in Subsection 4.5.7.

4.5.5 The Conditions for the Solvability of Problem 4.5.1

Separating linear system (4.5.7), (4.5.8) from system (4.5.1), (4.5.6) allows finding processes $\xi_i(t; t_0, \boldsymbol{x}_0)$ and $\eta_j(t; t_0, \boldsymbol{x}_0)$ ($i = 1, \ldots, q;\ j = 1, \ldots, r - q$) of system (4.5.1), (4.5.6) as the corresponding processes of the *linear* system.

Theorem 4.5.1. *Let the following conditions be valid:*

(1) *condition* (4.5.5) *in domain* (4.5.4);

(2) *for any* $\boldsymbol{x}_0 \in S$ *the levels* $\alpha^*_k(\boldsymbol{x}_0)$ *of control functions* u^*_k ($k = 1, \ldots, r$) *can be chosen so that, for any* $t \in [t_0, t_0 + \tau(\boldsymbol{x}_0)]$, *the following estimates hold*

$$\begin{gathered} \|\boldsymbol{x}(t; t_0, \boldsymbol{x}_0)\| \leq H_1, \\ \big|f_k[\boldsymbol{x}(t; t_0, \boldsymbol{x}_0), \boldsymbol{u}^*]\big| \leq \alpha_k \quad (k = 1, \ldots, r). \end{gathered} \tag{4.5.16}$$

Then, for all $\boldsymbol{x}_0 \in S$, *control rules* (4.5.9), (4.5.11), *and* (4.5.12) *solve Problem* 4.5.1; *the transfer of system* (4.5.1) *to the position* $\boldsymbol{y} = \boldsymbol{0}$ *is guaranteed by controls* $\boldsymbol{u} \in \mathcal{K}$ *in the finite time* $\tau = \tau(\boldsymbol{x}_0)$.

Proof. If condition (4.5.5) is satisfied in $\boldsymbol{\Lambda}_1$, the linear system (4.5.7), (4.5.8) is derived from the original nonlinear system (4.5.1), (4.5.6) by the corresponding nonlinear transformation of the variables. Under such a transformation, the state of variables ξ_i, η_j ($i = 1, \ldots, q;\ j = 1, \ldots, r - q$) of system (4.5.1), (4.5.6) in Λ_1 will be completely determined by the state of the same variables of linear system (4.5.7), (4.5.8). Then for all $\boldsymbol{x}_0 \in S_1 = \{\boldsymbol{x}_0 : \|\boldsymbol{x}(t)\| < H_1\}$, control rules (4.5.6), (4.5.11), and (4.5.12) guarantee that system (4.5.1) is transferred precisely to position (4.5.9) in the finite time τ.

Requirements (4.5.2) cause the admissible set of the initial perturbations $\boldsymbol{x}_0$ necessarily to satisfy the condition $\boldsymbol{x}_0 \in S_2 = \big\{\boldsymbol{x}_0 : \big|f_k[\boldsymbol{x}(t), \boldsymbol{u}^*]\big| \leq \alpha_k\big\}$.

Provided conditions (1)–(2) of the theorem are fulfilled, the given domain S of initial perturbations will belong to the intersection of sets S_1 and S_2. As a result, for all $\boldsymbol{x}_0 \in S$, control rules (4.5.6), (4.5.11), and (4.5.12) guarantee that system (4.5.1) is transferred precisely to position $\boldsymbol{y} = \boldsymbol{0}$ in the finite time τ and the control rules (4.5.6), (4.5.11), and (4.5.12) satisfy the given constraints (4.5.2). The Theorem is proved. □

4.5.6 Additional Conditions for the Solvability of Problem 4.5.1

Let us obtain conditions for the solvability of Problem 4.5.1 which are verified more constructively. To this end we note that in most cases the control rules solving the problems of control with respect to part of the variables and constructed in accordance with the feedback principle, evidently, are the functions of *all* the phase variables. Hence, to validate inequalities (4.5.16), the corresponding estimates are also to be obtained for "uncontrollable" (not included in $\boldsymbol{y}$) variables. This can be done, for example, by involving conditions additional to those of Theorem 4.5.1. If such additional conditions are fulfilled, the "uncontrollable" variables can be related to $\boldsymbol{y}$-variables of system (4.5.1).

Suppose that $r-q \geq n-r$. Along with F, let us also introduce the Jacobi matrix Q of functions $X_j^{(2)}(\boldsymbol{x})$ $(j=1,\ldots,r-q)$ with respect to $\boldsymbol{z}$; vector $\boldsymbol{z} \in R^{n-r}$ is composed of those phase variables of system (4.5.1) not included in vector $\boldsymbol{y}$. As a result, $\boldsymbol{x}=(\boldsymbol{y},\boldsymbol{z})$.

We assume that in the domain

$$\boldsymbol{\Lambda}_2 : \ \|\boldsymbol{x}\| \leq H_2 = \text{const} > 0 \quad (\boldsymbol{\Lambda}_2 \subseteq \boldsymbol{\Lambda}), \tag{4.5.17}$$

the following condition is satisfied:

$$\operatorname{rank} Q(\boldsymbol{x}) = m+p-r. \tag{4.5.18}$$

In the case $r \leq m$, the inequality $r-q \geq n-r$ and the inequality $p+q \geq r$ (which is the consequence of the inequality $p \geq r-q$ resulting from our choice of vector $\boldsymbol{y}$) hold simultaneously when

$$p+q=r=m. \tag{4.5.19}$$

Let us note that, if conditions (4.5.19) are satisfied, matrix Q is a *quadratic* $(m+p-r)\times(m+p-r)$ matrix.

Condition (4.5.19) takes place, for example, when system (4.5.1) is employed to simulate the angular velocity of a solid. Besides, in the framework of the proposed approach, this condition can be weakened. Under condition (4.5.18), in $\boldsymbol{\Lambda}_2$, the equalities $X_j^{(2)}(\boldsymbol{x}) = \mu_j$ $(j=1,\ldots,r-q)$ admit the solution (vector function $\boldsymbol{\Psi} \in R^{n-r}$ is continuous in $\boldsymbol{\Lambda}_2$)

$$\boldsymbol{z} = \boldsymbol{\Psi}(\boldsymbol{y},\boldsymbol{\mu}), \qquad \boldsymbol{\Psi}(\boldsymbol{0},\boldsymbol{0}) = \boldsymbol{0}. \tag{4.5.20}$$

Inequalities (4.5.20) relate the components of $\boldsymbol{x}$ not included in $\boldsymbol{y}$ to the components of $\boldsymbol{y}$, $\boldsymbol{\mu}$. As a result, in $\boldsymbol{\Lambda}_2$, the phase vector $\boldsymbol{x}$ of system (4.5.1) can be represented in the form $\boldsymbol{x} = [\boldsymbol{y}, \boldsymbol{z}(\boldsymbol{y},\boldsymbol{\mu})] \overset{\text{def}}{=} \boldsymbol{W}(\boldsymbol{y},\boldsymbol{\mu})$.

Therefore,

$$\boldsymbol{x}(t) = \boldsymbol{x}(t;t_0,\boldsymbol{x}_0) = \boldsymbol{W}[t;t_0,\boldsymbol{W}_0(\boldsymbol{y}_0,\boldsymbol{\mu}(\boldsymbol{x}_0))] = \boldsymbol{W}(t).$$

Thus, $\boldsymbol{x}(t)$ (not only the $\boldsymbol{y}(t)$-component of vector $\boldsymbol{x}$, as before) can be estimated through the estimates for the processes of *linear* (and not original nonlinear) system (4.5.7), (4.5.8), (4.5.11), and (4.5.12).

Theorem 4.5.2. *Let the following conditions be valid*:
(1) *condition* (4.5.5) *in domain* (4.5.4);

(2) *condition* (4.5.18) *in domain* (4.5.17);
(3) *condition* (4.5.19);
(4) *for any* $\boldsymbol{x}_0 \in S$ *the levels* $\alpha_k^*(\boldsymbol{x}_0)$ *of control functions* u_k^* $(k = 1, \dots, r)$ *can be chosen so that, for any* $t \in [t_0, t_0 + \tau(\boldsymbol{x}_0)]$, *the following estimates hold*:

$$\begin{aligned} \|\boldsymbol{W}(t)\| &\le \min(H_1, H_2), \\ \big|f_k[\boldsymbol{W}(t), \boldsymbol{u}^*]\big| &\le \alpha_k \quad (k = 1, \dots, r). \end{aligned} \tag{4.5.21}$$

Then, for all $\boldsymbol{x}_0 \in S$, *control rules* (4.5.6), (4.5.11), *and* (4.5.12) *solve Problem* 4.5.1; *the transfer of system* (4.5.1) *to the position* $\boldsymbol{y} = \boldsymbol{0}$ *is guaranteed by controls* $\boldsymbol{u} \in \mathcal{K}$ *in the finite time* $\tau = \tau(\boldsymbol{x}_0)$.

The proof scheme of Theorem 4.5.2 follows that for Theorem 4.5.1.

Remark 4.5.2.

1. Though seemingly alike, conditions of Theorems 4.5.1 and 4.5.5 essentially differ in that conditions (4.5.21) (as compared with condition (4.5.16)) are verified on trajectories (4.5.14), (4.5.15) of the auxiliary *linear* (not original *nonlinear*) system. This is achieved by introducing the additional condition (4.5.19).

2. Introducing conditions (4.5.19) allows one to represent control rules u_k $(k = 1, \dots, r)$ of type (4.5.6), (4.5.11), and (4.5.12), first obtained in the form $u_k = u_k(\boldsymbol{x}, \boldsymbol{x}_0)$, also in the *program* form $u_k = u_k(t)$, i.e., as *continuous* functions of t for given initial conditions $\boldsymbol{x} = \boldsymbol{x}_0$. Indeed, as $\boldsymbol{x}(t) = \boldsymbol{W}(t)$ under condition (4.5.19), then (for fixed initial conditions) control rules (4.5.6), (4.5.11), and (4.5.12) become *known* functions of t along the optimal trajectories (4.5.14), (4.5.15).

We indicate some possible ways of weakening the conditions of Theorem 4.5.2.

Let N first integrals $R_\varepsilon(\boldsymbol{x})$ $(\varepsilon = 1, \dots, N)$ of system (4.5.1) be known. (This means that there exist N functions $R_\varepsilon(\boldsymbol{x})$ such that $R_\varepsilon(\boldsymbol{x}) = \text{const}$ along the trajectories of this system.) Assuming $R_\varepsilon(\boldsymbol{0}) = 0$, $r - q + N \ge m + p - r$, instead of Q, we introduce the Jacobi matrix Q^* of functions $X_j^{(2)}, R_\varepsilon$ $(j = 1, \dots, r - q;\ \varepsilon = 1, \dots, N)$ with respect to $\boldsymbol{z}$.

Corollary 4.5.1. *Let the following conditions be valid*:
(1) $r \le p + q \le 2r - m + N$;
(2) $\operatorname{rank} Q^* = m + p - r$ *for* $\boldsymbol{x} \in \boldsymbol{\Lambda}_2$.

In addition, if conditions (1) *and* (4) *of Theorem* 4.5.2 *are satisfied, then, for all* $\boldsymbol{x}_0 \in S$, *control rules* (4.5.6), (4.5.11), *and* (4.5.12) *solve Problem* 4.5.1.

Another way of weakening the conditions of Theorem 4.5.2 consists of successively using *several* constructions of control rules of type (4.5.6). In this case system (4.5.1) is gradually transferred from $\boldsymbol{x}_0 \in S$ to $\boldsymbol{y} = \boldsymbol{0}$ (in a total finite time). In particular, the conditions of Theorem 4.5.1 can be weakened similarly.

4.5.7 Algorithm for Solving Problem 4.5.1

The proposed approach to solving Problem 4.5.1 comprises the following stages.

(1) Choosing construction (4.5.6) of control rules u_k in which u_k^* $(k = 1, \dots, r)$ have the form (4.5.11), (4.5.12).

(2) Estimating auxiliary controls u_k^*. Preliminarily choosing numbers $\alpha_k^* = \alpha_k^*(\boldsymbol{x}_0)$ from relationships (4.5.13). Numbers α_k^* predetermine the corresponding value of $\tau(\boldsymbol{x}_0)$.

(3) Validating inequalities (4.5.21) along trajectories (4.5.14), (4.5.15) of system (4.5.7), (4.5.8), (4.5.11), and (4.5.12) for all $\boldsymbol{x}_0 \in S$.

If these inequalities are violated or, vice versa, they are satisfied with a kind of "reserve," the search for the appropriate numbers α_k^* is continued. Otherwise, the control time equals $\tau(\boldsymbol{x}_0)$.

As a result, we get the *iterative* algorithm for solving Problem 4.5.1. The solvability of this algorithm depends on how the levels of the "original" controls u_k match the values of $\boldsymbol{x}_0, \tau$. The corresponding solvability conditions are presented in Subsections 4.5.5, 4.5.6.

Remark 4.5.3.

1. Though constructed as functions of the *current* phase vector $\boldsymbol{x}$, controls $\boldsymbol{u} \in \mathcal{K}$ still depend on the *initial* state $\boldsymbol{x}_0$, because the expressions for $\boldsymbol{u}$ contain constants α_k^* $(k = 1, \dots, r)$ depending on $\xi_{i0}, \eta_{j0}, \mu_{j0}(\boldsymbol{x}_0)$. However, under the conditions of Theorem 4.5.2, at any current moment when the phase vector $\boldsymbol{x} \in S$, it is possible to "*recalculate*" constants α_k^* for this current state. (In this case the current state $\boldsymbol{x}$ is taken as the "new initial state" $\boldsymbol{x}_0$.)

If the indicated adjustment is possible for *any admissible* (interferences taken into account) state of the system originating from domain $\boldsymbol{x}_0 \in S$, then, by their properties, control rules $\boldsymbol{u} \in \mathcal{K}$ are close to the control rules in the form of synthesis. To this end it is necessary that inequalities (4.5.21) are satisfied for possible "new" $\boldsymbol{x}_0$ (current $\boldsymbol{x}$) predetermined both by "original" and preceding "new" $\boldsymbol{x}_0$ and by possible deviations of the system from the current state from interferences.

2. The specified possibility of adjusting the controls agrees with a new view of the control *synthesis* problem which is being formed lately and which treats this problem as a *process of continuous successive adjustment* of the corresponding real-time *program* control problems. In other words, the synthesis problem is imbedded in a one-parameter family of program problems along a single trajectory being realized with real time as the parameter of the family. (For more detail, see Gabasov et al. [1992], where linear control systems are considered.)

4.5.8 The "Converse" of the Algorithm

(1) Choosing τ.

(2) Computing $\alpha_k^*(\boldsymbol{x}_0)$ by formulas (4.5.13).

(3) Estimating levels α_k of controls u_k (by formulas (4.5.6)) on trajectories (4.5.14), (4.5.15).

As a result, the capabilities of the control rules of type (4.6.6) can be estimated.

4.5.9 An Auxiliary Function of the Problem of Control with Respect to Part of the Variables in Control with Respect to All the Variables

Problem 4.5.2. Find control rules $\boldsymbol{u} \in \mathcal{K}^*$ *($\mathcal{K}^*$ is a class of functions piecewise-continuous with respect to $\boldsymbol{x}$, $\boldsymbol{x}_0$) transferring system (4.5.1) from a given domain S of initial perturbations $\boldsymbol{x}_0 \in S$ to the position $\boldsymbol{x} = \boldsymbol{0}$ in a finite time.*

The scheme for solving Problem 4.5.2 can be briefly described as follows:

(1) We first solve the problem of transferring system (4.5.1) to the position $\boldsymbol{x} = \boldsymbol{0}$ with respect to *part* of the variables, namely, variables ξ_i ($i = 1, \ldots, q$), η_j ($j = 1, \ldots, r-q$). This problem is reduced to the corresponding problem of control with respect to all the variables for the auxiliary linear $\boldsymbol{\mu}$-system (4.5.7), (4.5.8). As in the studies of partial stability (see Section 3.1), variables μ_j are "imposed" by a researcher into the number of controlled variables when the control with respect to part of the variables (to ξ_i, η_j) is carried out.

(2) Then we impose additional conditions under which the transfer of system (4.5.1) to the position $\boldsymbol{x} = \boldsymbol{0}$ with respect to the specified part of the variables is actually the solution of Problem 4.5.2 of control with respect to all the variables.

Following the scheme indicated, in the first stage we solve the time-optimal problem of transferring system (4.5.7), (4.5.8) to the position

$$\xi_i = 0 \quad (i = 1, \ldots, q), \qquad \eta_j = \mu_j = 0 \quad (j = 1, \ldots, r-q). \tag{4.5.22}$$

With α_k^* ($k = 1, \ldots, r$) in (4.5.10) fixed, the solution of this problem is given by the controls (Pontryagin et al. [1961])

$$u_i^*(\xi_i) = \begin{cases} -\alpha_i^* \operatorname{sgn} \xi_i, & \xi_i \neq 0 \\ 0, & \xi_i = 0 \ (i = 1, \ldots, q), \end{cases} \tag{4.5.23}$$

$$u_{q+j}^*(\eta_j, \mu_j) = \begin{cases} \alpha_{q+j}^* \operatorname{sgn} \psi_j(\eta_j, \mu_j), & \psi_j \neq 0 \\ \alpha_{q+j}^* \operatorname{sgn} \eta_j = -\alpha_{q+j}^* \operatorname{sgn} \mu_j, & \psi_j = 0, \end{cases} \tag{4.5.24}$$

$$\psi_j(\eta_j, \mu_j) = -\mu_j - 2\alpha_{q+j}^* \eta_j |\eta_j| \ (j = 1, \ldots, r-q).$$

Here ψ_j are the *switching functions.*

The value $\tau = \max(\tau_k)$ ($k = 1, \ldots, r$), where

$$\tau_i = (\alpha_i^*)^{-1} |\xi_i| \quad (i = 1, \ldots, q),$$

$$\tau_{q+j} = (\alpha_{q+j}^*)^{-1} \{ [\tfrac{1}{2}\mu_j^2 - \alpha_{q+j}^* \eta_j \operatorname{sgn} \psi_j]^{1/2} - \mu_j \operatorname{sgn} \psi_j \} \quad (j = 1, \ldots, r-q),$$

Table 4.5.1.

t	u_j^*	η_j	μ_j
$[0, \tau/2]$	α_j^*	$\eta_{j0} + 1/2\alpha_j^* t^2$	$\alpha_j^* t$
$(\tau/2, \tau]$	$-\alpha_j^*$	$\eta_{j1} - 1/2\alpha_j^*(t-\tau)^2$	$-\alpha_j^*(t-\tau)$

defines the minimum time to arrive at position (4.5.22) in the auxiliary linear problem.

The second stage of the above scheme of solving Problem 4.5.2 is realized in the same manner as in Subsection 4.5.6.

Corollary 4.5.2. *Let conditions* (1)–(4) *of Theorem* 4.5.2 *be satisfied. Then, for all* $\boldsymbol{x}_0 \in S$, *control rules* (4.5.6), (4.5.23), *and* (4.5.24) *guarantee that system* (4.5.1) *is transferred to the position* $\boldsymbol{x} = \mathbf{0}$ *in the finite time* $\tau(\boldsymbol{x}_0)$.

The time $\tau(\boldsymbol{x}_0)$ is found by *iteratively* assigning levels α_k^* $(k = 1, \dots, r)$ and subsequently verifying the given constraints (4.5.2) along the optimal trajectories of the closed system (4.5.7), (4.5.8), (4.5.23), and (4.5.24).

Example 4.5.1. Consider *Euler's dynamical equations* (4.4.1) together with *kinematic equations* (4.4.26), (4.4.27) in the variables which are components of Euler's quaternion.

Consider the problem of triaxially reorienting a solid (spacecraft) with $A = 4 \cdot 10^4$, $B = 8 \cdot 10^4$, and $C = 5 \cdot 10^4$ $kg{\cdot}m^2$ from the initial equilibrium position $\varphi_0 = \psi_0 = \pi/10$, $\theta_0 = -\pi/4$ into the final position $\varphi_1 = \psi_1 = \pi/6$, $\theta_1 = \pi/4$, also an equilibrium position of the solid. (Here ψ, θ, φ are *Euler's angles*: ψ is the *precession* angle, θ is the *nutation* angle, and ϕ is the angle of *pure rotation*.)

Let us take the original constraints on the controls in the form

$$|u_k| \le 300 \ (N{\cdot}m) \quad (k = 1, 2, 3). \tag{4.5.25}$$

Using the formulas for transforming Euler's angles into the components of Euler's quaternion (Lur'e [1961]),

$$\eta_1 = \sin 1/2\theta \cos 1/2(\psi - \varphi), \quad \eta_2 = \sin 1/2\theta \sin 1/2(\psi - \varphi),$$
$$\eta_3 = \cos 1/2\theta \sin 1/2(\psi + \varphi), \quad \eta_4 = \cos 1/2\theta \cos 1/2(\psi + \varphi),$$

following are the boundary conditions in the case under study:

$$\boldsymbol{x}_0 = \mathbf{0}, \quad \boldsymbol{\eta}_0 = (-0.383, 0.000, 0.286, 0.879)$$
$$\boldsymbol{x}_1 = \mathbf{0}, \quad \boldsymbol{\eta}_1 = (0.383, 0.000, 0.462, 0.800).$$

To solve this problem, we construct control rules of type (4.4.30). We take $u_2^* = 0$ and solve the time-optimal problem for the first and third equations of the linear controlled $\boldsymbol{\mu}$-system (4.4.31) under the corresponding boundary conditions. In solving this problem we take the constraints on u_1^* and u_3^* in the form $|u_j^*| \le \alpha_j^*$ $(j = 1, 3)$, where numbers α_j^* are iteratively selected in accordance with the algorithm described in Subsection 4.5.7.

Table 4.5.2.

τ	α_1^*	α_3^*	α_1	α_2	α_3
50.00	0.001225	0.000282	101.32	81.87	57.17
48.00	0.001329	0.000306	109.93	88.83	62.04
46.00	0.001447	0.000334	119.70	96.72	67.55
44.00	0.001581	0.000365	130.83	105.72	73.83
42.00	0.001736	0.000400	143.59	116.02	81.03
40.00	0.001913	0.000441	158.31	127.92	89.33
38.00	0.002120	0.000489	175.41	141.74	98.98
36.00	0.002362	0.000545	195.44	157.92	110.29
34.00	0.002648	0.000611	219.11	177.05	123.65
32.00	0.002990	0.000689	247.35	199.87	139.58
30.00	0.003402	0.000784	281.43	227.41	158.82
29.00	0.003640	0.000839	301.18	243.36	169.96
28.00	0.003905	0.000900	323.07	261.05	182.31
27.50	0.004048	0.000933	334.93	270.63	189.00
27.00	0.004200	0.000968	347.45	280.75	196.07
26.00	0.004529	0.001044	374.69	302.76	211.44
24.00	0.005315	0.001225	439.74	355.32	248.15
22.00	0.006325	0.001458	523.32	422.87	295.32
20.00	0.007654	0.001764	633.22	511.67	357.33
18.00	0.009449	0.002178	781.76	631.69	441.15
16.00	0.011959	0.002757	989.41	799.48	558.34
14.00	0.015620	0.003601	1292.29	1044.22	729.25
12.00	0.021260	0.004901	1758.95	1421.30	992.60
10.00	0.030615	0.007058	2532.89	2046.67	1429.34
8.00	0.047835	0.011028	3957.64	3197.92	2233.34

The relationships required to perform the calculations are presented in Table 4.5.1, where the control time τ (one and the same for η_1 and η_3) is obtained from the equations ($\eta_{11} = -\eta_{10}$ taken into account)

$$\tau = 2\big[2(\alpha_1^*)^{-1}\eta_{10}\big]^{1/2} = 2\big[(\alpha_3^*)^{-1}(\eta_{31} - \eta_{30})\big]^{1/2}.$$

The computations show that, in this case, $\tau = 29.05\ s$ and $\alpha_1^* = 3.640 \cdot 10^{-3}$, $\alpha_3^* = 8.390 \cdot 10^{-3}\ rad{\cdot}s^{-2}$.

Control rules u_k $(k = 1, 2, 3)$ of type (4.4.30) are *piecewise-continuous* functions with discontinuities at $t = \frac{1}{2}\tau$ (see Figures 4.5.2–4.5.4). Figure 4.5.5 shows the behavior of variables η_1, η_3, η_4.

We should stress that control function u_1 solely attains the limiting value while $|u_2| \leq 243.4$, $|u_3| \leq 169.9\ N{\cdot}m$.

Other results of numeric simulation are given in Table 4.5.2.

To compare and assess the performance of the construction of control rules of type (4.4.30), we note that the specified problem of optimal control was solved by a *numerical* method (Gulyaev et al. [1986]). It appeared that a *spatial* turn takes 27.5 s under constraints (4.5.25). (For comparison, it takes 38.5 s to perform a *planar* turn when the rotational axis of the body keeps a fixed position in space.)

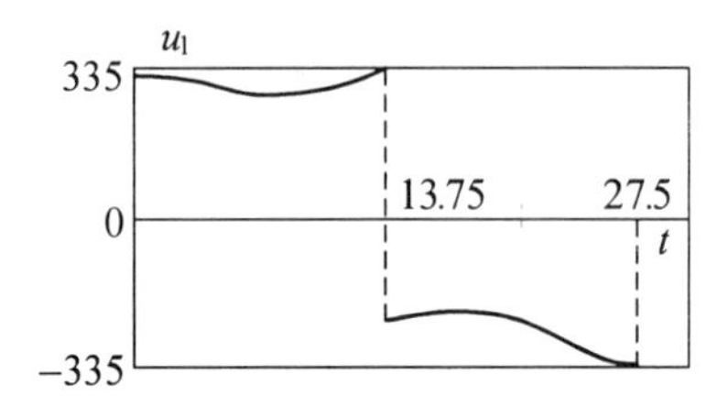

Figure 4.5.2. Control law u_1.

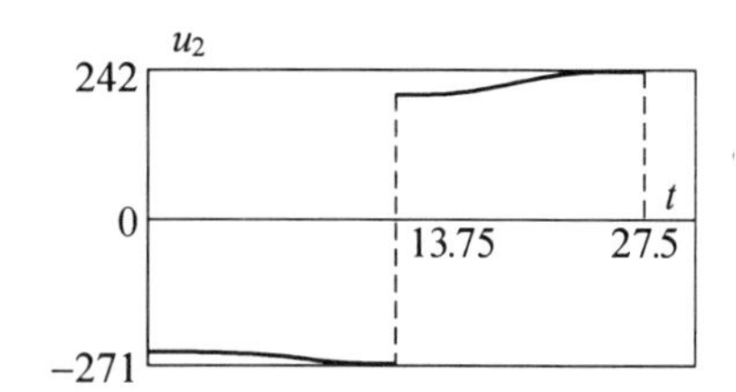

Figure 4.5.3. Control law u_2.

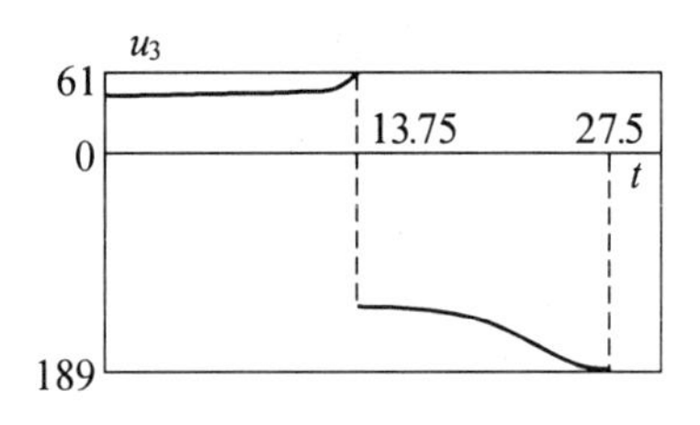

Figure 4.5.4. Control law u_3.

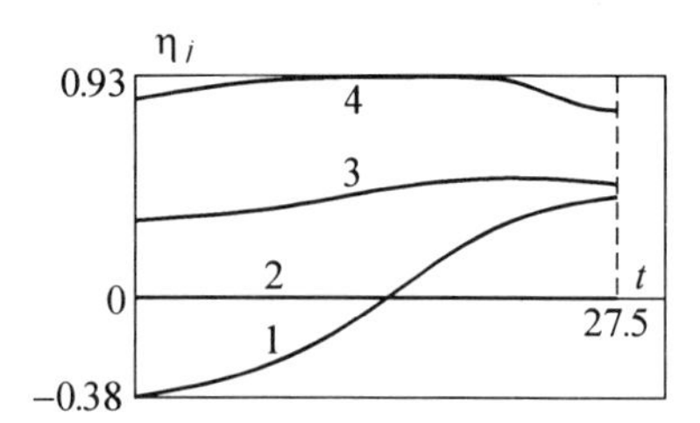

Figure 4.5.5. Functions η_j $(j = 1, \ldots, 4)$.

The corresponding optimal control rules are *piecewise-constant* (bang-bang control rules) and have one (for the planar turn) or up to four switches.

The comparison indicates that, in the case in question, control rules u_k $(k = 1, 2, 3)$ of type (4.4.30) are *suboptimal for speed of response.*

In conclusion, we note that, to solve the problem of reorientation considered, one can use control rules (4.4.30) and also control rules of the type

$$u_i = \frac{1}{\eta_3} f_i^*(\boldsymbol{x}, \boldsymbol{\eta}, \boldsymbol{u}^*) \quad (i = 1, 2, 3). \tag{4.5.26}$$

However, in this case the reorientation time (subject to constraints (4.5.25)) will sufficiently exceed 29.05 s because the range $[0.286, 0.462]$ of variable η_3 is closer to the "singularity" point $\eta_3 = 0$ of control rules (4.5.26) than the corresponding range [0.8,0.879] of η_4 to the "singularity" point $\eta_4 = 0$ of control rules (4.4.30).

4.6 The Nonlinear Problem of "Passage" of an Asymmetric Solid through a Given Angular Position

4.6.1 Preliminary Remarks

The control problem discussed in this section is typical of the subject matter of this book and is classified as essentially nonlinear problem of the "hard" encounter type in which the control is performed with respect to just *part* of the variables characterizing the state of a system, namely, we consider the turn of an asymmetric solid from a stationary state (or a state close to it) to a given angular position; it is not assumed that the turn takes the solid to a stationary state.

This approach to reorientation is compared with the conventional one which implies that the initial and final states are stationary. We demonstrate that the "unconventional" reorientation takes both significantly less time and energy, "geometric" constraints on the controls being the same. Besides, the controls thus obtained are easier to implement. They are *continuous*, as distinguished from the *piecewise-continuous* controls (see Example 4.5.1) in the traditional reorientation. Such an approach is not devoid of interest when the fastest possible reorientation of spacecraft is required to carry out *short-term* operations at the instant the desired position is attained. The examples of such operations are photographing, hitting a target (for military purposes), transmitting information, etc.

4.6.2 Formulation of the Problem

Consider *Euler's dynamical equations*

$$\begin{gathered} A\dot{\xi}_1 = (B-C)\xi_2\xi_3 + u_1, \\ B\dot{\xi}_2 = (C-A)\xi_1\xi_3 + u_2, \quad C\dot{\xi}_3 = (A-B)\xi_1\xi_2 + u_3, \end{gathered} \tag{4.6.1}$$

where the notation of the variables is chosen to conform to that used in the first group of equations of system (4.5.1). For uniformity, we also keep the indexing of variables ξ_i, u_k $(i,k = 1,2,3)$ unchanged, though in this case $l = r = 3$.

We shall consider equations (4.6.1) together with kinematic equations

$$\begin{gathered} 2\dot{\eta}_1 = \xi_1\eta_4 + \xi_3\eta_2 - \xi_2\eta_3, \\ 2\dot{\eta}_2 = \xi_2\eta_4 + \xi_1\eta_3 - \xi_3\eta_1, \quad 2\dot{\eta}_3 = \xi_3\eta_4 + \xi_2\eta_1 - \xi_1\eta_2 \end{gathered} \tag{4.6.2}$$

in terms of the Rodrigues–Hamilton variables which determine the orientation of a solid.

Variable η_4 in (4.6.2) is related to η_j $(j = 1,2,3)$ (with index j left unchanged to conform to the notation of Section 4.5) by the equation

$$\eta_4^2 + \sum_{j=1}^{3} \eta_j^2 = 1. \tag{4.6.3}$$

The equation for $\dot{\eta}_4$, if desired, can also be obtained.

For system (4.6.1)–(4.6.3), $l = p = r = 3$. In what follows we denote by $\boldsymbol{\eta}$ and $\boldsymbol{\xi}$ the vectors formed by the components η_j, η_4 and ξ_i, respectively, so $\boldsymbol{x} = (\boldsymbol{\xi}, \boldsymbol{\eta})$.

Vector $\boldsymbol{u}$ of the control rules is chosen in class $\mathcal{K}$ of vector functions $\boldsymbol{u}(\boldsymbol{x}, \boldsymbol{x}_0)$ subject to constraints (4.5.2). In this case inequalities (4.5.2) correspond to three pairs of engines fixedly attached to the solid.

Problem 4.6.1. Find control rules $\boldsymbol{u} \in \mathcal{K}$ *transferring a solid from the initial state* $\boldsymbol{\eta}(t_0) = \boldsymbol{\eta}_0$ *to the prescribed state* $\boldsymbol{\eta}(t_1) = \boldsymbol{\eta}_1$ *in a finite time. The initial state of the solid is a stationary state* $\boldsymbol{\xi}(t_0) = \boldsymbol{\xi}_0 = \mathbf{0}$. *The value* $\boldsymbol{\xi}_1 = \boldsymbol{\xi}(t_1)$ *of the angular velocity of the solid at the moment* $t = t_1$ *can be arbitrary. The time* t_1 *is not fixed.*

Remark 4.6.1.

1. The distinguishing feature of Problem 4.6.1 is that this is a problem of control with respect to *part* of the variables of the *"hard" encounter type.* The reorientation is performed without bringing the solid to a stationary state.

2. It is difficult to find rigorously optimal (in the sense of the minimum reorientation time, $t_1 \to \min$) control rules that ensure the solution of Problem 4.6.1. So, a natural effort should be made to devise a constructive method for finding suboptimal controls. This approach is taken as the keynote for further consideration of Problem 4.6.1.

Without loss of generality we assume that $\boldsymbol{\eta}_1 = (0,0,0,1)$. Thus the orientation is to be regulated relative to the coordinate system fixed at the initial time. Throughout this section, in accordance with the notation adopted, $i, j, k = 1, 2, 3$.

4.6.3 An Auxiliary Linear System

Let us differentiate both sides of each equation for $\dot{\eta}_j$ from (4.6.2) with respect to time and replace $\dot{\xi}_i$, $\dot{\eta}_j$ by their expressions from (4.6.1), (4.6.2). After reduction we get the equalities

$$\ddot{\eta}_j = f_j(\boldsymbol{x}, \boldsymbol{u}), \tag{4.6.4}$$

where

$$\begin{aligned} f_1 = {}& \frac{1}{2}\left[\eta_4(u_1 + M_1)A^{-1} + \eta_2(u_3 + M_3)C^{-1} - \eta_3(u_2 + M_2)B^{-1}\right] \\ & - \frac{1}{4}\eta_1 \sum_{i=1}^{3} \xi_i^2, \qquad M_1 = (B - C)\xi_2\xi_3 \qquad (123, ABC). \end{aligned}$$

We treat f_j as the auxiliary controls u_j^*. Then expressions (4.6.4) can be considered as the auxiliary *linear controlled* $\boldsymbol{\mu}$*-system*

$$\dot{\eta}_j = \mu_j, \qquad \dot{\mu}_j = u_j^* \tag{4.6.5}$$

of type (4.5.8).

The "original" controls u_k are expressed in terms of u_j^* through equalities of type (4.5.6) which, in the case in question, have the form

$$\begin{aligned} u_1 = {}& 2A\eta_4^{-1}\left[(\eta_4^2 + \eta_1^2)u_1^* + (\eta_1\eta_2 + \eta_4\eta_3)u_2^* + (\eta_1\eta_3 - \eta_4\eta_2)u_3^*\right] \\ & + \frac{1}{2}A\eta_1\eta_4^{-1}\sum_{i=1}^{3}\xi_i^2 + (C - B)\xi_2\xi_3 \qquad (ABC, 123). \end{aligned} \tag{4.6.6}$$

Remark 4.6.2.

1. Construction (4.6.6) can be regarded as the *general structural form* of the control rules in Problem 4.6.1. The parameters of this form, i.e., the auxiliary control rules u_j^* can be determined by solving the corresponding linear control problems.

2. Structure (4.6.6) of the control rules contains the factor η_4^{-1}, which formally implies a "*singularity*." However, thorough analysis reveals that in the case of $\boldsymbol{\eta}_1 = (0,0,0,1)$, the relationship $\eta_4 \in [\eta_{40}, 1]$ holds in the course of the control. Therefore, the specified "singularity" does not occur at all. If the value of η_{40} is small or $\boldsymbol{\eta}_1 \neq (0,0,0,1)$, it suffices to change the controls to those obtained from (4.6.6) by a permutation of the indices and coefficients A, B, and C (or to a combination of such controls). The final choice of "suboptimal" controls u_k is conducted in the iterative regime widely used in modern methods of applied control theory. In the context of the proposed algorithm for solving Problem 4.6.1, we note that the iterative search indicated for "suboptimal" controls is quite simple. It can be implemented in real time as a controlled object functions.

4.6.4 An Auxiliary Linear Control Problem

Let us solve the problem of time-optimal control with respect to part of the variables on the *simultaneous* transfer of system (4.6.5) to the position

$$\eta_j = 0. \tag{4.6.7}$$

The solution of this problem subject to the conditions of type (4.5.10) has the form

$$u_j^* = -\alpha_j^* \operatorname{sgn} \eta_j, \quad u_j^* = 0 \ \ (\eta_{j0} = 0). \tag{4.6.8}$$

Unlike in Section 4.5, we note that we set $u_j^* = 0$ for $\eta_{j0} = 0$ in (4.6.8), because $\boldsymbol{\xi}_0 = \mathbf{0}$ in Problem 4.6.1 and the condition $\eta_{j0} = 0$ means, in fact, that $\eta_{j0} = \mu_{j0} = 0$.

As in Section 4.5, the procedure for choosing levels α_j^* implies an iterative process which takes into account the necessity for *simultaneous* transfer of the system to position (4.6.7).

In view of $\mu_{j0} = 0$ (which is a consequence of $\boldsymbol{\xi}_0 = \mathbf{0}$), the value

$$\tau = \left[2|\eta_{j0}|(\alpha_j^*)^{-1}\right]^{1/2} \tag{4.6.9}$$

determines the *minimum* time τ in which (4.6.7) is achieved in the auxiliary linear problem.

The trajectories of the linear system (4.6.5), (4.6.8) corresponding to Problem 4.6.1 have the form

$$\mu_j = -(\alpha_j^* \operatorname{sgn} \eta_{j0})\, t, \quad \eta_j = \eta_{j0} - 1/2(\alpha_j^* \operatorname{sgn} \eta_{j0})\, t^2. \tag{4.6.10}$$

Analyzing equalities (4.6.10), we conclude that in the case $\boldsymbol{\eta}_1 = (0, 0, 0, 1)$, $\boldsymbol{\xi}_0 = \mathbf{0}$, the condition $\eta_4 > \gamma = \text{const} > 0$, once met at $t = t_0$, also remains valid for $t \in [t_0, t_0 + \tau]$. This fact rules out the "singularity" of controls u_k of type (4.4.6).

4.6.5 Algorithm for Solving Problem 4.6.1

(1) Choosing construction (4.6.6) of control rules u_k in which u_j^* have the form (4.6.8).

(2) "Assigning" levels $\alpha_j^*(\boldsymbol{\eta}_0)$ of auxiliary controls u_j^*, "*equalizing*" the control time with respect to each variable η_j taken into account. In so doing, α_j^* predetermine the corresponding value of $\tau(\boldsymbol{\eta}_0)$.

(3) Validating the original constraints (4.5.2) on controls u_k along trajectories (4.6.10) of the linear system (4.6.5), (4.6.8). In so doing, equalities of type (4.5.20) are employed which, in the case in question, take the form

$$\xi_1 = 2\eta_4^{-1}\big[(\eta_4^2+\eta_1^2)\mu_1+(\eta_1\eta_2+\eta_4\eta_3)\mu_2+(\eta_1\eta_3-\eta_4\eta_2)\mu_3\big] \ (123). \quad (4.6.11)$$

Besides, the relationship

$$\sum_{i=1}^{3} \xi_i^2 = 4\big[\eta_4^{-1}\sum_{j=1}^{3}(\xi_j\mu_j)\big]^2 + \sum_{i=1}^{3}\mu_j^2.$$

facilitates the calculation.

If estimates (4.5.2) fail to hold or, vice versa, they are satisfied with a kind of "reserve," the search for the appropriate α_j^* is to be continued. Otherwise, the reorientation time is determined by equality (4.6.9).

4.6.6 The Conditions for the Solvability of Problem 4.6.1

Taking into account (4.6.6), (4.6.11), we conclude that $u_k = u_k(\boldsymbol{z}^*, \boldsymbol{\mu}, \boldsymbol{\alpha}^*)$, where vectors $\boldsymbol{z}^*, \boldsymbol{\mu}$, and $\boldsymbol{\alpha}^*$ comprise η_j, μ_j, and α_j^*, respectively. Components η_j, μ_j are of the form (4.6.10), whereas α_j^* are *iteratively* chosen following the algorithm presented in Subsection 4.6.5. This means that control rules u_k in the form $u_k = u_k(\boldsymbol{x}, \boldsymbol{x}_0)$ can also be represented in the *program* form $u_k = u_k(t)$, i.e., as *continuous* functions of t for given initial conditions.

Theorem 4.6.1. 1. *Let* $\boldsymbol{\eta}_1 = (0, 0, 0, 1)$. *If levels* α_j^* *of the auxiliary control rules* u_j^* *are such that the inequalities*

$$\max_{t\in[t_0, t_0+\tau]} \big|u_k(t)\big| \le \alpha_k$$

are satisfied along trajectories (4.6.10) *of the linear system* (4.6.5), (4.6.8), *then control rules* (4.6.6), (4.6.8) *give a solution of Problem* 4.6.1 *in time* $\tau(\boldsymbol{\eta}_0)$. *The "passage" of the solid through the prescribed angular position* $\boldsymbol{\eta} = \boldsymbol{\eta}_1$ *in three-dimensional inertial space is guaranteed in time* τ. *The value* *of τ is determined by expression* (4.6.9) *and can be iteratively calculated by the algorithm presented in Subsection* 4.6.5.

2. *If* $\boldsymbol{\eta}_1 \neq (0,0,0,1)$ *or the value of* η_{40} *is small, then to solve Problem* 4.6.1, *it suffices to change to controls* u_k *obtained from* (4.6.6), (4.6.8) *by a permutation of the indices* (*or to a combination of such controls*).

Theorem 4.6.2. *For any values of* $\boldsymbol{\eta}_0$ *and* $\boldsymbol{\eta}_1$, *Problem* 4.6.1 *is solved by control rules* u_k *of type* (4.6.6), (4.6.8) *arbitrarily small in magnitude* (*or, by those obtained from* (4.6.6), (4.6.8) *by a permutation of the indices*), *where the reorientation takes a sufficiently long time.*

From the purely mathematical point of view, the formulations of Theorems 4.6.1 and 4.6.2 have a disadvantage: they require an iterative search for "time-suboptimal" controls u_k. However, this iterative process is easily implemented. Finally, Theorems 4.6.1 and 4.6.2 establish constructive conditions for solvability of the nonlinear Problem 4.6.1 which can be verified in real time.

4.6.7 Generalization of Problem 4.6.1

The proposed algorithm can also be carried over to the case when the initial state of a solid is not a stationary state $\boldsymbol{\xi}_0 \neq \mathbf{0}$. In this case in solving the auxiliary control problem for linear system (4.6.5), the constraint $\sum \eta_j^2 \leq 1$ imposed on variables η_j must be met. (When $\boldsymbol{\xi}_0 = \mathbf{0}$ such a constraint, which is a consequence of (4.6.3), is satisfied automatically.) With levels α_k of controls u_k fixed, this is possible up to a certain value $\Delta = |\boldsymbol{\xi}_0|$. (Otherwise, the angular velocity of the solid must be reduced in advance.) Besides, if $\eta_{j0} = 0$ for some values of j while $\mu_{j0} \neq 0$, one should set $\tau = 2(\alpha_j^*)^{-1}|\mu_{j0}|$ for the corresponding values of j.

4.6.8 Results of Computer Simulation

For a solid (spacecraft) with $A = 4{\cdot}10^4$, $B = 8{\cdot}10^4$, and $C = 5{\cdot}10^4$ $kg{\cdot}m^2$ we consider the triaxial reorientation from the position $\boldsymbol{\xi}_0 = \mathbf{0}$, $\boldsymbol{\eta}_0 = (0.353, 0.434, 0.432, 0.701)$ to $\boldsymbol{\eta}_1 = (0,0,0,1)$. The value of $\boldsymbol{\xi}_1$ is not assumed to be zero and can be arbitrary.

In the case when $\boldsymbol{\xi}_0 = \boldsymbol{\xi}_1 = \mathbf{0}$, the reorientation is attained (through one spatial turn) by control rules (4.6.6) with $\alpha_1 = 32.6$, $\alpha_2 = 80.1$, and $\alpha_3 = 68.0$ $N{\cdot}m$ in time $\tau = 70$ s. Control rules u_k are piecewise-continuous and have one switching moment $t = 35$ s.

The reorientation in the context of Problem 4.6.1 takes the same time $\tau = 70$ s (also one spatial turn) with $\alpha_1 = 29.9$, $\alpha_2 = 40.1$, $\alpha_3 = 24.9$ $N{\cdot}m$. The control rules are *continuous* and vary in the range $-29.9 \leq u_1 \leq -16.3$, $-40.1 \leq u_2 \leq -24.4$, and $-24.9 \leq u_3 \leq 2.4$ $N{\cdot}m$.

The same reorientation time is obtained in the context of Problem 4.6.1 with $\sum \alpha_k$ 41.6% lower than in the case when $\boldsymbol{\xi}_0 = \boldsymbol{\xi}_1 = \mathbf{0}$.

The computations also show that, for $\max \alpha_k \leq 80.1$ $N{\cdot}m$, the reorientation in the context of Problem 4.6.1 is attained in 49.5 s. Control u_2 alone achieves the boundary value, whereas $-32.6 \leq u_1 \leq -59.8$, $-49.9 \leq u_3 \leq 4.6$ $N{\cdot}m$.

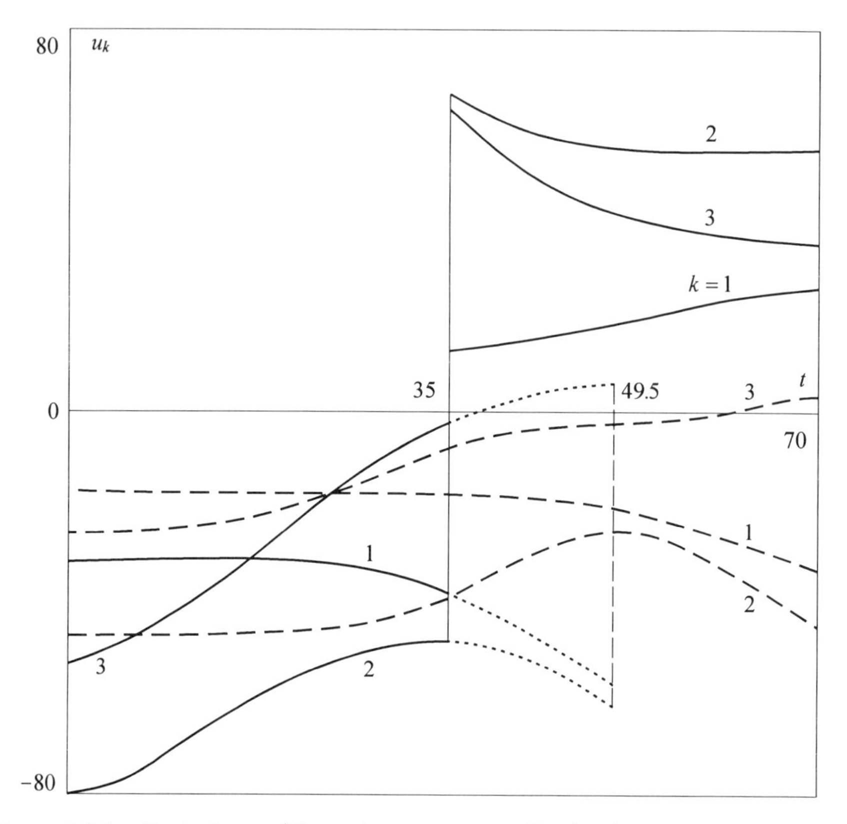

Figure 4.6.1. Controls u_k (discontinuous at $t = 35$ s) solving the reorientation problem for $\boldsymbol{\xi}_0 = \boldsymbol{\xi}_1 = \mathbf{0}$ in time $\tau = 70$ s are shown as solid lines. Dashed lines depict (continuous) controls u_k solving Problem 4.6.1 in time $\tau = 70$ s. Solid lines ($0 \leq t \leq 35$ s) continued by dotted line ($35 < t \leq 49.5$ s) depict (continuous) controls u_k solving Problem 4.6.1 for $\max \alpha_k \leq 80.1$ $N \cdot m$ (in time $\tau = 70$ s).

The gain in the reorientation time in the framework of Problem 4.6.1 is 29.3%, the "geometric" constraints on u_k being the same.

4.7 Overview of References

The approach to investigating nonlinear controlled systems proposed in Chapter 4 was outlined in the dissertation of Vorotnikov [1979a]. The essentials were discussed at Prof. V.V. Rumyantsev's seminar at Moscow State University in September, 1978.*

The quest for constructive methods for constructing the controls in problems of partial stabilization has stimulated the development of this approach. As a natural result of these efforts, the idea has been conceived of separating

*Vorotnikov, V.I. [1979] On a Method of Investigating Stability and Stabilization of Motion with Respect to Part of the Variables, *Izv. Akad. Nauk SSSR Mekh. Tverd. Tela*, 6, 166.

auxiliary linear controlled μ-systems which completely determine the state of a given part of the variables of an original nonlinear system. This idea turned out to be quite fruitful and made it possible to solve a number of nonlinear problems of controlling the angular motion of a solid (Vorotnikov [1979a], [1982b], [1985], [1986a]–[1986c], [1990], [1991a], [1997b], [1997c]). Nonlinear control rules obtained with this method contain terms that are *fractional functions* of the phase variables. We note that the method was also applied to the problems of stabilization with respect to *all* the variables (Vorotnikov [1988b], [1991a], [1991b]). In this case, the problem of partial stabilization plays the role of the auxiliary problem being analyzed in the *first stage* of the solution.

It should be mentioned that the approach under discussion is connected with the method of *exact feedback linearization* of nonlinear controlled systems, which has been studied intensively beginning with the report by Brockett [1978] at VII IFAC Congress (see monographs by Isidori [1989], Nijmejer and van der Schaft [1990]). A number of publications in this field of research consider the problem of controlling the angular motion of a solid (Hunt et al. [1983], Dwyer [1984], [1986], Monaco and Stornelli [1986]). These works and the relevant author's results (Vorotnikov [1985], [1986a], [1986b]; see also Vorotnikov [1984]) were published at about the same time. In essence, they all elaborate on (in different contexts and forms) *one and the same* very effective technique: an exact linearization of associated dynamical and kinematic equations describing the angular motion of a solid by means of a specially introduced feedback. The results of comparison with other methods, including *exact* numerical methods, show that the proposed technique of linearization of the original nonlinear systems allows obtaining *suboptimal* control laws.

A further principal step that extends the applicability of the method of constructing auxiliary linear controlled μ-systems is associated with solving nonlinear game-theoretic control problems (Vorotnikov [1994b], [1994c]), in particular, game-theoretic problems of controlling the angular motion of a solid (Vorotnikov [1994a], [1994c], [1997b], [1997e]). This line of investigation is discussed in Chapter 5.

We note that the control rules solving applied problems of controlling the angular motion of a solid discussed in Chapters 4 and 5 have the following structure:

$$\boldsymbol{u} = \varphi(\boldsymbol{x}) + \psi(\boldsymbol{x})\boldsymbol{u}^*, \tag{4.7.1}$$

were $\boldsymbol{u}^*$ is a vector of "auxiliary" controls. Here, an attempt to give a certain "physical" interpretation of this structural form is not devoid of interest. Thus, for example, the first term in expression (4.7.1) may be treated as a control which "*imposes*" a "*new dynamics*" on the original system (i.e., which alters its right-hand side). Expression $\psi(\boldsymbol{x})$ "aids" one to form an auxiliary linear controlled μ-system. In this case, the function of the "auxiliary" controls $\boldsymbol{u}^*$ reduces to controlling the newly formed linear μ-system. Because of this, most of the control laws constructed in this book cannot be classed among the controls obtained *by using natural, proper motions of an object* (see Bellman et al. [1958] and Krasovskii, A.A. [1987]). However, if preassigned restrictions (informational ones included) on the controls are met, the structural analysis of this kind may be of no importance. A well-known quip by Pasteur seems to fit the case: "Let our Stranger remain a stranger, we are content with her love."

The method proposed in Chapter 4 made it possible to study a number of applied problems of partial stabilization this book did not touch upon. We mention two of them: stabilization of a satellite in orbit with flywheels and stabilization of a solid containing cavity filled with an ideal liquid (Vorotnikov [1991a]).

An alternative way of investigation, which proved to be rather effective and allowed solving a number of applied problems of partial stabilization, rests on using the method of Lyapunov functions (Rumyantsev [1970], Krementulo [1977], Furasov [1977], Oziraner [1978], Rumyantsev and Oziraner [1987]).* This approach is a subject of in-depth analysis in the monograph by Rumyantsev and Oziraner [1987].

*See also: Oziraner, A.S. [1974] On Uniaxial Stabilization of a Dynamically Symmetric Satellite in a Circular Orbit, *Izv. Akad. Nauk SSSR Mekh. Tverd. Tela*, 3, 11–18; Oziraner, A.S. [1978] On Optimal Stabilization of Equilibrium Positions of a Solid Containing Cavity Filled with a Liquid, *Prikl. Mat. Mekh.*, **42**, 4, 749–753; Lisina, I.L. [1988] Optimal Stabilization of a Controlled Gyroscopic Compass on an Immobile Base, *Kosmich. Issledovaniya*, **26**, 5, 787–791.

Chapter 5

Nonlinear Game-Theoretic Problems of Control with Respect to Part of the Variables under Uncontrollable Interference

In this chapter the method of constructing auxiliary linear μ-systems (including controlled systems) proposed in the preceding chapters is applied to solving nonlinear control problems (with respect to all the variables and to part of them) under *uncontrollable interference.* The interference is identified with an intentionally acting *opponent* or with *uncertain factors* ("*nature*"). Such a formulation implies that only the minimum information about the interference, namely, their limits of variation, is available, so the systems of differential equations we shall deal with fall into the type of "*uncertain systems.*" The admissible controls must be chosen to ensure a *guaranteed result* for any possible action of interference, even though the most adverse. This formulation of the control problems is typical of the *differential games* theory.

We consider two groups of problems in the context of the formulation outlined of a game-theoretic control problem.

(1) Game-theoretic problems of control with respect to *part* of the variables which characterize the state of the original system.

(2) Game-theoretic problems of control with respect to *all* the variables whose solution technique implies that in the first (preliminary) stage the corresponding problems of control with respect to some initially chosen part of the variables are solved.

For both groups of problems we derive constructively verifiable sufficient conditions of solvability which embrace a relatively general class of nonlinear controlled systems subject to uncontrollable interference. These conditions guarantee the precise transfer of a system to the position in which a *specified part* of the variables (problems of group (1)) or *all* of them (problems of group (2)) are equal to zero in a *finite time*. The transfer is to be performed through controls that satisfy *prescribed* "geometric" *constraints*. The nonlinear game-theoretic control problem is reduced to the corresponding *linear* game-theoretic problems for specially constructed *auxiliary linear conflict-control* μ*-systems*.

At the same time, though fitting in well with the general scheme indicated, the solution technique for each of the two groups of problems has its own essential peculiarities. More specifically, the linear game-theoretic problems corresponding to the problems from the second group are those of control with respect to all the variables which we manage to reduce to optimal (nonconflict) control problems. This is not the case for the first group of problems because the corresponding linear game-theoretic problem is a problem of control with respect to part of the variables for which this reduction cannot be performed. As a result, the first group of problems turns out to be far more complicated.

To exemplify implementation of the proposed approach, we constructively solve a number of complicated nonlinear game-theoretic problems of controlling the angular motion of a solid (spacecraft) which are of independent theoretical and applied importance:

(1) *game-theoretic problem of the "passage" by a solid through a given angular position* (which relates to the first group of problems);

(2) *game-theoretic problems of uniaxial and triaxial reorientation of a solid* (which relate to the second group of problems).

The decisions obtained in these problems are time-suboptimal. Two methods for solving game-theoretic problem of reorientation are proposed depending on how the "auxiliary interference" is estimated in the auxiliary linear conflict-control systems.

The results of computer simulation are presented which confirm the effectiveness of the proposed approach.

5.1 Guaranteed Conditions for Controllability with Respect to Part of the Variables under Uncontrollable Interference

5.1.1 Preliminary Remarks

In this section the guaranteed conditions for controllability *with respect to part of the variables* for one sufficiently general class of nonlinear systems are obtained. If such conditions are satisfied, the system can be taken to the position in which the *specified part* of phase variables equals zero through the limited controls in a *finite* time. The control is realized under uncontrollable interference whose levels are not supposed to be small.

The original nonlinear game-theoretic control problem is reduced to the corresponding *linear* game-theoretic problems for specially constructed *auxiliary linear conflict-control $\boldsymbol{\mu}$-systems.* The solutions of the indicated linear game-theoretic problem can be constructively found in the framework of the discrete scheme of the Krasovskii *method of extremum aiming* (Krasovskii [1970]). As a result, the approach proposed in the previous chapters of the book is further developed as applied to the game-theoretic control problems.

5.1.2 Formulation of the Problem

Let the perturbed motion of a control object be described by a nonlinear finite-dimensional system of ordinary differential equations

$$\begin{aligned} \dot{\xi}_i &= X_i^{(0)}(\boldsymbol{\xi}, \boldsymbol{\eta}) + \sum_{k=1}^{r} X_{ik}^{(1)}(\boldsymbol{\xi}, \boldsymbol{\eta}) u_k + X_i^{(2)}(\boldsymbol{\xi}, \boldsymbol{\eta}) v_i, \\ \dot{\eta}_j &= X_j^{(3)}(\boldsymbol{\xi}, \boldsymbol{\eta}) \qquad (i = 1, \ldots, l;\ j = 1, \ldots, n - l). \end{aligned} \tag{5.1.1}$$

Here $\boldsymbol{\xi} = (\xi_1, \ldots, \xi_l)$, $\boldsymbol{\eta} = (\eta_1, \ldots, \eta_{n-l})$ are the vectors of the phase variables; $\boldsymbol{u} = (u_1, \ldots, u_r)$ is the vector of the controls; uncontrollable interference and perturbations constitute vector $\boldsymbol{v} = (v_1, \ldots, v_l)$. Let us denote $\boldsymbol{x} = (\boldsymbol{\xi}, \boldsymbol{\eta})$. We assume that, in the domain

$$\boldsymbol{\Lambda} : ||\boldsymbol{x}|| < H > 0,$$

functions $X_i^{(0)}$, $X_{ik}^{(1)}$, and $X_i^{(2)}$ are defined and continuous, and so are functions $X_j^{(3)}$ together with their partial derivatives with respect to ξ_i, η_j.

Functions $X_i^{(0)}$, $X_j^{(3)}$ vanish at $\boldsymbol{x} = \mathbf{0}$. For functions $X_{ik}^{(1)}$, on the contrary, we consider vanishing at $\boldsymbol{x} = \mathbf{0}$ impossible. Otherwise, we shall contradict the conditions given below.

From the standpoint of applications, the structural form (5.1.1) of a controlled object embraces, in particular, an important class of problems, such as the problems of controlling the angular motion of a solid in the presence of uncontrollable interference. Problems of this kind are closely related to those arising in the control of *aircraft* (*spacecraft*), *manipulators*, and objects of *biomechanics*. In this case the first group of equations of system (5.1.1) comprises *dynamical* equations and the second group (containing no controls and perturbations) some *kinematic* equations that characterize the orientation of the object considered.

The devices producing the controlling actions have limited capabilities. We shall assume that the "*geometric*" (nonintegral) constraints

$$|u_k| \leq \alpha_k = \text{const} > 0 \qquad (k = 1, \ldots, r). \tag{5.1.2}$$

are imposed on the controls. Here α_k are given numbers characterizing the maximum possible magnitudes (*levels*) of u_k.

Various assumptions about the interference v_i $(i = 1, \ldots, l)$ can be made. If the realizations of interference are a *stochastic process* (for example, *white noise*), then a *stochastic* (*probabilistic*) approach to the solution is possible.

However, in many cases, just the minimum information about the interference, namely, its limits of variation, is available and no probabilistic characteristics of the realizations of interference within these limits are known. In this case only constraints of the form

$$|v_i(t)| \le \beta_i = \text{const} > 0 \qquad (i = 1, \ldots, l). \tag{5.1.3}$$

are imposed.

The numbers β_i characterize the maximum possible interference magnitudes (levels), which can take the form of *any Lebesgue-integrable* time-dependent functions $\boldsymbol{v}(t)$ within the limits (5.1.3).

As opposed to the stochastic approach, *differential games* include problems on a type of control that guarantees the desired result for *any* possible action of interference, even the most adverse ("*worst*"). It is this approach (applied to the class of systems (5.1.1) under study) that is considered in the present chapter.

Let us make the control problems under consideration more specific. Suppose that the class L of control functions $\boldsymbol{u} = \boldsymbol{u}(t, \boldsymbol{x}, \boldsymbol{x}_0)$ is given (where $\boldsymbol{x}_0$ is the initial value of vector $\boldsymbol{x}$), with realizations $\boldsymbol{u}[t]$ being *Lebesgue integrable* functions. Controls $\boldsymbol{u} \in L$ satisfy the given constraints (5.1.2). Also specified is the class $L_1 = \{\boldsymbol{v} = \boldsymbol{v}(t)\}$ of interference satisfying constraints (5.1.3), which are *Lebesgue-integrable* over an arbitrary finite interval $t \in [t_0, t_1]$.

Let $\boldsymbol{y}$ be the vector of those phase variables of system (5.1.1), which are of interest in the course of controlling this system. We consider that $\boldsymbol{y} = (\eta_1, \ldots, \eta_r)$, i.e., vector $\boldsymbol{y}$ is composed of the first r components of vector $\boldsymbol{\eta}$. Note also that $\dim(\boldsymbol{y}) = \dim(\boldsymbol{u})$.

Problem 5.1.1. It is required to find control rules $\boldsymbol{u} \in L$ transferring the system from a given domain S of initial perturbations $\boldsymbol{x}(t_0) \in S$ to the position $\boldsymbol{y}(t_1) = \boldsymbol{0}$ in a finite time for any admissible realizations of interference $\boldsymbol{v} \in L_1$. The time $t_1 > t_0$ is not fixed. (The values of vector $\boldsymbol{x}$ components not entered into $\boldsymbol{y}$ can be arbitrary at $t = t_1$.)

5.1.3 An Auxiliary Linear Conflict-Control System

Consider vector function $\boldsymbol{\Phi} = (\Phi_1, \ldots, \Phi_r)$ whose components are defined as follows

$$\Phi_k(\boldsymbol{x}, \boldsymbol{u}) = \sum_{i=1}^{l} \frac{\partial X_k^{(3)}(\boldsymbol{x})}{\partial \xi_i} [X_i^{(0)}(\boldsymbol{x}) + \sum_{s=1}^{r} X_{is}^{(1)}(\boldsymbol{x}) u_s] + \sum_{j=1}^{n-l} \frac{\partial X_k^{(3)}(\boldsymbol{x})}{\partial \eta_j} X_j^{(3)}(\boldsymbol{x})$$

$$(k = 1, \ldots, r).$$

Let us introduce the *Jacobi matrix* $F(\boldsymbol{x}) = (\partial\Phi_s/\partial u_k)$ $(s, k = 1, \dots, r)$. Suppose that, in the domain

$$\boldsymbol{\Lambda}_1 : \|\boldsymbol{x}\| < H_1 \qquad (\boldsymbol{\Lambda}_1 \subseteq \boldsymbol{\Lambda}), \tag{5.1.4}$$

the following condition is satisfied:

$$\operatorname{rank} F(\boldsymbol{x}) = r. \tag{5.1.5}$$

By the *theory of implicit functions*, we conclude that in domain (5.1.4) the system of equations

$$\Phi_k(\boldsymbol{x}, \boldsymbol{u}) = u_k^* \qquad (k = 1, \dots, r) \tag{5.1.6}$$

(u_k^* are the control rules to be defined below, which constitute vector $\boldsymbol{u}^*$) has the solution

$$\boldsymbol{u} = \boldsymbol{f}(\boldsymbol{x}, \boldsymbol{u}^*). \tag{5.1.7}$$

The relationship $\boldsymbol{f}(\mathbf{0}, \mathbf{0}) = \mathbf{0}$ is true and components f_k of vector function $\boldsymbol{f} \in R^r$ are continuous in $\boldsymbol{x}$, $\boldsymbol{u}^*$ for $\boldsymbol{x} \in \boldsymbol{\Lambda}_1$ and for all $\boldsymbol{u}^*$.

Let us introduce the notation

$$v_j^* = \sum_{i=1}^{l} \frac{\partial X_j^{(3)}(\boldsymbol{x})}{\partial \xi_i} X_i^{(2)}(\boldsymbol{x}) v_i \qquad (j = 1, \dots, r) \tag{5.1.8}$$

and interpret v_j^* $(k = 1, \dots, r)$ as "*auxiliary interference.*"

As a result, under the assumptions made, from the closed nonlinear system (5.1.1), (5.1.7), one can segregate the *linear conflict-control* $\boldsymbol{\mu}$*-system*

$$\dot{\eta}_j = \mu_j, \quad \dot{\mu}_j = u_j^* + v_j^* \qquad (j = 1, \dots, r). \tag{5.1.9}$$

In $\boldsymbol{\mu}$-system (5.1.9), u_j^* $(j = 1, \dots, r)$ are the controls, and v_j^* $(j = 1, \dots, r)$ act as perturbations. We shall further use the designations $\boldsymbol{y} = (\eta_1, \dots, \eta_r)$, $\boldsymbol{\mu} = (\mu_1, \dots, \mu_r)$.

In what follows we construct the solution of the original nonlinear Problem 5.1.1 based on solving the corresponding *game-theoretic* control problems for the auxiliary linear conflict-control $\boldsymbol{\mu}$-system (5.1.9). As a result, the construction of control rules (5.1.7) can be regarded as the *general structural form* of the control rules in Problem 5.1.1. The parameters of this form, the auxiliary control rules u_j^* $(j = 1, \dots, r)$, are determined by solving the corresponding linear game-theoretic control problems.

5.1.4 An Auxiliary Linear Game-Theoretic Control Problem

The admissible constraints on "auxiliary interference" v_j^* $(j = 1, \dots, r)$ are defined by the given constraints (5.1.3) on v_i $(i = 1, \dots, l)$, namely, taking into account equalities (5.1.8), we get the following estimates

$$|v_j^*| \le \beta_j^* = \sum_{i=1}^{l} Y_{j,i}^* \beta_i, \quad (j = 1, \dots, r),$$
$$Y_{j,i}^* = \sup_{\boldsymbol{x} \in \bar{\boldsymbol{\Lambda}}_1} \left\{ \frac{\partial X_j^{(3)}(\boldsymbol{x})}{\partial \xi_i} X_i^{(2)}(\boldsymbol{x}) \right\}. \tag{5.1.10}$$

Here $\bar{\boldsymbol{\Lambda}}_1$ is the closure of the set $\boldsymbol{\Lambda}_1 \subseteq \boldsymbol{\Lambda}$.

Estimates (5.1.10) are obtained from the *reinforcing* inequalities. Consequently, the levels of the "auxiliary interference" v_j^* $(j = 1, \dots, r)$ are *overestimated* compared with the real values.

We note that if system (5.1.1) is a model of the angular motion of a solid (the kinematic equations are chosen in the Poisson or Rodrigues–Hamilton form), then obtaining the estimates of type (5.1.10) presents no difficulty (see Section 5.2).

In accordance with the objective of control in Problem 5.1.1, let us solve the control problem (via u_j^*) on bringing $\boldsymbol{\mu}$-system (5.1.9) to the position

$$\eta_j = 0 \qquad (j = 1, \dots, r) \tag{5.1.11}$$

in a *minimum time* for any *admissible perturbations* v_j^* that meet condition (5.1.10).

The terminal values of μ_j can be arbitrary. Taking this circumstance into account, Problem 5.1.1 implies that control u_j^* should *simultaneously* take η_j to position (5.1.11) for any admissible v_j^*.

We treat this problem as a *differential game.* In this game one of the players (controlling party) has at its disposal the auxiliary controls u_j^* $(j = 1, \dots, r)$ and aims at decreasing time τ for taking the system to the required position (5.1.11). The second player (opponent), aiming to increase τ or prevent attaining the position (5.1.11) at all, has at its disposal "auxiliary perturbations" v_j^* $(j = 1, \dots, r)$.

To solve this problem, we impose admissible constraints on u_j^*. We note that it *makes no sense* to study the auxiliary game-theoretic problem of attaining position (5.1.11) in a minimum time for each equation of $\boldsymbol{\mu}$-system (5.1.9) separately, subject to constraints (5.1.10) along with

$$|u_j^*| \le \alpha_j^* = \text{const} > 0 \qquad (j = 1, \dots, r). \tag{5.1.12}$$

Indeed, in solving this problem, the terminal values of η_j are arbitrary, and interference v_j^* is uncontrollable. As a result, the solutions of game-theoretic problems obtained *separately* for each equation of $\boldsymbol{\mu}$-system (5.1.9), subject to constraints (5.1.10), (5.1.12), *will not ensure simultaneous* passage of each variable η_j through position (5.1.11).

However, we shall demonstrate that one can avoid the above difficulty by considering all the equations of $\boldsymbol{\mu}$-system (5.1.9) in the *aggregate* and, instead of (5.1.12), by taking the admissible constraints on u_j^* in the form

$$\|\boldsymbol{u}^*\| \le \alpha^* = \text{const} > 0. \tag{5.1.13}$$

Here $\|\boldsymbol{u}^*\| = \left(\sum_{k=1}^{r}\left(u_k^*\right)^2\right)^{1/2}$ is the Euclidean norm of vector $\boldsymbol{u}^*$ in R^3.

Again, we impose the constraints on v_j^* in the form ($\beta^* = \left(\sum_{k=1}^{r}\left(\beta_k^*\right)^2\right)^{1/2}$)

$$\|\boldsymbol{v}^*\| \leq \beta^* = \text{const} > 0. \tag{5.1.14}$$

If $\alpha^* > \beta^*$, the solution of the auxiliary game-theoretic problem in question, subject to constraints (5.1.13), (5.1.14), is given by the *extremal deceleration strategy* $U^* = U^*(t, \boldsymbol{y}, \boldsymbol{\mu}, \theta)$ (Krasovskii [1970]) which ensures minimax time required to attain position (5.1.11) not later than by a moment $\tau = \theta(t_0, \boldsymbol{y}_0, \boldsymbol{\mu}_0)$. This will guarantee the simultaneous passage of position (5.1.11) by each variable η_j for any admissible v_j^*.

Controls u_j^* resulting from strategy U^* have the form (Krasovskii [1970])

$$u_j^*(t, \boldsymbol{y}, \boldsymbol{\mu}) = u_j^*(t, \boldsymbol{y}, \boldsymbol{\mu}, \theta(t, \boldsymbol{y}, \boldsymbol{\mu})) = \begin{cases} -\alpha^* L_j \left(\sum\limits_{k=1}^{r} L_k^2\right)^{-1/2}, & \omega > 0 \\ 0, & \omega \leq 0 \end{cases} \tag{5.1.15}$$

$$\omega = \left(\sum_{k=1}^{r} L_k^2\right)^{1/2} - 1/2\alpha^*(\theta - t)^2, \quad L_j = \eta_j + (\theta - t)\mu_j.$$

In expressions (5.1.15), variable θ is derived from the equation $\omega = 0$. The value of τ is determined as the *least* positive root of the following equation (we suppose that $t_0 = 0$):

$$\left(\sum_{k=1}^{r}\left(L_k^*\right)^2\right)^{1/2} - 1/2(\alpha^* - \beta^*)\tau^2 = 0, \quad L_k^* = \eta_{k0} + \tau\mu_{k0}. \tag{5.1.16}$$

Number τ specifies the *minimum guaranteed* (simultaneous with respect to all η_j) time of attaining position (5.1.11) in the linear problem under consideration. The transfer time corresponds to the case of the "*worst*" interference $v_j^* = -\beta^*(\alpha^*)^{-1}u_j^*$, i.e., optimal controls of the "opponent."

If v_j^* differ from the "worst" interferences, then position (5.1.11) will be attained (simultaneously in all η_j as well) in less than τ time. In this case position (5.1.11) will be attained at $v_j^* = \beta^*(\alpha^*)^{-1}u_j^*$ in time τ^* determined as the *least* positive root of the equation

$$\left(\sum_{k=1}^{r}\left(L_k^*\right)^2\right)^{1/2} - 1/2(\alpha^* + \beta^*)\tau^2 = 0, \quad L_k^* = \eta_{k0} + \tau\mu_{k0}. \tag{5.1.17}$$

(Here v_j^* act as other auxiliary controls additional to u_j^*.)

Remark 5.1.1.

1. Unlike the ordinary extremal strategy, the *decelerating* extremal strategy excludes the possibility of increasing the *absorption time* (the root of equation $\omega = 0$) during the game.

2. The *approximation scheme* for adjusting u_j^* at discrete times provides a constructive approach to constructing formal control rules (5.1.15) along with the motions of system (5.1.9), (5.1.15) corresponding to them. In this case one can regard controls (5.1.15) and the motions of system (5.1.9), (5.1.15) as the *limit representation* of the corresponding controls and motions generated by the discrete scheme mentioned. The motions of system (5.1.9), (5.1.15) thus obtained are *absolutely continuous* functions of time. In particular, when computer simulation of the control process is carried out, one can effectively construct controls (5.1.15) following this discrete scheme. Let us also note that the discrete scheme can be treated as an independent scheme if, from the very beginning, the auxiliary game-theoretic problem is posed in the class of controls corresponding to such a scheme.

3. If $\omega = 0$, in (5.1.15), instead of zero values, we can take any other values u_j^* that do not violate constraints (5.1.13). This can be done, for example, when it is desired to reduce the number of discontinuity points in realizations of u_j^* for each fixed set of initial conditions t_0, $\boldsymbol{y}_0$, $\boldsymbol{\mu}_0$. In this case both the "levels" of controls u_k and the time of control (within the interval from τ^* to τ, see expressions (5.1.16) and (5.1.17)) are changed correspondingly.

5.1.5 Conditions for the Solvability of Problem 5.1.1

Let $\boldsymbol{z}$ be the vector that comprises the phase variables ξ_i $(i = 1, \dots, l)$, η_j $(j = n - l - r + 1, \dots, n - l)$ of system (5.1.1) not included in $\boldsymbol{y}$. As a result, $\boldsymbol{x} = (\boldsymbol{y}, \boldsymbol{z})$. Along with the Jacobi matrix F introduced earlier, let us also introduce the Jacobi matrix Q of functions $X_j^{(3)}$ $(j = 1, \dots, r)$ with respect to components of vector $\boldsymbol{z}$. We assume that $2r \geq n$.

Suppose that, in the domain

$$\boldsymbol{\Lambda}_2 : \|\boldsymbol{x}\| < H_2, \tag{5.1.18}$$

the following condition holds:

$$\operatorname{rank} Q(\boldsymbol{x}) = n - r. \tag{5.1.19}$$

We note that, if $r \leq l$, the assumed inequality $2r \geq n$ together with the inequality $n - l \geq r$ (which is valid for vector $\boldsymbol{y}$ as chosen) are satisfied simultaneously, when

$$2r = 2l = n. \tag{5.1.20}$$

Condition (5.1.20) is met, for example, when system (5.1.1) is a model of the angular motion of a solid. In addition, this condition can be weakened. Denote by $\boldsymbol{X}_{\boldsymbol{v}} = \{\boldsymbol{x}_{\boldsymbol{v}} = \boldsymbol{x}(t; t_0, \boldsymbol{x}_0, \boldsymbol{v})\}$ the set of the processes $\boldsymbol{x} = \boldsymbol{x}(t; t_0, \boldsymbol{x}_0)$ of the closed system (5.1.1), (5.1.7), and (5.1.15) that correspond to all admissible realizations of interference $\boldsymbol{v} \in K_1$.

If condition (5.1.19) is satisfied, the equalities $X_j^{(3)}(\boldsymbol{x}) = \mu_j$ $(j = 1, \dots, r)$ admit the following solution in $\boldsymbol{\Lambda}_2$ (vector function $\Psi \in R^{n-r}$ is continuous in $\mathbf{\Lambda}_2$):

$$z = \Psi(\boldsymbol{y}, \boldsymbol{\mu}), \quad \Psi(\mathbf{0}, \mathbf{0}) = \mathbf{0}. \tag{5.1.21}$$

Equalities (5.1.21) relate components from $\boldsymbol{x}$ not entered into $\boldsymbol{y}$ to those from $\boldsymbol{y}$, $\boldsymbol{\mu}$. As a result, in $\mathbf{\Lambda}_2$, the phase vector $\boldsymbol{x}$ of system (4.5.1) can be represented in the form $\boldsymbol{x} = [\boldsymbol{y}, z(\boldsymbol{y}, \boldsymbol{\mu})] \overset{\text{def}}{=} \boldsymbol{W}(\boldsymbol{y}, \boldsymbol{\mu})$.

Consequently, the set $\boldsymbol{X}_{\boldsymbol{v}}$ can be expressed in the form

$$\boldsymbol{X}_{\boldsymbol{v}} = \{\boldsymbol{W} = \boldsymbol{W}[t; t_0, \boldsymbol{W}_0(\boldsymbol{y}_0, \boldsymbol{\mu}_0(\boldsymbol{x}_0)), \boldsymbol{v}]\} = \{\boldsymbol{W} = \boldsymbol{W}(t; t_0, \boldsymbol{x}_0, \boldsymbol{v})\}$$

and can be estimated through the corresponding estimates for the set of the processes of the *linear* system (5.1.9), (5.1.15) (rather than the original *nonlinear* system (5.1.1)).

Theorem 5.1.1. *Let the following conditions be satisfied*

(1) $2r = 2l = n$;
(2) *condition* (5.1.5) *in domain* (5.1.4);
(3) *condition* (5.1.19) *in domain* (5.1.18);
(4) *for any* $\boldsymbol{x}_0 \in S$, *the inequalities*

$$|\boldsymbol{W}(t; t_0, \boldsymbol{x}_0, \boldsymbol{v})| < \min(H_1, H_2), \tag{5.1.22}$$

$$\left| f_k[\boldsymbol{W}(t; t_0, \boldsymbol{x}_0, \boldsymbol{v}), \boldsymbol{u}^*] \right| \le \alpha_k \quad (k = 1, \dots, r) \tag{5.1.23}$$

are valid in the set $\boldsymbol{X}_{\boldsymbol{v}} = \{\boldsymbol{W} = \boldsymbol{W}(t; t_0, \boldsymbol{x}_0, \boldsymbol{v})\}$ *defined by the sets of possible trajectories of the linear system* (5.1.9), (5.1.15).

Then for all $\boldsymbol{x}_0 \in S$, *control rules* (5.1.7), (5.1.15) *solve Problem* 5.1.1. *It is guaranteed that system* (5.1.1) *is taken precisely to the position* $\boldsymbol{y} = \mathbf{0}$ *by bounded controls* $\boldsymbol{u} \in K$ *in a finite time* τ (*determined as the least positive root of equation* (5.1.16)) *for any admissible realizations of interference* $\boldsymbol{v} \in K_1$.

Proof. If condition (5.1.5) in domain (5.1.4) is satisfied, the linear system (5.1.9) is derived from the original nonlinear system (5.1.1), (5.1.7) by the appropriate nonlinear transformation of the variables. Under such a transformation, the state of variables η_j $(j = 1, \dots, r)$ of system (5.1.1), (5.1.7) in $\mathbf{\Lambda}_1$ will be completely determined by the state of the same variables of the linear system (5.1.9). Therefore, for all $\boldsymbol{x}_0 \in S_1 = \{\boldsymbol{x}_0 : \|\boldsymbol{x}_{\boldsymbol{v}}(t)\| < H_1\}$, control rules (5.1.7), (5.1.15) guarantee that system (5.1.1) is taken precisely to the position

$$\eta_j = 0 \qquad (j = 1, \dots, r) \tag{5.1.24}$$

in a finite time τ.

Besides, for all $\boldsymbol{x}_0 \in S_2 = \{\boldsymbol{x}_0 : \|\boldsymbol{x}_{\boldsymbol{v}}(t)\| < \min(H_1, H_2)\}$, the set $\boldsymbol{X}_{\boldsymbol{v}}$ can be estimated by using the corresponding estimates for the set of the processes of the *linear* system (5.1.9), (5.1.15) (rather than the original nonlinear system (5.1.1)). Requirements (5.1.2) result in the fact that the admissible set of the initial perturbations $\boldsymbol{x}_0$ should also satisfy the condition $\boldsymbol{x}_0 \in S_3 = \{\boldsymbol{x}_0 : |f_k[\boldsymbol{W}(t), \boldsymbol{u}^*]| \leq \alpha_k\}$.

Because the conditions of the theorem are satisfied, the given domain S of the initial perturbations will be a subset of the intersection of the sets S_1–S_3. The definition of the vector function $\boldsymbol{W} = \boldsymbol{W}(\boldsymbol{y}, \boldsymbol{\mu}) : R^{2r} \to R^n$ introduced implies that, for all $\boldsymbol{x}_0 \in S$, controls (5.1.7), (5.1.15) guarantee that system (5.1.1) is precisely taken to the position $\boldsymbol{y} = \boldsymbol{0}$ in a finite time τ for any $\boldsymbol{v} \in K_1$. Besides, control rules (5.1.7), (5.1.15) satisfy the given constraints (5.1.2). The theorem is proved. □

5.1.6 Algorithm for Solving Problem 5.1.1

The algorithm is divided into the following stages:

(1) Choosing construction (5.1.7) of control rules u_k $(k = 1, \ldots, r)$.

(2) Estimating levels β_i^* of the "auxiliary interference" v_i^* (by using formulas (5.1.10)). Evaluating number β^*.

(3) "Assigning" number α^* which determines the level of the auxiliary control rules u_j^*. Numbers α^*, β^* predetermine the corresponding value of the *guaranteed* time of control $\tau = t_1 - t_0$.

(4) Validating the original constraints (5.1.2) imposed on control rules u_k $(k = 1, \ldots, r)$. Since condition (5.1.19) is satisfied, this verification can be carried out with the help of the corresponding estimates for the set of the processes of the *linear* system (5.1.9), (5.1.15) (rather than the original nonlinear system (5.1.1)).

If estimates (5.1.2) fail to hold or, vice versa, they are satisfied with a kind of "reserve," the search for the appropriate number α^* is to be continued. Otherwise, the time of control equals τ.

As a result, we get an *iterative algorithm* for solving Problem 5.1.1.

5.2 The Nonlinear Game-Theoretic Problem of "Passage" of an Asymmetric Solid through a Given Angular Position

5.2.1 Preliminary Remarks

The problem considered below is typical of the game-theoretic problems of control *with respect to part of the variables* in the presence of uncontrollable interference. This problem is classified as an essentially nonlinear problem of the "hard" encounter type and is a development (for the case when interference is involved) of Problem 4.6.1. Its solution is based on the approach proposed in Section 5.1, namely, control rules are determined which ensure the guaranteed passage of an asymmetric solid through a given angular position for any admissible realizations of interference.

Apparently, the presence of uncontrollable interference significantly complicates both the purely mathematical aspect of solving the problem and the control rules obtained.

We note that ever growing needs of practice in such spheres as spacecraft dynamics, robotics, and biomechanics make it reasonable to study the problems of reorienting a solid in a *game-theoretic* formulation.

5.2.2 Formulation of the Problem

The angular motion of a solid is described by *Euler's dynamical equations*

$$\begin{gathered} A\dot{\xi}_1 = (B - C)\xi_2\xi_3 + u_1 + v_1, \\ B\dot{\xi}_2 = (C - A)\xi_1\xi_3 + u_2 + v_2, \quad C\dot{\xi}_3 = (A - B)\xi_1\xi_2 + u_3 + v_3, \end{gathered} \tag{5.2.1}$$

which are written here in the variables corresponding to the first group of equations of system (5.1.1). For notation to be uniform with that used in Section 5.1, we also leave unchanged both indices i, k $(i, k = 1, 2, 3)$ of variables ξ_i and u_k, respectively, though in this case $l = r = 3$.

In system (5.2.1) ξ_i, u_k are the projections of vector $\boldsymbol{\xi}$ of the solid's angular velocity and the controlling moment $\boldsymbol{u}$ onto the principal central axes of inertia, respectively; A, B, and C are the principal central moments of inertia. Moments v_i are the components of vector $\boldsymbol{v}$ which incorporates external forces and uncontrollable interference. Hereinafter, the sums under summation signs are taken over repeated indices from 1 to 3.

We shall consider equations (5.2.1) together with *kinematic equations*

$$\begin{gathered} 2\dot{\eta}_1 = \xi_1\eta_4 + \xi_3\eta_2 - \xi_2\eta_3, \\ 2\dot{\eta}_2 = \xi_2\eta_4 + \xi_1\eta_3 - \xi_3\eta_1, \quad 2\dot{\eta}_3 = \xi_3\eta_4 + \xi_2\eta_1 - \xi_1\eta_2 \end{gathered} \tag{5.2.2}$$

in terms of the Rodrigues–Hamilton variables which determine the solid's orientation.

Variables η_j $(j = 1, 2, 3)$, η_4 (with index j left unchanged to conform equations (5.2.2) to the second group of equations of system (5.1.1)) are related by

$$\sum_{j=1}^{3} \eta_j^2 + \eta_4^2 = 1. \tag{5.2.3}$$

The equation for $\dot{\eta}_4$, if desired, can also be obtained.

For system (5.2.1)–(5.2.3), $l = r = 3$, $n = 6$. In what follows we denote by $\boldsymbol{\eta}$ the vector formed by the variables η_j, η_4; $\boldsymbol{x} = (\boldsymbol{\xi}, \boldsymbol{\eta})$.

Controls $\boldsymbol{u}$ are chosen in class L of vector functions $\boldsymbol{u} = \boldsymbol{u}(t, \boldsymbol{x}, \boldsymbol{x}_0)$ (where $\boldsymbol{x}_0$ is the initial value of vector $\boldsymbol{x} = (\boldsymbol{\xi}, \boldsymbol{\eta})$) whose realizations $\boldsymbol{u}[t]$ are *Lebesgue-integrable.* Controls $\boldsymbol{u} \in L$ satisfy "geometric" constraints

$$|u_k| \leq \alpha_k = \text{const} > 0. \tag{5.2.4}$$

Inequalities (5.2.4) correspond to three pairs of engines "*fixedly*" *attached* to the solid.

Interference $\boldsymbol{v} \in L_1$ can take the form of any *Lebesgue-integrable* vector functions $\boldsymbol{v} = \boldsymbol{v}(t)$ subject to the constraints

$$|v_i| \leq \beta_i = \text{const} > 0. \tag{5.2.5}$$

Problem 5.2.1. Find control rules $\boldsymbol{u} \in L$ transferring a solid from the initial state $\boldsymbol{\eta}(t_0) = \boldsymbol{\eta}_0$ to the prescribed state $\boldsymbol{\eta}(t_1) = \boldsymbol{\eta}_1$ in a finite time for any $\boldsymbol{v} \in L_1$. The initial state of the solid is a stationary state $\boldsymbol{\xi}(t_0) = \boldsymbol{\xi}_0 = \mathbf{0}$. The value $\boldsymbol{\xi}_1 = \boldsymbol{\xi}(t_1)$ of the solid's angular velocity at the moment $t = t_1 > t_0$ can be arbitrary. The time t_1 is not fixed, but we shall seek to make it as short as possible.

Without loss of generality, we further suppose that $\boldsymbol{\eta}_1 = (0, 0, 0, 1)$. In fact, as has already been noted, the case in question corresponds to controlling the orientation relative to the coordinate system set at the initial time. In what follows, in accordance with the designations adopted, $i, j, k = 1, 2, 3$.

5.2.3 An Auxiliary Linear Conflict-Control System

Let us differentiate both sides of each equation for $\dot{\eta}_j$ in (5.2.2) with respect to time. Then we replace $\dot{\xi}_i$ with their expressions from (5.2.1) and $\dot{\eta}_j$ with the corresponding expressions from (5.2.2). Having completed the transformation, we arrive at the equalities

$$\begin{aligned} \ddot{\eta}_j &= f_j(\boldsymbol{\xi}, \boldsymbol{\eta}, \boldsymbol{u}) + \varphi_j(\boldsymbol{\eta}, \boldsymbol{v}), \\ f_1 &= \frac{1}{2}[\eta_4(u_1 + M_1)A^{-1} + \eta_2(u_3 + M_3)C^{-1} - \eta_3(u_2 + M_2)B^{-1}] \\ &\quad - \frac{1}{4}\eta_1 \sum_{i=1}^{3} \xi_i^2, \qquad M_1 = (B - C)\xi_2\xi_3 \qquad (123), \\ \varphi_1 &= \frac{1}{2}(\eta_4 v_1 A^{-1} + \eta_2 v_3 C^{-1} - \eta_3 v_2 B^{-1}) \qquad (ABC, 123). \end{aligned} \tag{5.2.6}$$

We interpret f_j and φ_j as the auxiliary control rules u_j^* forming vector $\boldsymbol{u}^*$ and the auxiliary interference v_j^* forming vector $\boldsymbol{v}^*$, respectively. As a result, expressions (5.2.6) can be considered a *linear conflict-control $\boldsymbol{\mu}$-system* of type (5.1.9)

$$\dot{\eta}_j = \mu_j, \quad \dot{\mu}_j = u_j^* + v_j^*; \tag{5.2.7}$$

the "original" control rules u_k are expressed in terms of u_j^* through the equalities of type (5.1.7) which in this case have the form

$$\begin{aligned} u_1 &= 2A\eta_4^{-1}[(\eta_4^2 + \eta_1^2)u_1^* + (\eta_1\eta_2 + \eta_4\eta_3)u_2^* + (\eta_1\eta_3 - \eta_4\eta_2)u_3^*] \\ &\quad + \frac{1}{2}A\eta_1\eta_4^{-1}\sum_{i=1}^{3}\xi_i^2 + (C - B)\xi_2\xi_3 \qquad (ABC, 123). \end{aligned} \tag{5.2.8}$$

Then we suggest that the solution of the original nonlinear game-theoretic Problem 5.2.1 be constructed based on solving the corresponding *game-theoretic* control problems for the linear system (5.2.7). As a result, one can regard contruction (5.2.8) as the *general structural form* of the control rules in Problem 5.2.1. The parameters of this form, the auxiliary control rules u_j^*, are determined by solving the corresponding linear game-theoretic problems.

Structure (5.2.8) of the control rules contains the factor η_4^{-1}, which formally implies a "*singularity.*" However, subsequent more thorough analysis reveals that, in the case $\boldsymbol{\eta}_1 = (0,0,0,1)$, the relationship $\eta_4 \in [\eta_{40}, 1]$ holds in the course of control. Therefore, the "singularity" indicated does not occur at all. If $\boldsymbol{\eta}_1 \neq (0,0,0,1)$ or the value of η_{40} is small, it suffices to change the controls to those obtained from (5.2.8) by a permutation of the indices and coefficients A, B, and C (or to a combination of such controls). The final choice of controls u_k is carried out in the *iterative* mode.

5.2.4 An Auxiliary Game-Theoretic Control Problem

By virtue of relationships (5.2.3), (5.2.5), and the *Cauchy–Schwarz inequality* $\left[\sum(a_i b_i)\right]^2 \le \sum a_i^2 \sum b_i^2$, the following relationships hold:

$$|v_j^*(t)| \le \beta_j^* = \frac{1}{2}\left[(\beta_1 A^{-1})^2 + (\beta_2 B^{-1})^2 + (\beta_3 C^{-1})^2\right]^{1/2}. \tag{5.2.9}$$

In line with the purposes of Problem 5.2.1, let us solve the game-theoretic control problem of bringing system (5.2.7) to the position

$$\eta_j = 0 \tag{5.2.10}$$

in the *shortest possible time.* The control is implemented through u_j^* for *any admissible* forms of v_j^* satisfying inequalities (5.2.9).

The terminal values of μ_j can be *arbitrary.* Taking this circumstance into account, Problem 5.2.1 implies that control u_j^* should *simultaneously* take η_j to position (5.2.10) for any admissible v_j^*.

We treat this problem as a *differential game.* In this game one of the players has at its disposal u_j^* and aims at decreasing the time τ of taking the system to position (5.2.10). The "opponent," aiming to increase τ (or prevent attaining the position (5.2.10) at all) has at its disposal "auxiliary perturbations" v_j^*.

To solve this problem, we impose admissible constraints on u_j^*. To this end, following the recommendations of Section 5.1, we shall take the admissible constraints on u_j^* in the form

$$\|\boldsymbol{u}^*\| \le \alpha^* = \text{const} > 0. \tag{5.2.11}$$

Here $\|\boldsymbol{u}^*\| = \left[\sum_{j=1}^{3} (u_j^*)^2\right]^{1/2}$ is the Euclidean norm of vector $\boldsymbol{u}^*$ in R^3.

Again, we impose the constraints on v_j^* in the form

$$\|\boldsymbol{v}^*\| \leq \beta^* = \text{const} > 0. \tag{5.2.12}$$

Here $\beta^* = \Big[\sum_{i=1}^{3}(\beta_i^*)^2\Big]^{1/2} = \dfrac{\sqrt{3}}{2}\big[(\beta_1 A^{-1})^2 + (\beta_2 B^{-1})^2 + (\beta_3 C^{-1})^2\big]^{1/2}$.

If $\alpha^* > \beta^*$, the solution of the auxiliary game-theoretic problem in question subject to constraints (5.2.11), (5.2.12) is given by the Krasovskii *extremal deceleration strategy* $U^* = U^*(t, \boldsymbol{\eta}, \boldsymbol{\mu}, \theta)$. Controls u_j^* resulting from this strategy are cited here for ease of reference

$$u_j^*(t, \boldsymbol{\eta}, \boldsymbol{\mu}) = \begin{cases} -\alpha^* L_j \Big(\sum\limits_{j=1}^{3} L_j^2\Big)^{-1/2}, & \omega > 0 \\ 0, & \omega \leq 0 \end{cases} \tag{5.2.13}$$

$$\omega = \Big(\sum_{j=1}^{3} L_j^2\Big)^{1/2} - 1/2\,\alpha^*(\theta - t)^2, \quad L_j = \eta_j + (\theta - t)\mu_j.$$

In expressions (5.2.13), variable θ is derived from the equation $\omega = 0$. Since, according to the formulation of Problem 5.2.1, the equality $\mu_{j0} = 0$ is true (as a consequence of $\boldsymbol{\xi}_0 = \mathbf{0}$), τ is determined as the positive root of the equation

$$\tau^2 = 2(\alpha^* - \beta^*)^{-1}\rho_0, \quad \rho_0 = \Big(\sum_{j=1}^{3} \eta_{j0}^2\Big)^{1/2}. \tag{5.2.14}$$

Number τ specifies the *minimum guaranteed* (simultaneous with respect to all η_j) time for attaining position (5.2.10) in the linear problem under consideration. This transfer time corresponds to the case of the "worst" interference $v_j^* = -\beta^*(\alpha^*)^{-1}u_j^*$, i.e., optimal controls of the "opponent."

If v_j^* differ from the "worst" interferences, then the set (5.2.10) will be attained (simultaneously also in all η_j) in less than time τ. In this case the set (5.2.10) will be attained at $v_j^* = \beta^*(\alpha^*)^{-1}u_j^*$ in time τ^* determined as the positive root of the equation $\tau^2 = 2(\alpha^* + \beta^*)^{-1}\rho_0$.

The trajectories of system (5.2.7), (5.2.13), when v_j^* are the "worst" auxiliary interference or auxiliary controls additional to u_j^*, are defined as follows

$$\mu_j = -(\alpha^* \pm \beta^*)\Gamma_j t, \quad \eta_j = \eta_{j0} - 1/2(\alpha^* \pm \beta^*)\Gamma_j t^2, \quad \Gamma_j = \eta_{j0}\rho_0^{-1}. \tag{5.2.15}$$

For any admissible realizations of v_j^*, the quantity $\rho = \Big(\sum_{j=1}^{3} \eta_j^2\Big)^{1/2}$ satisfies the estimates

$$\rho^-(t) \leq \rho(t) \leq \rho^+(t), \tag{5.2.16}$$

where $\rho^\pm$ determined from (5.2.15) are compiled in Table 5.2.1 (for brevity, $s^\pm = \alpha^* \pm \beta^*$, $t_0 = 0$).

The *approximation scheme* for adjusting u_j^* at discrete times provides a constructive approach to constructing control rules (5.2.13) together with the motions of system (5.2.8), (5.2.13) corresponding to them.

Table 5.2.1.

t	ρ^-	ρ^+
$[0, \tau^*]$	$\rho_0 - \frac{1}{2}s^+t^2$	$\rho_0 - \frac{1}{2}s^-t^2$
$(\tau^*, \tau]$	0	—— ॥ ——

5.2.5 Algorithm for Solving Problem 5.2.1

The algorithm is divided into the following stages:

(1) Choosing construction (5.2.8) of control rules u_k with u_j^* of type (5.2.13). (If $\boldsymbol{\eta}_1 \neq (0,0,0,1)$ or the value of η_{40} is small, it suffices to change the controls to those obtained from (5.2.8) by a permutation of the indices and coefficients A, B, and C.)

(2) Estimating levels β_j^* of the "auxiliary interference" v_j^* (by using formulas (5.2.9)). Evaluating number β^*.

(3) "Assigning" number α^* which determines the levels of the auxiliary control rules u_j^*. Numbers α^*, β^* predetermine the corresponding value of the *guaranteed* reorientation time $\tau = t_1 - t_0$.

(4) Verifying the original constraints (5.1.4) imposed on control rules u_k. To this end, we use equalities (4.6.11). As a result, this verification can be carried out on the trajectories of system (5.2.7), (5.2.13). See Subsection (5.2.6) regarding possible estimates for u_k.

If estimates (5.1.4) fail to hold or, vice versa, they are satisfied with a kind of "reserve," the search for the appropriate number α^* is to be continued. Otherwise, the reorientation time equals τ.

As a result, we get an *iterative algorithm* for solving Problem 5.2.1.

Analyzing inequalities (5.2.16), we note that in the case $\boldsymbol{\eta}_1 = (0,0,0,1)$, the condition $\eta_4 \geq \gamma = \text{const} > 0$, once met at $t = t_0$, remains valid for all $t \in [t_0, t_0 + \tau]$. It is implied in this case that the relationship $\eta_4 \in [\eta_{40}, 1]$ is valid in the course of the control. This fact rules out "singularity" of controls (5.2.8), (5.2.13).

5.2.6 Estimating Control Rules u_k

By the Cauchy–Schwarz inequality, under condition (5.2.11),

$$|\gamma_i u_1^* + \delta_i u_2^* + \varepsilon_i u_3^*| \leq \alpha^* (\gamma_i^2 + \delta_i^2 + \varepsilon_i^2)^{1/2}. \tag{5.2.17}$$

Relationships (5.2.17) can be used to estimate the first group of terms in (5.2.8) (that does not explicitly depend on ξ_i) if

$$\gamma_1 = \eta_4^{-1}(\eta_4^2 + \eta_1^2), \quad \delta_1 = \eta_4^{-1}(\eta_1\eta_2 + \eta_4\eta_3), \quad \varepsilon_1 = \eta_4^{-1}(\eta_1\eta_3 - \eta_4\eta_2) \quad (123).$$

We shall bear in mind that $\gamma_i^2 + \delta_i^2 + \varepsilon_i^2 = 1 + (\eta_4^{-1}\eta_i)^2$.

To estimate the remaining group of terms in (5.2.8), we use inequalities $|\xi_1\xi_2| \le 1/2 \sum \xi_i^2$ (123) and the relationship

$$\sum \xi_i^2 = G = 4\{[\eta_4^{-1} \sum_{j=1}^{3} (\eta_j \mu_j)]^2 + \sum_{j=1}^{3} \mu_j^2\}$$

which can be verified by using (4.6.11).

As a result, one can obtain the following estimates for u_k

$$|u_1| \le F_1 = 2A\alpha^* \big[1 + (\eta_4^{-1}\eta_1)^2\big]^{1/2} + 1/2 \big[A\eta_1\eta_4^{-1} + |C - B|\big] G^+ \quad (ABC, 123),$$

where G^+ is obtained from G by replacing μ_j with $|\mu_j|$.

Now we provide estimates for F_k. In so doing, it is necessary to take into account the constraints-inequalities (5.2.11) along with the inequalities

$$v^- \le v = \Big(\sum_{j=1}^{3} \mu_j^2\Big)^{1/2} \le v^+; \tag{5.2.18}$$

$v^\pm$ are compiled in Table 5.2.2.

Table 5.2.2.

t	v^-	v^+
$[0, \tau^*]$	$s^- t$	$s^+ t$
$(\tau^*, \tau]$	—— ॥ ——	$s^+ \tau^*$

It can be verified that $\sup F_k$ (for each k) under the constraint specified is attained on the *boundaries* of the set (5.2.16), (5.2.18).

Verifying alternatives, with matching $\rho^\pm \to v^\mp$ taken into account, we have the estimates

$$\begin{gathered} |u_k| \le F_k^+ = \max\big[F_k^*(\rho^-, v^+), F_k^*(\rho^+, v^-)\big], \\ F_1^* = 2A\alpha^* E + 1/2 (A\rho E + |C - B|) G^* \quad (ABC, 123), \\ G^*(\rho, v) = 4\big[(E\rho v)^2 + v^2\big] = 4(Ev)^2, \quad E^2(\rho) = (1 - \rho^2)^{-1}, \quad t \in [0, \tau]. \end{gathered} \tag{5.2.19}$$

The relationships

$$|\eta_j| \le \rho, \quad \big|1 + (\eta_4^{-1}\eta_j)^2\big| \le 1 + (E\rho)^2 = E^2.$$

were taken into account to obtain estimates (5.2.19).

5.2.7 The "Converse" of the Algorithm

(1) Choosing τ.
(2) Calculating number α^* by the formula

$$\alpha^* = \beta^* + 2\tau^{-2}\rho_0. \tag{5.2.20}$$

(3) Estimating levels α_k of controls u_k (by formulas (5.2.8)) over the state set of the linear system (5.2.7), (5.2.13).

As a result, one can assess the capabilities of control rules of type (5.2.8), (5.2.13) when the reorientation of a solid takes time τ.

The version of the algorithm indicated can be viewed as a principle, if Problem 5.2.1 is reformulated as follows.

Problem 5.2.2. Let the guaranteed reorientation time $\tau = t_1 - t_0$ *be fixed. Control rules belonging to class L are required that will take a solid from the initial state* $\boldsymbol{\xi}_0 = \mathbf{0}$, $\boldsymbol{\eta} = \boldsymbol{\eta}_0$ *to a given one* $\boldsymbol{\eta} = \boldsymbol{\eta}_1$ *in time* τ *for any admissible realizations of interference* $\boldsymbol{v} \in K_1$. *The value of* $\boldsymbol{\xi}_1$ *can be arbitrary. The values* α_k *of the levels of controls* u_k *are not given in advance but should be as small as possible.*

To solve Problem 5.2.2, we calculate the value of α^* by formula (5.2.20) and use control rules (5.2.8), (5.2.13) with the value of α^* obtained. In this case one can omit searching the accurate estimates for u_k, which presents the main difficulty when the proposed approach is applied. The fact that it is reasonable to use controls of type (5.2.8), (5.2.13) for solving Problem 5.2.2 can be backed up by some *indirect* considerations: for example, by the high efficiency of controls of type (5.2.8) in Problem 4.6.1 and by the results of computer simulation. Of course, the reasoning given does not replace the necessity of obtaining more accurate estimates for u_k which are to be worked through in the sequel.

5.2.8 Principal Results

Let us sum up the considerations presented.

Theorem 5.2.1. *If levels* α_k *of control rules* u_k *in system* (5.2.1)–(5.2.3) *are high enough, then, for any levels* β_i *of interference* v_i, *the rules* u_k *solving Problem* 5.2.1 *can be constructed in the form* (5.2.8), (5.2.13). *The precise guaranteed "passage" of a solid through a given angular position is ensured in a finite time* τ *for any* $\boldsymbol{v} \in K_1$. *The value of* τ *is determined iteratively as the positive root of equation* (5.2.14) *by using the algorithm given in Subsection* 5.2.5.

Theorem 5.2.1 states the possibility of solving Problem 5.2.1 by constructing control rules (5.2.8), (5.2.13). We can make this possibility more specific as follows (without loss of generality, we assume that $t_0 = 0$).

Theorem 5.2.2. *Let* $\boldsymbol{\eta}_1 = (0,0,0,1)$. *Suppose that one can find number* α^* *such that*

$$\alpha^* > \beta^* = \frac{\sqrt{3}}{2}\left[(\beta_1 A^{-1})^2 + (\beta_2 B^{-1})^2 + (\beta_3 C^{-1})^2\right]^{1/2}$$

$$F_k^+ \le \alpha_k, \quad t \in [0, \tau],$$

where F_k^+ *are defined by* (5.2.19). *Then control rules* (5.2.8), (5.2.13) *solve Problem* 5.2.1 *in a finite time* τ *found as the positive root of equation* (5.2.14).

Theorem 5.2.2 gives constructive conditions for solvability of Problem 5.2.1 which can be verified in real time. If the level of overestimates (5.2.19) is not acceptable, one can turn to Problem 5.2.2.

5.2.9 Generalization of Problem 5.2.1

The proposed algorithm can be carried over to the case when the initial state of a solid is not an equilibrium position ($\boldsymbol{\xi}_0 \neq 0$). In this case in the course of solving the auxiliary game-theoretic problem of control for the linear system (5.2.7), it is necessary to preserve the constraint $\sum \eta_j^2 \le 1$ imposed on variables η_j. (At $\boldsymbol{\xi}_0 = \mathbf{0}$ such a constraint resulting from (5.2.3) is automatically met.) For given levels α_k of controls u_k, this is only possible until some definite value of $|\boldsymbol{\xi}_0|$ is attained. Otherwise, the value of the solid's angular velocity should be reduced to the admissible limit in advance.

5.2.10 Results of Computer Simulation

For a solid (spacecraft) with $A = 4 \cdot 10^4$, $B = 8 \cdot 10^4$, and $C = 5 \cdot 10^4\ kg{\cdot}m^2$, we consider the triaxial reorientation from the position $\boldsymbol{\xi}_0 = \mathbf{0}$, $\boldsymbol{\eta}_0 = (0.353, 0.434, 0.432, 0.701)$ to the position $\boldsymbol{\eta}_1 = (0,0,0,1)$. The value of $\boldsymbol{\xi}_1$ is arbitrary, which corresponds to the nature of Problem 5.2.1.

We specify the levels of interference v_i through the value $\beta^* = \sqrt{3} \cdot 10^{-3}\ s^{-2}$.

Let $\tau = 70\ s$. We cite the results of simulation for various realizations of interference.

When control rules u_j^* and interference v_j^* vary in accordance with the extremal strategies, we obtain $\alpha_1 = 114.1$, $\alpha_2 = 280.5$, and $\alpha_3 = 174.5\ N{\cdot}m$. (Functions u_k for this case are shown in Figure 5.2.1.)

The results of simulation for some other possible forms of interference v_j^* are presented in Table 5.2.3.

Control rules u_k realized, following the discrete scheme given in Subsection 5.1.4 with the step $h = 0.1$ which corresponds to case (4) in Table 5.2.3, are presented in Figures 5.2.2–5.2.4.

On the whole the simulation demonstrates that the guaranteed reorientation time in Problem 5.2.1 is shorter than in the case of $\boldsymbol{\xi}_0 = \boldsymbol{\xi}_1 = \mathbf{0}$, where one and the same constraints are imposed on u_k. This conclusion agrees with the similar conclusion (see Section 4.6) for the case when interference is absent.

We also give (over)estimates for u_k computed by formula (5.2.19): $F_1^+ = 479.9$, $F_2^+ = 606.4$, and $F_3^+ = 626.4\ N{\cdot}m$.

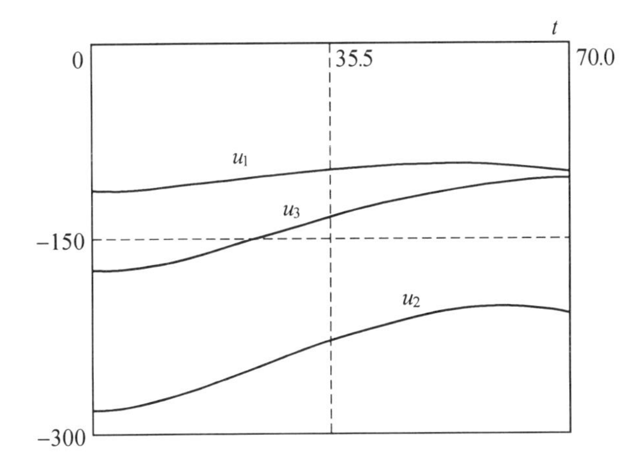

Figure 5.2.1. Controls u_k corresponding to the case when u_j^* and v_j^* vary in accordance with extremum strategies.

Table 5.2.3.

N	v_1^*	v_2^*	v_3^*	α_1	α_2	α_3
1	$-\beta^*$	0	0	166.8	243.1	180.1
2	0	$-\beta^*$	0	119.1	335.1	152.6
3	0	0	$-\beta^*$	96.0	282.1	206.7
4	β^*	0	0	149.6	358.3	128.1
5	0	β^*	0	53.9	314.7	174.0
6	0	0	β^*	129.1	139.1	223.5

5.3 An Auxiliary Function of the Problem of Control with Respect to Part of the Variables in Game-Theoretic Problems of Control with Respect to All the Variables

5.3.1 Preliminary Remarks

As the case with the problems of stability and stabilization, one of the important motives for investigating the game-theoretic problems of control with respect to *part of the variables* is the *auxiliary function* of such problems in the control process with respect to *all the variables* in the presence of uncontrollable interference.

In this section such an approach is applied to investigate the game-theoretic problem of control with respect to *all the variables* for one sufficiently general class of nonlinear controlled systems under the action of uncontrol-

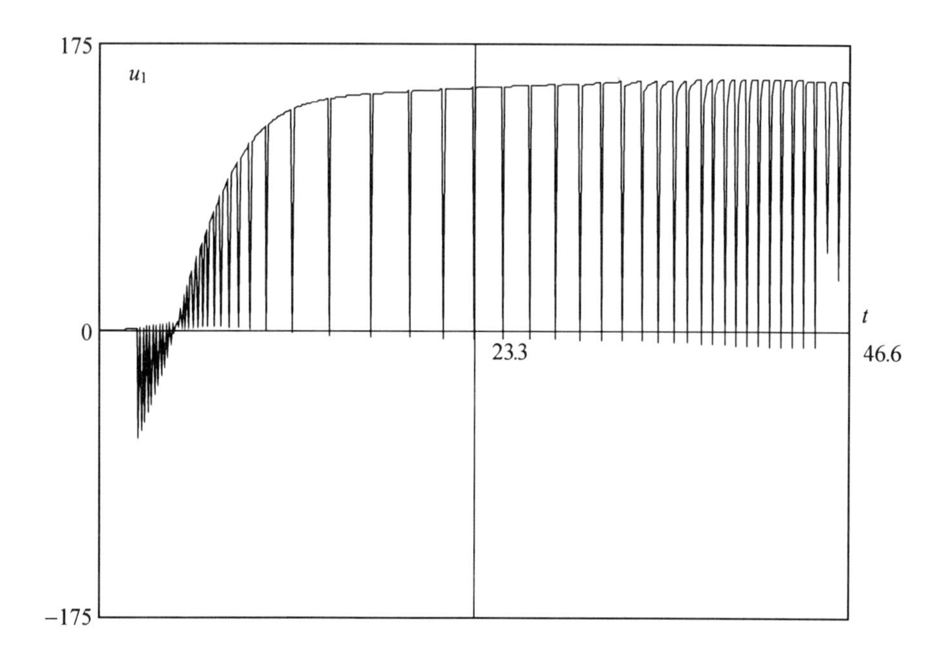

Figure 5.2.2. Control rule u_1 corresponding to case 4 from Table 5.2.3.

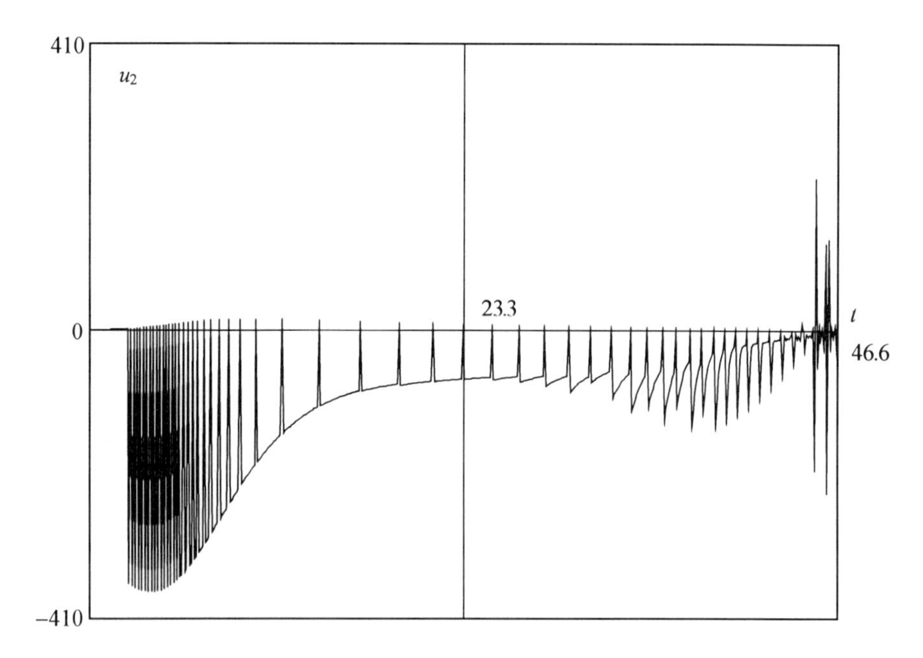

Figure 5.2.3. Control rule u_2 corresponding to case 4 from Table 5.2.3.

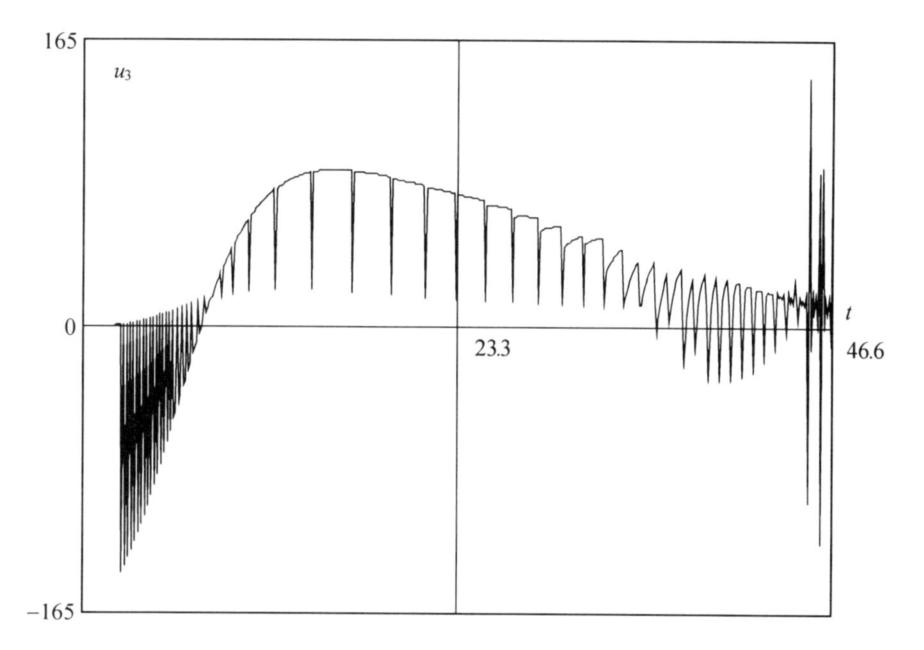

Figure 5.2.4. Control rule u_3 corresponding to case 4 from Table 5.2.3.

lable interference. This approach is based on the method of constructing *auxiliary linear conflict-control μ-systems* proposed in Section 5.1. But in this case, in contrast to Section 5.1, the corresponding problem of control with respect to *all the variables* is solved for the auxiliary linear μ-system. Such a game-theoretic problem can be reduced to the corresponding *optimal* (rather than conflict) *control problem.* The circumstance indicated allows one to simplify the solution process and also the control rules being obtained.

We also present the modification of the proposed approach which rests on the suggested *principle of assignment with subsequent confirmation of the levels of auxiliary interference.*

The applications illustrating the proposed approach are dealt with in Sections 5.4–5.6.

5.3.2 Formulation of the Problem

Consider the nonlinear controlled system (5.1.1) which is supposed to satisfy all the initial assumptions made in Subsection 5.1.2.

In contrast to Section 5.1, here L is not taken as a class of admissible controls but, from the standpoint of implementation, we choose them in a simpler class K^* of vector functions $\boldsymbol{u} = u(\boldsymbol{x}, \boldsymbol{x}_0)$ *piecewise-continuous* in $\boldsymbol{x}, \boldsymbol{x}_0$ (in the admissible domain of variation of $\boldsymbol{x}, \boldsymbol{x}_0$). Controls $\boldsymbol{u} \in K^*$ must also satisfy the prescribed "geometric" constraints (5.1.2).

Interference $\boldsymbol{v} \in K_1$ can take the form of any piecewise-continuous vector functions $\boldsymbol{v} = \boldsymbol{v}(t)$ subject to constraints (5.1.3).

Problem 5.3.1. Control rules $\boldsymbol{u} \in K^*$ are required to transfer system (5.1.1) *from a given domain S of initial perturbations $\boldsymbol{x}(t_0) \in S$ to the equilibrium position $\boldsymbol{x}(t_1) = \boldsymbol{0}$ in a finite time for any admissible realizations of interference $\boldsymbol{v} \in K_1$. The time $t_1 > t_0$ is not fixed.*

(1) First we solve the game-theoretic problem of the guaranteed transfer of system (5.1.1) to the position $\boldsymbol{x} = \boldsymbol{0}$ with respect to *part of the variables*, namely, to ξ_i $(i = 1, \ldots, q)$, η_j $(j = 1, \ldots, r - q)$. This problem is reduced to the corresponding game-theoretic problem of control with respect to all the variables for some specially constructed auxiliary linear conflict-control $\boldsymbol{\mu}$-system different from that considered in Section 5.1.
(2) Then additional conditions are imposed under which the guaranteed transfer of the system to the position $\boldsymbol{x} = \boldsymbol{0}$ with respect to the specified part of the variables actually turns out to be the solution of Problem 5.3.1 of control with respect to all the variables.

5.3.3 An Auxiliary Linear Conflict-Control System

Consider vector function $\boldsymbol{\Phi} = (\Phi_1, \ldots, \Phi_{q+n-l})$ whose first q components are defined as follows:

$$\Phi_i(\boldsymbol{x}, \boldsymbol{u}) = X_i^{(0)}(\boldsymbol{x}) + \sum_{k=1}^{r} X_{ik}^{(1)}(\boldsymbol{x}) u_k \qquad (i = 1, \ldots, q),$$

and the remaining $(n - l)$ components coincide with functions $\Phi_k(\boldsymbol{x}, \boldsymbol{u})$ introduced in Subsection 5.1.3 for $k = 1, \ldots, r$ only.

Let us introduce the *Jacobi matrix* $F(\boldsymbol{x}) = (\partial \Phi_s / \partial u_k)$ $(s = 1, \ldots, q + n - l;\ k = 1, \ldots, r)$. Suppose that condition (5.1.5) is satisfied in domain (5.1.4). We also assume, without loss of generality, that the first r rows of matrix F are linearly independent.

In this case by the *theory of implicit functions*, we conclude that in domain (5.1.4) the system of equations (5.1.6) has the solution

$$\boldsymbol{u} = \boldsymbol{f}^*(\boldsymbol{x}, \boldsymbol{u}^*) \tag{5.3.1}$$

of type (5.1.7).

Let us introduce notation

$$v_i^* = X_i^{(2)}(\boldsymbol{x}) v_i \qquad (i = 1, \ldots, q),$$
$$v_{q+j}^* = \sum_{s=1}^{l} \frac{\partial X_j^{(3)}(\boldsymbol{x})}{\partial \xi_s} X_s^{(2)}(\boldsymbol{x}) v_s \qquad (j = 1, \ldots, r - q)$$

and interpret v_k^* $(k = 1, \ldots, r)$ as "*auxiliary interference.*"

As a result, under the assumptions made, from the closed nonlinear system (5.1.1), (5.3.1) one can segregate the *linear conflict-control $\boldsymbol{\mu}$-system*

$$\dot{\xi}_i = u_i^* + v_i^* \qquad (i = 1, \dots, q < r < m), \tag{5.3.2}$$

$$\dot{\eta}_j = \mu_j, \quad \dot{\mu}_j = u_{q+j}^* + v_{q+j}^* \qquad (j = 1, \dots, r - q). \tag{5.3.3}$$

In $\boldsymbol{\mu}$-system (5.3.2), (5.3.3), u_k^* $(k = 1, \dots, r)$ are the controls, and v_k^* $(k = 1, \dots, r)$ act as perturbations. We shall further use the designations $y = (\xi_1, \dots, \xi_q, \eta_1, \dots, \eta_r)$, $\boldsymbol{\mu} = (\mu_1, \dots, \mu_r)$.

In what follows we construct the solution of the original nonlinear Problem 5.3.1 based on solving the corresponding *game-theoretic* control problems for the auxiliary linear conflict-control $\boldsymbol{\mu}$-system (5.3.2), (5.3.3). As a result, the contruction of control rules (5.3.1) (with vector function $\boldsymbol{\Phi}$ chosen above) can be regarded as the *general structural form* of the control rules in Problem 5.3.1. The parameters of this form, the auxiliary control rules u_k^* $(k = 1, \dots, r)$, are determined by solving the corresponding game-theoretic control problems.

5.3.4 An Auxiliary Linear Game-Theoretic Control Problem

The admissible constraints on "auxiliary interference" v_k^* $(k = 1, \dots, r)$ are defined by the given constraints (5.1.3) on v_i $(i = 1, \dots, l)$, namely, taking into account expressions for v_k^*, we get the following estimates:

$$|v_i^*| \le \beta_i^* = Y_i^* \beta_i, \quad |v_{q+j}^*| \le \beta_{q+j}^* = \sum_{s=1}^{l} Y_{q+j,s}^* \beta_s,$$

$$Y_i^* = \sup_{\boldsymbol{x} \in \bar{\boldsymbol{\Lambda}}_1} \left[X_i^{(2)}(\boldsymbol{x}) \right], \quad Y_{q+j,s}^* = \sup_{\boldsymbol{x} \in \bar{\boldsymbol{\Lambda}}_1} \left\{ \frac{\partial X_j^{(3)}(\boldsymbol{x})}{\partial \xi_s} X_s^{(2)}(\boldsymbol{x}) \right\} \tag{5.3.4}$$

$$(i = 1, \dots, q; \; j = 1, \dots, r - q).$$

Here $\bar{\boldsymbol{\Lambda}}_1$ is the closure of the set $\boldsymbol{\Lambda}_1 \subseteq \boldsymbol{\Lambda}$.

Estimates (5.3.4) are obtained using the *reinforcing* inequalities. Consequently, the levels of the "auxiliary interference" v_k^* $(k = 1, \dots, r)$ are *overestimated* compared with the real values.

We note that, if system (5.1.1) serves as a model of the angular motion of a solid (the kinematic equations are chosen in the Poisson or Rodrigues–Hamilton form), then obtaining the estimates of type (5.3.4) presents no difficulty (see Sections 5.4–5.5).

In accordance with the objective of control in Problem 5.3.1, let us solve the control problem (via u_k^*) of bringing $\boldsymbol{\mu}$-system (5.3.2), (5.3.3) to the position

$$\xi_i = 0, \quad \eta_j = \mu_j = 0 \qquad (i = 1, \dots, q; \; j = 1, \dots, r - q) \tag{5.3.5}$$

in a *minimum time* for *any* admissible perturbations v_k^*.

We treat this problem as a *differential game*. In this game one of the players (controlling participant) has at its disposal the auxiliary controls u_k^* ($k = 1, \dots, r$) and aims at decreasing time τ for taking the system to the required position (5.3.5). The second player (opponent), aiming to increase τ, has "auxiliary perturbations" v_k^* ($k = 1, \dots, r$) at its disposal.

The admissible levels of auxiliary controls u_k^* ($k = 1, \dots, r$) are to be taken so as not to violate the original constraints (5.1.2) on controls u_k ($k = 1, \dots, r$). This leads to the corresponding constraints on u_k^*, which we shall set in the form

$$|u_k^*| \le \alpha_k^* = \text{const} > 0 \qquad (k = 1, \dots, r). \tag{5.3.6}$$

In addition, to solve the problem, the admissible levels α_k^* of the auxiliary controls u_k^* ($k = 1, \dots, r$) *should be higher* than the obtained levels β_k^* of the auxiliary interference v_k^* ($k = 1, \dots, r$). As a result, in the auxiliary game-theoretic problem, the constraints take the form

$$|u_k^*| \le \alpha_k^*, \quad |v_k^*| \le \beta_k^* = \rho_k \alpha_k^*, \quad 0 < \rho_k < 1 \quad (k = 1, \dots, r). \tag{5.3.7}$$

The procedure for assigning numbers α_k^* can be constructively implemented, for example, by *iterative selection* with subsequent verification of inequalities (5.1.2) along the trajectories of system (5.1.1), (5.3.1). Naturally, to this end, one needs to know an explicit form of these trajectories or, at least, their appropriate estimates. The proposed approach, which implies reducing the original nonlinear Problem 5.3.1 to simple linear control problems, affords such an opportunity.

With values of α_k^* ($k = 1, \dots, r$) fixed, the differential game for each subsystem of system (5.3.2), (5.3.3) is a linear differential game of *one-type objects*. When $0 < \rho_k < 1$ ($k = 1, \dots, r$), its solution can be obtained by solving the *time-optimization* problem for the system

$$\dot{\xi}_i = (1 - \rho_i) u_i^* \quad (i = 1, \dots, q < r < m), \tag{5.3.8}$$

$$\dot{\eta}_j = \mu_j, \quad \dot{\mu}_j = (1 - \rho_{q+j}) u_{q+j}^* \quad (j = 1, \dots, r - q), \tag{5.3.9}$$

subject to the constraints $|u_k^*| \le \alpha_k^*$ ($k = 1, \dots, r$) (Krasovskii [1970]).

The boundary conditions are the same as for system (5.3.2), (5.3.3). System (5.3.8), (5.3.9) is derived from (5.3.2), (5.3.3) by setting the auxiliary perturbations to $v_k^* = -\rho_k u_k^*$. These are the "*worst*" v_k^*, which correspond to the optimal controls for the "opponent."

The solution of the time-optimization problem for systems of type (5.3.8), (5.3.9) has the form (Pontryagin et al. [1961])

$$u_i^*(\xi_i) = \begin{cases} -\alpha_i^* \operatorname{sgn} \xi_i, & \xi_i \ne 0 \\ 0, & \xi_i = 0 \quad (i = 1, \dots, q), \end{cases} \tag{5.3.10}$$

$$u_{q+j}^*(\eta_j, \mu_j) = \begin{cases} \alpha_{q+j}^* \operatorname{sgn} \psi_j^\rho(\eta_j, \mu_j), & \psi_j^\rho \ne 0 \\ \alpha_{q+j}^* \operatorname{sgn} \eta_j = -\alpha_{q+j}^* \operatorname{sgn} \mu_j, & \psi_j^\rho = 0, \end{cases} \tag{5.3.11}$$

$$\psi_j^\rho(\eta_j, \mu_j) = -\mu_j - \left[2\alpha_{q+j}^*(1 - \rho_{q+j})\right]^{-1} \eta_j |\eta_j| \quad (j = 1, \dots, r - q).$$

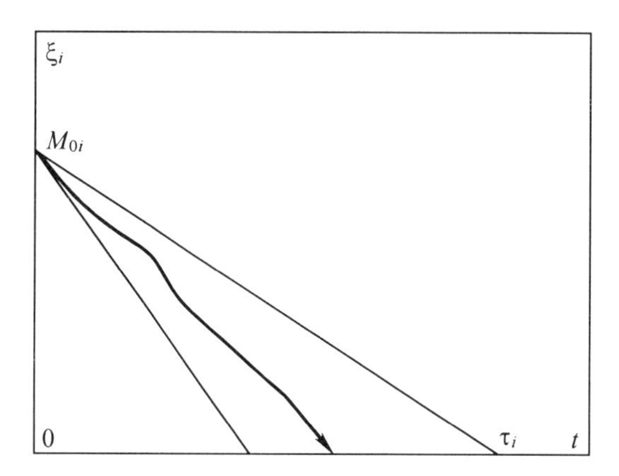

Figure 5.3.1. A phase trajectory of system (5.3.2), (5.3.10) for $v_k^* \not\equiv -\rho_k u_k^*$.

Here ψ_j^ρ are the *switching functions.*

The time τ_k $(k = 1, \ldots, r)$ necessary for attaining each position (5.3.5) along the optimal trajectories of system (5.3.8), (5.3.9) is defined by the equalities

$$\tau_i = [\alpha_i^*(1-\rho_i)]^{-1}|\xi_i| \qquad (i = 1, \ldots, q), \tag{5.3.12}$$

$$\begin{aligned}\tau_{q+j} = {}& [\alpha_{q+j}^*(1-\rho_{q+j})]^{-1}\{[{}^1\!/\!_2\mu_j^2 - \alpha_{q+j}^*(1-\rho_{q+j})\eta_j \operatorname{sgn}\psi_j^\rho]^{1/2} \\ & - \mu_j \operatorname{sgn}\psi_j^\rho\} \qquad (j = 1, \ldots, r-q).\end{aligned} \tag{5.3.13}$$

The value $\tau = \max(\tau_k)$ determines the *minimum guaranteed* time in the game-theoretic problem for the linear system (5.3.2), (5.3.3) under study. We note that, if v_k^* differ from the "worst," then the phase trajectories of system (5.3.2), (5.3.3) are not optimal and the time for moving the system to position (5.3.5) *will not exceed* τ.

Let us characterize the phase trajectories of system (5.3.2), (5.3.3), (5.3.10), and (5.3.11), when $v_k^* \not\equiv -\rho_k u_k^*$, to meet condition (5.1.2), which is crucial in solving Problem 5.3.1.

1. In the case of system (5.3.2), (5.3.10) for all admissible v_i^* $(i = 1, \ldots, q)$ the motion proceeds along the curves (or, for $v_i^* \equiv 0$ $(i = 1, \ldots, q)$, along the straight lines) located between the line segments which are the trajectories of this system for $v_i^* = \pm\rho_i u_i^*$ until the position $\xi_i = 0$ is attained. (See Figure 5.3.1, where M_{0i} is the initial point; for definiteness, $\xi_i(t_0) > 0$.)

2. We begin the consideration of possible trajectories of system (5.3.3), (5.3.11) with the case $v_{q+j}^* \equiv 0$ $(j = 1, \ldots, r-q)$. In this case the motion follows the arcs of parabolas which are the trajectories of the systems $\dot\eta_j = \mu_j$, $\dot\mu_j = (1-\rho_{q+j})u_{q+j}^*$ $(j = 1, \ldots, r-q)$ for u_{q+j}^* of type (5.3.11). Then, after arriving at the switching curves $\psi_j^\rho = 0$, the motion proceeds along them in the *sliding mode* until the desired final state $\eta_j = \mu_j = 0$ is attained. During this process, u_{q+j}^* takes the values $\pm\alpha_{q+j}^*$ with infinitely frequent sign changes, so that "on the average" $u_{q+j}^* = \pm(1-\rho_{q+j})\alpha_j^*$ on the corresponding branches of the switching curves.

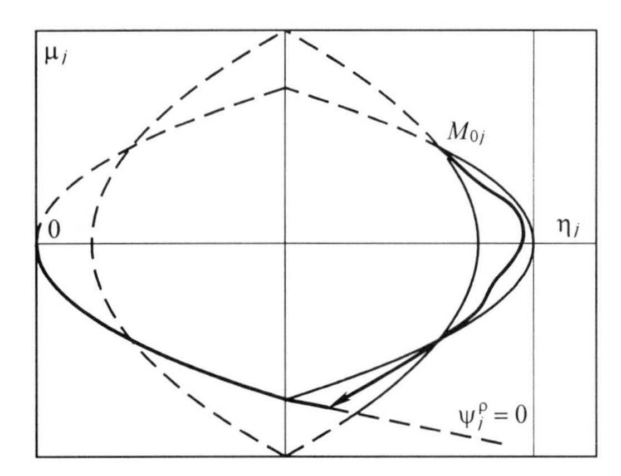

Figure 5.3.2. A phase trajectory of system (5.3.3), (5.3.11) for $v_k^* \not\equiv -\rho_k u_k^*$.

In the general case, the motion occurs initially (until the switching curves are reached) between the parabolic arcs of the systems $\dot{\eta}_j = \mu_j$ and $\dot{\mu}_j = (1 \pm \rho_{q+j})u_{q+j}^*$, respectively, where u_{q+j}^* has the the form (5.3.11). (Note that the system $\dot{\eta}_j = \mu_j$, $\dot{\mu}_j = (1 + \rho_{q+j})u_{q+j}^*$ corresponds to the case when the "auxiliary interference" v_{q+j}^* plays the role of additional auxiliary control rules and has the form $v_{q+j}^* = \rho_{q+j}u_{q+j}^*$.) Next, once the switching curves $\psi_j^\rho = 0$ are reached, the system will also move along these curves in the sliding mode until the state $\eta_j = \mu_j = 0$ is attained. (The process of motion originating from point M_{0j} in the phase plane of variables η_j, μ_j is represented by the solid line in Figure 5.3.2, where, to be specific, the initial values of variables η_j and μ_j have been taken as positive. The expressions for ψ_j^0 can be obtained from ψ_j^ρ by taking $\rho_j = 0$ and expressions for ψ_j^* by changing $(1 - \rho_j)$ to $(1 + \rho_j)$, respectively.)

5.3.5 Principal Results

Let $\boldsymbol{y}$ be a vector composed of variables ξ_i $(i = 1, \dots, q)$, η_j $(j = 1, \dots, r - q)$, and let $\boldsymbol{z}$ be a vector comprising the remaining phase variables of system (5.1.1). Consequently, $\boldsymbol{x} = (\boldsymbol{y}, \boldsymbol{z})$. Along with the Jacobi matrix F introduced earlier, let us introduce the Jacobi matrix Q of functions $X_j^{(3)}(\boldsymbol{x})$ $(j = 1, \dots, r - q)$ with respect to the variables that constitute vector $\boldsymbol{z}$. We assume that $r - q \geq n - r$.

Suppose that in the domain

$$\boldsymbol{\Lambda}_2 : \|\boldsymbol{x}\| < H_2, \tag{5.3.14}$$

the following condition is satisfied:

$$\operatorname{rank} Q(\boldsymbol{x}) = n - r. \tag{5.3.15}$$

Besides, we assume that one and the same rows of matrix Q are linearly independent for all $\boldsymbol{x}$ from domain (5.3.14).

Note that in the case $r \leq m$, the inequalities $n-l+q \geq r$ and $r-q \geq n-r$ introduced above can be simultaneously satisfied only when

$$n + q = 2r = 2l. \tag{5.3.16}$$

As already noted, condition (5.3.16) is observed, for example, in such important applied problems as the control of the angular motion of a solid. We shall also indicate some ways of weakening this condition.

Denote by $\boldsymbol{X_v} = \{\boldsymbol{x_v} = \boldsymbol{x}(t; t_0, \boldsymbol{x}_0, \boldsymbol{v})\}$ the set of the processes $\boldsymbol{x} = \boldsymbol{x}(t; t_0, \boldsymbol{x}_0)$ of the closed system (5.1.1), (5.3.1), (5.3.10), and (5.3.11) that correspond to all admissible realizations of interference $\boldsymbol{v} \in K_1$.

Under condition (5.3.16), the equalities $X_j^{(3)}(\boldsymbol{x}) = \mu_j$ $(j = 1, \dots, r-q)$ admit the solution

$$\boldsymbol{z} = \boldsymbol{\Psi}(\boldsymbol{y}, \boldsymbol{\mu}), \qquad \boldsymbol{\Psi}(\mathbf{0}, \mathbf{0}) = \mathbf{0} \tag{5.3.17}$$

in $\boldsymbol{\Lambda}_2$ (vector function $\boldsymbol{\Psi} \in R^{n-r}$ is continuous in $\boldsymbol{\Lambda}_2$).

Equalities (5.1.17) relate components of $\boldsymbol{x}$ not entered into $\boldsymbol{y}$ to components of $\boldsymbol{y}$, $\boldsymbol{\mu}$. As a result, in Λ_2 the phase vector $\boldsymbol{x}$ of system (5.1.1) can be represented in the form $\boldsymbol{x} = [\boldsymbol{y}, \boldsymbol{z}(\boldsymbol{y}, \boldsymbol{\mu})] \stackrel{\text{def}}{=} \boldsymbol{W}(\boldsymbol{y}, \boldsymbol{\mu})$.

Consequently, the set $\boldsymbol{X_v}$ can be represented in the form

$$\boldsymbol{X_v} = \{\boldsymbol{W} = \boldsymbol{W}[t; t_0, \boldsymbol{W}_0(\boldsymbol{y}_0, \boldsymbol{\mu}_0(\boldsymbol{x}_0), \boldsymbol{v}]\} = \{\boldsymbol{W} = \boldsymbol{W}(t; t_0, \boldsymbol{x}_0, \boldsymbol{v})\}$$

and estimated by employing the estimates for the sets of processes of the *linear* system (5.3.2), (5.3.3), (5.3.10), and (5.3.11).

Theorem 5.3.1. *Let the following conditions be satisfied:*

(1) *a constant* q $(0 \leq q < r)$ *can be found such that* $n + q = 2r = 2l$;
(2) *condition* (5.1.5) *in domain* (5.1.4);
(3) *condition* (5.3.15) *in domain* (5.3.14);
(4) *for any* $\boldsymbol{x}_0 \in S$, *the inequalities*

$$|\boldsymbol{W}(t; t_0, \boldsymbol{x}_0, \boldsymbol{v})| < \min(H_1, H_2), \tag{5.3.18}$$

$$\big|f_k[\boldsymbol{W}(t; t_0, \boldsymbol{x}_0, \boldsymbol{v}), \boldsymbol{u}^*]\big| \leq \alpha_k \quad (k = 1, \dots, r) \tag{5.3.19}$$

are valid in the set $\boldsymbol{X_v} = \{\boldsymbol{W} = \boldsymbol{W}(t; t_0, \boldsymbol{x}_0, \boldsymbol{v})\}$ *defined through the sets of possible trajectories of the linear system* (5.3.2), (5.3.3), (5.3.10), *and* (5.3.11).

Then, for all $\boldsymbol{x}_0 \in S$, *control rules* (5.3.1), (5.3.10), *and* (5.3.11) *solve Problem* 5.3.1. *It is guaranteed that system* (5.1.1) *is taken precisely to the equilibrium position* $\boldsymbol{x} = \mathbf{0}$ *by bounded controls* $\boldsymbol{u} \in K$ *in a finite time* τ *(determined as* $\max(\tau_k)$ $(k = 1, \dots, r)$ *from relationships* (5.3.12), (5.3.13)) *for any admissible realizations of interference* $\boldsymbol{v} \in K_1$.

Proof. If condition (5.1.5) is satisfied in domain (5.1.4), the linear system (5.3.2), (5.3.3) is derived from the original nonlinear system (5.1.1), (5.3.1) by the appropriate nonlinear transformation of the variables. Under such transformation, the state of variables η_j $(j = 1, \ldots, r)$ of system (5.1.1), (5.3.1) in $\boldsymbol{\Lambda}_1$ will be completely determined by the state of the same variables of the linear system (5.1.9). Therefore, for all $\boldsymbol{x}_0 \in S_1 = \{\boldsymbol{x}_0 : \|\boldsymbol{x}_{\boldsymbol{v}}(t)\| < H_1\}$, control rules (5.3.1), (5.3.10), and (5.3.11) guarantee that system (5.1.1) is taken precisely to the position

$$\eta_j = X_j^{(3)} = 0 \qquad (j = 1, \ldots, r) \tag{5.3.20}$$

in a finite time τ.

If condition (5.3.15) is satisfied in domain (5.3.14), relationship (5.3.17) is valid. Taking into account $\dot{\eta}_j = X_j^{(3)}(\boldsymbol{x})$ $(j = 1, \ldots, r)$, by discontinuity of vector function Ψ and condition $\Psi(\mathbf{0}, \mathbf{0}) = \mathbf{0}$, this function can be made arbitrarily small in magnitude by decreasing the absolute values of η_j and $X_j^{(3)}(\boldsymbol{x})$ $(j = 1, \ldots, r)$.

Thus, for all $\boldsymbol{x}_0 \in S_2 = \{\boldsymbol{x}_0 : \|\boldsymbol{x}_{\boldsymbol{v}}(t)\| < \min(H_1, H_2)\}$, control rules (5.3.1), (5.3.10), and (5.3.11) guarantee that system (5.1.1) will be precisely taken not only to (5.3.20) but to position $\boldsymbol{x}_0 = \mathbf{0}$.

From requirements (5.1.2), it follows that the admissible set of the initial perturbations $\boldsymbol{x}_0$ should meet the condition $\boldsymbol{x}_0 \in S_3 = \{\boldsymbol{x}_0 : |f_k[\boldsymbol{W}(t), \boldsymbol{u}^*]| \leq \alpha_k\}$.

Because the conditions of the theorem are satisfied, the given domain S of the initial perturbations will be a subset of the intersection of the sets S_1–S_3. The definition of the vector function $\boldsymbol{W} = \boldsymbol{W}(\boldsymbol{y}, \boldsymbol{\mu}) : R^{2r} \to R^n$ introduced implies that, for all $\boldsymbol{x}_0 \in S$, controls (5.3.1), (5.3.10), and (5.3.11) guarantee that system (5.1.1) is precisely taken to the position $\boldsymbol{x} = \mathbf{0}$ in a finite time τ for any $\boldsymbol{v} \in K_1$. Besides, control rules (5.3.1), (5.3.10), and (5.3.11) satisfy the given constraints (5.1.2). The theorem is proved. □

Let us indicate possible ways of weakening the conditions of Theorem 5.3.1.

Let N first integrals $R_\varepsilon(\boldsymbol{x}) = \text{const}$, $R_\varepsilon(\mathbf{0}) = 0$ of system (5.1.1) be known. Assuming $r - q + N \geq m + p - r$, instead of Q, we introduce the Jacobi matrix Q^* of functions $X_j^{(3)}$, R_ε with respect to the variables that constitute vector $\boldsymbol{z}$.

Corollary 5.3.1. *Let*

(1) q $(0 \leq q < r)$ *be found such that* $r \leq n + q \leq 2r + N$;

(2) $\operatorname{rank} Q^* = n - r$ *for* $\boldsymbol{x} \in \boldsymbol{\Lambda}_2$.

If conditions (2) *and* (4) *of Theorem* 5.3.1 *are satisfied, then, for all* $\boldsymbol{x}_0 \in S$, *control rules* (5.3.1), (5.3.10), *and* (5.3.11) *solve Problem* 5.3.1.

Another way of weakening the conditions of Theorem 5.3.1 consists of successively using *several* constructions of control rules of type (5.3.1). In this case system (5.1.1) is gradually transferred from $\boldsymbol{x}_0 \in S$ to $\boldsymbol{y} = \mathbf{0}$ (in a total finite time).

5.3.6 Algorithm for Solving Problem 5.3.1

The algorithm is divided into the following stages:

(1) Choosing construction (5.3.1) of control rules u_k $(k = 1, \ldots, r)$ (with vector function $\boldsymbol{\Phi}$ chosen according to Subsection 5.3.3).
(2) Estimating levels of the "auxiliary interference" v_k^* $(k = 1, \ldots, r)$. Computing numbers β_k^* $(k = 1, \ldots, r)$ in (5.3.4) from the values of β_i $(i = 1, \ldots, m)$.
(3) Estimating auxiliary controls u_k^* $(k = 1, \ldots, r)$. Preliminarily choosing numbers α_k^* $(k = 1, \ldots, r)$. In this case numbers α_k^*, β_k^* predetermine the corresponding value of the *guaranteed* reorientation time $\tau = t_1 - t_0$.
(4) Validating the inequalities (5.3.18), (5.3.19) for any $\boldsymbol{x}_0 \in S$. In so doing, one should rely on the estimates (two-sided) of the set of processes of system (5.3.2), (5.3.3), (5.3.10), and (5.3.11). Such estimates can be obtained based on analyzing possible trajectories of this system for all admissible realizations of auxiliary interference v_k^* $(k = 1, \ldots, r)$ given in Subsection 5.3.4.

If, with α_k^* $(k = 1, \ldots, r)$ thus chosen, inequalities (5.3.18), (5.3.19) fail to hold or, vice versa, they are satisfied with a kind of "reserve," the search for the appropriate numbers α_k^* is to be continued. Otherwise, the guaranteed control time equals τ.

As a result, we get an iterative algorithm for solving Problem 5.3.1. Its implementation as applied to the game-theoretic problem of reorienting an asymmetric solid is presented in Section 5.4.

5.3.7 A Modification of the Proposed Approach

The general structural form (5.3.1) of the control rules can be *simplified* by making the structure of "auxiliary interference" v_k^* $(k = 1, \ldots, r)$ *more complicated.*

To this end, in the vector function $\boldsymbol{\Phi}$ introduced in Subsection 5.3.3, we shall keep only expressions containing u_k $(k = 1, \ldots, r)$.

Suppose that condition (5.1.5) in domain (5.1.4) is satisfied. We assume, without loss of generality, that the first r rows in matrix F are linearly independent.

In this case by the theory of implicit functions we conclude that system (5.1.6) in domain (5.1.4) has the solution

$$\boldsymbol{u} = \boldsymbol{f}_*(\boldsymbol{x}, \boldsymbol{u}^*) \tag{5.3.21}$$

of type (5.3.1); expression (5.3.21) is *simpler* than (5.3.1).

Let us introduce the designations

$$v_i^* = X_i^{(0)}(\boldsymbol{x}) + X_i^{(2)}(\boldsymbol{x})v_i \qquad (i = 1, \ldots, q),$$

$$v_{q+j}^* = \sum_{s=1}^{l} \frac{\partial X_j^{(3)}(\boldsymbol{x})}{\partial \xi_s} \left(X_s^{(0)}(\boldsymbol{x}) + X_s^{(2)}(\boldsymbol{x})v_s \right)$$
$$+ \sum_{\varepsilon=1}^{n-l} \frac{\partial X_j^{(3)}(\boldsymbol{x})}{\partial \eta_\varepsilon} X_\varepsilon^{(3)}(\boldsymbol{x}) \qquad (j = 1, \ldots, r - q).$$

As a result, under assumptions made from the closed nonlinear system (5.1.1), (5.3.21), one can segregate the *linear conflict-control* μ*-system* (5.3.2), (5.3.3).

The "*direct*" estimates (5.3.4) of v_k^* $(k = 1, \ldots, r)$ were used in Subsection 5.3.4. Here, because of the complexity of the expressions for v_k^*, it is reasonable to use the *principle of assignment with subsequent confirmation* of levels of v_k^*. Levels of v_k^* having been "assigned," the auxiliary game-theoretic control problem for linear μ-system (5.3.2), (5.3.3) is solved as in Subsection 5.3.4.

As a result, we get the following algorithm for solving Problem 5.3.1.

(1) Choosing construction (5.3.21) of control rules u_k $(k = 1, \ldots, r)$.

(2) Preliminarily choosing the value τ of the guaranteed control time and and "assigning" levels β_k^* $(k = 1, \ldots, r)$ of "auxiliary interference" v_k^*, which predetermines values α_k^*, ρ_k $(k = 1, \ldots, r)$. The parameters in (5.3.10), (5.3.11) are thus specified.

(3) Verifying that the inequalities $|v_k^*| \leq \beta_k^*$ $(k = 1, \ldots, r)$ are actually valid over the state set Ω of linear system (5.3.2), (5.3.3).

(4) Validating the inequalities (5.3.18), (5.3.19) for any $x_0 \in S$.

If, with τ, β_k^* $(k = 1, \ldots, r)$ thus chosen, inequalities $|v_k^*| \leq \beta_k^*$ $(k = 1, \ldots, r)$ or (5.3.18), (5.3.19) fail to hold (or, vice versa, they are satisfied with a kind of "reserve"), the search for the appropriate numbers τ, β_k^* is to be continued. Otherwise, the guaranteed control time equals τ.

As a result, we get an iterative algorithm for solving Problems 5.3.1.

Its implementation as applied to the game-theoretic problem of reorienting an asymmetric solid is presented in Section 5.5.

5.4 The Nonlinear Game-Theoretic Problem of Triaxial Reorientation of an Asymmetric Solid (First Method of Solution)

5.4.1 Preliminary Remarks

Even without taking interference into account, it is hard to find rigorous solutions for the optimum problems of controlling the angular motion of a solid. So they are often divided into *two stages*:

(1) *damping of rotation*;

(2) *reorientation in space*.

In the first stage the study is based on Euler's dynamical equations; knowledge of the angular position of the body is unnecessary.

In the second stage the initial and final states are stationary states. The orientation can be altered in a number of ways:

(a) by three successive rotations about connected axes;

(b) in the class of *planar* rotations about Euler's axis (*extensive turn*), in which case the angular vector of the body keeps a fixed position in space;

(c) by several planar rotations;

(*d*) by one *spatial* turn without any additional restrictions on the character of the resulting motion.

The question of which of the schemes is appropriate for optimal control is solved separately in each particular case. As a rule, the prescribed speed and fuel consumption and the reliability and technological requirements are taken into account.

The optimization problems of controlling an extensive turn are very completely reflected in the literature. This method is particularly effective in those cases when the ellipsoid of inertia of the body is close to a sphere or the control is realized by small controlling moments. Problems concerned with the optimum reorientation of a solid by a single spatial rotation have been less extensively studied.

The actual conditions under which objects function make it necessary to account for constantly acting interference and perturbations. The control process becomes especially difficult if only the minimum information about the interference, namely their limits of variation, is available. Here one is concerned with control under *conditions of uncertainty*. In this case, however, as is shown in what follows, in the stage of altering the orientation of the body one can put forward a constructive method of solution subject to the constraints on the control, which, even if not rigorous, is completely acceptable from the viewpoint of applications.

This method employs the results of Section 5.3 and rests on choosing a structural form of contròl rules that enables one to reduce the original *nonlinear* problem to auxiliary *linear* control problems. The additive "*auxiliary interferences*" appearing in the auxiliary linear control systems (generated by an appropriate transformation of interference in the original system) are then treated as *player-opponent* control functions. As a result, the auxiliary linear control systems can be regarded as *conflict-control* systems. The parameters of the chosen structural form of the original control rules are determined as *player-partner* control strategies in the corresponding linear game-theoretic problems. In this way one obtains *robust* (*stable*) nonlinear control rules ensuring a *guaranteed* result when solving the original nonlinear control problem, namely, it is guaranteed that the solid will move *precisely* into the prescribed final state (by a *single spatial* rotation) for *any* admissible realization of interference.

The resulting rules for controlling the reorientation (making the most of the possibilities of the structural form proposed for them) are *time-suboptimal.*

The feasibility of extending the approach to the case of the problem of reorientation with simultaneous damping of the angular velocity of the body is discussed.

5.4.2 Formulation of the Problem

As a model of the angular motion of a solid, one can use a nonlinear controlled system of type (5.1.1) which includes *Euler's dynamical equations*

$$\begin{gathered} A\dot{\xi}_1 = (B-C)\xi_2\xi_3 + u_1 + v_1, \\ B\dot{\xi}_2 = (C-A)\xi_1\xi_3 + u_2 + v_2, \quad C\dot{\xi}_3 = (A-B)\xi_1\xi_2 + u_3 + v_3 \end{gathered} \tag{5.4.1}$$

and the *kinematic equations*

$$
\begin{gathered}
2\dot{\eta}_1 = \xi_1\eta_4 + \xi_3\eta_2 - \xi_2\eta_3, \\
2\dot{\eta}_2 = \xi_2\eta_4 + \xi_1\eta_3 - \xi_3\eta_1, \quad 2\dot{\eta}_3 = \xi_3\eta_4 + \xi_2\eta_1 - \xi_1\eta_2
\end{gathered} \tag{5.4.2}
$$

$$
\sum_{j=1}^{3} \eta_j^2 + \eta_4^2 = 1, \tag{5.4.3}
$$

in terms of the Rodrigues–Hamilton variables. Here, as in Section 5.2, $\boldsymbol{\eta}$ is the vector formed by variables η_j $(j = 1, 2, 3)$ and η_4; $\boldsymbol{x} = (\boldsymbol{\xi}, \boldsymbol{\eta})$. For uniformity of notation, we also leave the indices $i, j, k = 1, 2, 3$ of variables ξ_i, η_j and controls u_k unchanged.

But as distinguished from Section 5.2, from the standpoint of implementation, now we choose controls $\boldsymbol{u}$ in a *simpler* class $\boldsymbol{u} \in K^*$ of vector functions $\boldsymbol{u} = \boldsymbol{u}(\boldsymbol{x}, \boldsymbol{x}_0)$ *piecewise-continuous* in $\boldsymbol{x}$, $\boldsymbol{x}_0$, which satisfy the prescribed "geometric" constraints (5.2.4).

Class K_1 of admissible interference is *narrower* than that we dealt with in Section 5.2, namely, interference $\boldsymbol{v} \in K_1$ can take the form of arbitrary *piecewise-continuous* vector functions $\boldsymbol{v} = \boldsymbol{v}(t)$ within restrictions (5.2.5).

Numbers α_k and β_i $(i, k = 1, 2, 3)$ characterize the maximum possible magnitudes (*levels*) of the controls and interference, respectively.

Problem 5.4.1. Control rules $\boldsymbol{u} \in K^*$ are required that will take the body from the initial state $\boldsymbol{\eta}(t_0) = \boldsymbol{\eta}_0$ to the prescribed state $\boldsymbol{\eta}(t_1) = \boldsymbol{\eta}_1$ in a finite time for any admissible realizations of interference $\boldsymbol{v} \in K_1$. Both states are stationary: $\boldsymbol{\xi}(t_0) = \boldsymbol{\xi}_0 = \boldsymbol{\xi}(t_1) = \boldsymbol{\xi}_1 = \mathbf{0}$. The time $t_1 > t_0$ is not fixed, but we shall seek to make it as short as possible.

Without loss of generality, we further suppose that $\boldsymbol{\eta}_1 = (0, 0, 0, 1)$. In fact, as has already been noted, the case in question corresponds to controlling the orientation relative to the coordinate system set at the initial time. In what follows, in accordance with the designations adopted, $i, j, k = 1, 2, 3$.

5.4.3 An Auxiliary Linear Conflict-Control System

This system is constructed in just the same way as in Section 5.2 and has the form

$$
\dot{\eta}_j = \mu_j, \quad \dot{\mu}_j = u_j^* + v_j^*; \tag{5.4.4}
$$

the "original" control rules u_k are expressed in terms of u_j^* by formulas (5.2.8).

Next we suggest that the solution of the original nonlinear game-theoretic Problem 5.4.1 should be constructed based on the corresponding game-theoretic control problems for linear system (5.4.4).

Thus construction (5.2.8) can be regarded as the *general structural form* of the control rules in Problem 5.4.1, as done while solving Problem 5.2.1. But the parameters of this form, i.e., the auxiliary control rules u_j^*, are determined by solving linear game-theoretic problems *different* from those considered in Section 5.2.

Structure (5.2.8) of the control rules contains the factor η_4^{-1}, which formally implies a "*singularity.*" However, subsequent thorough analysis reveals, as in solving Problem 5.2.1, that in the case of $\boldsymbol{\eta}_1 = (0,0,0,1)$, the relationship $\eta_4 \in [\eta_{40}, 1]$ holds in the course of the control. Therefore, the specified "singularity" does not occur at all.

5.4.4 An Auxiliary Game-Theoretic Control Problem

By relationships (5.2.3), (5.2.5), and the Cauchy–Schwarz inequality, the following relationships hold:

$$|v_j^*(t)| \le \beta_j^* = \frac{1}{2}\left[(\beta_1 A^{-1})^2 + (\beta_2 B^{-1})^2 + (\beta_3 C^{-1})^2\right]^{1/2}. \tag{5.4.5}$$

In line with the purposes of Problem 5.4.1, let us solve the game-theoretic control problem of bringing system (5.4.4) to the position

$$\eta_j = \mu_j = 0 \tag{5.4.6}$$

in the *shortest* possible time. The control is implemented through u_j^* for *any* admissible realizations of v_j^* satisfying inequalities (5.4.5).

We shall interpret this problem as a *differential game* in which one of the players handles u_j^* and seeks to reduce the time τ for bringing the system to position (5.4.6). The "opponent" whose objective is to increase τ (or to avoid attaining position (5.4.6) at all) has "auxiliary perturbations" v_j^* at his disposal.

For this problem to be solvable, admissible levels u_j^* should be higher than levels v_j^*. So, we shall take the corresponding constraints in the form

$$|u_j^*| \le \alpha_j^*, \quad |v_j^*| \le \beta_j^* = \rho_j \alpha_j^*, \quad 0 < \rho_j < 1. \tag{5.4.7}$$

The procedure of assigning levels α_j^* will be considered later. Here we take them as prescribed, so that conditions (5.4.7) are satisfied.

Solving linear game-theoretic problem under study for system (5.4.4) is reduced to the time-optimal problem for the system

$$\dot{\eta}_j = \mu_j, \quad \dot{\mu}_j = (1-\rho_j)u_j^*, \quad |u_j^*| \le \alpha_j^*. \tag{5.4.8}$$

The boundary conditions are the same as for system (5.4.4).

The solution of time-optimal problems for systems of type (5.4.8) has the form (Pontryagin et al. [1961])

$$u_j^*(\eta_j, \mu_j) = \begin{cases} \alpha_j^* \operatorname{sgn} \psi_j^\rho(\eta_j, \mu_j), & \psi_j^\rho \ne 0 \\ \alpha_j^* \operatorname{sgn} \eta_j = -\alpha_j^* \operatorname{sgn} \mu_j, & \psi_j^\rho = 0, \end{cases} \tag{5.4.9}$$

$$\psi_j^\rho(\eta_j, \mu_j) = -\mu_j - \left[2\alpha_j^*(1-\rho_j)\right]^{-1} \eta_j |\eta_j|.$$

Here ψ_j^ρ are the *switching functions.*

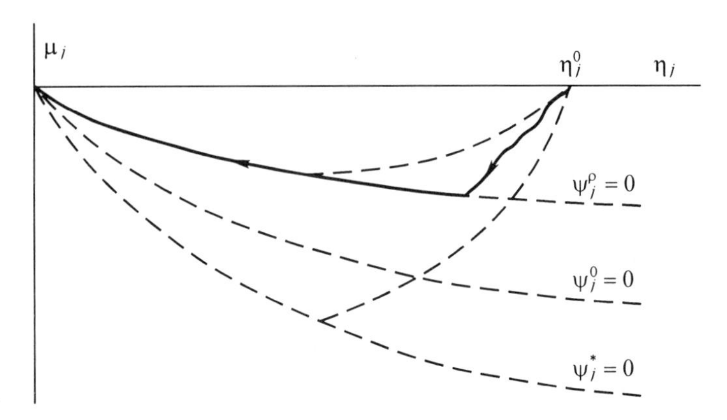

Figure 5.4.1. Phase trajectories of system (5.4.4), (5.4.9) for $v_j^* \not\equiv -\rho_j u_j^*$. Functions ψ_j^0 and ψ_j^* are defined in Subsection 5.3.4.

In the case of Problem 5.4.1 (when $\boldsymbol{\xi}_0 = \mathbf{0}$), the quantity

$$\tau = \max(\tau_j), \quad \tau_j = 2\{|\eta_{j0}|[\alpha_j^*(1-\rho_j)]^{-1}\}^{1/2} \tag{5.4.10}$$

determines the minimum guaranteed time τ for attaining position (5.4.6) in the auxiliary linear game-theoretic problem.

The characterization of phase trajectories of system (5.4.4), (5.4.9) for $v_j^* \not\equiv -\rho_j u_j^*$ is presented in Subsection 5.3.5. We note that in the case of Problem 5.4.1, the trajectories of this system originate from and terminate at the axes $\mu_j = 0$. (See Figure 5.4.1, where $\eta_{j0} > 0$, for definiteness).

5.4.5 Algorithm for Solving Problem 5.4.1

This algorithm consists of the following stages:

(1) Choosing construction (5.2.8) of control rules u_k in which u_j^* have the form (5.4.9). (If $\boldsymbol{\eta}_1 \neq (0,0,0,1)$ or the value of η_{40} is small, it suffices to change the controls to those obtained from (5.2.8) by permutation of the indices and coefficients A, B, and C.)

(2) Estimating levels β_j^* of the "auxiliary interference" v_j^* (by using formulas (5.4.5)).

(3) "Assigning" levels α_j^* of the auxiliary control rules u_j^*. Numbers α_j^*, β_j^* predetermine the corresponding value of the *guaranteed* reorientation time $\tau = t_1 - t_0$.

(4) Verifying the original constraints (5.2.4) imposed on control rules u_k. To this end we employ equalities (4.6.11). As a result, this verification is possible along trajectories of system (5.4.4), (5.4.9). See Subsection 5.4.6 regarding possible estimates for u_k.

If estimates (5.2.4) fail to hold or, vice versa, they are satisfied with a kind of "reserve," the search for the appropriate numbers α_j^* is to be continued. Otherwise, the reorientation time equals τ.

As a result, we get an *iterative algorithm* for solving Problem 5.4.1.

Analyzing possible trajectories of system (5.4.4), (5.4.9), we conclude that, in the case $\boldsymbol{\eta}_1 = (0,0,0,1)$, the condition $\eta_4 \geq \gamma = \text{const} > 0$, once met at $t = t_0$, remains valid for all $t \in [t_0, t_0 + \tau]$. It is implied in this case that, in the course of the control, the relationship $\eta_4 \in [\eta_{40}, 1]$ is valid and the "singularity" of controls (5.2.8), (5.4.9) is ruled out.

5.4.6 Estimating Control Rules u_k

For the proposed construction of control rules (5.2.8) not only the switching moments but also the quantities u_k explicitly depend on $\boldsymbol{\xi}, \boldsymbol{\eta}$.

The verification of constraints (5.2.4) can be constructively performed by estimating $|u_k|$ over the state set Ω of the linear system (5.4.4), (5.4.9). To this end, by rearranging the terms in expressions (5.2.8), we obtain the relationships

$$u_1 = 2A\{\eta_4^{-1}\eta_1\,[\Big(\sum_{j=1}^{3}\eta_j u_j^*\Big) + \frac{1}{4}\sum_{i=1}^{3}\xi_i^2] + \eta_4 u_1^* + \eta_3 u_2^* - \eta_2 u_3^*\}$$
$$+ (C - B)\xi_2\xi_3 \qquad (123, ABC).$$

Lemma 5.4.1. *Let* $\boldsymbol{\eta}_1 = (0,0,0,1)$, $\eta_{j0} > 0$. *The following (over)estimates are valid*

$$|u_1| \leq 2A(G_1 + P_1) + F_1 \quad (123, ABC),$$
$$P_1 = (\eta_4^-)^{-1}\eta_1^+(\eta_1^+\alpha_1^* + \eta_2^+\alpha_2^*) + \eta_4^+\alpha_1^* + \eta_3^+\alpha_2^* + l_1\alpha_3^* \quad (123),$$
$$l_1 = \max\big(\big|(\eta_4^-)^{-1}\eta_1^+\eta_3^+ - \eta_2^-\big|, \big|\eta_2^+ - (\eta_4^+)^{-1}\eta_1^-\eta_3^-\big|\big) \quad (123), \qquad (5.4.11)$$
$$G_1 = (\eta_4^-)^{-1}\eta_1^+ G, \quad G = \big[(\eta_4^-)^{-1}\sum_{i=1}^{3}(\eta_i^+\mu_i^-)\big]^2 + \sum_{i=1}^{3}(\mu_i^-)^2,$$
$$F_1 = 2\,|C - B|\,G \quad (123, ABC).$$

Here $\eta_j^{(+,-)}$ *are respective upper (lower) estimates for variables* η_j, μ_j^- *are lower estimates for* μ_j *over the set* Ω, *and*

$$\eta_4^{(+,-)} = \big[1 - \sum_{j=1}^{3}(\eta_j^{-,+})^2\big]^{1/2}.$$

Expressions for $\eta_j^{(+,-}$, μ_j^- *are compiled in Table* 5.4.1, *where*

$$T_j = 2(1-\rho_j)^{-1}T_j^*, \quad T_j^* = \big\{[\alpha_j^*(1+\rho_j)]^{-1}[\eta_{j0}(1-\rho_j)]\big\}^{1/2}, \quad t_0 = 0.$$

(*We note that* $T_j^* \leq \tau/2$, $T_j \leq \tau$ *for all* j.)

The specified estimates will be satisfied for any admissible realizations of interference v_i ("*auxiliary interference*" v_j^*).

Proof. The proof is based on the "*direct*" estimation of the expressions for u_k, using relationship (4.6.11). We note that the conditions $\boldsymbol{\eta}_1 = (0, 0, 0, 1)$, $\eta_{j0} > 0$ do not limit the generality of the procedure for obtaining estimates of type (5.4.11). With other boundary conditions imposed, the estimates of this type are derived by following the same scheme. □

Table 5.4.1.

t	η_j^+	η_j^-	μ_j^-
$[0, T_j^*]$	$\eta_{j0} - \frac{1}{2}(1-\rho_j)\alpha_j^* t^2$	$\eta_{j0} - \frac{1}{2}(1+\rho_j)\alpha_j^* t^2$	$-(1+\rho_j)\alpha_j^* t$
$(T_j^*, \tau/2]$	—‖—	$\frac{1}{2}(1-\rho_j)\alpha_j^*(t - T_j)^2$	$-(1+\rho_j)\alpha_j^* T_j^*$
$(\tau/2, T_j]$	$\frac{1}{2}(1-\rho_j)\alpha_j^*(t-\tau)^2$	—‖—	—‖—
$(T_j, \tau]$	—‖—	0	—‖—

Remark 5.4.1.

1. Computer simulation has demonstrated that the estimates of type (5.4.11) are acceptable enough for determining the guaranteed reorientation time. On the other hand, the development of an efficient computational procedure for solving the corresponding problem of *nonlinear programming* aimed at finding $\max|u_k|$ over the set Ω of possible states of a linear system of type (5.4.4), (5.4.9) is also desirable. This will allow one to estimate the guaranteed reorientation time more accurately.

2. Analyzing the structure of control rules (5.2.8), we also note that the values of $\max|u_k|$ over the set Ω are close enough to those of $\max|u_k|$ over the set Ω^* of those possible states of the linear system (5.4.4), (5.4.9), which correspond to various combinations of "edge" $v_j^* = \pm\rho_j u_j^*$ and "intermediate" $v_j^* \equiv 0$ realizations of v_j^*. Accurate calculation of $\max|u_k|$ over the set Ω^* presents no difficulties either.

Indeed, the structure of expressions (5.2.8) is such that the increase in values of $\max|u_k|$ is determined to a great extent by "nonsynchronism" of the switching moments of the auxiliary control rules u_j^*. The discussion of the set of possible states Ω^* of system (5.4.4), (5.4.9) takes into account (although not completely) the effect of "nonsynchronism" of switching u_j^*. Because of this, in the subsequent computation of $\max|u_k|$ over the set Ω, where the effect of "nonsynchronism" of switching u_j^* manifests itself fully, the values of $\max|u_k|$ will differ only slightly from their values over the set Ω^*.

3. The possibility indicated of solving the original nonlinear control problem under conditions of uncertainty with respect to interference by analyzing just the "edge" values of the "auxiliary interference" v_j^* is to some extent similar to that which arises in analyzing *robust stability* by using the "*edge*" *theorems* of the Kharitonov type [1978] (see also Bartlett et al. [1987]).

5.4.7 The Conditions for the Solvability of Problem 5.4.1

Let us sum up the considerations presented.

Theorem 5.4.1. *If levels α_k of control rules u_k in system* (5.4.1)–(5.4.3) *are high enough, then for any given levels β_i of interference v_i the rules u_k solving Problem* 5.4.1 *can be constructed in the form* (5.2.8) *with u_j^* of type* (5.4.9). *The precise reorientation of a solid is ensured in a finite time τ for any $\boldsymbol{v} \in K_1$. The value of τ is determined iteratively by using the algorithm given in Subsection* 5.4.5.

Theorem 5.4.1 states the possibility of solving Problem 5.2.1 by constructing control rules of type (5.2.8), (5.4.9). We can make this possibility more specific as follows.

Theorem 5.4.2. *Let $\boldsymbol{\eta}_1 = (0,0,0,1)$. Suppose that one can find numbers α_j^* such that*

$$\alpha_j^* > \beta_j^* = \frac{1}{2}\left[(\beta_1 A^{-1})^2 + (\beta_2 B^{-1})^2 + (\beta_3 C^{-1})^2\right]^{1/2},$$
$$2A(G_1 + P_1) + F_1 \le \alpha_1 \quad (123, ABC), \quad t \in [0, \tau],$$

where G_j, P_j, and F_j are defined by inequalities (5.4.11). *Then control rules* (5.2.8), (5.4.9) *solve Problem* 5.4.1 *in a finite time τ found in accordance with* (5.4.10).

5.4.8 Results of Computer Simulation

For a solid (spacecraft) with $A = 4 \cdot 10^4$, $B = 8 \cdot 10^4$, and $C = 5 \cdot 10^4$ $kg{\cdot}m^2$, we consider Problem 5.4.1 of the triaxial reorientation from the initial position $\boldsymbol{\xi}_0 = \mathbf{0}$, $\boldsymbol{\eta}_0 = (0.353, 0.434, 0.432, 0.701)$ to a given $\boldsymbol{\xi}_1 = \mathbf{0}$, $\boldsymbol{\eta}_1 = (0,0,0,1)$.

We define the admissible limits of variation of interference v_i by the inequalities

$$\frac{1}{2}\left|\eta_4 v_1 A^{-1} + \eta_2 v_3 C^{-1} - \eta_3 v_2 B^{-1}\right| \le \beta_1^* = 10^{-3}\ (s^{-2}) \quad (123, ABC). \qquad (5.4.12)$$

If reinforcing inequalities of type (5.4.5) are then used, we find from (5.4.12) that

$$\frac{1}{4}\left(\beta_1^2 A^{-2} + \beta_3^2 C^{-2} + \beta_2^2 B^{-2}\right) \le 10^{-6}\ (s^{-2}).$$

From this relationship we deduce that estimates (5.4.12) hold, for example, in the following cases:

(1) $\beta_1 = \beta_3 = 0, \quad \beta_2 = 160;$
(2) $\beta_1 = \beta_2 = 0, \quad \beta_3 = 100;$
(3) $\beta_2 = \beta_3 = 0, \quad \beta_1 = 80 \quad N{\cdot}m.$

However, as will be shown, the true values of v_i, subject to the constraints (5.4.12) may turn out to be much larger.

From the technological maneuverability requirements for a spacecraft, we shall assume that the guaranteed reorientation time τ must not exceed $\tau = 70$ s for the given admissible limits of variation of interference v_i. Let us evaluate the resources necessary to achieve this objective when using the construction of control rules u_k of the form (5.2.8), (5.4.9).

By (5.4.9), (5.4.10), the given value τ predetermines the values α_j^* of the maximum levels of the auxiliary control functions u_j^*. As the basis we take the equalities $\tau = \tau_j$, which mean that the minimum guaranteed reorientation time is "*levelled*" with respect to each variable η_j. As a result, we obtain the relationships

$$2\{\eta_{j0}[\alpha_j^*(1-\rho_j)]^{-1}\}^{1/2} = \tau. \tag{5.4.13}$$

Taking into account that $\alpha_j^*(1-\rho_j) = \alpha_j^* - \beta_j^*$, where $\beta_j^* = 10^{-3}$ in the case (5.4.12), we find from (5.4.13) that

$$\alpha_1^* = 1.289 \cdot 10^{-3}, \quad \alpha_2^* = 1.354 \cdot 10^{-3}, \quad \alpha_3^* = 1.353 \cdot 10^{-3} \ (s^{-2}). \tag{5.4.14}$$

Let us estimate the values $\alpha_k = \max|u_k|$ for the control rules u_k of the form (5.2.8), (5.4.9), and (5.4.14).

1° First we consider the case $v_j^* = -\rho_j u_j^*$ of the "*worst*" v_j^* slowing down, as much as possible, the process of taking the auxiliary linear system (5.4.4) to the desired state $\eta_j = \mu_j = 0$. In this case $\alpha_k = \max|u_k|$ can be found along the trajectories of system (5.4.8). The expressions necessary to carry out the computations are presented in Table 5.4.2.

Table 5.4.2.

t	u_j^*	μ_j	η_j
$[0, \tau/2]$	$-\alpha_j^*$	$-(1-\rho_j)\alpha_j^* t$	$\eta_{j0} - \frac{1}{2}(1-\rho_j)\alpha_j^* t^2$
$(\tau/2, \tau]$	α_j^*	$(1-\rho_j)\alpha_j^*(t-\tau)$	$\frac{1}{2}(1-\rho_j)\alpha_j^*(t-\tau)^2$

Computations indicate that, along the trajectories of system (5.4.8),

$$\alpha_1 = 137.8, \quad \alpha_2 = 300.6, \quad \alpha_3 = 203.2 \ (N\cdot m). \tag{5.4.15}$$

More precisely, for the "worst" interference v_j^*, the control rules u_k of the form (5.2.8), (5.4.9), and (5.4.14) are piecewise-continuous functions with discontinuities at $t = \tau/2$, which vary in the ranges $81.0 \leq |u_1| \leq 137.8$, $209.7 \leq |u_2| \leq 300.6$, and $120.6 \leq |u_3| \leq 203.2 \ N\cdot m$. Functions u_k are shown in Figure 5.4.2.

Solving the equations that define v_j^* for v_i, we obtain the expression

$$v_1 = \frac{2}{\eta_4}\left[(\eta_4^2 + \eta_1^2)v_1^* + (\eta_1\eta_2 + \eta_4\eta_3)v_2^* + (\eta_1\eta_3 - \eta_4\eta_2)v_3^*\right] \quad (123). \tag{5.4.16}$$

From formulas (5.4.16), one can find the program for varying the interference v_i corresponding to the "worst" realizations of v_j^*. In the case in question, v_i are piecewise-continuous functions with discontinuities at $t = \tau/2$ varying in the ranges $80.0 \leq |v_1| \leq 105.2$, $159.1 \leq |v_2| \leq 220.4$, and $100.0 \leq |v_3| \leq 153.3$. Note that in this case, the numbers $\gamma_i = |v_i||u_i|^{-1}$, which characterize the relationship between the levels of interference v_i and control rules u_k, vary in the ranges $0.611 \leq \gamma_1 \leq 1.030$, $0.688 \leq \gamma_2 \leq 0.944$, and $0.634 \leq \gamma_3 \leq 0.926$, which is, "on average," much higher than the "*golden section*" ratio ($\gamma_i > 0.618\ldots$).

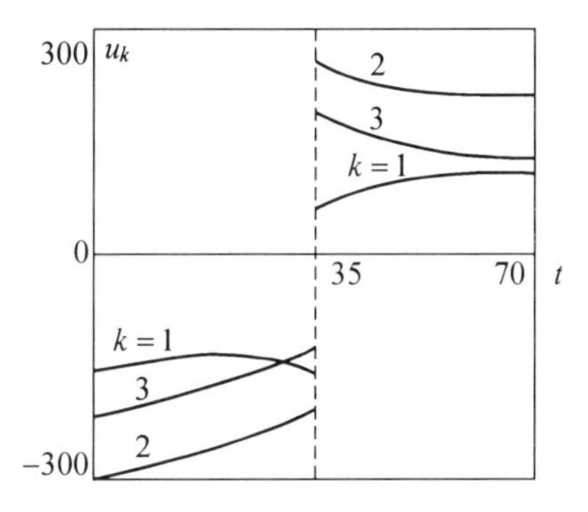

Figure 5.4.2. Controls u_k in the case when $v_j^* = -\rho_j u_j^*$.

2°. Next, we consider the case $v_j^* \equiv 0$ *without interference*. In this case the trajectories of system (5.4.4), (5.4.9), (5.4.14) are described in Subsection 5.4.4. The expressions necessary to estimate $|u_k|$ are given in Table 5.4.3. We have

$$T_j^{0*} = \{[\alpha_j^*(2-\rho_j)]^{-1}[2\eta_{j0}(1-\rho_j)]\}^{1/2}, \quad T_j^0 = (1-\rho_j)^{-1}(2-\rho_j)T_j^*.$$

Table 5.4.3.

t	u_j^*	μ_j	η_j
$[0, T_j^{0*}]$	$-\alpha_j^*$	$-\alpha_j^* t$	$\eta_{j0} - 1/2\alpha_j^* t^2$
$(T_j^{0*}, T^0]$	α_j^*	$(1-\rho_j)\alpha_j^*(t-T_j^0)$	$1/2(1-\rho_j)\alpha_j^*(t-T_j^0)^2$
$(T_j^0, \max T_j^0]$	0	0	0

Now, unlike the previous case, the reduction time T_j^0 will be *different for each variable* η_j. We have $T_j^0 < \tau$ and $T_1^0 = 54.73$, $T_2^0 = 55.63$, and $T_3^0 = 55.55$ s. Besides, control rules u_k given by (5.2.8), (5.4.9), and (5.4.14) will have *five* switching moments $t = 10.02$, 11.49, 11.53, 54.73, and 55.55 s and vary within the ranges $0 \le |u_1| \le 137.8$, $119.2 \le |u_2| \le 307.7$, and $0 \le |u_3| \le 290.5$ $N \cdot m$. Control rules u_k are shown in Figure 5.4.3, in which, for clarity, a different time scale is introduced in the sections between the points 0, 10.02, 11.49, 11.53, 54.73, 55.55, and 55.63 on the t-axis. Therefore, in the case $v_j^* \equiv 0$ when there is no interference, as distinguished from (5.4.15),

$$\alpha_1 = 137.8, \quad \alpha_2 = 307.7, \quad \alpha_3 = 290.5 \ (N \cdot m). \tag{5.4.17}$$

To assess the performance of the construction (5.2.8), (5.4.9) in the case $v_j^* \equiv 0$ in (5.4.9), we take the switching curves $\psi_j^0 = 0$ in place of $\psi_j^\rho = 0$. In this construction the reorientation time $T = 70$ s (equal to the same value τ in the "perturbed" problem) is attained for $\alpha_1 = 32.6$, $\alpha_2 = 80.1$, and $\alpha_3 = 68.0$ $N \cdot m$. However, we note that for the given levels (5.4.12) of interference v_i, the construction (5.2.8),

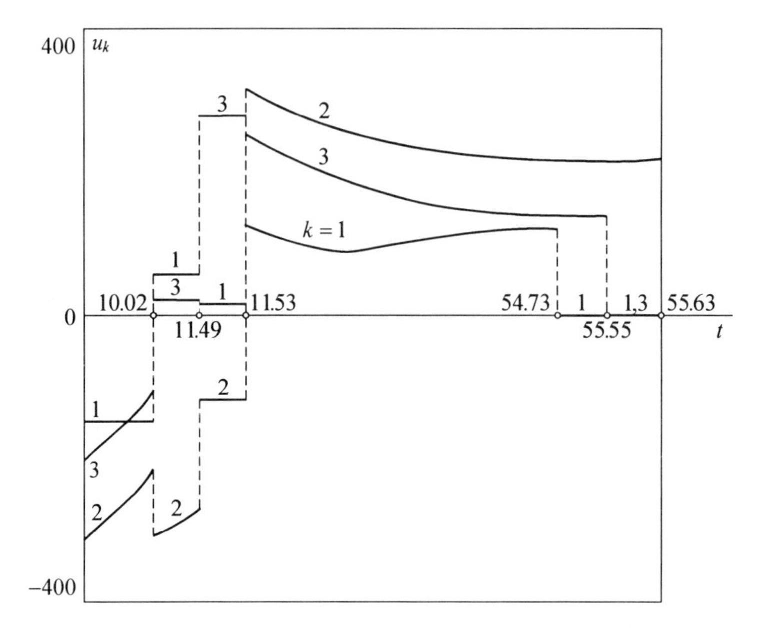

Figure 5.4.3. Controls u_k in the case when $v_j^* \equiv 0$.

(5.4.9) with the switching curves $\psi_j^0 = 0$ cannot ensure that the body will be reoriented in a *finite* time. Indeed, in this case $\alpha_1^* = 2.886 \cdot 10^{-4}$, $\alpha_2^* = 3.544 \cdot 10^{-4}$, and $\alpha_3^* = 3.527 \cdot 10^{-4}\ s^{-2}$ and the values of ρ_j will exceed the "golden section" value for at least one j.

According to the result of Chernous'ko [1990a], [1990b], this means that, for the index j in question, the auxiliary control functions u_j^* of the form (5.4.9) with switching curves $\psi_j^\rho = 0$ cannot transform variable η_j to the position $\eta_j = \mu_j = 0$ in a finite time. At the same time, as has been shown, control rules (5.2.8), (5.4.9) with switching curves $\psi_j^\rho = 0$ in (5.4.9) do ensure the required reorientation of the solid.

3° We also consider the case $v_j^* = \rho_j u_j^*$ when the "auxiliary interference" v_j^* *plays the role of the additional auxiliary control functions* u_j^*. The expressions necessary for computations are presented in Table 5.4.4. In this case

$$T_j^* = \{[\alpha_j^*(1+\rho_j)]^{-1}[\eta_{j0}(1-\rho_j)]\}^{1/2}, \quad T_j = 2(1-\rho_j)^{-1}T_j^*.$$

As in case 2° the reduction time T_j will be different for each variable η_j. Moreover, $T_j < T_j^0 < \tau$, and $T_1 = 52.49$, $T_2 = 53.12$, and $T_3 = 53.05\ s$. Besides, control rules (5.2.8), (5.4.9), and (5.4.14) will have *five* switching moments $t = 5.88$, 6.92, 6.94, 52.49, and 53.05 s and vary within the ranges $0 \le |u_1| \le 137.8$, $82.5 \le |u_2| \le 338.2$, and $0 \le |u_3| \le 322.5\ N \cdot m$. Thus, in the case $v_j^* = \rho_j u_j^*$, as distinguished from (5.4.15) and (5.4.17),

$$\alpha_1 = 137.8, \quad \alpha_2 = 338.2, \quad \alpha_3 = 322.5\ (N \cdot m) \tag{5.4.18}$$

4° Now let the "auxiliary interference" v_j^* *play different roles*: v_2^*, v_3^* are functions of the additional auxiliary control rules u_2^* and u_3^*, and $v_1^* \equiv 0$. The expressions

Table 5.4.4.

t	u_j^*	μ_j	η_j
$[0, T_j^*]$	$-\alpha_j^*$	$-(1+\rho_j)\alpha_j^* t$	$\eta_{j0} - 1/2(1+\rho_j)\alpha_j^* t^2$
$(T_j^*, T_j]$	α_j^*	$(1-\rho_j)\alpha_j^*(t-T_j)$	$1/2(1-\rho_j)\alpha_j^*(t-T_j)^2$
$(T_j, \max T_j]$	0	0	0

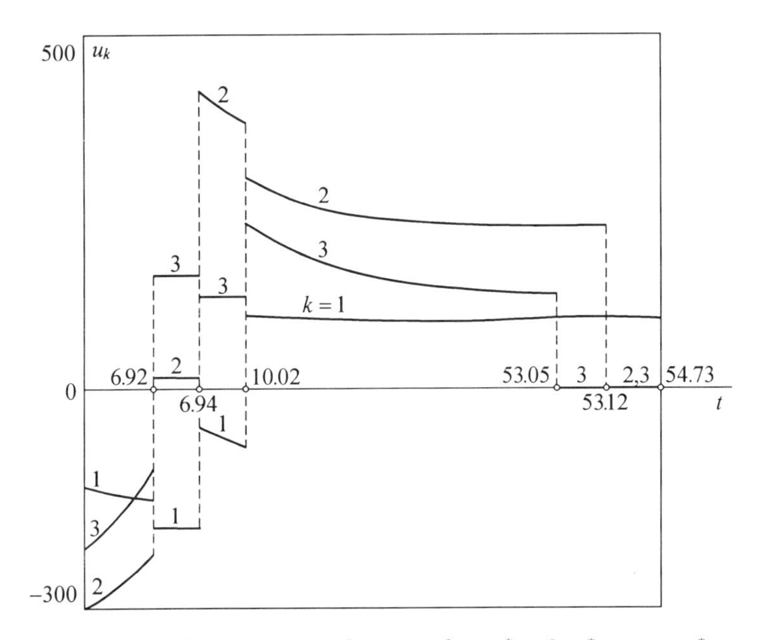

Figure 5.4.4. Controls u_k in the case when $v_1^* \equiv 0$, $v_{2,3}^* = \rho_{2,3} u_{2,3}^*$.

necessary for computations can be obtained by a suitable combination of the expressions in Tables 5.4.2 and 5.4.4.

In this case the reduction times $T_2 = 53.12$ and $T_3 = 53.05$ s for η_2 and η_3 are shorter than the reduction time $\tau = 54.73$ s for η_1. The rules (5.2.8), (5.4.9), and (5.4.14) will have *five* switching moments $t = 6.92$, 6.94, 10.02, 53.05, and 53.12 s and vary within the ranges $76.6 \le |u_1| \le 191.3$, $0 \le |u_2| \le 429.8$, and $0 \le |u_3| \le 234.8$ $N \cdot m$. The functions u_k are shown in Figure 5.4.4, in which, for clarity, we introduce a different time scale in the sections between the points 0, 6.92, 6.94, 10.02, 53.05, 53.12, and 54.73 on the t-axis. It follows that in the case in question,

$$\alpha_1 = 191.3, \quad \alpha_2 = 429.8, \quad \alpha_3 = 234.8\ (N \cdot m),$$

as distinguished (5.4.15), (5.4.17), and (5.4.18).

5° An analysis of cases 2°–4° indicates only that the control rules u_k of the form (5.2.8), (5.4.9), and (5.4.14) can change considerably when the "auxiliary interference" v_j^* differs from the "worst" scenario $v_j^* = -\rho_j u_j^*$ (case 1°). It is necessary to estimate numbers α_k for *any* admissible realizations of v_j^*. Computer modeling of estimates of type (5.4.11), embracing all admissible v_j^* at once, gives the following results: $\alpha_1 = 318.8$, $\alpha_2 = 514.5$, and $\alpha_3 = 415.9\ N\cdot m$.

Conclusions.

(1) For the given boundary conditions, values of A, B, and C, and interference levels v_i defined in accordance with (5.4.12), the *guaranteed* reorientation time $\tau = 70\ s$ can be attained in the construction of control rules (5.2.8), (5.4.9), and (5.4.14) for $\max \alpha_k = 514.5\ N\cdot m$. However, this is a "*cautious*" estimate and also a slight *overestimate* of the admissible levels of u_k.

(2) For the given levels of interference v_i, the "unperturbed" construction (5.2.8), (5.4.9) with switching curves $\psi_j^0 = 0$ in (5.4.9), in which we find α_j^* relying on the reorientation time $T = 70\ s$, *cannot* solve Problem 5.4.1 in a *finite* time.

(3) The "unperturbed" construction (5.2.8), (5.4.9) with switching curves $\psi_j^0 = 0$ in (5.4.9), in which we find α_j^* relying on the bounds $\max \alpha_k = 514.5\ N\cdot m$ (rather than on the prescribed reorientation time T), guarantees the reorientation time $T = 27.64\ s$ when $v_i \equiv 0$. Thus, the "*time loss*" $\gamma = \tau - T$ caused by the allowance for interference v_i amounts to $\gamma = 42.36\ s$ for the given levels of u_k and v_i and the "degree of accuracy" of the estimates involved.

5.4.9 Reorientation with Simultaneous Damping of the Angular Velocity

Problem 5.4.2. *It is required to find control rules $\mathbf{u} \in K^*$ to take the body from the initial state $\boldsymbol{\xi}(t_0) = \boldsymbol{\xi}_0 \neq \mathbf{0}$, $\boldsymbol{\eta}(t_0) = \boldsymbol{\eta}_0$ to the prescribed state $\boldsymbol{\xi}(t_1) = \boldsymbol{\xi}_1 = \mathbf{0}$, $\boldsymbol{\eta}(t_1) = \boldsymbol{\eta}_1$ in a finite time for any $\mathbf{v} \in K_1$. The time $t_1 > t_0$ is not fixed, but we shall seek to make it as short as possible.*

Without loss of generality, we further suppose that $\boldsymbol{\eta}_1 = (0, 0, 0, 1)$.

Now we shall highlight the features of solving Problem 5.4.2 by using construction of the control rules of type (5.2.8). For definiteness, let $\eta_{j0} > 0$, $\mu_{j0} > 0$, which does not entail any loss of generality either.

1. *Taking the constraints*

$$\sum_{j=1}^{3} \eta_j^2 \leq 1 - \varepsilon^2 \tag{5.4.19}$$

into consideration, which allows eliminating the "singularity" in control rules (5.2.8).

The condition $\boldsymbol{\xi}_0 = \mathbf{0}$ is not assumed in Problem 5.4.2. In this connection it is necessary, in general, to take into account the specified constraints, which are automatically satisfied in the case $\boldsymbol{\xi}_0 = \mathbf{0}$. This constraints will be satisfied only if the parabolic sections corresponding to the optimum trajectories of the system

$$\dot{\eta}_j = \mu_j, \quad \dot{\mu}_j = (1 + \rho_j)u_j^*, \quad |u_j^*| \leq \alpha_j^* \tag{5.4.20}$$

are "steep" enough.

These parabolic sections restrict the possible trajectories (for the admissible realizations of v_j^*) of system (5.4.4), (5.4.9) until they meet the switching curves $\psi_j^\rho = 0$. Such a choice of sufficiently "steep" parabolic sections imposes certain constraints on the levels α_j^* of the auxiliary control rules u_j^*. Constraints of this kind can contradict the original constraints (5.2.4) imposed on u_k.

Let us determine the levels α_j^* of the auxiliary control rules u_j^* that guarantee that the conditions (5.4.19) are satisfied. To this end, we observe that the parabolic sections of system (5.4.20), which are of interest, have the form

$$\eta_j = \eta_{j0} - [2\alpha_j^*(1+\rho_j)]^{-1}(\mu_j^2 - \mu_{j0}^2).$$

It follows that the points of intersection η_j^* of these trajectories with the axes $\eta_j = 0$ are given by

$$\eta_j^* = \eta_{j0} - [2\alpha_j^*(1+\rho_j)]^{-1}\mu_{j0}^2. \tag{5.4.21}$$

From (5.4.21), taking the inequalities $|\eta_j(t)| \leq \eta_j^*$ and equations (5.4.2) into account, we deduce that the desired range of values of α_j^* that guarantee that the inequalities (5.4.19) are satisfied can be found from the conditions

$$\begin{gathered}\sum\{\eta_{j0} + [2\alpha_j^*(1+\rho_j)]^{-1}(f_j^0)^2\}^2 \leq 1 - \varepsilon^2, \\ 2f_1^0 = \xi_{10}\eta_{40} + \xi_{30}\eta_{20} - \xi_{20}\eta_{30} \quad (123).\end{gathered} \tag{5.4.22}$$

For the values of α_j^* that do not contradict the original constraints (5.2.4), one can always ensure that the conditions (5.4.22) are met by reducing the angular velocity of the body in advance. The game-theoretic problem of reducing the angular velocity of the body can be solved based on Euler's equations (5.4.1) alone. It is not considered in this book (see Corless and Leitmann [1995]).

On the other hand, if levels α_k of the original control rules u_k are high enough, conditions (5.4.22) can always be satisfied by choosing appropriate values of α_j^* without violating (5.2.4).

2. *The nonuniqueness of the dependence of the maximum levels α_j^* of the auxiliary control rules u_j^* on τ.* Indeed, in the solution of Problem 5.4.2, even in the "unperturbed" case ($v_i \equiv 0$), the original constraints (5.2.4) on control functions (5.2.8) can be satisfied for two different sets of values of α_j^*. To the two sets of α_j^*, there correspond two sets of trajectories of system $\dot\eta_j = \mu_j$, $\dot\mu_j = u_j^*$, $|u_j^*| \leq \alpha_j^*$ leading to the position $\eta_j = \mu_j = 0$. It is as if the two tendencies to increase (reduce) α_k in (5.2.4) by reducing (increasing) η_4 in the denominator of expression (5.2.8) "balanced one another" in the above mentioned cases.

The nonuniqueness indicated is not an obstacle to getting the solution. It suffices to choose the set of α_j^* that corresponds to the shortest control time. One must, however, take this property into account when solving the problem, including also the "perturbed" case.

Theorem 5.4.3. *If levels α_k of control rules u_k in system* (5.4.1)–(5.4.3) *are high enough and levels α_j^* of "auxiliary" control rules are chosen in accordance with* (5.4.22), *then for any given levels of interference v_i the rules u_k in Problem* 5.4.2 *can be constructed based on relationships of type* (5.2.8). *Then it is ensured that Problem* 5.4.2 *can be solved precisely for any admissible realizations of interference v_i by reducing nonlinear Problem* 5.4.2 *to game control problems for the auxiliary linear control $\boldsymbol{\mu}$-systems of the form* (5.4.4).

In connection with the estimate of the guaranteed reorientation time, the conditions of Theorem 5.4.3 can be made more specific as in Subsection 5.4.7 (taking into account the above mentioned properties).

5.4.10 Results of Computer Simulation of Problem 5.4.2

For a spacecraft with the same values of A, B, and C, as in Subsection 5.4.8, we consider the triaxial reorientation problem with boundary conditions $\boldsymbol{\xi}_0 = (0.0101, 0.0111, 0.0176)$, $\boldsymbol{\eta}_0 = (0.353, 0.434, 0.432, 0.701)$ and $\boldsymbol{\xi}_1 = \mathbf{0}$, $\boldsymbol{\eta}_1 = (0, 0, 0, 1)$. Note that sufficiently large values of the components $K_1 = A\xi_{10}$ $(123, ABC)$ of the angular momentum vector $\boldsymbol{K}$ of the spacecraft $K_1 = 404$, $K_2 = 888$, $K_3 = 880$ $kg{\cdot}m^2{\cdot}s^{-1}$ correspond to the given initial value $\boldsymbol{\xi}_0$. As in Subsection 5.4.8, we choose the levels β_i of interference v_i in accordance with inequalities (5.4.12).

We shall use control rules of the form (5.2.8) to solve the problem. We confine ourselves to analyzing the case of the "*worst*" auxiliary interferences v_j^* that slow down, as much as possible, the process of bringing the system (5.4.4) to the position $\eta_j = \mu_j = 0$ desired for solving the problem in question. The expressions necessary for the computations are collected in Table 5.4.5, where

$$T_j^{**} = [\alpha_j^*(1-\rho_j)]^{-1}\{\mu_{j0} + [\tfrac{1}{2}\mu_{j0}^2 + \alpha_j^*\eta_{j0}(1-\rho_j)]\}^{1/2},$$
$$\alpha_j^* = \beta_j^* + \varepsilon_j + (\varepsilon_j^2 + \delta_j^2)^{1/2}, \quad \varepsilon_j = (2T^2)^{-1}(2T\mu_{j0} + 4\eta_{j0}), \quad \delta_j = T^{-1}\mu_{j0}. \tag{5.4.23}$$

Table 5.4.5.

t	u_j^*	μ_j	η_j
$[0, T_j^{**}]$	$-\alpha_j^*$	$\mu_{j0} - (1-\rho_j)\alpha_j^* t$	$\eta_{j0} + \mu_{j0}t - \frac{1}{2}(1-\rho_j)\alpha_j^* t^2$
$(T_j^{**}, T]$	α_j^*	$(1-\rho_j)\alpha_j^*(t-T)$	$\frac{1}{2}(1-\rho_j)\alpha_j^*(t-T)^2$

These expressions are composed to take into account the inequalities $\eta_{j0} > 0$ and $\mu_{j0} > 0$, which occur for the given boundary conditions, and are based on the assumption that $T = T_j$, where T_j is the reorientation time for each variable η_j. In particular, the expressions for α_j^* are obtained by solving the equations

$$T\alpha_j^*(1-\rho_j) = \mu_{j0} + 2\big[\tfrac{1}{2}\mu_{j0}^2 + \alpha_j^*\eta_{j0}(1-\rho_j)\big]^{1/2}$$

and using the equalities $\alpha_j^*(1-\rho_j) = \alpha_j^* - \beta_j^*$. In the course of finding the solution, the roots $\alpha_j^* = \beta_j^* + \varepsilon_j - (\varepsilon_j^2 + \delta_j^2)^{1/2}$ of the equation in question are discarded, because in this case $\alpha_j^* - \beta_j^* < 0$.

Computations indicate that, for example, for $|u_k| \leq 300\ N\cdot m$, the reorientation is impossible for any T in the framework of the construction of control rules (5.2.8) and (5.4.9). This is fully consistent with the first property mentioned in Subsection 5.4.9. At the same time, even for $\max \alpha_k = 300.9\ N\cdot m$, the reorientation is possible for the case $v_j^* = -\rho_j u_j^*$ of the "worst" v_j^*. In this case $T = 120\ s$. The functions u_k are piecewise-continuous and have *three* switching moments $t = 68.61$, 72.99, and $73.11\ s$. We emphasize that reducing and increasing T will result in a slow increase in α_k for a relatively long time interval. Thus, for $\max \alpha_k = 302.15\ N\cdot m$, the reorientation is achieved for $T = 108$ and $T = 130\ s$. This property also agrees fully with the second property mentioned in Subsection 5.4.9.

5.5 The Nonlinear Game-Theoretic Problem of Triaxial Reorientation of an Asymmetric Solid (Second Method of Solution)

5.5.1 Formulation of the Problem

The control rules obtained in Section 5.4 which solve Problem 5.4.1 are nonlinear functions of the variables determining the orientation of a solid and its angular velocity. They can be regarded as functions formed by *five impulses* of variable intensity for admissible realizations of interference (except for some specific cases).

The solution method proposed in Section 5.4 is modified below to obtain control rules which are *simpler* and, consequently, more convenient for implementation. In particular, they contain no constituents compensating *gyroscopic moments* of a solid. As in Section 5.4, the solution is obtained by handling basic linear game-theoretic problems; the control rules are simplified by complicating the structure of "auxiliary interference" in the linear systems formed.

Estimating the "auxiliary interference" requires performing a certain calculation over the state set of linear auxiliary conflict-control systems. To establish such estimates, it is necessary, as distinguished from Section 5.4, to use the *principle of "assignment with subsequent confirmation"* of the interference levels.

In spite of the indicated difference, the proposed modification of the method does not affect the complexity of calculations. Moreover, the calculation is simplified when constraints of some specific type are imposed on the control rules.

5.5.2 Auxiliary Linear Conflict-Control System

Let us differentiate both sides of each equation for $\dot{\eta}_j$ in (5.4.2) with respect to time. Then we replace $\dot{\xi}_i$ with their expressions from (5.4.1) and $\dot{\eta}_j$ with the corresponding expressions from (5.4.2). Having completed the transformation, we arrive at the equalities

$$\begin{aligned} \ddot{\eta}_1 &= f_1^*(\boldsymbol{\eta}, \boldsymbol{u}) + \varphi_1^*(\boldsymbol{\xi}, \boldsymbol{\eta}, \boldsymbol{v}), \\ f_1^* &= \frac{1}{2}(\eta_4 u_1 A^{-1} + \eta_2 u_3 C^{-1} - \eta_3 u_2 B^{-1}), \\ \varphi_1^* &= \frac{1}{2}[\eta_4(v_1 + M_1)A^{-1} + \eta_2(v_3 + M_3)C^{-1} \\ &\quad - \eta_3(v_2 + M_2)B^{-1}] - \frac{1}{4}\eta_1 \sum_{i=1}^{3} \xi_i^2 \qquad (123), \\ M_1 &= (B - C)\xi_2\xi_3 \quad (123). \end{aligned} \tag{5.5.1}$$

We interpret f_j^* and φ_j^* as the auxiliary control rules u_j^* and the interference v_j^*, respectively. In this case we note that the expressions for f_j^* and φ_j^* differ from the corresponding expressions for f_j and φ_j given in Section 5.4. As a result, expressions (5.5.1) can be considered a conflict-control $\boldsymbol{\mu}$-system of type (5.3.3):

$$\dot{\eta}_j = \mu_j, \quad \dot{\mu}_j = u_j^* + v_j^*, \tag{5.5.2}$$

where the "original" control rules u_k are expressed in terms of u_j^* by the equalities of type (5.3.1) which in this case have the form

$$u_1 = \frac{2A}{\eta_4}[(\eta_4^2 + \eta_1^2)u_1^* + (\eta_1\eta_2 + \eta_4\eta_3)u_2^* + (\eta_1\eta_3 - \eta_4\eta_2)u_3^*] \quad (123). \tag{5.5.3}$$

Construction (5.5.3) can be regarded as the *general structural form* of the control rules in Problem 5.4.1 along with construction (5.2.8) proposed earlier. The parameters of this form, i.e., the auxiliary control rules u_j^*, are determined by solving the corresponding linear game-theoretic problems.

Similar to structure (5.2.8), structure (5.5.3) of the control rules contains the factor η_4^{-1}, which formally implies a "*singularity.*" However, subsequent thorough analysis reveals that in this case the relationship $\eta_4 \in [\eta_{40}, 1]$ also holds for $\boldsymbol{\eta}_1 = (0,0,0,1)$. Therefore, the specified "singularity" does not occur at all. If the value of η_{40} is small or $\boldsymbol{\eta}_1 \neq (0,0,0,1)$, it suffices to change the controls to those obtained from (5.5.3) by permutation of the indices and coefficients A, B, and C (or to a combination of such controls).

5.5.3 Auxiliary Game-Theoretic Control Problem

Now we solve the problem of bringing the system (5.5.2) to the position

$$\eta_j = \mu_j = 0 \tag{5.5.4}$$

in the *shortest* possible time by the auxiliary control rules u_j^* for *any* admissible v_j^*.

We interpret this problem as a time-minimax *differential game* of the type considered in Subsection 5.3.4. Under the constraints of type (5.3.7), its solution can be obtained as that of the time-optimal control problem for a system of type (5.3.9) (the boundary conditions are the same).

However, as distinguished from Section 5.4, the procedure for choosing the levels α_j^*, β_j^* relies on the principle of "assignment with subsequent confirmation" of these levels. For now let us consider that they are prescribed, so that the conditions of type (5.3.7) are fulfilled.

By $\eta_{j0} = \mu_{j0} = 0$ (which results from $\boldsymbol{\xi}_0 = \boldsymbol{\xi}_1 = \mathbf{0}$), the value

$$\tau = \max(\tau_j), \quad \tau_j = 2\{|\eta_{j0}|[\alpha_j^*(1-\rho_j)]^{-1}\}^{1/2} \tag{5.5.5}$$

defines the minimum guaranteed time τ for attaining position (5.5.4) in the auxiliary linear game-theoretic problem in question. If $v_j^* \not\equiv -\rho_j u_j^*$, then the time for taking the system to (5.5.4) *does not exceed* τ.

The analysis of the phase portrait of system (5.5.2) for u_j^* of type (5.3.11) shows that the condition $\eta_4 > \gamma = \text{const} > 0$, satisfied at $t = t_0$, also holds for any admissible v_j^* for $t \in [t_0, t_0 + \tau]$. Thus "singularities" in control rules (5.5.2) are ruled out.

5.5.4 Algorithm for Solving Problem 5.4.1 through Control Rules (5.5.3)

(1) The choice of construction (5.5.3) of control rules u_k in which u_j^* are of type (5.3.11). At this stage not only α_j^* but also β_j^* (in contrast to Section 5.4) are undefined. We note that construction (5.5.3) of control rules u_k is *simpler* than that of (5.2.8) since it contains no terms explicitly dependent on ξ_i (in particular, it does not include components which compensate for the *gyroscopic moments* of the body).

(2) The preliminary choice of τ and the "assignment" of levels β_j^*. In accordance with (5.5.5), this predetermines α_j^*, ρ_j. Setting $\tau_j = \tau$ ("*equalizing*" the control time with respect to each z_j) leads to

$$\alpha_j^* = \beta_j^* + 4|\eta_{j0}|\tau^{-2}. \tag{5.5.6}$$

(3) The verification of whether the inequalities $|v_j^*| \leq \beta_j^*$ are actually satisfied in the state set Ω of the linear system (5.5.2) with u_j^* of type (5.3.11). Thus we employ the principle of "assignment with subsequent confirmation" of levels v_j^*.

(4) The verification of the original constraints on controls u_k in the set Ω. For structure (5.5.3) it is natural to anticipate that the values of $\max|u_k|$ are close in the state set Ω and its subset $\Omega^* \subset \Omega$ of the linear system (5.5.2) with u_j^* of type (5.3.11) for various combinations $v_j^* = \pm\rho_j u_j^*$ and $v_j^* \equiv 0$. Computation of $\max|u_k|$ over Ω^* does not present any difficulty.

As a result, we get an iterative algorithm for solving Problem 5.4.1.

5.5.5 Constructing Estimates for v_j^*

We use the inequalities $|\xi_1\xi_2| \leq ½\sum \xi_i^2$ (123) and the relationship

$$\sum_{i=1}^{3} \xi_i^2 = 4\{[\eta_4^{-1}\sum_{j=1}^{3}(\eta_j\mu_j)]^2 + \sum_{j=1}^{3}\mu_j^2\} \tag{5.5.7}$$

which can be verified by (4.6.11).

Thus, we arrive at the following (over)estimates:

$$\begin{aligned} |v_1^*| &\leq (\eta_1^+ + \eta_4^+ r_1 + \eta_2^+ r_3 + \eta_3^+ r_2)G + \Delta_1, \\ \Delta_1 &\leq \frac{1}{2}(\eta_4^+\beta_1 A^{-1} + \eta_2^+\beta_3 C^{-1} + \eta_3\beta_2 B^{-1}), \\ r_1 &= |B - C|A^{-1} \qquad (123, ABC), \\ G &= (\eta_4^-)^{-2}\Big[\sum_{j=1}^{3}(\eta_j^+\mu_j^-)\Big]^2 + \sum_{j=1}^{3}(\mu_j^-)^2, \\ \eta_4^{(-,+)} &= \Big[1 - \sum_{j=1}^{3}(\eta_j^{(+,-)})^2\Big]^{½}. \end{aligned} \tag{5.5.8}$$

The expressions for $\eta_j^{(+,-)}$, μ_j^- (overestimated μ_j^- at $t \geq t_0 + T_j^*$) are compiled in Table 5.4.1.

Overestimates (5.5.8) can be weakened. To this end, we note that, on numerous occasions, ξ_i of type (4.6.11) maintain the sign ($\xi_i \leq 0$ when $\eta_{j0} > 0$) for all admissible v_j^*. If, for example, $B \geq C \geq A$ (other cases obey this pattern), then the first group of inequalities in (5.5.8) can be replaced with the relationships

$$|v_1^*| \leq \big[\max(\eta_4^+ r_1 + \eta_3^+ r_2, \eta_1^+ + \eta_2^+ r_3)\big]G + \Delta_1 \quad (123). \tag{5.5.9}$$

5.5.6 Constructing Estimates for u_k

The computation is simplified if, instead of (5.2.4), the following inequality is used to estimate control rules (5.5.3):

$$E = \sum_{k=1}^{3}(u_k^2 A_k^{-2}) < \alpha = \text{const} > 0. \tag{5.5.10}$$

Indeed, by virtue of (5.5.3), E is determined by expression (5.5.7), if one replaces μ_j with u_j^* in it. Hence, inequality (5.5.10) holds if

$$E^* = \big[(\eta_{40})^{-1}\sum_{j=1}^{3}(|\eta_{j0}|\alpha_j^*)\big]^2 + \sum_{j=1}^{3}(\alpha_j^*)^2 < \frac{1}{4}\alpha. \tag{5.5.11}$$

In addition, one can use the inequality $E \leq 4\{[(\eta_4^-)^{-1}\sum(\eta_j^+\alpha_j^*)]^2 + \sum(\alpha_j^*)^2\}$ to *integrally* estimate $E = E(t)$ for $t \in [t_0, t_0+\tau]$.

5.5.7 Conditions for the Solvability of Problem 5.4.1 by Control Rules (5.5.3)

Let us sum up the considerations presented.

Theorem 5.5.1. *If levels α_k of control rules u_k in system* (5.4.1)–(5.4.3) *are high enough, then for any given levels β_i of interference v_i the rules u_k solving Problem* 5.4.1 *can be constructed in the form* (5.5.3) *with u_j^* of type* (5.3.11). *The precise reorientation of a solid is ensured in a finite time τ for any $v \in K_1$. The value of τ is determined iteratively by using the algorithm given in Subsection* 5.5.4.

Theorem 5.5.1 states the possibility of solving Problem 5.4.1 by constructing control rules of type (5.5.3), (5.3.11). We can make this possibility more specific by relying on the estimates for v_j^* obtained in Subsection 5.5.5.

Theorem 5.5.2. *Let $\boldsymbol{\eta}_1 = (0,0,0,1)$. Suppose that, for the "assigned" values of τ, β_j^* (and for the values of α_j^* which they predetermine),*

(1) *"confirmative" estimates $|v_j^*| \leq \beta_j^*$ are valid (derived from inequalities of type* (5.5.8) *or* (5.5.9));

(2) *inequality* (5.5.11) *is satisfied.*

Then control rules (5.5.3), (5.3.11) *solve Problem* 5.4.1 *in finite time τ.*

5.5.8 Results of Computer Simulation

To assess the algorithm's capability, consider the reorientation of a spacecraft with $A = 4\cdot 10^4$, $B = 8\cdot 10^4$, and $C = 5\cdot 10^4$ $kg{\cdot}m^2$ from the equilibrium position $\boldsymbol{\xi}_0 = \mathbf{0}$, $\boldsymbol{\eta}_0 = (0.353, 0.434, 0.432, 0.701)$ to the position $\boldsymbol{\xi}_1 = \mathbf{0}$, $\boldsymbol{\eta}_1 = (0,0,0,1)$.

We set $\tau = 70$ s. As computations show, one can set $\beta_j^* = 245\cdot 10^{-5}$ s^{-2} for

$$\tfrac{1}{2}|\eta_4 v_1 A^{-1} + \eta_2 v_3 C^{-1} - \eta_3 v_2 B^{-1}| \leq 10^{-3}\ s^{-2} \quad (123). \tag{5.5.12}$$

(Inequalities (5.5.9) which are satisfied in the case in question were used in the computations.) The corresponding value $4E^* = 185\cdot 10^{-6}$ $kg{\cdot}s^{-4}$.

Consequently, control rules (5.5.3) satisfy inequality (5.5.10) for $\alpha = 4E^*$. The "average" value of E is equal to $123\cdot 10^{-6}$ $kg{\cdot}s^{-4}$.

Let us also perform the comparative calculation following the technique described in Section 5.4 (the accuracy of the estimates is the same as of those obtained in Section 5.4). Taking the same constraints (5.5.12) and the same value of τ, we obtain control rules u_k of type (5.2.8) such that inequality (5.5.10) is satisfied for $\alpha = 174\cdot 10^{-6}$ $kg{\cdot}s^{-4}$. So, the comparison evidences the efficiency of the proposed approach.

5.6 The Nonlinear Game-Theoretic Problem of Uniaxial Reorientation of an Asymmetric Solid

5.6.1 Preliminary Remarks

The solution of the problem of *uniaxial* reorientation of a solid considered below is somewhat different from that of the problems of triaxial reorientation. The difference is that the auxiliary linear controlled μ-system in the case of uniaxial reorientation together with equations of the second order includes an equation of the first order. But the theory worked out in Section 5.3 also embraces this case.

5.6.2 Formulation of the Problem

Consider the problem of uniaxial reorientation of a unit vector $\boldsymbol{r}$ rigidly attached to a solid in the direction of a unit vector $\boldsymbol{\eta}$ immobile in the inertial space. To this end, along with *Euler's dynamical equations*

$$\begin{gathered} A\dot{\xi}_1 = (B-C)\xi_2\xi_3 + u_1 + v_1, \\ B\dot{\xi}_2 = (C-A)\xi_1\xi_3 + u_2 + v_2, \quad C\dot{\xi}_3 = (A-B)\xi_1\xi_2 + u_3 + v_3, \end{gathered} \tag{5.6.1}$$

consider the *Poisson kinematic equations*

$$\begin{gathered} \dot{\eta}_1 = \eta_2\xi_3 - \eta_3\xi_2, \\ \dot{\eta}_2 = \eta_3\xi_1 - \eta_1\xi_3, \quad \dot{\eta}_3 = \eta_1\xi_2 - \eta_2\xi_1, \end{gathered} \tag{5.6.2}$$

$$\eta_1^2 + \eta_2^2 + \eta_3^2 = 1. \tag{5.6.3}$$

In equations (5.6.2), (5.6.3), η_i $(i = 1, 2, 3)$ are the projections of vector $\boldsymbol{\eta}$ onto the principal central axes of inertia of the body.

For definiteness, let $\boldsymbol{r} = (0, 1, 0)$. In this case the direction of vector $\boldsymbol{r}$ coincides with that of one of the principal central axes of inertia of the solid.

Problem 5.6.1. It is required to find control rules $\boldsymbol{u} \in K^*$ *satisfying given constraints* (5.2.4) *to take the body from the initial state* $\boldsymbol{\eta}(t_0) = \boldsymbol{\eta}_0$ *to the prescribed state* $\boldsymbol{\eta}(t_1) = \boldsymbol{r}$ *in a finite time for any admissible realizations of interference* $\boldsymbol{v} \in K_1$. *Both states are stationary, and* $\boldsymbol{\xi}(t_0) = \boldsymbol{\xi}_0 = \boldsymbol{\xi}(t_1) = \boldsymbol{\xi}_1 = \mathbf{0}$. *The time* $t_1 > t_0$ *is not fixed, but we shall seek to make it as short as possible.*

5.6.3 Solution of Problem 5.6.1

We shall define the structural form of the desired game strategies based on the possibility of solving Problem 5.6.1 by using the corresponding linear game-theoretic control problems.

For this purpose we shall employ the control rules u_k $(k = 1, 2, 3)$ of the type considered in Section 5.3 which in this case take the form

$$
\begin{aligned}
u_1 &= \frac{A}{\eta_2}\left[\varphi_1(\boldsymbol{\xi}, \boldsymbol{\eta}) + \frac{\eta_1 u_2}{B} + u_3^*\right], \\
u_2 &= B\varphi_2(\boldsymbol{\xi}) + u_2^*, \\
u_3 &= \frac{C}{\eta_2}\left[\varphi_3(\boldsymbol{\xi}, \boldsymbol{\eta}) + \frac{\eta_3 u_2}{B} - u_1^*\right].
\end{aligned}
\tag{5.6.4}
$$

Functions φ_i $(i = 1, 2, 3)$ are such that, from the closed system (5.6.1)–(5.6.4), one can segregate the *linear conflict-control* $\boldsymbol{\mu}$*-system*

$$\dot{\xi}_2 = u_2^* + v_2^*; \quad \dot{\eta}_j = \mu_j, \quad \dot{\mu}_j = u_j^* + v_j^* \quad (j = 1, 3) \tag{5.6.5}$$

of type (5.3.2), (5.3.3); v_k^* $(k = 1, 2, 3)$ are the "auxiliary interferences" formed by appropriately transforming interference v_i $(i = 1, 2, 3)$ in the original system

$$Bv_2^* = v_2, \quad v_1^* = \eta_2 v_3 C^{-1} - \eta_3 v_2 B^{-1}, \quad v_3^* = \eta_1 v_2 B^{-1} - \eta_2 v_1 A^{-1}. \tag{5.6.6}$$

(We note that the agreement in indices of η_j and u_j^* in (5.6.5) leads, however, to mismatching them for u_j and u_j^* in (5.6.4).)

The relationships of type (5.3.17) in the case in question take the form

$$\xi_1 = \frac{1}{\eta_2}(-\mu_3 + \xi_2\eta_1), \quad \xi_3 = \frac{1}{\eta_2}(\mu_1 + \xi_2\eta_3). \tag{5.6.7}$$

By using control rules of type (5.6.4), Problem 5.6.1 can be solved in two stages.

(1) First the solid is transformed to the position

$$\boldsymbol{\xi} = \mathbf{0}, \quad \eta_1 = \eta_3 = 0, \quad \eta_2 = 1 \tag{5.6.8}$$

which meets the requirements of Problem 5.6.1 only with respect to *part* of the variables, i.e., variables ξ_2, η_1, and η_3. This problem of control with respect to part of the variables is solved as the corresponding control problem of taking the auxiliary linear system (5.6.5) to the position $\xi_2 = 0$, $\eta_j = \mu_j = 0$ $(j = 1, 3)$ with respect to *all* the variables.

(2) By (5.6.7), we conclude that the transformation of the system to position (5.6.8) with respect to variables ξ_2, η_1, η_3 actually solves Problem 5.6.1.

Then the procedure for solving Problem 5.6.1 can be worked through along the same lines as in Sections 5.4 and 5.5.

Together with construction (5.6.4), we can also consider constructions of control rules obtained from (5.6.4) by permutation of indices and coefficients A, B, and C. The corresponding expressions of type (5.3.17) are also derived from (5.6.7) by the indicated permutation of indices and coefficients.

This set of the control rules of type (5.6.4) (their successive use included, see Section 4.4) can be applied to solving the game-theoretic problem of uniaxial reorientation of a solid under arbitrary boundary conditions.

5.6.4 Results of Computer Simulation

We shall consider Problem 5.6.1 of uniaxial reorientation from the initial equilibrium position $\boldsymbol{\xi}_0 = \mathbf{0}$, $\boldsymbol{\eta}_0 = (0.4, 0.69, 0.6)$ to the prescribed one $\boldsymbol{\xi}_1 = \mathbf{0}$, $\boldsymbol{\eta}_1 = (0, 1, 0)$ for a spacecraft with $A = 4 \cdot 10^4$, $B = 8 \cdot 10^4$, and $C = 5 \cdot 10^4$ $kg{\cdot}m^2$.

For simplicity, setting $v_2 \equiv 0$, we shall define the admissible boundaries of varying interference v_i $(i = 1, 2, 3)$ by the inequalities

$$|\eta_2 v_3 C^{-1}| \leq \beta^*, \quad |\eta_2 v_1 A^{-1}| \leq \beta^* = 10^{-3}\ s^{-2}. \tag{5.6.9}$$

If reinforcing inequalities are used, then, taking into account the inequality $0.69 \leq s_2(t) \leq 1$, $t \in [t_0, t_1]$, we find from (5.6.9) the relationships $\beta_3 C^{-1} = \beta^*$, $\beta_1 A^{-1} = \beta^* = 10^{-3}$ s^{-2}. As a result, $\beta_1 = 40.0$, $\beta_2 = 0$, and $\beta_3 = 50.0$ $N{\cdot}m$. However, real values v_j $(j = 1, 3)$ in the framework of constraints (5.6.9) may turn out to be significantly larger, which will be shown below.

Suppose that, based on the technical requirements for the maneuverability of a spacecraft, the guaranteed reorientation time τ under the prescribed admissible boundaries of varying interference v_j $(j = 1, 3)$ should not exceed $\tau = 70$ s. We shall assess what resources this will require if the construction of control rules u_k $(k = 1, 2, 3)$ of type (5.6.4) are used, where u_k^* $(k = 1, 2, 3)$ have the form (5.3.10), (5.3.11). Taking into account the structure of expressions (5.6.4) and the initial assumption $v_2 \equiv 0$, from $v_2^* \equiv 0$, $\boldsymbol{\xi}_0 = \boldsymbol{\xi}_1 = \mathbf{0}$, we conclude that $u_2^* \equiv 0$ over the entire control interval.

By (5.3.11), the given value of τ predetermines values α_j^* $(j = 1, 3)$ of maximum levels of auxiliary controls u_j^*. We shall proceed from the equalities $\tau = \tau_j$ which imply "*equalizing*" the guaranteed reorientation time with respect to each variable η_j $(j = 1, 3)$. As a result, we arrive at the relationships

$$2\{\eta_{j0}[\alpha_j^*(1 - \rho_j)]^{-1}\}^{1/2} = \tau. \tag{5.6.10}$$

Taking into account that $\alpha_j^*(1 - \rho_j) = \alpha_j^* - \beta_j^*$, where $\beta_j^* = 10^{-3}$ in the case (5.6.9), from (5.6.10) we find

$$\alpha_1^* = 1.49 \cdot 10^{-3}, \quad \alpha_3^* = 1.33 \cdot 10^{-3}\ s^{-2}. \tag{5.6.11}$$

Now we estimate values $\alpha_k = \max|u_k|$ $(k = 1, 2, 3)$ of control rules u_k of type (5.6.4), (5.3.11) when $u_2^* \equiv 0$.

In what follows $(i, k = 1, 2, 3)$ (both the indices are kept for uniformity of notation adopted in Chapter 5); $j = 1, 3$. In accordance with the assumption made, from now on we set $v_2^* \equiv 0$.

1º Let us first consider the case $v_j^* = -\rho_j u_j^*$ of the "*worst*" v_j^* slowing down, as much as possible, the process of taking the auxiliary linear system of type (5.3.3) to the desired state $\eta_j = \mu_j = 0$. In this case $\alpha_k = \max|u_k|$ can be found along the trajectories of linear system (5.3.3). The expressions necessary to carry out the computations are presented in Table 5.4.1.

Computations indicate that in this case

$$\alpha_1 = 76.6, \quad \alpha_2 = 2.3, \quad \alpha_3 = 107.5\ N{\cdot}m. \tag{5.6.12}$$

More precisely, for the "worst" interference v_j^*, the control rules u_k defined by relationships (5.6.4), (5.3.11) have the following form: u_1, u_3 are piecewise-continuous functions with discontinuities at $t = \tau/2$, which vary in the ranges

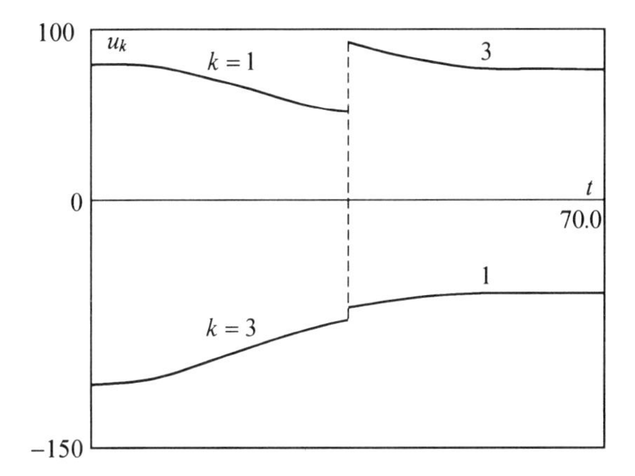

Figure 5.6.1. Controls u_1, u_3 in the case when $v_j * = -\rho_j u_j^*$.

$-60.1 \leq u_1 \leq 76.6$, $-107.5 \leq u_3 \leq 86.1$ $N\cdot m$; u_2 is a *continuous* function varying within $0 \leq u_2 \leq 2.3$ $N\cdot m$. It is obvious that control u_2, in contrast to u_1, u_3, performs just a *corrective* function. Functions u_1, u_3 are shown in Figure 5.6.1.

From formulas (5.6.6), one can also find the program for varying the interference v_j corresponding to the "worst" realizations of v_j^*. In the case in question, interference v_j are piecewise-continuous functions with discontinuities at $t = \tau/2$ varying in the ranges $-57.4 \leq v_1 \leq 42.5$, $-53.2 \leq v_3 \leq 72.2$. In this case we note that the quantities $\gamma_j = |v_j||u_j|^{-1}$, which characterize the relationship between the levels of the corresponding "original" interference v_i and the "original" control rules u_k, vary in the ranges $0.708 \leq \gamma_1 \leq 0.817$, $0.884 \leq \gamma_3 \leq 1.021$, which is, "on average," much higher than the "*golden section*" ratio ($\gamma_j > 0.618\ldots$).

2° Next, we consider the case $v_j^* \equiv 0$ *without interference.* In this case the trajectories of system (5.6.5), (5.3.11) are described in Subsection 5.3.5. The expressions necessary to estimate $|u_k|$ are given in Table 5.4.2. We have

$$T_j^{0*} = \{[\alpha_j^*(2-\rho_j)]^{-1}[2\eta_{j0}(1-\rho_j)]\}^{1/2}, \quad T_j^0 = (1-\rho_j)^{-1}(2-\rho_j)T_j^{0*}.$$

Now, unlike the previous case, the reduction time T_j^0 will be *different* for each variable η_j. We have $T_j^0 < \tau$ and $T_1^0 = 57.05$, $T_3^0 = 55.01$ s. Besides, control rules u_1, u_3 given by (5.6.4), (5.3.11), and (5.6.11) will have *three* switching moments $t = 10.94$, 14.12, and 55.01 s and vary within the ranges $-81.4 \leq u_1 \leq 76.8$, $-107.5 \leq u_3 \leq 106.4$ $N\cdot m$. Control u_2 is *continuous* and varies within $0 \leq u_2 \leq 3.9$ $N\cdot m$. Control rules u_1, u_3 are shown in Figure 5.6.2, in which, for clarity, a different time scale is introduced in the intervals $[0, 55.01]$ and $(55.01, 57.05]$ on the t-axis. Therefore, in the case $v_j^* \equiv 0$ when there is no interference, as distinguished from (5.6.12),

$$\alpha_1 = 81.4, \quad \alpha_2 = 3.9, \quad \alpha_3 = 107.5\ N\cdot m. \tag{5.6.13}$$

To assess the performance of the construction (5.6.4), (5.3.11), and (5.6.11) in the case $v_j^* \equiv 0$ in (5.3.11), we take the switching curves $\psi_j^0 = 0$ in place of $\psi_j^\rho = 0$. In this construction the reorientation time $T = 70$ s (equal to the same value τ in the "perturbed" problem) is attained for $\alpha_1 = 18.9$, $\alpha_2 = 2.3$, and $\alpha_3 = 35.3$ $N\cdot m$. However, we note that, for the given levels (5.6.9) of interference v_i, the construction

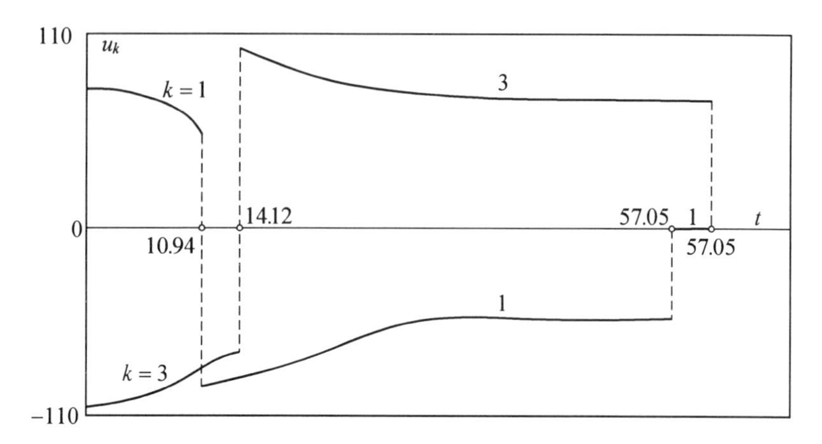

Figure 5.6.2. Controls u_1, u_3 in the case when $v_j* \equiv 0$.

(5.6.4), (5.3.11), and (5.6.11) with the switching curves $\psi_j^0 = 0$ cannot ensure that the body will be reoriented in a *finite* time. Indeed, in this case $\alpha_1^* = 4.9 \cdot 10^{-4}$, $\alpha_3^* = 3.3 \cdot 10^{-4}\ s^{-2}$ and the values of ρ_j will exceed the "golden section" value for at least one j.

As already noted in Subsection 5.4.8, this means that, for the index j in question, the auxiliary control functions u_j^* of the form (5.3.11) with switching curves $\psi_j^\rho = 0$ cannot transform variable η_j to the position $\eta_j = \mu_j = 0$ in a finite time. At the same time, as has been shown, control rules (5.6.4), (5.3.11), and (5.6.11) with switching curves $\psi_j^\rho = 0$ in (5.3.11) do ensure the required reorientation of the solid.

3° We also consider the case $v_j^* = \rho_j u_j^*$ when the "auxiliary interference" v_j^* *plays the role of the additional auxiliary control functions* u_j^*. The expressions necessary for computations are presented in Table 5.4.3. In this case

$$T_j^* = \left\{[\alpha_j^*(1+\rho_j)]^{-1}[\eta_{j0}(1-\rho_j)]\right\}^{1/2}, \quad T_j = 2(1-\rho_j)^{-1}T_j^*.$$

As in case 2° the reduction time T_j will be different for each variable η_j. Moreover, $T_j < T_j^0 < \tau$ and $T_1 = 54.14$, $T_3 = 52.61\ s$. Besides, control rules u_1, u_3 defined by relationships (5.6.4), (5.3.11), and (5.6.11) will have *three* switching moments $t = 6.53$, 8.90, and 52.61 s and vary within the ranges $-86.6 \le u_1 \le 76.8$, $-107.5 \le u_3 \le 123.6\ N\cdot m$. Control u_2 is *continuous* and varies within $0 \le u_2 \le 4.9$ $N\cdot m$. Thus, in the case $v_j^* = \rho_j u_j^*$, as distinguished from (5.6.12) and (5.6.13),

$$\alpha_1 = 86.6, \quad \alpha_2 = 4.9, \quad \alpha_3 = 123.6\ N\cdot m. \tag{5.6.14}$$

Now let "auxiliary interference" v_j^* perform different functions.

4° Let $v_1^* = \rho_1 u_1^*$, $v_3^* \equiv 0$. The expressions necessary for computation can be obtained by a suitable combination of the expressions in Tables 5.4.2 and 5.4.4.

In this case the reduction time $T_1^* = 54.14\ s$ for η_1 are shorter than the reduction time $T_3^0 = 55.01\ s$ for η_3. Controls u_1, u_3 defined by relationships (5.6.4), (5.3.11), and (5.6.11) will have *three* switching moments $t = 8.90$, 10.94, and 54.14 s and vary within the ranges $-78.5 \le u_1 \le 76.8$, $-107.5 \le u_3 \le 127.3\ N\cdot m$. Control u_2 is *continuous* and varies within $0 \le u_2 \le 4.5\ N\cdot m$. The functions u_1, u_3 are shown in Figure 5.6.3, in which, for clarity, we introduce a different time scale in the intervals

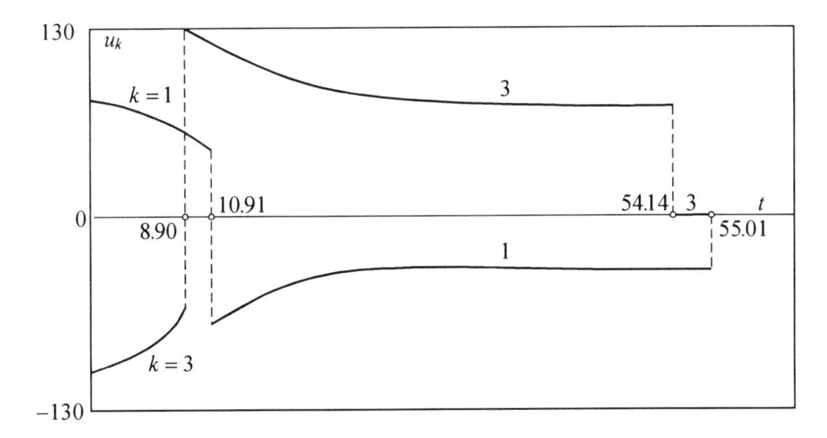

Figure 5.6.3. Controls u_1, u_3 in the case when $v_1* = \rho_1 u_1^*$, $v_3* \equiv 0$.

$[0, 54.14]$ and $(54.14, 55.01]$ on the t-axis. It follows that in the case in question, as distinguished from (5.6.12)–(5.6.14),

$$\alpha_1 = 78.5, \quad \alpha_2 = 4.5, \quad \alpha_3 = 127.3 \ N \cdot m. \tag{5.6.15}$$

5° Let $v_3^* = \rho_3 u_3^*$, $v_1^* \equiv 0$. Control rules u_1, u_3 defined by relationships (5.6.4), (5.3.11), and (5.6.11) will have *three* switching moments $t = 6.53$, 14.12, and 52.61 s and vary within the ranges $-86.5 \le u_1 \le 76.8$, $-107.5 \le u_3 \le 104.1 \ N \cdot m$. Control u_2 is *continuous* and varies within $0 \le u_2 \le 3.6 \ N \cdot m$. In this case, compared with (5.6.12)–(5.6.15),

$$\alpha_1 = 86.5, \quad \alpha_2 = 3.6, \quad \alpha_3 = 107.5 \ N \cdot m. \tag{5.6.16}$$

An analysis of cases 2°– 5° indicates only that the control rules u_k of the form (5.6.4), (5.3.11), and (5.6.11) can change considerably when the "auxiliary interference" v_j^* differs from the "worst" scenario $v_j^* = -\rho_j u_j^*$ (case 1°). It is necessary to estimate numbers α_k for *any* admissible realizations of v_j^*.

Taking into account relationships (5.6.12)–(5.6.16), we conclude that the values of $\alpha_k = \max|u_k|$ on the subset Ω^* of the set Ω of those possible states of the linear system (5.6.4), (5.3.11), and (5.6.11), which correspond to various combinations of "edge" $v_j^* = \pm\rho_j u_j^*$ and "intermediate" $v_j^* \equiv 0$ realizations of v_j^*, are as follows:

$$\alpha_1 = 86.6, \quad \alpha_2 = 4.9, \quad \alpha_3 = 127.5 \ N \cdot m. \tag{5.6.17}$$

As already noted, these values should be close to those of $\alpha_k = \max|u_k|$ over the entire set Ω.

Let us also estimate the values $\alpha_k = \max|u_k|$ taking as the basis the (upper and lower) estimates, presented in Table 5.4.1, for the variables which determine the state of the linear system of type (5.3.2), (5.3.3) for all admissible realizations v_j^* at once.

To do so, let us make the construction of control rules (5.6.4) more specific by using relationships (5.6.7). In the case in question, $v_2 \equiv v_2^* \equiv u_2^* \equiv 0$, $\xi_2 \equiv 0$, and control rules (5.6.4) take the form

$$u_1 = \frac{A}{\eta_2}(-u_3^* + \eta_1\xi_1\xi_3 - \eta_3\xi_1^2), \quad u_3 = \frac{C}{\eta_2}(u_1^* - \eta_3\xi_1\xi_3 + \eta_1\xi_3^2),$$
$$u_2 = (A - C)\xi_1\xi_3, \tag{5.6.18}$$
$$\xi_1 = -\mu_3\eta_2^{-1}, \quad \xi_3 = \mu_1\eta_2^{-1}.$$

Using the inequalities

$$0 \le \eta_j(t) \le \eta_j^+(t), \quad 0 \le \eta_2^-(t) \le \eta_2(t) \le 1,$$
$$0 \le \xi_1(t) \le \xi_1^+(t) = -[\mu_3^-(t)][\eta_2^-(t)]^{-1},$$
$$[\mu_1^-(t)][\eta_2^-(t)]^{-1} = \xi_3^-(t) \le \xi_3(t) \le 0, \quad t \in [t_0, t_1],$$

which are valid in the case under consideration, and estimating expressions (5.6.18) for u_k by the reinforcing inequalities, we arrive at the relationships

$$|u_1| \le \frac{A}{\eta_2^-}[\alpha_3^* - \eta_1^+\xi_1^+\xi_3^- + \eta_3^+(\xi_1^+)^2],$$
$$|u_2| = (A - C)\xi_1^+\xi_3^-, \tag{5.6.19}$$
$$|u_3| \le \frac{C}{\eta_2^-}[\alpha_1^* - \eta_3^+\xi_1^+\xi_3^- + \eta_1^+(\xi_1^-)^2], \quad t \in [t_0, t_1].$$

Calculations by formulas (5.6.19) lead to the estimates

$$\alpha_1 = 105.2, \quad \alpha_2 = 6.53, \quad \alpha_3 = 160.1\ N\cdot m. \tag{5.6.20}$$

Comparison of (5.6.20) with (5.6.17) shows that the values of $\alpha = \max \alpha_k$ differ by approximately 25%. Note that in the test example of the problem of triaxial reorientation presented in Subsection 5.4.8 the difference is nearly the same (on the order of 20%).

Conclusions.

(1) For the given boundary conditions, values of A, B, C, and interference levels v_j defined in accordance with (5.6.9), the guaranteed reorientation time $\tau = 70$ s can be attained in the construction of control rules (5.6.4), (5.3.11), and (5.6.11) for $\max \alpha_k = 160.1\ N\cdot m$. However, this is a "*cautious*" estimate and also a slight *overestimate* of the admissible levels of u_k obtained by employing the reinforcing inequalities (5.6.19).

(2) For the given levels of interference v_j, the "unperturbed" construction (5.6.4), (5.3.11), and (5.6.11) with switching curves $\psi_j^0 = 0$ in (5.3.11), in which we find α_j^* relying on the reorientation time $T = 70$ s, cannot solve Problem 5.6.1 in a *finite* time.

(3) The "unperturbed" construction (5.6.4), (5.3.11), and (5.6.11) with switching curves $\psi_j^0 = 0$ in (5.3.11), in which we find α_j^* relying on the bounds $\max \alpha_k = 160.1\ N\cdot m$ (rather than on the prescribed reorientation time T), guarantees the reorientation time $T = 32.9$ s when $v_i \equiv 0$. Thus, the "*time loss*" $\gamma = \tau - T$ caused by the allowance for interference v_j amounts to $\gamma = 37.1$ s for the given levels of u_k and v_j and the "degree of accuracy" of the estimates involved.

5.7 Overview of References

The term "*differential games*" is comparatively recent. It was coined by Isaacs in the 50s. By now, the bibliography on the subject is quite extensive (see, for example, monographs by Blaquière et al. [1969], Friedman [1971], Hajek [1975], Kurzhanskii [1977], Chernous'ko and Melikyan [1978], Subbotin and Chentsov [1981], Krasovskii and Subbotin [1988], A.N. Krasovskii and N.N. Krasovskii [1995], and the survey by Leitmann [1993]).

The elaboration of constructive methods for investigating complex essentially nonlinear systems is one of the most important problems which govern advances in the theory of games. Naturally, the current state of the theory gives no effective ways of constructing game-theoretic control laws even with the use of high-performance modern computers.

In a number of recent works an attempt was made to achieve this purpose for certain classes of nonlinear controlled systems (Vorotnikov [1994a]–[1994c], [1995c], [1997b], [1997e]). Chapter 5 presents these results.

In connection with the problems Chapter 5 deals with, we indicate two lines of investigation in modern control theory:

(1) the problem of *decomposition of a nonlinear controlled system* (Chernous'ko [1990a], [1990b], [1992a], [1992b], [1993], Anan'evskii, Dobrynina, and Chernous'ko [1995]);

(2) the problem of *aircraft motion control in an uncertain environment* (see Miele et al. [1986], Bryson and Zhao [1987], Miele [1989], Chen and Pandey [1989], Botkin et al. [1989], [1993], Leitmann and Pandey [1990], Zhao and Bryson [1990], [1992], Bulirsch et al. [1991]).

Also worthy of mention is the problem of damping the *rotational motions* of an asymmetric solid in the presence of uncontrollable interference, which was solved by Corless and Leitmann [1995]. The damping of rotational motions is performed until the angular velocity of the body attains a *desired* value (the zero value included). This problem is considered based solely on Euler's dynamical equations and does not include the problem of reorientation. However, as was already noted, it is often possible to subdivide the process of controlling the angular motion of a solid into two stages: rotation damping and reorientation in space.* Therefore, the results of Chapter 5 together with the above result provide one of possible ways of *completely solving* the game-theoretic problem of controlling the angular motion of a solid in the presence of uncontrollable interference. Such a solution can be used, for example, in designing attitude control systems for spacecraft which serve various purposes.

We note that constructing the control rules which solve the problems of reorienting a solid considered in Chapter 5 requires information on the current state of the solid. To be specific, the components of the angular velocity

*The latest studies showed (Zelikin [1996]) that even in the absence of interference the time-optimal reorientation of a solid without taking this assumption as the starting point leads to rather complicated control rules with *infinite* number of switching points (so called "*chattering*" controls).

vector and also the Rodrigues–Hamilton variables (or the Poisson variables in the case of uniaxial reorientation) must be available. In space engineering the current values of the angular velocity vector of a body are measured by special gyroscopes. On-board computers allow real-time integration of kinematic equations. In this respect the form of kinematic equations employed in Chapter 5 appears to be well-chosen.

Further development of the approach to solving the problem of reorienting a solid proposed in Chapter 5 seems reasonable from the viewpoint of theory and practice. It might be interesting, for example, to obtain solutions

(1) *alternatively interpreting* the differential games for auxiliary linear conflict-control μ-systems;
(2) choosing controls in the class of *piecewise-constant* functions (*bang-bang* or *relay* controls) that simulate engines operating on chemical propellant;*
(3) using *incomplete information* on the current state (the angular velocity and position) of the body;
(4) using control by *rotating masses* (flywheels, gimballed gyroscope).

Chapter 6

Stability and Stabilization of Functional-Differential Equations with Respect to Part of the Variables

The theory of systems of *functional-differential equations* is a significant and rapidly developing sphere of modern mathematics which finds extensive application in complex systems of automatic control and in economic, ecological, and biological models. Naturally, the problem arises of the stability of the processes described by the class of equations mentioned.

In line with the subject matter of this book, in this chapter we consider the problem of *stability with respect to a given part of the variables* characterizing the state of a system of functional-differential equations. The principal results of the chapter can be summarized as follows.

(1) The formulation of the problem of stability with respect to part of the variables is made more specific.

(2) Theorems on stability and asymptotic stability with respect to part of the variables are established based on the *Lyapunov direct method.* Here we use both the *Lyapunov functions* and *Lyapunov–Krasovskii functionals.* Investigation is carried out using the *method of introducing additional μ-functions* suggested in Chapter 3. However, following this path we encounter crucial peculiarities, which results in the theorems obtained differing from their analogs for the case of ordinary differential equations.

(3) Sufficient (in a number of cases, necessary and sufficient) conditions for stability and asymptotic stability with respect to part of the variables are obtained for *linear* systems with constant coefficients and constant holdover of an argument and conditions for asymptotic stability with respect to part of the variables *in a linear approximation.* The *method*

of constructing auxiliary μ-systems proposed in Chapters 1–2 is further developed.

(4) Linear equations with a retarded argument of a *neutral type* are considered which are embraced by the method of constructing auxiliary μ-systems.

(5) A method for constructing controls in the problem of partial stabilization of linear systems is proposed.

6.1 Formulation of the Problem of Stability with Respect to Part of the Variables

6.1.1 System of Equations with Holdover (Delay)

Let there be given a system of differential equations *with holdover* (in the vector form)

$$\begin{gathered} \dot{\boldsymbol{x}}(t) = \boldsymbol{X}[t, \boldsymbol{x}(t+\theta)], \\ \boldsymbol{x}^{\mathsf{T}} = (x_1, \ldots, x_n) = (y_1, \ldots, y_m, z_1, \ldots, z_p) = (\boldsymbol{y}^{\mathsf{T}}, \boldsymbol{z}^{\mathsf{T}}), \\ m > 0, \quad p \geq 0, \quad n = m + p. \end{gathered} \tag{6.1.1}$$

Here $\boldsymbol{X} = (X_1, \ldots, X_n)^{\mathsf{T}}$ is a given mapping (operator), $-\tau \leq \theta \leq 0$.

A particular case of system (6.1.1) is a system of equations with a *retarded argument* (with *delay*) which is often termed a system of *difference-differential* equations

$$\dot{\boldsymbol{x}}(t) = \boldsymbol{X}[t, \boldsymbol{x}(t), \boldsymbol{x}(t-\tau)], \tag{6.1.2}$$

where $\boldsymbol{X}$ is a given vector function of variables t, $\boldsymbol{x}(t)$, $\boldsymbol{x}(t-\tau)$ and τ is a delay, which can be constant, dependent on time and on the desired function $\boldsymbol{x}(t)$ and its derivatives.

The most simple and worked through is the case of *constant* delay τ. This assumption fits a wide range of problems.

Another important case of system (6.1.1) is a broad class of the *Volterra integro-differential equations*, which fall into the type of equations with a *distributed* delay (see Lakshmikantham and Rao [1994]).

System (6.1.1) enables one to embrace cases of an arbitrary number of concentrated and distributed delays and delays dependent on the desired function. This form of equation has been systematically employed lately in many works.

In recent years the intensively developing theory of functional-differential equations with an *abstract Volterra operator* has stepped up its ever increasing role in investigations (see Corduneanu [1991], [1993], Azbelev et al. [1991] and the bibliography therein).

Let t_0 be an initial point in time, and let $\varphi_0(\theta)$, $-\tau \leq \theta \leq 0$, $\tau = \text{const} > 0$, be a given initial continuous vector function. A vector function $\boldsymbol{x}(t)$ that is continuous for all $t \geq t_0 - \tau$, continuously differentiable for all $t \geq t_0$, satisfies system (6.1.1) for all $t > t_0$, and meets the initial condition $\boldsymbol{x}(t_0+\theta) = \boldsymbol{\varphi}_0(\theta)$ for $t \in [t_0 - \tau, t_0]$ is understood as a solution $\boldsymbol{x}(t) = \boldsymbol{x}(t; t_0, \boldsymbol{\varphi}_0(\theta))$ of system (6.1.1). We assume that $\dot{\boldsymbol{x}}(t_0) = \boldsymbol{X}[t_0, \boldsymbol{\varphi}_0(\theta)]$ at the time $t = t_0$.

6.1.2 Definitions of Stability with Respect to Part of the Variables

We shall assume that system (6.1.1) is a *system of equations of perturbed motion* composed for the process being analyzed for stability. In this case $\boldsymbol{X}(t, \mathbf{0}) \equiv \mathbf{0}$, and the process under study will be the zero equilibrium position $\boldsymbol{x} = \mathbf{0}$ or, in other words, the *unperturbed motion* of this system.

Consider the *Banach space* $C_{[0,\tau]}$ of continuous vector functions $\boldsymbol{\varphi}(\theta)$: $R^1 \to R^n$ and represent $\boldsymbol{\varphi}(\theta)$ in the form

$$\boldsymbol{\varphi}(\theta) = [\boldsymbol{\varphi}_y(\theta), \boldsymbol{\varphi}_z(\theta)],$$
$$\boldsymbol{\varphi}_y(\theta) = [\varphi_1(\theta), \ldots, \varphi_m(\theta)], \quad \boldsymbol{\varphi}_z(\theta) = [\varphi_{m+1}(\theta), \ldots, \varphi_n(\theta)].$$

Suppose that

(1) the operator $\boldsymbol{X}[t, \boldsymbol{\varphi}(\theta)]$ defining the right-hand side of system (6.1.1) is continuous and satisfies the conditions for existence and uniqueness of solutions in the domain

$$\begin{gathered} t \geq 0, \quad \|\boldsymbol{\varphi}_y(\theta)\| \leq H, \quad \|\boldsymbol{\varphi}_z(\theta)\| < \infty, \\ \|\boldsymbol{\varphi}_y(\theta)\| = \sup|\varphi_i(\theta)|, \quad \|\boldsymbol{\varphi}_z(\theta)\| = \sup|\varphi_{m+j}(\theta)|, \\ (-\tau \leq \theta \leq 0;\ i = 1, \ldots, m;\ j = 1, \ldots, p); \end{gathered} \tag{6.1.3}$$

(2) solutions of system (6.1.1) are *$\boldsymbol{z}$-continuable* (i.e., all the solutions are defined for all $t \geq t_0$ such that $\|\boldsymbol{y}(t; t_0, \boldsymbol{\varphi}_0(\theta))\| \leq H$).

Further we shall also use the following representation of system (6.1.1)

$$\dot{\boldsymbol{y}}(t) = \boldsymbol{Y}[t, \boldsymbol{y}(t+\theta), \boldsymbol{z}(t+\theta)], \quad \dot{\boldsymbol{z}}(t) = \boldsymbol{Z}[t, \boldsymbol{y}(t+\theta), \boldsymbol{z}(t+\theta)].$$

Definition 6.1.1. The unperturbed motion $\boldsymbol{x} = \mathbf{0}$ of system (6.1.1) is said to be

(1) *stable with respect to* $y_1, \ldots, y_m$ (briefly, *$\boldsymbol{y}$-stable*) if for any numbers $\varepsilon > 0$, $t_0 \geq 0$, there is a number $\delta(\varepsilon, t_0) > 0$ such that from $\|\boldsymbol{\varphi}_0(\theta)\| < \delta$ it follows that $\|\boldsymbol{y}(t; t_0, \boldsymbol{\varphi}_0(\theta))\| < \varepsilon$ for all $t \geq t_0$;

(2) *uniformly $\boldsymbol{y}$-stable* if in definition (1) the number δ does not depend on t_0;

(3) *asymptotically $\boldsymbol{y}$-stable* if it is $\boldsymbol{y}$-stable and, in addition, for each $t_0 \geq 0$, there exists a number $\Delta(t_0) > 0$ such that solutions $\boldsymbol{x}(t; t_0, \boldsymbol{\varphi}_0(\theta))$ with $\|\boldsymbol{\varphi}_0(\theta)\| < \Delta$ satisfy the condition

$$\lim \|\boldsymbol{y}(t; t_0, \boldsymbol{\varphi}_0)\| = 0, \quad t \to \infty \tag{6.1.4}$$

(the domain $\|\boldsymbol{\varphi}_0(\theta)\| < \Delta$ is contained in the *domain of $\boldsymbol{y}$-attraction* of the point $\boldsymbol{x} = \mathbf{0}$ for the initial time t_0);

(4) *uniformly asymptotically* $\boldsymbol{y}$*-stable* if in definition (3) the number Δ does not depend on t_0 and relationship (6.1.4) holds uniformly with respect to t_0, $\varphi_0(\theta)$ from the domain $t_0 \geq 0$, $\|\varphi_0(\theta)\| < \Delta$;

(5) *exponentially asymptotically* $\boldsymbol{y}$*-stable* if there exist positive constants Δ, M, and α such that solutions $\boldsymbol{x}(t; t_0, \varphi_0(\theta))$ with $\|\varphi_0(\theta)\| < \Delta$ satisfy the condition

$$\|\boldsymbol{y}(t; t_0, \varphi_0(\theta))\| \leq M(\|\varphi_0(\theta)\|)e^{-\alpha(t-t)}, \quad t \geq t_0.$$

Remark 6.1.1.

1. Definition 6.1.1 is a development of the Lyapunov–Rumyantsev definitions as applied to the problem of partial stability for systems of functional-differential equations. Such a problem was, evidently, considered for the first time by Halanay [1963], Corduneanu [1975], and Yudaev [1975].

2. In Definition 6.1.1 we take the class C of *continuous* initial vector functions. However, in real systems the class of natural initial perturbations can be *narrower*: for example, the class C^1 of continuously differentiable functions. In this case, the solution $\boldsymbol{y}$-unstable in class C may turn out to be $\boldsymbol{y}$-stable in class C^1 (Krasovskii [1959]). This circumstance shows that the conditions for instability with respect to part of the variables are of relatively lower value.

3. In systems (6.1.1), (6.1.2), the value of τ is assumed to be *finite*. The case of *infinite* delay value is also considered in the literature (Corduneanu and Lakshmikantham [1980], Hino et al. [1991]).

6.2 Using the Method of Lyapunov–Krasovskii Functionals

6.2.1 Preliminary Remarks

Elsgoltz* was the first to notice the possibility of formally using the method of Lyapunov functions in problems of stability of systems with holdover (delay). As distinguished from systems of ordinary differential equations, the principal difficulty in the way of using this method is that the derivative of the Lyapunov V-function is a *functional* even for the simplest systems of type (6.1.1) or (6.1.2).

To overcome this difficulty, two approaches were proposed almost simultaneously:

(1) Krasovskii's approach [1956] which uses *functionals*;

(2) the approach (Razumikhin [1956], Krasovskii [1956]) which extends the possibility of using the *Lyapunov functions* of a finite number of variables.

The analysis of these approaches demonstrates their *significant difference* though the function is a particular case of the functional.

---

*Elsgoltz, L.E. [1954] Stability of Solutions of Difference-Differential Equations, *Uspekhi Mat. Nauk*, **9**, 4, 95–112 (in Russian).

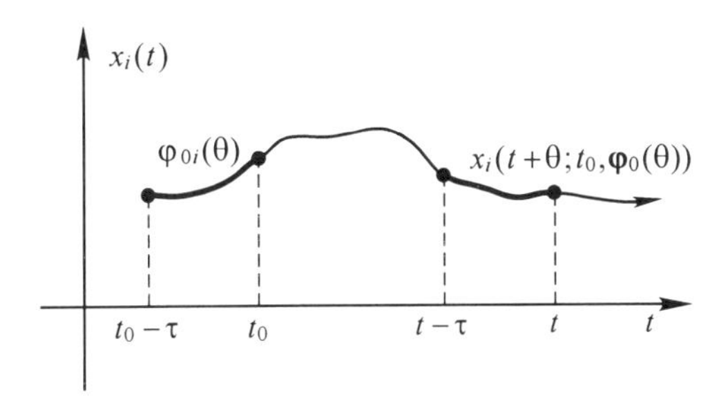

Figure 6.2.1. A component $x_i(t+\theta;t_0,\varphi_0(\theta))$ of a vector interval $\boldsymbol{x}_t = \boldsymbol{x}(t+\theta;t_0,\varphi_0(\theta))$; $\varphi_{0i}(\theta)$ is the ith component of the initial vector function $\varphi_0(\theta)$.

Krasovskii's approach [1956], [1959] is based on using, instead of the Lyapunov V-functions, the corresponding *V-functionals* which are often called the *Lyapunov–Krasovskii functionals.* In this case, one takes, as an element of the trajectory $\boldsymbol{x}(t) = \boldsymbol{x}(t;t_0,\varphi_0(\theta))$ of systems of type (6.1.1) or (6.1.2) at the current time t, a *vector interval* $\boldsymbol{x}_t = \boldsymbol{x}(t+\theta;t_0,\varphi_0(\theta))$ of this trajectory for $-\tau \leq \theta \leq 0$ rather than a point $\boldsymbol{x}(t)$ of R^n. This interval is a point of the corresponding *functional* space $C_{[0,\tau]}$, and a *unique* value of functional $\dot{V}$ (by virtue of systems (6.1.1) or (6.1.2)) in an interval of the trajectory virtue of systems (6.1.1) or (6.1.2)) in an interval of the trajectory corresponds to the value of a given V-functional in the same interval.

The idea of extending the problem of stability to the functional space made it possible to formulate in a natural way (Krasovskii [1956], [1959]) the analogs of the principal theorems of the method of Lyapunov functions as applied to the systems of type (6.1.1) and (6.1.2). The fact that the functional analogs of the Lyapunov theorems of such a type are invertible is successfully used to substantiate the stability in the first approximation and the stability in the presence of constantly acting perturbations. However, the constructive construction of functionals in specific problems is extremely complicated: it is, in fact, required to know the solutions of the original system. That is why the efforts of many researchers (see, for example, Krasovskii [1959], Hale [1977], Burton [1978], [1985], Burton and Hatvani [1989], Gapalsamy [1992], Kim [1992], Kolmanovskii and Myshkis [1992], Hale and Lunel [1993], Kuang [1993]) are focused on the further development of this approach. A number of constructively verifiable conditions for stability and asymptotic stability have been obtained in the context of the method of Lyapunov functionals.

Another line of investigation deals with the formulation of new ideas allowing one to constructively use the method of Lyapunov functions of a finite number of variables in its application to the class of functional-differential equations. The specific feature of this approach is that the conditions for stability are verified not for all admissible initial vector functions; as a result, the behavior of the Lyapunov function along the integral curves can be nonmonotonic. This approach is also being intensively developed (see, for example, Razumikhin [1956], [1988], Lakshmikantham and Leela [1969], Hale [1977], Kato [1980], Lakshmikantham and Martynyuk [1993]).

Efforts have been undertaken of late to treat these two approaches based on common ground (Gaishun and Knyazhishche [1994]).

We stress that, as in the works by Lyapunov, the problem of stability (asymptotic stability) was analyzed mainly *with respect to all the variables* which characterize the state of a sufficiently general system of functional-differential equations of perturbed motion.

In line with the subject matter of this book, in this section we focus on the problem of stability *with respect to part of the variables* for systems (6.1.1) and (6.1.2) in the context of the method of Lyapunov–Krasovskii functionals. In this case, there are peculiarities which distinguish the corresponding theorems from their analogs for ordinary differential equations. Section 6.3 studies the problem of partial stability in the framework of the method of Lyapunov functions.

6.2.2 Conditions for Stability (Asymptotic Stability) with Respect to Part of the Variables

Consider single-valued continuous functionals $V = V[t, \boldsymbol{\varphi}(\theta)]$ that satisfy the condition $V[t, \mathbf{0}] \equiv 0$ and are defined on an arbitrary piecewise-continuous vector function $\boldsymbol{\varphi}(\theta)$, $-\tau \le \theta \le 0$ in domain (6.1.3).

On substituting vector function $\boldsymbol{x}_t = \boldsymbol{x}(t+\theta; t_0, \boldsymbol{\varphi}_0(\theta))$, which defines an element of the trajectory at a time $t \ge t_0$, in the functional $V[t, \boldsymbol{\varphi}(\theta)]$, we obtain the value of functional $V[t, \boldsymbol{x}_t]$ on solutions of system (6.1.1).

By the *derivative* $\dot{V}$ of a functional along the solutions of system (6.1.1) we shall understand the value (Yoshizawa [1966])

$$\dot{V} = \overline{\lim_{\delta \to 0+}}\, \delta^{-1}\{V[t+\delta, \boldsymbol{x}_{t+\delta}] - V[t, \boldsymbol{x}_t]\}.$$

If V satisfies the local Lipschitz condition with respect to the second argument, then the indicated limit is uniquely defined.

Along with the earlier introduced norm $||\boldsymbol{\varphi}(\theta)||$, we shall also further use the following norm (Krasovskii [1959])

$$||\boldsymbol{\varphi}(\theta)||_\tau = \left(\int_{-\tau}^{0} \sum_{i=1}^{n} \varphi_i^2(\theta)\, d\theta \right)^{1/2}.$$

Functions $a(r)$, $b(r)$, and $c(r)$ have the same meaning as in Section 0.4.

Theorem 6.2.1.

1. *Suppose that for system* (6.1.1) *it is possible to specify a functional* $V = V[t, \boldsymbol{\varphi}(\theta)]$ *that is defined, continuous, and locally Lipschitzian in domain* (6.1.3) *such that the conditions*

$$a(||\boldsymbol{\varphi}_y(0)||) \le V[t, \boldsymbol{\varphi}(\theta)], \quad \dot{V}[t, \boldsymbol{x}_t] \le 0,$$

are satisfied in domain (6.1.3).

Then the unperturbed motion $\boldsymbol{x} = \mathbf{0}$ *of system* (6.1.1) *is* $\boldsymbol{y}$*-stable.*

2. *Suppose that the following conditions are satisfied in domain* (6.1.3):

$$a(\|\boldsymbol{\varphi}_y(0)\|) \leq V[t, \boldsymbol{\varphi}(\theta)] \leq b(\|\boldsymbol{\varphi}(\theta)\|), \quad \dot{V}[t, \boldsymbol{x}_t] \leq 0.$$

Then the unperturbed motion $\boldsymbol{x} = \mathbf{0}$ *of system* (6.1.1) *is uniformly* $\boldsymbol{y}$*-stable.*

3. *Suppose that the following conditions are satisfied in domain* (6.1.3):

$$a(\|\boldsymbol{\varphi}_y(0)\|) \leq V[t, \boldsymbol{\varphi}(\theta)] \leq b(\|\boldsymbol{\varphi}_y(\theta)\|), \quad \dot{V}[t, \boldsymbol{x}_t] \leq -c(\|\boldsymbol{y}_t\|).$$

Then the unperturbed motion $\boldsymbol{x} = \mathbf{0}$ *of system* (6.1.1) *is uniformly asymptotically* $\boldsymbol{y}$*-stable (on the whole with respect to* $\boldsymbol{\varphi}_{z0}$*).*

4. *Suppose that the following conditions are satisfied in domain* (6.1.3):

$$a(\|\boldsymbol{\varphi}_y(0)\|) \leq V[t, \boldsymbol{\varphi}(\theta)] \leq b_1(\|\boldsymbol{\varphi}_y(0)\|) + b_2(\|\boldsymbol{\varphi}_y(\theta)\|_\tau),$$
$$\dot{V}[t, \boldsymbol{x}_t] \leq -c(\|\boldsymbol{y}(t)\|).$$

Then the unperturbed motion $\boldsymbol{x} = \mathbf{0}$ *of system* (6.1.1) *is uniformly asymptotically* $\boldsymbol{y}$*-stable (on the whole with respect to* $\boldsymbol{\varphi}_{z0}$*).*

5. *Suppose that the following conditions are satisfied in domain* (6.1.3):

$$a(\|\boldsymbol{\varphi}_y(0)\|) \leq V[t, \boldsymbol{\varphi}(\theta)] \leq b(\|\boldsymbol{\varphi}(\theta)\|),$$
$$\dot{V}[t, \boldsymbol{x}_t] \leq -c(\|\boldsymbol{y}(t)\|),$$
$$\|\boldsymbol{Y}[t, \boldsymbol{\varphi}(\theta)]\| \leq M = \text{const} > 0.$$

Then the unperturbed motion $\boldsymbol{x} = \mathbf{0}$ *of system* (6.1.1) *is uniformly asymptotically* $\boldsymbol{y}$*-stable.*

Proof.

1. Functional $V[t, \boldsymbol{\varphi}(\theta)]$ is defined and continuous in domain (6.1.3) and the condition $V[t, \mathbf{0}] \equiv 0$ is valid. Therefore, for any numbers $\varepsilon > 0$, $t_0 \geq 0$, one can find a number $\delta(\varepsilon, t_0) > 0$ such that $\|\boldsymbol{\varphi}_0(\theta)\| < \delta$ yields the inequality $V[t_0, \boldsymbol{\varphi}_0(\theta)] \leq a(\varepsilon)$. By virtue of $\dot{V}[t, \boldsymbol{x}_t] \leq 0$, for a solution $\boldsymbol{x}(t; t_0, \boldsymbol{\varphi}_0(\theta))$ of system (6.1.1) with $\|\boldsymbol{\varphi}_0(\theta)\| < \delta$,

$$\begin{aligned} a(\|\boldsymbol{y}(t; t_0, \boldsymbol{\varphi}_0(\theta))\|) &\leq V[t, \boldsymbol{x}(t+\theta; t_0, \boldsymbol{\varphi}_0(\theta))] \\ &\leq V[t_0, \boldsymbol{\varphi}_0(\theta)] \leq a(\varepsilon) \end{aligned} \tag{6.2.1}$$

for all $t \geq t_0$.

These inequalities, with the properties of function $a(r)$ taken into account, give the estimate $\|\boldsymbol{y}(t; t_0, \boldsymbol{\varphi}_0(\theta))\| < \varepsilon$ for all $t \geq t_0$. The first part of Theorem 6.2.1 is proved.

2. If $V[t, \boldsymbol{\varphi}(\theta)] \leq b(\|\boldsymbol{\varphi}(\theta)\|)$, number δ appearing in the proof of the first part of the theorem can be chosen as independent of t_0. Indeed, let us take $\delta(\varepsilon) = b^{-1}(a(\varepsilon))$. Then for $t_0 \geq 0$, $\|\boldsymbol{\varphi}_0(\theta)\| < \delta$, by virtue of $V[t_0, \boldsymbol{\varphi}_0(\theta)] \leq b(\|\boldsymbol{\varphi}_0(\theta)\|) \leq b(\delta) = a(\varepsilon)$, we have relationships (6.2.1) for all $t \geq t_0$. The second part of Theorem 6.2.1 is proved.

3. Again let us take $\delta(\varepsilon) = b^{-1}(a(\varepsilon))$. Then for $t_0 \geq 0$, $\|\boldsymbol{\varphi}_{y0}(\theta)\| < \delta$, $\|\boldsymbol{\varphi}_{z0}(\theta)\| < \infty$, by virtue of $V[t_0, \boldsymbol{\varphi}_0(\theta)] \leq b(\|\boldsymbol{\varphi}_{y0}(\theta)\|) \leq b(\delta) = a(\varepsilon)$, we

have relationships (6.2.1) for all $t \geq t_0$. The uniform $\boldsymbol{y}$-stability (on the whole with respect to $\boldsymbol{\varphi}_{z0}(\theta)$) is proved.

To proceed with the proof, for a given number H, let us choose $\Delta(H) > 0$, so that $b(\Delta) < a(H)$. The uniform asymptotic $\boldsymbol{y}$-stability (on the whole with respect to $\boldsymbol{\varphi}_{z0}(\theta)$) will be proved if, for each $\varepsilon > 0$, there can be found a number $T = T(\varepsilon) > 0$ such that $\|\boldsymbol{y}(t; t_0, \boldsymbol{\varphi}_0(\theta))\| < \varepsilon$ for all $t > t_0 + T(\varepsilon)$ if $\|\boldsymbol{\varphi}_{y0}(\theta)\| < \Delta$, $\|\boldsymbol{\varphi}_{z0}(\theta)\| < \infty$, $t_0 \geq 0$.

Let $\varepsilon > 0$ be given and $\delta = \delta(\varepsilon)$ be chosen in the same way as in the proof of the uniform $\boldsymbol{y}$-stability (on the whole with respect to $\boldsymbol{\varphi}_{z0}(\theta)$). Suppose that $\|\boldsymbol{\varphi}_{y0}(\theta)\| < \Delta$, $\|\boldsymbol{\varphi}_{z0}(\theta)\| < \infty$ yields

$$\delta(\varepsilon) \leq \|\boldsymbol{y}(t+\theta; t_0, \boldsymbol{\varphi}_0(\theta))\| \leq H \tag{6.2.2}$$

for all $t \geq t_0$.

Then, by the condition $\dot{V}[t, \boldsymbol{x}_t] \leq -c(\|\boldsymbol{y}_t\|)$, the inequality $\dot{V}[t, \boldsymbol{x}_t] \leq -c(\delta(\varepsilon)) \stackrel{\text{def}}{=} \alpha = \text{const} < 0$ is valid, and, consequently,

$$V[t, \boldsymbol{x}_t] \leq V[t_0, \boldsymbol{\varphi}_0(\theta)] - \alpha(t - t_0) \leq b(\Delta) - \alpha(t - t_0). \tag{6.2.3}$$

For $t > t_0 + T$, where $T = \alpha^{-1} b(\Delta)$, by (6.2.3), we obtain $V[t, \boldsymbol{x}_t] < 0$, which contradicts the condition $V[t, \boldsymbol{x}_t] \geq a(\|\boldsymbol{y}(t)\|)$.

Therefore, for a certain $t^* = [t_0, t_0 + T]$, $\|\boldsymbol{y}(t^* + \theta; t_0, \boldsymbol{\varphi}_0(\theta))\| < \delta$. On the basis of uniform $\boldsymbol{y}$-stability (on the whole with respect to $\boldsymbol{\varphi}_{z0}(\theta)$), we conclude that $\|\boldsymbol{y}(t; t_0, \boldsymbol{\varphi}_0(\theta))\| < \varepsilon$ for all $t \geq t^*$ (the more so for $t > t_0 + T$). The third part of Theorem 6.2.1 is proved.

4. For any $\varepsilon > 0$, we choose $\delta(\varepsilon) > 0$ so that

$$b_1(\delta) + b_2(\delta(m\tau)^{1/2}) < a(\varepsilon).$$

Then for $t_0 \geq 0$, $\|\boldsymbol{\varphi}_{y0}(\theta)\| < \delta$, $\|\boldsymbol{\varphi}_{z0}(\theta)\| < \infty$, by virtue of $V[t_0, \boldsymbol{\varphi}_0(\theta)] \leq b_1(\|\boldsymbol{\varphi}_{y0}(0)\|) + b_2(\|\boldsymbol{\varphi}_{y0}(\theta)\|_\tau) \leq b_1(\delta) + b_2(\delta(m\tau)^{1/2}) < a(\varepsilon)$, we have relationships (6.2.1) for all $t \geq t_0$. The uniform $\boldsymbol{y}$-stability (on the whole with respect to $\boldsymbol{\varphi}_{z0}(\theta)$) is proved.

We shall prove the uniform asymptotic $\boldsymbol{y}$-stability following the scheme proposed in Burton [1978].

For $H^* = \min(H, 1)$, let us choose $\Delta(H^*) > 0$ so that $b_1(\Delta) + b_2(\Delta(m\tau)^{1/2}) < a(H^*)$. We shall demonstrate that, for each $\varepsilon > 0$, a number $T = T(\varepsilon) > 0$ can be found such that $\|\boldsymbol{y}(t; t_0, \boldsymbol{\varphi}_0(\theta))\| < \varepsilon$ for all $t > t_0 + T(\varepsilon)$, provided $\|\boldsymbol{\varphi}_{y0}(\theta)\| < \Delta$, $\|\boldsymbol{\varphi}_{z0}(\theta)\| < \infty$, $t_0 \geq 0$.

Given $\varepsilon > 0$, let us choose $\varepsilon_1 < \varepsilon$ and $\varepsilon_2 < \varepsilon$ so that
(1) $b_1(\varepsilon_1) < \frac{1}{2}a(\varepsilon)$,
(2) $b_2(\|\boldsymbol{y}_t\|_\tau) < \frac{1}{2}a(\varepsilon)$, if $\|\boldsymbol{y}_t\| < \varepsilon_2$.

The inequality $b_2(\|\boldsymbol{y}_t\|_\tau) \leq b_2(\varepsilon_2(m\tau)^{1/2})$ for $\|\boldsymbol{y}_t\| < \varepsilon_2$ ensures the possibility of the indicated choice of ε_2.

Let us take $\varepsilon_3 = \min(\varepsilon_1, \varepsilon_2)$ and show that if $\|\boldsymbol{\varphi}_{y0}(\theta)\| < \Delta$, $\|\boldsymbol{\varphi}_{z0}(\theta)\| < \infty$, then there is a number $T_1 > \tau$ such that the inequality $\|\boldsymbol{y}(t; t_0, \boldsymbol{\varphi}_0(\theta))\| \geq \varepsilon_3$ is violated for some value of t in each interval of length T_1.

Indeed, if $\|\boldsymbol{y}(t;t_0,\boldsymbol{\varphi}_0(\theta))\| \geq \varepsilon_3$ over the interval $[t_1,t_2]$, then

$$V[t_2,\boldsymbol{x}_{t_2}] \leq V[t_1,\boldsymbol{x}_{t_1}] - c(\varepsilon_3)(t_2-t_1) \leq \alpha^* - c(\varepsilon_3)(t_2-t_1);$$
$$\alpha^* = b_1(\Delta) + b_2(\Delta(m\tau)^{1/2}).$$

Therefore, if $(t_2-t_1) > \alpha^* c^{-1}(\varepsilon_3)$, then $V[t_2,\boldsymbol{x}_{t_2}] < 0$, which contradicts the condition $V[t_2,\boldsymbol{x}_{t_2}] \geq a(\|\boldsymbol{y}(t_2)\|) \geq 0$. As a result, there exists a monotone sequence $\{t_n\} \to \infty$ such that $\|\boldsymbol{y}(t_n)\| < \varepsilon_3$. Passing to the subsequence $\{t_k\}$, we shall require that $t_k+\tau < t_{k+1}$. As a result, if $\|\boldsymbol{\varphi}_{y0}(\theta)\| < \Delta$, $\|\boldsymbol{\varphi}_{z0}(\theta)\| < \infty$, then one can find a number $T_1(\varepsilon) > \tau$ such that $\|\boldsymbol{y}(t_n)\| < \varepsilon_3$ for $t_k \in [t_0+kT_1, t_0+(k+1)T_1]$.

Let us consider the sequence of functions $\{\boldsymbol{y}_{t_k}\}$ corresponding to the chosen subsequence $\{t_k\}$ and show that

$$b_2(\|\boldsymbol{y}_{t_k}\|_\tau) < 1/2\, a(\varepsilon). \tag{6.2.4}$$

Assuming the contrary, we have $\|\boldsymbol{y}_{t_k}\|_\tau \geq \beta$ for some number $\beta > 0$, which yields $\sum\limits_{i=1}^{m}\int\limits_{-\tau}^{0} y_i^2(t_k+\theta)d\theta \geq \beta^2$. Therefore,

$$\sum_{i=1}^{m}\frac{1}{m}\int_{t_k-\tau}^{t_k} y_i^2(s)\,ds \geq \frac{\beta^2}{m} \overset{\text{def}}{=} \beta^* \tag{6.2.5}$$

and, consequently, there exists a number $\gamma > 0$ such that

$$\int_{t_k-\tau}^{t_k} c\Big(\frac{1}{m}\sum_{i=1}^{m} y_i^2(t;t_0,\boldsymbol{\varphi}_0(\theta))\Big)dt \geq \gamma. \tag{6.2.6}$$

However, for $t > t_k$, the following inequalities are valid

$$\begin{aligned}
V[t,\boldsymbol{x}_t] &\leq \alpha^* - \int_{t_0}^{t} c(\|\boldsymbol{y}(u;t_0,\boldsymbol{\varphi}_0(\theta))\|)\,du \\
&\leq \alpha^* - \sum_{i=2}^{k}\int_{t_i-\tau}^{t_i} c(\|\boldsymbol{y}(t;t_0,\boldsymbol{\varphi}_0(\theta))\|)\,dt \\
&\leq \alpha^* - \sum_{i=2}^{k}\int_{t_i-\tau}^{t_i} c\Big(\frac{1}{m}\sum_{i=1}^{m} y_i^2(t;t_0,\boldsymbol{\varphi}_0(\theta))\Big)dt,
\end{aligned}$$

since $y_i^2(t;t_0,\boldsymbol{\varphi}_0(\theta)) \leq H^2 < 1$. Therefore, if inequalities (6.2.5), (6.2.6) hold for $\boldsymbol{y}_{t_1},\ldots,\boldsymbol{y}_{t_k}$, then

$$V[t,\boldsymbol{x}_t] \leq \alpha^* - (k-1)\gamma,$$

and there exists a number N such that for $k > N$ $\alpha^*(k-1)\gamma < 0$. Since $V[t,\boldsymbol{x}_t] \geq 0$, we conclude that inequality (6.2.4) is satisfied for some function $\boldsymbol{y}_{t_i}$ and for $1 \leq i \leq N$.

Then for $t \geq t_i$,

$$\begin{aligned}
a(\|\boldsymbol{y}(t)\|) &\leq V[t,\boldsymbol{x}_t] \leq V[t_i,\boldsymbol{x}_{t_i}] \\
&\leq b_1(\|\boldsymbol{y}(t_i)\|) + b_2(\|\boldsymbol{y}_{t_i}\|_\tau) \leq b_1(\varepsilon_3) + 1/2\, a(\varepsilon) \leq a(\varepsilon)
\end{aligned}$$

and, consequently, $\|\boldsymbol{y}(t;t_0,\boldsymbol{\varphi}_0(\theta))\| < \varepsilon$ for all $t \geq t_i$.

Bearing in mind that $t_N < t_0 + 2NT_1$, we take $T = 2NT_1$. Then for all $t > t_0 + T$, the inequality $\|\boldsymbol{y}(t; t_0, \boldsymbol{\varphi}_0(\theta))\| < \varepsilon$ is valid, provided $\|\boldsymbol{\varphi}_{y0}(\theta)\| < \Delta$, $\|\boldsymbol{\varphi}_{z0}(\theta)\| < \infty$, $t_0 \geq 0$.

The fourth part of Theorem 6.2.1 is proved.

5. The uniform $\boldsymbol{y}$-stability follows from the second part of Theorem 6.2.1.

To proceed with the proof of the theorem, we shall choose $\Delta(H^*) > 0$ in the same way as in the proof of the fourth part of Theorem 6.2.1. To prove the uniform asymptotic $\boldsymbol{y}$-stability, let us demonstrate that for each $\varepsilon > 0$ there exists a number $T = T(\varepsilon) > 0$ such that $\|\boldsymbol{y}(t; t_0, \boldsymbol{\varphi}_0(\theta))\| < \varepsilon$ for all $t > t_0 + T(\varepsilon)$, provided $\|\boldsymbol{\varphi}_0(\theta)\| < \Delta$, $t_0 \geq 0$.

Let $\varepsilon > 0$ be given and $\delta = \delta(\varepsilon)$ be chosen as in the proof of the uniform $\boldsymbol{y}$-stability. Suppose that if $\|\boldsymbol{\varphi}_0(\theta)\| < \Delta$, then, for all $t \geq t_0$, relationship (6.2.2) is valid. Then there exists a sequence $\{t_k\}$ such that

$$t_0 + (2k-1)\tau \leq t_k \leq t_0 + 2k\tau \quad (k = 1, 2, \ldots),$$
$$\delta(\varepsilon) \leq \|\boldsymbol{y}(t_k; t_0, \boldsymbol{\varphi}_0(\theta))\| \leq H^*.$$

Under the conditions of the theorem, there exists a number M such that $\|\dot{\boldsymbol{y}}(t)\| \leq M$ for all $t \geq t_0$. Consequently, $\|\boldsymbol{y}(t)\| > \frac{1}{2}\delta$ over the intervals $t_k - \rho \leq t \leq t_k + \rho$, $\rho = (2M)^{-1}\delta$.

By choosing, if required, number M sufficiently large, we can assume that these intervals do not intersect. Thus,

$$V[t_k, \boldsymbol{x}_{t_k}] = V[t_0, \boldsymbol{\varphi}_0(\theta)] + \int_{t_0}^{t_k} \dot{V} dt$$
$$\leq b(\Delta) - c(\tfrac{1}{2}\delta) \sum_{i=1}^{k-1} \int_{t_i-\rho}^{t_i+\rho} dt = b(\Delta) - 2\rho(k-1)c(\tfrac{1}{2}\delta).$$

When $k > 1 + \frac{1}{2}b(\Delta)\rho^{-1}c^{-1}(\frac{1}{2}\delta)$, $V[t_k, \boldsymbol{x}_{t_k}] < 0$, which contradicts the condition $V[t_k, \boldsymbol{x}_{t_k}] \geq a(\|\boldsymbol{y}(t_k)\|) \geq 0$. The further proof follows that of the third part of Theorem 6.2.1. The fifth part of Theorem 6.2.1 is proved.

The theorem is proved. □

Remark 6.2.1.

1. Theorem 6.2.1 is an extension of theorems by Krasovskii [1956], [1959] and Burton [1978] (to the problem of partial stability) and by Rumyantsev [1957a], [1970] (to the systems of functional-differential equations). We note that Theorem 6.2.1 (except for its fifth part) does not require that $\|\boldsymbol{Y}(t, \boldsymbol{\varphi})\|$ be *bounded.*

2. Following the scheme proposed by Krasovskii [1959] and Rumyantsev and Oziraner [1987], one can prove that if the unperturbed motion $\boldsymbol{x} = \boldsymbol{0}$ of system (6.1.1) is uniformly asymptotically $\boldsymbol{y}$-stable (on the whole with respect to $\boldsymbol{\varphi}_{z0}(\theta)$), then there exists a functional V satisfying the conditions of the third part of Theorem 6.2.1.

6.2.3 On Construction of V-Functionals

In solving the problems of partial stability of the unperturbed motion $\boldsymbol{x} = \mathbf{0}$ of systems (6.1.1) and (6.1.2) by using V-functionals, the question of how they can be constructed naturally arises.

We would like to indicate one of possible ways which is expected to be rather effective in certain situations.

In this case, as applied to the problem on $\boldsymbol{y}$-stability (asymptotic $\boldsymbol{y}$-stability), we seek a V-functional in the form

$$V[t, \boldsymbol{\varphi}(\theta)] \equiv V^*[t, \boldsymbol{\varphi}_y(\theta), \boldsymbol{\mu}[\boldsymbol{\varphi}(\theta)]], \tag{6.2.7}$$

where $\boldsymbol{\mu}(\boldsymbol{\varphi})$ is a certain continuous vector function of $\boldsymbol{\varphi}$ such that $\boldsymbol{\mu}(\mathbf{0}) = \mathbf{0}$.

With this approach, one can use standard V^*-functionals which are sign-definite with respect to *all* the variables (with respect to $\boldsymbol{y}$, $\boldsymbol{\mu}$) and, in particular, *quadratic* V^*-functionals.

Let us introduce the following designations:

$$\boldsymbol{\xi}(\theta) = \{\boldsymbol{\varphi}_y(\theta), \boldsymbol{\mu}[\boldsymbol{\varphi}(\theta)]\},$$

$$||\boldsymbol{\xi}(\theta)|| = (||\boldsymbol{\varphi}_y(\theta)||^2 + ||\boldsymbol{\mu}[\boldsymbol{\varphi}(\theta)]||^2)^{1/2},$$

$$||\boldsymbol{\xi}(\theta)||_\tau = \left(\int_{-\tau}^{0} \sum_{i=1}^{m+k} \xi_i^2(\theta)\, d\theta \right)^{1/2}, \quad k = \dim(\boldsymbol{\mu}),$$

where $\xi_i(\theta)$ are the components of vector function $\boldsymbol{\xi}(\theta)$.

To substantiate the proposed approach to the construction of V-functionals, we shall modify Theorem 6.2.1 in the following manner.

Corollary 6.2.1. *Suppose that for system* (6.1.1), *along with a V-functional, it is possible to specify a continuous vector function* $\boldsymbol{\mu} = \boldsymbol{\mu}(\boldsymbol{\varphi})$, $\boldsymbol{\mu}(\mathbf{0}) = \mathbf{0}$ *such that in the domain*

$$t \geq 0, \quad ||\boldsymbol{\xi}(\theta)|| \leq H, \quad ||\boldsymbol{\varphi}_z(\theta)|| < \infty \tag{6.2.8}$$

the following conditions are satisfied:

(1) $a(||\boldsymbol{\xi}(0)||) \leq V[t, \boldsymbol{\varphi}(\theta)] \leq b_1(||\boldsymbol{\xi}(0)||) + b_2(||\boldsymbol{\xi}(\theta)||_\tau)$,

(2) $\dot{V}[t, \boldsymbol{x}_t] \leq -c((||\boldsymbol{y}(t)||^2 + ||\boldsymbol{\mu}[\boldsymbol{x}(t)]||^2)^{1/2})$.

Then the unperturbed motion $\boldsymbol{x} = \mathbf{0}$ of system (6.1.1) *is uniformly asymptotically $\boldsymbol{y}$-stable.*

Proof. Taking into account the continuity of $\boldsymbol{\mu}$ with respect to $\boldsymbol{\varphi}$ and the condition $\boldsymbol{\mu}(\mathbf{0}) = \mathbf{0}$, for any $\delta_1 > 0$, one can choose $\delta(\delta_1) > 0$ so that $||\boldsymbol{\varphi}_0(\theta)|| < \delta$ yields $||\boldsymbol{\mu}[\boldsymbol{\varphi}_0(\theta)]|| < \delta_1$. For any number $\varepsilon > 0$, we take $\delta = \delta(\varepsilon)$ so that

$$b_1\big((\delta^2 + \delta_1^2)^{1/2}\big) + b_2\big((\tau(\delta^2 m + \delta_1^2 k))^{1/2}\big) < a(\varepsilon).$$

Then, for all $||\boldsymbol{\varphi}_0(\theta)|| < \delta$ and all $t \geq t_0$, the following relationships are valid:

$$a((\|\boldsymbol{y}(t)\|^2 + \|\boldsymbol{\mu}[\boldsymbol{x}(t)]\|^2)^{1/2}) \le V[t, \boldsymbol{x}_t] \le V[t_0, \boldsymbol{\varphi}_0(\theta)]$$
$$\le b_1((\|\boldsymbol{\varphi}_{y0}(0)\|^2 + \|\boldsymbol{\mu}[\boldsymbol{\varphi}_0(0)]\|^2)^{1/2}) + b_2((\|\boldsymbol{\varphi}_{y0}(\theta)\|_\tau^2 + \|\boldsymbol{\mu}[\boldsymbol{\varphi}_0(\theta)]\|_\tau^2)^{1/2})$$
$$\le b_1((\delta^2 + \delta_1^2)^{1/2}) + b_2((\tau(\delta^2 m + \delta_1^2 k))^{1/2}) < a(\varepsilon).$$

From these inequalities and from the properties of function $a(r)$, it follows that $\|\boldsymbol{y}(t; t_0, \boldsymbol{\varphi}_0(\theta))\| < \varepsilon$ for all $t \ge t_0$. Thus the uniform $\boldsymbol{y}$-stability of the unperturbed motion $\boldsymbol{x} = \mathbf{0}$ of system (6.1.1) is proved.

The uniform asymptotic $\boldsymbol{y}$-stability can be proved in the same manner as in the proof of the fourth part of Theorem 6.2.1.

The corollary is proved. □

Remark 6.2.2.

1. Even in the case $\dim(\boldsymbol{y}) = \dim(\boldsymbol{z}) = 1$, the fact that the condition $V[t, \boldsymbol{\varphi}(\theta)] \ge a(\|\boldsymbol{\xi}(0)\|)$ is satisfied in domain (6.2.8) does not, generally, imply that the condition $V[t, \boldsymbol{\varphi}(\theta)] \ge a(\|\boldsymbol{\varphi}_y(0)\|)$ holds in domain (6.1.3). The latter is known as the condition for *$\boldsymbol{y}$-sign-definiteness* of a V-functional (in Rumyantsev's sense [1957a]).

2. The choice of a $\boldsymbol{\mu}$-function in (6.2.7) and the choice of a V-functional itself depend on specificity of the problem under investigation. In this regard, the $\boldsymbol{\mu}$-function can be treated as an *auxiliary Lyapunov vector function* to be used alongside the main V-functional.

The introduction of the auxiliary $\boldsymbol{\mu}$-function (together with the main V-functional) allows weakening the conditions of Theorem 6.2.1.

Corollary 6.2.2. *Suppose that for system* (6.1.1), *along with a V-functional, it is possible to specify a continuous vector function $\boldsymbol{\mu} = \boldsymbol{\mu}(\boldsymbol{\varphi})$, $\boldsymbol{\mu}(\mathbf{0}) = \mathbf{0}$ such that in domain* (6.1.3) *the following conditions are satisfied*:
(1) $a(\|\boldsymbol{\varphi}_y(0)\|) \le V[t, \boldsymbol{\varphi}(\theta)] \le b_1(\|\boldsymbol{\xi}(0)\|) + b_2(\|\boldsymbol{\xi}(\theta)\|_\tau)$,
(2) $\dot{V}[t, \boldsymbol{x}_t] \le -c((\|\boldsymbol{y}(t)\|^2 + \|\boldsymbol{\mu}[\boldsymbol{x}(t)]\|^2)^{1/2})$.

Then the unperturbed motion $\boldsymbol{x} = \mathbf{0}$ of system (6.1.1) *is uniformly asymptotically $\boldsymbol{y}$-stable.*

Remark 6.2.3.

1. The condition $V[t, \boldsymbol{\varphi}(\theta)] \le b_1(\|\boldsymbol{\xi}(0)\|) + b_2(\|\boldsymbol{\xi}(\theta)\|_\tau)$ is weaker than the corresponding condition $V[t, \boldsymbol{\varphi}(\theta)] \le b_1(\|\boldsymbol{\varphi}_y(0)\|) + b_2(\|\boldsymbol{\varphi}_y(\theta)\|_\tau)$, in the fourth part of Theorem 6.2.1.

2. In the absence of the delay, conditions (1) and (2) of Corollary 6.2.2 lead to a differential inequality in V. However, in systems with holdover, the question is wholly unclear whether conditions (1) and (2) can be interpreted as a kind of differential inequality.

Example 6.2.1.

1. Let the system of equations of perturbed motion (6.1.2) have the form

$$\begin{aligned}
\dot{y}_1(t) &= -y_1(t) + z_1^2(t) z_2(t) + Y_1[y_1(t), y_1(t-\tau)],\\
\dot{z}_1(t) &= z_1(t) + y_1(t) z_1(t) + z_1(t) Z_1[y_1(t), y_1(t-\tau)],\\
\dot{z}_2(t) &= -3z_2(t) - 2y_1(t) z_2(t) + z_2(t) Z_2[y_1(t), y_1(t-\tau)],
\end{aligned} \tag{6.2.9}$$

where Y_1, Z_1, and Z_2 are nonlinear perturbations, $\tau = \text{const} > 0$.

Let us examine the unperturbed motion $y_1 = z_1 = z_2 = 0$ of system (6.2.9) for asymptotic stability with respect to y_1.

To this end, following (6.2.7), we take the following expressions as a V-functional and $\boldsymbol{\mu}$-function (ε_1 and ε_2 are some positive constants)

$$\begin{gathered} V[\boldsymbol{x}_t] = V^*[y_{1t}, \mu_{1t}[\boldsymbol{x}_t]] = \tfrac{1}{2}y_1^2(t) + \tfrac{1}{2}y_1(t)\mu_1(t) + \tfrac{3}{4}\mu_1^2(t) \\ + \varepsilon_1 \int_{-\tau}^{0} y_1^2(t+\theta)\,d\theta + \varepsilon_2 \int_{-\tau}^{0} \mu_1^2(t+\theta)\,d\theta, \\ \mu_1 = z_1^2 z_2. \end{gathered} \tag{6.2.10}$$

Taking into account the equalities

$$\int_{-\tau}^{0} y_1^2(t+\theta)\,d\theta = \int_{t-\tau}^{t} y_1^2(s)\,ds$$

and the analogous equalities for μ_1, we arrive at the following relationships:

$$\begin{aligned} \dot V[\boldsymbol{x}_t] &= -y_1^2(t) - \mu_1^2(t) + \varepsilon_1[y_1^2(t) - y_1^2(t-\tau)] + \varepsilon_2[\mu_1^2(t) - \mu_1^2(t-\tau)] \\ &+ y_1(t)Y_1 + \tfrac{1}{2}\mu_1(t)\{Y_1 + [y_1(t) + 3\mu_1(t)](2Z_1 + Z_2)\}. \end{aligned}$$

As a result, constants ε_1 and ε_2 can be chosen so that, for a sufficiently small H, the estimates

$$V[\boldsymbol{x}_t] \geq l_1[y_1^2(t) + \mu_1^2(t)], \quad \dot V[\boldsymbol{x}_t] \leq -l_2[y_1^2(t) + \mu_1^2(t)]$$

are valid in domain (6.2.8), with $l_1 > 0$, $l_2 > 0$ constants.

Consequently, the V-functional and $\boldsymbol{\mu}$-function considered satisfy conditions of Corollary 6.2.1, and the unperturbed motion $y_1 = z_1 = z_2 = 0$ of system (6.2.9) is uniformly asymptotically stable with respect to y_1.

2. Consider *Euler's dynamical equations*

$$A\dot x_1 = (B - C)x_2x_3 + u_1 \quad (ABC, 123), \tag{6.2.11}$$

which describe the angular motion of a solid (spacecraft) about its center of mass. Suppose that the control rules are formed as a linear feedback with retardation of the state

$$\begin{gathered} u_i = \alpha_i x_i(t) + \beta_i(t)x_i(t-\tau) \quad (i = 1, 2), \quad u_3 \equiv 0, \\ \alpha_i, \tau = \text{const}, \quad \tau < 0. \end{gathered} \tag{6.2.12}$$

It is required to find the domain of variation for α_i, $\beta_i(t)$ in which the equilibrium position $x_1 = x_2 = x_3 = 0$ of system (6.2.11), (6.2.12) is uniformly asymptotically stable with respect to $\boldsymbol{y} = (x_1, x_2)$. This case corresponds to dampening the angular velocity of the solid with respect to two given principal central axes of its ellipsoid of inertia. We assume that the perturbed motion of the solid is caused by an initial perturbation of its angular velocity; the perturbation is characterized by an arbitrary vector function $\varphi_0(\theta)$ that is continuous in the interval $[t_0 - \tau, t_0]$.

Suppose that $C < A, B$ and consider the functional

$$V[\boldsymbol{y}_t] = \sum_{i=1}^{2}[\gamma_i x_i^2(t) + 2\varepsilon_i \int_{-\tau}^{0} x_i^2(t+\theta)\, d\theta)], \tag{6.2.13}$$

$$\varepsilon_i = \text{const} > 0 \quad (i = 1, 2), \quad \gamma_1 = B^{-1}(A - C), \quad \gamma_2 = A^{-1}(B - C),$$

whose derivative satisfies the relationship

$$\dot{V}[\boldsymbol{y}_t] = -2\sum_{i=1}^{2}[-(\varepsilon_i + \gamma_i\alpha_{i1})x_i^2(t) - \gamma_i\beta_{i1}(t)x_i(t)x_i(t-\tau) + \varepsilon_i x_i^2(t-\tau)],$$

$$\alpha_{11} = A^{-1}\alpha_1, \quad \alpha_{21} = B^{-1}\alpha_2, \quad \beta_{11} = A^{-1}\beta_1, \quad \beta_{21} = B^{-1}\beta_2.$$

The bracketed quadratic forms will be positive-definite on condition that

$$-(\varepsilon_i + \gamma_i\alpha_{i1})\varepsilon_i - {}^{1}\!/_{4}\gamma_i^2\beta_{i1}^2 \geq \varepsilon_i^* = \text{const} > 0 \quad (i = 1, 2).$$

Since the maximum in the left-hand side of the inequalities obtained is attained at $\varepsilon_i = -{}^{1}\!/_{2}\gamma_i\alpha_{i1}$, we have the relationships

$$\alpha_i < 0, \quad |\beta_{i1}(t)| < -\alpha_{i1} \quad (i = 1, 2). \tag{6.2.14}$$

Thus, if conditions (6.2.14) are satisfied, the inequality $\dot{V}[\boldsymbol{y}_t] \leq -c(\|\boldsymbol{y}(t)\|)$ is valid and the V-functional obeys the conditions of the fourth part of Theorem 6.2.1.

6.3 Using the Method of Lyapunov Functions

6.3.1 The Structure of Lyapunov Functions to be Used in Problems of Stability with Respect to Part of the Variables

As indicated earlier, the seminal idea of using the method of Lyapunov V-functions of a *finite* number of variables in problems of stability for systems with delay was put forward by Razumikhin and Krasovskii in 1956.

Here we consider the application of this method to the problem of partial stability: asymptotic $\boldsymbol{y}$-stability of the unperturbed motion $\boldsymbol{x} = \mathbf{0}$ of system (6.1.2).

In this case we shall seek a Lyapunov V-function in the form

$$V(t, \boldsymbol{x}) \equiv V^*(t, \boldsymbol{y}, \boldsymbol{\mu}(\boldsymbol{x})), \tag{6.3.1}$$

where $\boldsymbol{\mu}(\boldsymbol{x})$ is a continuous vector function of $\boldsymbol{x}$ such that $\boldsymbol{\mu}(\mathbf{0}) = \mathbf{0}$. A definite choice of such a $\boldsymbol{\mu}$-function depends on specificity of the problem under study.

As a result, we can use V^*-functions that are sign-definite with respect to *all* the variables (with respect to $\boldsymbol{y}$, $\boldsymbol{\mu}$) and, in particular, *quadratic* V^*-functions.

6.3.2 Conditions for Asymptotic Stability with Respect to Part of the Variables

Suppose that the right-hand sides of system (6.1.2) are bounded along trajectories originating from a sufficiently small neighborhood of the position $\boldsymbol{x} = \mathbf{0}$.

Theorem 6.3.1. *Suppose that for system* (6.1.2), *along with a V-function, it is possible to specify a continuous vector* $\boldsymbol{\mu}(\boldsymbol{x})$*-function,* $\boldsymbol{\mu}(\mathbf{0}) = \mathbf{0}$ *such that in the domain*

$$t \geq 0, \quad (\|\boldsymbol{y}\|^2 + \|\boldsymbol{\mu}(\boldsymbol{x})\|^2)^{1/2} \leq H, \quad \|\boldsymbol{z}\| < \infty, \tag{6.3.2}$$

the following conditions are satisfied

(1) $a\big((\|\boldsymbol{y}\|^2+\|\boldsymbol{\mu}(\boldsymbol{x})\|^2)^{1/2}\big) \leq V(t,\boldsymbol{x}) = V^*\big(t,\boldsymbol{y},\boldsymbol{\mu}(\boldsymbol{x})\big) \leq b\big((\|\boldsymbol{y}\|^2+\|\boldsymbol{\mu}(\boldsymbol{x})\|^2)^{1/2}\big)$;

(2) $\dot{V}[t,\boldsymbol{x}_t] \leq -c\big((\|\boldsymbol{y}(t)\|^2 + \|\boldsymbol{\mu}[\boldsymbol{x}(t)]\|^2)^{1/2}\big)$

along integral curves of system (6.1.2) *that meet the condition*

$$V^*\big(\theta,\boldsymbol{y}(\theta),\boldsymbol{\mu}(\theta)\big) \leq f\Big(V^*\big(t,\boldsymbol{y}(t),\boldsymbol{\mu}(t)\big)\Big) \tag{6.3.3}$$

for all $t-\tau \leq \theta < t$, $t \geq T > t_0 + \tau$ *and for some continuous monotone increasing function* $f(r) > r > 0$, $f(0) = 0$.

Then the unperturbed motion $\boldsymbol{x} = \mathbf{0}$ *of system* (6.1.2) *is uniformly asymptotically* $\boldsymbol{y}$*-stable.*

Remark 6.3.1. Theorem 6.3.1 is an extension of theorems by Razumikhin [1956] and Krasovskii [1956], [1959] (to the problem of partial stability) and Rumyantsev [1957a], [1970] (to the systems of functional-differential equations).

Example 6.3.1. Let us consider (as we have done in Example 6.2.1) the problem of finding the domain of variation for α_i and $\beta_i(t)$ in which the equilibrium position $x_1 = x_2 = x_3 = 0$ of system (6.2.11), (6.2.12) is uniformly asymptotically stable with respect to $\boldsymbol{y} = (x_1, x_2)$.

We assume that $C < A, B$ and take V-function in the form

$$\begin{aligned} V(\boldsymbol{y}) &= \gamma_1 x_1^2(t) + \gamma_2 x_2^2(t), \\ \gamma_1 &= B^{-1}(A-C), \quad \gamma_2 = A^{-1}(B-C). \end{aligned} \tag{6.3.4}$$

The derivative $\dot{V}$ is determined as follows:

$$\begin{gathered} \dot{V}[\boldsymbol{y}_t] = 2\sum_{i=1}^{2} \gamma_i[\alpha_{i1}x_i^2(t) + \beta_{i1}x_i(t)x_i(t-\tau)], \\ \alpha_{11} = A^{-1}\alpha_1, \quad \alpha_{21} = B^{-1}\alpha_2, \\ \beta_{11} = A^{-1}\beta_1, \quad \beta_{21} = B^{-1}\beta_2. \end{gathered}$$

When the inequalities

$$|x_i(\theta)| \leq q|x_i(t)|, \quad q = \text{const} > 1 \tag{6.3.5}$$

hold true, we get

$$\dot{V} \leq 2\sum_{i=1}^{2} \gamma_i[\alpha_{i1} + q^2|\beta_{i1}|]x_i^2(t).$$

Taking $f(r) = q^2 r$ in (6.3.3), one can choose constant q so that if the strengthened conditions (6.2.14) are met, V-function (6.3.4) satisfies all the conditions of Theorem 6.3.1 (when $\boldsymbol{\mu} = \mathbf{0}$).

We point out that, when using V-function (6.3.4) but not V-functional (6.2.13), one can assume that delay τ is *variable.*

6.4 The Stability of Linear Delayed Systems with Respect to Part of the Variables

6.4.1 The Linear Delayed System. Formulation of the Problem

Consider a linear system of differential equations of perturbed motion with constant coefficients and a constant delay of the argument

$$\frac{dx_i(t)}{dt} = \sum_{j=1}^{n} A_{ij}x_j(t) + \sum_{j=1}^{n} A'_{ij}x_j(t-\tau) \qquad (i = 1, \ldots, n),$$

$$\boldsymbol{x}^{\mathsf{T}} = (x_1, \ldots, x_n) = (y_1, \ldots, y_m, z_1, \ldots, z_p) = (\boldsymbol{y}^{\mathsf{T}}, \boldsymbol{z}^{\mathsf{T}}),$$

$$m > 0, \qquad p \geq 0, \qquad n = m + p.$$

In $\boldsymbol{y}, \boldsymbol{z}$ variables the system takes the form

$$\frac{dy_i(t)}{dt} = \sum_{k=1}^{m} a_{ik}y_k(t) + \sum_{l=1}^{p} b_{il}z_l(t) + \sum_{k=1}^{m} a'_{ik}y_k(t-\tau) + \sum_{l=1}^{p} b'_{il}z_l(t-\tau),$$

$$\frac{dz_j(t)}{dt} = \sum_{k=1}^{m} c_{jk}y_k(t) + \sum_{l=1}^{p} d_{jl}z_l(t) + \sum_{k=1}^{m} c'_{jk}y_k(t-\tau) + \sum_{l=1}^{p} d'_{jl}z_l(t-\tau)$$

$$(i = 1, \ldots, m;\ j = 1, \ldots, p), \tag{6.4.1}$$

where a_{ik}, a'_{ik}, b_{il}, b'_{il}, c_{jk}, c'_{jk}, d_{jl}, d'_{jl}, and τ are constants.

We shall study the problem of stability (asymptotic stability) of the unperturbed motion $\boldsymbol{x} = \mathbf{0}$ of system (6.4.1) with respect to variables $y_1, \ldots, y_m$.

We note that for the linear system (6.4.1) the *global Lipschitz condition* is satisfied, which guarantees, for each t_0 and for any continuous initial function $\varphi_0(\theta)$, the existence of a unique solution defined for all $t \geq t_0$.

6.4.2 Possible Number of Asymptotically Stable Variables

The behavior of the variables that characterize the state of system (6.4.1) is determined by the roots of its *characteristic quasi-polynomial*

$$\det(A^* + e^{-\lambda\tau} A'^* - \lambda I_n) = 0,$$
$$A^* = (A_{ij}), \qquad A'^* = (A'_{ij}) \qquad (i, j = 1, \ldots, n).$$

It is known that only a *finite* number of roots with nonnegative real parts can be among the roots of the characteristic quasi-polynomial. We denote this number by N. Let us demonstrate how the number R of the variables with respect to which the unperturbed motion $\boldsymbol{x} = \mathbf{0}$ of system (6.4.1) is asymptotically stable is related to numbers n and N.

Lemma 6.4.1. 1. *If $N \geq n$, the unperturbed motion $\boldsymbol{x} = \mathbf{0}$ of system* (6.4.1) *cannot be stable with respect to any variable.*

2. *If $N < n$, the number R of asymptotically stable variables satisfies the inequality $R \leq n - N$.*

Proof. Let $\lambda_1, \ldots, \lambda_N$ be the roots of the characteristic quasi-polynomial of system (6.4.1) with *nonnegative* real parts. The general solution of system (6.4.1) can be represented in the form (Shimanov [1965], Hale [1977])

$$\boldsymbol{x}(t + \theta) = \sum_{j=1}^{N} \xi_j \boldsymbol{x}_j(\theta) + \boldsymbol{\eta}(t + \theta), \tag{6.4.2}$$

where $\boldsymbol{x}_j(\theta)$ are the eigenvectors of system (6.4.1) corresponding to the roots $\lambda_1, \ldots, \lambda_N$; $\|\boldsymbol{\eta}(t + \theta)\| \leq Ce^{-\beta t}\|\varphi_0(\theta)\|$; C and β are positive numbers independent of the choice of the initial function $\varphi_0(\theta)$; ξ_j are the solutions of the canonical system of ordinary differential equations constructed on the roots $\lambda_1, \ldots, \lambda_N$.

Let us rewrite (6.4.2) in matrix form: $\boldsymbol{x}(t + \theta) = L\boldsymbol{\xi} + \boldsymbol{\eta}(t + \theta)$, where L is a matrix of dimensions $n \times N$ and $\boldsymbol{\xi}$ is a vector of dimensions $N \times 1$.

1. Suppose that $N \geq n$. In this case, the asymptotically stable coordinates can occur in the general solution (6.4.2) if and only if matrix L contains zero rows. However, in the case $N \geq n$, matrix L cannot contain any zero row. Indeed, any n vectors from $\boldsymbol{x}_j(\theta)$ $(j = 1, \ldots, N)$ are linearly independent and, if there are zero rows in matrix L, there are no minors of order n different from zero in this matrix. This contradiction proves the first part of the lemma.

2. Suppose that $N < n$. Now the linear independence of N vectors $\boldsymbol{x}_1(\theta), \ldots, \boldsymbol{x}_N(\theta)$ does not contradict the possibility of the existence of zero rows in matrix L, but their number is limited by the value $n - N$. The statement is proved. □

6.4.3 An Auxiliary Linear System of Equations

Now we demonstrate that the problem of stability (asymptotic stability) of the unperturbed motion of system (6.4.1) with respect to $y_1, \ldots, y_m$ can be reduced to the problem of stability (asymptotic stability) of a specially constructed auxiliary linear system of the same type as system (6.4.1) with respect to all the variables. The dimension of this auxiliary system does not exceed that of the original system (6.4.1).

Consider the vectors $\boldsymbol{b}_i = (b_{i1}, \ldots, b_{ip})^\top$ and $\boldsymbol{b}'_i = (b'_{i1}, \ldots, b'_{ip})^\top$ $(i = 1, \ldots, m)$ composed of the corresponding coefficients of system (6.2.1). Without loss of generality, we assume that the first m_1 vectors $\boldsymbol{b}_1, \ldots, \boldsymbol{b}_{m_1}$ $(m_1 \leq m)$ and the first m_2 vectors $\boldsymbol{b}'_1, \ldots, \boldsymbol{b}'_{m_2}$ $(m_2 \leq m)$ constitute a linearly independent basis of the system of vectors $\boldsymbol{b}_j$, $\boldsymbol{b}'_j$ $(j = 1, \ldots, m)$.

To construct the auxiliary system of equations, we introduce the new variables

$$\mu_i = \sum_{l=1}^{p} b_{il} z_l, \quad \mu_{m_1+j} = \sum_{l=1}^{p} b'_{jl} z_l \quad (i = 1, \ldots, m_1;\ j = 1, \ldots, m_2). \tag{6.4.3}$$

With the new variables thus introduced, two cases are possible.

The first case. The following system can be formed:

$$\begin{aligned}
\frac{dy_i(t)}{dt} &= \sum_{k=1}^{m} a_{ik} y_k(t) + \sum_{l=1}^{m_1+m_2} \alpha_{il} \mu_l(t) + \sum_{k=1}^{m} a'_{ik} y_k(t-\tau) + \sum_{l=1}^{m_1+m_2} \alpha'_{il} \mu_l(t-\tau),\\
\frac{d\mu_j(t)}{dt} &= \sum_{k=1}^{m} a^*_{jk} y_k(t) + \sum_{l=1}^{m_1+m_2} \alpha^*_{jl} \mu_l(t) + \sum_{k=1}^{m} a'^*_{jk} y_k(t-\tau) + \sum_{l=1}^{m_1+m_2} \alpha'^*_{jl} \mu_l(t-\tau)\\
&\qquad (i = 1, \ldots, m;\ j = 1, \ldots, m_1 + m_2),
\end{aligned} \tag{6.4.4}$$

where a^*_{jk}, a'^*_{jk}, α_{il}, α'_{il}, α^*_{jl}, and α'^*_{jl} are constants. The behavior of variables $y_1, \ldots, y_m$, with respect to which we study the problem of stability of the unperturbed motion $\boldsymbol{x} = \boldsymbol{0}$ of system (6.4.1), is completely determined by system (6.4.4). This system is called the ***μ*-*system*** relative to the original system (6.4.1).

The second case. Suppose that all $m_1 + m_2$ equalities in the second group of equations of system (6.4.4) cannot be simultaneously satisfied, i.e., system (6.4.1) has the form

$$\begin{aligned}
\frac{dy_i(t)}{dt} &= \sum_{k=1}^{m} a_{ik} y_k(t) + \sum_{l=1}^{m_1+m_2} \alpha_{il} \mu_l(t) + \sum_{k=1}^{m} a'_{ik} y_k(t-\tau) + \sum_{l=1}^{m_1+m_2} \alpha'_{il} \mu_l(t-\tau),\\
\frac{d\mu_j(t)}{dt} &= \sum_{k=1}^{m} a^*_{jk} y_k(t) + \sum_{l=1}^{p} b^*_{jl} z_l(t) + \sum_{k=1}^{m} a'^*_{jk} y_k(t-\tau) + \sum_{l=1}^{p} b'^*_{jl} z_l(t-\tau),\\
\frac{dz_s(t)}{dt} &= \sum_{k=1}^{m} c_{sk} y_k(t) + \sum_{l=1}^{p} d_{sl} z_l(t) + \sum_{k=1}^{m} c'_{sk} y_k(t-\tau) + \sum_{l=1}^{p} d'_{sl} z_l(t-\tau)\\
&\qquad (i = 1, \ldots, m;\ j = 1, \ldots, m_1 + m_2;\ s = 1, \ldots, p),
\end{aligned} \tag{6.4.5}$$

where b^*_{jl} and b'^*_{jl} are constants.

In this case, we introduce the new variables once more:

$$\mu_{m_1+m_2+i} = \sum_{l=1}^{p} b^*_{il} z_l, \quad \mu_{2(m_1+m_2)+i} = \sum_{l=1}^{p} b'^*_{il} z_l \quad (i = 1, \ldots, m_1 + m_2).$$

Again system (6.4.1) will reduce to a system of type (6.4.4) or (6.4.5).

Suppose that, by successively repeating the procedure of introducing new variables, we get $p = n - m$ linearly independent new variables $\mu_1, \ldots, \mu_p$. Then, further entering new variables loses its meaning, since each successive newly introduced variable can be represented as a linear combination of $\mu_1, \ldots, \mu_p$. Consequently, the maximum order of the $\boldsymbol{\mu}$-system equals n.

Thus the following lemma is true.

Lemma 6.4.2. *By introducing no more than p groups of new variables, for system* (6.4.1), *it is possible to construct an auxiliary $\boldsymbol{\mu}$-system whose order does not exceed that of the original system. The equations and variables of the $\boldsymbol{\mu}$-system split into subsets so that each equation of the corresponding subset except those of the last one can involve variables from all the previous and the single following subsets and does not include variables from the subsequent subsets.*

6.4.4 The Order of an Auxiliary System and its Spectrum

Let us represent system (6.4.1) in matrix form

$$\begin{aligned} \dot{\boldsymbol{y}}(t) &= A\boldsymbol{y}(t) + A'\boldsymbol{y}(t-\tau) + B\boldsymbol{z}(t) + B'\boldsymbol{z}(t-\tau), \\ \dot{\boldsymbol{z}}(t) &= C\boldsymbol{y}(t) + C'\boldsymbol{y}(t-\tau) + D\boldsymbol{z}(t) + D'\boldsymbol{z}(t-\tau), \end{aligned}$$

where A, A', B, B', C, C', D, and D' are constant matrices of appropriate sizes. We introduce the notations $B_0 = (B^{\mathsf{T}}, B'^{\mathsf{T}})$, $D_0 = D^{\mathsf{T}}$, and $D_1 = D'^{\mathsf{T}}$ and consider the matrices

$$K_0 = B_0, \quad K_i = (D_0 K_{i-1}, D_1 K_{i-1}) \qquad (i = 1, \ldots, p).$$

Lemma 6.4.3. 1. *Let s be the minimum number such that* $\operatorname{rank} K_{s-1} = \operatorname{rank} K_s$. *Then the order of the $\boldsymbol{\mu}$-system equals $m + \operatorname{rank} K_{s-1}$.*

2. *The set of roots of the characteristic quasi-polynomial of the $\boldsymbol{\mu}$-system is a subset* (*in a particular case, coinciding with the whole set*) *of the characteristic quasi-polynomial roots of system* (6.4.1).

Proof. 1. Let $\operatorname{rank} K_{s-1} = h$. We denote h linearly independent vector columns of matrix K_{s-1} by $\boldsymbol{g}_i$, $\boldsymbol{g}_i = (g_{i1}, \ldots, g_{ip})^{\mathsf{T}}$ $(i = 1, \ldots, h)$. The new variables $\mu_i = g_{i1}z_1 + \cdots + g_{ip}z_p$ $(i = 1, \ldots, h)$ correspond to these linearly independent vectors. Since $\operatorname{rank} K_{s-1} = \operatorname{rank} K_s$, it is meaningless to further introduce new variables in constructing the $\boldsymbol{\mu}$-system for (6.4.1). Indeed, each of the subsequent newly introduced variables can be represented as a linear combination of variables $\mu_1, \dots, \mu_h$. This means that the order of the $\boldsymbol{\mu}$-system equals $m + h$.

2. Consider $n \times n$ matrix G

$$G = \begin{pmatrix} I_m & 0 \\ 0 & G_1 \\ 0 & G_* \end{pmatrix}, \quad G_1 = (g_{ij}) \qquad (i = 1, \dots, h;\ j = 1, \dots, p), \qquad (6.4.6)$$

where I_m is the identity matrix of order m and G_* is an arbitrary constant $(n-m-h) \times p$ matrix such that matrix G is nonsingular. Then, by the change of variables $\boldsymbol{w} = G\boldsymbol{x}$ in system (6.4.1), we obtain the system of equations $\dot{\boldsymbol{w}} = GA^*G^{-1}\boldsymbol{w}(t) + GA'^*G^{-1}\boldsymbol{w}(t-\tau)$. The first $m + h$ equations of this system form the $\boldsymbol{\mu}$-system for (6.4.1).

The following equalities are valid

$$\begin{aligned} &\det(GA^*G^{-1} + e^{-\lambda\tau}GA'^*G^{-1} - \lambda I_n) \\ &\quad = \det[GA^*G^{-1} + e^{-\lambda\tau}GA'^*G^{-1} - G(\lambda I_n)G^{-1}] \\ &\quad = \det[G(A^* + e^{-\lambda\tau}A'^* - \lambda I_n)G^{-1}] = \det(A^* + e^{-\lambda\tau}A'^* - \lambda I_n). \end{aligned}$$

Consequently, the set of characteristic quasi-polynomial roots of the $\boldsymbol{\mu}$-system is a subset of characteristic quasi-polynomial roots of the original system (6.4.1). The statement is proved. □

6.4.5 The Form of the Characteristic Quasi-Polynomial of an Auxiliary System

Suppose that $\operatorname{rank} K_{s-1} = h$, where s is the minimum number such that $\operatorname{rank} K_{s-1} = \operatorname{rank} K_s$, and consider that matrices G_i $(i = 1, \dots, 5)$ have the same structure as matrices L_i $(i = 1, \dots, 5)$ introduced in Subsection 1.1.4.

Lemma 6.4.4. *The equation*

$$\det(G_4A^*G_5 + e^{-\lambda\tau}G_4A'^*G_5 - \lambda I_{m+h}) = 0, \qquad (6.4.7)$$

which is equivalent to the equation

$$\det \begin{bmatrix} A + e^{-\lambda\tau}A' - \lambda I_m & (B + e^{-\lambda\tau}B')G_3 \\ G_1(C + e^{-\lambda\tau}C') & G_1DG_3 + e^{-\lambda\tau}G_1DG_3 - \lambda I_h \end{bmatrix} = 0,$$

is the characteristic quasi-polynomial of the $\boldsymbol{\mu}$-system.

Proof. The auxiliary $\boldsymbol{\mu}$-system for the original system (6.4.1) comprises (in the case $\operatorname{rank} K_{s-1} = h$) the first $m + h$ equations of the system

$$\dot{\boldsymbol{w}}(t) = GA^*G^{-1}\boldsymbol{w}(t) + GA'^*G^{-1}\boldsymbol{w}(t-\tau),$$

where G is matrix (6.4.6). Therefore, coefficients g_{ij} and g'_{ij} $(i, j = 1, \dots, m + h)$ of the auxiliary $\boldsymbol{\mu}$-system

$$\dot{\xi}_i(t) = \sum_{j=1}^{m+h} g_{ij}\xi_j(t) + \sum_{j=1}^{m+h} g'_{ij}\xi_j(t-\tau) \qquad (i = 1, \dots, m + h)$$

are the coefficients with indices $i = 1, \ldots, m+h$ of matrices $GA^*G^{-1} = (g_{ij})$ and $GA'^*G^{-1} = (g'_{ij})$, respectively.

The following equalities are valid:

$$\bar{g}_{ij} = \frac{(-1)^{i+j} \det G_{ji}}{\det G} \qquad (i,j = 1, \ldots, n),$$

where matrix G_{ji} is obtained from G by deleting its jth row and ith column. Following the scheme of proof for Theorem 1.1.1, we conclude that $G_5 = (\bar{g}_{ij})$ $(i = 1, \ldots, n;\ j = 1, \ldots, m+h)$.

Consequently,

$$G_4 A^* G_5 = (g_{ij}), \quad G_4 A'^* G_5 = (g'_{ij}) \qquad (i,j = 1, \ldots, m+h),$$

and the $\boldsymbol{\mu}$-system for (6.4.1) is formed by the equations

$$\dot{\boldsymbol{\xi}}(t) = G_4 A^* G_5 \boldsymbol{\xi}(t) + G_4 A^* G_5 \boldsymbol{\xi}(t-\tau).$$

Taking into account the structure of matrices G_4 and G_5, we can also write the auxiliary $\boldsymbol{\mu}$-system in the equivalent form

$$\begin{aligned}
\dot{\boldsymbol{y}}(t) &= A\boldsymbol{y}(t) + A'\boldsymbol{y}(t-\tau) + BG_3\boldsymbol{\mu}(t) + B'G_3\boldsymbol{\mu}(t-\tau),\\
\dot{\boldsymbol{\mu}}(t) &= G_1C\boldsymbol{y}(t) + G_1C'\boldsymbol{y}(t-\tau) + G_1DG_3\boldsymbol{\mu}(t) + G_1D'G_3\boldsymbol{\mu}(t-\tau).
\end{aligned}$$

This means that equation (6.4.7) and the equation equivalent to it are indeed the characteristic quasi-polynomial of the $\boldsymbol{\mu}$-system for the original system (6.4.1). The statement is proved. □

6.4.6 The Conditions for Asymptotic Stability with Respect to Part of the Variables

Theorem 6.4.1. *For the unperturbed motion of system* (6.4.1) *to be asymptotically stable with respect to* $y_1, \ldots, y_m$, *it is sufficient that the zero solution of the auxiliary* $\boldsymbol{\mu}$*-system be Lyapunov asymptotically stable* (*all the roots of equation* (6.4.7) *have negative real parts*). *This condition is necessary and sufficient in the cases*

(1) $(B = 0) \wedge \{[(D = 0) \vee (B'D = 0)] \vee [(D' = 0) \vee (B'D' = 0)]\}$,

(2) $(B' = 0) \wedge \{[(D = 0) \vee (BD = 0)] \vee [(D' = 0) \vee (BD' = 0)]\}$

(*where the zeros are understood as the zero matrices of appropriate sizes*).

Proof. The *sufficiency* is obvious.

Necessity. First consider case (2).

The asymptotic stability of the unperturbed motion $\boldsymbol{y} = \mathbf{0}$, $\boldsymbol{z} = \mathbf{0}$ of system (6.4.1) is known to be *exponential* asymptotic stability. As can be verified by directly integrating the first m equations of system (6.4.1), this necessitates that the following inequalities be satisfied on trajectories of the system

$$|\mu_i(t)| = \left|\sum_{l=1}^{p} b_{il} z_l(t)\right| \leq A_{i1} e^{-\alpha_{i1}(t-t_0)}, \quad t \geq t_0 \quad (i = 1, \ldots, m),$$

where A_{i1} and α_{i1} are positive constants.

Therefore, for the unperturbed motion $\boldsymbol{y} = \mathbf{0}$, $\boldsymbol{z} = \mathbf{0}$ of system (6.4.1) to be asymptotically stable with respect to $y_1, \ldots, y_m$, it is necessary that this motion also be asymptotically stable with respect to variables $\mu_1, \ldots, \mu_m$. In the case when it is possible to segregate $\boldsymbol{\mu}$-system (6.4.4) from system (6.4.1), the necessity of the conditions of the theorem is proved.

When system (6.4.1) is not reduced to the form (6.4.4) (and, consequently, has the form (6.4.5) with $b'^{*}_{jl} = 0$ ($j = 1, \ldots, m_1+m_2$; $l = 1, \ldots, p$)), we repeat the above reasoning for system (6.4.5). Since the reduction to a $\boldsymbol{\mu}$-system is always possible, the necessity of the conditions of the theorem in case (2) is proved.

Case (1) is handled similarly.

The theorem is proved. □

Corollary 6.4.1. *For the unperturbed motion of system* (6.4.1) *to be stable with respect to* $y_1, \ldots, y_m$, *it is sufficient and, in cases* (1) *and* (2), *it is necessary and sufficient that the zero solution of the* $\boldsymbol{\mu}$*-system be Lyapunov stable.*

Example 6.4.1.

1. Let the equations of the perturbed motion (6.4.1) have the form

$$\begin{aligned} \dot{y}_1(t) &= -y_1(t-\tau) + z_1(t-\tau) - 2z_2(t-\tau), \\ \dot{z}_1(t) = 4y_1(t) + z_1(t-\tau), &\quad \dot{z}_2(t) = 2y_1(t) + z_1(t-\tau) - z_2(t-\tau). \end{aligned} \tag{6.4.8}$$

The auxiliary $\boldsymbol{\mu}$-system in this case comprises two equations

$$\dot{y}_1(t) = -y_1(t-\tau) + \mu_1(t-\tau), \quad \dot{\mu}_1(t) = -\mu_1(t-\tau). \tag{6.4.9}$$

In the interval $\tau \in (0, \pi/2)$, all the roots of the characteristic quasi-polynomial $\Phi(\lambda) = (e^{-\lambda\tau} + \lambda)^2 = 0$ of the auxiliary $\boldsymbol{\mu}$-system (6.4.9) have negative real parts. Consequently, when $\tau \in (0, \pi/2)$, the unperturbed motion $y_1 = z_1 = z_2 = 0$ is asymptotically stable with respect to y_1.

2. Let the equations of the unperturbed motion (6.4.1) have the form

$$\dot{y}_1(t) = -y_1(t) + 0.1z_1(t) - z_1(t-\tau), \quad \dot{z}_1(t) = z_1(t). \tag{6.4.10}$$

Now we show that, if the assumptions (1) and (2) of Theorem 6.4.1 do not hold, the sufficient conditions for asymptotic stability with respect to $y_1, \ldots, y_m$, generally speaking, will not be necessary. Indeed, system (6.4.10) is, at the same time, the $\boldsymbol{\mu}$-system, whose zero solution is unstable (with respect to both the variables) for any $\tau > 0$. However, there is a τ^* for which the unperturbed motion of system (6.4.10) is asymptotically stable with respect to y_1. Bearing in mind that $z_1(t) = z_1(t_0)e^{(t-t_0)}$, we can choose this particular value of τ^* from the condition

$$0.1z_1(t) - z_1(t-\tau^*) = z_1(t_0)e^{(t-t_0)}(0.1 - e^{-\tau^*}) = 0.$$

Thus, for $z_1(t_0) \neq 0$, we arrive at the equation $0.1 - e^{-\tau^*} = 0$ relative to τ^*, and by solving it, we find $\tau^* = \ln 10$. This means that when $\tau^* = \ln 10$, the unperturbed motion of system (6.4.10) is asymptotically stable with respect to y_1, though among the roots of the characteristic quasi-polynomial of the $\boldsymbol{\mu}$-system there are those with positive real parts.

3. Suppose that three neutral subsystems $\dot{y}_1 = 0$, $\dot{z}_1 = 0$ $(i = 1, 2)$ are interacting by linear links among them so that the consolidated system has the form

$$\begin{gathered} \dot{y}_1(t) = b_{11} z_1(t) + b_{12} z_2(t), \\ \dot{z}_1(t) = c_{11} y_1(t) + c_{11}' y_1(t-\tau), \quad \dot{z}_1(t) = c_{21} y_1(t) + c_{21}' y_1(t-\tau). \end{gathered} \tag{6.4.11}$$

Here b_{11}, b_{12}, c_{11}, c_{11}', c_{21}, c_{21}', and $\tau > 0$ are constants. Let us find the domain of variation of these constants where the equilibrium position $y_1 = z_1 = z_2 = 0$ of the consolidated system (6.4.11) is asymptotically stable with respect to y_1.

Entering the new variable $\mu_1(t) = b_{11} z_1(t) + b_{12} z_2(t)$ into the original system (6.4.11), we construct the $\boldsymbol{\mu}$-system

$$\begin{gathered} \dot{y}_1(t) = \mu_1(t), \quad \dot{\mu}_1(t) = c_1^* y_1(t) + c_2^* y_1(t-\tau), \\ c_1^* = b_{11} c_{11} + b_{12} c_{11}', \quad c_2^* = b_{11} c_{21} + b_{12} c_{21}'. \end{gathered}$$

Under the conditions (Kolmanovskii and Nosov [1981])

$$c_1^* < 0, \quad c_2^* > 0, \quad c_1^* > -2c_2^* > 2c_1^*, \quad 0 < \tau < \sqrt{c_2^*/(2c_2^* - c_1^*)},$$

all the roots of the characteristic quasi-polynomial of the $\boldsymbol{\mu}$-system have negative real parts. Consequently, in this case the equilibrium position $y_1 = z_1 = z_2 = 0$ of system (6.4.11) is asymptotically stable with respect to y_1.

6.4.7 The Stability of Linear Systems with Delay of a Neutral Type with Respect to Part of the Variables

Consider a linear system of differential equations of perturbed motion with delay of *neutral type*

$$\begin{gathered} \frac{dx_i(t)}{dt} = \sum_{j=1}^{n} A_{ij} x_j(t) + \sum_{j=1}^{n} A_{ij}' x_j(t-\tau) + \sum_{j=1}^{n} A_{ij}'' \frac{dx_j(t-\tau)}{dt} \quad (i = 1, \ldots, n), \\ \boldsymbol{x}^{\mathsf{T}} = (x_1, \ldots, x_n) = (y_1, \ldots, y_m, z_1, \ldots, z_p) = (\boldsymbol{y}^{\mathsf{T}}, \boldsymbol{z}^{\mathsf{T}}), \\ m > 0, \quad p \geq 0, \quad n = m + p. \end{gathered} \tag{6.4.12}$$

We assume that coefficients A_{ij}, A_{ij}', A_{ij}'', and the delay $\tau > 0$ are constants.

As particular cases of system (6.4.12) we can list the following:
(1) systems with *retarded argument* (for $A_{ij}'' = 0$);
(2) systems of *ordinary* differential equations (for $A_{ij}' = A_{ij}'' = 0$);
(3) systems of *difference* (*discrete*) equations (for $A_{ij} = A_{ij}' = 0$).

These circumstances are quite enough to arouse interest in the class of equations (6.4.12) as such. Besides, there are a number of applied problems that lead to the study of systems of type (6.4.12) (see, for example, Hale [1977]).

In $\boldsymbol{y}$, $\boldsymbol{z}$ variables, system (6.4.12) takes the form

$$\begin{aligned} \frac{dy_i(t)}{dt} &= \sum_{k=1}^{m} a_{ik} y_k(t) + \sum_{l=1}^{p} b_{il} z_l(t) + \sum_{k=1}^{m} a_{ik}' y_k(t-\tau) \\ &\quad + \sum_{l=1}^{p} b_{il}' z_l(t-\tau) + \sum_{k=1}^{m} a_{ik}'' \frac{dy_k(t-\tau)}{dt} + \sum_{l=1}^{p} b_{il}'' \frac{dz_l(t-\tau)}{dt}, \end{aligned}$$

$$\frac{dz_j(t)}{dt} = \sum_{k=1}^{m} c_{jk} y_k(t) + \sum_{l=1}^{p} d_{jl} z_l(t) + \sum_{k=1}^{m} c'_{jk} y_k(t-\tau) \tag{6.4.13}$$
$$+ \sum_{l=1}^{p} d'_{jl} z_l(t-\tau) + \sum_{k=1}^{m} c''_{jk} \frac{dy_k(t-\tau)}{dt} + \sum_{l=1}^{p} d''_{jl} \frac{dz_l(t-\tau)}{dt},$$
$$(i = 1, \ldots, m;\ j = 1, \ldots, p),$$

where a_{ik}, a'_{ik}, a''_{ik}, b_{il}, b'_{il}, b''_{il}, c_{jk}, c'_{jk}, c''_{jk}, d_{jl}, d'_{jl}, d''_{jl}, and $\tau > 0$ are constants.

Provided that $|A''_{ij}| < 1$ $(i, j = 1, \ldots, n)$, for any *absolutely continuous* initial vector function $\varphi_0(\theta)$, $t_0 - \tau \leq \theta \leq t_0$, there exists (Kolmanovskii and Myshkis [1992]) an *absolutely continuous* vector function $\boldsymbol{x}(t) = \boldsymbol{x}(t; t_0, \varphi(\theta))$, $t \geq t_0$, that almost everywhere satisfies equations (6.4.13) and the initial conditions

$$\boldsymbol{x}(t_0 + \theta) = \varphi_0(\theta), \quad \dot{\boldsymbol{x}}(t_0 + \theta) = \dot{\varphi}_0(\theta), \quad t_0 - \tau \leq \theta \leq t_0$$

and at the same time is a solution of system (6.4.13).

To evaluate the deviation of system (6.4.13) from the unperturbed motion $\boldsymbol{x} = \mathbf{0}$, we introduce the following norms:
(1) $\|\boldsymbol{y}\|^*$—the Euclidean norm of the components of vectors $\boldsymbol{y}$ and $\dot{\boldsymbol{y}}$;
(2) $\|\varphi(\theta)\|^* = \sup\{|\varphi_s(\theta)|, |\dot{\varphi}_s(\theta)|\}$ $(t_0 - \tau \leq \theta \leq t_0,\ s = 1, \ldots, n)$.

Definition 6.4.1. The unperturbed motion $\boldsymbol{x} = \mathbf{0}$ of system (6.4.13) is said to be

(1) *stable with respect to* $y_1, \ldots, y_m$ (*$\boldsymbol{y}$-stable*) if for any numbers $\varepsilon > 0$, $t_0 \geq 0$, there is a number $\delta(\varepsilon, t_0)$ such that from $\|\varphi_0(\theta)\|^* < \delta$ it follows that $\|\boldsymbol{y}(t; t_0, \varphi_0(\theta))\|^* \leq \varepsilon$ for $t \geq t_0$, where $\varphi_0(\theta)$ is an arbitrary *absolutely continuous* initial function;
(2) *asymptotically stable with respect to* $y_1, \ldots, y_m$ (*asymptotically $\boldsymbol{y}$-stable*) if it is $\boldsymbol{y}$-stable and, in addition, for any $t_0 \geq 0$ there exists $\Delta(t_0) > 0$ such that a solution $\boldsymbol{x}(t) = \boldsymbol{x}(t; t_0, \varphi_0(\theta))$ with $\|\varphi_0(\theta)\|^* \leq \Delta$ satisfies the condition $\lim \|\boldsymbol{y}(t; t_0, \varphi_0(\theta))\|^* = 0$, $t \to \infty$.

Because system (6.4.13) is autonomous, $\boldsymbol{y}$-stability is *uniform* (i.e., δ in definition (1) does not depend on t_0). At the same time, the peculiarity of system (6.4.13) (as compared to systems (1.1.1) or (6.4.1)) is that asymptotic $\boldsymbol{y}$-stability may be observed and *uniform* asymptotic $\boldsymbol{y}$-stability may be absent (for this issue, see Hale [1977]).

As is the case with systems with delay, for system (6.4.13) one can construct the auxiliary linear $\boldsymbol{\mu}$-system of the same class. The dimension of this auxiliary $\boldsymbol{\mu}$-system will not exceed that of the original system (6.4.13).

To construct the auxiliary $\boldsymbol{\mu}$-system, instead of (6.4.3), we introduce the new variables which are linear independent among the variables

$$\mu_i = \sum_{l=1}^{p} b_{il} z_l, \quad \mu_{m+i} = \sum_{l=1}^{p} b'_{il} z_l, \quad \mu_{2m+i} = \sum_{l=1}^{p} b''_{il} z_l \tag{6.4.14}$$
$$(i = 1, \ldots, m).$$

Omitting all the details of the construction, we cite the final result. To this end, we represent system (6.4.13) in matrix form:

$$\dot{\boldsymbol{y}}(t) = A\boldsymbol{y}(t) + A'\boldsymbol{y}(t-\tau) + A''\dot{\boldsymbol{y}}(t-\tau) + B\boldsymbol{z}(t) + B'\boldsymbol{z}(t-\tau) + B''\dot{\boldsymbol{z}}(t-\tau),$$
$$\dot{\boldsymbol{z}}(t) = C\boldsymbol{y}(t) + C'\boldsymbol{y}(t-\tau) + C''\dot{\boldsymbol{y}}(t-\tau) + D\boldsymbol{z}(t) + D'\boldsymbol{z}(t-\tau) + D''\dot{\boldsymbol{z}}(t-\tau).$$

Let us introduce the designations

$$K_0 = (B^{\mathsf{T}}, B'^{\mathsf{T}}, B''^{\mathsf{T}}),$$
$$K_i = (D^{\mathsf{T}}K_{i-1}, D'^{\mathsf{T}}K_{i-1}, D''^{\mathsf{T}}K_{i-1}) \quad (i = 1, \ldots, p)$$

and, guided by rules from Subsection 6.4.5, compose matrices G_4 and G_5.

Theorem 6.4.2. *Suppose that the zero solution $\boldsymbol{\xi} = \mathbf{0}$ of the auxiliary $\boldsymbol{\mu}$-system*

$$\dot{\boldsymbol{\xi}}(t) = G_4 A^* G_5 \boldsymbol{\xi}(t) + G_4 A'^* G_5 \boldsymbol{\xi}(t-\tau) + G_4 A''^* G_5 \dot{\boldsymbol{\xi}}(t-\tau) \qquad (6.4.15)$$

is Lyapunov stable (asymptotically stable).

Then the unperturbed motion $\boldsymbol{x} = \mathbf{0}$ of the original system (6.4.13) *is stable (asymptotically stable) with respect to $y_1, \ldots, y_m$.*

Proof. Using the proof scheme for Lemma 6.4.4, one can show that the introduction of the new variables of the form (6.4.14) leads to the auxiliary $\boldsymbol{\mu}$-system for the original system (6.4.13) which comprises equations (6.4.15).

Since $\xi_i = y_i$ $(i = 1, \ldots, m)$, the conditions of the theorem ensure that the unperturbed motion $\boldsymbol{x} = \mathbf{0}$ of system (6.4.13) is stable (asymptotically stable) with respect to $y_1, \ldots, y_m$.

The theorem is proved. □

Example 6.4.2. Let system (6.4.13) have the form

$$\begin{aligned} \dot{y}_1(t) &= -y_1(t-\tau) + 0.1[\dot{z}_1(t-\tau) - 2\dot{z}_2(t-\tau)], \\ \dot{z}_1(t) &= 0.4\dot{y}_1(t-\tau) + z_1(t-\tau), \\ \dot{z}_2(t) &= 0.2\dot{y}_1(t-\tau) + z_1(t-\tau) - z_2(t-\tau). \end{aligned} \qquad (6.4.16)$$

The auxiliary $\boldsymbol{\mu}$-system comprises the equations

$$\begin{aligned} \dot{y}_1(t) &= -y_1(t-\tau) + 0.1\dot{\mu}_1(t-\tau), \\ \dot{\mu}_1(t) &= -\mu_1(t-\tau). \end{aligned} \qquad (6.4.17)$$

In the interval $\tau \in (0, \pi/2)$, it is possible to specify an everywhere dense set M in which all the roots of the characteristic quasi-polynomial $(e^{-\lambda\tau} + \lambda)^2 = 0$ of the auxiliary $\boldsymbol{\mu}$-system (6.4.17) satisfy the condition $\operatorname{Re}\lambda_j \le \alpha = \text{const} < 0$.

Therefore, for $\tau \in M$, the unperturbed motion $y_1 = z_1 = z_2 = 0$ of system (6.4.16) is asymptotically stable with respect to y_1 by Theorem 6.4.2.

6.5 The Stabilization of Linear Delayed Systems with Respect to Part of the Variables

6.5.1 Formulation of the Problem

Consider a controlled object whose dynamics of the perturbed motion are described by a linear system of differential equations with retarded argument

$$\dot{\boldsymbol{x}}(t) = A^*\boldsymbol{x}(t) + A'^*\boldsymbol{x}(t-\tau) + B^*\boldsymbol{u},$$
$$\boldsymbol{x}^{\mathsf{T}} = (y_1, \ldots, y_m, z_1, \ldots, z_p) = (\boldsymbol{y}^{\mathsf{T}}, \boldsymbol{z}^{\mathsf{T}}), \quad m > 0, \quad p \geq 0, \quad n = m + p.$$

In $\boldsymbol{y}$, $\boldsymbol{z}$ variables, this system takes the form

$$\begin{aligned} \dot{\boldsymbol{y}}(t) &= A\boldsymbol{y}(t) + A'\boldsymbol{y}(t-\tau) + B\boldsymbol{z}(t) + B'\boldsymbol{z}(t-\tau) + P\boldsymbol{u}, \\ \dot{\boldsymbol{z}}(t) &= C\boldsymbol{y}(t) + C'\boldsymbol{y}(t-\tau) + D\boldsymbol{z}(t) + D'\boldsymbol{z}(t-\tau) + Q\boldsymbol{u}, \end{aligned} \tag{6.5.1}$$

where A, A', B, B', C, C', D, D', P, and Q are constant matrices of appropriate sizes, $\boldsymbol{x}$ is the vector characterizing the state of the system, $\boldsymbol{u} = (u_1, \ldots, u_r)^{\mathsf{T}}$ is the vector of controls, and $\tau = \text{const} > 0$ is a value of the delay.

The problem of stabilizing dynamic systems with delay was first considered by Krasovskii and Osipov (see Osipov [1965a]). To solve this problem they used the control rules in the form of a linear feedback of *integral type* constructed by using the method of the Lyapunov–Krasovskii functionals. Implementing this method requires finding eigenvalues and eigenfunctions of the system, which is a very difficult task.

In this connection another approach to the problem of stabilization was shaped in the middle 70s.* This approach uses *discrete control rules with delay in the state* (see Borkovskaya and Marchenko [1996]). These control rules have the form

$$\boldsymbol{u}(t) = \sum_{j=0}^{N} F_j \boldsymbol{x}(t - j\tau), \tag{6.5.2}$$

where F_j are constant matrices of appropriate sizes and N is a natural number. We shall express this case as $\boldsymbol{u} \in U_D$.

The implementation of control rules $\boldsymbol{u} \in U_D$ is appreciably simpler than control rules of integral type.

Of particular interest is the special case of control rules (6.5.2) with $N = 2$, which leaves the closed system within the class being considered.

Problem 6.5.1. Find control rules $\boldsymbol{u} \in U_D$ such that the unperturbed motion $\boldsymbol{y} = \boldsymbol{0}$, $\boldsymbol{z} = \boldsymbol{0}$ of system (6.5.1) *is asymptotically $\boldsymbol{y}$-stable.*

---

*Morse, A.S [1976] Ring Models for Delay-Differential Systems, *Automatica*, **12**, 5, 529–531. Asmykovich, I.K. and Marchenko, V.M. [1976] Control of a Spectrum of Systems with Delay, *Avtomat. i Telemekh.*, 7, 5–14. Datko, R. [1985] Remarks Concerning the Asymptotic Stability and Stabilization of Linear Delay Differential Equations, *J. Math. Anal. Appl.*, **111**, 2, 571–584.

6.5.2 The General Idea of Constructing Control Rules

We shall construct the control rules $\boldsymbol{u} \in U_D$ solving Problem 6.5.1 in the form

$$\boldsymbol{u} = \Gamma \boldsymbol{z}(t) + \Gamma_\tau \boldsymbol{z}(t-\tau) + \boldsymbol{u}^*, \quad \boldsymbol{u}^* \in U_D. \tag{6.5.3}$$

Constant matrices Γ and Γ_τ are determined by a transformation that allows some auxiliary μ-system (of the same class as system (6.5.1)) to be segregated from system (6.5.1). Control $\boldsymbol{u}^* \in U_D$ solves the problem of stabilizing this auxiliary μ-system with respect to *all* the variables.

6.5.3 Condition for Stabilizability with Respect to Part of the Variables

We denote by G_4^0, G_5^0 the matrices constructed for the closed system (6.5.1), (6.5.3) with $\boldsymbol{u}^* = \boldsymbol{0}$ in accordance with the rules from Subsection 6.4.5.

Theorem 6.5.1. *Let* $r = 1$, *and let there exist matrices* Γ *and* Γ_τ *such that for any* α *the following condition is satisfied*:

$$\det(B^0, A^0 B^0, \ldots, (A^0)^{m+h-1} B^0) \neq 0,$$
$$A^0 = G_4^0 A^* G_5^0 + \alpha G_4^0 A'^* G_5^0, \quad B^0 = G_4^0 B^*.$$

Then control rules (6.5.3) *solve Problem* 6.5.1 *for system* (6.5.1).

Proof. The auxiliary controlled μ-system corresponding to matrices Γ and Γ_τ has the form

$$\dot{\boldsymbol{\xi}}(t) = G_4^0 A^* G_5^0 \boldsymbol{\xi}(t) + G_4^0 A'^* G_5^0 \boldsymbol{\xi}(t-\tau) + G_4^0 B^* \boldsymbol{u}_1^*.$$

Under the conditions of the theorem, this μ-system is stabilizable with respect to all the variables by a control $\boldsymbol{u}_1^* \in U_D$ (Borkovskaya and Marchenko [1996]). Therefore, one can construct a control $\boldsymbol{u}_1 \in U_D$ of the form (6.5.3) that solves Problem 6.5.1. The theorem is proved. □

Example 6.5.1. Consider an *automatic rheostatic voltage regulator* whose dynamics are described by the system of equations with delay (Vicker [1934]*)

$$\begin{aligned} \dot{y}_1(t) &= -\frac{\varepsilon}{\theta_1} z_1(t) + \frac{\varepsilon}{\theta_1} z_2(t) + \frac{B}{\theta_1} z_1(t-\tau) + u_1, \\ \dot{y}_2(t) &= -\frac{k}{\theta_2} y_2(t) + \frac{\varepsilon}{\theta_2} z_1(t) - \frac{\varepsilon}{\theta_2} z_2(t), \\ \dot{z}_i(t) &= y_i(t) \quad (i = 1, 2), \end{aligned} \tag{6.5.4}$$

where θ_1 (θ_2) and z_1 (z_2) are the moments of inertia and rotational angles of the regulator armature (of the damper sector and the washer), respectively, k is a damping coefficient, B and ε are constants, u_1 is the controlling action (a voltage at terminals of the armature), and τ is a constant delay.

*Vicker, D.A. [1934] Dynamics of Automatic Rheostat Voltage Regulators, *Electricity*, 9, 26–30 (in Russian).

Let us solve the problem of stabilizing the unperturbed motion $y_1 = y_2 = z_1 = z_2 = 0$ of system (6.5.4) with respect to y_1, y_2 by a control rule $u_1 \in U_D$. Following (6.5.3), we shall seek control rule u_1 in the form

$$u_1 = \Gamma \boldsymbol{z}(t) + \Gamma_\tau \boldsymbol{z}(t-\tau) + u_1^*,$$
$$\Gamma = \left(\frac{\varepsilon}{\theta_1}, -\frac{\varepsilon}{\theta_1}\right) \quad \Gamma_\tau = \left(-\frac{B}{\theta_1}, 0\right), \quad \boldsymbol{z} = (z_1, z_2)^{\mathsf{T}}. \tag{6.5.5}$$

The auxiliary $\boldsymbol{\mu}$-system for the closed system (6.5.4), (6.5.5) comprises the equations

$$\dot{y}_1(t) = u_1^*, \quad \dot{y}_2(t) = -\frac{k}{\theta_2} y_2(t) + \mu_1(t), \tag{6.5.6}$$
$$\dot{\mu}_1(t) = \frac{\varepsilon}{\theta_2} y_1(t) - \frac{\varepsilon}{\theta_2} y_2(t).$$

System (6.5.6) is completely controllable with respect to all the variables. Thus, the problem of stabilizing the zero solution $y_1 = y_2 = \mu_1 = 0$ of system (6.5.6) (with respect to all the variables) can be solved by control rule u_1^* expressed as a linear combination of variables $y_1(t)$, $y_2(t)$, and $\mu_1(t)$. This means that control rule (6.5.5) solves the problem of stabilizing the unperturbed motion $y_1 = y_2 = z_1 = z_2 = 0$ of system (6.5.4) with respect to variables y_1, y_2, and $u_1 \in U_D$ for $N = 2$.

We note that the unperturbed motion $y_1 = y_2 = z_1 = z_2 = 0$ of system (6.5.4), (6.5.5) is exponentially asymptotically stable with respect to variables y_1, y_2 and is also Lyapunov (nonasymptotically) stable. Besides, from $\mu_1(t) \to 0$, it follows that with time $z_1(t) \to z_2(t)$.

6.6 The Stability of Nonlinear Delayed Systems in a Linear Approximation

6.6.1 Original System of Equations

Let the equations of perturbed motion (6.1.2) have the form

$$\begin{aligned}
\frac{dy_i(t)}{dt} &= \sum_{k=1}^{m} a_{ik} y_k(t) + Y_i^0\big[\boldsymbol{y}(t), \boldsymbol{y}(t-\tau)\big] Y_i^*\big[\boldsymbol{z}(t)\big] + Y_i^{**}\big[t, \boldsymbol{x}(t), \boldsymbol{x}(t-\tau)\big], \\
\frac{dz_j(t)}{dt} &= \sum_{l=1}^{p} d_{jl} z_l(t) + Z_j\big[t, \boldsymbol{x}(t), \boldsymbol{x}(t-\tau)\big] \\
&\qquad (i = 1, \dots, m; \ j = 1, \dots, p),
\end{aligned} \tag{6.6.1}$$

where a_{ik}, d_{jl}, and $\tau > 0$ are constants; $Y_i^0(\mathbf{0}, \mathbf{0}) = 0$, Y_i^{**}, and Z_j are *nonlinear* perturbations (Y_i^0 can also be linear). Functions Y_i^* satisfy the condition $Y_i^* = \sum_{s=2}^{r} U_s^{(i)}\big[\boldsymbol{z}(t)\big]$, where $U_s^{(i)}$ are homogeneous forms of variables $z_1, \dots, z_p$ of order $s \le r$ and r is a finite number.

6.6.2 Construction of an Auxiliary System

Now we introduce the new variables

$$\mu_i = \sum_{s=2}^{r} U_s^{(i)}(z) \quad (i = 1, \ldots, m).$$

Without loss of generality, we assume that the vectors corresponding to these variables (see Subsection 2.2.3) are linearly independent.

With the new variables thus introduced, two cases are possible.

The first case. System (6.6.1) takes the form

$$\begin{aligned}
\frac{dy_i(t)}{dt} &= \sum_{k=1}^{m} a_{ik} y_k(t) + Y_i^0\big[\boldsymbol{y}(t), \boldsymbol{y}(t-\tau)\big]\mu_i(t) + Y_i^{**}\big[t, \boldsymbol{x}(t), \boldsymbol{x}(t-\tau)\big], \\
\frac{d\mu_j(t)}{dt} &= \sum_{k=1}^{m} L_{jk}\mu_k(t) + Z_j^*\big[t, \boldsymbol{x}(t), \boldsymbol{x}(t-\tau)\big], \qquad (6.6.2)\\
\frac{dz_v(t)}{dt} &= \sum_{l=1}^{p} d_{vl} z_l(t) + Z_v\big[t, \boldsymbol{x}(t), \boldsymbol{x}(t-\tau)\big],
\end{aligned}$$

$$Z_j^* = \sum_{l=1}^{p} \frac{\partial \mu_j}{\partial z_l} Z_l \qquad (i, j = 1, \ldots, m;\ v = 1, \ldots, p),$$

where L_{jk} are constants. We shall term the linear part of the first two group of equations from (6.6.2) the ***μ**-system of linear approximation* for the original system (6.6.1).

The second case. System (6.6.1) is not reduced to (6.6.2) and has the form

$$\begin{aligned}
\frac{dy_i(t)}{dt} &= \sum_{k=1}^{m} a_{ik} y_k(t) + Y_i^0\big[\boldsymbol{y}(t), \boldsymbol{y}(t-\tau)\big]\mu_i(t) + Y_i^{**}\big[t, \boldsymbol{x}(t), \boldsymbol{x}(t-\tau)\big], \\
\frac{d\mu_j(t)}{dt} &= \bar{Y}_j^*(\boldsymbol{z}(t)) + Z_j^*\big[t, \boldsymbol{x}(t), \boldsymbol{x}(t-\tau)\big], \quad \bar{Y}_j^* = \sum_{s=2}^{r} \bar{U}_s^{(j)}\big[\boldsymbol{z}(t)\big], \\
\frac{dz_v(t)}{dt} &= \sum_{l=1}^{p} d_{vl} z_l(t) + Z_v\big[t, \boldsymbol{x}(t), \boldsymbol{x}(t-\tau)\big],
\end{aligned}$$

$$(i, j = 1, \ldots, m;\ v = 1, \ldots, p),$$

where $\bar{U}_s^{(j)}$ are homogeneous forms of variables $\boldsymbol{z}(t)$ of order s. Suppose that none of the following equalities can be valid:

$$\bar{Y}_i^* = \sum_{k=1}^{m} L_{ik}\mu_k(t) \qquad (i = 1, \ldots, m).$$

In this case, we introduce the new variables once more

$$\mu_{m+i} = \bar{Y}_i^*(\boldsymbol{z}) \qquad (i = 1, \ldots, m).$$

Since the new variables are forms of a *finite* order in variables $z_1, \dots, z_p$, then, following the above reasoning, one can construct a finite-dimensional system of type (6.6.2) for system (6.6.1). The linear part of this system (except for the last group of equations) is the $\boldsymbol{\mu}$-system of linear approximation for the original system (6.6.1).

6.6.3 The Condition for Stability with Respect to Part of the Variables

Without loss of generality, we assume that the auxiliary $\boldsymbol{\mu}$-system of linear approximation for the original system (6.6.1) can already be constructed in the first stage of introducing the new variables.

Let us introduce the following designations:

(1) $\boldsymbol{\xi}$ is the vector whose components are variables $y_i(t)$, $y_i(t-\tau)$, $\mu_j(t)$, and $\mu_j(t-\tau)$ $(i,j = 1,\dots,m)$;

(2) $\boldsymbol{Y}_*$ is the vector whose components are functions Y_i^{**} and Z_j^* $(i,j = 1,\dots,m)$.

Suppose that the following inequality is satisfied in the domain $t \geq 0$, $\|\boldsymbol{\xi}\| \leq H$,

$$\|\boldsymbol{Y}_*[t, \boldsymbol{x}(t), \boldsymbol{x}(t-\tau)]\| \leq \alpha \|\boldsymbol{\xi}\|, \tag{6.6.3}$$

where α and H are sufficiently small positive constants.

Theorem 6.6.1. *Let the zero solution* $y_i = 0$, $\mu_j = 0$ $(i,j = 1,\dots,m)$ *of the* $\boldsymbol{\mu}$*-system of linear approximation*

$$\frac{dy_i(t)}{dt} = \sum_{k=1}^{m} a_{ik} y_k, \qquad \frac{d\mu_j(t)}{dt} = \sum_{k=1}^{m} L_{jk}\mu_k \tag{6.6.4}$$

be Lyapunov asymptotically stable.

Then, under condition (6.6.3), *the unperturbed motion* $\boldsymbol{x} = \boldsymbol{0}$ *of system* (6.6.1) *is uniformly asymptotically stable with respect to* $y_1, \dots, y_m$.

Proof. If the zero solution of system (6.6.4) is Lyapunov asymptotically stable, then there exists the Lyapunov function $v = v(\boldsymbol{y}, \boldsymbol{\mu})$ that satisfies the conditions

$$c_1(\|\boldsymbol{y}\|^2 + \|\boldsymbol{\mu}\|^2) \leq v(\boldsymbol{y}, \boldsymbol{\mu}) \leq c_2(\|\boldsymbol{y}\|^2 + \|\boldsymbol{\mu}\|^2),$$
$$\dot{v}(\boldsymbol{y}, \boldsymbol{\mu}) = -(\|\boldsymbol{y}\|^2 + \|\boldsymbol{\mu}\|^2).$$

Consider the V-functional

$$V[\boldsymbol{x}_t] = v(\boldsymbol{y}, \boldsymbol{\mu}) + \sum_{i=1}^{m} \varepsilon_i \int_{-\tau}^{0} y_i^2(t+\theta)\, d\theta + \sum_{j=1}^{m} \varepsilon_{m+j} \int_{-\tau}^{0} \mu_j^2(t+\theta)\, d\theta. \tag{6.6.5}$$

If condition (6.6.3) is satisfied, constants ε_i and ε_{m+j} in (6.6.5) can be chosen so that the estimate

$$\dot V[\boldsymbol{x}_t] \le -c_3(\|\boldsymbol{y}(t)\|^2 + \|\boldsymbol{\mu}(t)\|^2), \quad c_3 = \text{const} > 0$$

is valid on trajectories of system (6.6.2) (and, consequently, also on trajectories of the original system (6.6.1)).

Therefore, V-functional (6.6.5) satisfies all the conditions of Corollary 6.2.2. The theorem is proved. □

Example 6.6.1.

1. Let the equations of perturbed motion (6.6.1) have the form

$$\begin{aligned} \dot y_1(t) &= a_1 y_1(t) + b_1 y_1(t-\tau) z_1(t) z_2(t) + Y_1^{**}\big[t, \boldsymbol{x}(t), \boldsymbol{x}(t-\tau)\big], \\ \dot z_i(t) &= d_i z_i(t) + Z_i\big[t, \boldsymbol{x}(t), \boldsymbol{x}(t-\tau)\big] \quad (i = 1, 2), \end{aligned} \tag{6.6.6}$$

where a_1, b_1, and d_i are constants and Y_1^{**} and Z_i are nonlinear perturbations.

By introducing the new variable $\mu_1 = z_1 z_2$, we construct the auxiliary system

$$\begin{aligned} \dot y_1(t) &= a_1 y_1(t) + b_1 y_1(t-\tau)\mu_1(t) + Y_1^{**}\big[t, \boldsymbol{x}(t), \boldsymbol{x}(t-\tau)\big], \\ \dot\mu_1(t) &= (d_1 + d_2)\mu_1(t) + Z_1^*\big[t, \boldsymbol{x}(t), \boldsymbol{x}(t-\tau)\big], \\ \dot z_i(t) &= d_i z_i(t) + Z_i\big[t, \boldsymbol{x}(t), \boldsymbol{x}(t-\tau)\big] \quad (i = 1, 2). \end{aligned}$$

Let the following conditions be satisfied:

(1) $a_1 < 0$, $d_1 + d_2 < 0$;

(2) condition (6.6.3) with $\boldsymbol{\xi} = (y_1(t), y_1(t-\tau), \mu_1(t), \mu_1(t-\tau))$ and $\boldsymbol{Y}_* = (Y_1^{**}, Z_1^*)$.

Then, for any constant number b_1 and sufficiently small numbers α and H, the unperturbed motion $y_1 = z_1 = z_2 = 0$ of system (6.6.6) is uniformly asymptotically stable with respect to y_1 by Theorem 6.6.1.

2. Consider the *Lur'e controlled system* with delay in the final control element and the system itself in the critical case of two zero roots:

$$\begin{aligned} \dot{\boldsymbol{y}}(t) &= A\boldsymbol{y}(t) + A'\boldsymbol{y}(t-\tau) + \boldsymbol{b}^{(1)}\varphi[\sigma(t-\tau)], \\ \dot z_1 &= b_1^{(2)}\varphi[\sigma(t-\tau)], \quad \dot z_2 = b_2^{(2)}\varphi[\sigma(t-\tau)], \\ \sigma(t) &= \boldsymbol{\beta}^{(1)}\boldsymbol{y}(t) + \beta_1^{(2)} z_1(t) + \beta_2^{(2)} z_2(t). \end{aligned} \tag{6.6.7}$$

In this case, the auxiliary $\boldsymbol{\mu}$-system comprises the equations

$$\begin{aligned} \dot{\boldsymbol{y}}(t) &= A\boldsymbol{y}(t) + A'\boldsymbol{y}(t-\tau) + \boldsymbol{b}^{(1)}\varphi[\sigma(t-\tau)], \\ \dot\mu_1 &= b^*\varphi[\sigma(t-\tau)], \quad \sigma(t) = \boldsymbol{\beta}^{(1)}\boldsymbol{y}(t) + \gamma\mu_1(t), \\ b^* &= \frac{1}{\gamma}(\beta_1^{(2)} b_1^{(2)} + \beta_2^{(2)} b_2^{(2)}). \end{aligned} \tag{6.6.8}$$

The conditions for absolute stability of $\boldsymbol{\mu}$-system (6.6.8) with respect to *all* the variables* will be sufficient for the original system (6.6.7) to be absolutely $\boldsymbol{y}$-stable.

---

*Razvan, V. [1975] *Stabilitatea Absoluta a Systemelor Automatĕ cu Intirziere*, Ed. Acad. Rep. Soc. Romania, Bucharest.

6.7 Overview of References

The problem of *partial* stability for systems of functional-differential equations goes back to the work by Halanay [1963], Corduneanu [1975], and Yudaev [1975].

Halanay [1963] studied the system

$$\dot{\boldsymbol{y}}(t) = \boldsymbol{Y}\big[t, \boldsymbol{y}(t+\theta), \boldsymbol{z}(t+\theta)\big], \quad \dot{\boldsymbol{z}}(t) = \boldsymbol{Z}\big[t, \boldsymbol{y}(t+\theta), \boldsymbol{z}(t+\theta)\big], \\ \boldsymbol{Y}(t, \mathbf{0}, \mathbf{0}) \equiv \mathbf{0}, \quad \boldsymbol{Z}(t, \mathbf{0}, \mathbf{0}) \equiv \mathbf{0} \tag{6.7.1}$$

and proved the following result: if the zero solution $\boldsymbol{y} = \boldsymbol{z} = \mathbf{0}$ of system (6.7.1) is uniformly $\boldsymbol{y}$-stable (uniformly asymptotically $\boldsymbol{y}$-stable) and the zero solution of the "*truncated*" system

$$\dot{\boldsymbol{z}}(t) = \boldsymbol{Z}\big[t, \mathbf{0}, \boldsymbol{z}(t+\theta)\big]$$

is Lyapunov uniformly asymptotically stable, then the zero solution of system (6.7.1) is Lyapunov uniformly stable (Lyapunov uniformly asymptotically stable).

The conditions for exponential asymptotic $\boldsymbol{y}$-stability of the zero solution of system (6.7.1) were obtained by Yudaev [1975] based on analyzing another "truncated" system

$$\dot{\boldsymbol{y}}(t) = \boldsymbol{Y}\big[t, \boldsymbol{y}(t+\theta), \mathbf{0}\big].$$

The problem of partial asymptotic stability also arises in the context of the "*reduction principle*" for systems with delay (Osipov [1965b]).

The method of the Lyapunov–Krasovskii functionals as applied to the problem of *partial* stability of systems of functional-differential equations is considered in the following publications: Corduneanu [1975], Vorotnikov [1979a], [1980b], [1991a], Il'yasov [1982], and Kalistratova [1986].

The partial stability of autonomous linear systems with holdover and the partial stability in the linear approximation are considered in the works by Vorotnikov [1980a], [1991a]. These results are cited in the present chapter. Terjeki [1983a] suggested another approach to investigating partial stability of autonomous linear systems.

A number of works (see, for example, Liu [1989], Makay [1991], Lakshmikantham and Martynyuk [1993]) deal with the problem of stability (asymptotic stability) *with respect to two measures* for systems with delay.

We also mention the work by Karnishin* which studies the partial stability of functional-differential equations using *Azbelev's W-method* (Azbelev et al. [1991]).

*Karnishin, S.G. [1987] On Stability of Linear Functional-Differential Equations with Respect to Part of the Variables, in *Functional-Differential Equations*, 48–52, Perm': Izdat. Permsk. Politekhn. Inst. (in Russian).

Chapter 7

Stability and Stabilization of Stochastic Systems with Respect to Part of the Variables

In formulating applied problems of partial stability and stabilization and problems of control with respect to part of the variables, one should often take into account possible *uncertainty* concerning description of a system and an external agency. A source of uncertainty might also appear as interference in the control or observation channels.

Mathematically, it is possible to formalize uncertain problems in a number of ways. Thus, a *game-theoretic approach* to the problems of control with respect to part of the variables in the presence of uncontrollable interference has been presented in Chapter 5. Another way of allowing for uncertainty is interpreting it as the action of *random interference* possessing given *statistical* parameters.

This chapter extends the approach proposed in Chapters 1–4 to the problems of partial stability and stabilization for *stochastic systems of differential equations in Ito's form*. These problems are equivalent (under the corresponding passage to the limit) to those of partial stability and stabilization for dynamic systems subject to random perturbations of the "*white noise*" *type*.

Sufficient conditions for partial asymptotic stability of linear systems with constant coefficients are obtained. Conditions for partial asymptotic stability in the linear approximation and conditions for partial stabilization are derived for nonlinear systems. Also considered is a stochastic version of the method of Lyapunov functions as applied to the problem of partial stability.

As an illustration, we work through the problem of damping the angular rotations of a solid (with respect to part of the variables) in the presence of "white noise" in control channels.

7.1 Formulation of the Problem of Stability with Respect to Part of the Variables

7.1.1 Ito's Stochastic System of Differential Equations

Consider a *stochastic* system of differential equations in *Ito's form*

$$dx = X(t, x) + \sum_{k=1}^{r} \sigma_k(t, x) dw_k(t),$$

$$X(t, 0) \equiv 0, \quad \sigma_i(t, 0) \equiv 0 \quad (i = 1, \dots, r), \tag{7.1.1}$$

$$x^{\mathsf{T}} = (y_1, \dots, y_m, z_1, \dots, z_p) = (y^{\mathsf{T}}, z^{\mathsf{T}}), \quad m > 0, \quad p \geq 0, \quad n = m + p.$$

Here $X(t, x)$, $\sigma_k(t, x)$ $(k = 1, \dots, r)$ are n-dimensional vector functions

$$X = (X_1, \dots, X_n)^{\mathsf{T}}, \quad \sigma_k = (\sigma_{k1}, \dots, \sigma_{kn})^{\mathsf{T}},$$

and w_k $(k = 1, \dots, r)$ are *independent Wiener's processes* such that $\mathbf{E}\, w_k = 0$, $\mathbf{E}\, w_k^2 = t$ ($\mathbf{E}$ stands for the expected value).

Suppose that the following conditions are satisfied.

(1) In domain (0.3.2) vector functions $X(t, x)$, $\sigma_k(t, x)$ $(k = 1, \dots, r)$ are continuous, regarding the arguments in the aggregate, and satisfy the conditions for existence and uniqueness of solutions $x(t) = x(t; t_0, x_0)$. Here a solution $x(t)$ is understood as a *Markov stochastic process* continuous with probability one with a Feller transition function.

(2) Solutions of system (7.1.1) are *z-continuable*, which means that any solution $x(t)$ is defined for all $t \geq 0$ such that $\|y(t)\| \leq H$ in a corresponding probabilistic sense.

Differential equations whose right-hand sides contain a *stochastic process* were evidently first considered by Langevin.* Then, starting in the 30s, mathematical publications on the theory of stochastic differential equations began to appear. A sufficiently simple and convenient theory of such equations was proposed by Ito [1951]. This theory is described in detail in a number of monographs (see, for example, Khas'minskii [1969]).

A radically different form of describing stochastic systems is provided by the *Fokker–Planck equations* which involve density of distribution of a representative point in a Euclidean state space. These are partial differential equations.

We note that the approaches to describing stochastic systems which we mentioned appeared after the first publications on the *theory of Brownian motion* by Einstein and Smolukhovskii.

7.1.2 Definitions of Stability with Respect to Part of the Variables

Definition 7.1.1. The unperturbed motion $x = 0$ of system (7.1.1) is said to be

*Langevin, P. [1908] Sur la Théorie du Mouvement Brownien, *C. R. Acad. Sci. Paris*, **146**, 530–533.

(1) *stable with respect to* $y_1, \ldots, y_m$ (*$\boldsymbol{y}$-stable*) *in probability* if, for any $\varepsilon > 0$, $t_0 \geq 0$, and $\gamma > 0$, there exists $\delta(\varepsilon, t_0) > 0$ such that for $\|\boldsymbol{x}_0\| < \delta$

$$\mathsf{P}\,\{\sup\|\boldsymbol{y}(t; t_0, \boldsymbol{x}_0)\| > \varepsilon\} < \gamma, \quad t \geq t_0$$

(P is a probability measure);

(2) *asymptotically $\boldsymbol{y}$-stable in probability* if it is $\boldsymbol{y}$-stable in probability and, besides,

$$\lim_{t\to\infty} \mathsf{P}\,\{\lim_{\boldsymbol{x}_0\to\boldsymbol{0}}\|\boldsymbol{y}(t; t_0, \boldsymbol{x}_0)\| = 0\} = 1;$$

(3) *exponentially asymptotically $\boldsymbol{y}^N$-stable* if there exist positive constants Δ, A, α, and N such that for $\|\boldsymbol{x}_0\| < \Delta$ the following inequalities are satisfied:

$$\mathsf{E}\,\|\boldsymbol{y}(t; t_0, \boldsymbol{x}_0)\|^N \leq A\|\boldsymbol{x}_0\|^N e^{-\alpha(t-t_0)}, \quad t \geq t_0; \tag{7.1.2}$$

(4) *globally uniformly asymptotically $\boldsymbol{y}^N$-stable* if

$$\sup \mathsf{E}\,\|\boldsymbol{y}(t; t_0, \boldsymbol{x}_0)\|^N \to 0 \text{ as } \delta \to 0,\ t_0 \geq 0, \text{ and } \|\boldsymbol{x}_0\| < \delta,\ t \geq t_0$$

and, besides, $\mathsf{E}\,\|\boldsymbol{y}(t; t_0, \boldsymbol{x}_0)\|^N \to 0$ as $t \to \infty$ uniformly for any $t_0 \geq 0$ and $\|\boldsymbol{x}_0\| < \infty$;

(5) *globally exponentially asymptotically $\boldsymbol{y}^N$-stable* if inequality (7.1.2) holds for any $\|\boldsymbol{x}_0\| < \infty$.

Remark 7.1.1.

1. In cases (4) and (5), the assumptions concerning the right-hand sides of system (7.1.1) are valid in the domain $t \geq 0$, $\|\boldsymbol{x}\| < \infty$.

2. In cases (3)–(5), for $N = 1$ we have *asymptotic (exponential) $\boldsymbol{y}$-stability in the mean* and for $N = 2$, *in the quadratic mean.*

7.2 Using the Method of Lyapunov Functions

7.2.1 The Class of Lyapunov Functions under Consideration

We shall consider Lyapunov functions $V = V(t, \boldsymbol{x})$, $V(t, \boldsymbol{0}) \equiv 0$ that are twice continuous differentiable with respect to $\boldsymbol{x}$ and once with respect to t except, possibly, at the point $\boldsymbol{x} = \boldsymbol{0}$ in domain (0.3.2). Functions $a(r)$, $b(r)$, and $c(r)$ have the same meaning as in Section 0.4.

We shall further consider a "*differential generator*" LV of a function V along the trajectories of system (7.1.1):

$$LV(t, \boldsymbol{x}) = \frac{\partial V}{\partial t} + \sum_{i=1}^{n} \frac{\partial V}{\partial x_i} X_i(t, \boldsymbol{x}) + \tfrac{1}{2} \sum_{k=1}^{r} \Big[\langle \boldsymbol{\sigma}_k(t, \boldsymbol{x}) \cdot \frac{\partial}{\partial \boldsymbol{x}} \rangle\Big]^2 V,$$

$$\frac{\partial}{\partial \boldsymbol{x}} = \left(\frac{\partial}{\partial x_1}, \ldots, \frac{\partial}{\partial x_n}\right)^{\mathsf{T}}.$$

7.2.2 Conditions for Stability with Respect to Part of the Variables

Theorem 7.2.1. *Suppose that in the domain $t \geq 0$, $\|\boldsymbol{x}\| < \infty$ there exists a V-function such that, for certain positive constants c_i ($i = 1, 2, 3$), N, and $\boldsymbol{x} \neq \boldsymbol{0}$, the following inequalities are satisfied:*

$$c_1 \|\boldsymbol{y}\|^N \leq V(t, \boldsymbol{x}) \leq c_2 \|\boldsymbol{y}\|^N, \quad LV(t, \boldsymbol{x}) \leq -c_3 \|\boldsymbol{y}\|^N.$$

Then the unperturbed motion $\boldsymbol{x} = \boldsymbol{0}$ of system (7.1.1) is globally exponentially $\boldsymbol{y}^N$-stable.

Proof. Under the conditions of the theorem the Dynkin* formula is valid:

$$\mathbf{E}\, V(t, \boldsymbol{x}(t; t_0, \boldsymbol{x}_0)) - V(t_0, \boldsymbol{x}_0) = \int_{t_0}^{t} \mathbf{E}\, LV(s, \boldsymbol{x}(s, t_0, \boldsymbol{x}_0))\, ds.$$

Differentiating this equality with respect to t and taking into account the conditions imposed on function $V(t, \boldsymbol{x})$, we obtain the relationship

$$\frac{d}{dt} \mathbf{E}\, V(t, \boldsymbol{x}(t; t_0, \boldsymbol{x}_0)) = -\frac{c_3}{c_2} \mathbf{E}\, V(t, \boldsymbol{x}(t; t_0, \boldsymbol{x}_0)). \tag{7.2.1}$$

From (7.2.1), with $V(t, \boldsymbol{x}) \geq c_1 \|\boldsymbol{y}\|^N$ for all $t \geq t_0$ taken into account, we obtain the estimate

$$\begin{aligned} \mathbf{E}\, \|\boldsymbol{y}(t; t_0, \boldsymbol{x}_0)\|^N &\leq \frac{1}{c_1} \mathbf{E}\, V(t, \boldsymbol{x}(t; t_0, \boldsymbol{x}_0)) \\ &\leq \frac{1}{c_1} V(t_0, \boldsymbol{x}_0) e^{[-{c_3}/{c_2}(t - t_0)]}. \end{aligned} \tag{7.2.2}$$

Since $V(t_0, \boldsymbol{x}_0) \leq c_2 \|\boldsymbol{y}_0\|^N$, from (7.2.2) it follows that, for all $t \geq t_0$,

$$\mathbf{E}\, \|\boldsymbol{y}(t; t_0, \boldsymbol{x}_0)\|^N \leq \frac{c_2}{c_1} \|\boldsymbol{y}_0\|^N e^{[-{c_3}/{c_2}(t - t_0)]}. \tag{7.2.3}$$

Setting $A = {c_2}/{c_1}$ and $\alpha = {c_3}/{c_2}$, we conclude that the unperturbed motion $\boldsymbol{x} = \boldsymbol{0}$ of system (7.1.1) is globally exponentially asymptotically $\boldsymbol{y}^N$-stable. The theorem is proved. □

Remark 7.2.1.

1. Theorem 7.2.1 is an extension of theorems by Krasovskii [1959] and Nevel'son and Khas'minskii (see Khas'minskii [1969]) to the problem of partial stability of stochastic systems.

2. Inequality (7.2.3) is more general than (7.1.2). Indeed, estimates of $\boldsymbol{y}$-components of solutions of system (7.1.2) do not depend on $\boldsymbol{z}_0$. Therefore, the conditions of Theorem 7.2.1 also guarantee exponential asymptotic $\boldsymbol{y}^N$-stability *on the whole with respect to $\boldsymbol{z}_0$.*

3. When $N < 1$, a *smooth* function V for $\boldsymbol{x} = \boldsymbol{0}$ may not exist.

---

*Dynkin, E.B. [1963], *The Markov Processes*, Moscow: Fizmatgiz (in Russian).

Corollary 7.2.1. *Suppose that in the domain $t \geq 0$, $\|\boldsymbol{x}\| < \infty$ along with a V-function, it is possible to specify a continuous vector $\boldsymbol{\mu}(\boldsymbol{x})$-function, $\boldsymbol{\mu}(\mathbf{0}) = \mathbf{0}$ such that*

$$\begin{aligned} c_1\|\boldsymbol{y}\|^2 \leq V(t,\boldsymbol{x}) &\leq c_2\big(\|\boldsymbol{y}\|^2 + \|\boldsymbol{\mu}(\boldsymbol{x})\|^2\big), \\ LV(t,\boldsymbol{x}) &\leq -c_3\big(\|\boldsymbol{y}\|^2 + \|\boldsymbol{\mu}(\boldsymbol{x})\|^2\big). \end{aligned} \tag{7.2.4}$$

Then the unperturbed motion $\boldsymbol{x} = \mathbf{0}$ of system (7.1.1) is globally uniformly asymptotically $\boldsymbol{y}$-stable in the quadratic mean. In this case, for arbitrary initial conditions t_0, $\boldsymbol{x}_0$, solutions of system (7.1.1) satisfy the following estimate for all $t \geq t_0$

$$\mathbf{E}\,\|\boldsymbol{y}(t;t_0,\boldsymbol{x}_0)\|^2 \leq \frac{c_2}{c_1}\big(\|\boldsymbol{y}_0\|^2 + \|\boldsymbol{\mu}(\boldsymbol{x}_0)\|^2\big)e^{[-c_3/c_2(t-t_0)]}. \tag{7.2.5}$$

Corollary 7.2.2. *Suppose that conditions (7.2.4) hold in domain (0.3.2). Then the unperturbed motion $\boldsymbol{x} = \mathbf{0}$ of system (7.1.1) is exponentially asymptotically $\boldsymbol{y}$-stable in the quadratic mean ("in the small").*

Proof. Since $\boldsymbol{\mu}$ is continuous and the condition $\boldsymbol{\mu}(\mathbf{0}) = \mathbf{0}$ is true, one can find a number $\gamma > 0$ such that $\|\boldsymbol{\mu}(\boldsymbol{x}_0)\| \leq \gamma\|\boldsymbol{x}_0\|$ for a sufficiently small $\|\boldsymbol{x}_0\|$. From (7.2.5) we conclude that, for $A = c_2(\gamma+1)/c_1$ and $\alpha = c_3/c_2$, the relationship (7.1.2) is valid. The statement is proved. □

Corollary 7.2.1 justifies the *possibility of constructing Lyapunov V-functions* of the form

$$V(t,\boldsymbol{x}) \equiv V^*(t,\boldsymbol{y},\boldsymbol{\mu}(\boldsymbol{x})), \tag{7.2.6}$$

where V^* are functions sign-definite with respect to *all* the variables (with respect to $\boldsymbol{y}$, $\boldsymbol{\mu}$). In particular, one can use *quadratic* V^*-functions whose coefficients satisfy the *generalized Sylvester criterion.*

Now we formulate the conditions for $\boldsymbol{y}$-stability (asymptotic $\boldsymbol{y}$-stability) *in probability* of the unperturbed motion $\boldsymbol{x} = \mathbf{0}$ of system (7.1.1).

Theorem 7.2.2. *Suppose that in domain (0.3.2) there exists a V-function satisfying, for $\boldsymbol{x} \neq \mathbf{0}$, the inequalities*

$$V(t,\boldsymbol{x}) \geq a\big(\|\boldsymbol{y}\|\big), \quad LV(t,\boldsymbol{x}) \leq 0.$$

Then the unperturbed motion $\boldsymbol{x} = \mathbf{0}$ of system (7.1.1) is $\boldsymbol{y}$-stable in probability.

In addition, if

$$V(t,\boldsymbol{x}) \leq b\big(\|\boldsymbol{y}\|\big), \quad LV(t,\boldsymbol{x}) \leq -c\big(\|\boldsymbol{y}\|\big),$$

then this motion is asymptotically $\boldsymbol{y}$-stable in probability.

Remark 7.2.2.

1. Theorem 7.2.2 is an extension of theorems by Rumyantsev [1957a], [1970] (to the case of stochastic systems), Kushner [1967] and Khas'minskii [1969] (to the problem of partial stability).
2. The proof of Theorem 7.2.2 can be found in the work by Sharov [1978].

Example 7.2.1. Let the equations of perturbed motion (7.1.1) have the form

$$\begin{gathered} dy_1 = (-2y_1 + z_1^2 z_2)dt + y_1 dw_1(t), \\ dz_1 = (z_1 + y_1 z_1)dt, \quad dz_2 = (-6z_2 - 2y_1 z_2)dt + z_2 dw_2(t). \end{gathered} \tag{7.2.7}$$

Following (7.2.6), we take the Lyapunov V-function of the form

$$V(\boldsymbol{x}) = y_1^2 + 2y_1\mu_1 + 2\mu_1^2, \quad \mu_1 = z_1^2 z_2. \tag{7.2.8}$$

Let us evaluate the operator LV of function (7.2.8) on trajectories of system (7.2.7). To this end, we use the expressions

$$\begin{gathered} \boldsymbol{\sigma}_k = (\sigma_{k1}, \sigma_{k2}, \sigma_{k3})^{\mathsf{T}} \quad (k = 1, 2), \\ \sigma_{11} = y_1, \quad \sigma_{23} = z_2, \quad \sigma_{12} = \sigma_{13} = \sigma_{21} = \sigma_{22} = 0, \\ \sum_{k=1}^{2} \Big(\boldsymbol{\sigma}_k^{\mathsf{T}} \frac{\partial}{\partial \boldsymbol{x}} \Big)^2 V = \sigma_{11}^2 \frac{\partial^2 V}{\partial y_1^2} + \sigma_{23}^2 \frac{\partial^2 V}{\partial z_2^2} = 2y_1^2 + 4\mu_1^2. \end{gathered}$$

As a result,

$$LV(\boldsymbol{x}) = -2(y_1^2 + 2y_1\mu_1 + 5\mu_1^2).$$

Thus, V-function (7.2.8) satisfies the conditions of Corollary 7.2.1 and, consequently, the unperturbed motion $y_1 = z_1 = z_2 = 0$ of system (7.2.7) is (globally) uniformly asymptotically stable in the quadratic mean with respect to y_1. For arbitrary t_0, $\boldsymbol{x}_0$ and for all $t \geq t_0$, the following estimate is valid:

$$\mathsf{E}\, y_1^2(t; t_0, \boldsymbol{x}_0) \leq A(y_{10}^2 + z_{10}^4 z_{20}^2) e^{-\alpha(t - t_0)}.$$

In this case

$$c_1 = \tfrac{1}{2}(3 - \sqrt{5}), \quad c_2 = \tfrac{1}{2}(3 + \sqrt{5}), \quad c_3 = 6 - \sqrt{20},$$

and we can take $A = c_2/c_1$, $\alpha = c_3/c_2$.

We also note that by Corollary 7.2.2 we have *exponential* asymptotic stability in the quadratic mean with respect to y_1 for sufficiently small values of z_{10} and z_{20}.

7.3 Damping of Rotational Motion of a Solid (with Respect to Part of the Variables) with Random Interference in Control Channels Taken into Account

7.3.1 Formulation of the Problem

Consider *Euler's dynamical equations*

$$\begin{gathered} A\dot{x}_1 = (B - C)x_2 x_3 + u_1, \\ B\dot{x}_2 = (C - A)x_1 x_3 + u_2, \quad C\dot{x}_3 = (A - B)x_1 x_2 + u_3 \end{gathered} \tag{7.3.1}$$

describing the angular motion of a solid (spacecraft) about its center of mass.

As already noted, the control rules

$$u_i = \alpha_i x_i, \quad \alpha_i = \text{const} < 0 \quad (i = 1, 2), \quad u_3 \equiv 0 \tag{7.3.2}$$

damp the angular motion of the solid with respect to its *two* (of three) principal center axes of inertia.

Suppose that coefficients α_1 and α_2 represent just *mean values* of the factors, whereas their true values which occur when generating control rules (7.3.2) by special engines are *stochastic processes*

$$\eta_i(t) = \alpha_i + \beta_i \dot{w}_i(t) \quad (i = 1, 2),$$

where $\dot{w}_i(t)$ are "*white noise*" processes (independent *Gaussian processes* such that $\mathbf{E}\, \dot{w}_i(t) = 0$, $\mathbf{E}\, [\dot{w}_i(s)\dot{w}_i(t)] = \delta(t - s)$ and δ is *Dirac's delta function*); constants β_i characterize the noise intensities.

In this case, control rules (7.3.2) actually have the form

$$u_i = [\alpha_i + \beta_i \dot{w}_i(t)] x_i \quad (i = 1, 2), \quad u_3 \equiv 0. \tag{7.3.3}$$

Problem 7.3.1. Suppose that numbers α_1 and α_2 are preassigned. Find a domain of admissible values of constants β_i characterizing the intensity of noises for which the equilibrium position $x_1 = x_2 = x_3 = 0$ of system (7.3.1), (7.3.3) *is (globally) exponentially asymptotically stable with respect to x_1, x_2 in the quadratic mean.*

7.3.2 Constructing Ito's Stochastic System

When analyzing applied problems in which "white noise" is an idealized model of some real process with a small correlative time, it is naturally to understand the corresponding stochastic equation as a stochastic equation in *Stratonovich's form.**

We shall hold this viewpoint to make the meaning of the closed system (7.3.1), (7.3.3) more specific and consider it as a stochastic system in the Stratonovich form

$$\begin{aligned} A dx_1 &= [\alpha_1 x_1 + (B - C) x_2 x_3] dt + \beta_1 x_1 d^* w_1(t), \\ B dx_2 &= [\alpha_2 x_2 + (C - A) x_1 x_3] dt + \beta_2 x_2 d^* w_2(t), \\ C dx_3 &= (A - B) x_1 x_2 dt. \end{aligned} \tag{7.3.4}$$

A Stratonovich stochastic equation

$$d\boldsymbol{x} = \boldsymbol{X}(t, \boldsymbol{x}) dt + \sum_{k=1}^{r} \boldsymbol{\sigma}_k(t, \boldsymbol{x}) d^* w_k(t)$$

---

*Stratonovich, R.L. [1966] *Conditional Markov Processes and Their Application to Optimal Control Theory,* Moscow: Izdat. Moskov. Gos. Univ. (in Russian).

is equivalent to Ito's stochastic equation in which vector function $\tilde{\boldsymbol{X}}$ with components

$$\tilde{X}_i = X_i(t, \boldsymbol{x}) + \frac{1}{2}\sum_{k=1}^{r}\langle \boldsymbol{\sigma}_k(t, \boldsymbol{x}) \cdot \frac{\partial \sigma_{ki}}{\partial \boldsymbol{x}}\rangle \quad (i = 1, \ldots, n)$$

is substituted for $\boldsymbol{X}$.

Bearing in mind this dependency, we pass to *Ito's stochastic system*

$$\begin{aligned} A dx_1 &= \left[\left(\alpha_1 + \frac{\beta_1^2}{2A}\right) x_1 + (B - C) x_2 x_3\right] dt + \beta_1 x_1 dw_1(t), \\ B dx_2 &= \left[\left(\alpha_2 + \frac{\beta_2^2}{2B}\right) x_2 + (C - A) x_1 x_3\right] dt + \beta_2 x_2 dw_2(t), \\ C dx_3 &= (A - B) x_1 x_2 dt. \end{aligned} \tag{7.3.5}$$

7.3.3 Solution of Problem 7.3.1

Theorem 7.3.1. *If the conditions*

$$\beta_1^2 < -\alpha_1 A, \quad \beta_2^2 < -\alpha_2 B, \quad C < A, B \tag{7.3.6}$$

are satisfied, then the equilibrium position $x_1 = x_2 = x_3 = 0$ of system (7.3.1), (7.3.3) *is (globally) exponentially asymptotically stable with respect to x_1, x_2 in the quadratic mean.*

Proof. To prove the theorem, we consider a Lyapunov V-function of the form

$$V(\boldsymbol{x}) = (AB)^{-1}\left[A(A - C)x_1^2 + B(B - C)x_2^2\right]. \tag{7.3.7}$$

Let us evaluate the differential generator LV of function V along the trajectories of system (7.3.5). To this end, we use the expressions

$$\boldsymbol{\sigma}_k = (\sigma_{k1}, \sigma_{k2}, \sigma_{k3})^{\mathsf{T}} \quad (k = 1, 2),$$

$$\sigma_{11} = \beta_1 A^{-1} x_1, \quad \sigma_{22} = \beta_2 B^{-1} x_2, \quad \sigma_{12} = \sigma_{13} = \sigma_{21} = \sigma_{23} = 0,$$

$$\begin{aligned} \sum_{k=1}^{2}\left(\boldsymbol{\sigma}_k^{\mathsf{T}} \frac{\partial}{\partial \boldsymbol{x}}\right)^2 V &= \sigma_{11}^2 \frac{\partial^2 V}{\partial x_1^2} + \sigma_{22}^2 \frac{\partial^2 V}{\partial x_2^2} \\ &= 2(AB)^{-1}\left[A^{-1}(A - C)\beta_1^2 x_1^2 + B^{-1}(B - C)\beta_2^2 x_2^2\right]. \end{aligned}$$

As a result,

$$LV = \frac{A - C}{B}\left[\frac{2}{A}\left(\alpha_1 + \frac{\beta_1^2}{2A}\right) + \frac{\beta_1^2}{A^2}\right] x_1^2 + \frac{B - C}{A}\left[\frac{2}{B}\left(\alpha_2 + \frac{\beta_2^2}{2B}\right) + \frac{\beta_2^2}{B^2}\right] x_2^2.$$

If conditions (7.3.6) are satisfied, V-function (7.3.7) satisfies all the conditions of Theorem 7.2.1 for $\boldsymbol{y} = (x_1, x_2)$ and $N = 2$. This means that the zero solution $x_1 = x_2 = x_3 = 0$ of system (7.3.5) is (globally) exponentially asymptotically stable with respect to x_1, x_2 in the quadratic mean.

Because system (7.3.5) is "equivalent" to system (7.3.1), (7.3.3), Theorem 7.3.1 is proved. □

Remark 7.3.1.

1. Under conditions (7.3.6), the Lyapunov function

$$W(\boldsymbol{x}) = \tfrac{1}{2}(Ax_1^2 + Bx_2^2 + Cx_3^2) \tag{7.3.8}$$

(*kinetic energy* of the solid) satisfies the condition $LW(\boldsymbol{x}) \leq 0$ for all x_i. This means that the zero solution $x_1 = x_2 = x_3 = 0$ of system (7.3.5) is (nonasymptotically) stable *with respect to all* the variables in *probability* by Theorem 7.2.2.

2. The equilibrium position $x_1 = x_2 = x_3 = 0$ of system (7.3.1), (7.3.3) is also exponentially asymptotically stable with respect to x_1, x_2 in the quadratic mean *on the whole with respect to* x_{30}.

3. We call the reader's attention to the fact that the *smooth* V-functions (7.2.8) and (7.3.7) were employed in Example 7.2.1 and in the proof of Theorem 7.3.1. Also the W-function (7.3.8) is smooth.

7.4 The Stability of Linear Systems with Respect to Part of the Variables

7.4.1 The Linear Stochastic System. Formulation of the Problem

Consider the case when equations (7.1.1) are linear equations with constant coefficients (for simplicity, we take $r = 1$)

$$dx_i = \left[\sum_{j=1}^{n} A_{ij}x_j\right]dt + \left[\sum_{j=1}^{n} A'_{ij}x_j\right]dw_1(t) \qquad (i = 1, \dots, n),$$

$$\boldsymbol{x}^{\mathsf{T}} = (x_1, \dots, x_n) = (y_1, \dots, y_m, z_1, \dots, z_p) = (\boldsymbol{y}^{\mathsf{T}}, \boldsymbol{z}^{\mathsf{T}}),$$

$$m > 0, \quad p \geq 0, \quad n = m + p.$$

In $\boldsymbol{y}, \boldsymbol{z}$ variables, the system takes the form

$$\begin{aligned} dy_i &= \left[\sum_{k=1}^{m} a_{ik}y_k + \sum_{l=1}^{p} b_{il}z_l\right]dt + \left[\sum_{k=1}^{m} a'_{ik}y_k + \sum_{l=1}^{p} b'_{il}z_l\right]dw_1(t), \\ dz_j &= \left[\sum_{k=1}^{m} c_{jk}y_k + \sum_{l=1}^{p} d_{jl}z_l\right]dt + \left[\sum_{k=1}^{m} c'_{jk}y_k + \sum_{l=1}^{p} d'_{jl}z_l\right]dw_1(t) \\ &\qquad (i = 1, \dots, m;\ j = 1, \dots, p), \end{aligned} \tag{7.4.1}$$

where a_{ik}, a'_{ik}, b_{il}, b'_{il}, c_{jk}, c'_{jk}, d_{jl}, and d'_{jl} are constants.

We shall study the problem of exponential asymptotic stability in the quadratic mean of the unperturbed motion $\boldsymbol{x} = \boldsymbol{0}$ of system (7.4.1) with respect to variables $y_1, \dots, y_m$.

We note that for the linear system (7.4.1) the *global Lipschitz condition* is satisfied, which guarantees, for all t_0, $\boldsymbol{x}_0$, the existence of a unique solution defined for all $t \geq t_0$.

7.4.2 Constructing an Auxiliary Linear System

Consider vectors $\boldsymbol{b}_i = (b_{i1}, \dots, b_{ip})^\top$ and $\boldsymbol{b}'_i = (b'_{i1}, \dots, b'_{ip})^\top$ $(i = 1, \dots, m)$ composed of the corresponding coefficients of system (7.4.1). Without loss of generality, suppose that the first m_1 vectors $\boldsymbol{b}_1, \dots, \boldsymbol{b}_{m_1}$ $(m_1 \leq m)$ and the first m_2 vectors $\boldsymbol{b}'_1, \dots, \boldsymbol{b}'_{m_2}$ $(m_2 \leq m)$ constitute a linearly independent basis of the system $\boldsymbol{b}_j$, $\boldsymbol{b}'_j$ $(j = 1, \dots, m)$.

To construct an auxiliary linear system, we introduce the new variables

$$\mu_i = \sum_{l=1}^{p} b_{il} z_l, \quad \mu_{m_1+j} = \sum_{l=1}^{p} b'_{jl} z_l \qquad (i = 1, \dots, m_1,\ j = 1, \dots, m_2).$$

With new variables thus introduced, two cases are possible.

The first case. The following system can be formed:

$$\begin{aligned} dy_i &= \left[\sum_{k=1}^{m} a_{ik} y_k + \sum_{l=1}^{m_1+m_2} \alpha_{il} \mu_l\right] dt + \left[\sum_{k=1}^{m} a'_{ik} y_k + \sum_{l=1}^{m_1+m_2} \alpha'_{il} \mu_l\right] dw_1(t), \\ d\mu_j &= \left[\sum_{k=1}^{m} a^*_{jk} y_k + \sum_{l=1}^{m_1+m_2} \alpha^*_{jl} \mu_l\right] dt + \left[\sum_{k=1}^{m} a'^*_{jk} y_k + \sum_{l=1}^{m_1+m_2} \alpha'^*_{jl} \mu_l\right] dw_1(t) \\ &\qquad (i = 1, \dots, m;\ j = 1, \dots, m_1 + m_2), \end{aligned} \tag{7.4.2}$$

where a^*_{jk}, a'^*_{jk}, α_{il}, α'_{il}, α^*_{jl}, and α'^*_{jl} are constants. The behavior of variables $y_1, \dots, y_m$ of system (7.4.1), with respect to which we study the problem of stability of the unperturbed motion, is completely determined by system (7.4.2). This system is referred to as the ***μ*-*system*** relative to the original system (7.4.1).

The second case. Suppose, without loss of generality, that all $m_1 + m_2$ equalities in the second group of equations of system (7.4.2) cannot be simultaneously satisfied. In this case system (7.4.1) comprises the equations

$$\begin{aligned} dy_i &= \left[\sum_{k=1}^{m} a_{ik} y_k + \sum_{l=1}^{m_1+m_2} \alpha_{il} \mu_l\right] dt + \left[\sum_{k=1}^{m} a'_{ik} y_k + \sum_{l=1}^{m_1+m_2} \alpha'_{il} \mu_l\right] dw_1(t), \\ d\mu_j &= \left[\sum_{k=1}^{m} a^*_{jk} y_k + \sum_{l=1}^{p} b^*_{jl} z_l\right] dt + \left[\sum_{k=1}^{m} a'^*_{jk} y_k + \sum_{l=1}^{p} b'^*_{jl} z_l\right] dw_1(t), \\ dz_s &= \left[\sum_{k=1}^{m} c_{sk} y_k + \sum_{l=1}^{p} d_{sl} z_l\right] dt + \left[\sum_{k=1}^{m} c'_{sk} y_k + \sum_{l=1}^{p} d'_{sl} z_l\right] dw_1(t) \\ &\qquad (i = 1, \dots, m;\ j = 1, \dots, m_1 + m_2;\ s = 1, \dots, p), \end{aligned}$$

where b^*_{jl} and b'^*_{jl} are constants.

Assuming that the first $m_3 \leq m_1 + m_2$ vectors $\boldsymbol{b}^*_j = (b^*_{j1}, \dots, b^*_{jp})$ and the first $m_4 \leq m_1 + m_2$ vectors $\boldsymbol{b}'^*_j = (b'^*_{j1}, \dots, b'^*_{jp})$ form a basis of the system $\boldsymbol{b}^*_j$,

$b_j'^*$ $(j = 1, \ldots, m_1 + m_2)$, we introduce the new variables once more

$$\mu_{m_1+m_2+i} = \sum_{l=1}^{p} b_{il}^* z_l, \quad \mu_{m_1+m_2+m_3+j} = \sum_{l=1}^{p} b_{jl}'^* z_l$$
$$(i = 1, \ldots, m_3;\ j = 1, \ldots, m_4).$$

Successively repeating the indicated process of introducing new variables, for system (7.4.1) we can always construct a $\boldsymbol{\mu}$-system of a dimension not exceeding that of the original system (7.4.1).

7.4.3 The Condition for Stability with Respect to Part of the Variables

Let us introduce the designations $B_0 = \{(b_{il})^\top, (b_{il}')^\top\}$, $D_0 = (d_{jl})^\top$, and $D_1 = (d_{jl}')^\top$. Consider the matrices

$$K_0 = B_0, \quad K_i = (D_0 K_{i-1}, D_1 K_{i-1}) \quad (i = 1, \ldots, p)$$

and matrices G_4, G_5 constructed in the same way as matrices L_4, L_5 from Subsection 1.1.4.

Theorem 7.4.1. *Suppose that the zero solution* $\boldsymbol{\xi} = \mathbf{0}$ *of the auxiliary* $\boldsymbol{\mu}$*-system*

$$\begin{gathered} d\boldsymbol{\xi} = (G_4 A^* G_5 \boldsymbol{\xi})dt + (G_4 A'^* G_5 \boldsymbol{\xi}) dw_1(t) \\ A^* = (A_{ij}), \quad A'^* = (A_{ij}') \quad (i, j = 1, \ldots, n) \end{gathered} \tag{7.4.3}$$

is exponentially asymptotically stable in the quadratic mean with respect to all the variables.

Then the unperturbed motion $\boldsymbol{x} = \mathbf{0}$ *of the original system* (7.4.1) *is exponentially asymptotically stable in the quadratic mean with respect to variables* $y_1, \ldots, y_m$.

Proof. The auxiliary $\boldsymbol{\mu}$-system is derived by applying the linear transformation $\boldsymbol{w} = G\boldsymbol{x}$ to the original system (7.4.1). In this case matrix G has the form (6.4.6) and the $\boldsymbol{\mu}$-system is composed of the first $m + h$ equations of the system

$$d\boldsymbol{\eta} = (GA^* G^{-1} \boldsymbol{\eta})dt + (GA'^* G^{-1} \boldsymbol{\eta}) dw_1(t). \tag{7.4.4}$$

Following the proof scheme for Theorem 1.1.1, we can demonstrate that the first $m + h$ equations of system (7.4.4) have the form (7.4.3).

Since $\xi_i = y_i$ $(i = 1, \ldots, m)$ and there is a constant α such that $\|\boldsymbol{\xi}_0\| \leq \alpha \|\boldsymbol{x}_0\|$, then under the conditions of the theorem the unperturbed motion $\boldsymbol{x} = \mathbf{0}$ of system (7.4.1) is exponentially asymptotically stable in quadratic mean with respect to variables $y_1, \ldots, y_m$. The theorem is proved. □

Example 7.4.1. Let system (7.4.1) have the form

$$\begin{gathered} dy_1 = (a_{11}y_1 + z_1 - z_2)dt + a'_{11}y_1dw_1(t), \\ dz_1 = z_2dt + z_1dw_1(t), \quad dz_2 = (-y_1 + z_2)dt + z_1dw_1(t). \end{gathered} \tag{7.4.5}$$

The auxiliary $\boldsymbol{\mu}$-system comprises the equations

$$\begin{aligned} dy_1 &= (a_{11}y_1 + \mu_1)dt + a'_{11}y_1dw_1(t), \\ d\mu_1 &= y_1dt. \end{aligned} \tag{7.4.6}$$

If the condition $2a_{11} < -(a'_{11})^2$ is satisfied, the zero solution $y_1 = \mu_1 = 0$ of the auxiliary $\boldsymbol{\mu}$-system (7.4.6) is exponentially asymptotically stable in the quadratic mean with respect to y_1, μ_1. Therefore, the unperturbed motion $y_1 = z_1 = z_2 = 0$ of system (7.4.5) is exponentially asymptotically stable in the quadratic mean with respect to y_1.

7.5 Stability with Respect to Part of the Variables in a Linear Approximation

7.5.1 The Original System of Equations

Let the equations of perturbed motion (7.1.1) have the form (for simplicity, we take $r = 1$)

$$\begin{gathered} dx_i = \left[\sum_{j=1}^{n} A_{ij}x_j + X_i(t, \boldsymbol{x})\right]dt + \left[\sum_{j=1}^{n} A'_{ij}x_j + X'_i(t, \boldsymbol{x})\right]dw_1(t) \quad (i = 1, \ldots, n), \\ \boldsymbol{x}^{\mathsf{T}} = (x_1, \ldots, x_n) = (y_1, \ldots, y_m, z_1, \ldots, z_p) = (\boldsymbol{y}^{\mathsf{T}}, \boldsymbol{z}^{\mathsf{T}}), \\ m > 0, \quad p \geq 0, \quad n = m + p, \end{gathered} \tag{7.5.1}$$

where A_{ij} and A'_{ij} are constants and X_i and X'_i are nonlinear perturbations.

In $\boldsymbol{y}, \boldsymbol{z}$ variables, the system can be written as

$$\begin{aligned} dy_i &= \left[\sum_{k=1}^{m} a_{ik}y_k + Y_i(t, \boldsymbol{x})\right]dt + \left[\sum_{k=1}^{m} a'_{ik}y_k + Y'_i(t, \boldsymbol{x})\right]dw_1(t), \\ dz_j &= \left[\sum_{k=1}^{m} c_{jk}y_k + \sum_{l=1}^{p} d_{jl}z_l + Z_j(t, \boldsymbol{x})\right]dt \\ &\quad + \left[\sum_{k=1}^{m} c'_{jk}y_k + \sum_{l=1}^{p} d'_{jl}z_l + Z'_j(t, \boldsymbol{x})\right]dw_1(t) \\ &\qquad (i = 1, \ldots, m;\ j = 1, \ldots, p), \end{aligned} \tag{7.5.2}$$

where a_{ik}, a'_{ik}, c_{jk}, c'_{jk}, d_{jl}, and d'_{jl} are constants and Y_i, Y'_i, Z_j, and Z'_j are nonlinear perturbations.

We note that the first group of equations of system (7.5.2) contains no terms linear in z_j. When analyzing the $\boldsymbol{y}$-stability of the unperturbed motion of system (7.5.1), such an assumption can be made if the linear part of system (7.5.1) is transformed as indicated in Section 7.4.

Let us single out those terms of the nonlinear expressions which are *finite-order homogeneous forms* in variables $x_1, \ldots, x_n$, i.e., let us represent Y_i, Y_i' in the form

$$Y_i = Y_i^*(\boldsymbol{x}) + Y_i^{**}(t, \boldsymbol{x}), \quad Y_i' = Y_i'^*(\boldsymbol{x}) + Y_i'^{**}(t, \boldsymbol{x}),$$

$$Y_i^* = \sum_{l=2}^{r} U_l^{(i)}(\boldsymbol{x}), \quad Y_i'^* = \sum_{l=2}^{s} U_l'^{(i)}(\boldsymbol{x}) \quad (i = 1, \ldots, m),$$

where $U_l^{(i)}$ and $U_l'^{(i)}$ are homogeneous forms in variables $x_1, \ldots, x_n$ of order l, $l < \max(r, s)$ and r and s are finite numbers.

7.5.2 The Transformation of System (7.5.2)

Now we introduce the new variables

$$\mu_i = Y_i^*(\boldsymbol{x}), \quad \mu_{m+i} = Y_i'^*(\boldsymbol{x}) \quad (i = 1, \ldots, m). \tag{7.5.3}$$

Without loss of generality we assume that the vectors corresponding to these variables are linearly independent (see Subsection 2.2.3).

Applying *Ito's stochastic differentiation formula*, we obtain

$$d\mu_j = \left[Y_i^{(1)}(\boldsymbol{x}) + Y_i^{(2)}(t, \boldsymbol{x})\right]dt + \left[Y_i'^{(1)}(\boldsymbol{x}) + Y_i'^{(2)}(t, \boldsymbol{x})\right]dw_1(t),$$

$$Y_i^{(1)} = \sum_{k=1}^{n} \frac{\partial Y_i^*(\boldsymbol{x})}{\partial x_k} \sum_{s=1}^{n} A_{ks} x_s + \frac{1}{2} \sum_{k,l=1}^{n} \frac{\partial^2 Y_i^*(\boldsymbol{x})}{\partial x_k \partial x_l} \left(\sum_{s=1}^{n} A'_{ks} x_s\right)\left(\sum_{s=1}^{n} A'_{ls} x_s\right),$$

$$Y_i^{(2)} = \sum_{k=1}^{n} \frac{\partial Y_i^*(\boldsymbol{x})}{\partial x_k} X_k(t, \boldsymbol{x}) + \frac{1}{2} \sum_{k,l=1}^{n} \frac{\partial^2 Y_i^*(\boldsymbol{x})}{\partial x_k \partial x_l} X'_k(t, \boldsymbol{x}) X'_l(t, \boldsymbol{x}), \tag{7.5.4}$$

$$Y_i'^{(1)} = \sum_{k=1}^{n} \frac{\partial Y_i^*(\boldsymbol{x})}{\partial x_k} \sum_{s=1}^{n} A'_{ks} x_s, \quad Y_i'^{(2)} = \sum_{k=1}^{n} \frac{\partial Y_i^*(\boldsymbol{x})}{\partial x_k} X'_k(t, \boldsymbol{x}) \quad (i = 1, \ldots, m),$$

where expressions for $Y_i^{(1)}$ and $Y_i'^{(1)}$ are the aggregate of forms of the same order which appear in expression for Y_i^*. By analogy, we can derive expressions for $d\mu_{m+i}$ $(i = 1, \ldots, m)$.

Having introduced the new variables (7.5.3), we consider two possible cases.

The first case. The expressions for $Y_i^{(1)}$ and $Y_i'^{(1)}$ from (7.5.4) and the analogous expressions in formulas for $d\mu_{m+i}$ $(i = 1, \ldots, m)$ have the form

$$Y_j^{(1)} = \sum_{l=1}^{2m} e_{jl} \mu_l, \quad Y_j'^{(1)} = \sum_{l=1}^{2m} e'_{jl} \mu_l, \quad (j = 1, \ldots, 2m). \tag{7.5.5}$$

In this case, system (7.5.2) is transformed as follows:

$$\begin{aligned} dy_i &= \left[\sum_{k=1}^{m} a_{ik} y_k + \mu_i + Y_i^{**}(t,\boldsymbol{x})\right] dt \\ &\quad + \left[\sum_{k=1}^{m} a'_{ik} y_k + \mu_{m+i} + Y_i^{\prime **}(t,\boldsymbol{x})\right] dw_1(t), \\ d\mu_j &= \left[\sum_{l=1}^{2m} e_{jl}\mu_l + Y_j^{(2)}(t,\boldsymbol{x})\right] dt + \left[\sum_{l=1}^{2m} e'_{jl}\mu_l + Y_j^{\prime(2)}(t,\boldsymbol{x})\right] dw_1(t), \\ dz_s &= \left[\sum_{k=1}^{m} c_{sk} y_k + \sum_{l=1}^{p} d_{sl} z_l + Z_s(t,\boldsymbol{x})\right] dt \\ &\quad + \left[\sum_{k=1}^{m} c'_{sk} y_k + \sum_{l=1}^{p} d'_{sl} z_l + Z'_s(t,\boldsymbol{x})\right] dw_1(t) \\ &\qquad (i = 1,\dots,m;\ j = 1,\dots,2m;\ s = 1,\dots,p), \end{aligned} \tag{7.5.6}$$

where e_{jl} and e'_{jl} are constants. The linear part of the first two groups of equations of system (7.5.6) will be called the *$\boldsymbol{\mu}$-system of linear approximation* for the original system (7.5.2).

The second case. System (7.5.2) is not reduced to the form (7.5.6). For example, let none of the equalities (7.5.5) be possible. In this case we introduce the new variables once more:

$$\mu_{2m+i} = Y_i^{(1)}(\boldsymbol{x}), \quad \mu_{3m+i} = Y_i^{\prime(1)}(\boldsymbol{x}), \quad (i = 1,\dots,m). \tag{7.5.7}$$

This will result in reducing system (7.5.2) to the form (7.5.6) or further introducing new variables of type (7.5.3), (7.5.7).

Since functions Y_i^*, $Y_i^{\prime *}$ are *finite-order* forms in variables $x_1,\dots,x_n$, using the proof scheme for Lemma 2.2.1, we can show that, at some finite step of introducing new variables, it is possible to construct a *finite-dimensional* system of type (7.5.6) for system (7.5.2). The linear part of such a system (except for the last group of equations) forms the $\boldsymbol{\mu}$-system of linear approximation for the original system (7.5.2).

7.5.3 The Condition for Stability with Respect to Part of the Variables

Suppose, without loss of generality, that the $\boldsymbol{\mu}$-system of linear approximation for the original system (7.5.2) is formed at the first step of introducing new variables.

We shall use the following designations:

(1) $\boldsymbol{\xi}$ is a vector composed of variables y_i, μ_j $(i = 1,\dots,m;\ j = 1,\dots,2m)$, which determine the state of the $\boldsymbol{\mu}$-system of linear approximation;

(2) $\boldsymbol{Y}_*$ is a vector function composed of functions Y_i^{**}, $Y_i^{\prime **}$, $Y_j^{(2)}$, and $Y_j^{\prime(2)}$ $(i = 1,\dots,m;\ j = 1,\dots,2m)$.

Suppose that the following equation is valid in the domain $t \geq 0$, $||\boldsymbol{\xi}|| \leq H$, $||\boldsymbol{z}|| < \infty$:

$$||\boldsymbol{Y}_*(t, \boldsymbol{x})|| \leq \alpha ||\boldsymbol{\xi}||, \tag{7.5.8}$$

where α and H are sufficiently small positive constants.

Theorem 7.5.1. *Let the zero solution $\boldsymbol{\xi} = \boldsymbol{0}$ of the auxiliary $\boldsymbol{\mu}$-system of linear approximation for system* (7.5.2) *be asymptotically stable in probability with respect to all the variables. Then, if condition* (7.5.8) *is satisfied, the unperturbed motion $\boldsymbol{x} = \boldsymbol{0}$ of system* (7.5.2) *is asymptotically $\boldsymbol{y}$-stable in probability.*

Proof. Under the conditions of the theorem, for the $\boldsymbol{\mu}$-system of linear approximation, there is a function $V(\boldsymbol{\xi})$ such that, for some $l > 0$, $k_i > 0$, the following estimates are valid (Khas'minskii [1969]):

$$k_1 ||\boldsymbol{\xi}||^l \leq V(\boldsymbol{\xi}) \leq k_2 ||\boldsymbol{\xi}||^l, \quad L_\mu V \leq -k_3 ||\boldsymbol{\xi}||^l,$$
$$\left| \frac{\partial V}{\partial \xi_i} \right| \leq k_4 ||\boldsymbol{\xi}||^{l-1}, \quad \left| \frac{\partial^2 V}{\partial \xi_i \partial \xi_j} \right| \leq k_4 ||\boldsymbol{\xi}||^{l-2},$$

where L_μ is a differential generator by virtue of the system of linear approximation.

One can demonstrate that under condition (7.5.8) the following estimate is valid on the trajectories of the original nonlinear system (7.5.2):

$$LV(\boldsymbol{\xi}) \leq k ||\boldsymbol{\xi}||^l, \quad k = \text{const} > 0.$$

By Theorem 7.2.2, the unperturbed motion $\boldsymbol{x} = \boldsymbol{0}$ of system (7.5.1) is asymptotically $\boldsymbol{y}$-stable in probability. The theorem is proved. □

Remark 7.5.1.

1. Conditions for stability in the linear approximation for Ito's stochastic systems in the problem of stability with respect to *all* the variables were obtained by Khas'minskii [1969]. Theorem 7.5.1 is an extension of these conditions to the problem of *partial* stability.

2. As in the deterministic case (see Remark 2.2.3), in constructing auxiliary systems of type (7.5.6) one can use the method of "*splitting*" the nonlinear terms Y_i and Y_i' $(i = 1, \ldots, m)$.

Example 7.5.1.

1. Let the equations of perturbed motion (7.5.1) have the form

$$\begin{aligned} dy_1 &= (a y_1 + y_1^2 z_1) dt, \\ dz_1 &= [b z_1 + Z_1(t, y_1, z_1)] dt + b' z_1 dw_1(t), \end{aligned} \tag{7.5.9}$$

where a, b, and b' are constants. We consider the problem of asymptotic stability in probability of the unperturbed motion $y_1 = z_1 = 0$ with respect to y_1.

By introducing the new variable $\mu_1 = y_1 z_1$ system (7.5.9) is transformed in the following way:

$$\begin{aligned} dy_1 &= (ay_1 + y_1\mu_1)dt, \\ d\mu_1 &= [(2a+b)\mu_1 + \mu_1^2 + y_1 Z_1(t, y_1, z_1)]dt + b'\mu_1 dw_1(t), \\ dz_1 &= [bz_1 + Z_1(t, y_1, z_1)]dt + b' z_1 dw_1(t). \end{aligned}$$

If the conditions

$$a < 0, \qquad 2a + b < \tfrac{1}{2}(b')^2 \tag{7.5.10}$$

are satisfied, the zero solution $y_1 = \mu_1 = 0$ of the auxiliary $\boldsymbol{\mu}$-system of linear approximation

$$\begin{aligned} dy_1 &= ay_1 dt, \\ d\mu_1 &= (2a+b)\mu_1 dt + b'\mu_1 dw_1(t) \end{aligned}$$

is asymptotically stable (with respect to y_1, μ_1) in probability.

In the domain $t \geq 0$, $\|\boldsymbol{\xi}\| \leq H$, $|z_1| < \infty$ ($\boldsymbol{\xi} = (y_1, \mu_1)$), let the following condition also be satisfied:

$$|y_1 Z_1(t, y_1, z_1)| \leq \gamma_1 |y_1| + \gamma_2 |\mu_1|, \tag{7.5.11}$$

where γ_1, γ_2, and H are sufficiently small positive constants.

If conditions (7.5.10), (7.5.11) are satisfied, the unperturbed motion $y_1 = z_1 = 0$ of system (7.5.9) is asymptotically stable in probability with respect to y_1 by Theorem 7.5.1.

2. Let the equations of perturbed motion (7.5.1) have the form

$$\begin{aligned} dy_1 &= [-y_1 + y_1(z_1 z_2 - 2z_1 z_3)]dt, \\ dz_i &= [\Sigma_i + Z_i(t, y_1, z_1, z_2, z_3)]dt + z_i dw_1(t) \quad (i = 1, 2, 3), \\ \Sigma_1 &= ay_1 - z_1, \quad \Sigma_2 = by_1 + 4z_1 + z_2, \quad \Sigma_3 = cy_1 + 2z_1 + z_2 - z_3, \end{aligned} \tag{7.5.12}$$

where a, b, and c are constants. Introducing the new variable $\mu_1 = z_1 z_2 - 2z_1 z_3$, we obtain the equations

$$\begin{aligned} dy_1 &= (-y_1 + y_1\mu_1)dt, \\ d\mu_1 &= (-\mu_1 + y_1\mu_2 + Z_4)dt + 2\mu_1 dw_1(t), \\ d\mu_2 &= (-\mu_2 + Z_5)dt + \mu_2 dw_1(t), \\ dz_i &= (\Sigma_i + Z_i)dt + z_i dw_1(t) \quad (i = 1, 2, 3), \\ \mu_2 &= (b - 2c)z_1 - a(z_2 - 2z_3), \\ Z_4 &= Z_1(z_1 - 2z_3) + z_1(Z_2 - 2Z_3), \\ Z_5 &= (b - 2c)Z_1 - a(Z_2 - 2Z_3). \end{aligned} \tag{7.5.13}$$

The zero solution $y_1 = \mu_1 = \mu_2 = 0$ of the linear part of the first three equations of system (7.5.13) (which is the auxiliary $\boldsymbol{\mu}$-system of linear approximation for the original system (7.5.12)) is asymptotically stable in probability for arbitrary values of a, b, and c.

Consequently, if condition (7.5.8) with $\boldsymbol{Y}_* = (Z_4, Z_5)$, $\boldsymbol{\xi} = (y_1, \mu_1, \mu_2)$ is satisfied, the unperturbed motion $y_1 = z_1 = z_2 = 0$ of system (7.5.12) is asymptotically stable in probability with respect to y_1.

7.6 The Stabilization of Nonlinear Controlled Systems with Respect to Part of the Variables

7.6.1 Formulation of the Problem

Suppose that the perturbed motion of a controlled object is described by a nonlinear system of stochastic differential equations in Ito's form

$$\begin{gathered} d\boldsymbol{x} = \boldsymbol{A}(t,\boldsymbol{x}) + \boldsymbol{F}(t,\boldsymbol{x},\boldsymbol{u})dt + \boldsymbol{\sigma}(t,\boldsymbol{x})dw_1(t), \\ \boldsymbol{x}^{\top} = (y_1,\ldots,y_m,z_1,\ldots,z_p) = (\boldsymbol{y}^{\top},\boldsymbol{z}^{\top}), \quad \boldsymbol{u} = (u_1,\ldots,u_l), \\ \boldsymbol{A}(t,\boldsymbol{0}) \equiv \boldsymbol{0}, \quad \boldsymbol{F}(t,\boldsymbol{0},\boldsymbol{0}) \equiv \boldsymbol{0} \quad \boldsymbol{\sigma}(t,\boldsymbol{0}) \equiv \boldsymbol{0}, \end{gathered} \tag{7.6.1}$$

where $\boldsymbol{A}(t,\boldsymbol{x})$, $\boldsymbol{F}(t,\boldsymbol{x},\boldsymbol{u})$, and $\boldsymbol{\sigma}(t,\boldsymbol{x})$ are vector functions of dimension $n\times 1$. We assume that the right-hand sides of equations (7.6.1) are defined and continuous in domain (0.3.2).

We shall seek the control $\boldsymbol{u}$ in the form $\boldsymbol{u} = \boldsymbol{u}(t,\boldsymbol{x})$, $\boldsymbol{u}(t,\boldsymbol{0}) \equiv \boldsymbol{0}$ and assume that $\boldsymbol{u}(t,\boldsymbol{x})$ is defined and continuous in the domain $t\geq 0$, $||\boldsymbol{y}||\leq H$, $||\boldsymbol{z}||<\infty$, whereas system (7.6.1) at $\boldsymbol{u}=\boldsymbol{u}(t,\boldsymbol{x})$ satisfies all the conditions imposed on system (7.1.1). (In this case, $\boldsymbol{u}\in\mathfrak{U}$.)

Problem 7.6.1. Find control rules $\boldsymbol{u} = \boldsymbol{u}^0(t,\boldsymbol{x}) \in \mathfrak{U}$ that ensure asymptotic $\boldsymbol{y}$-stability in probability of the unperturbed motion $\boldsymbol{x}=\boldsymbol{0}$ of system (7.6.1).

7.6.2 Theorem on Stabilization with Respect to Part of the Variables

Theorem 7.6.1. *Let there exist control rules $\boldsymbol{u} = \boldsymbol{u}^0(t,\boldsymbol{x}) \in \mathfrak{U}$ and N continuously differentiable functions $\mu_i = \mu_i(\boldsymbol{x})$, $\mu_i(\boldsymbol{0}) = \boldsymbol{0}$ such that from system* (7.6.1) *it possible to segregate the equations*

$$\begin{gathered} d\boldsymbol{y} = \boldsymbol{A}^{(1)}(t,\boldsymbol{y},\boldsymbol{\mu})dt + \boldsymbol{\sigma}^{(1)}(t,\boldsymbol{y},\boldsymbol{\mu})dw_1(t), \\ d\boldsymbol{\mu} = \boldsymbol{A}^{(2)}(t,\boldsymbol{y},\boldsymbol{\mu})dt + \boldsymbol{\sigma}^{(2)}(t,\boldsymbol{y},\boldsymbol{\mu})dw_1(t), \\ \boldsymbol{A}^{(i)}(t,\boldsymbol{0},\boldsymbol{0}) \equiv \boldsymbol{0}, \quad \boldsymbol{\sigma}^{(i)}(t,\boldsymbol{0},\boldsymbol{0}) \equiv \boldsymbol{0} \quad (i=1,2), \quad \boldsymbol{\mu} = (\mu_1,\ldots,\mu_N)^{\top}, \\ \boldsymbol{A}^{(1)} = (A_1+F_1,\ldots,A_m+F_m)^{\top}, \quad \boldsymbol{\sigma}^{(1)} = (\sigma_1,\ldots,\sigma_m)^{\top}, \\ \boldsymbol{A}^{(2)} = \Bigg[\sum_{i=1}^{n}\frac{\partial\mu_1}{\partial x_i}(A_i+F_i) + \frac{1}{2}\sum_{i,j=1}^{n}\frac{\partial^2\mu_1}{\partial x_i\partial x_j}\sigma_i\sigma_j,\ldots, \\ \sum_{i=1}^{n}\frac{\partial\mu_N}{\partial x_i}(A_i+F_i) + \frac{1}{2}\sum_{i,j=1}^{n}\frac{\partial^2\mu_N}{\partial x_i\partial x_j}\sigma_i\sigma_j\Bigg]^{\top}, \\ \boldsymbol{\sigma}^{(2)} = \Bigg[\sum_{i=1}^{n}\frac{\partial\mu_1}{\partial x_i}\sigma_i,\ldots,\sum_{i=1}^{n}\frac{\partial\mu_N}{\partial x_i}\sigma_i\Bigg]^{\top}, \\ \boldsymbol{A} = (A_1,\ldots,A_n)^{\top}, \quad \boldsymbol{F} = (F_1,\ldots,F_n)^{\top}, \quad \boldsymbol{\sigma} = (\sigma_1,\ldots,\sigma_n)^{\top} \end{gathered} \tag{7.6.2}$$

and in the domain $t \geq 0$, $\|\boldsymbol{y}\| \leq H$, $\|\boldsymbol{\mu}\| \leq H$, *conditions of the existence and uniqueness theorem are satisfied.*

If the solution $\boldsymbol{y} = \mathbf{0}$, $\boldsymbol{\mu} = \mathbf{0}$ *of system* (7.6.2) *is asymptotically stable in probability, then control rules* $\boldsymbol{u} = \boldsymbol{u}^0(t, \boldsymbol{x}) \in \mathcal{U}$ *solve Problem* 7.6.1 *on* $\boldsymbol{y}$*-stabilization of system* (7.6.1).

The proof follows the proof scheme for Theorem 4.1.1, taking the stochastic analog of the procedure for constructing the $\boldsymbol{\mu}$-system into account.

Remark 7.6.1.

1. Theorem 7.6.1 allows analyzing the problem of stabilization with respect to part of the variables to be reduced to the problem of *stability* with respect to *all* the variables for system (7.6.2).

2. Another approach to solving Problem 7.6.1 can be devised based on the stochastic version of the method of Lyapunov functions, considered in Section 7.2.

Example 7.6.1. Consider the problem of *damping the angular rotations of a solid* (*spacecraft*) about its center of mass in a field of linear dissipative, accelerating, and random forces. In this case, let Ito's stochastic system have the form

$$\begin{aligned} A dy_1 &= [\alpha y_1 + (B - C) z_1 z_2] dt, \\ B dz_1 &= [\beta z_1 + (C - A) y_1 z_2 + u_1] dt + z_1 dw_1(t), \\ C dz_2 &= [\gamma z_2 + (A - B) y_1 z_1 + u_2] dt + z_2 dw_1(t), \end{aligned} \tag{7.6.3}$$

where A, B, and C are central moments of inertia of the solid and coefficients α, β, and γ are constants that characterize the total effect of the forces.

We consider Problem 7.6.1 on stabilization of the equilibrium position $y_1 = z_1 = z_2 = 0$ of the solid with respect to y_1 by using controls u_1, u_2.

Taking u_1, u_2 and an auxiliary $\boldsymbol{\mu}$-function in the form

$$u_1 = (A - C) y_1 z_2, \quad u_2 = (B - A) y_1 z_1, \quad \mu_1 = z_1 z_2, \tag{7.6.4}$$

we arrive at the $\boldsymbol{\mu}$-system

$$\begin{aligned} A dy_1 &= [\alpha y_1 + (B - C) \mu_1] dt, \\ BC d\mu_1 &= (C\beta + B\gamma + 1) \mu_1 dt + (B + C) \mu_1 dw_1(t). \end{aligned} \tag{7.6.5}$$

By introducing the new variable $\eta_1 = [\alpha y_1 + (B - C)\mu_1] A^{-1}$, we reduce equations (7.6.5) to the form more convenient for analysis:

$$\begin{gathered} dy_1 = \eta_1 dt, \quad d\eta_1 = (k_1 y_1 + k_2 \eta_1) dt + (k_3 y_1 + k_4 \eta_1) d\eta_1(t), \\ k_1 = -\frac{\alpha k^*}{ABC}, \quad k_2 = \frac{A k^* + \alpha BC}{ABC}, \\ k_3 = -\frac{\alpha k_4}{A}, \quad k_4 = \frac{B\gamma + C\beta}{BC}, \quad k^* = C\beta + (B + \beta)\gamma. \end{gathered} \tag{7.6.6}$$

If the conditions

$$k_1 < 0, \quad k_2 < 0, \quad 2 k_1 k_2 > k_3^2 - k_1 k_4^2 \tag{7.6.7}$$

are satisfied, the solution $y_1 = \eta_1 = 0$ of system (7.6.6) and, consequently, the solution $y_1 = \mu_1 = 0$ of system (7.6.5) are asymptotically stable in probability.

Therefore, in the case (7.6.7), the equilibrium position $y_1 = z_1 = z_2 = 0$ of system (7.6.3), (7.6.4) is asymptotically stable in probability with respect to y_1 by Theorem 7.6.1.

7.7 Overview of References

The publications by Bertram and Carachik,* Katz and Krasovskii [1960] gave an impetus to investigation of the problem of stochastic stability. These works showed the way for a constructive use of the method of Lyapunov functions as applied to stochastic systems.

Monographs by Kushner [1967] and Khas'minskii [1969] are intended for systematic study of the problems of stochastic stability. These books deal with the analysis of stability with respect to *all* the variables characterizing the state of a system.

As regards the problems of control in stochastic systems, see the works by Kushner [1967], Khas'minskii [1969], Åström [1970], Chernous'ko and Kolmanovskii [1978].

As to the problems of partial stability and stabilization of stochastic systems, they are still paid too little attention. Formulation of the problem and some conditions for partial stability and stabilization in the context of Lyapunov's direct method were given by Sharov [1978].

The matter of Chapter 7 is associated with another approach to this problem (Vorotnikov [1983b], [1991a]). Certain development of this approach as applied to stochastic discrete models is presented in Phillis [1984].

We also note the following applied problems of control with respect to part of the variables for stochastic systems:*

(1) the time-optimal problem of damping the rotational motions of an asymmetric solid by flywheels under random perturbations of the "white noise" type;

(2) the time-optimal problem of the "hard encounter" type considering the passage of a mass point in a line through the zero coordinate (the velocity can be arbitrary at this time).

In the context of this book it is reasonable to obtain solutions of the problems of reorienting a solid discussed in Chapter 5 using a stochastic model of perturbations. In this case, instead of an auxiliary linear *conflict-control* μ-system, we should use a corresponding auxiliary linear *stochastic* controlled μ-system.

In conclusion, just a few words about the problem related to a new concept** of randomness in the theory of dynamic systems. The modern view of the world around us is such*** that the reality is far from being an arena

---

*Bertram, J.E. and Carachik, P.E. [1959] Stability of Circuits with Randomly Time-Varying Parameters, *Proc. of the Intern. Sympos. on Circuit and Inform. Theory*, Los Angeles, CA: IRE Trans., CT-6, 809–823.

*Kolmanovskii, V.B. [1976] On a Problem of Controlling the Motion of a Gyrostat under Random Perturbations, *Avtomat. i Telemekh.*, 11, 48–55; Afanas'ev, V.N., Kolmanovskii, V.B., and Nosov, V.R. [1989] *The Mathematical Theory of the Control Systems Design*, Moscow: Vysshaya Shkola (in Russian).

**See, for example: Neimark, Yu.I. [1978] *Dynamic Systems and Controlled Processes*, Moscow: Nauka (in Russian).

***See foreword by O. Toffler to the book: Prigogine, I. and Stengers, I. [1984] *Order out of Chaos. Man's New Dialogue with Nature*, London: Heinemann.

where order, stability, and equilibrium dominate; on the contrary, *instability* and *nonequilibrium* prevail all around.

In the context of this book it is not devoid of interest to analyze whether (and in which manner) this instability can be interpreted as a kind of *partial* stability and nonequilibrium—as a kind of *partial* equilibrium. This could enable the theory of partial stability to play a proper part in understanding the reality.

References

Ackermann, J.

[1993] *Robust Control. Systems with Uncertain Physical Parameters*, London: Springer-Verlag.

Akulenko, L.D.

[1980] Approximate Synthesis of Optimal Control of Motion with Respect to Part of the Variables, *Izv. Akad. Nauk SSSR Mekh. Tverd. Tela*, 5, 3–13.

[1987] *Asymptotic Methods of Optimal Control*, Moscow: Nauka (in Russian).

[1994a] *Problems and Methods of Optimal Control*, Dordrecht: Kluwer Acad. Publ.

[1994b] Perturbed Time-Optimal Problem of Controlling the Terminal State of a Mass Point by a Constrained Force, *Prikl. Mat. Mekh.*, **58**, 2, 12–21.

Aleksandrov, A.Yu.

[1996] On the Stability of Equilibrium of Unsteady Systems, *Prikl. Mat. Mekh.*, **60**, 2, 205–209.

Aminov, A.B. and Sirazetdinov, T.K.

[1987] The Method of Lyapunov Functions in the Problem of Polystability of Motion, *Prikl. Mat. Mekh.*, **51**, 5, 709–716.

Anan'evskii, I.M., Dobrynina, I.S., and Chernous'ko, F.L.

[1995] The Method of Decomposition in the Problem of Control of Mechanical System, *Izv. Akad. Nauk Teor. i Sistemy Upravleniya*, 2, 3–14.

Anapol'skii, L.Yu. and Chaikin, S.V.

[1993] On Partial Stability of Relative Equilibriums of an Elastic Satellite, *Prikl. Mat. Mekh.*, **57**, 2, 67–76.

Anashkin, O.V. and Khapaev, M.M.

[1995] On Partial Stability of Nonlinear Systems of Ordinary Differential Equations with Small Parameter, *Differentsial'nye Uravneniya*, **31**, 3, 371–381.

Anchev, A.A. and Rumyantsev, V.V.

[1979] On the Dynamics and Stability of Gyrostats, *Uspekhi Mekhaniki*, **2**, 3, 3–45.

Andreev, A.S.

[1979] On Asymptotic Stability and Instability of Nonautonomous Systems, *Prikl. Mat. Mekh.*, **43**, 5, 796–805.

[1982] On Asymptotic Stability and Instability with Respect to Part of the Variables, *Dokl. Akad. Nauk UzSSR*, 5, 9–12.

[1984] On Asymptotic Stability and Instability of the Zero Solution of a Nonautonomous System with Respect to Part of the Variables, *Prikl. Mat. Mekh.*, **48**, 5, 707–713.

[1987] Investigation of Partial Asymptotic Stability and Instability Based on Limiting Equations, *Prikl. Mat. Mekh.*, **51**, 2, 253–259.

[1991] Investigation of Partial Asymptotic Stability, *Prikl. Mat. Mekh.*, **55**, 4, 539–547.

[1996] On Stability of Equilibrium Positions of a Nonautonomous Mechanical System, *Prikl. Mat. Mekh.*, **60**, 3, 388–396.

Arnold, V.I.

[1990] *The Catastrophe Theory*, 3rd ed., Moscow: Nauka (in Russian).

Artstein, Z.

[1977] Topological Dynamics of an Ordinary Differential Equation, *J. Diff. Equat.*, **33**, 2, 216–223.

Åström, K.J.

[1970] *Introduction to Stochastic Control Theory*, New York: Academic Press.

Atanasov, V.A. and Willems, P.I.

[1993] On NonLinear Control of Gyrostat in Orbit, *Izv. Akad. Nauk Tekhn. Kibernet.*, 1, 16–23.

Athans, M. and Falb, P.L.

[1966] *Optimal Control*, New York: McGraw–Hill.

Azbelev, N.V., Maksimov, V.P., and Rakhmatullina, L.F.

[1991] *Introduction to the Theory of Functional-Differential Equations*, Moscow: Nauka (in Russian).

Bajic, V.B.

[1986] Partial Exponential Stability of Semi-State Systems, *Intern. J. Contr.*, **44**, 5, 1383–1394.

Ballieu, R.J. and Peiffer, K.

[1978] Attractivity of the Origin for the Equation $\ddot{x} + f(t, x, \dot{x})|\dot{x}|^{\alpha}\dot{x} + g(x) = 0$, *J. Math. Anal. Appl.*, **65**, 321–332.

Barbashin, E.A.

[1951] The Method of Sections in the Theory of Dynamic Systems, *Mat. Sb.*, **29**, 2, 233–280.

Barbashin, E.A. and Krasovskii, N.N.

[1952] On Global Stability of Motion, *Dokl. Akad. Nauk SSSR*, **86**, 3, 453–456.

Bartlett, A.C., Hollot, C.V., and Lin, H.

[1987] Root Locations of an Entire Polytope of Polynomials: It Suffices to Check the Edges, *Math. Contr., Signals Syst.*, **1**, 61–71.

Beletskii, V.V.
[1965] *The Motion of an Earth's Artificial Satellite about Its Center of Mass*, Moscow: Nauka (in Russian).

Bellman, R.
[1953] *Stability Theory of Differential Equations*, New York: McGraw–Hill.
[1962] Vector Lyapunov function, *SIAM J. Control*, **1**, 1, 32–34.

Bellman, R., Glicksberg, I., and Gross, O.
[1958] *Some Aspects of the Mathematical Theory of Control Processes*, Rand Corporation. Santa Monica, CA.

Blaquiére, A., Gerard, F., and Leitmann, G.
[1969] *Quantitative and Qualitative Games*, New York: Academic Press.

Bondi, P., Fergola, P., Gambardella, L., and Tenneriello, C.
[1981] Partial Asymptotic Stability via Limiting Equations, *Math. Meth. Appl. Sci.*, **3**, 4, 516–522.

Borkovskaya, I.M. and Marchenko, V.M.
[1996] On One Approach to the Problem of Stabilizing Delayed Systems, *Prikl. Mat. Mekh.*, **60**, 4.

Botkin, N.D., Kein, V.M., Patsko, V.S., and Turova, V.L.
[1989] Aircraft Landing Control in the Presence of Windshear, *Problems of Contr. Inform. Theory*, **18**, 4, 223–235.

Botkin, N.D., Zarkh, M.A., Kein, V.M., Patsko, V.S., and Turova, V.L.
[1993] Differential Games and Aircraft Control in the Presence of Windshear, *Izv. Akad. Nauk Tekhn. Kibernet.*, 1, 68–76.

Brockett, R.W.
[1978] Feedback Invariants for Nonlinear Systems, *Proc. IFAC Congress. Helsinki, 1978*, 1115–1120.

Bryson, A.E., Jr. and Zhao, Y.
[1987] Feedback Control for Penetrating a Downburst, *AIAA Paper*, AIAA-87-2343.

Bulirsch, R., Montrone, F., and Pesch, H.J.
[1991] Abort Landing in the Presence of a Windshear as Minimax Optimal Control Problem. Part 1, 2, *J. Optim. Theory Appl.*, **70**, 1, 1–23, **70**, 2, 223–254.

Burton, T.A.
[1978] Uniform Asymptotic Stability in Functional Differential Equations, *Proc. Amer. Math. Soc.*, **68**, 3, 195–199.
[1985] *Stability and Periodic Solutions of Ordinary and Functional Differential Equations*, Orlando: Academic Press.
[1993] Generalization of Krasovskii's Theory of Stability, *Izv. Akad. Nauk Tekhn. Kibernet.*, 1, 44–51.

Burton, T.A. and Hatvani, L.
[1989] Stability Theorems for Nonautonomous Functional Differential Equations by Liapunov Functionals, *Tohoku Math. J.*, **41**, 1, 65–104.

Campbell, S.L.

[1982] *Singular Systems of Differential Equations; I, II (1983)*, Marshfield, Mass.: Pitman Publishing Co.

Cantarelli, G.

[1995] Global Existence and Boundedness for Quasi-Variational Ordinary Differential Systems, *Proc. VI Intern. Coll. on Diff. Equations, Plovdiv (Bulgaria), August 18–23.*

Cantarelli, G. and Risito, C.

[1992] Criteri di Esistenza Globale e di Limitatezza per i Sistemi Olonomi Scleronomi, *Ann. Math. Pure Appl. (IV)*, **162**, 383–394.

Carrington, C.K. and Junkins, J.L.

[1986] Optimal Nonlinear Feedback Control for Spacecraft Attitude Maneuvers, *J. Guidance, Control, and Dynamics*, **9**, 99–107.

Cesari, L.

[1959] *Asymptotic Behavior and Stability Problems in Ordinary Differential Equations*, Berlin: Springer-Verlag.

Chaplygin, S.A.

[1919] A new Method of Approximate Integration of Differential Equations, *Selected Works on Mechanics and Mathematics*, 490–538, Moscow: Gostekhizdat, 1954; Collected Works, Vol. 1, 347–419, Moscow: GITTL, 1948.

Chen, Y.H. and Pandey, S.

[1989] Robust Control Strategy for Take-Off Perfomance in a Windshear, *Opt. Contr. Appl. and Methods*, 10, 65–79.

Cheremenskii, A.G.

[1987] Observability and Stability with Respect to Only Some of the Variables, *Differentsial'nye Uravneniya*, **23**, 4, 680–685.

Chernous'ko, F.L.

[1990a] Decomposition and Suboptimal Control in Dynamic Systems, *Prikl. Mat. Mekh.*, **54**, 6, 883–893.

[1990b] Decomposition and Control Synthesis in Dynamic Systems, *Izv. Akad. Nauk SSSR Tekhn. Kibernet.*, 6, 64–82.

[1992a] Control Synthesis for Nonlinear Dynamic System, *Prikl. Mat. Mekh.*, **56**, 2, 179–191.

[1992b] Constrained Control in Distributed System, *Prikl. Mat. Mekh.*, **56**, 5, 810–826.

[1993] Decomposition and Suboptimal Control in Dynamic Systems, *Izv. Akad. Nauk SSSR Tekhn. Kibernet.*, 1, 209–214.

[1994] *State Estimation for Dynamic Systems*, Boca Raton: CRC Press.

Chernous'ko, F.L. and Kolmanovskii, V.B.

[1978] *Optimal Control of Stochastic Processes*, Moscow: Nauka (in Russian).

Chernous'ko, F.L. and Melikyan, A.A.

[1978] *Game-Theoretic Problems of Control and Search*, Moscow: Nauka (in Russian).

Chetayev, N.G.

[1946] *The Stability of Motion*, 1st ed., Moscow: Gostekhizdat ; 4th ed., Moscow: Nauka, 1990; Oxford: Pergamon Press, 1961.

Claude, D.

[1986] Everything You Always Wanted to Know about Linearization but were Afraid to Ask, in: *Algebraic and Geometric Methods in Nonlinear Control Theory. Proc. Conf.* (M. Fliess and M. Hazewinkel, eds.), 381–438, Paris.

Corduneanu, C.

[1960] Application of Differential Inequalities to Stability Theory, *Anal. Stiintifice Univ. A.J. Cusa Din Jasi. Ser.1*, **6**, 1, 47–58, **7**, 2, 247–252.

[1964] Sur La Stabilité Partielle, *Revue Roumaine De Math. Pure Et. Appl.*, **9**, 3, 229–236.

[1971] Some Problems Concerning Partial Stability, *Symp.Math. V. 6, Meccanica Non-Lineare e Stabilitá. 23-26 Febbrario, 1970*, 243–265, New York: Academic Press.

[1975] On Partial Stability for Delay Systems, *Ann. Polon. Math.*, **29**, 377–381.

[1991] *Integral Equations and Applications*, Cambridge University Press.

[1993] Functional Differential Equations with Abstract Volterra Operators and Their Control, *Intern. Meeting on Ordinary Differential Equations and Their Applications. Firenze (Italy), Sept. 20–24, 1993*, 61–81.

Corduneanu, C. and Lakshmikantham, V.

[1980] Equations with Unbounded Delay: a Survey, *Nonlinear Analysis: TMA*, **4**, 5, 831–877.

Corless, M.

[1993] Control of Uncertain Nonlinear Systems, *ASME J. Dynamic Systems, Measur., Control*, **115**, 362–372.

Corless, M. and Leitmann, G.

[1995] Exponential Convergence for Uncertain Systems with Component-Wise Bounded Controllers, *Engineering and Automation*, **3–4**, 14–28.

Demin, V.G.

[1964] Application of a Theorem of V.V. Rumyantsev on Stability with Respect to Part of the Variables in Problems on Stellar Mechanics, *Kosmich. Issledovaniya*, **2**, 5, 416–418.

[1968] *The Motion of an Artificial Satellite in a Noncentral Gravitational Field*, Moscow: Nauka (in Russian).

Demin, V.G. and Furasov, V.D.

[1976] On Stabilization of Control Systems with Respect to Part of the Variables, *Prikl. Mat. Mekh.*, **40**, 2, 355–359.

Deng Zongqi and Ruan Shigui

[1991] Oscillations with Respect to Part Variables of Linear Second-Order Differential Systems, *Proc. Amer. Math. Soc.*, **113**, 3, 777–783.

Desoer, C.A. and Vidyasagar, M.

[1975] *Feedback Systems: Input-Output Properties*, New York: Academic Press.

Dwyer, T.A.W., III

[1984] Exact Nonlinear Control of Large Angle Rotational Maneuvers, *IEEE Trans. Autom. Contr.*, **AC-29**, 9, 769–774.

[1986] Exact Nonlinear Control of Spacecraft Slewing Maneuvers with Internal Momentum Transfer, *J. Guidance, Control, and Dynamics*, 2, 240–247.

Dykhman, E.I.

[1950] On Reduction Principle, *Izv. Akad. Nauk Kazakh. SSR Ser. Fiz.-Mat.*, **97**, 4, 372–390.

Ehricke, K.A.

[1962] *Space Flight. V.2: Dynamics*, New York: Van Nostrand Reinhold.

Emelyanov, S.V.

[1967] *Variable-Structure Control Systems*, Moscow: Nauka (in Russian).

Emelyanov, S.V., Korovin, S.K., Nersisian, A.L., and Nisenzon, Yu.E.

[1992] Output Feedback Stabilization of Uncertain Plants: A Variable Structure Systems Approach, *Intern. J. Contr.*, **55**, 61–82.

Falb, P.L. and Wolovich, W.A.

[1967] Decompling In the Design and Synthesis of Multivariables Control Systems, *IEEE Trans. Autom. Contr.*, **AC-12**, 6, 651–659.

Fergola, P. and Moauro, V.

[1970] On Partial Stability, *Ricerche Mat.*, **19**, 185–207.

Floquet, G.

[1883] Sur les Equations Differentielles Lineaires a Coefficients Periodiques, *Ann. De L'Ecole Normale, 2-e Series*, **12**, 47–88.

Friedman, A.

[1971] *Differential Games*, New York: Wiley–Interscience.

Furasov, V.D.

[1977] *Stability of Motion, Estimates and Stabilization*, Moscow: Nauka (in Russian).

[1982] *Stability and Stabilization of Discrete Processes*, Moscow: Nauka (in Russian).

Gabasov, R., Kirillova, F.M., and Kostyukova, O.I.

[1992] Real-Time Optimization of a Linear Control System, *Izv. Akad. Nauk Tekhn. Kibernet.*, 4, 3–19.

Gaishun, I.V. and Knyazhishche, L.B.

[1994] Nonmonotone Lyapunov Functionals. Conditions for Stability of Equations with Delay, *Differentsial'nye Uravneniya*, **30**, 8, 1291–1298.

Galiullin, A.S., Mukhametzyanov, I.A., Mukharlyamov, R.G., and Furasov, V.D.

[1971] *Construction of Programmed Motion Systems*, Moscow: Nauka (in Russian).

Gapalsamy, K.

[1992] *Equations of Mathematical Ecology*, Dordrecht: Kluwer Acad. Publ.

Grammel, R.

[1920] Die Stabilitat der Standeschen Kreiselbewegungen, *Math. Zeitschrift*, **6**.

Guelman, M.

[1989] On Gyrostat Dynamics and Recovery, *J. Astronaut. Sci.*, **37**, 2, 109–119.

Gulyaev, V.P., Koshkin, V.L., and Savilova, I.V.

[1986] Time-Optimal Control of Triaxial Orientation of a Solid with Constrained Control Parameters, *Izv. Akad. Nauk SSSR Mekh. Tverd. Tela*, 5, 11–15.

Habets, P. and Peiffer, K.

[1973] Classification of Stability-Like Concepts and Their Study Using Vector Liapunov Functions, *J. Math. Anal. Appl.*, **43**, 537–570.

Hacker, T.

[1961] Stability of Partially Controlled Motion of an Aircraft, *J. Aerospace Sci.*, **28**, 1, 15–27.

Hagedorn, P.

[1982] Steady Motions in Controlled Gyrostat Systems, *Ing. Arch.*, **52**, 183–204.

Hahn, W.

[1967] *Stability of Motion*, Berlin: Springer-Verlag.

Hajek, O.

[1975] *Pursuit Games*, New York: Academic Press.

Halanay, A.

[1963] *Differential Equation: Stability, Oscillations, Time Lags*, Bucharest; New York: Academic Press, 1966.

Hale, J.K.

[1969] *Ordinary Differential Equations*, Wiley–Interscience.

[1977] *Theory of Functional Differential Equations*, New York: Springer-Verlag.

Hale, J.K. and Lunel, S.M.V.

[1993] *Introduction to Functional Differential Equations*, Berlin: Springer-Verlag.

Hatvani, L.

[1975a] On Some Criteria for Stability with Two Lyapunov Functions, *Prikl. Mat. Mekh.*, **39**, 1, 172–177 (Erratum: *Prikl. Mat. Mekh.*, **40**, 2, 251).

[1975b] On Application of Differential Inequalities to Stability Theory, *Vestnik Moskov. Univ. Ser. I Mat. Mekh.*, 3, 83–89.

[1976] On the Absence of Asymptotic Stabilty with Respect to Part of the Variables, *Prikl. Mat. Mekh.*, **40**, 2, 245–251.

[1979a] On Stability and Partial Stability of Non-Autonomous Systems of Differential Equation, *Alkalmaz. Mat. Lapok.*, 5, 1–48 (in Hungarian).

[1979b] A Generalization of the Barbashin–Krasovskij Theorems to Partial Stability in Nonautonomous Systems, *Coll. Math. Soc. Janos Bolyai, Qualitative Theory of Diff. Equations, Szeged (Hungary)*, 381–409.

[1982] On the Partial Asymptotic Stability by Ljapunov Function with Semidefinite Derivative, *MTA Szám. és Autom. Kutató Intézet. Közlemények*, **26**, 85–88.

[1983a] On Partial Asymptotic Stability and Instability, I (Autonomous Systems), *Acta Sci. Math.*, **45**, 219–231.

[1983b] On Partial Asymptotic Stability and Instability, II (The Method of Limiting Equations), *Acta Sci. Math.*, **46**, 143–156.

[1984] On the Asymptotic Stability by Nondecrescent Ljapunov Function, *Nonlinear Analysis: TMA*, **8**, 1, 67–77.

[1985a] On Partial Asymptotic Stability and Instability, III (Energy-Like Ljapunov Functions), *Acta Sci. Math.*, **49**, 157–167.

[1985b] On Partial Asymptotic Stability by the Method of Limiting Equation, *Ann. Mat. Pura Appl. (IV)*, **139**, 65–82.

[1991] On the Stability of the Solutions of Ordinary Differential Equations with Mechanical Applications, *Alkalmaz. Mat. Lapok.*, **15**, 1/2, 1–90.

Hatvani, L. and Terjeki, I.

[1985] Stability Properties of the Equilibrium under the Influence of Unbounded Damping, *Acta Sci. Math.*, **48**, 1–4, 187–200.

Hill, G.W.

[1886] On the Part of the Motion of the Lunar Perigee which is a Function of the Mean Motion of the Sun and Moon, *Acta Math.*, 8, 1–36.

Hino, Y., Murakami, S., and Naito, T.

[1991] Functional Differential Equations with Infinite Delay, *Springer Lecture Notes, 1473*, Berlin: Springer-Verlag.

Hunt, L.R., Su, R., and Meyer, G.

[1983] Global Transformations of Nonlinear Systems, *IEEE Trans. Autom. Contr.*, **AC-28**, 1, 24–31.

Ignat'ev, A.O.

[1988] Stability of Motion with Respect to Part of the Variables in the Presence of Constantly Acting Perturbations, *Mat. Fiz. Nelinein. Mekh.*, 10, 20–25.

[1989a] On Stability of Almost Periodic Systems with Respect to Part of the Variables, *Differentsial'nye Uravneniya*, **25**, 8, 1446–1448.

[1989b] On Preserving of Uniform Asymptotically Stability with Respect to Part of the Variables, *Prikl. Mat. Mekh.*, **53**, 1, 167–171.

Ikeda, M., Ohta, Y., and Siljak, D.D.

[1990] Parametric Stability, in *New Trends in System Theory*, Boston: Birkhauser, pp. 1–20.

Il'yasov, K.

[1982] On Stability of Delay Systems with Respect to Part of the Variables, *Matem. Fiz. (Kiev)*, 31, 32–36.

[1984] On Stability of Solutions of Linear Difference Equations with Respect to Part of the Variables, *Matem. Fiz. (Kiev)*, 35, 30–32.

Isaacs, R.

[1965] *Differential Games*, New York: John Wiley & Sons.

Isidori, A.

[1989] *Nonlinear Control Systems*, 2nd ed., Berlin: Springer-Verlag.

Ito, K.

[1951] On Stochastic Differential Equations, *Mem. Amer. Math. Soc.*, 4, 1–51.

Jacubczyk, B. and Recpondek, W.

[1980] On Linearization of Control Systems, *Bull. Acad. Pol. Sci. Ser.Sci. Math.*, **28**, 9–10, 517–522.

Jia-Xiang, Z.

[1990] On Partial Asymptotic Stability and Limit Equations, *J. Nanjing Univ. Math. Biquart.*, **7**, 1, 100–108.

Kalistratova, M.A.

[1986] On Stability of Delay Systems with Respect to Part of the Variables, *Avtomat. i Telemekh.*, 5, 32–37.

Kalman, R.E.

[1960a] Contributions to the Theory of Optimal Control, *Bol. Soc. Mat. Mexic. Seg. Ser.*, 1, 102–119.

[1960b] On the General Theory of Control Systems, in: *Proc. I IFAC Congress (Moscow, 1960)*, 481–492, London: Butterworth.

Kamenkov, G.V.

[1971] *Selected Works Vol. 1 (1972, Vol. 2)*, Moscow: Nauka (in Russian).

Karapetyan, A.V.

[1981] On Permanent Rotations of a Massive Solid on an Absolutely Rough Horizontal Plane, *Prikl. Mat. Mekh.*, **45**, 5, 808–814.

Karapetyan, A.V. and Rumyantsev, V.V.

[1983] Stability of Conservative and Dissipative Systems, *Itogi Nauki i Tekhniki. Obshchaya Mekhanika*, **6**, 3–127, Moscow: VINITI (in Russian); *Appl. Mech. Soviet Reviews. Stability and Analytical Mech.*, **1**, New York: Hemisphere Publ. Co., 1990.

Karimov, A.U.

[1973] On Stability with Respect to Part of the Variables in the Presence of Constantly Acting Perturbations, in: *Mat. Fizika i Elektrodinamika*, 3–10, Moscow: Izd. Moskov. Gos. Univ. (in Russian).

Kato, J.

[1980] Liapunov's Second Method in Functional Differential Equations, *Tohoku Math. J. Ser.2*, **32**, 4, 487–497.

Katz, I.Ya. and Krasovskii, N.N.

[1960] On Stability of Systems with Random Parameters, *Prikl. Mat. Mekh.*, **24**, 5, 809–823.

Kertész, V.

[1991] Partial Stability Investigations of Differential Equation of Second Order, *Comput. and Math. Appl.*, **21**, 1, 95–102.

Khapaev, M.M.

[1986] *Averaging in the Theory of Stability: Investigation of Resonance Multifrequency Systems*, Moscow: Nauka (in Russian).

Khapaev, M.M. and Shinkin, V.N.

[1983] On Analysis of Resonant Almost Periodic Systems for Partial Stability, *Prikl. Mat. Mekh.*, **47**, 2, 334–337.

Kharitonov, V.L.

[1978] Asymptotic Stability of a Family of Systems of Linear Differential Equations, *Differentsial'nye Uravneniya*, **14**, 11, 2086–2088.

Khas'minskii, R.Z.

[1969] *Stability of Systems of Differential Equations Whose Parameters are Randomly Perturbed*, Moscow: Nauka (in Russian).

Kholostova, O.V.

[1992] On Stability of a Particular Motion of a Solid with Elastic Membrane in a Circular Orbit, *Prikl. Mat. Mekh.*, **56**, 1, 24–33.

Kim, A.V.

[1992] *The Lyapunov Direct Method in the Theory of Stability of Delay Systems*, Ekaterinburg: Izd. Ural. Gos. Univ. (in Russian).

Kolmanovskii, V.B. and Myshkis, A.D.

[1992] *Applied Theory of Functional Differential Equations*, Dordrecht: Kluwer Acad. Publ.

Kolmanovskii, V.B. and Nosov, V.R.

[1981] *Stability and Periodic Modes of Control Systems with Holdover*, Moscow: Nauka (in Russian).

Kosov, A.A.

[1988] On the Problem of Stability of Motion with Respect to Part of the Variables, in *Problems in the Qualitative Theory of Differential Equations*, 195–203, Novosibirsk: Nauka (in Russian).

Kovalev, A.M.

[1993] Controllability of Dynamic Systems with Respect to Part of the Variables, *Prikl. Mat. Mekh.*, **57**, 6, 41–50.

[1994] Control and Stabilization Problems with Respect to a Part of the Variables, *ZAMM*, **74**, 7, 59–60.

Kozlov, V.V.

[1991a] On the Stability of Equilibrium Positions in Nonstationary Force Fields, *Prikl. Mat. Mekh.*, **55**, 1, 12–19.

[1991b] The Stability of Periodic Trajectories and Chebyshev Polynomials, *Vestnik Moskov. Univ. Ser. I Mat. Mekh.*, 5, 7–13.

Krasovskii, A.A.

[1992] Nonclassical Objective Functionals and Problems in Optimal Control Theory, *Izv. Akad. Nauk Tekhn. Kibernet.*, 1, 3–41.

Krasovskii, A.A., (ed.)
[1987] *Handbook on the Theory of Automatic Control*, Moscow: Nauka (in Russian).

Krasovskii, A.A., Bukov, V.N., and Shendrik, V.S.
[1977] *Universal Algorithms for Optimal Control of Continuous Systems*, Moscow: Nauka (in Russian).

Krasovskii, A.N. and Krasovskii, N.N.
[1995] *Control under Lack of Information*, Boston: Birkhauser.

Krasovskii, N.N.
[1956] On Application of the Method of Lyapunov Functions to the Equations with Delay, *Prikl. Mat. Mekh.*, **20**, 2, 315–327; **20**, 4, 513–518.
[1959] *Stability of Motion*, Moscow: Fizmatlit (in Russian); Stanford: Stanford University Press, 1963.
[1966] Problems on Stabilizaton of Controlled Motion, in *Malkin I.G. The Theory of Stability of Motion, Appendix 4*, 475–514, Moscow: Nauka (in Russian).
[1968] *The Theory of Motion Control*, Moscow: Nauka (in Russian).
[1970] *Game-Theoretic Problems on Rendezvous of Motions* (in Russian).

Krasovskii, N.N. and Subbotin, A.I.
[1988] *Game-Theoretical Control Problems*, New York: Springer-Verlag.

Krementulo, V.V.
[1977] *Stabilization of Stationary Motion of Rigid Bodies by Means of Rotating Masses*, Moscow: Nauka (in Russian).

Krivosheev, Yu.A. and Lutsenko, A.V.
[1980] On Stability of Motion with Respect to Part of the Variables for Linear Systems with Constant and Almost Constant Matrices, *Prikl. Mat. Mekh.*, **44**, 2, 205–210.

Kuang, Y.
[1993] *Delay Differential Equations with Applications in Population Dynamics*, San Diego: Academic Press.

Kukhtenko, A.I.
[1984] Basic Stages in Creation of the Theory of Invariance, I, II, *Avtomatika*, 2, 3–13; 1985, 2, 3–14.

Kurakin, L.G.
[1994] On Stability of a Regular Vortex n-gon, *Dokl. Akad. Nauk*, **335**, 6, 729–731.

Kurzhanskii, A.B.
[1977] *Control and Observation under Conditions of Uncertainty*, Moscow: Nauka (in Russian).

Kushner, H.
[1967] *Stochastic Stability and Control*, New York: Academic Press.

Lakshmikantham, V. and Leela, S.
[1969] *Differential and Integral Inequalities. Theory and Applications, Vol. 1, 2*, New York: Academic Press.

Lakshmikantham, V., Leela, S., and Martynyuk, A.A.

[1989] *Stability Analysis of Nonlinear Systems*, New York: Marcel Dekker, Inc.

Lakshmikantham, V. and Martynyuk, A.A.

[1993] Extension of the Lyapunov Method to the Systems with Holdover. A Survey, *Prikl. Mekh.*, **29**, 2, 3–16.

Lakshmikantham, V., Matrosov, V.M., and Sivasundaram, S.

[1991] *Vector Lyapunov Function and Stability Analysis of Nonlinear Systems*, Dordrecht: Kluwer Acad. Publ.

Lakshmikantham, V. and M. Rama Mohana Rao

[1994] *Theory of Integro-Differential Equations*, Gordon and Breach Sci. Publ.

Lefschetz, S.

[1965] *Stability of Nonlinear Control Systems*, New York: Academic Press.

Leimanis, E.

[1965] *The General Problem of the Motion of Coupled Rigid Bodies about a Fixed Point*, Berlin: Springer-Verlag.

Leitmann, G.

[1993] On the Approach to the Control of Uncertain Systems, *ASME J. Dynamic Systems, Measur., Control*, **115**, 373–380.

Leitmann, G. and Pandey, S.

[1990] Aircraft Control under Conditions of Windshear, in *Control and Dynamic Systems* (C.T. Leondes, ed.), **34**, 1–79, New York: Academic Press.

[1991] Aircraft Control for Flight in an Uncertain Environment: Takeoff in Windshear, *J. Optim. Theory Appl.*, **70**, 1, 25–55.

Letov, A.M.

[1960] The Analytical Design of Control Systems, *Avtomat. i Telemekh.*, **21**, 4–6, 436–441, 561–568, 661–665.

[1981] *The Mathematical Theory of Control Processes* (N.N. Krasovskii, ed.), Moscow: Nauka (in Russian).

Liao, Xiao, Xin

[1989] On Asymptotic Stability of Motion Relative to Part of the Variables for Linear Systems, *Prikl. Mat. Mekh.*, **53**, 6, 1034–1035.

Lilov, L.K.

[1972] On Stabilization of Stationary Motions of Mechanical Systems Relative to Part of the Variables, *Prikl. Mat. Mekh.*, **36**, 6, 977–985.

Lilov, L.K. and Kalkovski, A.G.

[1986] Partial Stability, *Proc. Fifteenth Annual Bulgarian Mathematical Conference. April 6–9, 1986*, 429–432, Sofia.

Lim, Y.S. and Kazda, L.F.

[1964] A Study of Second Order Linear Systems, *J. Math. Anal. Appl.*, **8**, 3, 423–444.

Liu Xinzhi
[1989] Stability in Terms of Two Measures for Functional Differential Equations, *J. Diff. Integr. Equat.*, 3, 257–261.
[1993] On (h_0, h)-Stability of Autonomous Systems, *J. Appl. Math. Stoch. Anal.*, **5**, 4, 331–337.

Liu Xinzhi and Sivasundaram, S.
[1992] On the Direct Method of Lyapunov in Terms of Two Measures, *J. Math. Phys. Sci.*, **26**, 4, 389–400.
[1995] Stability of Nonlinear Systems under Constantly Acting Perturbations, *Int. J. Math. and Math. Sci.*, **18**, 2, 273–278.

Lizunova, M.G.
[1991] On Stability with Respect to Part of the Variables in the Critical Case of a Pair of Pure Imaginary Roots, in *Stability and Nonlinear Oscillations*, 59–65, Sverdlovsk: Izd. Ural. Gos. Univ. (in Russian).

Lorenz, E.N.
[1963] Deterministic Nonperiodic Flow, *J. Atmos. Sci.*, **20**, 130–141.

Lotka, A.J.
[1920] Analytical Note on Certain Rhythmic Relations in Organic Systems, *Proc. Nat. Acad. Sci. USA*, **6**, 410–415.

Lur'e, A.I.
[1951] *Some Nonlinear Problems of Automatic Control Theory*, Moscow-Leningrad: Gostekhizdat (in Russian).
[1961] *Analytical Mechanics*, Moscow: Fizmatlit (in Russian); *Mécanique Analytique*, Louvain: Librarie Universitaire, 1968.

Lutsenko, A.V. and Stadnikova, L.V.
[1973] On Partial Stability in the First Approximation, *Differentsial'nye Uravneniya*, **9**, 8, 1530–1533.

Lyapunov, A.M.
[1892] *The General Problem of Stability of Motion*, Kharkov: Izd. Khark. Univ. (in Russian).
[1893] Analysis of One of the Singular Cases of the Problem of Stability of Motion, in *Collected Works, Vol. 2*, 272–331, Moscow-Leningrad: Izd. Akad. Nauk SSSR, 1956.

Magnus, K.
[1955] Beitrage Sur Dynamik Des Kreftefrein Kordansc Gelagerten Kreisels, *ZAMM*, **35**, 1/2.
[1971] *Kreisel. Theorie und Anwendungen*, Berlin: Springer-Verlag.
[1980a] Aktive Kreiselsysteme, *Ing. Arch.*, **49**, 295–308.
[1980b] Die Stabilitat von Schwingungen in Rotorsystemen Mit Synchronantieb, in *Karl - Marguerre - Gedachtnis - Kolloquium. TH Darmstadt. 27–28 Mai, 1980*, **16**, 179–188, Darmstadt: THD-Schriftenreihe Wissenschaft und Technik.

Makay, G.
[1991] On the Asymptotic Stability in Terms of Two Measures for Functional Differential Equations, *Nonlinear Analysis: TMA*, **16**, 7/8, 721–727.

Malkin, I.G.

[1938] On Stability of Motion in Lyapunov's Sense, *Mat. Sb.*, **3(45)**, 47–100.

[1966] *The Theory of Stability of Motion*, Moscow: Nauka (in Russian).

Mamedova, T.F.

[1995] Asymptotic Equivalence of Differential Equations and Stability of Solutions with Respect to Part of the Variables, in *Mater. Mezdunar. Konf. "Differentsial'nye Uravneniya i ikh Prilozheniya," Saransk, Russia, 20–22 Dec. 1994*, 216–227, Saransk: Izd. Sar. Gos. Univ. (in Russian).

Marachkov, V.P.

[1940] On a Stability Theorem, *Izv. Fiz.-Mat. Ob. i Nauchno-Issled. Inst. Mat. i Mekh. pri Kazanskom Univ. Ser. 3*, **12**, 171–174.

Markeev, A.P.

[1992] *Dynamics of a Body Touching a Rigid Surface*, Moscow: Nauka (in Russian).

Martynyuk, A.A.

[1972] On Technical Stability of Motion with Respect to Individually Specified Coordinates, *Prikl. Mekh.*, **8**, 3, 87–91.

[1983] *Practical Stability of Motion*, Kiev: Naukova Dumka (in Russian).

[1991a] A New Trend in the Method of Lyapunov Matrix Functions, *Dokl. Akad. Nauk SSSR*, **312**, 3, 554–557.

[1991b] A Theorem on Polystability, *Dokl. Akad. Nauk SSSR*, **318**, 4, 808–811.

[1994] A New Trend in the Theory of Stabilty: The Problem on Polystability, *Prikl. Mekh.*, **30**, 5, 3–17.

Martynyuk, A.A. and Chernetskaya, L.N.

[1993a] On Polystability of Linear Autonomous Systems, *Dokl. Akad. Nauk Ukrainy*, 8, 17–18.

[1993b] On Polystability of Linear Systems with Periodic Coefficients, *Dokl. Akad. Nauk Ukrainy*, 11, 61–65.

Massera, I.L.

[1949] On Liapounoff's Condition of Stability, *Ann. Math.*, **50**, 3, 705–721.

[1956] Contributions to Stability Theory, *Ann. Math.*, **64**, 182–206 (Erratum: *Ann. Math.*, 1958, **68**, 202).

Matrosov, V.M.

[1962a] On Stability of Motion, *Prikl. Mat. Mekh.*, **26**, 5, 885–895.

[1962b] Toward a Theory of Stability of Motion, *Prikl. Mat. Mekh.*, **26**, 6, 992–1002.

[1965] Development of the Method of Liapunov Function in Stability Theory, in *Proc. II All-Union Conf. Theor. Appl. Mech.* (L.I. Sedov, ed.), 112–125; Israel Program for Scientific Translation, Jerusalem, 1968.

[1969] The Comparison Principle with Lyapunov Vector-Function IV, *Differentsial'nye Uravneniya*, **5**, 12, 2129–2143.

[1989] The Method of Lyapunov Vector Functions in Nonlinear Mechanics, *Uspekhi Mekhaniki*, **12**, 3, 59–82.

Merkin, D.R.

[1987] *Introduction to the Theory of Stability of Motion*, Moscow: Nauka (in Russian).

Michel, A.N.

[1969] On the Bounds of the Trajectories of Differential Systems, *Intern. J. Contr.*, **10**, 593–600.

Miele, A.

[1989] Final Report on NASA Grant No. NAG-1-156: Optimization and Guidance of Flight Trajectories in Presence of Windshear,1984-1989, *Aero-Aeronautics Report No. 244*, Houston, Texas: Rice University.

Miele, A., Wang, T., and Melvin, W.W.

[1986] Optimal Take-Off Trajectories in the Presence of Windshear, *J. Optim. Theory Appl.*, **49**, 1, 1–45.

Miki, K.

[1990] Partially Integral Stability Theorems by Comparison Principle, *Res. Rept. Akita Techn. Coll.*, **25**, 84–89.

Miki, K., Masamichi, A., and Shoichi, S.

[1985] On the Partially Total Stability and Partially Total Boundedness of a System of Ordinary Differential Equations, *Res. Rept. Akita Techn. Coll.*, **20**, 105–109.

Mlodzeevsky, B.K.

[1894] On Permanent Axes in the Motion of a Massive Solid About a Fixed Point, *Trudy Otdeleniya Fiz. Nauk Obshch. Lyubit. Estestvoznan.*, **VII**.

Moiseev, N.N. and Rumyantsev, V.V.

[1965] *The Dynamics of Bodies with Liquid-Filled Cavities*, Moscow: Nauka (in Russian); Berlin: Springer-Verlag, 1968.

Monaco, S. and Stornelli, S.

[1986] A Nonlinear Feedback Control Law for Attitude Control, *Algebraic and Geometric Methods in Nonlinear Control Theory. Proc. Conf.* (M. Fliess and M. Hazewinkel, eds.), 573–595, Paris.

Movchan, A.A.

[1960] Stability of Processes with Respect to Two Metrics, *Prikl. Mat. Mekh.*, **24**, 6, 988–1001.

Mukharlyamov, R.G.

[1989] Control of Programmed Motion with Respect to Part of the Variables, *Differentsial'nye Uravneniya*, **25**, 8, 938–942.

Muller, P.S.

[1977] *Stabilitat und Matrizen*, Berlin: Springer-Verlag.

[1982] Zum Problem Der Partiellen Asymptotichen Stabilitat, *Fest. Zum 70 Celebr. von Herr Prof. Dr. K. Magnus, TU Munchen, 15.11.1982.*

[1984] Kriterien sur Untersuchung der Partiellen Asymptotichen Stabilitas, *ZAMM*, **64**, 4, 71–72.

Nabiullin, M.K.

[1990] *Stationary Motions and Stability of Elastic Satellites*, Novosibirsk: Nauka (in Russian).

Nemytskii, V.V. and Stepanov, V.V.

[1949] *Qualitative Theory of Differential Equations*, Moscow: Gostekhizdat (in Russian); Princeton: Princeton University Press, 1960.

Nijmeijer, H. and A.J. van der Schaft

[1990] *Nonlinear Dynamic Control Systems*, Berlin: Springer-Verlag.

Nita, M.M.

[1973] *Teorie Zborului Spatial*, Bucurest: Acad. RSR.

Nosov, V.R. and Furasov, V.D.

[1979] On the Stability of Discrete Processes Relative to Given Variables and Convergence of Certain Optimization Algorithms, *Zh. Vychisl. Mat. i Mat. Fiz.*, **19**, 2, 316–328.

Ohta, Y. and Siljak, D.D.

[1994] Parametric Quadratic Stabilizability of Uncertain Nonlinear Systems, *Syst. Contr. Letters*, **22**, 437–444.

Osipov, Yu.S.

[1965a] Stabilization of Controlled Systems with Delay, *Differentsial'nye Uravneniya*, **1**, 5, 605–618.

[1965b] On Reduction Principle in Critical Cases of Stability of Motion of Systems with Delay, *Prikl. Mat. Mekh.*, **29**, 5, 810–820.

Oziraner, A.S.

[1971] On the Problem of Stability of Motion Relative to Part of the Variables, *Vestnik Moskov. Univ. Ser. I Mat. Mekh.*, 1, 92–100.

[1972a] On Asymptotic Stability of Motion Relative to Part of the Variables for Linear Systems, *Vestnik Moskov. Univ. Ser. I Mat. Mekh.*, 1, 73–80.

[1972b] On Some Theorems Concerning Lyapunov's Second Method, *Prikl. Mat. Mekh.*, **36**, 3, 396–404.

[1972c] On the Stability of Equilibrium Positions of Rigid Bodies with Liquid-Filled Cavities, *Prikl. Mat. Mekh.*, **36**, 5, 522–528.

[1973] On Asymptotic Stability and Instability Relative to Part of the Variables, *Prikl. Mat. Mekh.*, **37**, 4, 659–665.

[1975] On the Stability of Motion in Critical Cases, *Prikl. Mat. Mekh.*, **39**, 3, 415–421.

[1978] On Optimal Stabilization of the Motion Relative to Part of the Variables, *Prikl. Mat. Mekh.*, **42**, 2, 272–276.

[1979] On a Theorem of Malkin and Massera, *Prikl. Mat. Mekh.*, **43**, 6, 975–979.

[1981] On Stability with Respect to Part of the Variables in the Presence of Constantly Acting Perturbations, *Prikl. Mat. Mekh.*, **45**, 3, 419–427.

[1986] On Analysis of Stability with Respect to Part of the Variables by Means of Quadratic Forms, *Prikl. Mat. Mekh.*, **50**, 1, 163–167.

Oziraner, A.S. and Rumyantsev, V.V.

[1972] The Method of Lyapunov Functions in the Problem of Stability of Motion Relative to Part of the Variables, *Prikl. Mat. Mekh.*, **36**, 2, 364–384.

Peiffer, K.

[1968] *La Méthode de Liapunoff Appliquée à l'Étude de la Stabilité Partielle*, Université Catholique de Louvain. Faculté des Sciences.

Peiffer, K. and Rouche, N.

[1969] Liapounov's Second Method Applied to Partial Stability, *J. Mécanique*, **8**, 2, 323–334.

Petrov, B.N.

[1960] The Principle of Invariance and Conditions for its Use in Designing Linear and Nonlinear Systems, in *Proc. I IFAC Congress (Moscow, 1960)*, 259–275, London: Butterworth.

Petrov, B.N., Rutkovskii, V.Yu., and Zemlyakov, S.D.

[1980] *Adaptive Coordinate-Parametric Control of Nonstationary Objects*, Moscow: Nauka (in Russian).

Phillis, Y.

[1984] y-Stability and Stabilizability in the Mean of Discrete-Time Stochastic Systems, *Intern. J. Contr.*, **40**, 1, 149–160.

Plymale, B.T. and Goodstein, R.

[1955] Nutation of a Free Gyro Subjected to an Impulse, *J. Appl. Mech. Trans. Amer. Soc. Mech. Eng.*, **22**, 3.

Pontryagin, L.S., Boltyanskii, V.G., Gamkrelidze, R.V., and Mishchenko, E.F.

[1961] *The Mathematical Theory of Optimal Processes*, Moscow: Fizmatlit (New York: Interscience, 1962).

Prokop'ev, V.P.

[1975] On Stability of Motion Relative to Part of the Variables in the Critical Case of One Zero Root, *Prikl. Mat. Mekh.*, **39**, 3, 422–426.

Pyatnitskii, E.S.

[1970] Absolute Stability of Nonstationary Nonlinear Systems, *Avtomat. i Telemekh.*, 1, 5–15.

[1993] Synthesis of Systems of Stabilizing Programmed Motions of Nonlinear Systems, *Avtomat. i Telemekh.*, 7, 19–37.

Pyatnitskii, E.S. and Rapoport, L.B.

[1991] Periodic Motion and Criteria for Absolute Stability of Nonlinear Nonstationary Systems, *Avtomat. i Telemekh.*, 10, 63–73.

Raushenbakh, B.V. and Tokar', E.N.

[1974] *Spacecraft Attitude Control*, Moscow: Nauka (in Russian).

Razumikhin, B.S.

[1956] On Stability of Systems with Delay, *Prikl. Mat. Mekh.*, **20**, 4, 500–512.

[1988] *The Stability of Hereditary Systems*, Moscow: Nauka (in Russian).

Risito, C.

[1970] Sulla Stabilita Asintotica Parziale, *Ann. Math. Pura Appl. (IV)*, **84**, 279–292.

[1971] Some Theorems on the Stability and the Partial Asymptotic Stability of Systems with Known First Integrals, *Comptes Rendus des Journess Nationales du C.B.R.M. 24-26 Mai*, 53–56.

[1974] The Comparison Method Applied to the Stability of Systems with Known First Integrals, in *Proc. VI Int. Conf. on Non-Linear Oscillations, Poznan, 1972. Non-Linear Vibration Problems*, **15**, 25–45.

[1976] Metodi per lo Studio della Stabilita di Sistemi con Integrali Primi Noti, *Ann. Math. Pura Appl. (IV)*, **107**, 49–94.

Roitenberg, Ya.N.

[1987] *Automatic Control*, 2nd ed., Moscow: Nauka (in Russian).

Rouche, N. and Peiffer, K.

[1967] Le Théorème de Lagrange-Dirichlet et la Deuxième Méthode de Liapunoff, *Ann. Soc. Scient. Bruxelles, Ser. I*, **81**, 1, 19–33.

Rouche, N., Habets, P., and Peiffer, K.

[1977] *Stability Theory by Liapunov's Direct Method*, New York: Springer-Verlag.

Rubanovskii, V.N.

[1982a] Stability of Steady-State Motion of Complex Mechanical Systems, in *Itogi Nauki i Tekhniki. Obshchaya Mekhanika*, **5**, 62–134, Moscow: VINITI (in Russian).

[1982b] Stability of Stationary Motion of a Rigid Body with an Elastic Shell Partially Filled with Liquid, *Prikl. Mat. Mekh.*, **46**, 4, 543–552.

Rubanovskii, V.N. and Rumyantsev, V.V.

[1979] On the Stability of Motion of Complex Mechanical Systems, *Uspekhi Mekhaniki*, 2, 53–79.

Rumyantsev, V.V.

[1956] Stability of Permanent Rotation of a Heavy Rigid Body, *Prikl. Mat. Mekh.*, **20**, 1, 50–66.

[1957a] On the Stability of Motion with Respect to Part of the Variables, *Vestnik Moskov. Univ. Ser. Mat. Mekh. Fiz. Astron. Khim.*, 4, 9–16.

[1957b] On the Stability of Permanent Rotations of a Solid about a Fixed Point, *Prikl. Mat. Mekh.*, **21**, 3, 339–346.

[1960] A Theorem on Stability of Motion, *Prikl. Mat. Mekh.*, **24**, 1, 47–54.

[1967] *On the Stability of Stationary Motions of Satellites*, Moscow: Izd. VTs Akad. Nauk SSSR (in Russian).

[1968] The Method of Lyapunov Functions in the Theory of Stability of Motion, *Fifty Years of Soviet Mechanics, Vol. I*, 7–66, Moscow: Nauka (in Russian).

[1970] On Optimal Stabilization of Control Systems, *Prikl. Mat. Mekh.*, **34**, 3, 440–456.

[1971a] On Asymptotic Stability and Instability of Motions Relative to Part of the Variables, *Prikl. Mat. Mekh.*, **35**, 1, 147–152.

[1971b] On the Stability with Respect to a Part of the Variables, *Symp. Math. Vol. 6. Meccanica Non-Lineare e Stabilitá, 243-265, 23-26 Febbrario, 1970*, New York: Academic Press.

[1972a] On Control and Stabilization of Systems with Cyclic Coordinates, *Prikl. Mat. Mekh.*, **36**, 6, 966–976.

[1972b] Problems on Stability of Motion with Respect to Part of the Variables, in *Problems on Solid Media and Related Problems in Analysis*, 429–436, Moscow: Nauka (in Russian).

[1983] The Development of Motion-Stability Theory in the USSR, *Differentsial'nye Uravneniya*, **19**, 5, 739–776.

[1987] On Stability and Stabilization of Motion Relative to Part of the Variables, in *Contemporary Problems in Mathematical Physics: Proc. All-Union Symposium, Tbilisi, April 22-25, 1987, Vol. I*, 85–100, Tbilisi: Izd. Tbiliss. Gos. Univ. (in Russian).

[1993] On Optimal Stabilization of Motion with Respect to Part of the Variables, *Izv. Akad. Nauk Tekhn. Kibernet.*, 1, 184–189.

Rumyantsev, V.V. and Karapetyan, A.V.

[1976] Stability of Motion of Nonholonomic Systems, *Itogi Nauki i Tekhniki. Obshchaya Mekhanika*, **3**, 5–42, Moscow: VINITI (in Russian).

Rumyantsev, V.V. and Oziraner, A.S.

[1987] *Stability and Stabilization with Respect to Part of the Variables*, Moscow: Nauka (in Russian).

Rusinov, I.

[1988] Asymptotic Stability and Instability with Respect to Part of the Variables in Nonautonomous Systems, *Nauch. Tr. Mat. Plovdiv. Univ*, **26**, 3, 67–73.

Salehi, S.V. and Ryan, E.P.

[1985] A Nonlinear Feedback Attitude Regulator, *Intern. J. Contr.*, **41**, 1, 281–287.

Salvadori, L.

[1969] Sulla Stabilita del Movimento, *Matematiche*, **24**, 218–239.

[1972] Sul Problema della Stabilita Asimptotica, *Rend. Acc. Naz. Lincei*, **53 (2)**, 35–38.

Savchenko, A.Ya. and Ignat'ev, A.O.

[1989] *Problems in the Theory of Stability of Nonautonomous Systems*, Kiev: Naukova Dumka (in Russian).

Sell, G.R.

[1967] Nonautonomous Differential Equations and Topological Dynamics, *Trans. Amer. Math. Soc.*, **127**, 2, 241–283.

Sharov, V.F.

[1978] Stability and Stabilization of Stochastic Systems with Respect to Part of the Variables, *Avtomat. i Telemekh.*, 11, 63–71.

Shchennikov, V.N.

[1984] Analysis of Partial Stability for Differential Systems with Homogeneous Right Sides, *Differentsial'nye Uravneniya*, **20**, 9, 1645–1649 (in Russian only).

[1985] On Partial Stability in the Critical Case of $2k$ Purely Imaginary Roots, in *Differential and Integral Equations: Methods of Topological Dynamics*, 46–50, Gorky: Izd. Gork. Gos. Univ. (in Russian).

Shestakov, A.A.

[1990] *Generalization of Lyapunov's Direct Method for Systems with Distributed Parameters*, Moscow: Nauka (in Russian).

Shiehlen, W.

[1986] *Technishe Dynamik*, Stuttgart: Teubner.

Shimanov, S.N.

[1965] On the Theory of Linear Differential Equations with Delay, *Differentsial'nye Uravneniya*, **1**, 1, 102–116.

Shklyar, V.N. and Malyshenko, A.M.

[1975] On the Problem of Optimal Spatial Turn of Spacecraft about its Center of Mass, *Kosmich. Issledovaniya*, **13**, 4, 473–480.

Shoichi, S.

[1990] On the Partially Iniformly Integral Stability of Solutions or Ordinary Differential Equation, *Res. Rept. Akita Techn. Coll.*, **25**, 78–83.

Shoichi, S., Miki, K., and Masamichi, A.

[1982] Partial Stability Theorems by Comparison Principle, *Res. Repts. Akita Tech.*, **17**, 95–99.

Silakov, V.P. and Yudaev, G.S.

[1975] On Stability of Difference Systems Relative to Part of the Variables, *Differentsial'nye Uravneniya*, **11**, 5, 909–913.

Siljak, D.D.

[1978] *Large-Scale Dynamic Systems*, New York: North Holland.

[1989] Parameter Space Methods for Robust Control Design: A Guided Tour, *IEEE Trans. Autom. Contr.*, **AC-34**, 7, 674–688.

[1991] *Decentralized Control of Complex Systems*, Cambridge: Academic Press.

Simenov, P.S. and Bainov, D.D.

[1986] Stability with Respect to a Part of Variables in Systems with Impulse Effect, *J. Math. Anal. Appl.*, **117**, 1, 247–263; 1987, **124**, 2, 547–560.

Singh, S.N. and Bossart, T.C.

[1993] Exact Feedback Linearization and Control of Space Station Using CMG, *IEEE Trans. Autom. Contr.*, **AC-38**.

Sirazetdinov, T.K.

[1987] *Stability of Systems with Distributed Parameters*, Novosibirsk: Nauka (in Russian).

Skimel', V.N.

[1978] On the Property of Stiffness of Motion, *Prikl. Mat. Mekh.*, **42**, 3, 407–414.

[1992] Application of the Method of Lyapunov Functions to Some Problems of Acceptability of Approximate Solutions of Differential Equations, *Prikl. Mat. Mekh.*, **56**, 6, 918–925.

Smets, H.B.

[1961] Stability in the Large of Heterogeneous Power Reactors, *Acad. R. Belg. Bull. Cl. Sci.*, **47**, 382–405.

Smirnov, E.Ya. and Ermolina, M.V.

[1988] Stabilization of Programmed Motion for Mechanical Systems Relative to Part of the Variables, in *Problems in the Qualitative Theory of Differential Equations*, 251–260, Novosibirsk: Nauka (in Russian).

Smirnov, E.Ya., Pavlikov, V.Yu., Shcherbakov, P.P., and Yurkov, A.V.

[1985] *Controlling the Motion of Mechanical Systems*, Leningrad: Izd. Leningr. Gos. Univ. (in Russian).

Staude, O.

[1894] Uber Permanente Rotationsachsen bei der Bewegung Eines Schweren Korpers um Einen Festen, *Punkt. J. fur die Reine und Angewandte Mathematik*, **113**, 318–334.

Strogaya, G.V.

[1990] Stability with Respect to Some of the Variables of Systems Splitting in the Absence of Perturbations, *Differentsial'nye Uravneniya*, **26**, 1, 1949–1955.

Subbotin, A.I. and Chentsov, A.G.

[1981] *Guarantee Optimization in Control Problems*, Moscow: Nauka (in Russian).

Talpalary, P. and Stefancu, D.

[1987] On Partial Stability of Multivalued Differential Equations, *Bull. Inst. Politech. Iasi. Ser. I*, **33**, 1–4, 41–46.

Tamm, I.E.

[1976] *The Basics of the Electricity Theory*, 9th ed., Moscow: Nauka (in Russian).

Terjeki, J.

[1982] Analysis of Stability and Convergence with Lyapunov Functions, *MTA Szám. és Autom. Kutató Intézete. Közlemétnyek*, **26**, 125–129.

[1983a] On Partial Stability of Solutions to Linear Autonomous Differential-Difference Equations, *Stud. Sci. Math. Hung.*, **18**, 2–4, 143–152 (in Russian).

[1983b] On the Stability and Convergence of Solutions of Differential Equations by Ljapunov's Direct Method, *Acta Sci. Math.*, **46**, 1–4, 157–171.

Terjeki, J. and Hatvani, L.

[1981] On Partial Stability and Covergence of Motion, *Prikl. Mat. Mekh.*, **45**, 3, 428–435.

[1982] On Asymptotic Halting in the Presence of Viscous Friction, *Prikl. Mat. Mekh.*, **46**, 1, 20–26.

[1985] Lyapunov Functions of Mechanical Energy Type, *Prikl. Mat. Mekh.*, **49**, 6, 894–899.

Tertychnyi, V.Yu.

[1989] Stochastic Stabilization of Controlled Rotational Motion of a Solid, *Izv. Akad. Nauk SSSR Mekh. Tverd. Tela*, 2, 9–14.

[1993] *Synthesis of Controlled Mechanical Systems*, St. Petersburg: Politekhnika (in Russian).

Tisserand, F.

[1891] *Traité de Mécanique Céleste, Vol. 2*, Paris.

Vadali, S.R. and Junkins, J.L.

[1983] Spacecraft Large Angle Rotational Maneuvers with Optimal Momentum Transfer, *J. Astronaut. Sci.*, **31**, 217–235.

[1984] Optimal Open-Loop and Stable Feedback Control of Rigid Spacecraft Attitude Maneuvers, *J. Astronaut. Sci.*, **32**, 105–122.

Veretennikov, V.G.

[1984] *Stability and Oscillation of Nonlinear Systems*, Moscow: Nauka (in Russian).

Volterra, V.

[1931] *Leçons sur la Théorie Mathématique de la Lutte Pour la Vie*, Paris: Gauthier–Villars.

Vorotnikov, V.I.

[1979a] One Method for Analyzing Stability and Stabilization of Motion with Respect to Part of the Variables, *Dissertation, Moscow State University*, Moscow (in Russian).

[1979b] On the Stability of Motion Relative to Part of the Variables for Certain Nonlinear Systems, *Prikl. Mat. Mekh.*, **43**, 3, 441–450; *J. Appl. Math. Mech.*, **43**, 3, 476–485.

[1980a] Stability and Stabilization of Motion with Respect to Some of the Variables for Linear Delayed Systems, *Avtomat. i Telemekh.*, 8, 36–47; *Autom. Remote Contr.*, 8, 1068–1077.

[1980b] Stability with Respect to Part of the Variables for Systems with Holdover, *Abstracts of the Fourth Conf. of Young Scientists*, 25–26, Sverdlovsk: NTO (in Russian).

[1982a] On the Stability and Stabilization of Motion with Respect to a Part of the Variables, *Prikl. Mat. Mekh.*, **46**, 6, 914–923; *J. Appl. Math. Mech.*, **46**, 6, 733–740.

[1982b] On Complete Controllability and Stabilization of Motion with Respect to Some of the Variables, *Avtomat. i Telemekh.*, 2, 15–21; *Autom. Remote Contr.*, 2, 277–283.

[1983a] On Motion Stability Relative to a Part of the Variables under Persistent Perturbations, *Prikl. Mat. Mekh.*, **47**, 2, 291–301; *J. Appl. Math. Mech.*, **47**, 2, 244–252.

[1983b] Stability of Motion with Respect to Some of the Variables for Stochastic Systems, *Avtomat. i Telemekh.*, 7, 76–86; *Autom. Remote Contr.*, 7, 890–898.

[1984] Stabilization of Permanent Rotations of a Massive Solid, *Mater. Nauchn. Tekhn. Konfer. Ural'sk. Politekhn. Inst.*, Sverdlovsk: Izd. Ural'sk. Politekhn. Inst. (in Russian).

[1985] Stabilization of Permanent Rotations of a Heavy Rigid Body Attached at a Fixed Point, *Izv. Akad. Nauk SSSR Mekh. Tverd. Tela*, **20**, 3, 16–18; *Mech. Solids*, **20**, 3, 15–17.

[1986a] Stability with Respect to a Specified Number of Variables, *Prikl. Mat. Mekh.*, **50**, 3, 353–359; *J. Appl. Math. Mech.*, **50**, 3, 266–271.

[1986b] Stabilization of Orientatiton of a Gyrostat in a Circular Orbit in a Newtonian Force Field, *Izv. Akad. Nauk SSSR Mekh. Tverd. Tela*, **21**, 3, 25–30; *Mech. Solids*, **21**, 3, 22–27.

[1986c] On Controlling the Angular Motion of a Rigid Body, *Izv. Akad. Nauk SSSR Mekh. Tverd. Tela*, **21**, 6, 38–43; *Mech. Solids*, **21**, 6.

[1988a] The Partial Stability of Motion, *Prikl. Mat. Mekh.*, **52**, 3, 372–385; *J. Appl. Math. Mech.*, **52**, 3, 289–300.

[1988b] Optimal Stabilization of Motion, *Izv. Akad. Nauk SSSR Mekh. Tverd. Tela*, **23**, 2, 22–31; *Mech. Solids*, **23**, 2, 18–27.

[1990] Optimal Stabilization of Motion with Respect to Some of the Variables, *Prikl. Mat. Mekh.*, **54**, 5, 726–736; *J. Appl. Math. Mech.*, **54**, 5, 597–605.

[1991a] *Stability of Dynamic Systems with Respect to Part of the Variables*, Moscow: Nauka (in Russian).

[1991b] Optimal Stabilization of Nonlinear Systems, *Avtomat. i Telemekh.*, 3, 22–34; *Autom. Remote Contr.*, 3, 314–324.

[1993] Stability and Stabilization of Motion: Research Approaches, Results, Distinctive Characteristics, *Avtomat. i Telemekh.*, 3, 3–62; *Autom. Remote Contr.*, 3, 339–397.

[1994a] The Control of the Angular Motion of a Solid with Interference. A Game-Theoretic Approach, *Prikl. Mat. Mekh.*, **58**, 3, 82–103; *J. Appl. Math. Mech.*, **58**, 3, 457–476.

[1994b] On Nonlinear Synthesis of Bounded Control in the Presence of Disturbances, *Dokl. Akad. Nauk*, **337**, 1, 44–47; *Physics-Doklady*, **39**, 7, 519–522.

[1994c] The Problem of the Nonlinear Synthesis of Speed-of-Response-Wise Suboptimal Bounded Controls When There is Interference, *Izv. Akad. Nauk Tekhn. Kibernet.*, 6, 88–116; *Journal of Computer and Systems Sciences International*, **34**, 2, 57–81.

[1995a] On the Theory of Partial Stability, *Prikl. Mat. Mekh.*, **59**, 4, 553–561; *J. Appl. Math. Mech.*, **59**, 4, 525–531.

[1995b] The Theory of Stability with Respect to a Part of Variables, *Dokl. Akad. Nauk*, **341**, 3, 334–337; *Physics-Doklady*, **40**, 3, 146–149.

[1995c] On Bounded Control Synthesis in a Game Theory Problem of Reorientation of an Asymmetric Solid, *Dokl. Akad. Nauk*, **343**, 5, 630–634; *Physics-Doklady*, **40**, 8, 421–425.

[1997a] On Theorems of Barbashin–Krasovskii's Type in the Problems on Stability with Respect to Part of the Variables, *Dokl. Akad. Nauk*, **354**, 4, 481–484; *Physics-Doklady*, **42**, 6, 315–318.

[1997b] On the Nonlinear Problem of Reorientating an Asymmetric Solid, *Dokl. Akad. Nauk*, **352**, 3, 331–334; *Physics-Doklady*, **42**, 1, 47–51.

[1997c] The Construction of Bounded Game-Theoretic Control for Nonlinear Dynamical Systems, *Prikl. Mat. Mekh.*, **61**, 1, 63–74; *J. Appl. Math. Mech.*, **61**, 1, 59–69.

[1997d] On Null-Controllability of Nonlinear Dynamic Systems with Respect to Part of the Variables, *Avtomat. i Telemekh.*, 6, 50–64; *Autom. Remote Contr.*, 6 (to appear).

[1997e] On Nonlinear Game-Theoretic Problem of "Passing" through a Given Angular Position by an Asymmetric Solid, *Dokl. Akad. Nauk*, **357**, 1; *Physics-Doklady* (to appear).

Vorotnikov, V.I. and Prokop'ev, V.P.

[1978] On Stability of Motion of Linear Systems with Respect to a Part of Variables, *Prikl. Mat. Mekh.*, **42**, 2, 268–271; *J. Appl. Math. Mech.*, **42**, 2, 280–283.

Vu Tuan

[1980] On Stability in the First Approximation Relative to Part of the Variables, *Prikl. Mat. Mekh.*, **44**, 2, 211–220.

Vuyichich, V.A. and Kozlov, V.V.

[1991] On Lyapunov's Problem on Stability with Respect to Given Functions of State, *Prikl. Mat. Mekh.*, **55**, 4, 555–559.

Wada, T., Ohta, Y., Ikeda, M., and Siljak, D.D.

[1995] Parametric Absolute Stability of Lur'e Systems, *Proc. 34 Conf. Decis. Contr.:New Orleans, Dec. 1995*, 1449–1454.

Wazcwski, T.

[1950] Systèmes des Équations et des Inequalités Differentialles Ordinaires aux Deuxièmes Members Monotones et leurs Applications, *Ann. Soc. Polon. Math.*, **23**, 112–166.

Wei-Xuan, C.

[1984] The Connective Stability with Respect to a Part of Variables of Dynamic Systems with Delay, *J. Xiamen Univ. Nat. Sci.*, **23**, 3, 291–299.

Weiss, L. and Infante, E.F.

[1967] Finite Time Stability under Perturbing Forces and on Product Spaces, *IEEE Trans. Autom. Contr.*, **AC-12**, 1, 54–59.

Yakubovich, V.A. and Starzhinskii, V.M.

[1972] *Linear Differential Equations with Periodic Coefficients*, Moscow: Nauka (in Russian); New York: John Wiley & Sons, 1975.

Yoshizawa, T.

[1959] Liapunov's Function and Boundedness of Solutions, *Funkcialaj Ekvacioj*, **2**, 95–142.

[1966] *Stability Theory by Liapounov's Second Method*, Tokyo: Math. Soc. Japan.

[1975] *Stability Theory and Existence of Periodic Solutions and Almost Periodic Solutions*, New York: Springer-Verlag.

Yudaev, G.S.

[1975] Stability with Respect to Some (Not All) of the Variables, *Differentsial'nye Uravneniya*, **11**, 6, 1023–1029.

[1977] On Stability of Stochastic Differential Equations, *Prikl. Mat. Mekh.*, **41**, 3, 430–435.

Yurkov, A.V.

[1988] Stabilization of the Motion of Mechanical Systems with Indirect Control Relative to Part of the Variables, in *Problems in the Qualitative Theory of Differential Equations*, 270–272, Novosibirsk: Nauka (in Russian).

Zaitsev, V.V.

[1993] An Existence Criterion and Estimates for Invariant Bounded Sets of a System of Autonomous Differential Equations, *Differentsial'nye Uravneniya*, **29**, 5, 766–772.

Zelikin, M.I.

[1996] Optimal Control of the Rotational Motion of a Solid, *Dokl. Akad. Nauk*, **346**, 3, 334–336.

Zelikin, M.I. and Borisov, V.F.

[1994] *Theory of Chattering Control with Applications to Astronautics, Robotics, Economics and Engineering*, Boston: Birkhauser.

Zhao, Y. and Bryson, A.E., Jr.

[1990] Control of an Aircraft in Downbursts, *J. Guidance, Control and Dynamics*, **13**, 5, 819–823.

[1992] Approach Guidance in a Downburst, *J. Guidance, Control and Dynamics*, **15**, 4, 893–900.

Zhu Hai-ping and Mei Fang-xiang

[1995] On the Stability of Nonholonomic Mechanical Systems with Respect to Part Variables, *Chin. J. Appl. Math. Mech.*, **16**, 3, 225–233.

Zubov, V.I.

[1957] *The Methods of Liapunov and Their Applications*, Leningrad: Izd. Leningr. Gos. Univ. (in Russian); Groningen: Noordhoff, 1964.

[1959] *Mathematical Methods for the Study of Automatic Control Systems*, 1st ed., Leningrad: Sudpromgiz (in Russian); 2nd ed., Moscow: Mashinostroenie, 1974; Oxford: Pergamon Press, Jerusalem Acad. Press, 1962.

[1975] *The Control Theory*, Moscow: Nauka (in Russian); *Théorie de la Commande*, Moscow: Mir, 1978.

[1980] *The Problem of Stability of Control Processes*, Leningrad: Sudostroenie.

[1982] *The Dynamics of Controlled Systems*, Moscow: Vysshaya Shkola.
[1996] Asymptotic Stability in the First, in the Broad Sense, Approximation, *Dokl. Akad. Nauk*, **346**, 3, 295–296.

Zubov, V.I., Ermolin, V.S., Sergeyev, S.L., and Smirnov E.Ya.
[1978] *Controlling the Rotational Motion of a Solid*, Leningrad: Izd. Leningr. Gos. Univ. (in Russian).

Index

M

O

P

Q

R

T

U

V

W

Y

Z